Repair of Vehicle Bodies

Repair of Vehicle Bodies

Fifth edition

The late A. ROBINSON
RTechEng, MInstBE, MIBCAM, AWeldI, MISME
Formerly Section Leader for Vehicle Bodywork and Vehicle Body Repair
Course Coordinator for CGLI 398 Vehicle Body Competences
Course Coordinator for CGLI 385 Vehicle Bodywork
at Gateshead College

Updated by

ANDREW LIVESEY, BEd, FIMI, MIMechE
Head of School of Engineering
North West Kent College

AMSTERDAM • BOSTON • HEIDELBERG • LONDON • NEW YORK • OXFORD
PARIS • SAN DIEGO • SAN FRANCISCO • SINGAPORE • SYDNEY • TOKYO

Butterworth-Heinemann is an imprint of Elsevier

Butterworth-Heinemann
An imprint of Elsevier
Linacre House, Jordan Hill, Oxford OX2 8DP, UK
84 Theobald's Road, London WC1X 8RR, UK

First published by Heinemann Educational Books Ltd 1973
Reprinted 1977, 1979, 1981, 1982
Second edition published by Butterworth-Heinemann Ltd 1989
Reprinted 1990, 1991, 1992, 1993
Third edition 1993
Reprinted 1994
Fourth edition 2000
Fifth edition 2005

NOTICE

No responsibility is assumed by the publisher for any injury and/or damage to
persons or property as a matter of products liability, negligence or otherwise,
or from any use or operation of any methods, products, instructions or ideas
contained in the material herein. Because of rapid advances in the medical
sciences, in particular, independent verification of diagnoses and drug dosages
should be made

British Library Cataloguing in Publication Data
A catalogue record for this book is available from the British Library

Library of Congress Cataloguing in Publication Data
A catalogue record for this book is available from the Library of Congress

ISBN 13: 978-0-75-066753-1
ISBN 10: 0-75-066753-2

For information on all Elsevier Butterworth-Heinemann publications
visit our website at books.elsevier.com

Typeset by Integra Software Services Pvt. Ltd, Pondicherry, India
www.integra-india.com
Printed and bound in Great Britain

To my wife Jean for her help and support whilst writing,
and to my children Rachel, Rebecca and Samuel for
accompanying me to shows and rallies to collect information.

Contents

Foreword

Since 1914 the Vehicle Builders and Repairers Association has been intimately involved in the history and development of the motor vehicle and repair techniques. The onward march of technical progress increases each year with vehicles becoming more complex and where the ancillary equipment, especially the electrics, require delicate handling and treatment as part of the repair process.

It has never been more necessary to encourage young people into the industry, to equip them with the necessary technical skills and tools plus an ability to work quickly, confidently and efficiently to maximize profit whilst safeguarding their own and their colleagues' health and safety.

Vehicle Body repair is both art and science. *Repair of Vehicle Bodies* addresses both sides of that equation.

Qualified technicians able and willing to apply themselves diligently and accurately to their task are a scarce resource, commanding increasing respect and greater recognition than ever before.

Everyone has to begin somewhere. This book is an excellent starting point providing the basic essentials as well as the more advanced skills and knowledge needed to equip students for a successful career in the body repair industry.

Like all businesses, body repairers rely heavily on their most valuable asset, their staff, for survival and success in their specialist field. Pride in the job is still very much in evidence but it is essential that everyone involved in the repair process performs efficiently and consistently. This book underpins this process and goes on to become a significant source of reference for the future.

Malcolm Tagg
Director General
Vehicle Builders and Repairers Association Limited

Preface

This book is written for a wide range of students and professional practitioners in the field of vehicle body repair and re-finishing. It is a text book and reference work which covers the concepts, principles and skills needed to carry out effective vehicle body repair and re-finishing.

It will be useful for the professional practitioner whether working on current models or restoring vintage or classic cars. The DIY enthusiast will find it easy to read and useful as a work-shop manual when carrying out tasks, or projects, related to the vehicle body.

Students who are studying for NVQ, SNVQ, National Diploma, Higher National Diploma or Foundation Degree qualifications in motor vehicle engineering will find the book an ideal text and reference book for both the practical skills and the examinable content of those courses.

For NVQ/SNVQ students, Tables P1 and P2 will help in portfolio planning. Table P1 shows the qualification framework for Vehicle and Body Paint Operations Units from Automotive Skills Limited. Table P2 links the qualification framework with the contents of the book, showing which sections in the book relate to specific units. This table will help guide readers who are using the book as a course text or reference, with links to the underpinning knowledge and skills as well as to the material included in the NVQ assessment.

The book deals with the craft techniques needed for working with ferrous, non-ferrous metals and composite materials as well as with materials joining technologies and the use of body- and re-finishing shop equipment and tools. It is written to be as comprehensive as possible. I hope that the reader will enjoy it. If you have comments on this book, mail me at

Andrew Livesey
Andrew@AndrewLivesey.co.uk

Table P1 Automotive skills S/NVQ qualification framework: Vehicle body and paint operations

Units		Level 2 MET/Body Fitting	Level 2 Body repair	Level 2 Refinishing	Level 3 MET/Body fitting	Level 3 Body repair	Level 3 Refinishing
				Routes			
G1	Contribute to workplace good housekeeping	M	M	M	M	M	M
G2	Ensure your own actions, reduce risks to health and safety	M	M	M	M	M	M
G3	Maintain positive working relationships	M	M	M	M	M	M
BP01	Remove and fit basic Mechanical, Electrical and Trim (MET) components to vehicles	M					
BP02	Remove and fit non-welded non-structural vehicle body panels	M	M				
BP03	Remove and fit non-welded non-structural motorcycle body panels**						
BP04	Remove, renew and refit Mechanical, Electrical and Trim (MET) units within vehicle systems	M					
BP05	Remove and replace vehicle non-structural body panels		M				
BP06	Repair vehicle non-structural body panels		M				
BP07	Prepare vehicle panels to accept foundation and topcoats			M			
BP08	Prepare and apply foundation materials to vehicles			M			
BP09	Repair minor vehicle paint defects			M			
BP10	Carry out complete vehicle refinishing operations						M
BP11	Mix and match vehicle paint colours						M
BP12	Identify and rectify vehicle paint defects and faults						M
BP13	Remove and replace vehicle body panels					M	
BP14	Repair vehicle body panels					M	
BP15	Remove and reinstate vehicle mechanical and electrical systems and assemblies following accident damage				M		
BP16	Remove and reinstate vehicle trim fitments following accident damage				M		
BP17	Rectify vehicle misalignment					M	
BP18	Repair glass reinforced panels and vehicle bodies*	A					
MR06	Inspect vehicles				M		
MR09	Valet vehicles*	A					
Note: **6 mandatory units to be completed for each route within each level.** * These units are available for accreditation in addition to those necessary for the full NVQ/SVQ, if required. ** This unit appears as an optional unit for motorcycle technicians within the M & R Level 2 S/NVQ framework.		Mandatory framework	Mandatory framework	Mandatory framework	Mandatory framework	Mandatory framework	Mandatory framework

Table P2 Automotive skills: S/NVQ units and related sections

	Units	Related sections
G1	Contribute to workplace good housekeeping	2.1–2.8; 3.5; 12.6; 13.5
G2	Ensure your own actions, reduce risks to health and safety	2.1–2.8; 9.13; 11.8; 12.6; 15.6; 16.11
G3	Maintain positive working relationships	N/A
BP01	Remove and fit basic Mechanical, Electrical and Trim (MET) components to vehicles	1.3–1.5; 13.18
BP02	Remove and fit non-welded non-structural vehicle body panels	14.4.4–14.4.6
BP03	Remove and fit non-welded non-structural motorcycle body panels	N/A
BP04	Remove, renew and refit Mechanical, Electrical and Trim (MET) units within vehicle systems	1.13–1.5; 13.18
BP05	Remove and replace vehicle non-structural body panels	1.3–1.5; 3.1; 3.17; 4.13; 7.1–7.8; 14.4.9–14.4.11
BP06	Repair vehicle non-structural body panels	13.1–13.16
BP07	Prepare vehicle panels to accept foundation and topcoats	1.5; 16.9; 17.12–17.13
BP08	Prepare and apply foundation materials to vehicles	17.13; 17.16
BP09	Repair minor vehicle paint defects	17.5; 17.11
BP10	Carry out complete vehicle refinishing operations	17.1–17.20
BP11	Mix and match vehicle paint colours	17.1–17.6
BP12	Identify and rectify vehicle paint defects and faults	17.11; 17.12; 17.15; 17.19
BP13	Remove and replace vehicle body panels	1.1–1.5; 3.1; 7.1–7.8
BP14	Repair vehicle body panels	13.1–13.17; 14.1–14.5; 16.1–16.11
BP15	Remove and reinstate vehicle mechanical and electrical systems and assemblies following accident damage	13.18
BP16	Remove and reinstate vehicle trim fitments following accident damage	4.13; 13.16.10; 14.4.9–14.4.11; 3.24.3
BP17	Rectify vehicle misalignment	14.1–14.5
BP18	Repair glass reinforced panels and vehicle Bodies	16.1–16.11
MR06	Inspect vehicles	N/A
MR09	Valet vehicles	14.4.14

Acknowledgements

I wish to express my appreciation of assistance given by my colleagues in the Body Work and Painting Section at Gateshead College, and especially that of Mr. O. Carr (NCTEC Final, CGLI Final Motor Vehicle Painting and Industrial Finishing), who is a lecturer in motor vehicle spray painting and industrial finishing and has compiled the information and illustrations in Chapter 17. I am also grateful for the efficient help given by the library staff of the college.

I wish to thank my wife Norma for the many hours spent working at the computer to assist me again throughout the revision of this edition. Also I would like to take this opportunity to thank my son Andrew for the production of some of the photographs taken especially for this edition.

I would like to thank ICI's Autocolor International Bodyshop Planning Consultant, Mr Ernest Godfrey, a partner at Pickles Godfrey Design Partnership; Martin Ferguson, Planning Manager of Dana Distribution Limited; and Gill Nichol, Editor of *Bodyshop Magazine*; for helping me compile the chapter on bodyshop planning. I would also like to take this opportunity of offering my sincere thanks to the following firms and/or their representatives who have so readily cooperated with me by permitting the reproduction of data, illustrations and photographs:

AGA Ltd
Al Welders Ltd
Akzo Coatings PLC
Alcan Design and Development
Alkor Plastics (UK) Ltd
Aluminium Federation Ltd
Ambi-Rad Ltd
ARO Welding Ltd
Aston Martin Lagonda Ltd
Auchard Development Co. Ltd
Autoglym
Autokraft Ltd
Auto-Quote
Avdel Ltd
Bayer UK Ltd
Berger Industrial Coatings
Black and Decker Ltd
Blackburn College of Technology
Blackhawk Automotive Ltd
BMW
BOC Ltd
Bodyshop Magazine
Bodymaster UK
Bondaglass-Voss Ltd
Bostick Ltd
BP Chemicals (UK) Ltd
British Motor Industry Heritage Trust
British Standards Institution
British Steel Stainless

Brooklands College
Brooklands Museum
BSC Strip Product Group
Callow & Maddox Bros Ltd
Car Bench
Car-O-Liner UK Ltd
Celette UK Ltd
Cengar Universal Tool Company Ltd
Chicago Pneumatic Tool Co. Ltd
Chief Automotive Ltd
Chubb Fire Ltd
Citroen UK Ltd
Dana Distribution Ltd, Martyn Ferguson
Dataliner, Geotronics Ltd
Department for Transport
Department of Transport (USA)
Desoutter Automotive Ltd
DeVilbiss Automotive Refinishing Products
Dinol-Protectol Ltd
Dr Ahmed
Dr Barnard
Dr Ramnefors
DRG Kwikseal Products
Du Pont (UK) Ltd
Duramix
Dunlop Adhesives
East Coast Traders
Edwards Pearson Ltd
EEVC

ESAB Ltd
European Industrial Services
EuroNCAP
Express Garage
Facom Tools Ltd
Farécla Products Ltd
Farnborough College of Technology
Fifth Generation Technology Ltd
Fire Extinguishing Trades Association
Ford Motor Company Ltd
Forest Fasteners
Frost Auto Restoration Techniques Ltd
Fry's Metals Ltd
George Marshall (Power Tools) Ltd
GE Plastics Ltd
GKN Screws and Fasteners Ltd
Glass's Guide Service Ltd
Glasurit Automotive Refinish
Glas-Weld Systems (UK) Ltd
Go-Jo Industries Europe Ltd
Gramos Chemicals International Ltd
Gray Campling Ltd
Harboran Ltd
Herberts
Hooper & Co. (Coachbuilders) Ltd
Huck UK Ltd
Ian Williamson – IHMS Ltd
IBCAM Journal
ICI Autocolor
ICI Chemicals and Polymers Group
ICI Paints
Industrial and Trade Fairs Ltd
Institute of the Motor Industry
Institution of Engineering Designers
Institution of Mechanical Engineers
Jack Sealey Ltd
Jaguar Cars Ltd
James Holyfield – Automotive Skills
John Cotton (Colne) Ltd
Kärcher
Kenmore Computing Services Ltd
Kroll (UK) Ltd
Land Rover Ltd
Lotus Engineering
McLaren Cars
McLaren Racing
Mercedes Benz
Mick Ellender
Migatronic Welding Equipment Ltd
Mig Tig Arc Online

Minden Industrial Ltd
Morgan Motor Company Ltd
Motor Insurance Repair Centre (Thatcham)
Motorspart Industry Association
Murex Welding Products Ltd
3M Automotive Trades
National Adhesives & Resins Ltd
Nederman Ltd
Neill Tools Ltd
Nettlefolds Ltd
North West Kent College
Olympus Welding & Cutting Technology Ltd
Owens-Corning Fiberglas (GB) Ltd
Oxford Brookes University
Partco Engineering
Permabond
Performance Racing Industry (USA)
Pickles Godfrey Design Partnership, Ernest W. Godfrey
Plastics and Rubber Institute
Power-Tec Body Shop Solutions
Proton
Racal Safety Ltd
Rachel Margetts – Graphic Designs
Rally Speed
Rebecca Maturin – AVC
Reliant Motors Ltd
Renault UK Ltd
Rolls-Royce Motor Car Ltd
Rover Cars
Saab-Scania
Schlegel (UK) Ltd
Scott Bader Co. Ltd
Selson Machine Tool Co. Ltd
SIP (Industrial Products) Ltd
SL Kars
Spraybake Ltd
SPSystems
Standard Forms Ltd
Stanners Ltd
Strode College, Street
Sun Electric UK Ltd
Sykes-Pickavant Ltd
Teroson UK
The Institute of Materials
The Motor Insurance Repair Research Centre
The National Motor Museum
Thornley and Knight Ltd
THATCHAM
Triplex Safety Glass

Tri-Sphere Ltd
TRW United-Car Ltd
Tucker Fasteners Ltd
UK Fire International Ltd
United Continental Steels Ltd
USLB – AVC
Vauxhall Motors Ltd
VCA
Vehicle Builders and Repaires Association
Vitamol Ltd

Vines Guildford (BMW and MINI)
Volkswagen-Audi
Volvo Concessionaires Ltd
VOSA
W. David and Sons Limited
Welding & Metal Fabrication
Welwyn Tool Co. Ltd
Wheelforce V. L. Churchill
Wholesale Welding Supplies Ltd

A. Robinson
(List updated by W. A. Livesey)

Glossary

Abrasive A substance used for wearing away a surface by rubbing.

Acceleration Acceleration is the rate of change of speed.

Accelerator A constituent of synthetic resin mix which hastens a reaction.

Accessories Optional extras not essential to the running of a vehicle, e.g. radio, heater.

Acetone A liquid hydrocarbon capable of dissolving twenty-five times its own volume of acetylene gas at atmospheric pressure.

Acetylene A combustible gas which is mixed with oxygen and used in oxy-acetylene welding.

Adhesion The ability to adhere to a surface.

Adhesive A substance that allows two surfaces to adhere together.

Adjustments Necessary alterations to improve tolerances in fit.

Air bags A passive safety restraint system that inflates automatically on vehicle impact to protect the driver.

Alignment The operation of bringing into line two or more specified points on a vehicle structure.

All-metal construction Generally this applies to those body shells of both private cars and light commercial vehicles in which the construction is in the form of steel pressings assembled by welding, thus forming a fabricated unit.

Alloy A mixture of two or more metals with, or without, other metallic or non-metallic elements.

A-post A structural pillar on which the front door is hung.

Backfire In gas welding, a momentary return of gases indicated in the blowpipe by a pop or loud bang, the flame immediately recovering and burning normally at the blowpipe.

Backhand welding Sometimes classified as 'rightward welding'. A technique in which the flame is directed backwards against the completed part of the weld.

Back light A central window in the rear panel of the driving cab, or the rear window of a saloon body.

BC-post Central pillar acting as a central roof and side support between the rear and the front of the car.

Bevel angle The angle of a prepared edge creating a bevel prior to welding.

Billet An oblong piece of metal having a square section.

Binder A resin or cementing constituent of a compound.

Blowpipe A tool used for welding, known as a welding torch.

Body The structured part of a vehicle which encompasses the passenger, engine and luggage compartments.

Body hardware Functional accessories of vehicle body, e.g. door handles.

Body lock pillar A body pillar that incorporates a lock striker plate.

Body mounting Conventional body mounted on car chassis in composite method of body construction.

Body panels Pressed metal panels, or plastic moulded composite panels, which are fastened together to form the skin of a car body.

Body side moulding The exterior trim moulding fastened to the exterior of the body in a horizontal position.

Body sill The panel directly below the bottom of the doors.

Body spoon A body repairer's hand tool.

Body trim The materials which are used in the interior of the body for lining and upholstery.

Bonnet The metal cover over the engine compartment.

Boot A compartment provided in a car body which takes the luggage and often the spare wheel and fuel tank. It may be at the front or rear of the body depending upon the location of the engine.

Boot lid Door covering luggage compartment.

Bottom side   The frame member of the base of the body extending along the full of the main portion of the body.

Brazing   A non-fusion process in which the filler metal has a lower melting point than the parent metal(s).

Buckles   The resulting distortion of body panels after collision.

Bulkhead   A transverse support in a body structure.

Bumping   Reshaping metal with a hammer and dolly.

Burr   The resulting condition left on a metal edge after cutting or filing.

Bursting disc   A type of pressure relief device which consists of a disc, usually of metal, which is so held that it confines the pressure of the cylinder under normal conditions. The disc is intended to rupture between limits of over-pressure due to abnormal conditions, particularly when the cylinder is exposed to fire.

Butt joint   A welded joint in which the ends or edges of two pieces of metal directly face each other.

Calibrate   To check irregularities in measuring instruments.

Cant panel   The curved section of the roof top running between the comparatively flat top and the rain drip or gutter.

Cantrail   The longitudinal framing of the roof at the joint.

Carbon dioxide   A heavy colourless and incombustible gas which results from the perfect combustion of carbon.

Carbon fibre   An extremely strong, though expensive, reinforcement which can be used in conjunction with fibreglass. It gives increased rigidity to the laminate.

Carbonizing flame   An oxy-acetylene flame adjustment created by an excess of acetylene over oxygen, resulting in an excess of carbon in the flame.

Case hardening   This is the process of hardening the outer case or shell articles, which is accomplished by inducing additional carbon into the case of the steel by a variety of methods.

Catalyst   A chemical substance which brings about a chemical change to produce a different substance.

Catalyst dispenser   A purpose-designed container for measuring and dispensing liquid catalyst without splashing.

Centre pillar   The centre vertical support of a four-door saloon.

Chassis   The base frame of a motor vehicle of composite construction to which the body is attached.

Check value   A safety device that controls the passage of gas or air in one direction, in order to prevent the reversal of gas flow and a consequent accident.

Chemical reaction   The resulting change when two or more chemical substances are mixed.

Chopped strand mat   Chopped strands bonded into a mat to produce a popular economical general-purpose reinforcement.

Chopped strands   As the name suggests, glass fibre strands chopped into short (about 12 mm) lengths. They can be used as fillers. Useful for bodywork repairs.

Circuit   The path along which electricity flows. When the path is continuous, the circuit is closed and the current flows. When the path is broken, the circuit is open and no current flows.

Cold curing   Generic term for materials which harden at room temperatures, after the addition of catalyst.

Collapsible steering column   A safety feature in the form of an energy-absorbing steering column designed to collapse on impact.

Compartment shelf panel   The horizontal panel situated between the rear seat back and the back window.

Compression ratio (*CR*)   This is the ratio between the volume of the gas above the piston when it is at BDC compared to that at TDC.

Compressive strength   The ability of a material to withstand being crushed. It is found by testing a sample to failure: the load applied, divided by the cross section of the sample, gives the compressive strength.

Condensation   A change of state from a gas to a liquid caused by temperature or pressure changes. It may also be formed by moisture from the air being deposited on a cool surface.

Conductor   Any material or substance that allows current or heat to flow through it.

Copper acetylide   A spontaneously explosive and inflammable substance which forms when acetylene is passed through a copper tube.

Corrosion   The wearing away or gradual destruction of a substance, e.g. rusting of metal.

Curing   The change of a binder from soluble fusible state to insoluble infusible state by chemical action.

Curing time   The time needed for liquid resin to reach a solid state after the catalyst has been added.

Cutting tip   A torch especially adapted for cutting.

Cylinder   Steel containers used for storage of compressed gases.

Dash panel   A panel attached to the front bulkhead assembly and which provides a mounting for all instruments necessary to check the performance of the vehicle.

Deposited metal   Filler metal from a welding rod or electrode which has been melted by a welding process and applied in the form of a joint or built up.

Diagnosis   The determination, by examination, of the cause of a problem.

Dinging   Straightening damaged metal with spoons, hammers or dollies. In the early days the dingman was the tradesman who worked on completed bodies to remove minor imperfections without injury to the high gloss lacquer or varnish.

Dinging hammer   A special hammer used for dinging or removal of dents.

Direct damage   Primary damage which results from an impact on the area in actual contact with the object causing the damage.

Dolly block   A hand tool, made from special steel, shaped to suit the contour of various panel assemblies and used in conjunction with a planishing hammer to smooth out damaged panel surfaces.

Door skins   Outside door panels.

Door trim   The interior lining of a door.

D-post   The rear standing pillar providing a shut face for the rear door and forming the rear quarter panel area.

Drip moulding   A roof trough to direct water from door openings.

Electrode   The usual term for the filler rod which is deposited when using the electric arc welding process.

Electrolyte   A substance which dissolves in water to give a solution capable of conducting an electric current.

Epoxy   Based on an epoxy resin which is mixed with a hardener.

Evaporation   A change of state from solid or liquid into vapour.

Expansion   The increase in the dimensions of metals due to heat.

Extrude   To draw into lengths.

Fender   American term for wing.

Filler   Inorganic types used to extend low-pressure resins, usually polyesters.

Filler metal   Metal added to a weld in the form of a rod, electrode or coil.

Fillet weld   A weld in which two surfaces at right angles to one another are welded together.

Firewall   Panel dividing engine compartment from interior of body.

Flange   A reinforcement on the edge of a panel formed at approximately right angles to the panel.

Flashback   Occurs when the flame disappears from the end of the welding tip and the gases burn in the torch.

Flat   A panel is said to be flat when insufficient shaping has caused uneven contours and so flat areas are obvious.

Floor pan   Main floor of the passenger compartment of an underbody assembly.

Flux   A chemical material or gas used to dissolve and prevent the formation of surface oxides when soldering, brazing or welding.

Foams (flexible) A resin which is often used for cushioning in the automobile industries. These foams are usually urethanes.

Foams (rigid) A resin with a higher modulus than the flexible foams. These are also normally urethanes and are used in more structural applications such as cores in sandwich constructions.

Force Mass is to do with the number of atoms or molecules in a material.

Four-door Denotes the type of saloon body having four doors.

Frame gauges Self-centring alignment gauges which are hung from a car's underbody.

Friction Pressing the brake pads against the discs causes friction which converts the dynamic energy of the moving vehicle into heat energy.

Friction The resistance to motion that a body meets when moving over another.

Fusion welding A process in which metals are welded together by bringing them to the molten state at the surface to be joined, with or without the addition of filler metal, and without the application of mechanical pressure or blows.

Gas welding A fusion welding process which uses a gas flame to provide the welding heat.

Gel Resin takes on a gel-like consistency (gels) usually within 10–15 minutes of being catalyzed. At this point it is impossible to spray, paint or pour. Stored resin which has passed its shelf life may gel without being catalyzed.

Gelcoat A thixotropic resin normally used without reinforcement and applied first to the mould. It forms the smooth shiny surface of the finished article.

Glass fibre Glass filaments drawn together into fibres and treated for use as reinforcement.

Hardener A chemical curing or hardening agent.

Hardening Heating to a critical temperature followed by a relatively rapid rate of cooling.

Headlining The cloth or other material used to cover the inner surface of the car roof.

Heat and temperature Temperature is the hotness or coldness of a body measured in degrees centigrade (C, also called Celsius) or absolute temperature in Kelvin (K). Heat is energy measured in joules (J).

Heelboard The vertical board or panel under the rear seat which forms the support for the seat cushion.

Hinge pillar A pillar on which a door is swung.

Hood American term for bonnet.

Hydraulic pressure Pressure transmitted by a liquid.

Hydraulics The use of pressurized liquid to transfer force.

Impregnated The particles of one substance infused into that of another.

Independent front suspension Suspension system in which each wheel is independently supported by a spring.

Indirect damage Secondary damage found in the area surrounding the damage which caused it.

Inertia Inertia is the resistance of a body to stop doing whatever it is doing.

Inertia Property of an object by which it continues in its existing state of rest or motion in a straight line, unless that state is changed by an external force.

Insulation A material which is non-conductive of either heat or electricity.

Integral A necessary part to complete a whole unit.

Interchangeability The ability to substitute one part for another.

Kerb weight The weight of an empty vehicle without passengers and luggage.

Kevlar A synthetic aramid fibre used as a reinforcement for resins. It is noted for its high impact resistance and is used in racing car bodywork.

Kinetic energy The energy of motion.

Laminates A material composed of a number of layers.

Lap joint A form of joint obtained by overlapping the edges of two pieces of metal. The overlapping parts must be in the sample plane.

Latex A natural rubber used for making flexible moulds. It is a liquid which solidifies in contact with air.

Lay-up Layers of glass fibres are laid on top of wet resin and then pressed down into the liquid resin.

Leftward welding   This is known as forehand welding.

Levers   Levers are used to increase the force exerted by pivoting about a point.

Mass production   Large-scale, high-speed manufacture.

Metal fatigue   A metal structural failure, resulting from excessive or repeated stress, finally resulting in a crack.

Molecule   A minute particle into which a substance can be divided and retain its properties.

Monomer   A simple molecule capable of combining with itself, or a compatible similar chemical, to form a chain (polymer).

Moulding   The resulting shape of a plastics material when it is removed from its mould.

Mould release   A substance used to coat the mould to prevent sticking of the resin that will be used to make a part. It facilitates the removal of that part from the mould.

Near side   The left-hand side of the vehicle as viewed from the driver's seat.

Neutral flame   A balanced flame, indicating perfect combustion of both oxygen and acetylene gases.

Non-ferrous metals   Metals which do not contain any ferrite or iron.

Normalizing   Heating to a high temperature to produce a refinement of the grain structure of a metal or alloy.

Off side   The right-hand side of a vehicle as viewed from the driver's seat.

Original finish   The paint applied at the factory by the vehicle manufacturer.

Oxidation   Chemical reaction between oxygen and some other element resulting in oxides.

Oxidizing flame   A gas welding flame which has an excess of oxygen when burning.

Paddle   A wooden tool shaped for spreading body solder.

Parent metal   The material of a part to be welded.

Pascal's law   Pascal discovered that the pressure in a body is equal in all directions.

Penetration   Depth of fusion or weld penetration.

Pickle   To soak metal in an acid solution in order to free the surface of rust or scale.

Pillar   A vertical support of a body frame.

Pillar face   The front of a pillar visible when the door is opened.

Pinch weld   Two metal flanges butted together and spot welded along the join of the flat surfaces.

Polyurethane   A versatile material used for adhesives, paints, varnishes, resins and foam materials. These are often used ion conjunction with polyester-based GRP.

Porosity   The presence of gas pockets or inclusions within a weld.

Power   Power is usually expressed in horse power (HP), that is, the number of horses which would be needed to do the equivalent amount of work in the same time.

Pressure   The amount force per unit area

Prototype   An original model.

Puddle   The small body of molten metal created by the flame of a welding torch.

Quarter light   The window directly above the quarter panel.

Quarter panel   The side panel extending from the door to the rear end of the body (including rear wing).

Reinforcement   Filler material added to plastics (resin) in order to strengthen the finished product.

Relative density (RD)   Relative density is the relationship which other materials have with this.

Resin   Resins occur in nature as organic compounds, insoluble in water, e.g. amber, shellac. Synthetic resins have similar properties and are normally converted to solids by ploymerization.

Return sweep   A reverse curve.

Saloon   An enclosed body not having a partition between the front and rear seats.

Scuttle panel   The panel between the bonnet and windscreen.

Self-tapping screw   A screw that cuts its own threads into a predrilled hole.

Silicone rubbers   Used, amongst other applications, for sealants and flexible mould compounds. They are usually cold curing.

Solvent   A chemical fluid which will dissolve, dilute or liquefy another material.

Specifications   Information provided by the manufacturer on vehicle data in the form of dimensions.

Squab   The rear seat back construction.

Stream lining   The shaping of a vehicle body to minimize air resistance.

Subframe   Members to which the engine and front-end assembly are attached.

Swage   A raised form of moulding pressed into a piece of metal in order to stiffen it.

Swage line   A design line on a vehicle body, caused by a crease or step in a panel.

Sweating   Uniting two or more metal surfaces by the use of heat and soft solder.

Synthetic   A substance produced artificially.

Temperature   The measurement, in degrees, of the intensity of heat.

Template   A form or pattern made so that other parts can be formed to exactly the same shape.

Tensile strength   The resistance to breaking which metal offers when subject to a pulling stress.

Thermoplastic   Plastic which can be softened by heating, and which still retains its properties after it has been cooled and hardened. Typical thermoplastics are polythene and PVC.

Thermosetting   Plastic which hardens by non-reversible chemical reaction, initiated by heat and/or curing agents. Once hardened, it cannot be melted down without being destroyed.

Thixotropic   Generally used to describe substances which have a very high viscosity when stable, but low viscosity when stirred or brushed. 'Non-drip' paint is an obvious example; another is gelcoat resin.

Torque   Torque is the turning moment about a point expressed in newton-metres (Nm).

Tunnel   A raised floor panel section for drivershaft clearance.

Turret   American term for roof.

Weld bead   One single run of an electrode welding rod in manual metal arc welding.

Weld deposit   Metal which has been added to a joint by one of the welding processes.

Wheel alignment   The adjustment of a vehicle's wheels so that they are positioned to drive correctly.

Wheel arch   Panel forming inner housing for rear wheels.

Wheelbase   The distance between the centre lines of the front and rear wheels of a vehicle.

Abbreviations and symbols

ABS	acrylonitrile butadiene styrene
ABS	anti-lock braking system
AF	across flats (bolt size)
C_d	aerodynamic drag coefficient
AC	alternating current
A	ampere
AFFF	aqueous film forming foam (fire fighting)

bar	10^6 dyn/cm^2; 10^5 N/m^2; 0.98682 atm; 14.505 psi
BATNEEC	best available techniques not enabling excessive costs
BS	British Standard
BSI	British Standards Institution
BCF	bromochlorodifluoromethane (fire extinguishers)

CO	carbon monoxide
CO_2	carbon dioxide
C	Centigrade (Celsius)
cm	centimetre
CAD	computer-aided design
CAE	computer-aided engineering
CAM	computer-aided manufacturing
CIM	computer-integrated manufacturing
COSHH	Control of Substances Hazardous to Health (Regulations)
cm^3	cubic centimetres
in^3	cubic inches

deg. or °	degree (angle or temperature)
DTI	dial test indicator
dia.	diameter
DC	direct current
dB	decibel

ECU	Electronic control unit
EPA	Environmental Protection Act
EC	European Community

F	Fahrenheit
ft	foot
ft/min	feet per minute

gal	gallon (imperial)
GLS	general lighting service (lamp)
GRP	glass fibre reinforced plastic
g	gram (mass)
HASAWA	Health and Safety at Work Act
HSE	Health and Safety Executive
HT	high tension (electrical)
HVLP	high velocity low pressure (spray guns)
in	inch
IFS	independent front suspension
IR	infrared
ID	internal diameter
IED	Institution of Engineering Designers
IMI	Institute of the Motor Industry
ISO	International Organization for Standardization
kg	Kilogram (mass)
kW	kilowatt
LH	left-hand
LHD	left-hand drive
LHThd	left-hand thread
l	litre
LT	Low tension
lumen	light energy radiated per second per unit solid angle by a uniform point source of l candela intensity
lux	unit of illumination equal to l lumen/m^2
max.	maximum
MAG	metal active gas (welding)
MIG	metal inert gas (welding)
m	metre
mm	millimetre
min.	minimum
−	minus (of tolerance)
′	minute (of angle)
(−)	negative (electrical)
Nm	newton metre
dB(A)	noise level at the ear
$L_{ep,d}$	noise exposure level, personal (daily)
$L_{ep,w}$	noise exposure level, personal (weekly)
no.	Number
ozf	ounce (force)
oz	ounce (mass)
OD	outside diameter

part no.	part number
%	percentage
PPE	Personal Protection Equipment (Regulations)
pt	pint (imperial)
+	plus (tolerance)
±	plus or minus
PVA	polyvinyl acetate
PVC	polyvinyl chloride
+	positive (electrical)
lbf	pound force
lb	pound (mass)
lbf ft	pound force foot (torque)
lbf/in^2	pound force per square inch
r	radius
ref.	reference
rev/min	revolutions per minute
RH	right-hand
RHD	right-hand drive
″	second (angle)
SAE	Society of Automobile Engineers
cm^2	square centimetres
in^2	square inches
std	standard
TIG	tungsten inert gas (welding)
VIN	vehicle identification number
VOCs	volatile organic compounds
V	volt

The history, development and construction of the car body

1.1 Development of the motor car body

1.1.1 Brief history

The first motor car bodies and chassis frames, made between 1896 and 1910, were similar in design to horse-drawn carriages and, like the carriages, were made almost entirely of wood.

The frames were generally made from heavy ash, and the joints were reinforced by wrought iron brackets which were individually fitted. The panels were either cedar or Honduras mahogany about 9.5 mm thick, glued, pinned or screwed to the framework. The tops, on cars which had them, were of rubberized canvas or other fabrics. Some bodies were built with closed cabs, and the tops were held in place by strips of wood bent to form a solid frame. About 1921 the Weymann construction was introduced, in which the floor structure carried all the weight of the seating, and the body shell, which was of very light construction, was attached to the floor unit. Each joint in the shell and between the shell and the floor was made by a pair of steel plates, one on each side of the joint and bolted through both pieces of timber, leaving a slight gap between the two pieces. The panelling was of fabric, first canvas, then a layer of wadding calico and finally a covering of leather cloth. This form of construction allowed flexibility in the framing and made a very light and quiet body frame, but the outer covering had a very short life.

As the demand for vehicles increased it became necessary to find a quicker method of production.

Up to that time steel had been shaped by hand, but it was known that metal in large sheets could be shaped using simple die tools in presses, and machine presses were introduced to the steel industry to form steel sheets into body panels. Initially the sheets were not formed into complex shapes or contours, and the first bodies were very square and angular with few curves. The frame and inner construction was still for the most part made of wood, as shown in Figure 1.1. About 1923 the first attempts were made to build all-steel bodies, but these were not satisfactory as the design principles used were similar to those which had been adopted for the timber-framed body. The real beginning of the all-steel body shell came in 1927, when presses became capable of producing a greater number of panels and in more complex shapes; this was the dawn of the mass production era. During the 1930s most of the large companies who manufactured motor vehicles adopted the use of metal for the complete construction of the body shell, and motor cars began to be produced in even greater quantities.

Owing to the ever-increasing demand for private transport, competition increased between rival firms, and in consequence their body engineers began to incorporate features which added to the comfort of the driver and passenger. This brought about the development of the closed cars or saloons as we know them today. The gradual development of the shape of the motor car body can be clearly seen in Figure 1.2, which shows a

Figure 1.1 Timber constructed bodies: (a) De Dion Works body shop, Finchley, *c.* 1923 (b) Gordon England Ltd, 1922 (*National Motor Museum, Beaulieu*)

selection of Austin vehicles from 1909 to 1992. That is, from Edwardian to modern times.

The inner construction of the head roof of these saloons was concealed by a headlining. Up to and including the immediate post-war years, this headlining was made from a woollen fabric stitched together and tacked into position on wooden frames. However, the more recently developed plastic and vinyl materials were found to be more suitable than fabric, being cheaper and easier to clean and fit. They are fitted by stretching over self-tensioning frames which are clipped into position for easy removal, or alternatively the headlining is fastened into position with adhesives.

Comfort improved tremendously with the use of latex foam rubber together with coil springs in the seating, instead of the original plain springing. The general interior finish has also been improved by the introduction of door trim pads, fully trimmed

dash panels and a floor covering of either removable rubber or carpeting.

Then came the general use of celluloid for windows instead of side curtains, and next a raising and lowering mechanism for the windows. Nowadays the windscreen and door glasses are made of laminated and/or toughened safety glass. The window mechanism in use today did not begin to develop until well into the 1920s.

Mudguards, which began as wooden or leather protections against splattered mud, grew into wide splayed deflectors in the early part of the twentieth century and then gradually receded into the body work, becoming gracefully moulded into the streamlining of the modern motor car and taking the name of wings. Carriage steps retained on earlier models gave place to running boards which in their turn disappeared altogether.

Steering between 1890 and 1906 was operated by a tiller (Figure 1.3). This was followed by the steering wheel which is in current use. The position of the gear lever made an early change from the floor to the steering column, only to return to some convenient place on the floor.

Some of the first vehicles, or horseless carriages as they were known, carried no lights at all; then carriage candle lamps made their appearance. Later came oil lamps, acetylene lamps and finally the electric lighting system, first fitted as a luxury extra and ultimately becoming standard and finally obligatory equipment which must conform with legislation of the day.

When windscreens were first introduced such accessories as windscreen wipers and washers were unknown. Then came the single hand-operated wiper, followed by the suction wiper and finally electrically driven wipers.

The design of the wheels was at first dictated by fashion. It was considered necessary for the rear wheels to be larger than the front, a legacy from the elegant horse-drawn carriages. Wooden spokes and iron tyres were the first wheels to appear, and with both rear and front wheels of the same dimensions. Then came the wooden-spoked artillery wheel with pneumatic tyre (Figure 1.4). The artillery wheel gave way to the wire-spoked wheel, and this in turn to the modern disc wheel with tubeless tyres.

Great strides have been made in the evolution of the motor car since 1770, when Cugnot's steam wagon travelled at 3 mile/h (4.8 km/h), to the modern

1909 The first Baby Austin

1922 Austin Tourer

1932 Austin Saloon

1946 Austin 16

Figure 1.2   Development of the Austin car body 1909 to 1992, from Edwardian to modern construction methods

1948 Austin A–70

1952 Austin Seven

1959 Austin Princess

1960 Austin Mini

Figure 1.2 (*continued*)

1963 Austin 1100

1964 Austin 1800

1970 Austin Maxi 1800

1973 Austin Allegro

Figure 1.2   (*continued*)

1975 Austin Princess

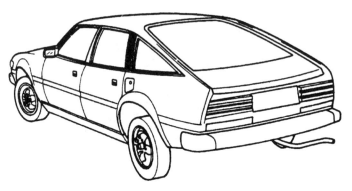

1976 Austin Rover SDI

1980 Austin Metro

1983 Austin Maestro

Figure 1.2 (*continued*)

1984 Austin Montego

1986 Austin Rover 200

1987 Austin Rover Sterling

1988 Rover 820 Fastback

1989 Mini Flame

Figure 1.2   (*continued*)

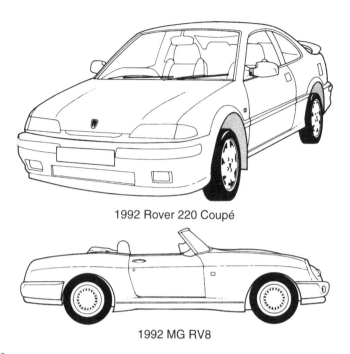

1992 Rover 220 Coupé

1992 MG RV8

Figure 1.2 (*continued*)

vehicle which can carry driver and passengers in silence, comfort and safety at speeds which at one time were thought to be beyond human endurance; indeed, special vehicles on prepared tracks are now approaching the speed of sound.

It must be borne in mind that the speed of the vehicle is governed by (*a*) the type of power unit, (*b*) its stability and manoeuvring capabilities and (*c*) its shape, which is perhaps at present one of the most important features in high-speed travel. Whatever the

mechanical future of the car, we may rest assured that the shape of the motor car body will continue to change as technical progress is made (Figure 1.5).

1.1.2 Highlights of motor vehicle history

The idea of a self-propelled vehicle occurs in Homer's *Iliad*. Vulcan, the blacksmith of the gods, in one day made 20 tricycles which 'self-moved obedient to the beck of the gods'. The landmarks in more modern motor vehicle history are as follows:

1688   Ferdinand Verbiest, missionary in China, made a model steam carriage using the steam turbine principle.
1740   Jacques de Vaucansen showed a clockwork carriage in Paris.
1765   Watt developed the steam engine.
1765   Nicholas Joseph Cugnot, a French artillery officer, built a steam wagon which carried four people at a speed of 2.25 mile/h. It overturned in the streets of Paris and Cugnot was thrown into prison for endangering the populace.
1803   Richard Trevithick built a steam carriage and drove it in Cornwall.
1831   Sir Charles Dance ran a steam coach (built by Sir Goldsworthy Gurney) on a regular

Figure 1.3 First Vauxhall, tiller operated, 1903 (*Vauxhall Motors Ltd*)

Figure 1.4 1905 and 1909 Vauxhalls with wooden spoked artillery wheels with pneumatic tyres (*Vauxhall Motors Ltd*)

Figure 1.5 Twin concept car with interchangeable engines (petrol/electric motor drive module) (*Vauxhall Motors Ltd*)

service from Gloucester to Cheltenham. Sometimes they did four round trips a day, doing 9 miles in 45 minutes. The steam coaches were driven off the road by the vested interests of the stage coach companies, who increased toll charges and piled heaps of stones in the roads along which the steam coaches passed. This, combined with the problems of boilers bursting and mechanical breakdowns and the advent of the railways, contributed to the withdrawal of the steam coaches.

1859 Oil was discovered in USA.

1865 The Locomotive Act of 1865 (the Red Flag Act) was pushed through by the railway and coach owners. One of the stipulations was that at least three people must be employed to conduct the locomotive through the streets, one of whom had to walk 60 yards in front carrying a red flag. Speeds were restricted to 2–4 mile/h. This legislation held back the development of the motor vehicle in Great Britain for 31 years, allowing the continental countries to take the lead in this field.

1885   Karl Benz produced his first car. This is recognized as being the first car with an internal combustion engine as we know it.

1886   Gottlieb Daimler also produced a car.

1890   Panhard and Levasser began making cars in France.

1892   Charles and Frank Duryea built the first American petrol-driven car, although steam cars had been in use long before this.

1895   First motor race in Paris.
First Automobile Club formed in Paris.

1896   The repeal of the Red Flag Act. This is commemorated by the London to Brighton veteran car run. The speed limit was raised to 12 mile/h and remained at that until 1903, when the 20 mile/h limit in built-up areas was introduced. There was much persecution of motorists by police at this time, which led to the formation of the RAC and the AA.

1897   The RAC was formed, largely through the efforts of F. R. Simms, who also founded the SMMT in 1902.

1899   Jenatzy set world speed record of 66 mile/h.

1900   Steering wheel replaces tiller.
Frederick Lanchester produced his first car, a 10 hp model. He had built an experimental phaeton in 1895.

1901   Front-mounted engine.
Mercedes car produced.

1902   Running board.
Serpollet did a speed of 74 km/h in a steamcar.

1903   Pressed steel frames.
First windshield.
The Motor Car Act resulted in considerable persecution of the motorist for speeding, number plates and lights, so much so that the motoring organizations paid cyclists to find police speed traps.

1904   Folding windshield.
Closed saloon-type body.
A petrol car reached 100 mile/h and, in the same year, a Stanley steam car achieved a speed of 127 mile/h. Stanley steam cars used paraffin in a multitube boiler and had a chassis made from hickory.
Rolls-Royce exhibited their first car in Paris. The motoring press were impressed with its reliability.
Veteran cars are cars up to and including this year.

1.1.3 Terms used to describe early vehicle body styles

In the history of the motor car there has been some ambiguity in the names used to describe various types of body styles, built by coach builders from different countries. The following terms relate to the vehicles produced during the period 1895 to 1915, and show the derivation of the terminology used to describe the modern vehicle.

Berlina   Rarely used before the First World War. A closed luxury car with small windows which allowed the occupants to see without being seen.

Cab   A term taken directly from the days of the horse-drawn carriages. Used to describe an enclosed vehicle which carried two passengers, while the driver was situated in front of this compartment and unprotected.

Cabriolet   Used towards the end of the period. Describes a car with a collapsible hood and seating two or four people.

Coupé   A vehicle divided by a fixed or movable glass partition, behind the front seat. The driver's position was only partially protected by the roof whilst the rear compartment was totally enclosed and very luxurious.

Coupé cabriolet or double cabriolet   A long vehicle having the front part designed as a coupé and the rear part designed as a cabriolet. There were often two supplementary seats.

Coupé chauffeur   A coupé with the driving position completely covered by an extension of the rear roof.

Coupé de ville   A coupé having the driving position completely open.

Coupé limousine   A vehicle having a totally enclosed rear compartment and the front driving position closed on the sides only.

Double Berlina   A longer version of the Berlina but having the driving position separated from the rear part of the vehicle.

Double landaulet A longer version of the landaulet. It had two permanent seats plus two occasional seats in the rear and a driving position in front.

Double phaeton A phaeton which had two double seats including the driver's seat.

Double tunneau A longer version of the tonneau in which the front seats were completely separated from the rear seats.

Landau A cabriolet limousine having only the roof behind the rear windows collapsible.

Landaulet or landaulette A small landau having only two seats in the closed collapsible roof portion.

Limousine A longer version of the coupé with double side windows in the rear compartment.

Limousine chauffeur A limousine with an extended rear roof to cover the driving position.

Phaeton A term from the days of the horse-drawn carriage. In early motoring it was used to describe a lightweight car with large spoked wheels, one double seat and usually a hood.

Runabout An open sporting type of vehicle with simple bodywork and two seats only.

Tonneau An open vehicle having a front bench seat and a semicircular rear seat which was built into the rear doors.

Glass saloon A large closed vehicle similar to a double Berlina but with enlarged windows.

Saloon A vehicle having the driving seat inside the enclosed car but not separated from the rear seat by a partition.

Torpedo A long sports vehicle having its hood attached to the windscreen.

Victoria Another term derived from the era of horses. The Victoria was a long, luxurious vehicle with a separate driving position and a large rear seat. It was equipped with hoods and side screens.

Wagon saloon A particularly luxurious saloon used for official purposes.

1.1.4 Vehicle classification

There are many ways in which motor vehicles may be classified into convenient groups for recognition. Much depends on such factors as the manufacturer, the make of the car, the series and the body type or style. Distinctive groups of passenger vehicle bodies include the following:

1 Small-bodied mass-produced vehicles
2 Medium-bodied mass-produced vehicles
3 Large-bodied mass-produced vehicles
4 Modified mass-produced bodywork to give a standard production model a more distinctive appearance
5 Specially built vehicles using the major components of mass-produced models
6 High-quality coach-built limousines (hand made)
7 Sports and GT bodywork (mass-produced)
8 Specially coach-built sports cars (hand made).

Styling forms include the following:

Saloon The most popular style for passenger vehicles is the two-door or four-door saloon. It has a fully enclosed, fixed-roof body for four or more people. This body style also has a separate luggage or boot compartment (Figure 1.6a).

Hatchback This body style is identified by its characteristics sloping rear tailgate, which is classed as one of the three or five doors. With the rear seats down there is no division between the passenger and luggage compartments and this increases the luggage carrying capacity of the vehicle (Figure 1.6b).

Estate This type of vehicle is styled so that the roof extends to the rear to give more luggage space, especially when the rear seats are lowered (Figure 1.6c).

Sports coupé and coupé A sports coupé is a two-seater sports car with a fixed roof and a high-performance engine. A coupé is a two-door, fixed-roof, high-performance vehicle with similar styling but with two extra seats at the rear, and is sometimes referred to as a '2-plus-2' (Figure 1.6d).

Convertible or cabriolet This can have either two or four doors. It has a soft-top folding roof (hood) and wind-up windows, together with fully enclosed or open bodywork (Figure 1.6e).

Sports This is a two-seater vehicle with a high-performance engine and a folding or removable roof (hood) (Figure 1.6f).

Limousine This vehicle is characterized by its extended length, a high roofline to allow better headroom for seating five passengers comfortably behind the driver, a high-quality finish and luxurious interiors (Figure 1.6g).

1.1.5 The evolution of design

When the first motor cars appeared, little attention was paid to their appearance; it was enough that they ran. Consequently the cars initially sold

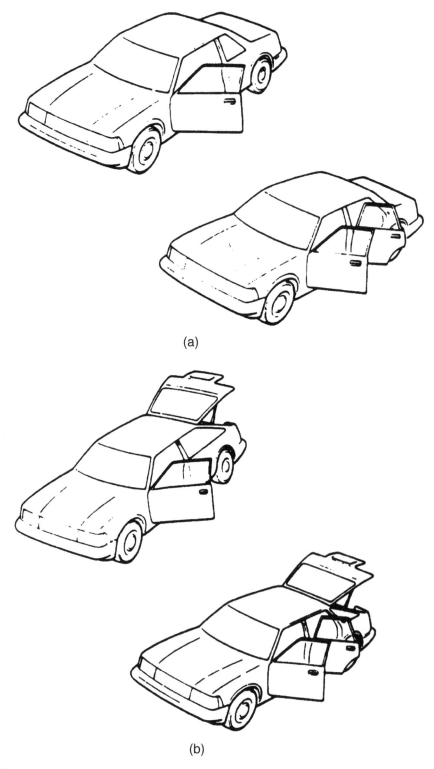

(a)

(b)

Figure 1.6 Vehicle styling forms: (a) saloon (b) hatchback

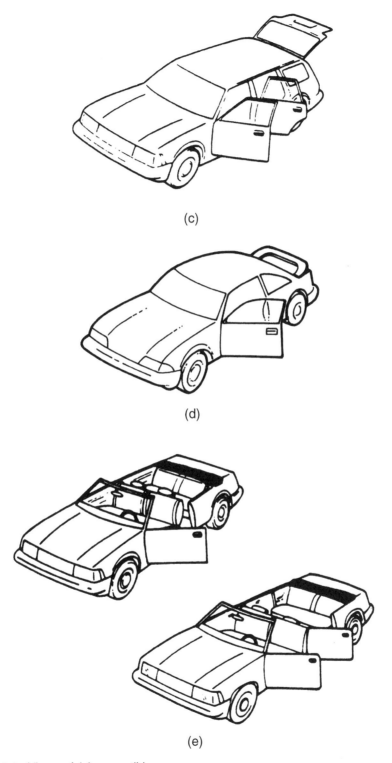

(c)

(d)

(e)

Figure 1.6   (c) estate (d) coupé (e) convertible

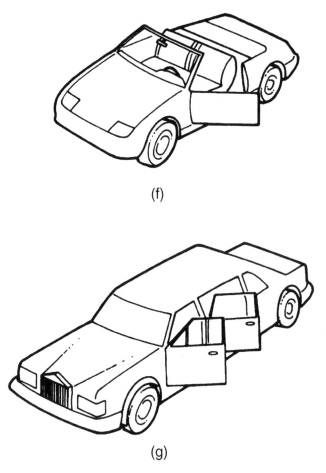

(f)

(g)

Figure 1.6 (f) sports (g) limousine

to the public mostly resembled horse-drawn car-riages with engines added. Henry Ford launched his Model T in 1908, and it sold on its low price and utility rather than its looks. However, the body design of this car had to be changed over its 19 year production span to reflect changes in customer taste.

The 1930s saw greater emphasis on stream-lining design. Manufacturers began to use wind tunnels to eliminate unnecessary drag-inducing projections from their cars. One of the dominant styling features of the 1950s and 1960s was the tail fin, inspired by the twin tail fins of the wartime Lockhead Lightning fighter aircraft. Eventually a reaction set in against such excesses and the trend returned to more streamlined styling.

In creating cars for today's highly competitive car market, designers have to do far more than just achieve a pleasing shape. National legal require-ments determine the positions of lamps, direction indicators and other safety-related items, while the buying market has become much more sophisti-cated than before. Fuel economy, comfort, function and versatility are now extremely important.

1.2 Creation of a new design from concept to realization

The planning, design, engineering and develop-ment of a new motor car is an extremely complex process. With approximately 15 000 separate parts, the car is the most complicated piece of equipment built using mass production methods.

Every major design project has its own design team led by a design manager, and they stay with the project throughout. The size of the team varies according to the progress and status of the project. The skill and judgement of the trained and experienced automotive designer is vital to the creation of any design concept.

To assist in the speed and accuracy of the ensuing stages of the design process (the implementation), some of the most advanced computer-assisted design equipment is used by the large vehicle manufacturers. For example, computer-controlled measuring bridges that can automatically scan model surfaces, or machines that can mill surfaces, are linked to a computer centre through a highly sophisticated satellite communication network. The key terms in computer equipment are as follows:

Computer-aided design (CAD)   Computer-assisted design work, basically using graphics.
Computer-aided engineering (CAE)   All computer-aided activities with respect to technical data processing, from idea to preparation for production, integrated in an optimum way.
Computer-aided manufacturing (CAM)   Preparation of production and analysis of production processes.
Computer-integrated manufacturing (CIM)   All computer-aided activities from idea to serial production.

The use of CAE is growing in the automotive industry and will probably result in further widespread changes. Historically, the aerospace industry was the leader in CAE development. The three major motor companies of GM, Ford and Chrysler started their CAE activities as soon as computers became readily available in the early 1960s. The larger automotive companies in Europe started CAE activities in the early 1970s – about the same time as the Japanese companies.

Each new project starts with a series of detailed paper studies, aimed at identifying the most competitive and innovative product in whichever part of the market is under review. Original research into systems and concepts is then balanced against careful analysis of operating characteristics, features performance and economy targets, the projected cost of ownership and essential dimensional requirements. Research into competitors' vehicles, market research to judge tastes in future years, and possible changes in legislation are all factors that have to be taken into account by the product planners when determining the specification of a new vehicle.

The various stages of the design process are as follows:

1 Vehicle styling, ergonomics and safety
2 Production of scale and full-size models
3 Engine performance and testing
4 Wind tunnel testing
5 Prototype production
6 Prototype testing
7 Body engineering for production

1.2.1 Vehicle styling

Styling
Styling has existed from early times. However, the terms 'stylist' and 'styling' originally came into common usage in the automotive industry during the first part of the twentieth century.

The automotive stylist needs to be a combination of artist, inventor, craftsman and engineer, with the ability to conceive new and imaginative ideas and to bring these ideas to economic reality by using up-to-date techniques and facilities. He must have a complete understanding of the vehicle and its functions, and a thorough knowledge of the materials available, the costs involved, the capabilities of the production machinery, the sources of supply and the directions of worldwide changes. His responsibilities include the conception, detail, design and development of all new products, both visual and mechanical. This includes the exterior form, all applied facias, the complete interior, controls, instrumentation, seating, and the colours and textures of everything visible outside and inside the vehicle.

Styling departments vary enormously in size and facilities, ranging from the individual consultant stylist to the comprehensive resources of major American motor corporations like General Motors, who have more than 2000 staff in their styling department at Detroit. The individual consultant designer usually provides designs for organizations which are too small to employ full-time stylists. Some act as an additional brain for organizations who want to inject new ideas into their own production. Among the famous designers are the Italians Pininfarina (Lancia, Ferrari, Alfa), Bertone (Lamborghini), Ghia (Ford) and Issigonis (Mini).

The work of the modern car stylist is governed by the compromise between his creativity and the world of production engineering. Every specification, vehicle type, payload, overall dimensions, engine power and vehicle image inspire the stylist and the design proposals he will make. Initially he makes freehand sketches of all the fundamental components placed in their correct positions. If the drawing does not reduce the potential of the original ideas, he then produces more comprehensive sketches of this design, using colours to indicate more clearly to the senior executives the initial thinking of the design (Figure 1.7). Usually the highly successful classic designs are the work of one outstanding individual stylist rather than of a team.

The main aim of the designer is to improve passenger comfort and protection, vision, heating and ventilation. The styling team may consider the transverse engine as a means of reducing the space occupied by the mechanical elements of the car. Front-wheel drive eliminates the driveshaft and tunnel and the occupants can sit more comfortably. Certain minimum standards are laid down with regard to seat widths, kneeroom and headroom. The interior dimensions of the car are part of the initial specifications and not subject to much modification. Every inch of space is considered in the attempt to provide the maximum interior capacity for the design. The final dimensions of the interior and luggage space are shown in a drawing, together with provision for the engine and remaining mechanical assemblies.

Ergonomics

Ergonomics is a fundamental component of the process of vehicle design. It is the consideration of human factors in the efficient layout of controls in the driver's environment. In the design of instrument panels, factors such as the driver's reach zones and his field of vision, together with international standards, all have to be considered. Legal standards include material performance in relation to energy absorption and deformation under impact. The vision and reach zones are geometrically defined, and allow for the elimination of instrument reflections in the windshield.

Basic elements affecting the driver's relationship to the instrument panel controls, instruments, steering wheel, pedals, seats and other vital elements in the car are positioned for initial evaluation using the 'Manikin', which is a two- and three-dimensional measuring tool developed as a result of numerous anthropometric surveys and representing the human figure. Changes are recorded until the designer is satisfied that an optimum layout has been achieved.

1.2.2 Safety

With regard to bodywork, the vehicle designer must take into account the safety of the driver, passengers and other road users. Although the vehicle cannot be expected to withstand collision with obstacles or other vehicles, much can be done to reduce the effects of collision by the use of careful design of the overall shape, the selection of suitable materials and the design of the components. The chances of injury can be reduced both outside and inside the vehicle by avoiding sharp-edged, projecting elements.

Every car should be designed with the following *crash safety principles* in mind:

1   The impact from a collision is absorbed gradually by controlled deformation of the outer parts of the car body.

Figure 1.7   Style artist at work (*Ford Motor Company Ltd*)

2 The passenger area is kept intact as long as possible.

3 The interior is designed to reduce the risk of injury.

Safety-related vehicle laws cover design, performance levels and the associated testing procedures; requirements for tests, inspections, documentation and records for the process of approval; checks that standards are being maintained during production; the issue of safety-related documentation; and many other requirements throughout the vehicle's service life.

Primary or active safety

This refers to the features designed into the vehicle which reduce the possibility of an accident. These include primary design elements such as dual-circuit braking systems, anti-lock braking systems, high aerodynamic stability and efficient bad-weather equipment, together with features that make the driver's environment safer, such as efficient through ventilation, orthopaedic seating, improved all-round vision, easy to read instruments and ergonomic controls.

An anti-lock braking system (ABS) enhances a driver's ability to steer the vehicle during hard braking. Sensors monitor how fast the wheels are rotating and feed data continuously to a microprocessor in the vehicle to signal that a wheel is approaching lockup. The computer responds by sending a signal to apply and release brake pressure as required. This pumping action continues as long as the driver maintains adequate force on the brake pedal and impending wheel lock condition is sensed.

The stability and handling of the vehicle are affected by the width of the track and the position of the centre of gravity. Therefore the lower the centre of gravity and the wider the track the more stable is the vehicle.

Secondary or passive safety

If a crash does happen, secondary safety design should protect the passengers by

1 Making sure that, in the event of an accident, the occupants stay inside the car

2 Minimizing the magnitude and duration of the deceleration to which they are subjected

3 Restraining the occupants so that they are not injured by secondary impacts within the car,

and, if they do strike parts of the inside of the vehicle, making sure that there is sufficient padding to prevent serious injury

4 Designing the outside of the vehicle so that the least possible injury is caused to pedestrians and others who may come into contact with the outside of the vehicle.

The primary concern is to develop efficient restraint systems which are comfortable to wear and easy to use. Manufacturers are now fitting automatic seatbelt tensioners. These automatic 'body lock' front seatbelt tensioners reduce the severity of head injuries by 20 per cent with similar gains in chest protection. In impacts over 12 mile/h (20 km/h) the extra tension in the seatbelt buckle triggers a sensor which tightens the lap and diagonal belts in 22 milliseconds, that is before the occupant even starts to move. In addition, because it operates at low speeds, it covers a broad spectrum of accident situations. Anti-submarining ramps built into the front seats further aid safety by reducing the possibility of occupants sliding under the belt (Figure 1.8).

There are also engineering features such as impact energy-absorbing steering columns, head restraints, bumpers, anti-burst door locks, and self-aligning steering wheels. Anti-burst door locks are to prevent unrestrained occupants from falling out of the vehicle, especially during roll-over. The chances of survival are much reduced if the occupant is thrown out. Broad padded steering wheels are used to prevent head or chest damage. Collapsible steering columns also prevent damage to the chest and abdomen and are designed to

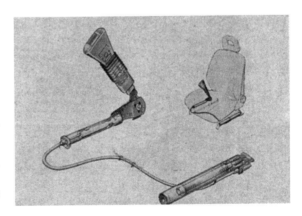

Figure 1.8 Automatic seatbelt tensioner (*Vauxhall Motors Ltd*)

prevent the steering column being pushed back into the passenger compartment whilst the front end is crumpling. The self-aligning steering wheel is designed to distribute force more evenly if the driver comes into contact with the steering wheel during a crash. This steering wheel has an energy-absorbing hub which incorporates six deformable metal legs. In a crash, the wheel deforms at the hub and the metal legs align the wheel parallel to the chest of the driver to help spread the impact and reduce chest, abdomen and facial injuries.

Body shells are now designed to withstand major collision and rollover impacts while absorbing shock by controlled deformation of structure in the front and rear of the vehicle. Vehicle design and accident prevention is based on the kinetic energy relationship of damage to a vehicle during a collision. Energy is absorbed by work done on the vehicle's materials by elastic deformation. This indicates that, to be effective, bumpers and other collision-absorbing parts of a vehicle should be made of materials such as foam-filled plastics and heavy rubber sections. Data indicates that long energy-absorbing distances should be provided in vehicle design, and the panel assemblies used for this purpose should have a lower stiffness than the central section or passenger compartment of the vehicle. The *crumple zones* are designed to help decelerate the car by absorbing the force of collision at a controlled rate, thereby cushioning the passengers and reducing the risk of injury (Figure 1.9). The *safety cage* (or safety cell) is the central section of the car body which acts as the passenger compartment. To ensure passenger safety, all body apertures around the passenger area should be reinforced by box-type profiles; seats should be secured rigidly to the floor; and heavy interior padding should be used around the dashboard areas. A strengthened roof construction, together with an anti-roll bar, afford additional protection in case of overturning (Figure 1.10).

To counteract side impact manufacturers are now fitting, in both front and rear doors, lateral side supports in the form of twin high-strength steel tubular beams, which are set 90 mm apart to reduce the risk of the vehicle riding over the beams during side collision. These beams absorb the kinetic energy produced when the vehicle is struck from the side. To further improve the body structure the BC-pillars are being reinforced at the points of attachment to the sill and roof, again giving more strength to the safety cage and making it stronger and safer when the vehicle is involved in collision (Figure 1.11a, b).

Visibility in design is the ability to see and be seen. In poor visibility and after dark, light sources must be relied upon. The lights on vehicles now are much more efficient than on earlier models. The old tungsten filament lamp has given way to quartz-halogen lamps which provide much better illumination. The quartz-halogen lamp is able to produce a more powerful beam because the filament can be made hotter without shortening its lifespan. Hazard, reversing and fog lights are now fitted to most vehicles to improve safe driving.

In daylight, colour is probably the most important factor in enabling cars to be seen. If a vehicle is coloured towards the red end of the spectrum, it can be less obvious to other road users than a yellow one, especially in sodium vapour street lights: a red car absorbs yellow light from the street light and

Figure 1.9 Crumple zones (*Volvo Concessionaires Ltd*)

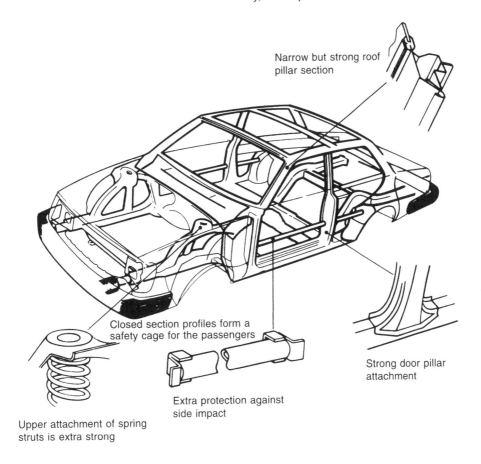

Narrow but strong roof pillar section

Closed section profiles form a safety cage for the passengers

Strong door pillar attachment

Extra protection against side impact

Upper attachment of spring struts is extra strong

Safety cage protects the occupants of the car

Figure 1.10   Safety cage (*Volvo Concessionaires Ltd*)

reflects little, and so appears to be dark in colour, whereas a yellow car reflects the yellow light and appears more obvious. Silver vehicles will blend into mist and fog and become difficult to see.

Blind spots can be diminished firstly by good design of front pillars, making them slim and strong, and secondly by reducing the area of rear quarter sections. This elimination of blind spots is now being achieved by using bigger windscreens which wrap round the front A-post, and rear windows which wrap round the rear quarter section, giving a wider field of vision.

Many automotive manufacturers now believe that a seatbelt/airbag combination provides the best possible interior safety system. Airbags play an important safety role in the USA since the wearing of seatbelts is not compulsory in many of the states. As competition to manufacture Europe's safest car

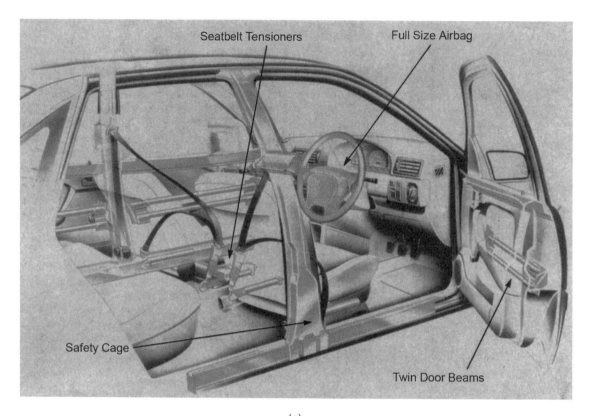

(a)

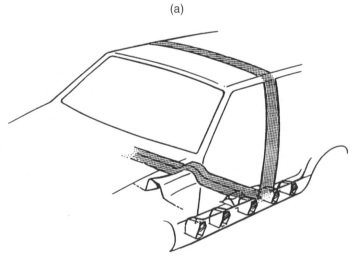

Improvements to a conventional uni-body.
Shaded areas indicate reinforcements

(b)

Figure 1.11 (a) Safety features included in the safety cage (*Vauxhall Motors Ltd*) (b) Reinforced BC-pillar and anti-roll bar (*Volvo Concessionaires Ltd*)

Figure 1.12 Driver's airbag system (*Du Pont (UK) Ltd*)

increases, more manufacturers including those in the UK are starting to fit airbags. These Eurobags, or facebags as they are now called, since their main function in Europe and the UK is to protect the face rather than the entire body in the event of collision, are less complex than their USA counterparts.

The first automotive airbags were made more than 20 years ago using nylon-based woven fabrics, and these remain the preferred materials among manufacturers. Nylon fabrics for airbags are supplied in two basic designs depending on whether the airbag is to protect the driver or the front passenger. The driver's airbag is housed in the steering wheel and requires special attention because of the confined space (Figure 1.12). The passenger's airbag system has a compartment door, located in front of the passenger in the dash area, which must open within 10 milliseconds and deploy the airbag within 30 milliseconds. The vehicle has a crash sensor which signals the airbags to deploy on impact (Figure 1.13).

1.2.3 Production of models

Scale models

Once the initial designs have been accepted, scale models are produced for wind tunnel testing to determine the aerodynamic values of such a design. These models are usually constructed of wood and clay to allow for modifications to be made easily. At the same time, design engineering personnel construct models of alternative interiors so that locations of instruments can be determined.

A $\frac{1}{4}$ or $\frac{3}{8}$ scale model is produced from the stylist's drawings to enable the stylist designer to evaluate the

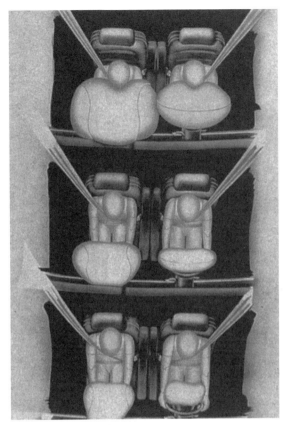

Figure 1.13 Driver and front passenger airbag systems in use (*Du Pont (UK) Ltd*)

three-dimensional aspect of the vehicle. These scale models can look convincingly real (Figure 1.14). The clay surfaces are covered with thin coloured plastic sheet which closely resembles genuine painted metal. Bumpers, door handles and trim strips are all cleverly made-up dummies, and the windows are made of Plexiglass. The scale models are examined critically and tested. Changes to the design can be made at this stage.

Full-size models

A full-size clay model is begun when the scale model has been satisfactorily modified. It is constructed in a similar way to the scale model but uses a metal, wood and plastic frame called a *buck*. The clay is placed on to the framework by professional model makers, who create the final outside shape of the body to an accuracy of 0.375 mm. The high standard of finish and detail results in an exact replica of the future full-size vehicle (Figure 1.15).

Figure 1.14 Scale model maker at work (*Ford Motor Company Ltd*)

This replica is then evaluated by the styling management and submitted to top management for their approval. The accurate life-size model is used for further wind tunnel testing and also to provide measurements for the engineering and production departments. A scanner, linked to a computer, passes over the entire body and records each and every dimension (Figure 1.16). These are stored

Figure 1.15 Full-size clay model (*Ford Motor Company Ltd*)

Figure 1.16 Checking dimensional accuracy of the full-size model (*Ford Motor Company Ltd*)

Figure 1.17 Interior styling model
(*Ford Motor Company Ltd*)

and can be produced on an automatic drafting machine. The same dimensions can also be projected on the screen of a graphics station; this is a sophisticated computer-controlled video system showing three-dimensional illustrations, allowing design engineers either to smooth the lines or to make detail alterations. The use of computers or CAD allows more flexibility and saves a lot of time compared with the more conventional drafting systems.

At the same time as the exterior model is being made, the interior model is also being produced accurately in every detail (Figure 1.17). It shows the seating arrangement, instrumentation, steering wheel, control unit location and pedal arrangements. Colours and fabrics are tried out on this mock-up until the interior styling is complete and ready for approval.

1.2.4 Engine performance and testing

Development engineers prepare to test an engine in a computer-linked test cell to establish the optimum settings for best performance, economy and emission levels. With the increasing emphasis on performance with economy, computers are used to obtain the best possible compromise. They are also used to monitor and control prolonged engine testing to establish reliability characteristics. If current engines and transmissions are to be used for a new model, a programme of refining and adapting for the new installation has to be initiated. However, if a completely new engine, transmission or driveline configuration is to be adopted, development work must be well in hand by this time.

1.2.5 Aerodynamics and wind tunnel testing

Aerodynamics is an experimental science whose aim is the study of the relative motions of a solid body and the surrounding air. Its application to the design of a car body constitutes one of the chief lines of the search for energy economy in motor vehicles.

In order to move over flat ground, a car must overcome two forces:

1 Resistance to tyre tread motion, which varies with the coefficient of tyre friction over the ground and with the vehicle's mass.
2 Aerodynamic resistance, which depends on the shape of the car, on its frontal area, on the density of the air and on the square of the speed.

One of the objects of aerodynamic research is to reduce the latter: in other words to design a shape that will, for identical performance, require lower energy production. An aerodynamic or streamlined body allows faster running for the same consumption of energy, or lower consumption for the same speed. Research for the ideal shape is done on reduced-scale models of the vehicle. The models are placed in a *wind tunnel*, an experimental installation producing wind of a certain quality and fitted with the means for measuring the various forces due to the action of the wind on the model or the vehicle. Moreover, at a given cruising speed, the more streamlined vehicle has more power left available for acceleration: this is a safety factor.

The design of a motor car body must, however, remain compatible with imperatives of production, of overall measurements and of inside spaciousness. It is also a matter of style, for the coachwork must be attractive to the public. This makes it impossible to apply the laws of aerodynamics literally. The evolution of the motor car nevertheless tends towards a gradual reduction in aerodynamic resistance.

Aerodynamic drag

The force which opposes the forward movement of an automobile is aerodynamic drag, in which air rubs against the exterior vehicle surfaces and forms disturbances about the body, thereby retarding forward movement. Aerodynamic drag increases with speed; thus if the speed of a vehicle is doubled, the corresponding engine power must be increased by eight times. Engineers express the magnitude of aerodynamic drag using the *drag coefficient* C_d. The coefficient expresses the aerodynamic efficiency of the vehicle: the smaller the value of the coefficient, the smaller the aerodynamic drag.

Figure 1.18 illustrates the improvements in aerodynamic drag coefficient achieved by alterations to the shape of vehicles. Over the years, the value of C_d has been reduced roughly as follows:

1910	0.95	1960	0.40
1920	0.82	1970	0.36
1930	0.56	1980	0.30
1940	0.45	1990	0.22
1950	0.42	1993	0.20

During the wind tunnel test all four wheels of the car rest on floating scales connected to a floor balance, which has a concrete foundation below the main floor area. The vehicle is then subjected to an air stream of up to 112 mile/h; the sensitive balances register the effect of the headwind on the vehicle as it is either pressed down or lifted up from the floor, pushed to the left or right, or rotated about its longitudinal axis. The manner in which the forces affect the vehicle body and the location at which the forces are exerted depends upon the body shape, underbody contours and projecting parts. The fewer disturbances which occur as air moves past the vehicle, the lower its drag. Threads on the vehicle exterior as well as smoke streams indicate the air flow, and enable test engineers to see where disturbance exists and where air flows are interrupted or redirected, and therefore where reshaping of the body is necessary in order to produce better aerodynamics (Figures 1.19 and 1.20).

1.2.6 Prototype production

The new model now enters the prototype phase. The mock-ups give way to the first genuine road going vehicle, produced with the aid of accurate drawings and without complex tooling and machinery. The prototype must accurately reproduce the exact shape, construction and assembly conditions of the final production body it represents if it is to be of any value in illustrating possible manufacturing problems and accurate test data. The process begins with the issue of drawing office instructions to the experimental prototype workshop. Details of skin panels and other large pressings are provided in the form of tracings or as photographic reproductions of the master body drafts. As the various detailed parts are made, by either simple press tools or traditional hand methods, they are spot welded into minor assemblies or subassemblies; these later become part of a major assembly to form the completed vehicle body.

1.2.7 Prototype testing

Whilst still in the prototype stage, the new car has to face a number of arduous tests. For these tests a mobile laboratory is connected to the vehicle by a cable, which transmits signals from various sensors on the vehicle back to the onboard computer for collation and analysis. The prototype will also be placed on a computer-linked simulated rig to monitor, through controlled vibrations, the stresses and strains experienced by the driveline, suspension and body.

Crash testing (Figure 1.21) is undertaken to establish that the vehicle will suffer the minimum of damage or distortion in the event of an impact and that the occupants are safely installed within the strong passenger compartment or safety cell. The basic crash test is a frontal crash at 30 mile/h (48 km/h) into a fixed barrier set perpendicularly

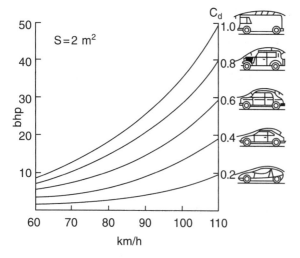

Figure 1.18 Theoretical drag curves for four types of vehicle, all reduced for comparison purposes to a front section of 2 m². Since air resistance increases in proportion to the square of the speed, a truck with C_d 1.0 requires 35 bhp at 100 km/h, whereas a coupé with C_d 0.2 requires only 7 bhp

Figure 1.19 Wind tunnel testing of a prototype: front view (*Ford Motor Company Ltd*)

Figure 1.20 Wind tunnel testing of a prototype: side view (*Ford Motor Company Ltd*)

to the car's longitudinal axis. The collision is termed 100 per cent overlap, as the complete front of the car strikes the barrier and there is no offset (Figure 1.22). The main requirement is that the steering wheel must not be moved back by more than 120 mm (5 in), but there is no requirement to measure the force to which the occupants will be subject in collision. The manufacturers use anthropometric dummies suitably instrumented with decelerometers and strain gauges which collect

(a)

(b)

(c)

(d)

Figure 1.21 Basic frontal crash and side impact (angled side swipe) tests (*Vauxhall Motors Ltd*)

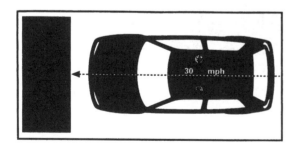

Figure 1.22 Standard frontal impact test

relevant data on the effect of the collision on the dummies. A passenger car side impact test aimed at reducing chest and pelvic injuries will be legal in the USA from 1993. This stricter standard requires that a new vehicle must pass a full-scale crash test designed to simulate a collision at an intersection in which a car travelling at 15 mile/h is hit in the side by another car travelling at 30 mile/h. This test is called an angled side-swipe: the displacement is 27 degrees forward from the perpendicular of the test vehicle's main axis. The test is conducted by propelling a movable deformable barrier at 33.5 mile/h into the side of a test car occupied by dummies in the front and rear seats. The dummies are wired with instruments to predict the risk potential of human injury. Volvo do a very unusual promotional crash test which involves propelling a car from the top of a tall building (Figure 1.23).

Figure 1.23 Volvo crash test
(*Volvo Concessionaires Ltd*)

Extensive durability tests are undertaken on a variety of road surfaces in all conditions (Figure 1.24). Vehicles are also run through water tests (Figure 1.25) and subjected to extreme climatic temperature changes to confirm their durability.

Figure 1.24 Road testing a prototype
(*Ford Motor Company Ltd*)

The final stages are now being reached; mechanical specifications, trim levels, engine options, body styles and the feature lists are confirmed.

1.2.8 Body engineering for production

The body engineering responsibilities are to simulate the styling model and overall requirements laid down by the management in terms of drawings and specification. The engineering structures are designed for production, at a given date, at the lowest possible tooling cost and to a high standard of quality and reliability.

As competition between the major car manufacturers increases, so does the need for lighter and more effective body structures. Until recently the choice of section, size and metal gauges was based upon previous experience. However, methods have now been evolved which allow engineers to solve problems with complicated geometry on a graphical display computer which can be constructed to resemble a body shape (Figure 1.26). The stiffness and stress can then be computed from its geometry, and calculations made of the load bearing of the structures using finite-element methods (Figure 1.27).

With the final specifications approved, the new car is ready for production. At this stage an initial batch of cars is built (a *pilot run*) to ensure that the plant facilities and the workforce are ready for the start of full production. When the production line begins to turn out the brand new model, every stage of production is carefully scrutinized to ensure quality in all the vehicles to be built.

1.2.9 EuroNCAP

The governments in most countries have some form of regulations covering vehicle safety. These regulations are aimed at giving both the occupants of the vehicle protection in the case of an accident, and ensuring that pedestrians and cyclists are not subject to unnecessary injury if they come into contact with a car. The regulations are in most cases very minimal. In the UK the Department for Transport (DfT) works with a number of bodies on vehicle safety, much of the DfT work is sub-contracted to Transport Research Laboratory (TRL) Ltd – formerly a wholly government funded institution. In America there is the United States Department of Transportation (DOT). There is also the EEVC (European Enhanced Vehicle-safety Committee).

The most pro-active of vehicle safety organisations is EuroNCAP. The full title is European New Car Assessment Programme. This programme is jointly funded and supported by it members which includes:

- Allgemeiner Deutscher Automobil-Club e V (ADAC), motoring organisation – Germany
- Bundesministerium fur Verker, Bau- und Wohnungswesen, government department – Germany

Figure 1.25 Water testing a prototype (*Ford Motor Company Ltd*)

Figure 1.26 Three-dimensional graphics display of a scale model (*Ford Motor Company Ltd*)

- Department for Transport (DfT), government department – UK
- Dutch Ministry of Transport, Public Works and Water Management, government department – Holland

- European Commission – Belgium
- FIA Foundation for the Automobile and Society, motoring organisation – UK
- Government of Catalonia, government department – Spain
- International Consumer Research and Testing, consumer group – UK
- Ministere de l'Equipment, government department – France
- Swedish Road Administration, government department – Sweden
- Thatcham – representing British Motor Insurers – GB

EuroNCAP have a number of tests which vehicles are subjected to, the results of the tests are then subjected to a number of calculations which lead to star ratings. Basically the more stars the safer the vehicle (Figure 1.28). The tests appear simple; but the recording of results is quite complex. Each test has a 50-page operating manual. Readings are taken from the dummies inside the vehicles as well as the photographs of the vehicle as it deforms under impact. The dummies contain electrical sensing equipment, mainly measuring acceleration rates. Each dummy costs the same amount of money as a super car such as a Ferrari.

The EuroNCAP tests are designed to encourage vehicle manufacturers to consider safety standards above and beyond those required by the government regulations.

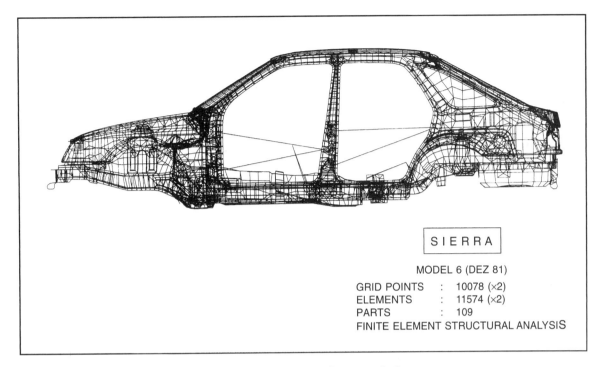

SIERRA

MODEL 6 (DEZ 81)
GRID POINTS   :   10078 (×2)
ELEMENTS   :   11574 (×2)
PARTS   :   109
FINITE ELEMENT STRUCTURAL ANALYSIS

Figure 1.27   Finite-element structural analysis (*Ford Motor Company Ltd*)

Front impact test

Frontal impact takes place at 64 kph (40 mph) when a car strikes a deformable barrier that is offset (Figure 1.28a). This test is similar to many road accidents where one car hits another car, or another object, offset to one side.

Side impact test

This is similar to accidents where the car is hit by another on the side. The impact takes place at 50 kph (30 mph) when a trolley with a deformable front is towed into the driver's side of the car to simulate a side-on crash (see Figure 1.28c).

Pole test

Accident patterns vary from country to country within Europe, but approximately a quarter of all serious-to-fatal injuries happen in side impact collisions. Many of these injuries occur when one car runs into the side of another. To encourage manufacturers to fit head protection devices, an optional pole or head protection test may be performed, where such safety features are fitted to the vehicle. Side impact airbags help protect the head by providing a padding effect and by preventing the head from passing through the window opening (see Figure 1.28e).

In this test, the car being tested is propelled sideways at 29 kph (18 mph) into a rigid pole. The pole is relatively narrow, like a telegraph pole of lamp post, so there is major penetration into the side of the car. In an impact without the head protecting airbag, a driver's head could hit the pole with sufficient force to cause a fatal head injury.

Pedestrian impact test

A series of tests (Figure 1.28b) are carried out to replicate accidents involving child and adult pedestrians where impact occurs at 40 kph (25 mph) – maximum speed in build up areas in France. Impact sites are then assessed and rated fair, weak and poor. As with the other tests, these are based on EEVC guidelines.

Star ratings

Each vehicle tested is given a star rating for its protection of:

- adult occupant
- child occupant
- pedestrians

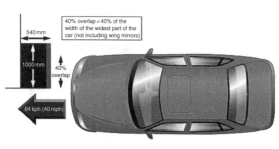

(a) Frontal impact test

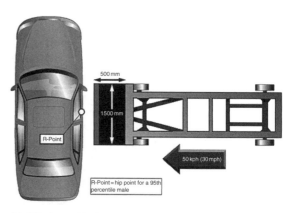

(b) Side impact test

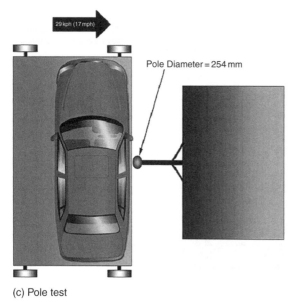

(c) Pole test

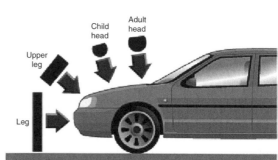

(d) Pedestrain impact test

Figure 1.28 EuroNCAP

The Star Ratings are based on calculations carried out after the tests, the latest figures can be found on the EuroNCAP website: www.euroncap.com.

1.2.10 Computational fluid dynamics

Definition

Any material which flows, such as air or water, can be referred to as a fluid; 'dynamics' simply means moving and 'computational' is about calculations. Aerodynamics was for many years about observing the air flow over a vehicle, with sample calcula-

tions for specific areas of the car; the advent of computers meant that calculations could be done many times faster than by long hand. So, it became possible to carry out calculations for large sections of the car very quickly.

In the 1970s engineers became interested in the aerodynamics which were taking place both underneath and inside the car, places which could not be seen. More recently software has been developed, such as AutoCAD, 3D Studio and Pro/ENGINEER, which allows solid modelling and the facility to virtually walk through a design. This means that there is now no longer a need to make a buck, or a mock-up, of a car to be able to visualize the design. As an example, vehicle body engineers used to use a wooden buck of an engine to help them to design the body work, and see if it could be fitted to and removed from the initial body design. The Rover 75 is the first car to be designed without this; a solid modelling package was used for a virtual engine and body design, allowing an onscreen test fit.

Having designed the virtual car it is possible to observe it in a virtual wind tunnel and carry out calculations both internally and externally, this is what computational fluid dynamics (CFD) is all about. As you can appreciate, the cost of a virtual design and aerodynamic testing without a wind tunnel is only a fraction of the cost of building a buck and using a real wind tunnel.

The calculations

The following are some of the calculations which an aerodynamist may be concerned with; it should be remembered that these calculations are often carried out on about 10 000 000 (ten million) individual grid squares, or cells, on the car body, a slight change of design will need a new set of calculations. Even using the latest computer software, the slightest change may take several days; without CFD it would take months, and the level of accuracy would be much less. If you choose to use any of these formulae, remember to use SI units, metres, newtons and seconds where appropriate.

Dynamic pressure, which is also a kinetic energy of unit volume in terms of cubic metres, comes from the Bernoulli equations. Bernoulli was a scientist whose fluid flow theories were first used in the design of ships' hulls.

Dynamic pressure
$$= 1/2 \times \text{air density} \times \text{vehicle velocity squared}$$
$$= 1/2\,\rho V^2$$

As you can see, the speed (velocity) of the vehicle is important for these calculations. Of course velocity is a vector quantity, it is related to the direction of the wind. Wind is very rarely a straight-on head wind, so calculations can be done for any of the 360 degree possible wind directions

for each of the ten million grid squares. Yes, that is 3.6 billion calculations for each speed and of course the air density varies with altitude; at sea level the value is 1.226 kilogrammes per cubic metre.

Reynolds Number is a ratio which gives a good guide to the air flow pattern and is an important consideration of what is called scale effect.

Reynolds Number
$$= \text{air density} \times \text{air velocity}$$
$$\times \text{ length of flow/air Viscosity}$$

$$\text{Re} = \rho v l / \mu$$

Drag is the aerodynamic resistance of the vehicle, its resistance to pass through air. Drag in newtons force is found by the formula:

Drag $= 1/2$ air density $\times$ velocity squared
$$\times \text{ frontal area} \times \text{coefficient of drag}$$
$$= 1/2\rho V^2 A C_D$$

You will see that part of the formula is familiar, and part of it is the same as dynamic pressure, therefore:

Drag $=$ dynamic pressure $\times$ frontal area
$$\times \text{ coefficient of drag}$$

The coefficient of drag is a number which indicates the resistance of the car to pass through the air, typical values are between 0.25 and 0.35.

Lift is the force generated by an aerofoil section normal to the direction of fluid flow. In other words it is the upward lifting force which is generated when passing horizontally through air. For road vehicles wings are used to hold a vehicle on to the road, this can be called downthrust or negative lift.

Lift $=$ dynamic pressure $\times$ wing area
$$\times \text{ coefficient of lift}$$

When working with road vehicles the frontal area is often used for the wing area figure. On aircraft, the wing plan area is more appropriate. With advanced aerodynamic work the plan area is related to a reference area. For most road vehicles the frontal area and the plan area are proportional; also the coefficient of lift and the coefficient of drag are also proportional.

$$\text{Lift} = 1/2\rho V^2 A C_L$$

Grid system

The vehicle body shape is broken into grid squares, or cells, see Figure 1.29. Depending on the shape of the panel, the grid cells may be of different shapes between square and oblong.

The squares are then considered as imaginary cube shapes, see Figure 1.30. A set of calculations called the Navier–Stokes equations gives a relationship between pressure, momentum and viscous forces in three-dimensional space. There is also a similar set called the Euler equations. The above calculations covered some of these concepts. The computer is used to calculate the amount of energy which is entering each cube and in turn leaving it. Obviously the two figures should balance and there will be flow between adjacent cubes.

Ahmed model

For benchmark testing of CFD systems the simplified vehicle shape, known as the Ahmed model, is used, see Figure 1.31. This is a simplified model of a hatchback car. The Ahmed model can be made from a wooden block and used in any wind tunnel.

1.3 Methods of construction

The steel body can be divided into two main types: those which are mounted on a separate chassis frame, and those in which the underframe or floor forms an integral part of the body. The construction of today's mass-produced motor car has

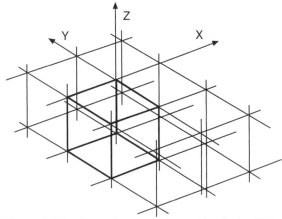

Figure 1.30 One cube shape in a grid cell, the CFD calculates the energy of entering and leaving each cube (*Dr Barnard 1996*)

changed almost completely from the composite, that is conventional separate chassis and body, to the integral or mono unit. This change is the result of the need to reduce body weight and cost per unit of the total vehicle.

1.3.1 Composite construction (conventional separate chassis)

The chassis and body are built as two separate units (Figure 1.32). The body is then assembled on to the chassis with mounting brackets, which

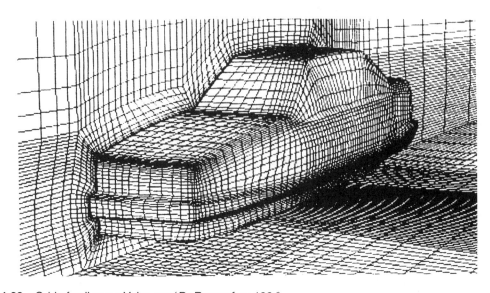

Figure 1.29 Grid of cells on a Volvo car (*Dr Ramnefors 1994*)

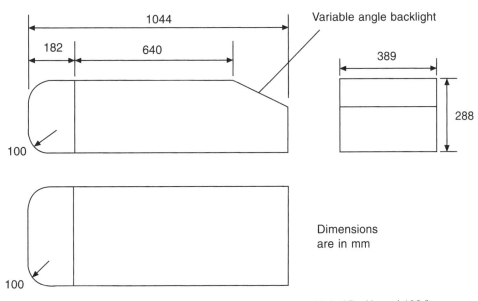

Figure 1.31 The Ahmed Model, a simplified shape of a hatchback vehicle (*Dr Ahmed 1984*)

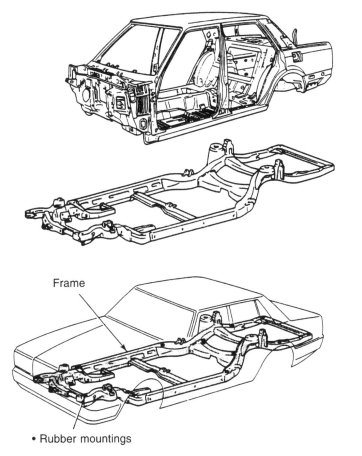

Figure 1.32 Composite construction (conventional separate chassis)

have rubber-bushed bolts to hold the body to the rigid chassis. These flexible mountings allow the body to move slightly when the car is in motion. This means that the car can be dismantled into the two units of the body and chassis. The chassis assembly is built up of engine, wheels, springs and transmission. On to this assembly is added the body, which has been pre-assembled in units to form a complete body shell (Figure 1.33).

1.3.2 Integral (mono or unity) construction

Integral body construction employs the same principles of design that have been used for years in the aircraft industry. The main aim is to strengthen without unnecessary weight, and the construction does not employ a conventional separate chassis frame for attachment of suspension, engine and other chassis and transmission components (Figure 1.34). The major difference between composite and integral construction is hence the design and construction of the floor (Figure 1.35). In integral bodies the floor pan area is generally called the underbody. The underbody is made up of

formed floor sections, channels, boxed sections, formed rails and numerous reinforcements. In most integral underbodies a suspension member is incorporated in both the front and rear of the body. The suspension members have very much the same appearance as the conventional chassis frame from the underside, but the front suspension members end at the cowl or bulkhead and the rear suspension members end just forward of the rear boot floor. With the floor pan, side rails and reinforcements welded to them, the suspension members become an integral part of the underbody, and they form the supports for engine, front and rear suspension units and other chassis components. In the integral body the floor pan area is usually of heavier gauge metal than in the composite body, and has one or more box sections and several channel sections which may run across the floor either from side to side or from front to rear; this variety of underbody construction is due largely to the difference in wheelbase, length and weight of the car involved. A typical upper body for an integral constructed car is very much the same as the conventional composite body shell; the major differences lie in the rear seat area and the construction which joins the front wings to the front

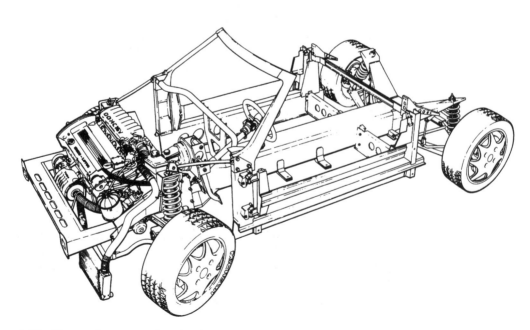

Figure 1.33 Composite construction showing a Lotus Elan chassis before fitting the body (*Lotus Engineering*)

bulkhead or cowl assembly. The construction in the area to the rear of the back seat is much heavier in an integral body than in a composite body. The same is true of the attaching members for the front wings, front bulkhead and floor assembly, as these constructions give great strength and stability to the overall body structure.

1.3.3 Semi-integral methods of construction

In some forms of integral or mono assemblies, the entire front end or subframe forward of the bulkhead is joined to the cowl assembly with bolts. With this construction, the bolts can be easily removed and the entire front (or in some cases rear) subframe can be replaced as one assembly in the event of extensive damage.

1.3.4 Glass fibre composite construction

This method of producing complex shapes involves applying layers of glass fibre and resin in a prepared mould. After hardening, a strong moulding is produced with a smooth outer surface requiring little maintenance. Among the many shapes available in this composite material are lorry cabs, bus front

canopies, container vehicles, and the bodies of cars such as the Reliant Scimitar. The Italian designer, Michelotti, styled the Scimitar body so that separately moulded body panels could be used and overlapped to hide the attachment points. This allows the panels to be bolted directly to the supporting square-section steel tube armatures located on the main chassis frame. The inner body, which rests directly on the chassis frame and which forms the base for all internal trim equipment, is a complex GRP moulding. The windscreen aperture is moulded as a part of the inner body, and incorporates steel reinforcing hoops which are braced directly to the chassis. The boot compartment is also a separate hand-laid GRP moulding, as are the doors and some of the other panels. Most of the body panels are secured by self-tapping bolts which offer very positive location and a useful saving in assembly time (see Figures 1.36 and 1.37).

1.3.5 Galvanized body shell clad entirely with composite skin panels

Renault have designed a high-rise car which has a skeletal steel body shell (Figure 1.38, clad entirely with composite panels. After assembly the complete body shell is immersed in a bath of

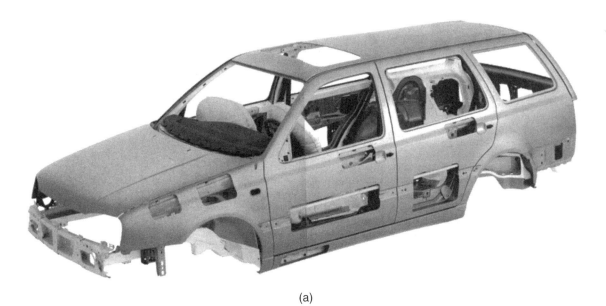

(a)

Figure 1.34 (a) VW Golf Estate body assembly

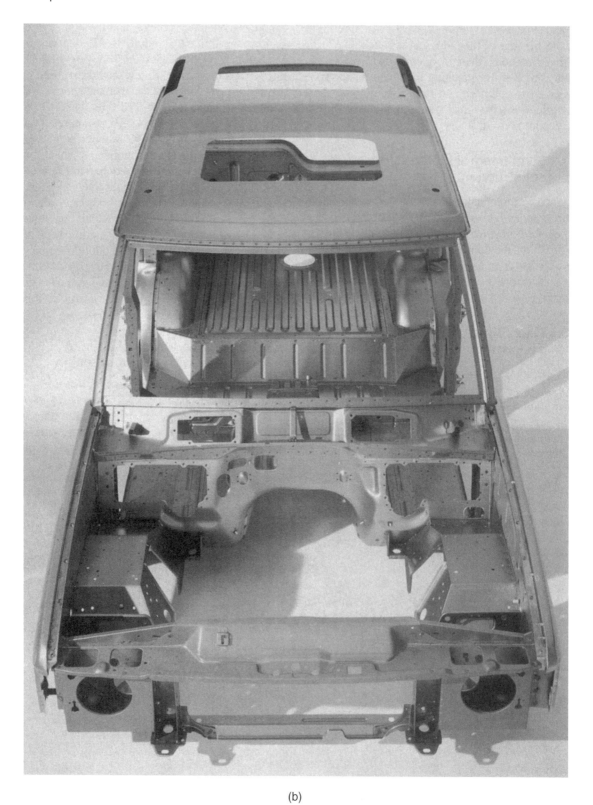

Figure 1.34 (b) Land Rover Discovery body assembly

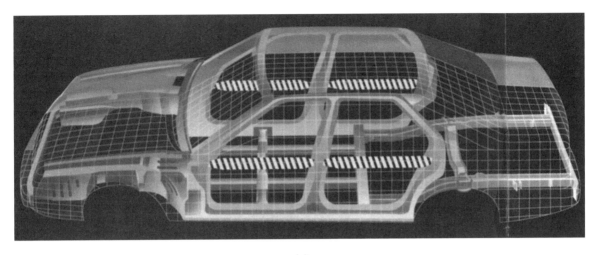

(c)

Figure 1.34   (c) Rover 800 load levelling body parts

molten zinc, which applies an all-over 6.5 micron (millionth of a metre) coating. The process gives anti-rust protection, while the chemical reaction causes a molecular change in the steel which strengthens it. Lighter-guage steel can therefore be used without sacrificing strength, resulting in a substantial weight saving even with the zinc added.

Skin panels are formed in reinforced polyester sheet, made of equal parts of resin, fibreglass and mineral filler. The panels are joined to the galvanized frame and doors by rivets or bonding as appropriate. The one-piece high-rise tailgate is fabricated entirely from polyester with internal steel reinforcements (Figure 1.39). Damage to panels through impact shocks is contained locally and absorbed through destruction of the material, unlike the steel sheet which transmits deformation. Accident damage and consequent repair costs are thus reduced.

1.3.6 Variations in body shape

Among the motor car manufacturers there are variations in constructional methods which result in different body types and styles. Figure 1.40 illustrates four types of body shell – a saloon with a boot, a hatchback, an estate car and a light van. Figure 1.41 shows a coach-built limousine of extremely high quality, built on a Rolls-Royce Silver Spirit chassis by the coach-builders Hooper & Co.

This vehicle has been designed for the use of heads of state and world-ranking VIPs.

1.4 Basic body construction

A typical four-door saloon body can be likened to a hollow tube with holes cut in the sides. The bulkhead towards the front and rear completes the box-like form and assists in providing torsional stability. The roof, even if it has to accommodate a sunshine roof, is usually a quite straightforward and stable structure; the curved shape of the roof panel prevents lozenging (going out of alignment in a diamond shape). The floor is a complete panel from front to rear when assembled, and is usually fitted with integral straightening ribs to prevent lozenging. With its bottom sides or sill panels, wheel arches, cross members and heelboard, it is the strongest part of the whole body. The rear bulkhead, mainly in the form of a rear squab panel, is again a very stable structure. However, the scuttle or forward bulkhead is a complex structure in a private motor car. Owing to the awkward shape of the scuttle and the accommodation required for much of the vehicle's equipment, it requires careful designing to obtain sufficient strength. Body sides with thin pillars, large windows and door openings are inherently weak, requiring reinforcing with radiusing corners to the apertures to give them sufficient constructional strength.

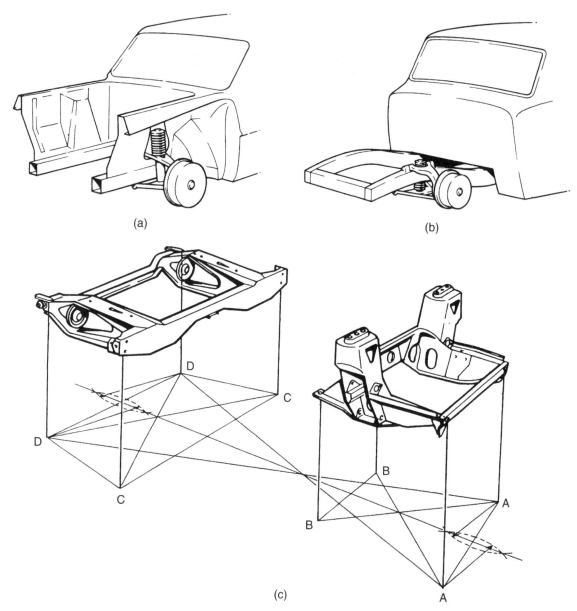

(a)

(b)

(c)

Figure 1.35 Front end construction (a) integral or mono (b) composite and (c) front sub-frame, this is bolted separately to body assembly

A designer in a small coach building firm will consider methods necessary to build the body complete with trim and other finishing processes. The same job in a mass production factory may be done by a team of designers and engineers all expert in their own particular branch of the project. The small manufacturer produces bodies with skilled labour and a minimum number of jigs, while the mass producer uses many jigs and automatic processes to achieve the necessary output. However, the problems are basically the same: to maintain strength and stability, a good standard of finish and ease of production.

Figure 1.42 shows the build-up details of a four-door saloon, from the main floor assembly to the complete shell assembly. In the figure the main floor

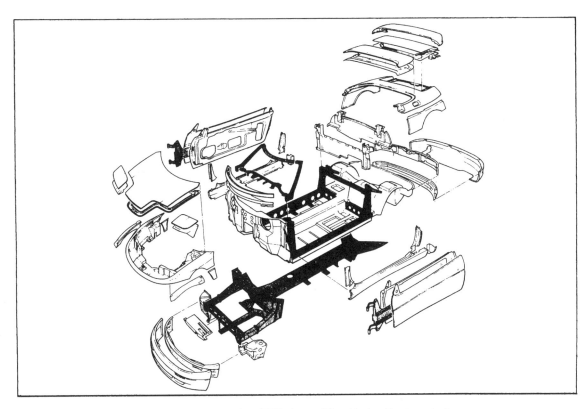

Figure 1.36   Motor body panel assembly using GRP: Lotus Elan (*Lotus Engineering*)

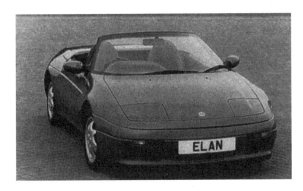

Figure 1.37   Complete Lotus Elan SE body shell (*Lotus Engineering*)

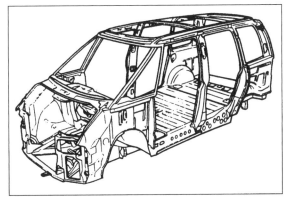

Figure 1.38   Espace high-rise car with galvanized skeletal steel body shell (*Renault UK Ltd*)

unit (1), commencing at the front, comprises a toe-board or pedal panel, although in some cases this may become a part of the scuttle or bulkhead. Apart from providing a rest for the front passengers' feet, it seals off the engine and gearbox from the body

and connects the scuttle to the main floor. The main centre floor panel (2) should be sufficiently reinforced to carry the weight of the front seats and passengers. It may be necessary to have a tunnel running the length of the floor in the centre to clear

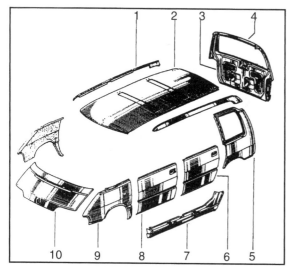

Plastic structure

Figure 1.39 Espace high-rise car showing composite panel cladding (*Renault UK Ltd*). Plastic parts are made from a composite material based on polyester resin: pre-impregnated type (SMC) for parts 1, 4, 5, 6, 7, 8, 9, 10; injected resin type for parts 2, 3. Parts bonded to chassis: Detachable parts:

1	Body top	3	Tailgate lining
2	Roof	4	Tailgate outer panel
5	Rear wing	6	Rear door panel
9	Front wing	8	Front door panel
7	Sill	10	Bonnet

the transmission system from the engine to the rear axle, and holes may have to be cut into the floor to allow access to the gearbox, oil filler, and dipstick, in which case removable panels or large grommets would be fitted in these access holes (3).

The front end of the main floor is fixed to the toe-board panel and the sides of the main centre floor are strengthened by the bottom sills (4) and/or some form of side members which provide the necessary longitudinal strength. The transverse strength is provided by the cross members. The floor panel itself prevents lozenging, and the joints between side members and cross members are designed to resist torsional stresses.

The rear end of the floor is stiffened transversely by the rear seat heelboard (5). This heelboard also stiffens the front edges of the rear seat panel. In addition it often provides the retaining lip for the rear seat cushion, which is usually made detachable

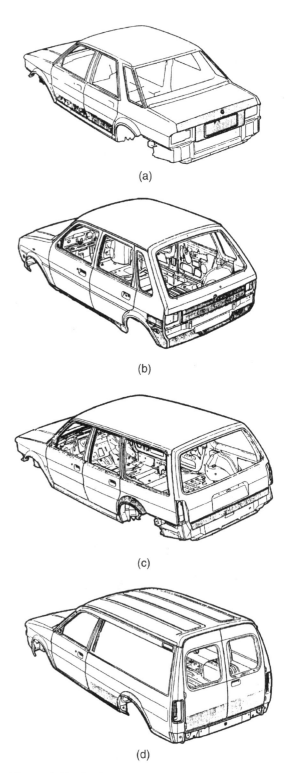

Figure 1.40 Body shell variations: (a) saloon with boot (b) hatchback (c) estate car (d) light van (*Rover Group Ltd*)

Figure 1.41 Coach-built limousine: Emperor State Landaulette (*Hooper & Co (Coach-builders) Ltd*)

from the body. The heelboard, together with the rear panel and rear squab panel, forms the platform for the rear seat.

The rear seat panel (6) is reinforced or swaged if necessary to gain enough strength to support the rear passengers. Usually the rear seat panel has to be raised to provide sufficient clearance for the deflection of the rear axle differential housing. The front edge of the rear seat panel is stiffened by the rear seat heelboard, and the rear edge of the seat panel is stiffened by the rear squab panel. The rear squab panel completes this unit and provides the rear bulkhead across the car. It seals off the boot or luggage compartment from the main body or passenger compartment.

The boot floor (7), which extends from the back of the rear squab panel to the extreme back of the body, completes the floor unit. In addition to the luggage, the spare wheel has to be accommodated here. The front edge of the boot floor is reinforced by the rear squab panel and the rear end by a cross member of some form (8). The sides of the floor are stiffened by vertical boot side panels at the rear, while the wheel arch panels complete the floor structure by joining the rear end of the main floor and its side members. The wheel arch panels (9) themselves seal the rear road wheels from the body.

In general the floor unit is made up from a series of panels with suitable cross members or reinforcements. The edges of the panels are stiffened either by flanging reinforcing members, or by joining to the adjacent panels. The boot framing is joined at the back to the rear end of the boot floor, at the sides to the boot side panels and at the top to the shelf panel behind the rear squab (10). It has to be sufficiently strong at the point where the boot lid hinges are fitted to carry the weight of the boot lid when this is opened. Surrounding the boot lid opening there is a gutter to carry away rain and water to prevent it entering the boot; opposite the hinges, provision is made for the boot lid lock striking plate (11) to be fixed. From the forward edge of the boot, the next unit is the back light and roof structure (12), and this extends to the top of the windscreen or canopy rail (13). The roof is usually connected to the body side frames, which comprise longitudinal rails or stringers and a pair of cantrails which form the door openings (14). Provision in the roof should be made for the interior lights and wiring and also the fixing of the interior trimming. The scuttle and windscreen unit, including the front standing pillar or A-post (15), provides the front bulkhead and seals the engine from the passenger compartment.

Accommodation has to be made for the instrumentation of the car, the wiring, radio, windscreen wipers and driving cable, demisters and ducting, steering column support, handbrake support and pedals. The scuttle (16) is a complicated structure which needs to be very strong. When the front door is hinged at the forward edge, provision has to be made in the front pillar for the door hinges, door check and courtesy light switches.

The centre standing pillar or BC-post (17) is fixed to the side members of the main floor unit and supports the cantrails of the roof unit. It provides a shut face for the front door, a position for the door lock striking plate and buffers or dovetail, and also a hinge face for the rear door; as with the front standing pillar, provision is made for the door hinges and door check. The rear standing pillar or D-post (18) provides the shut face for the rear end of the floor side members at the bottom, whilst the top is fixed to the roof cantrails and forms the front of the quarters.

The quarters (19) are the areas of the body sides between the rear standing pillars and the back light and boot. If the body is a six-light saloon there will be a quarter window here with its necessary surrounding framing, but in the case of a four-light saloon this portion will be more simply constructed. Apart from the doors, bonnet, boot lid and

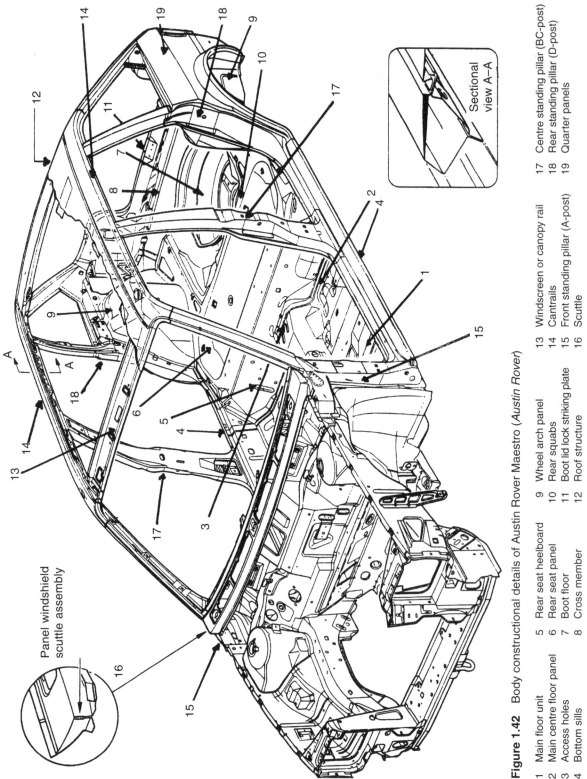

Figure 1.42 Body constructional details of Austin Rover Maestro (*Austin Rover*)

1 Main floor unit	5 Rear seat heelboard	9 Wheel arch panel	13 Windscreen or canopy rail	17 Centre standing pillar (BC-post)
2 Main centre floor panel	6 Rear seat panel	10 Rear squabs	14 Cantrails	18 Rear standing pillar (D-post)
3 Access holes	7 Boot floor	11 Boot lid lock striking plate	15 Front standing pillar (A-post)	19 Quarter panels
4 Bottom sills	8 Cross member	12 Roof structure	16 Scuttle	

Sectional view A–A

Panel windshield scuttle assembly

front wings this completes the structure of the average body shell.

1.5 Identification of major body pressings

The passenger-carrying compartment of a car is called the body, and to it is attached all the doors, wings and such parts required to form a complete body shell assembly (Figure 1.43).

1.5.1 Outer construction

This can be likened to the skin of the body, and is usually considered as that portion of a panel or panels which is visible from the outside of the car.

1.5.2 Inner construction

This is considered as all the brackets, braces and panel assemblies that are used to give the car strength (Figure 1.44). In some cases the entire panels are inner construction on one make of car and a combination of inner and outer on another.

1.5.3 Front-end assembly including cowl or dash panel

The front-end assembly (Figure 1.45) is made up from the two front side member assemblies which are designed to carry the weight of the engine, suspension, steering gear and radiator. The suspension system used will affect the design of the panels, but whatever system is used the loads must be transmitted to the wing valances and on to the body panels. The front cross member assembly braces the front of the car and carries the radiator and headlamp units. The side valance assemblies form a housing for the wheels, a mating edge for the bonnet and a strong box section for attachment of front wings. Both the side frames and valance assemblies are connected to the cowl or dash panel, The front-end assembly is attached to the main floor at the toe panel.

The cowl or dash panel forms the front bulkhead of the body (Figure 1.45) and is usually formed by joining smaller panels (the cowl upper panel and the cowl side panel) by welds to form an integral unit. In some cases the windscreen frame is integral with the cowl panel. The cowl extends upwards

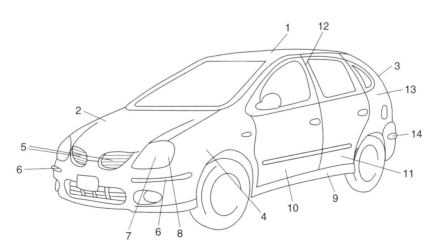

Figure 1.43   Major body panels

1	Roof panel	8	Sidelamps
2	Bonnet panel	9	Sill panel
3	Boot lid	10	Front door
4	Front wing	11	Rear door
5	Radiator grille	12	Centre pillar
6	Front bumper bar	13	Rear quarter panel
7	Headlamps	14	Rear bumper bar

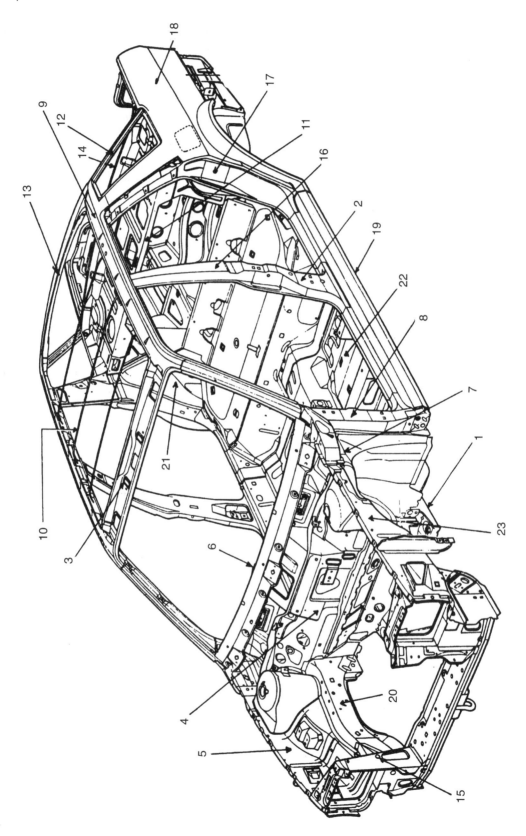

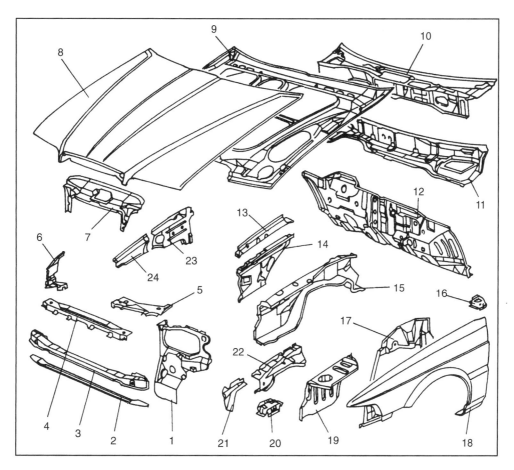

(Facing page)

Figure 1.44 Body shell assembly (*Austin Rover Group Ltd*)

1　Underbody assembly
2　Body side frame assembly
3　Windscreen upper rail assembly
4　Cowl and dash panel assembly
5　Front wheel house complete panel
6　Instrument panel assembly
7　Cowl side lower brace
8　Front body hinge pillar (A-post)
9　Roof panel assembly
10　Roof bow assembly
11　Bulkhead brace assembly
12　Rear quarter centre panel assembly (back window)
13　Back window upper rail panel assembly
14　Rear-end upper panel assembly
15　Radiator panel complete assembly
16　Centre pillar (BC-post)
17　D-post
18　Rear quarter assembly
19　Sill panel
20　Front side member assembly
21　Rear wheel arch assembly
22　Main floor assembly
23　Front valance complete assembly

Figure 1.45 Complete front-end assemblies (*Rover Group Ltd*)

1　Headlamp panel RH and LH
2　Front cross member closing panel
3　Front cross member
4　Bonnet lock panel
5　Headlamp panel reinforcement RH and LH
6　Front wing corner piece RH and LH
7　Bonnet frame extension
8　Bonnet skin
9　Bonnet frame
10　Dash panel
11　Scuttle panel
12　Front bulkhead
13　Chassis leg reinforcement RH and LH
14　Front inner wing RH and LH
15　Front chassis leg RH and LH
16　Subframe mounting RH and LH
17　Front wheel arch RH and LH
18　Front wing RH and LH
19　Battery tray
20　Chassis leg gusset RH and LH
21　Bumper mounting reinforcement RH and LH
22　Chassis leg extension RH and LH
23　A-post rear reinforcement RH and LH
24　A-post front reinforcement RH and LH

around the entire windscreen opening so that the upper edge of the cowl panel forms the front edge of the roof panel. In this case the windscreen pillars, i.e. the narrow sloping construction at either side of the windscreen opening, are merely part of the cowl panel. In other constructions, only a portion of the windscreen pillar is formed as part of the cowl. The cowl is sometimes called the fire wall because it is the partition between the passenger and engine compartments, and openings in the cowl accommodate the necessary controls, wiring and tubing that extend from one compartment to the other. The instrument panel, which is usually considered as part of the cowl panel although it is a complex panel in itself, provides a mounting for the instruments necessary to check the performance of the vehicle during operation. Cowl panels usually have both inner and outer construction, but in certain constructions only the upper portion of the cowl around the windscreen is visible. On many vehicles the front door hinge pillar is also an integral part of the cowl.

1.5.4 Front side member assembly

This is an integral part of the front-end assembly; it connects the front wing valances to the cowl or dash assembly. It is designed to strengthen the front end; it is part of the crumple zone, giving lateral strength on impact and absorbing energy by deformation during a collision. It also helps to support the engine and suspension units (see Figure 1.45; key figure references 13, 15, 16, 20, 22).

1.5.5 A-post assembly

This is an integral part of the body side frame. It is connected to the front end assembly and forms the front door pillar or hinge post. It is designed to carry the weight of the front door and helps to strengthen the front bulkhead assembly (Figure 1.45).

1.5.6 Main floor assembly

This is the passenger-carrying section of the main floor. It runs backwards from the toe panel to the heelboard or back seat assembly. It is strengthened to carry the two front seats, and in some cases may have a transmission tunnel running through its centre. Strength is built into the floor by the transmission tunnel acting like an inverted channel section. The body sill panels provide extra reinforcement in the form of lateral strength. Transverse strength is provided by box sections at right angles to the transmission tunnel, generally in the areas of the front seat and in front of the rear seat. The remaining areas of flat floor are ribbed below the seats and in the foot wells to add stiffness (Figure 1.46).

1.5.7 Boot floor assembly

This is a section of the floor between the seat panel and the extreme back of the boot. It is strengthened by the use of cross members to carry the rear seat passengers. This area forms the rear bulkhead between the two rear wheel arches, forming the rear seat panel or heelboard, and in a saloon body shell can incorporate back seat supports and parcel shelf. The boot floor is also strengthened to become the luggage compartment, carrying the spare wheel and petrol tank. At the extreme back it becomes the panel on to which the door or tailgate closes (Figure 1.46).

1.5.8 Complete underbody assembly

This is commonly called the floor pan assembly, and is usually composed of several smaller panels welded together to form a single floor unit. All floor panels are reinforced on the underside by stiffening members or cross members. Most floor pans are irregular in shape for several reasons. They are formed with indentations or heavily swaged areas to strengthen the floor sections between the cross members, and foot room for the passengers is often provided by these recessed areas in the floor. Figure 1.46 shows a complete underbody assembly.

1.5.9 Body side frame assembly

On a four-door saloon this incorporates the A-post, the BC-post, the D-post and the rear quarter section. The side frames reinforce the floor pan along the sill sections. The hinge pillar or A-post extends forward to meet the dash panel and front bulkhead to provide strength at this point. The centre pillars or BC-posts connect the body sills to the roof cantrails. They are usually assembled as box sections using a top-hat section and flat plate. These are the flanges which form the attachments for the door weather seals and provide the four

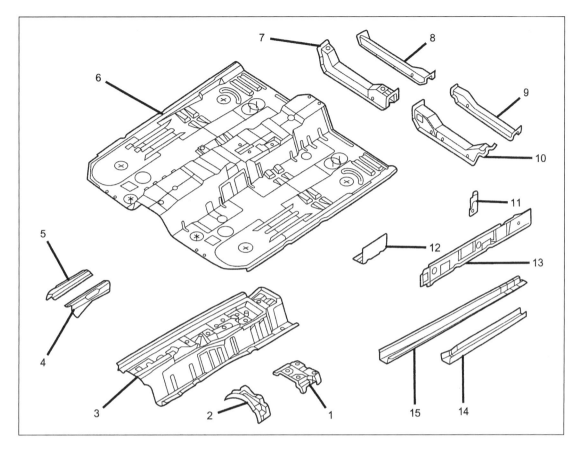

Figure 1.46   Main floor assemblies and boot floor assemblies (*Proton*)

1 Reinf. parking brake lever	9 Crossmember assy front floor rear LH
2 Crossmember assy backbone	10 Crossmember assy front floor front LH
3 Reinf. assy backbone	11 Reinf. seat belt side LH/RH
4 Bracket A-Frame LH	12 Bracket anti zipper
5 Bracket A-Frame RH	13 Sill front floor side inner LH/RH
6 Pan front floor	14 Reinf. sidemember front floor LH/RH
7 Crossmember assy front floor front RH	15 Sidemember front floor LH/RH
8 Crossmember assy front floor rear RH	

NOTE:
A large reinforcement has been added to the front floor backbone (Reinforcement assy backbone), and this is coupled with the side sill, sidemember and crossmember to provide increased rigidity to the total floor.

door openings. The D-post and rear quarter section is integral with the rear wheel arch and can include a rear quarter window (Figure 1.47).

1.5.10 Roof panel

The roof panel is one of the largest of all major body panels, and it is also one of the simplest in construction. The area which the roof covers varies between different makes and models of cars. On some cars, the roof panel ends at the windscreen. On others it extends downwards around the windscreen so that the windscreen opening is actually in the roof. On some cars the roof ends above the rear window, while on others it extends downwards so that the rear window opening is in the lower rear roof. When this is the case the roof panel forms the top panel

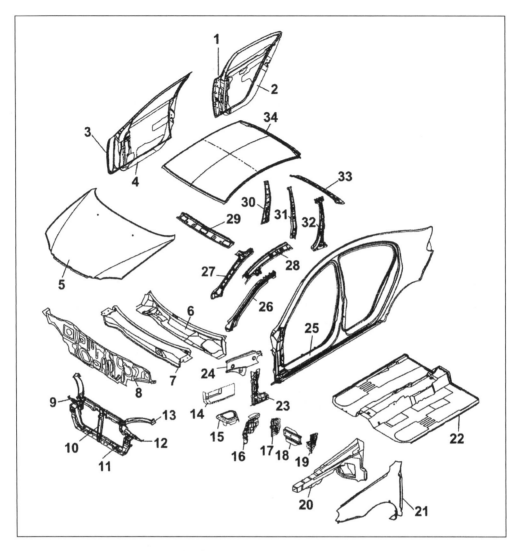

Figure 1.47 Body side assemblies, roof, BC-post, front and rear door of a hatchback (*Proton*)

1	Panel outer rear door	18	Reinf. pillar front inner lower
2	Panel inner rear door	19	Pillar front inner lower
3	Panel outer front door	20	Sidemember front
4	Panel inner rear door	21	Panel front fender
5	Panel hood	22	Pan front floor
6	Panel cowl top inner	23	Reinf. front pillar lower
7	Panel cowl top outer	24	Extension upper frame outer
8	Panel assy dash	25	Side structure
9	Reinf. radiator side RH	26	Reinf. front pillar centre
10	Stay hood lock	27	Pillar front inner upper
11	Crossmember front end	28	Rail roof side inner
12	Reinf. radiator side LH	29	Rail roof front
13	Bar front end upper	30	Pillar centre inner
14	Bulkhead front pillar lower	31	Pillar centre outer
15	Bulkhead front pillar side sill	32	Pillar rear inner
16	Bracket crossmember front	33	Rail roof rear
17	Bracket crossmember	34	Panel roof

around the rear boot opening. Some special body designs incorporate different methods of rear window construction, which affects the roof panel; this is particularly true for estate cars, hatchbacks and hardtop convertibles. Alternatively the top is joined to the rear quarter panel by another smaller panel which is part of the roof assembly.

The stiffness of the roof is built in by the curvature given to it by the forming presses, while the reinforcements, consisting of small metal strips placed crosswise to the roof at intervals along the inside surface, serve to stiffen the front and rear edges of the windscreen and rear window frames. In some designs the roof panel may have a sliding roof built in (Figure 1.47) or a flip-up detachable sunroof incorporated.

1.5.11 Rear quarter panel or tonneau assembly

This is integral with the side frame assembly and has both inner and outer construction. The inner construction comprises the rear wheel arch and the rear seat heelboard assembly. This provides the support for the rear seat squab in a saloon car; if the vehicle is a hatchback or estate car, the two back seats will fold flat and the seat squabs will not need support. This area is known as the rear bulkhead of the car; it gives additional transverse strength between the wheel arch sections and provides support for the rear seat. The rear bulkhead also acts as a partition between the luggage and passenger compartments (Figure 1.47).

1.5.12 Rear wheel arch assembly

This assembly is constructed as an integral part of the inner construction of the rear quarter panel. It is usually a two-piece construction comprising the wheel arch and the quarter panel, which are welded together (Figure 1.46).

1.5.13 Wings

A wing is a part of the body which covers the wheel. Apart from covering the suspension construction, the wing prevents water and mud from being thrown up on to the body by the wheels. The front wings (or the fender assembly) are usually attached to the wing valance of the front end assembly (see Figure 1.45) by means of a flange the length of the wing, which is turned inwards from the outer surface and secured by either welding or bolts. Adjustment for the front wing is usually provided for by slotting the bolt holes so that the wing can be moved either forwards or backwards by loosening the attaching bolts. This adjustment cannot be made if the wing is welded to the main body structure.

In some models the headlights and sidelights are recessed into the front wing and fastened in place by flanges and reinforcement rims on the wing. Any trim or chrome which appears on the side of the wing is usually held in place by special clips or fasteners which allows easy removal of the trim.

The unsupported edges of the wing are swaged edges known as beads. The bead is merely a flange which is turned inwards on some cars and then up to form a U-section with a rounded bottom. It not only gives strength but prevents cracks developing in the edges of the wing due to vibration, and it provides a smooth finished appearance to the edge of the wing.

In general the rear wing is an integral part of the body side frame assembly and rear quarter panel. When the wing forms an integral part of the quarter panel, the inner construction is used to form part of the housing around the wheel arch. The wheel arch is welded to the rear floor section and is totally concealed by the rear quarter panel, while the outer side of the wheel arch is usually attached to the quarter panel around the wheel opening. This assembly prevents road dirt being thrown upwards between the outer panel and inner panel construction.

1.5.14 Doors

Several types of door are used on each vehicle built, although the construction of the various doors is similar regardless of the location of the door on the vehicle, as indicated on Figure 1.47. The door is composed of two main panels, an outer and an inner panel, both being of all-steel construction. The door derives most of its strength from the inner panel since this is constructed mainly to act as a frame for the door. The outer panel flanges over the inner panel around all its edges to form a single unit, which is then spot welded or, in some cases, bonded with adhesives to the frame.

The inner panel has holes or apertures for the attachment of door trim. The trim consists of the window regulator assembly and the door locking mechanism. These assemblies are installed through the large apertures in the middle of the inner panel. Most of the thickness of the door is due to the depth of the inner panel which is necessary to accommodate the door catch and window mechanism. The inner panel forms the lock pillar and also the hinge pillar section of the door. Small reinforcement angles are usually used between the outer and inner panel, both where the lock is inserted through the door and where the hinges are attached to the door. The outer panel is either provided with an opening through which the outside door handle protrudes, or is recessed to give a more streamlined effect and so to create better aerodynamics.

The upper portion of the door has a large opening which is closed by glass. The glass is held rigidly by the window regulator assembly, and when raised it slides in a channel in the opening between the outer and inner panels in the upper portion of the door. When fully closed the window seats tightly in this channel, effectively sealing out the weather.

1.5.15 Boot lid or tailgate

This is really another door which allows access to the luggage compartment in the rear of the car (Figure 1.46). A boot lid is composed of an outer and an inner panel. These panels are spot welded along their flanged edges to form a single unit in the same manner as an ordinary door. The hatchback and estate car have a rear window built into the boot lid, which is then known as a tailgate. Some manufacturers use external hinges, while others use concealed hinges attached to the inner panel only. A catch is provided at the lower rear edge of the boot lid or tailgate and is controlled by an external handle or locking mechanism. This mechanism may be concealed from the eye under a moulding or some type of trim. In some models there is no handle or external locking mechanism; instead the hinges are spring loaded or use gas-filled piston supports, so that when

the lid is unlocked internally it automatically rises and is held in the open position by these mechanisms.

1.5.16 Bonnet

The bonnet (Figure 1.45) is the panel which covers the engine compartment where this is situated at the front of the vehicle, or the boot compartment of a rear-engined vehicle. Several kinds of bonnets are in use on different makes of cars. The bonnet consists of an outer panel and an inner reinforcement constructed in the H or cruciform pattern, which is spot welded to the outer skin panel at the flanged edges of the panels. The reinforcement is basically a top-hat section, to give rigidity to the bonnet. In some cases the outer panel is bonded to the inner panel using epoxy resins. This system avoids the dimpling effect on the outer surface of the bonnet skin which occurs in spot welding.

Early models used a jointed type of bonnet which was held in place by bolts through the centre section of the top of the bonnet into the body of the cowl and into the radiator. A pianotype hinge was used where the bonnet hinged both at the centre and at the side.

The most commonly used bonnet on later constructions is known as the mono or one-piece type, and can be opened by a variety of methods. On some types it is hinged at the front so that the rear end swings up when the bonnet is open. Others are designed so that they can be opened from either side, or unlatched from both sides and removed altogether. Most bonnets, however, are of the alligator pattern, which is hinged at the rear so that the front end swings up when opened.

The type of bonnet catch mechanism depends on the type of bonnet used. When a bonnet opens from the rear the catch mechanism is also at the rear. When it opens from either side the combination hinge and catch are provided at each side. The alligator bonnets have their catches at the front, and in most cases the catches are controlled from inside the car.

Bonnets are quite large, and to make opening easier the hinges are usually counterbalanced by means of tension or torsion springs. Where smaller bonnets are used the hinges are not counterbalanced and the bonnet is held in place by a bonnet stay from the side of the wing to the bonnet. Adjustment of the bonnet position is sometimes possible by moving the hinges.

1.5.17 Trims

Some details of exterior and interior trims are shown in Figures 1.48 and 1.49.

1.5.18 Complete body shell

A contemporary vehicle embracing all the latest techniques of panel assembly is shown in Figures 1.50 and 1.51. Figure 1.50 illustrates the completed structure with all panel assemblies in place. Figure 1.51 shows the completely finished vehicle ready for the road.

1.5.19 Comparative terms in common use by British, American and European car manufacturers

As manufacturers use differing terms for the various body panel assemblies and individual panels, difficulties may arise when identifying specific panels. The following are the terms in most common use:

Bonnet, hood
Boot lid, deck lid, trunk lid, tailgate
Cantrail, roof side rail, drip rail
Centre pillar, BC-post
Courtesy light, interior light
Cowl, scuttle, bulkhead, fire wall
Dash panel, facia panel
Door opening plates, scuff plates
Door skin, outside door panel

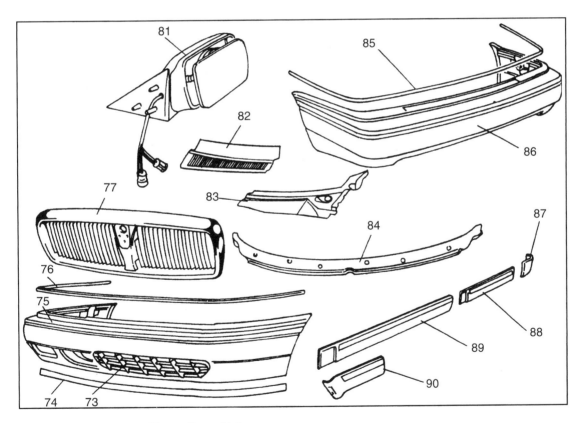

Figure 1.48 Exterior trim (*Rover Group Ltd*)

73 Lower front grille	77 Front grille	84 Lower screen moulding	88 Rear door waist moulding
74 Front spoiler	81 Door mirror assembly	85 Rear bumper insert	89 Front door waist moulding
75 Front bumper	82 Scuttle grille	86 Rear bumper	90 Front wing waist moulding
76 Front bumper insert	83 Scuttle moulding	87 Rear wing waist moulding	

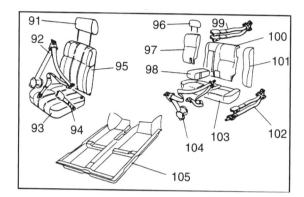

Figure 1.49 Interior trim (*Rover Group Ltd*)

91 Front seat headrest
92 Front seatbelt assembly
93 Front seat cushion
94 Front seatbelt centre stalk
95 Front seat back rest
96 Front seat head rest
97 Rear seat back rest small section
98 Rear seat centre arm rest
99 Rear seatbelt buckle assembly
100 Rear seat back rest large section
101 Rear seat side bolster
102 Rear seatbelt lap assembly
103 Rear seat cushion
104 Rear seatbelt assembly
105 Main floor carpet

Face bar, bumper bar
Front pillar, A-post, windscreen pillar
Light, window
Quarter panel, tonneau assembly
Roof, turret
Roof lining, headlining
Sill panel, rocker panel
Squab, seat back
Underbody, floor pan assembly
Valance of front wing, fender side shield
Vent window, flipper window
Waist rail, belt rail
Wheel arch, wheel house
Windscreen, windshield
Wing, fender.

1.5.20 Vehicle identification numbers

The vehicle identification number (VIN) is stamped on a plate located typically inside the engine compartment or on a door pillar. A VIN system is shown in Figure 1.52. The figure also shows the paint and trim codes which are usually included on the VIN plate.

The car body number is provided separately in the engine or boot compartments.

1.6 Vehicle type approval

Many industrial sectors are subject to some form of approval or certification system but road vehicles are a special case, because of their importance to and impact upon society, and have been subject to specific technical standards almost from their first invention. Within Europe, two systems of type approval have been in existence for over 20 years. One is based around EC Directives and provides for the approval of whole vehicles, vehicle systems and separate components. The other is based around ECE (United Nations) Regulations and provides for approval of vehicle systems and separate components, but not whole vehicles.

Type approval is the confirmation that production samples of a design will meet specified performance standards. The specification of the product is recorded and only that specification is approved.

Automotive EC Directives and ECE Regulations require third party approval – testing, certification and production conformity assessment by an independent body. Each member state is required to appoint an Approval Authority to issue the approvals and a Technical Service to carry out the testing to the Directives and Regulations. An approval issued by one Authority will be accepted in all the member states. Vehicle Certification Agency (VCA) is the designated UK Approval Authority and Technical Service for all type of approvals to automotive EC Directives and ECE Regulations. The VCA also has offices in Europe, North America and the far East.

Approved parts carry the E-mark. That is a letter 'e' or 'CE' followed by a number which indicates the country of approval.

1.6.1 Single vehicle approval scheme

The Single Vehicle Approval (SVA) scheme is a pre-registration inspection for cars and light goods vehicles that have not been type-approved to British or European standards. The main purpose

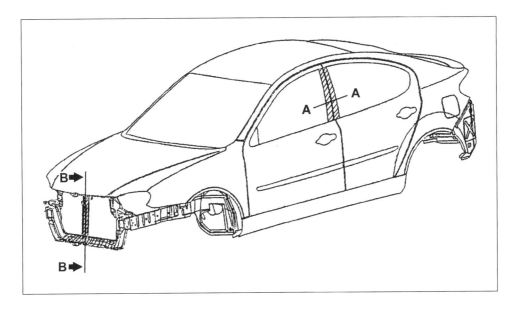

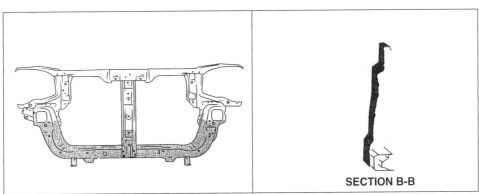

SECTION B-B

Figure 1.50   Complete body shell (*Proton*)

Figure 1.51   Proton

of the scheme is to ensure that these vehicles have been designed and constructed to modern safety and environmental standards before they can be used on public roads.

Single Vehicle Approval checks that vehicles constructed for non-European Economic Area markets comply with British law. Even vehicles outwardly similar to European-specification models, but intended for other markets, can often be unsuitable for use in Britain without, at least some, modification. SVA recognises certain non-European technical standards as acceptable alternatives to the SVA requirements.

Vehicle Information Code Plate — for UK/EC market only

Location

Vehicle information code plate is riveted on the toe board inside the engine compartment.

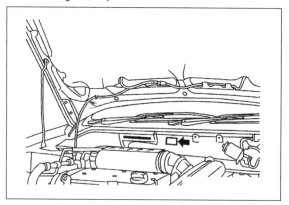

CODE PLATE DESCRIPTION

The plate shown model code, engine model, transmission model and colour code.

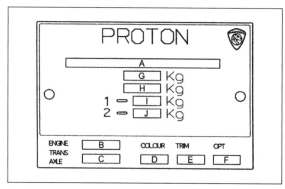

Position	Description
A	VIN CODE
B	ENGINE
C	TRANS AXLE
D	EXTERIOR CODE
E	INTERIOR CODE
F	OPTION CODE
G	GROSS VEHICLE WEIGHT
H	GROSS COMBINATION WEIGHT
I	FRONT AXLE WEIGHT
J	REAR AXLE WEIGHT

Figure 1.52 Vehicle identification number

Single Vehicle Approval also checks that the construction of amateur-built vehicles, rebuilt vehicles and vehicles using parts from a previously registered vehicle meet modern safety and environmental standards. It also provides an alternative to type-approval for vehicles manufactured in very low volume; vehicles converted for the disabled prior to registration, as well as hearses and armoured vehicles for civilian use.

Most of the items inspected in the SVA scheme are those that are tested when manufacturers apply for type-approval of mass-produced vehicles. Table 1.1 lists the items inspected for cars and light goods vehicles for the two levels of SVA. All the items for Standard SVA are checked at the SVA test station. However, items for ESVA (other than seatbelts) cannot be tested at the SVA test station and therefore documentary evidence of compliance as to be produced. The SVA test station will then check that the vehicle aligns to the documentation presented.

Acceptable alternatives are listed in the SVA Inspection Manual which is produced by VOSA.

For standard SVA, all the inspections are undertaken at selected testing stations operated by the Vehicle and Operator Services Agency (VOSA), an executive agency of The Department for Transport. These tests can also be conducted by VOSA Examiners at 'designated premises' if certain conditions are met. Designated premises are privately owned premises that have been authorized by VOSA for their examiners to use to conduct SVA tests.

Although the same items are tested as in type-approval they are not tested in the same way. To keep the fee to a level affordable by individuals the examiner will in the case of certain items conduct an engineering assessment. This check will be in the form of either visual inspection or a simple test on the vehicle to ensure that it complies with the regulations.

The examination will be limited to parts of the vehicle which can be readily seen without dismantling. However, the presenter may be asked to open lockable compartments and remove engine covers, inspection/access panels, trims or carpeting. This is to allow access to items subject to examination. Also, because vehicles are inspected individually the tests are not destructive or damaging in any way.

A Minister's Approval Certificate (MAC) under SVA is issued when the examiner is satisfied that the vehicle would meet the requirements of the regulations in relation to the design and construction of the vehicle. Unlike the MOT, the examination is not primarily concerned with vehicle

Table 1.1

Item	Standard SVA		Additional items for enhanced SVA	
	Cars	Light goods vehicles	Cars	Light goods vehicles
Doors, their latches and hinges	Yes	No	No	No
Radio interference suppression	Yes	Yes	No	No
Protective steering	Yes	No	Evidence of compliance with type-approval standard or acceptable alternative	No
Exhaust emissions	Yes*	Yes*	Evidence of compliance with type-approval standard or acceptable alternative May require independent test	Evidence of compliance with type-approval standard or acceptable alternative May require independent test
Smoke emissions (diesels only)	Yes	Yes	No	No
Lamps, reflectors and devices	Yes	No	No	No
Rear-view mirrors	Yes	No	No	No
Anti-theft devices	Yes*	No	Evidence of compliance with type-approval standard or acceptable alternative May require independent test	No
Seat belts	Yes	No	Evidence of compliance or visual assessment at test site	No
Seat belt anchorages	Yes*	No	Evidence of compliance with full type-approval standard or acceptable alternative	No
Installation of seat belts	Yes	No	Evidence of compliance with type-approval standard or acceptable alternative	No
Brakes	Yes	Yes	Evidence of compliance with type-approval standard or acceptable alternative	Evidence of compliance with type-approval standard or acceptable alternative

Table 1.1   (*continued*)

Item	Standard SVA		Additional items for enhanced SVA	
	Cars	Light goods vehicles	Cars	Light goods vehicles
Noise and silencers	Yes*	Yes*	Evidence of compliance with type-approval standard or acceptable alternative May require independent test	Evidence of compliance with type-approval standard or acceptable alternative May require independent test
Glass: windscreen and other windows outside	Yes	No	No	No
Seats and their anchorages	Yes	No	No	No
Tyres	Yes	No	No	No
Interior fittings	Yes	No	No	No
External projections	Yes	No	No	No
Speedometers	Yes	No	No	No
Wiper and washer system	Yes	No	No	No
Defrosting and demisting system	Yes	No	No	No
Fuel input	Yes	Yes	No	No
Design weights	Yes	No	No	No
General vehicle construction	Yes	Yes	No	No
CO_2 emissions and fuel consumption (vehicles manufactured after 1 January 1997)	No	No	Evidence of compliance with full type-approval standard or acceptable alternative	No Does not apply to LGVs
CO_2 emissions and fuel consumption (vehicles manufactured after 1 January 1997)	No	No	Evidence of compliance with full type-approval standard or acceptable alternative	No Does not apply to LGVs
Front impact protection (vehicles manufactured after 1 October 2003)	No	No	Evidence of compliance with type-approval standard or acceptable alternative N.B. May require comparison test against EC approved vehicle	No

Table 1.1   (*continued*)

Item	Standard SVA		Additional items for enhanced SVA	
	Cars	Light goods vehicles	Cars	Light goods vehicles
Plate for goods vehicles	No	Yes	No	No
Side Impact Protection (vehicles manufactured after 1 October 2003)	No	No	Evidence of compliance with type-approval standard or acceptable alternative N.B. May require comparison test against EC approved vehicle	No

* Compliance with the enhanced requirements shall exempt the vehicles from these SVA items.

condition, although poor condition may make it impossible to assess whether some construction requirements are met.

Questions

1   Why were the earliest motor vehicle bodies made almost entirely of wood?

2   When and why did manufacturers commence to use metal for the construction of vehicle bodies?

3   Give a brief history of the development of the vehicle body style, illustrating the significant changes which have taken place.

4   What is meant by monocoque construction, and why has it become so popular in motor vehicle manufacture?

5   With the aid of sketches, describe the general principles of monocoque construction.

6   Describe, with the aid of sketches, the general principles of composite and integral methods of body construction.

7   Draw a sketch of a vehicle body shell and name all the major body panels.

8   State the location and function on a vehicle body of the following sections: (a) BC-post (b) quarter panel (c) wheel arch (d) bonnet.

9   What is the most common form of vehicle body construction?

10   What are the alternatives to integral construction?

11   What is a load-bearing stressed panel assembly? Give examples.

12   What is a non-load-bearing panel assembly? Give examples.

13   Explain how rigidity and strength are achieved in monoconstruction.

14   Describe the location and function of the front and rear bulkheads.

15   Give a brief description of the following early vehicle body styles: coupé, cabriolet, limousine, saloon.

16   What is meant by a veteran vehicle? Name and describe three such vehicles.

17   Name two people who were associated with the early development of the motor vehicle, and state their involvement.

18   Explain what is meant by the semi-integral method of construction.

19   Explain why it is difficult to mass produce composite constructed vehicles.

20   In integral construction, what section of the body possesses the greatest amount of strength?

21   What is the front section of the body shell called, and what are its principal panel assemblies?

22   Explain the role of the stylist in the design organization.

23 Name one vehicle design stylist who has become well known during the last 25 years.

24 List the stages of development in the creation of a new vehicle body design.

25 State the definition of the symbol C_d.

26 Define the term CAD-CAM.

27 Explain the role of the clay modeller in the structure of the styling department.

28 With the aid of a sketch, explain what is meant by profile aerodynamic drag.

29 Explain the necessity for prototype testing.

30 Explain the use of dummies in safety research and testing.

31 Explain the difference in manufacture between a medium-bodied mass-produced vehicle and a high-quality coach-built limousine.

32 Describe the body work styling of a Sports or GT vehicle.

33 What is the difference in design between a saloon and a hatchback vehicle?

34 With the aid of a sketch, explain the body styling of a coupé vehicle.

35 Explain ABS as an active safety feature on a vehicle.

36 How are vehicles made safe against side impact involvement?

37 Explain how the airbag system works in a vehicle.

38 Explain the VIN number and why it is used on a vehicle.

39 Name the two main types of seatbelt arrangement which are fitted to a standard saloon vehicle.

40 State the letters used in design to identify the body pillars on a four-door saloon.

41 State the main purpose of a vehicle subframe.

42 Explain why seatbelt anchorages must be reinforced on a vehicle body.

43 State why GRP bodywork is normally associated with separate body construction.

44 List the design features that characterize a vehicle body as a limousine.

45 Explain the necessity for a hydraulic damper in the suspension of a motor vehicle.

46 Why is GRP not used in the mass production of vehicle body shells on an assembly line?

47 Name one of the persons who was associated with the early development of the motor vehicle and state his involvement.

48 State the purposes of the inner reinforcement members of a bonnet panel and say how they are held in place.

49 State the reasons for swaging certain areas of a vehicle floor pan.

50 Explain the importance of the use of scale models in vehicle design.

51 Why are current body shapes more rounded than previous designs?

52 Why are radiator grilles shaped differently on different makes of cars?

2

Health and safety

The main responsibility for occupational health and safety lies with the employer. It is the employer who must provide a safe working environment, safe equipment and safety protection and must also ensure that all work methods are carried out safely.

The Health and Safety at Work Act (HASAWA) 1974 is a major piece of occupational legislation, which requires the employer to ensure, as far as is reasonably practicable, the health and safety of all staff and any other personnel who may be affected by the work carried out. The other two important regulations affecting bodyshops are the Control of Substances Hazardous to Health (COSHH) Regulations 1988 and the Environmental Protection Act (EPA) 1990. Chapter 15 provides further information about these regulations.

2.1 Personal safety and health practices

2.1.1 Skin care (personal hygiene) systems

All employees should be aware of the importance of personal hygiene and should follow correct procedures to clean and protect the skin in order to avoid irritants causing skin infections and dermatitis. All personnel should use a suitable barrier cream before starting work and again when recommencing work after a break. There are waterless hand cleaners available which will remove heavy dirt on skin prior to thorough washing. When the skin has been washed, after-work creams will help to restore its natural moisture.

Many paints, refinishing chemicals and bodyshop materials will cause irritation on contact with the skin and must be removed promptly with a suitable cleansing material. Paint solvents may cause dermatitis, particularly where skin has been in contact with peroxide hardeners or acid catalysts: these have a drying effect which removes the natural oils in the skin. There are specialist products available for the bodyshop which will remove these types of materials from the skin quickly, safely and effectively.

2.1.2 Hand protection

Body technicians and painters are constantly handling substances which are harmful to health. The harmful effect of liquids, chemicals and materials on the hands can be prevented, in many cases, by wearing the correct type of gloves. To comply with COSHH Regulations, vinyl disposable gloves must be used by painters to give skin protection against toxic substances. Other specialist gloves available are: rubber and PVC gloves for protection against solvents, oil and acids; leather gloves for hard wear and general repair work in the bodyshop; and welding gauntlets, which are made from specially treated leather and are longer than normal gloves to give adequate protection to the welder's forearms.

2.1.3 Protective clothing

Protective clothing is worn to protect the worker and his clothes from coming into contact with dirt, extremes of temperature, falling objects and chemical substances. The most common form of protective clothing for the body repairer is the overall a one-piece boiler suit made from good quality cotton, preferably flame-proof. Worn and torn materials should be avoided as they can catch in moving machinery. Where it is necessary to protect the skin, closely fitted sleeves should be worn down to the wrist with the cuffs fastened. All overall buttons must be kept fastened, and any loose items such as ties and scarves should not be worn.

Protective clothing worn in the paint shop by the spray painters should be either good quality washable nylon garments, anti-static, and complete with hood, elasticated wrists and ankles; or low-linting disposable coveralls, which offer a liquid barrier protection from splashes, airborne dusts and paint overspray. The coveralls must withstand continuous exposure to a variety of chemicals and must be suitable for protection when using isocyanate-based two-pack paints; they also prevent the environment being contaminated by particles from the operator's clothing and hair. They can be of the one-piece variety or can have separate disposable hoods (Figure 2.1a, b).

(a)

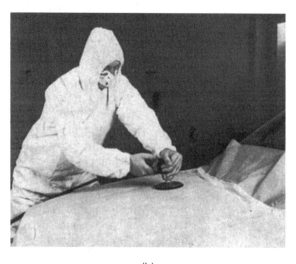

(b)

Figure 2.1 (a) Disposable protective coverall (b) protective coverall and face mask in use (*Gramos Chemicals International Ltd*)

2.1.4 Head protection

Head protection is very important to the body worker when working underneath a vehicle or under its bonnet while it is being repaired. A light safety helmet, normally made from aluminium, fibreglass or plastic, should be worn if there is any danger from falling objects, and will protect the head from damage when working below vehicles. Hats and other forms of fabric headwear keep out dust, dirt and overspray and also prevent long hair (tied back) becoming entangled in moving equipment.

2.1.5 Eyes and face protection

Eye protection is required when there is a possibility of eye injury from flying particles when using a grinder, disc sander, power drill or pneumatic chisel, or when removing glass windscreens or working underneath vehicles. Many employers are now requiring all employees to wear some form of safety glasses when they are in either the repair or the paint areas of the bodyshop, because in any bodyshop location there is always the possibility of flying objects, dust particles, or splashing liquids entering the eyes. Not only is this painful but it can, in extreme cases, cause loss of sight. Eyes are irreplaceable: therefore it is advisable to wear safety goggles, glasses or face shields in all working areas.

The following types of eye protection are available:

Lightweight safety spectacles with adjustable arms and with side shields for extra protection. There is a choice of impact grades for the lenses (Figure 2.2). *General-purpose safety goggles* with a moulded PVC frame which is resistant to oils, chemicals and water. These have either a clear acetate or a polycarbonate lens with BS impact grades 1 and 2 (Figure 2.3).
Face shields with an adjustable head harness and deep polycarbonate brow guard with replaceable swivel-up clear or anti-glare polycarbonate visor BS grade 1, which gives protection against sparks, molten metal and chemicals (Figure 2.4a, b).
Welding helmet or welding goggles with appropriate shaded lens to BS regulations. These must be worn at all times when welding. They will protect the eyes and face from flying molten particles of steel when gas welding and brazing, and from the harmful light rays generated by the arc when MIG/MAG, TIG or MMA welding (Figure 2.5a, b).

Figure 2.2 Lightweight safety spectacles (*Racal Safety Ltd*)

Figure 2.3 General-purpose safety goggles (*Racal Safety Ltd*)

2.1.6 Foot protection

Safety footwear is essential in the bodyshop environment. Boots or shoes with steel toecaps will protect the toes from falling objects. Rubber boots will give protection from acids or wet conditions. Never wear defective footwear as this becomes a hazard in any workshop environment.

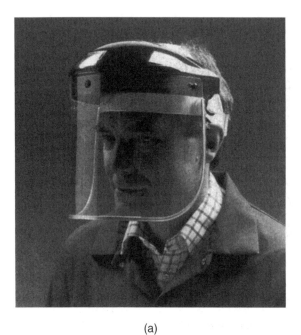

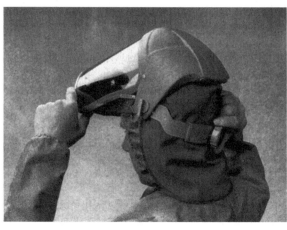

(a)   (b)

Figure 2.4 (a) Face shield (*Racal Safety Ltd*) (b) Face shield with protective hood (*DeVilbiss Automotive Refinishing Products*)

(a)   (b)

Figure 2.5 (a) Standard visor-type welding helmet (b) Standard welding goggles with hinged lenses (*Racal Safety Ltd*)

2.1.7 Respiratory protection (lungs)

One of the most important hazards faced by the body-shop worker is that of potential damage to the lungs. Respirators are usually needed in body repair shops even though adequate ventilation is provided for the working areas. During welding, metal or paint preparation, or spraying, some form of protection is necessary. Under the COSHH Regulations, respiratory protection is essential and therefore must be used.

Respirators give protection against abrasive dusts, gases, vapours from caustic solutions and solvents, and spray mist from undercoats and finishing paint, by filtering the contaminated atmosphere before it is inhaled by the wearer. They may be either simple filtering devices, where the operator's lungs are used to draw air through the filter, or powered devices incorporating a battery-driven fan to draw contaminated air through the filters and deliver a flow of clean air to the wearer's face. There are four primary types of respirator available to protect the bodyshop technicians: dust respirators, cartridge filter respirators, powered respirators, and constant-flow air line breathing apparatus.

Dust respirators (masks)

The most basic form of respiratory protection is the disposable filtering half-mask, typically used when preparing or finishing bodywork such as by rubbing down or buffing, and where dust, mist and fumes are a problem. This face mask provides an excellent face seal while at the same time allowing the wearer to speak freely without breaking the seal. Breathing resistance is minimal, offering cool and comfortable use. Various types of mask are available for use in a variety of environments where contaminants vary from nuisance dust particles to fine dusts and toxic mists. These masks can only be used in atmospheres containing less than the occupational exposure limit of the contaminant (Figure 2.6a, b).

Cartridge filter mask

The cartridge filter or organic vapour type of respirator, which covers the nose and mouth, is equipped with a replacement cartridge that removes the organic vapours by chemical absorption. Some of these are also designed with a pre-filter to remove solid particles from the air before the air passes

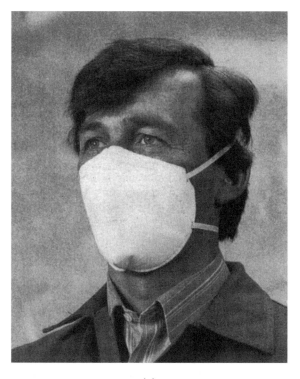

(a)

(b)

Figure 2.6 (a) Dust mask (b) Dust mask in use (*Racal Safety Ltd*)

through the chemical cartridge. They are used in finishing operations with non-toxic paints, but not with isocyanate paints. For the vapour/particle respirator to function correctly it is essential that it fits properly against the face. Follow the manufacturer's instructions for changing the cartridges when spraying over a continuous period (Figure 2.7a, b).

(a)

(b)

Figure 2.7 (a) Standard cartridge mask with filters (*Racal Safety Ltd*) (b) Cartridge mask and protective coveralls being used while spraying (*Gramos Chemicals International Ltd*)

Powered respirators

Powered respirators using canister filters offer protection against toxic dusts and gases. The respirator draws contaminated air through filters with a motor fan powered by a rechargeable battery and supplies clean air to the wearer's face. This avoids discomfort and fatigue caused by the effort of having to inhale air through filters, permitting longer working periods. These devices find great use both in the spray shop and in the repair shop when carrying out welding (Figure 2.8a, b).

Constant-flow air line breathing apparatus

The constant-flow compressed air line breathing apparatus is designed to operate from an industrial compressed air system in conjunction with the spray gun. Using a waist-belt-mounted miniature fixed-pressure regulator and a pre-filter, the equipment supplies breathing quality air through a small-bore hose to a variety of face masks and visors to provide respiratory protection for paint spraying (such as with isocyanates), cleaning and grinding (Figure 2.9a, b). The COSHH Regulations have made it mandatory for all respiratory protection equipment to be both approved and suitable for the purpose, for the operatives to be correctly trained in the equipment's use and maintenance, and for proper records to be kept.

2.1.8 Ear protection

The Noise at Work Regulations 1989 define three action levels for exposure to noise at work:

1 A daily personal exposure of up to 85 dB(A). Where exposure exceeds this level, suitable hearing protection must be provided on request (Figure 2.10).
2 A daily personal exposure of up to 90 dB(A). Above this second level of provision, hearing protection is mandatory.
3 A peak sound pressure of 200 pascals (140 dB).

Where the second or third levels are reached, employers must designate ear protection zones and require all who enter these zones to wear ear protection. Where the third level is exceeded, steps must be taken to reduce noise levels as far as is reasonably practicable. In every case where there is

(a) (b)

Figure 2.8 (a) Airstream welding helmet (powered respirator) (b) Powered respirator in use in welding (*Racal Safety Ltd*)

a risk of significant exposure to noise, assessment must be carried out and action taken to minimize hearing damage.

The first two noise action levels relate to exposure over a period (one day) and are intended to cater for the risks of prolonged work in noisy surroundings. The third level is related to sudden impact noises like those occurring in metal working procedures.

2.2 Fire precautions

The Fire Precautions (Places of Work) Regulations 1992 replaced and extended the old Fire Precautions Act 1971 as from 1 January 1993. These Regulations are aligned with standard practice in EC Directives in placing the responsibility for compliance on the employers. They require employers not only to assess risks from fire, but now to include the

preparation of an evacuation plan, to train staff in fire precautions, and to keep records. Workplaces with fewer than 20 employees may require emergency lighting points and fire warning systems. The self-employed who do not employ anyone but whose premises are regularly open to the public may only require fire extinguishers and warning signs; they will, however, need to be able to demonstrate that there is a means of escape in case of fire. Where five or more persons work on the premises as employees, all assessments need to be recorded in writing.

Most of these requirements were already covered by existing legislation. The prime differences are the recording of assessments, the provision of training, and the requirement that means of fighting fire, detecting fire and giving warning in case of fire, be maintained in good working order.

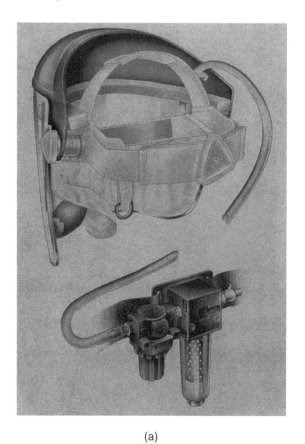

(a)

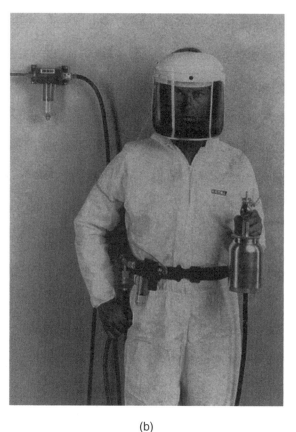

(b)

Figure 2.9   (a) Visionair constant-flow breathing apparatus (b) Operator wearing complete constant-flow breathing apparatus (*Racal Safety Ltd*)

Figure 2.10   Ear protectors (*Racal Safety Ltd*)

2.2.1 What is fire?

Fire is a chemical reaction called combustion (usually oxidation resulting in the release of heat and light). To initiate and maintain this chemical reaction, or in other words for an outbreak of fire to occur and continue, the following elements are essential (Figure 2.11):

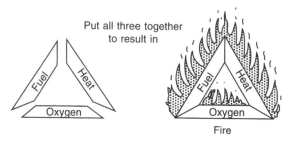

Figure 2.11   The fire triangle (*Chubb Fire Ltd*)

Fuel A combination substance, either solid, liquid or gas.
Oxygen Usually air, which contains 21 per cent oxygen.
Heat The attainment of a certain temperature (once a fire has started it normally maintains its own heat supply).

2.2.2 Methods of extinction

Because three ingredients are necessary for fire to occur, it follows logically that if one or more of these ingredients is removed, fire will be extinguished. Basically three methods are employed to extinguish a fire: removal of heat (cooling); removal of fuel (starving); and removal or limitation of oxygen (blanketing or smothering).

Removal of heat
If the rate of heat generation is less than the rate of dissipation, combustion cannot continue. For example, if cooling water can absorb heat to a point where more heat is being absorbed than generated, the fire will go out.

Removal of fuel
This is not a method that can be applied to fire extinguishers. The subdividing of risks can starve a fire, prevent large losses and enable portable extinguishers to retain control; for example, part of a building may be demolished to provide a fire stop.

The following advice can contribute to a company's fire protection programme:

1 What can cause fire in this location, and how can it be prevented?
2 If fire starts, regardless of cause, can it spread?
3 If so, where to?
4 Can anything be divided or moved to prevent such spread?

Removal or limitation of oxygen
It is not necessary to prevent the contact of oxygen with the heated fuel to achieve extinguishment. It will be found that where most flammable liquids are concerned, reducing the oxygen in the air from 21 to 15 per cent or less will extinguish the fire. Combustion becomes impossible even though a considerable proportion of oxygen remains in the atmosphere. This rule applies to most solid fuels although the degree to which oxygen content must be reduced

may vary. Where solid materials are involved they may continue to burn or smoulder until the oxygen in the air is reduced to 6 per cent. There are also substances which carry within their own structures sufficient oxygen to sustain combustion.

2.2.3 Fire risks in the workshop

Fire risks in the vehicle body repair shop cover all classes of fire: class A, i.e. paper, wood and cloth; class B, i.e. flammable liquids such as oils, spirits, alcohols, solvents and grease; class C, i.e. flammable gases such as acetylene, propane, butane; and also electrical risks. It is essential that fire is detected and extinguished in the early stages. Workshop staff must know the risks involved and should be aware of the procedures necessary to combat fire. Bodyshop personnel should be aware of the various classes of fire and how they relate to common workshop practice.

Class A fires: wood, paper and cloth
Today wood is not used in cars, although there are exceptions. Cloth materials are used for some main trim items and are therefore a potential fire hazard. The paper used for masking purposes is a prime area of concern. Once it has done its job and is covered in overspray it is important that it is correctly disposed of, ideally in a metal container with a lid, and not scrunched up and thrown on the floor to form the potential start of a deep seated fire.

Class B fires: flammable liquids
Flammable liquids are the stock materials used in the trade for all body refinishing processes: gun cleaner to clear finish coats, cellulose to the more modern finishes, can all burn and produce acrid smoke.

Class C fires: gases
Not many cars run on liquid propane gas (LPG), but welding gases or propane space heaters not only burn but can be the source of ignition for A or B fires.

Electrical hazards
Electricity is not of itself a class of fire. It is, however, a potential source of ignition for all of the fire classes mentioned above.

The Electricity at Work Regulations cover the care of cables, plugs and wiring. In addition, in the bodyshop the use of welding and cutting equipment produces sparks which can, in the absence of good housekeeping, start a big fire. Training in how to use fire fighting equipment can stop a fire in its early stages. Another hazard is the electrical energy present in all car batteries. A short-circuit across the terminals of a battery can produce sufficient energy to form a weld and in turn heating, a prime source of ignition. When tackling a car fire a fireman will always try to disconnect the battery, as otherwise any attempt to extinguish a fire can result in the reignition of flammable vapours.

Body filler

A further possible source of ignition to be aware of in general use in the body repair business is the mixing of two materials to use as a body filler. The result of mixing in the wrong proportions can give rise to an exothermic (heat releasing) reaction; in extreme cases the mix can ignite.

2.2.4 General precautions to reduce fire risk

(a) Good housekeeping means putting rubbish away rather than letting it accumulate.
(b) Read the manufacturer's material safety data sheets so that the dangers of flammable liquids are known.
(c) Only take from the stores sufficient flammable material for the job in hand.
(d) Materials left over from a specific job should be put back into a labelled container so that not only you but anyone (and this may be a fireman) can tell what the potential risk may be.
(e) Take care when welding that sparks or burning underseal do not cause a problem, especially when working in confined areas of vehicles.
(f) Be extremely careful when working close to plastic fuel lines.
(g) Petrol tanks are a potential hazard: supposedly empty tanks may be full of vapour. To give some idea of the potential problem, consider one gallon of petrol: it will evaporate into 33 ft^3 of neat vapour, which will mix with air to form 2140 ft^3 of flammable vapour. Thus the average petrol tank needs only a small amount of petrol to give a tank full of vapour waiting to ignite and explode.

The key to fire safety is:

1 Take care.
2 Think.
3 Train staff in the correct procedures before things go wrong.
4 Ensure that these procedures are written down, understood and followed by all personnel within the workshop.

2.2.5 Portable extinguishers: types and uses

The colour codes for each type of appliance are as follows:

Red for water.
Cream for foam.
Black for CO_2.
Blue for powder.
Green for halon (BCF).

Figures 2.12a–c show various types of extinguisher. Although older fire extinguishers are colour coded, new ones are all red in colour with labelling to identify their contents and applications.

Water

Water is the most widely used extinguisher agent. With portable extinguishers, a limited quantity of water can be expelled under pressure and its direction controlled by a nozzle.

There are basically two types of water extinguishers. The gas (CO_2) cartridge operated extinguisher, when pierced by a plunger, pressurizes the body of the extinguisher, thus expelling the water and producing a powerful jet capable of rapidly extinguishing class A fires. In stored pressure extinguishers the main body is constantly under pressure from dry air or nitrogen, and the extinguisher is operated by opening the squeeze grip discharge valve. These extinguishers are available with 6 litre or 9 litre capacity bodies and thus provide alternatives of weight and accessibility.

Foam

Foam is an agent most suitable for dealing with flammable liquid fires. Foam is produced when a solution of foam liquid and water is expelled under pressure through a foam-making branch pipe at which point air is entrained, converting the solution into a foam.

Types of modern fire extinguishers
B.S.I. DD48:1976

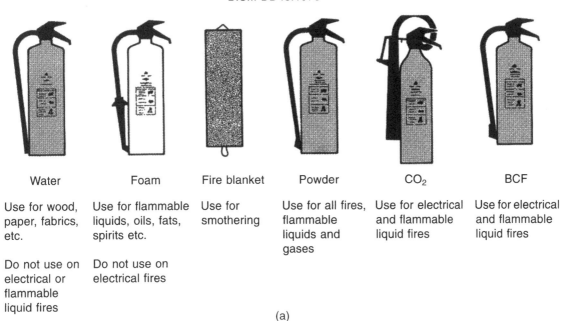

Water	Foam	Fire blanket	Powder	CO$_2$	BCF
Use for wood, paper, fabrics, etc.	Use for flammable liquids, oils, fats, spirits etc.	Use for smothering	Use for all fires, flammable liquids and gases	Use for electrical and flammable liquid fires	Use for electrical and flammable liquid fires
Do not use on electrical or flammable liquid fires	Do not use on electrical fires				

(a)

(b)

(c)

Figure 2.12   (a) Types of portable fire extinguisher (b) Types of fire fighting equipment (*UK Fire International Ltd*) (c) Portable fire extinguishers suitable for a bodyshop (*Chubb Fire Ltd*)

Foam extinguishers can be pressurized either by a CO_2 gas cartridge or by stored pressure. The standard capacities are 6 and 9 litres.

Spray foam

Unlike conventional foams, aqueous film forming foam (AFFF) does not require to be fully aspirated in order to extinguish fires. Spray foam extinguishers expel an AFFF solution in an atomized form which is suitable for use on class A and class B fires. AFFF is a fast and powerful means of tackling a fire and seals the surfaces of the material, preventing re-ignition. The capacity can be 6 or 9 litres, and operation can be by CO_2 cartridge or stored pressure.

Carbon dioxide

Designed specifically to deal with class B, class C and electrical fire risks, these extinguishers deliver a powerful concentration of carbon dioxide gas under great pressure. This not only smothers the fire very rapidly, but is also non-toxic and is harmless to most delicate mechanisms and materials.

Dry powder

This type of extinguisher is highly effective against flammable gases, open or running fires involving flammable liquids such as oils, spirits, alcohols, solvents and waxes, and electrical risks. The powder is contained in the metal body of the extinguisher from which it is supplied either by a sealed gas cartridge, or by dry air or nitrogen stored under pressure in the body of the extinguisher in contact with the powder.

Dry powder extinguishers are usually made in sizes containing 1 to 9 kg of either standard powder or (preferably and more generally) all-purpose powder, which is suitable for mixed risk areas.

Vaporizing liquid (halon 1211, BCF)

Portable extinguishers of this type are manufactured in sizes ranging from 1 to 15 kg. They are particularly effective for dealing with class B fires and with fires started by an electrical source.

Halon 1211 (bromochlorodifluoromethane, BCF) has a low toxicity level, is considered to be non-corrosive and has a long storage life. It is clean to use and leaves no residue, thus rendering it harmless to delicate fabrics and machinery. However, owing to the contribution of halons to atmospheric ozone depletion most companies have decided to cease production of halon 1211.

Choosing and siting portable extinguishers

Because there is such a variety of fire risks in bodyshops, it is important to analyse these risks separately and (with the help of experts such as fire officers) to choose the correct fire fighting medium to deal with each possible fire situation. It should be noted that portable fire extinguishers are classified as first-aid fire fighting and are designed for ease of operation in an emergency. It is important to realize that because they are portable they have only a limited discharge. Therefore their siting, together with an appreciation of their individual characteristics, is fundamental to their success in fighting fire (Figure 2.13).

2.3 Safety signs in the workshop

It is a legal requirement that all safety signs used in a bodyshop comply with BS 5378: Part 1. Each of these signs is a combination of colour and design, within which the symbol is inserted. If additional information is required, supplementary text may be used in conjunction with the relevant symbol, provided that it does not interfere with the symbol. The text can be in an oblong or square box of the same colour as the sign, with the text in the relevant contrasting colour, or the box can be white and the text black.

BS 5378 divides signs into four categories (Figure 2.14):

Prohibition Prohibition signs have a red circular outline and crossbar running from top left to bottom right on a white background (Figure 2.15a). The symbol displayed on the sign must be black and placed centrally on the background, without obliterating the crossbar. The colour red is associated with 'stop' or 'do not'.

Warning Warning signs have a yellow triangle with a black outline (Figure 2.15b). The symbol or text used on the sign must be black and placed centrally on the background. This combination of black and yellow identifies caution.

Mandatory Mandatory signs have a blue circular background (Figure 2.15c). The symbol or text used must be white and placed centrally on the background. Mandatory signs indicate that a specific course of action is to be taken.

Class of fire	Water	Foam (AFFF)	CO$_2$ gas	Powder
A Paper Wood Textile Fabric	✔	✔		✔
B Flammable liquids		✔	✔	✔
C Flammable gases			✔	✔
Electrical hazards			✔	✔
Vehicle protection		✔		✔

Figure 2.13 Which extinguisher to use (*Chubb Fire Ltd*)

Key to British and European Standard safety signs

| Prohibition
Don't do | Warning
Risk of danger | Safe condition
The safe way | Mandatory
Must do |

Figure 2.14 Standard safety signs

Safe condition The safe condition signs provide information for a particular facility (Figure 2.15d) and have a green square or rectangular background to accommodate the symbol or text, which must be in white. The safety colour green indicates 'access' or 'permission'.

Fire safety signs are specified by BS 5499, which gives the characteristics of signs for fire equipment, precautions and means of escape in case of fire (Figure 2.16). It uses the basic framework concerning safety colours and design adopted by BS 5378.

2.4 General safety precautions in the workshop

The Health and Safety at Work Act imposes on employers a statutory duty to ensure safe working conditions and an absence of risk in the use of equipment and the handling of materials, and to comply with Regulations regarding safe working practices in order to reduce to a minimum the hazards to health and safety associated with vehicle body repair work. To skilled and experienced operators this does not mean that any additional restrictions are imposed on their activities, but merely that they should carry out their tasks with constant regard for the health and safety of themselves and their fellow workers.

Prohibition signs

No smoking

Smoking and naked flames prohibited

Do not extinguish with water

Not drinking water

Do not operate

(a)

Warning signs

Caution, risk of fire

Caution, risk of explosion

Caution, toxic hazard

Caution, laser beam

Caution, corrosive substance

(b)

Figure 2.15   (a) Prohibition signs (b) Warning signs

Mandatory signs

 Eye protection must be worn

 Head protection must be worn

 Hearing protection must be worn

 Respiratory protection must be worn

 Foot protection must be worn

 Hand protection must be worn

(c)

Safe condition signs

 First aid

 Indication of direction (may be used in conjunction with A.4.1)

 Emergency eye wash

 Emergency telephone

 Emergency stop push-button

(d)

Figure 2.15 (c) Mandatory signs (d) Safe condition signs

Figure 2.16 Fire signs

Particular hazards may be encountered in the bodyshop, and safety precautions associated with them are as follows:

1 Do wash before eating, drinking or using toilet facilities to avoid transferring the residues of sealers, pigments, solvents, filings of steel, lead and other metals from the hands to the inner parts and other sensitive areas of the body.

2 Do not use kerosene, thinners or solvents to wash the skin. They remove the skin's natural protective oils and can cause dryness and irritation or have serious toxic effects.

3 Do not overuse waterless hand cleaners, soaps or detergents, as they can remove the skin's protective barrier oils.

4 Always use barrier cream to protect the hands, especially against fuels, oils, greases, hydrocarbon solvents and solvent-based sealers.

5 Do follow work practices that minimize the contact of exposed skin and the length of time liquids or substances stay on the skin.

6 Do thoroughly wash contaminants such as used engine oil from the skin as soon as possible with soap and water. A waterless hand cleaner can be used when soap and water are not avail-able. Always apply skin cream after using waterless hand cleaner.

7 Do not put contaminated or oily rags in pockets or tuck them under a belt, as this can cause continuous skin contact.

8 Do not dispose of dangerous fluids by pouring them on the ground, or down drains or sewers.

9 Do not continue to wear overalls which have become badly soiled or which have acid, oil, grease, fuel or toxic solvents spilt over them. The effect of prolonged contact from heavily soiled overalls with the skin can be cumulative and life threatening. If the soilants are or become flammable from the effect of body temperature, a spark from welding or grinding could envelop the wearer in flames with disastrous consequences.

10 Do not clean dusty overalls with an air line: it is more likely to blow the dust into the skin, with possible serious or even fatal results.

11 Do wash contaminated or oily clothing before wearing it again.

12 Do disguard contaminated shoes.

13 Wear only shoes which afford adequate protection to the feet from the effect of dropping tools and sharp and/or heavy objects on them, and also from red hot and burning materials. Sharp or hot objects could easily penetrate unsuitable footwear such as canvas plimsolls or trainers. The soles of the shoes should also be maintained in good condition to guard against upward penetration by sharp or hot pieces of metal.

14 Ensure gloves are free from holes and are clean on the inside. Always wear them when handling materials of a hazardous or toxic nature.

15 Keep goggles clean and in good condition. The front of the glasses or eyepieces can become obscured by welding spatter adhering to them. Renew the glass or goggles as necessary. Never use goggles with cracked glasses.

16 Always wear goggles when using a bench grindstone or portable grinders, disc sanders, power saws and chisels.

17 When welding, always wear adequate eye protection for the process being used. MIG/MAG welding is particularly high in ultraviolet radiation which can seriously affect the eyes.

18 Glasses, when worn, should have 'safety' or 'splinter-proof' glass or plastic lenses.

19 Always keep a suitable mask for use when dry flatting or working in dusty environments and when spraying adhesive, sealers, solvent carried waxes, and paints.

20 In particularly hostile environments such as when using volatile solvents or isocyanate materials, respirators or fresh air fed masks must be worn.

21 Electric shock can result from the use of faulty and poorly maintained electrical equipment or misuse of equipment. All electrical equipment must be frequently checked and maintained in good condition. Flexes, cables and plugs must not be frayed, cracked, cut or damaged in any way. Equipment must be protected by the correctly rated fuse.

22 Use low-voltage equipment wherever possible (110 volts).

23 **In case of electric shock:**
    (a) Avoid physical contact with the victim.
    (b) Switch off the electricity.
    (c) If this is not possible, drag or push the victim away from the source of the electricity using non-conductive material.
    (d) Commence resuscitation if trained to do so.
    (e) Summon medical assistance as soon as possible.

2.5 Electrical hazards

The Electricity at Work Act 1989 fully covers the responsibilities of both the employee and the employer. As a body repairer you are obliged to follow these regulations for the protection of yourself and your colleagues. Some of the important points to be aware of are given below.

2.5.1 Voltages

The normal mains electricity voltage via a three-pin socket outlet is 240 volts; heavy duty equipment such as vehicle hoists use 415 volts in the form of a three-phase supply. Both 240 volt and 415 volt supplies are likely to kill anybody who touches them. Supplies of 415 volts must be used through a professionally installed system. If 240 volts is used for power tools, then a safety circuit breaker should be used. A safer supply for power tools is 110 volts; this may be wired into the workshop as a separate circuit or provided through a safety transformer.

Inspection hand-lamps are safest with a 12 volt supply; but for reduced current flow 50 volt hand-lamp systems are frequently used.

2.5.2 Check list

Before using electrical equipment the body repairer is advised to check the following:

1 Cable condition – check for fraying, cuts or bare wires.

2 Fuse rating – the fuse rating should be correct for the purpose as recommended by the equipment manufacturer.

3 Earth connection – all power tools must have sound earth connections.

4 Plugs and sockets – do not overload plugs and sockets, ensure that only one plug is used in one socket.

5 Water – do not use any electrical equipment in any wet conditions.

6 PAT testing – it is a requirement of the Electricity at Work Regulations that all portable electrical appliances are tested regularly, they should be marked with approved stickers and the inspection recorded in a log.

2.6 COSHH

The Control of Substances Hazardous to Health regulations require that assessments are made of all substances used in the body repair shop, for instance paint and body filler. This assessment must state the hazards of using the materials and how to deal with accidents arising from misuse. Your wholesale supplier will provide you with this information as set out by the manufacturer in the form of either single sheets on individual substances, or a small booklet covering all the products in a range.

2.7 RIDDOR

The Reporting of Injuries, Diseases and Dangerous Occurrences Regulations 1995 require that certain information is reported to the Health and Safety Executive (HSE). This includes the following:

1 Death or major injury – if an employee or member of the public is killed or suffers major injury the HSE must be notified immediately by telephone.

2 Over-three-day injury – if as the result of an accident connected with work an employee is

absent for more than three days an accident form must be sent to the HSE.

3 Disease – if a doctor notifies an employer that an employee suffers from a reportable work-related disease then this must be reported to the HSE.

4 Dangerous occurrence – if an explosion or other dangerous occurrence happens, this must be reported to the HSE, it does not need to involve a personal injury.

2.8 Maintain the health, safety and security of the work environment

2.8.1 Guidelines, statutory regulations and safe systems for health and safety protection are followed

It is the duty of every employee and employer in the motor industry to comply with the statutory regulations relating to health and safety and the associated guidelines which are issued by the various government offices. That means you must work in a safe and sensible manner. A body repairer is expected to follow the health and safety recommendations of his/her employer; employers are expected to provide a safe working environment and advise on suitable safe working methods. The current regulations which affect those who work in the motor repair industry are given in this long list which you are not expected to remember:

Factories Act 1961
Offices, Shops and Railway Premises Act 1963
Abrasive Wheels Regulations 1970
Fire Precautions Act 1971
Highly Flammable Liquids and Liquefied Petroleum
  Gas Regulations 1972
Health and Safety at Work Act 1974
Eye Protection Regulations 1974
Control of Lead at Work Regulations 1980
Health and Safety (First Aid) Regulations 1981
Reporting of Injuries, Diseases and Dangerous
  Occurrences Regulations 1985
Control of Substances Hazardous to Health
  Regulations 1988
Electricity at Work Regulations 1989
Noise at Work Regulations 1989
Pressure Systems and Transportable Gas Containers
  Regulations 1989

A trainee body repairer is expected to appreciate the principal requirements of the three main regulations. So you will need to remember the names of the following regulations and say how they affect you at work.

Factories Act This sets out specific regulations relating to: working temperature and heating of buildings, fitting of machine guards, lighting of the working area, control of dust, fire escape provision, washing and toilet facilities and rest rooms. The Factories Act was aimed at employers, it was designed to make factories, including garages, better places to work. The trainee body repairer can help the employer comply with this Act by working in a clean and tidy manner and reporting any breakages, or shortages, immediately to the chargehand or senior technician.

For example, if the barrier cream dispenser is empty, see the chargehand for permission to install a refill from stores.

Health and Safety at Work Act 1974 (HSWA) This states that it is the duty of every employee to work in a safe and secure manner that will not cause any harm or injury to the individual or anybody else, also to take care of any safety equipment and to cooperate with employers to comply with any related regulations. The HSWA was designed to cover the areas of work which the Factories Act did not cover, for instance schools and colleges, especially their motor vehicle workshops. You will not find any specific statements in the HSWA, but the guidelines issued by government bodies and trade associations interpret how the HSWA should be read.

As a trainee you must follow the health and safety advice issued by your employer, unless you know that it is wrong, in which case you should point this out and seek advice from your union safety representative. You must not damage any equipment which is provided for your use, nor knowingly break any health and safety regulations. It is your duty in the eyes of the law to know about the regulations which affect you; in practice you will learn them as you learn your trade. But using common sense and thinking before you act is always good policy, as is asking questions about things you do not know.

Control of Substances Hazardous to Health Regulations 1974 (COSHH) These require that substances used in the workplace are correctly labelled and instructions for their safe storage and use are available.

This means almost any substance which you are likely to find in a garage, but particularly oils, greases, paints, brake fluid, battery acid and cleaning materials. As a trainee mechanic you should take time to read the information which your employer has provided on the substances, these are known as COSHH sheets and will be available from either the service department office or the parts department workshop counter.

2.8.2 Identified hazards in the immediate working environment are removed where possible

You must work in a safe manner or you are breaking the HSWA and are liable to a fine of up to £2000 and/or imprisonment. This means following the safe working practices which are normally used within the industry. The guidelines published by the Health and Safety Executive (HSE) and motor vehicle textbooks usually identify industry accepted safe working practices. Examples of important procedures are:

1 Always use axle stands when a vehicle is jacked up.
2 Always use an exhaust extractor when running an engine in the garage.
3 Always wear overalls, safety boots and any other personal protective equipment (PPE) when it is needed, for example safety goggles when grinding or drilling and a breathing mask when working in dusty conditions.
4 Always use the correct tools for the job.

Even if you are working in a safe and careful manner you are still likely to spill the odd small amount of fluid or snag the airline, this will then create a hazard. The procedure here is always to remove the hazard, no matter how it was created, immediately.

If you spill petrol or oil when you disconnect a pipe from the engine you should clean it up immediately or else you, or a colleague, may slip and fall. Absorbent granules should be used for this job, as they will soak up the liquid without causing a fire hazard or making the floor more slippery.

Brake fluid is a special hazard because if it is spilled on the vehicle's paint work it will soften the paint and may cause it to peel off, just like paint stripper. Therefore any spilled brake fluid should be wiped off immediately and the paint surface washed and polished if needed. Antifreeze spilled on paintwork will soften the paint surface and cause discoloration, so it too must be wiped off immediately with absorbent paper roll or towel and washed down if needed.

When working on any system which contains fluids it is good practice to use a drip-tray to catch any possible spillages, this saves having to clean the floor as well as ensuring that all the used oils and fluids are disposed off safely, that is, you can pour them from your drip-tray into your disposal container. The Environment Protection Act requires that you dispose of used oils and other fluids in a way which will not cause pollution. In practice many vehicle manufacturers now collect used oil and brake fluid for recycling or safe disposal. Smaller garages without franchises will dispose of these liquids either through a private waste collection company or through a scheme in conjunction with the local authority. No waste oil, petrol, brake fuid or similar chemicals must be allowed to enter the drainage system.

Exhaust fumes are very dangerous, they can kill you. Small intakes of exhaust fumes will give you bad headaches, and over time can cause lung and/or brain diseases so ensure that you do not run a vehicle in a workshop without an exhaust extractor. Also ensure that the extractor pipe is correctly connected and is not leaking.

The airline used in most garages operates at between 100 and 150 psi (7 and 10 bar), this is a very high pressure, so it must be handled with great care. When you are using an airline always wear safety goggles to prevent dust entering your eyes. You must not use an airline for dusting off components, especially brake and clutch parts as the very fine dust can cause damage to the throat and lungs. Before you use an airline ensure that the coupling is fitted firmly into the socket and that the pipe is not leaking anywhere along its length. Any damage or leaks should be immediately reported to your supervisor or manager so that they can be repaired. The high pressure of the air can quickly turn a small leak in an airline into a large gash which in turn may make the airline wip around and cause damage to colleagues or customers' vehicles.

Another area of potential danger is when using electrical equipment such as an electric drill, handlamp or grinder. Most mains operated equipment runs at 240 volts; an electric shock from such a voltage is most likely to kill you straight away. Some companies use 110 volt equipment which is operated through a transformer, this is much safer, especially if the transformer is fitted with an overload cutout. Hand-lamps should operate at 50 volts, or preferably 12 volts to give the highest level of safety. Plugs should only be fitted to electrical equipment by skilled persons, at the same time a fuse of the correct amperage rating should be fitted and the equipment tested and logged in accordance with the Portable Appliance Testing Regulations (PAT testing). PAT tested equipment should be numbered and carry a test date label. Before you use any electrical equipment visually check it for signs of damage and check that the cable is not frayed or split. Then ensure that you plug it into the correct voltage outlet. Do not attempt to use any electrical equipment which you suspect may be faulty; report the fault immediately to your supervisor.

2.8.3 Where identified hazards cannot be removed, appropriate action is taken immediately to minimize risk to own and others' health and safety

This section is about those situations where the hazards cannot be readily removed, that is, how do you behave in accident situations, or when equipment malfunctions and you can see an accident about to happen? Most mechanics are only likely to encounter such problems every few years, but the professional is the person who can save the day. The following is an example of where a service manager colleague came to the rescue. The central locking and the car alarm on a six-month-old vehicle malfunctioned. A small child was trapped in the vehicle; it was a very hot day at a local car boot sale. The mother and child were hysterical, the father had gone for the fire brigade; other members of the public just watched. The mother shouted for help. My quick-witted friend grabbed a screwdriver off one of the stalls, inserted it behind the rear quarterlight rubber and levered out the glass, put his hand inside the car, opened the door and released the child. The panic was over.

There were other ways in which this situation could have been dealt with, but this one was acceptable because it provided a very quick solution and caused the minimum amount of damage to the vehicle. The important point is that people come first and property second, although the amount of damage to the property should be kept to the minimum.

Working by the roadside is always hazardous, but you can minimize the risks by following a few simple rules:

1 Always wear a reflective safety vest (bright green or orange).
2 Use the warning triangle and hazard lights.
3 Use the flashing lamp on your recovery vehicle.
4 Only work on hard and level ground.
5 If you are on the hard shoulder get as far over to the left as possible.
6 Always use props when vehicle cabs or other panels are raised.
7 Think through the possible hazards, never take risks.
8 Have the vehicle towed to the workshop if necessary.

Often it is better to do nothing than cause damage, this is referred to as preserving the situation. Many times things look different after a cup of tea, or you have had time to check it out with a colleague.

2.8.4 Dangerous situations are reported immediately and accurately to authorized persons

As a trainee in the motor industry your company will require you to report any dangerous situations to your supervisor; this will be a person that you know as the chargehand, foreperson or service manager. Any internal matter should in the first instance be reported to one of these people – you will know who this is from your induction training. However, if you are working alone or the matter is not a company one, then you must inform the relevant authority. The four emergency services in the UK are Police, Fire, Ambulance and Coast Guard. To call them use any telephone and dial **999**.

2.8.5 Suppliers' and manufacturers' instructions relating to safety and safe use of all equipment are followed

Many pieces of garage equipment are marked 'only to be used by authorized personnel'. This is mainly because incorrect use can cause damage to the equipment, the workpiece or the operator. Do not operate equipment which you have not been properly trained to use and have not been given specific permission to use.

The suppliers of garage equipment issue operating instructions, and as part of your training you must read these instruction booklets so that you will understand the job better. You will also find that certain safety instructions are marked on the equipment. The vehicle hoist (ramp), hydraulic jack and other lifting equipment are marked with the Safe Working Load (SWL) in either tonnes or kilogrammes. You must ensure that you do not exceed these maximum load figures.

Some items of equipment have two-handed controls or deadman grips – do not attempt to operate these items incorrectly.

2.8.6 Approved/safe methods and techniques are used when lifting and handling

Do not attempt to manually carry a load which you cannot easily lift and which you cannot see above and around. The maximum weight of load that you should lift is 20 kilogrammes, but as a trainee this may still be too heavy for you.

When you are lifting items from the floor always keep your back straight and bend your knees. Bending your back whilst lifting can cause back injury. If you keep your feet slightly apart this will improve your balance. It is always a good idea to wear safety gloves when manually lifting.

Hoists and jacks are available for lifting vehicles; hydraulic or chain-operated equipment is available for lifting engines; hydraulic devices are available to lift gearboxes. For moving equipment and heavy components you should have either a trolley or a sack-truck.

You are advised to seek the assistance of a colleague when moving a heavy load even when you are using lifting equipment.

2.8.7 Required personal protective clothing and equipment are worn for designated activities and in designated areas

The following table lists typical items of personal protective equipment (PPE) and states when they must be worn.

PPE	Usage
cotton overalls (boiler suit)	all the time
safety footwear	all the time
disposable gloves	dealing with dirty or oily items
'rubber gloves'	operating the parts cleaning bath
reinforced safety gloves	lifting heavy/sharp edged items
dust mask	rubbing down body filler or dusty items
breathing apparatus and paper coveralls	certain types of spray painting
goggles	using a grinder or drill
waterproof overalls and boots	steam or pressure cleaning

You will often see safety notices requiring you to wear certain PPE in some areas at all times, this is because other people are working in the area and you may be at risk. Hard hats are sometimes required when working underneath vehicles on a hoist.

2.8.8 Injuries involving individuals are reported immediately to competent first aiders and/or appropriate authorized persons and appropriate interim support is organized to minimize further injury

Should there be an accident the first thing to do is call for help. Either contact your supervisor or a known first aid person. Should any of these not be available, and it is felt appropriate, call for your local doctor or an ambulance.

You are not expected to be a first aid expert, nor are you advised to attempt to give first aid unless you are properly qualified. However, as a professional in the motor industry you should be able to preserve the scene, that is, prevent further injury and make the injured person comfortable. The following points are suggested as ones worth remembering:

1   Switch off any vehicle or power source.
2   Do not move the person if injury to the back or neck is suspected.
3   In the case of electric shock turn off the electricity supply.
4   In the case of a gas leak, turn off the gas supply.
5   Do not give the person any drink or food, especially alcohol, in case surgery is needed.
6   Keep the person warm with a blanket or coat.

7 If a wound is bleeding heavily, apply pressure to the wound with a clean bandage to reduce the loss of blood.

8 If a limb has been trapped, use a safe jack to free the limb.

2.8.9 Visitors are alerted to potential hazards

The best policy is not to let customers into the workshop – many garages have a notice to this effect on the workshop door. For MOT purposes garages must have a customer viewing area. However, it is not always possible to keep people out of the workshop. Insurance company assessors and RAC/AA engineers will probably also require entry to the workshop as well as some customers who are concerned about their vehicles. So, before allowing them into the workshop you should warn them of potential hazards. For instance the dangers of oil and grease and the requirement to wear a hard hat.

It is always a good idea to accompany customers when they are in the workshop, this way you can advise them in the event that they may do something potentially dangerous or if there is a hazard of which they may not be readily aware.

2.8.10 Injuries resulting from accidents or emergencies are reported immediately to a competent first aider or appropriate authority

If a person is injured the first action must be to ensure that first aid is given by a competent first aider or other suitable person. Most companies have a designated first aider who is trained to deal with accidents and emergencies. If your company has no such a person on the staff then you will have a designated person who you must contact in the event of a colleague being injured. That person may be your supervisor or another senior member of the staff. If no manager or other senior person is available you should either dial 999 for an ambulance or telephone your company doctor, then inform the garage manager.

2.8.11 Incidents and accidents are reported in an accident book

By law all companies are required to maintain records of accidents which take place at work. These records are usually kept in an accident book. Accident books may be inspected by HSE inspectors (generally referred to as factory inspectors); they must be kept for a period of at least three years from the date of the last entry.

The information which is required to be recorded in the accident book is:

Name and address of injured person

Date, time and place of accident/dangerous occurrence

Name of person making the report and date of entry Brief account of accident and details of any equipment/substances which were involved.

It is always a good idea to keep a notepad to help remind you which way round things go when working on unfamiliar vehicles, this would also be useful for making any other notes, such as those about an accident.

2.8.12 Where there is a conflict over limitation of damage priority is always given to the person's safety

You can always buy a new wing for a car, but you cannot buy a new arm for a mechanic. In the event of an accident people come first. For instance if a building is on fire, do not re-enter to retrieve your belongings, wait until the fire is out and there is no risk before going back into the building. If a car is about to fall off a jack, get out of the way, do not try to catch the car with your hands or some such other dangerous action.

2.8.13 Professional emergency services are summoned immediately by authorized persons in the event of a fire/disaster

An authorized person is somebody who has the task of carrying out a specific job. Anybody may call the emergency services if they are needed.

The four emergency services are:

Police
Fire
Ambulance
Coastguard.

All are called by dialling **999** on an outside-line telephone. The emergency services operator will ask you which service you require. In certain cases the police will automatically be called, for instance in the case of a severe fire.

All emergency telephone calls are recorded on tape at the telephone exchange. You will be asked for your name, the place where the emergency is and where you are calling from. With the introduction of electronic telephone exchanges the number which you are calling from is automatically recorded, and you will be asked for the number to help confirm that your call is not a hoax.

Many companies have a direct telephone line to the fire station, and these automatically call the fire service if a fire is detected by sensors or by breaking the glass of a fire alarm. In such cases, if the fire service is called out and there is not a fire, they may charge the company a large fee. So, do not tamper with such a device unless you are authorized to do so or there is a dangerous fire.

2.8.14 Alarm/alert/evacuation systems are activated immediately by authorized persons

Generally only senior staff (managers) are allowed to operate alarms and other forms of alert/evacuation systems. This is because of the costs which may be involved if the fire service is called out wrongly and the damage which may occur if staff and/or customers panic.

In the case of a fire the normal alarm is a form of siren or bell. For other emergencies say a serious injury, an audible warning from a speaker announcement system (often called a tannoy) may be used. In a garage these are usually operated from the service department office.

2.8.15 Selection of fire extinguishers is appropriate for a given type of fire

There are five different types of fire extinguishers in common use in garage premises, these are identified by their colour, as given in the following table:

Colour	Type
red	water
cream	foam
green	vaporizing liquid
blue	dry chemical powder
black	carbon dioxide

In addition there are 'fire buckets' full of sand, and 'fire blankets'.

Before you use any fire extinguisher you must ask yourself three questions:

1 Can the fire be put out easily?
2 If it cannot be put out easily can the spread of the fire be slowed down or stopped without risk by using an extinguisher?
3 Which fire extinguisher should be used?

To help you to choose the correct fire extinguisher, fires are classified into four classes:

Class	Description	Colour of extinguisher
class A	fires of solid material, such as wood, paper, cloth, rubber	red
class B	fire of liquids, such as petrol, paraffin, brake fluid	black, blue or green
class C	fires involving leaking gas such as acetylene, Calor or natural gas	cream or blue
class D	fires of metals which burn, such as magnesium and nickel	blue, but it must be an inert dry powder

Liquid and gas fires are easily spread by using water, water can also conduct electricity.

Let us look at a few typical examples of fires which sometimes occur in garages:

1 Petrol spillage fire – if this is on the forecourt the need is for quick action. The black carbon dioxide (CO_2) extinguisher will put out the fire and not leave a mark anywhere.
2 Fire under the car bonnet, cause unknown – blue extinguisher using dry chemical powder is safe on both petrol and electrical fires, it is also easily cleaned off and will not damage the engine.
3 Fire in a rubbish bin, cause unknown – the use of a fire bucket full of sand or a fire blanket spread over the top of the bin should extinguish this fire.

Fires need fuel, oxygen from the air and heat. Remove any one of these and the fire will go out. Most fire extinguishers tend to both starve the fire of oxygen and lower the temperature of the fire so that it goes out.

2.8.16 In the event of warnings, procedures for isolating machines and evacuating premises are followed

If you hear a fire/emergency warning you must follow the company's evacuation procedure – this is usually stated on the workshop wall. If you hear a fire alarm a typical evacuation procedure is:

1 Shut off the electricity by pressing the emergency stop button.
2 Leave the premises by the nearest route, go to assembly point (AP) 1 which is in the customers' car park.
3 Do NOT re-enter the building until your supervisor tells you that it is safe to do so.

In the event of discovering a fire raise the alarm by breaking the glass of the alarm button.

2.8.17 Reports/records are available to authorized persons and are complete and accurate

The Social Security (Claims and Payments) Regulations 1979 require employers to maintain an accident book as well as regulation 7 of the HSWA. This book requires brief details of any accident or dangerous occurrence to be recorded. An approved book BI 510 is available from the HSE direct or through most good book stores. For more detailed information HSE Form 2508 should be completed. HSE staff have a statutory right to see a completed accident book or Form 2508, and they may also ask for further information. If you are personally involved in an accident you are advised to keep a copy of the book entry and any completed forms as well as your own notes on the event. These may be useful in the event of legal proceedings.

2.8.18 Machines and equipment are isolated, where appropriate, from the mains prior to cleaning and routine maintenance operations

You must always isolate an electrical machine from the mains supply before either cleaning it or carrying out any maintenance or repairs. There are two reasons for this: first, if you touch an electrically live part you may get an electric shock; second, the machine may be accidentally started which could cause injury or damage.

With portable electrical appliances this simply means switching off and taking the plug out of the socket.

With fixed machinery, for instance a pillar drill, you will need to switch off the power supply at the isolator switch. This is usually found on the wall near the machine. Isolating this way is fine while cleaning the machine, but for carrying out maintenance or repair work it is advisable to remove the supply fuse from the isolator box. With the fuse removed the machine cannot be restarted if the isolator is accidentally turned on by a colleague who confuses the isolator for the one on an adjacent machine.

2.8.19 Safe and approved methods for cleaning machines/equipment are used

There are three main items of cleaning equipment used in the garage: the cleaning bath (or tank), the pressure washer and the steam cleaner.

The cleaning bath uses a chemical solvent, this is usually used for cleaning dirty/oily components. The components are submerged in the solvent and dirt is loosened with a stiff-bristled brush.

The pressure washer is used for cleaning the mud off the underside of vehicles; water at very high pressure will clean off mud. For hard-to-remove dirt detergent can be added to the pressure washer.

The steam cleaner, often referred to as a steam jenny (jenny = generator), produces hot pressurized water with the option of detergent. This is used for removing very stubborn grease and dirt, like that found on the underside of high mileage goods vehicles.

If you are cleaning an engine or electrical equipment it is important not to get water inside. Before cleaning inside an engine bay with a pressure cleaner the electrical components and engine inlets should be covered over with polythene.

When cleaning portable electrical appliances be careful not to get water on the plug, this could cause a short circuit.

The mechanical parts of fixed machines may be cleaned with solvents, then dried with absorbent paper towel.

2.8.20 Appropriate cleaning and sanitizing agents are used according to manufacturer's instructions

Before using any solvent, detergent or sanitizing agent such as bleach you must read both the label on the container and the COSHH sheet which the manufacturer or your company has prepared.

Solvent should only be used in the cleaning bath for which it is designed.

The pressure washer or steam jenny should only be used with the recommended detergent.

Electrical items can be cleaned with one of the many aerosol sprays which are available for this purpose, but the volatile fumes which are given off must not be breathed in.

You should remember that all cleaning agents should be kept away from your mouth and eyes, and contact with your skin may cause irritation or a more serious skin disease. Always wash your hands and any other exposed areas of skin with toilet soap after carrying out a cleaning task.

2.8.21 Used agents are safely disposed of according to local and statutory regulations

The Environmental Protection Act (EPA) and local by-laws in most areas require that used cleaning solvents must be disposed of safely. This means that they must be put into drums and either collected by a refuse disposal firm or taken to a local authority amenity site where they are put into a large tank for bulk incineration. Several local authorities, for instance Surrey and Hampshire, are now looking at ways of using the energy produced by burning waste material to produce electricity. Emptying used solvents into the drain can lead to a heavy fine or even imprisonment.

Detergents are by their nature biodegradable, that is, they break down, do not build up sludge and will not explode, unlike solvents. However, if you use large quantities of detergents, wash bays which are fitted with the correct type of drainage system should be used.

2.8.22 Machinery, equipment and work areas are cleaned according to locally agreed schedules

In your company's Health and Safety Policy document there will be reference to the cleaning of the floors and equipment in the garage and general amenities such as toilets and rest areas. Also there will be maintenance and repair records for the workshop equipment which will include a regular schedule of cleaning and inspection.

Most companies work on the basis of sweeping down fixed machinery and floors at the end of each day, unless the generation of dirt requires more frequent attention.

On a weekly basis there will be a more thorough cleaning programme which may include window cleaning and wet cleaning certain areas.

Workshop equipment is usually cleaned and inspected on a monthly basis unless there is reason, such as a fault, for a more regular treatment.

2.8.23 Appropriate safety clothing and equipment is used when working with hazardous cleansing agents and equipment

To protect yourself from the cleaning agents which you are using you must, where appropriate, wear personal protective equipment (PPE). Most cleaning agents are poisonous and cause irritation or more serious complaints if allowed to come into contact with your eyes or skin.

Whenever you are working on a motor vehicle it is expected that you wear cotton overalls and safety footwear. In addition the HSWA requires that employers provide and employees wear the appropriate PPE for hazardous jobs such as using cleaning equipment. The general requirements are as follows:

1 Cleaning bath – rubber protective gloves which extend over the user's wrists, goggles and plastic apron. Avoid getting solvent on your overalls as this can lead to skin irritation, be especially careful not to put solvent soaked or oily rags in your overall pockets.
2 Pressure washer – rubber protective gloves and goggles, waterproof (plastic) over-trousers and jacket, and finally rubber boots (wellingtons). The idea is to be able to take the waterproof gear off and be dry underneath.
3 Steam cleaning plant – the hazard here is that as well as being wet the water is scolding hot. So the waterproof clothes must be of such a manufacture that they will protect the wearer from

the high temperature, high pressure steam. This means thick and strong over-trousers, coat, boots, gloves and a hat. A full-face mask is used to give complete protection.

Health and Safety issues are further discussed in chapter 15.

Questions

1 State five basic rules concerning dress and behaviour which demonstrate personal safety in the workshop environment.

2 List five necessary precautions for safety in the workshop and describe each one briefly.

3 What is meant by a skin care system as used in the workshop?

4 Explain the importance of eye and face protection in the workshop environment.

5 Explain the importance of protective clothing for a body repairer and a paint sprayer.

6 Explain the significance of headwear and footwear while working in the workshop.

7 Name the four types of respirator used in a bodyshop.

8 Why have the COSHH Regulations made the use of respiratory equipment mandatory?

9 State the minimum noise level at which ear protection must be used.

10 With the aid of a diagram, explain the fire triangle.

11 Name the three methods of fire extinction.

12 Explain the three classifications of fire.

13 Identify the correct colour code for the following fire extinguishers: water, foam, CO_2, powder, halon.

14 Name the four categories of safety signs used in the workshop.

15 Sketch and identify a safety sign used in a bodyshop.

16 Name the items of personal safety equipment that should be used when operating the following power tools: power saw, power chisel, disc sander.

17 State the essential personal safety precautions to be taken before working under a vehicle which is on a hoist.

18 Give practical reasons for wearing safety gloves in the workshop.

19 Identify the type of fire extinguisher that must be used when dealing with a solvent fire.

20 Explain the precautions which must be taken when handling toxic substances in a workshop environment.

21 When dealing with a petrol fire, which would be the correct type of fire extinguisher to use?

22 Explain the importance of the use of a barrier cream.

23 Name three important Acts of Parliament which influence the working procedures in a bodyshop.

24 Explain the following abbreviations: COSHH, EPA, HASAWA.

25 State the health hazards associated with the use of GRP for repairs.

26 What is meant by duty of care?

27 Describe how to safely store paint.

28 Describe how you might account for waste paint products.

29 What is PAT testing?

30 Why should you not smoke or drink alcohol when working as a vehicle repairer or refurbisher?

Hand and power tools

3.1 Hand tools used in body repair work

Mass production methods have made the present-day motor car such that it requires special techniques, skills and tools for the rectification of body damage. Specialist tools have been designed to suit the varying contours and shapes of the present all-steel bodies and panels. These tools are made of high-carbon tool steel, which is forged and then heat treated to give long service in the hands of a skilled body repair worker. In a body repair toolkit the basic tools are the hammer and dolly. All other tools have been developed around these, giving us the specialist tools which are now currently available. A repair job cannot be successfully carried out before one has completely mastered the skill of using the planishing hammer and dolly in coordination with each other, as this skill is the basis of all body repair work involving the use of hand tools.

Sections 3.2 to 3.8 describe body repair hand tools.

3.2 Hammers

Planishing hammer

The planishing or panel hammer is used more than any other tool in the body repair trade, and for this reason the best hammer available should be obtained. The principal purpose of the panel hammer is for the smoothing and finalizing of a panel surface after it has been roughed out to the required shape. The planishing hammer should have a true and unmarked face, and it must be kept polished and free from road tar, underseal and paint, which readily adhere to its working faces during use. This tool is designed solely to be used in conjunction with a dolly block; it must never be used for chiselling or any other work which might

mark or impair the faces of the hammer, for if the face of the hammer became marked the marks would be transferred to the surface of the panel.

These particular hammers are generally made with one face square. This face is usually flat, while the other, round end is slightly domed or crowned. However, hammers are available with the square end domed and the round end flat, or alternatively with both faces flat or both faces crowned. In practice the square end, which is usually flat, is used for planishing on a curved surface of a panel, or in corners, or against swaged recessed sections, and the crowned end for reasonably flat panels. The difference in the faces stops the panel surface being marked with the edge of the hammer when used in conjunction with the dolly block. The weight of the planishing hammer for general and new work ranges from 12 oz (340 g) to 16 oz (450 g), and the handles, which are usually very thin at the neck of the shaft for balance purposes, are made of hickory or ash to give the hammer a good rebound action when used with a dolly block.

Standard bumping hammer

This hammer (Figure 3.1a, b) is used for initial roughing out of work on damaged panels. It is also used for finalizing and finishing. The round face is $1\frac{3}{8}$ in (35 mm) in diameter and the square face is $1\frac{1}{2}$ in (38 mm) square, and the total weight is 14 oz (395 g). The hammer is made with either flat or crowned faces.

Light bumping hammer

This (Figure 3.1c, d) tool is ideal for work on light gauge materials. It is used in the same manner as the standard bumping hammer. The squared face is 1 in (25 mm) square and the round face is $1\frac{1}{4}$ in (32 mm) in diameter, and it has a weight of 12 oz

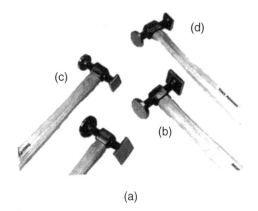

Figure 3.1 (a) Standard bumping hammer (crowned faces) (b) standard bumping hammer (flat faces) (c) light bumping hammer (crowned faces) (d) light bumping hammer (flat faces) (*Sykes-Pickavant Ltd*)

(340 g). This hammer is also obtainable in flat or crowned faces.

Dinging hammer

This hammer (Figure 3.2a, b) is a long-reach planishing hammer and is designed for careful, controlled finishing work. It is extremely well

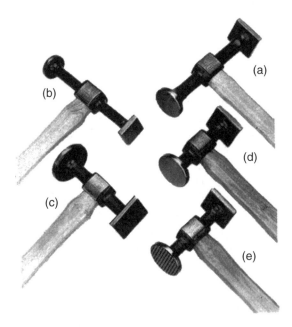

Figure 3.2 (a) Heavy dinging hammer (flat faces) (b) light dinging hammer (crowned faces) (c) heavy shrinking hammer (serrated flat face) (d) light shrinking hammer (milled flat face) (e) light shrinking hammer (shallow milled flat faces) (*Sykes-Pickavant Ltd*)

balanced and gives a very good finish when used correctly. Hammers can be obtained for light or heavy work with weights from 12 oz (340 g) to 18 oz (510 g).

Shrinking hammers

Shrinking hammers (Figure 3.2c, d, e) are similar in design to a normal planishing hammer but have faces which, instead of being smooth, are serrated, giving a cross-milled effect like a file. The purpose of these serrations is to achieve a shrinking effect when the hammer is used in conjunction with a dolly block. This is caused by the fact that the contact area between hammer and metal is greatly reduced by the serrations on the face. This tool is used largely when beating the surface on over-stretched panel areas which have to be hot-shrunk in order to return them to their normal contours. Hammers are available for light or heavy shrinking, according to the depth of the serrations.

Pick and finishing hammer

This tool (Figure 3.3a, b) is used in place of, or in conjunction with, the planishing hammer. Its main use is to pick up small, low areas on the surface of a panel which is in the process of being repaired by planishing. On panels that are reasonably flat, such as door panels, parts of roof panels and bonnets, this method of raising low areas is quick and, if carried out correctly, does not unduly stretch the metal.

To lift a low area with a pick hammer, one or two taps with the pick end of the hammer are directed from underneath the panel under repair to the centre of the low area. The blows stretch the metal sufficiently to raise the surface surrounding the point of the low spots where the blows were struck. This slightly raised area is next tapped down lightly with a planishing hammer or the finishing end of the pick hammer on to a suitably shaped dolly block, and the panel is finished off by filing with a panel file. When one becomes proficient in using this tool it is possible to raise the surface with light blows and finish off by filing only. However, without sufficient experience there is a danger of over-stretching the metal owing to the inability to direct the blow accurately on to the low area under repair.

The pick and finishing hammer has a pointed end which is suitable for removing low spots and is also a useful finishing hammer, having a crowned

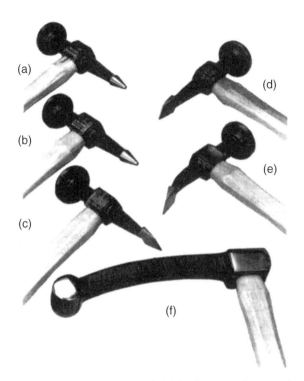

Figure 3.3   (a) Pick and finishing hammer (crowned face) (b) pick and finishing hammer (flat face) (c) straight pein and finishing hammer (crowned face) (d) straight pein and finishing hammer (flat face) (e) curved pein and finishing hammer (crowned face) (f) fender bumping hammer (*Sykes-Pickavant Ltd*)

surface on a round face of $1\frac{1}{2}$ in (38 mm) diameter; it is well balanced and weighs 14 oz (395 g).

Straight pein and finishing hammers

These hammers (Figure 3.3c, d) are used in a similar manner to the pick and finishing hammer, but are designed with either a straight or a curved peined end which acts like a chisel, and a domed round end which is used for planishing. They are suitable for roughing out prior to planishing, or in the finishing stages of planishing for stretching small low areas. These hammers can be used to dress out sections which are difficult to work on owing to their awkward shape or position, such as around lamp openings and in recessed and moulded sections on panels.

Curved pein and finishing hammer

This tool (Figure 3.3e) is identical in use to the straight pein and finishing hammer except that its curved pein end allows for greater flexibility in dressing out sections which are difficult to work on owing to their awkward shape or position.

Fender bumping hammer

This tool (Figure 3.3f) has a long curved head with one face which is circular to reduce the effect of stretching the metal when in use. This hammer is used for roughing out and dressing out damaged sections on panels to restore them to their correct shape and curvature before planishing begins. The heavy weight of this hammer, together with the curve, makes it very effective for hammering out difficult and inaccessible sections.

File hammer or beating file

This tool (Figure 3.4) is designed to be used like a hammer in conjunction with a dolly block, although it is actually a file with a serrated face and is suitably shaped for holding in the hand. The milling on the file blade tends to shrink the panel as well as leaving a regular rough patterned surface ideal for locating low spots on the panel under repair and for finishing with a body file. The tool is used in conjunction with a hand dolly, and with a glancing blow. It is most effective on large flat sections, where it will be found ideal for smoothing and levelling out wavy panels. Two types of beating files are available; one is flat for use on low- and high-crowned surfaces, while the other is half-round in shape and is used on convex or reverse-curved panel sections.

Mallets

Mallets can be of the round or pear-shaped type made from boxwood or lignum vitae, or can be

Figure 3.4   Beating files (*Sykes-Pickavant Ltd*)

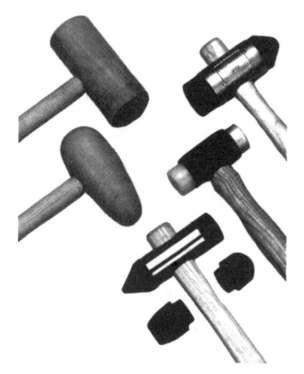

Figure 3.5   Mallets (*Sykes-Pickavant Ltd*)

rubber, aluminium or plastic faced (Figure 3.5). Some mallets have interchangeable heads so that the correct head can be used for the material being worked. A mallet is greatly used in the initial stages of smoothing and roughing out of a panel prior to planishing. When hot shrinking, the mallet is the tool to achieve a most successful shrink because a normal planishing hammer would tend to stretch the metal rather than shrink it. Without a mallet, aluminium work would be most difficult as this metal is so easily marked and stretched. The working faces of the mallet must be kept in first-class order, or marks on the surface of the metal will result.

3.3 Hand Dollies

These are either cast or drop forged steel blocks, heat treated to provide the correct degree of hardness. The shapes of the dolly blocks have been designed to provide a working surface that is highly polished and suitable for use on the many contours found on motor vehicle bodies. They are used in conjunction with the planishing hammer or beating file and act as a support or anvil to smooth out the surface area of panels that have been damaged.

These dollies, together with the planishing hammers, are the most essential tools for the panel beating trade. Obviously one dolly block will not be suitable for all shapes requiring planishing; therefore it is advisable to have a set of these dollies which would be suitable for a wide range of the shapes and contours encountered on the ever-changing body styles of the modern motor vehicle.

When selecting a dolly block for a particular job, it should always be remembered that as flat a dolly block should be used as possible for the job in hand; then the dolly will not only cover the panel area quicker because of its bigger face area, but will smooth out the metal without excessive stretching. When working on panel contours it must be borne in mind that a dolly block having a high-crowned surface will tend to stretch the metal much quicker than one having a low-crowned surface; hence the choice of block depends on whether the particular section under repair needs to be stretched quickly or just smoothed and planished without stretching.

It is very common for the faces of the dolly blocks to become coated with paint, road tar, or anti-drum compounds which it picks up from the underside of panels under repair. This coating must be removed from the surface of the dolly block so that when it is used with the hammer there is a metallic contact (metal to metal) between dolly block and work and hammer. This contact should be heard as a ringing noise if successful planishing is to be achieved.

Various dollies are shown in Figures 3.6 and 3.7. Some details are as follows.

Double-ended hand dolly is conveniently designed for a good grip with two useful surfaces, one high crowned and one low crowned. It is ideal for general planishing and roughing out. Its weight is 3 lb 4 oz (1473 g).

Utility dolly forms the basis of every body kit. It offers a variety of useful faces, and is found ideal for working in confined spaces.

General-purpose dolly is very well shaped and easy to hold and offers wide, low-crowned faces which are essential when working on the new body styles. Its weight is 2 lb 12 oz (1247 g).

Heel dolly has a low flat face with radiused corners. It is suitable for corner and angle work and is easy to handle because of its fairly small weight of 2 lb 14 oz (1020 g).

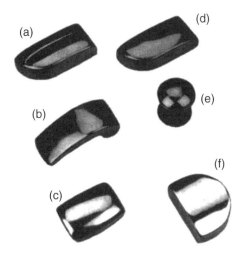

Figure 3.6 (a) Shrinking dolly (b) lightweight curved dolly or comma block (c) regular dolly (d) toe dolly (e) round dolly (f) heel dolly (*Sykes-Pickavant Ltd*)

Toe dolly combines a large flat face with a low crown on other faces and is long and thin for easy handling in narrow sections. The bottom face is ground flat which adds to the versatility of this block. Its weight is 3 lb (1360 g).

Thin toe dolly has similar surfaces to the toe dolly but is thinner and is ideal for working in awkward

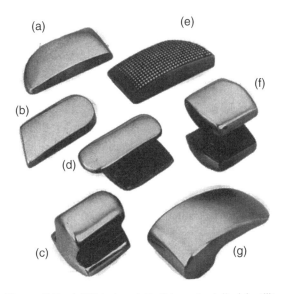

Figure 3.7 (a) Thin toe dolly (b) angle dolly (c) utility dolly (d) general-purpose dolly (e) grid dolly (f) double-ended dolly (g) curved dolly or comma block (*Sykes-Pickavant Ltd*)

places where other dollies cannot be used. Its weight is 2 lb 4 oz (1020 g).

Shrinking dolly block is designed for shrinking welded seams and reducing stretched areas prior to filling. It has a groove down the centre of the block into which the stretched or welded seams can be dressed. Its weight is 3 lb (1360 g).

Grid dolly is similar in shape to the toe dolly but has a large crowned grid face on the upper surface. The base has a plain flat face for normal finishing work. It has been designed to act as a shrinking dolly when used in combination with a shrinking hammer and the application of heat. The weight is 2 lb 4 oz (1130 g).

Curved dolly or comma block has a long curved face, combined with the high- and low-crowned areas and the tapering face; it is extremely well suited to the modern body design, and is comfortable to hold in difficult areas and narrow corners. The weight is 3 lb 6 oz (1530 g).

Round dolly is light and small and very easy to hold. It offers high- and low-crowned faces for work on small areas of damage. The weight is 1 lb 10 oz (736 g).

3.4 Body spoons

These tools are made from a high-grade steel which has been drop forged and heat treated. They are sometimes called prying spoons because the spoon end is used in the same manner as a dolly in conjunction with a hammer. The body spoon really does the same job as a dolly block but is designed for use in confined spaces where a normal dolly block cannot be held in the hand, e.g. between door frames and outer door panels. The spoon end, which acts as the dolly, must be kept in good condition and free from anti-drum compound so that it gives a metal-to-metal contact when used in conjunction with a hammer. The spoon can also be used for roughing or easing out by wedging the spoon in between panels.

Surface spoon (Figure 3.8a) has an extra large working area which is very slightly crowned, and it is ideal for working in between and round struts and brackets without much dismantling.

General-purpose spoon (Figure 3.8b) is a double-ended body spoon and is also a basic item in most panel beaters' toolkits. The applications for a spoon

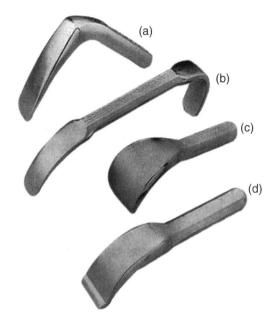

Figure 3.8 (a) Surface spoon (b) general-purpose spoon (c) high-crowned spoon (d) drip moulding spoon (*Sykes-Pickavant Ltd*)

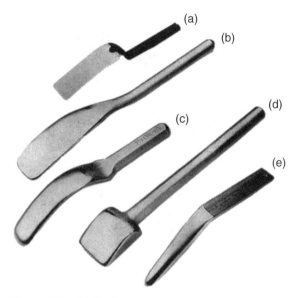

Figure 3.9 (a) Spring hammering spoon (b) heavy-duty pry spoon (c) pry and surfacing spoon (d) long-reach dolly spoon (e) thin spoon (*Sykes-Pickavant Ltd*)

offering such different curves are many and varied and meet the needs of modern body shapes.

High-crowned spoon (Figure 3.8c) has been designed to offer a broad working blade with a high crown to work in positions where other dollies and spoons cannot be used.

Drip moulding spoon (Figure 3.8d) has a special lip which can be hooked under the drip moulding of the roof section and hence simplifies the repair. The curved surface is fully finished to enable the tool to be used as a short standard spoon.

Spring hammering spoon (Figure 3.9a) is a light pressed steel spoon which is designed for spring hammering on panels which have minor blemishes in their finishing stages. The broad blade spreads the blow evenly over a larger contact area, thereby reducing the possibility of sinking the metal. This spoon is not made for prying or levering.

Heavy-duty pry spoon (Figure 3.9b) is ideal for heavy prying and roughing-out work. It has a long low curved blade which can be inserted into very thin sections.

Pry and surfacing spoon (Figure 3.9c) has a short handle which gives easy access in limited spaces and can be used efficiently for prying behind brackets and panel edges.

Long-reach dolly spoon (Figure 3.9d) is a special long-handled body spoon designed to be used in restricted spaces between double-skinned panels such as doors and quarter panels.

Thin spoon (Figure 3.9e) is a very special thin-bladed spoon, the blade having a very slow taper which permits entry into restricted spaces between double-skinned panels which would not be possible by using a normal spoon.

3.5 Body files

Flexible panel file

This tool (Figure 3.10a) is designed with a two-position handle, and has a 14 in (30 cm) spring steel backing plate to give adequate support over the whole blade. Positioned between the two hand grips is a turnscrew threaded left and right hand for adjusting the blade to concave or convex positions to suit the user's requirements. The main use of this stool is to assist in the final planishing of the work. First and most important, it locates areas which are low on the surface of the panel under repair; second, it files out small marks or defects on the panel surface. It can be adapted to file the surface of almost any shaped panel by setting the blade

Half-round body file

This tool (Figure 3.10c) is similar in design to the flat file but the blade is half-round. It is very useful for curved surfaces.

Abrasive file

This (Figure 3.10d) is not a normal file but an abrasive holder with a wood handle having spring clips at either end on to which can be attached an abrasive grit paper. This tool is of greatest use when rubbing down plastic body fillers to their final finish.

Body file blades

These flexible, double-sided blades are produced from a special alloy steel and heat treated. The milled teeth allow a smooth filing operation and are specially shaped to reduce clogging. The blades are either flat or radiused according to the surface to be filed.

Supercut blades are 8 tpi (teeth per inch) general-purpose standard blades.
Standard cut blades are 9 tpi and used for soft metals such as lead and solder fillers.
Fine cut blades are 13 tpi and used for aluminium, copper and brass.
Extra fine cut clades are 17 tpi and suitable for cast iron, steel or any narrow section of metal.
Plasticut blades are 6.5 tpi and used for plastic body fillers only.

3.6 Hand snips

The offset combination of universal snips is preferred by the panel beater when cutting thin gauge metal. Universal snips (Figure 3.11a, b) are suitable for cutting straight lines, outside and inside curves. A right- and left-hand pair of combination snips will be suitable for most of the sheet metal cutting that will be encountered by the panel beater, and there is no need for any curved-blade snips. When the more popular right-hand snips are used, the waste metal forms a coil to the left of the cutting blades, thus causing little distortion to the surface of the sheet or panel being cut; similarly, with the left-hand pair of snips the waste metal passes to the right of the cutting blades, leaving the undistorted sheet or panel on the left. Snips can be obtained in

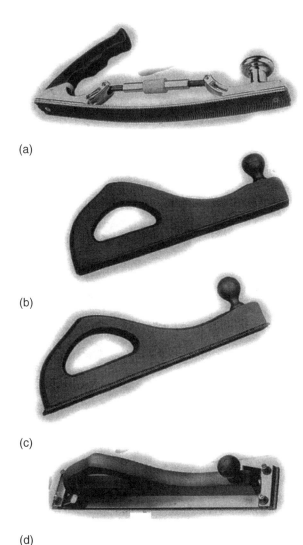

(a)

(b)

(c)

(d)

Figure 3.10   (a) Flexible panel file (b) flat file (c) half-round body file (d) abrasive file (*Sykes-Pickavant Ltd*)

either straight, concave or convex. The file blades are specially designed so that they do not remove too much metal, and the milled serrations are wide apart and curved to prevent clogging when filing metal which has been painted, soldered or plastic filled. It is important to release the tension on the file blade after use in order to reduce the risk of breakage if it is dropped or struck by accident.

Flat file

This file (Figure 3.10b) has a solid wood holder designed to take a standard 14 in (30 cm) blade, which must be used in a rigid position.

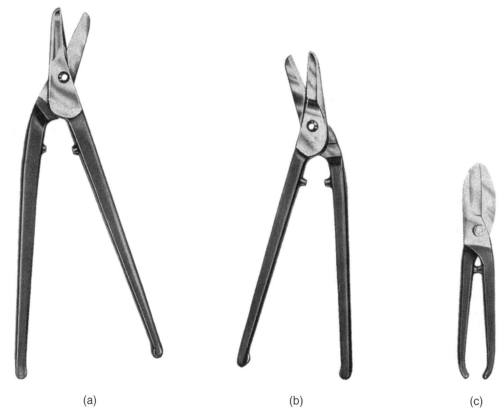

(a) (b) (c)

Figure 3.11 (a) Universal combination snips RH (b) universal combinations snips LH (c) standard pattern straight snips (*Sykes-Pickavant Ltd*)

varying sizes with either straight or crank handles. A straight pair of snips is often necessary for long straight cuts (Figure 3.11c).

Section 3.15.1 gives further information on hand snips.

3.7 Specialist panel beating tools

Panel puller

This tool (Figure 3.12) comprises a long steel rod with a cross T-piece at the top which acts as a hand grip. At the other end a fixed hexagonal nut holds a strong self-tapping screw, while a heavy cylindrical weight can slide up and down the shaft against a stop which is near the handle. This tool has been designed to pull out dents and creases from the face side of the panel without the necessity of removing the trim and lining materials to gain access behind the dents. It can also be used on double-skinned panels where access is impossible with conventional tools.

The panel puller is used by drilling a $\frac{1}{8}$ in (3.175 mm) diameter hole at the deepest part of the dent, then inserting the self-tapping screw in the

Figure 3.12 Panel pullers (*Sykes-Pickavant Ltd*)

hole and screwing it until a firm grip is obtained. One hand holds the T-bar while the other pulls the sliding weight towards the stop; there the weight rebounds, forcing the dented panel out under the impact of the blow.

Zipcut spot-weld remover

The Zipcut spot-weld remover (Figure 3.13) is used with an electric or air drill, and is ideal for removing spot welds on all areas of bodywork and subframes. The cutter blade A is reversible with two cutting edges. Adjustment B provides for varying depth of cut so that only the upper panel is released (c), leaving the original spot weld behind on the lower panel.

Cone drills or variable hole cutters

These are special alloy steel drill bits which have been hardened to give a fine cutting edge (Figure 3.14a). They are shaped like a cone and fit in the chuck of a standard power drill. The hole diameter can vary from 6 to 40 mm depending on the hole cutter in use. These cutters are used for fitting wing mirrors, aerials and rubber grommets into body structures.

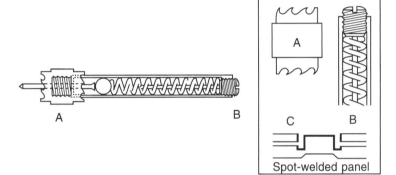

Figure 3.13   Zipcut spot-weld remover (*Sykes-Pickavant Ltd*)

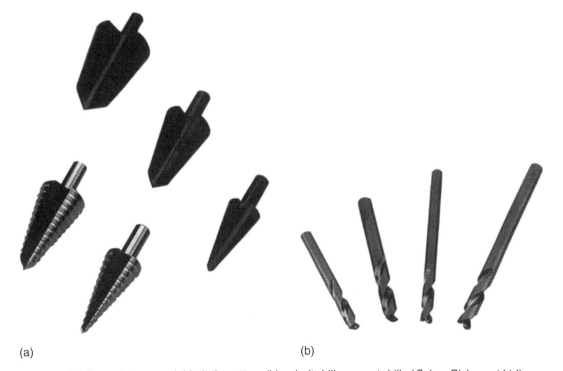

(a)   (b)

Figure 3.14   (a) Cone drills or variable hole cutters (b) cobalt drills or spot drills (*Sykes-Pickavant Ltd*)

Cobalt drills or spot drills

These are special alloy steel drills to which cobalt has been added to give a very hard sharp cutting edge (Figure 3.14b). The design of the cutting edge allows spot welds to be drilled out of panel assemblies without creating any panel distortion.

Impact driver

The impact driver (Figure 3.15) will loosen or tighten the most stubborn screws, nuts and bolts. The tool is supplied with a complete range of interchangeable bits.

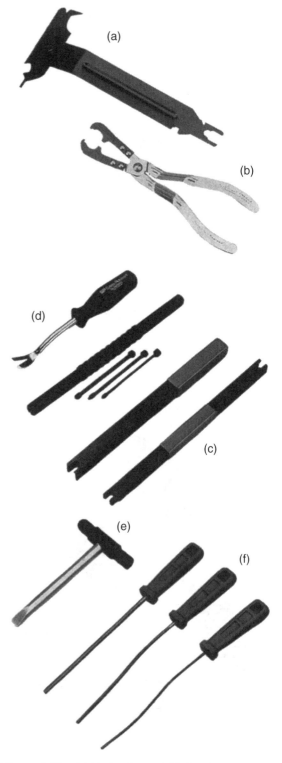

Figure 3.15 Impact driver (*Sykes-Pickavant Ltd*)

Body trim tool

This is a universal spring steel tool for removing body trims and clips, and also weather strips, door trims, headlamp fittings, windscreen clips, plastic mouldings, motifs and badges (Figure 3.16a).

Door handle spring clip removers

These are specially designed pliers for fitting or removing the spring clips used on many modern door handles. One jaw sets in the neck of the spring whilst the other grips the other end, and the whole spring is then levered out (Figure 3.16b).

Trim panel remover

This tool is designed to remove vehicle trim panels, upholstery and roof liners without damage (Figure 3.16c). It also fits a wide variety of button-type fasteners used on vehicle trim.

Door hinge pin remover and replacer

This is designed for the removal and replacement of hollow hinge pins used on vehicle door hinges (Figure 3.16d).

Figure 3.16 (a) Body trim tool (b) door handle spring clip remover (c) trim panel remover (d) door hinge pin remover and replacer (e) corrosion assessment tool (f) bendable files (*Sykes-Pickavant Ltd*)

Corrosion assessment tool

This tool has been designed for vehicle inspection (Figure 3.16e). It acts as a combined tapping hammer, blunt scraper and short lever. It has been introduced to standardize the methods of corrosion assessment employed by MOT testers, and is ideal for bodyshops conducting pre-MOT checks and assessment for corrosion.

Bendable file

This tool is a standard round file but it can be bent to the desired shape, which allows it to be used in difficult locations where a normal straight file would not have easy access (Figure 3.16f).

Clamps

Sheet metal clamp (Figure 3.17a) is a general-purpose clamp which can be locked on to sheet metal with a powerful quick-release grip, making panel assembly and tack welding procedures much easier operations.

Welding clamp (Figure 3.17b) has a powerful grip which holds parts in alignment while leaving both hands free for the welding operation. The deep-throated jaws and centre opening provide the operator with maximum visibility and full access to the welding area.

C clamp (Figure 3.17c, d) has a wide jaw opening with a relatively small gripping area, which produces pressure without permanent jaw damage to panel surfaces. It allows accurate close-up working in restricted areas and on awkward shapes, e.g. sill panels.

Punches

Hole punches (Figure 3.18a) have interchangeable heads to punch holes of either $\frac{3}{16}$ in (5 mm) or $\frac{1}{4}$ in (6 mm) diameter, and enable joining panels to be accurately aligned for welding. The $\frac{3}{16}$ in (5 mm)

(a)   (b)

Figure 3.18   (a) Hole punch (b) wing punch (*Sykes-Pickavant Ltd*)

diameter hole is for gas welding or brazing and the $\frac{1}{4}$ in (6 mm) hole is for MIG welding.

Wing punch (Figure 3.18b) is a hole punch with a specially designed head which allows it to be used on wing panels and channel sections and also fit over roof gutters and wheel arches.

Edge setters

The edge setter (Figure 3.19a) is a portable, hand-operated tool designed to provide a 'joggled' joint or stepped edge on a repaired or new panel, thus

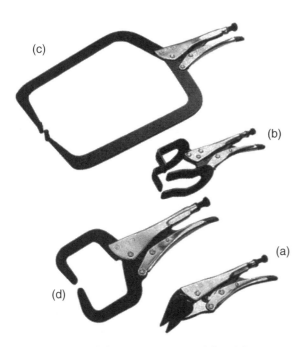

Figure 3.17   (a) Sheet metal clamp (b) welding clamp (c) extra large C clamp (d) C clamp (*Sykes-Pickavant Ltd*)

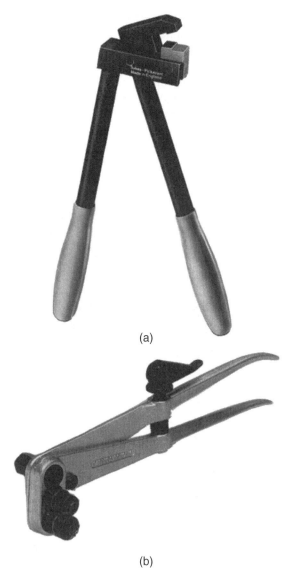

(a)

(b)

Figure 3.19 (a) Edge setter (b) Rolastep edge setter (*Sykes-Pickavant Ltd*)

Door skinner

This is a special tool (Figure 3.20) for crimping flanges tightly on replacement door skins. It is used by first bending the flange to an acute angle and then tightening or crimping the flange to the door frame. During this operation, interchangeable tough nylon pads prevent damage to the surface of the door skin.

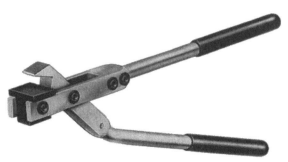

Figure 3.20 Door skinner (*Sykes-Pickavant Ltd*)

Inter-grip welding clamp

The inter-grip clamp is approximately 25 mm cube shaped, and uses a wing nut and toggle bar to firmly clamp pieces of metal together edge to edge and perfectly level, leaving only a small gap which will allow full penetration of the weld and filler rod (Figure 3.21). Once the two sections are tack welded together the wing nut is slackened, the toggle bar removed and the clamp lifted from the

creating a flush-fitting lap joint. The joggled panel edge provides a stiff joint and helps prevent distortion when welding. It can be used on mild steel up to 1.2 mm (18 gauge) and applied in many situations which require this type of joint.

The rolastep edge setter (Figure 3.19b) is a portable tool used on the vehicle or a panel assembly. Its rollers produce a smooth, uniform stepped panel edge to create a flush-fitting lap joint, allowing a panel replacement to be inserted.

Figure 3.21 Inter-grip welding clamps (*Frost Auto Restoration Techniques Ltd*)

job, prior to final welding. The end result is a first-class butt weld requiring very little dressing. The clamps are ideal when welding patch repairs in floors, wings and door skinning, or when assembling fabricated or new panels.

Temporary sheet metal fastener system

This sheet metal holding system is a set of pins, installed with special pliers, which quickly and easily clamps panels together before final welding, bonding or riveting (Figure 3.22). The clamps will securely hold together body panels and sections where no other clamp will reach because there is no rear access. The 3 mm blind fasteners have an expanding pin which, when inserted through the panel and released, exerts over 8 kg (18 lb) holding pressure. The edge grips have 12 mm deep jaws to clamp flanges and other edge work. To install or remove either, the pliers are needed to overcome the powerful spring pressure.

3.8 Recommended basic toolkits for panel beaters

The pride of the craftsman is the set of tools he possesses. Care of tools throughout their working life is important. All bright surfaces should be kept clean and free from scours and blemishes that could be transferred to the body panels. Stowage of tools when not in use is also important, but this will depend on the working conditions in the bodyshop. Wall boards with the necessary clips and tool silhouettes can be used with a specified set of tools. This is ideal for general usage from a central store, as missing tools can be quickly identified. Metal toolboxes are most useful when each worker maintains his own toolkit.

The apprentice, improver and tradesman will each collect his or her own basic toolkit: suggested sets are shown in Figures 3.23, 3.24 and 3.25 respectively.

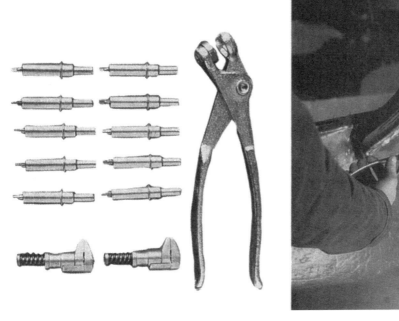

(a)

(b)

Figure 3.22   Temporary sheet metal fastener: (a) system (b) in use (*Frost Auto Restoration Techniques Ltd*)

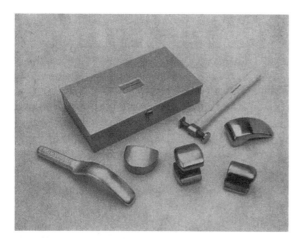

Figure 3.23 Apprentice basic toolkit (*Sykes-Pickavant Ltd*)

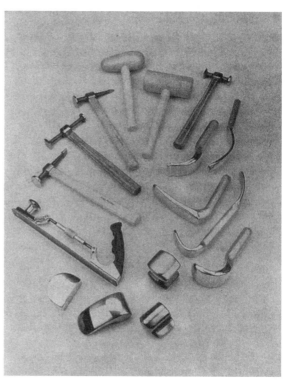

Figure 3.25 Tradesman basic toolkit (*Sykes-Pickavant Ltd*)

replacement panels and sections are not readily available. These specialist tools are also frequently used in the coach building industry during the construction of new vehicles.

Sections 3.10 to 3.16 cover fabrication tools.

3.10 Hammers and mallets

3.10.1 Hammers

Hammers (Figure 3.26) are still the most important forming tools as far as the body worker is concerned, and from the commencement of his career he should attempt to acquire a skilful hammering technique. The essential points to look for when hammering are first, to hold the hammer correctly, gripping the shaft firmly but not tightly and towards the shank end in order to take full advantage of its length. Second, the action should be produced by the wrist rather than the forearm, and as strength is developed in the wrist it will be found that it is possible to control the hammer under all circumstances. The hammer head should only be swung

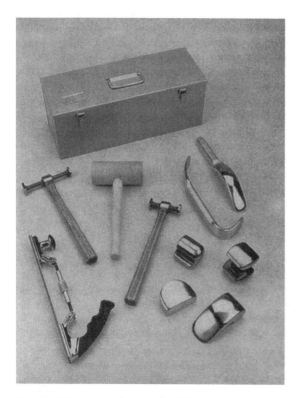

Figure 3.24 Improver basic toolkit (*Sykes-Pickavant Ltd*)

3.9 Hand tools used in the fabrication of sheet metal

The panel beater often finds the need to fabricate articles or components in mild steel, using specialist hand tools. This need usually arises when

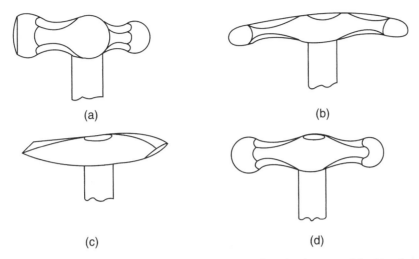

Figure 3.26 (a) Engineer's ball-pein hammer (b) stretching and flanging hammer (c) wiring hammer (d) hollowing or blocking hammer

sufficiently to ensure that the metal is struck where required using the correct amount of force.

Engineer's ball-pein hammer

This is a general purpose hammer. Its main uses are for riveting, chiselling, forming angles and brackets, planishing heavy welds to refine their structure and flattening heavy-gauge plate. The hammer has one end round and flat and the other a rounded ball-pein, and it usually has an ash handle fitted into the head. It is made in a variety of weights to suit individual needs.

Stretching and flanging hammer

This is used to form external and internal flanges on curved surfaces by hand. The hammer is curved with two rounded cross-peins which allow blows to be struck at right angles in the direction of stretching. By tilting the hammer very slightly it is possible to gain maximum stretching on the outside of curved work. This hammer is designed for turning small flanges on highly curved work and is normally used in conjunction with a stake.

Wiring hammer

This is designed for closing thin-gauge metal over wire to form wired edges. The head of the hammer is curved and has two very sharp chisel-like cross-peins which allow it access to tuck down the metal around the wired edge. It can be used for either straight or curved wiring, and is used extensively in the manufacture of wings made for heavy transport.

Hollowing or blocking hammer

This is a shaping hammer which has two ball-peined ends and is rather heavy. It is designed to be used on mild steel in conjuctions with a sandbag or wood hollowing block, and to create a double-curved shape by hand. The hammer is used on the initial stages of shaping metal, the component normally being planished or wheeled to its final shape.

3.10.2 Mallets

Mallets (Figure 3.27) come in two shapes.

Figure 3.27 Round-faced mallet, pear-shaped mallet and sandbags (*F. J. Edwards Ltd*)

Round-faced mallets

These are used in conjunction with steel stakes for forming and shaping. There are two types of mallet – boxwood and rawhide. The latter is made in strips which are rolled and pinned in position. Usually rawhide is used on aluminium and the lighter metals as it gives a softer blow than boxwood and is less liable to mark the metal. The boxwood type is an all-purpose mallet and can be used on either mild steel or aluminium with equal success.

When using a round-faced mallet the metal should be struck as squarely as possible to avoid wear on one side of the face. Care should also be taken to avoid striking the raw edges of sheet metal, as these could split the mallet head or damage the faces. At all times the faces of these mallets must be kept smooth, true and free from marks as these could be automatically transferred on to the workpiece.

Pear-shaped mallets

These are usually made in boxwood or lignum vitae, which is a very hard wood and hence is ideally suitable for this type of work. The mallet is used in conjunction with a sandbag or wood block for either hollowing or raising when shaping metal by hand to a double curvature shape. The high-crowned end of the mallet is used for hollowing into a sandbag and the small-radiused end of the pear shape is used for glancing blows as used in raising. The mallet is normally used on aluminium and thingauge mild steel. Care should be taken not to allow the faces of the mallet to become damaged when in use.

3.11 Sandbags and hardwood blocks

Sandbags are leather bags made from two pieces of the finest quality leather, filled with very fine sand and stitched together. They are available in a variety of weights and sizes but are usually round or square in shape (Figure 3.27). The sandbag is used with the pear-shaped mallet or the blocking hammer to create panels by hand to a double curvature shape. The sandbag is resilient and at each blow of the mallet allows the metal to shape into the sandbag.

In some instances hardwood blocks are hollowed out to different depths and diameters. These blocks are used in the same manner as the sandbag for shaping metal, although the blocks are more solid and do not 'give' as much as the resilient sandbags.

3.12 Sheet metal bench stakes

Quite often when making panels and parts of panels in sheet metal, it is necessary to perform some operation such as bending, flanging, seaming, shaping and planishing which cannot be done by machine. In such cases these operations will have to be carried out by hand using bench stakes. There are many different types of stake in use; the smaller types are usually made from cast steel, the heavier are of cast iron, and for special work stainless steel stakes may be used.

All these stakes must have highly polished working surfaces and should be kept in a dry atmosphere to reduce the possibility of corrosion on the working surface. Certain stakes are used only occasionally, and to maintain a bright, smooth surface they must be given a coating of light oil from time to time. The condition of the stakes has much to do with the workmanship of the finished job, for if a stake has been roughened by hammer or chisel marks the completed job will look rough and lack finish. For this reason stakes should not be used to 'back up' a sheet of metal or a panel directly when cutting it with a chisel; hammer blows should be aimed with care; and a mallet should be used whenever possible when forming sheet metal over the stakes.

Special reinforced square holes of various sizes are made at intervals in the bench for insertion of the bases of the stakes, which are usually square in cross-section and taper in shape to facilitate a good grip in the bench. There are many different types of stake available, and those most commonly used are as follows (Figure 3.28):

Pipe stake consists of one or two cylindrical arms having different diameters. This stake is used for forming pipes and cylindrical workpieces.
Grooving stake has a grooved slot cut along the top face. The slots are used for wiring and swaging.
Funnel stake has a thick tapered head and is used in forming, riveting and seaming tapered articles such as funnels.
Anvil stake has a flat, square-shaped head with a short shank and is used for general working operations.

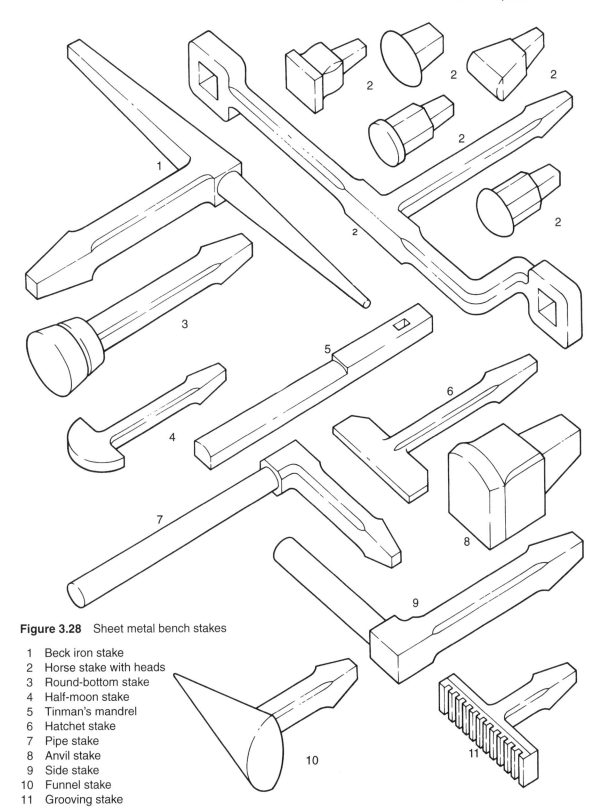

Figure 3.28 Sheet metal bench stakes

1 Beck iron stake
2 Horse stake with heads
3 Round-bottom stake
4 Half-moon stake
5 Tinman's mandrel
6 Hatchet stake
7 Pipe stake
8 Anvil stake
9 Side stake
10 Funnel stake
11 Grooving stake

Side stake consists of one long elliptical shaped head. It is used for bending, seaming and forming curved radii.

Beck iron stake has a square tapering flat head on one side and a round tapered head on the other side. It serves as a general purpose anvil for riveting and shaping round and flat surfaces, straight bending and corner seams.

Horse stake with heads has holders (the horse part of the stake) which can take various shaped small heads. This stake is useful in many operations for which other stakes are not suitable.

Round-bottom stake consists of a single vertical piece with a flat round head on top. It is used for flanging circular and curved work.

Tinman's mandrel is a single horizontal metal bar, one section of which has a flat surface and the other section of which is rounded. The flat length has a slot cut in it which permits the stake to be fastened directly to the bench. The stake is used for riveting, seaming and forming.

Half-moon stake consists of a single vertical shank with a half-rounded head on top. The stake is used for forming or shaping curved flanges.

Hatchet stake consists of a horizontal, sharp, straight edge, and is used for making straight sharp bends.

Panel head stake is shaped like a mushroom and is made in various sizes. It is used for raising, tucking and planishing.

Ball stake is shaped like a ball and is used for the planishing of hollowed or raised articles. It is also used for raising domes or hemispherical shapes by mallet or raising hammer.

3.13 Hand grooving tools

The hand groover is used when making a seam joint by hand between two pieces of sheet metal. The end of the tool is recessed to fit over the joint formed when the edges of the two pieces of metal are prebent into a U-shape and interlocked. The grooving tool is then placed over the interlocked edges and hammered. One edge of the tool acts as a guide while the other forces the metal down into a crease, thus forming a groove. This crease or set in the metal should be made gradually, the tool being hammered while tilted very slightly in the direction in which the crease is to be made. This will avoid marking the metal on the opposite side of the joint to the crease. A tool a size larger than that of the

grooved joint should be chosen to give clearance in use. The tools are manufactured in various sizes from $\frac{1}{8}$ in to $\frac{3}{4}$ in (3–19 mm). These grooved joints can be used to advantage when fastening large flat panels together as used on commercial vehicles.

3.14 Rivet sets

A rivet set is made of tool steel. A deep hole in the bottom of the set is used to draw the rivet through the metal sheet before riveting, and a cup-shaped hole is used to form the finished head on the rivet. The drawing-up hole and the cup hole are offset slightly to the direction of the vertical hammer blows, and therefore great care should be taken to hold the set in an upright position when riveting. If the tool is allowed to tilt it is liable to make indentations on the area of the metal which has been riveted. Hand rivet sets are made in various sizes ranging from $\frac{1}{8}$ in to $\frac{3}{4}$ in (3.1–19 mm).

3.15 Cutting tools

3.15.1 Hand snips

These are used to a great extent in the fabrication or repair side of the panel beating trade, for cutting thin-gauge mild steels and aluminium alloys. The limits to which these snips can cut satisfactorily, leaving a good edge, is a thickness of 20 SWG (1 mm) in mild steel and slightly greater thicknesses in aluminium. Snips are available in a variety of types and sizes and are made from the finest tool steel forgings. The blades are ground and set for the accurate cutting of all sheet material.

The most popular hand snips are the *universal combination snips,* which are made in sizes 10 in to 13 in (254–280 mm) with narrow or broad blades, with straight or cranked handles, and left or right handed. The handles are offset on the cranked variety so that straight and curved cuts can be made in any direction in one operation with the sheet kept flat and the operator's hand clear of the metal, the cut being followed from above. The word 'universal' means that this single pair of snips can be used for cutting straight lines and both external and internal curved edges. *Standard pattern straight snips* are also very popular, although limited in their use to straight line cutting or large external curves. Sizes vary from 6 in

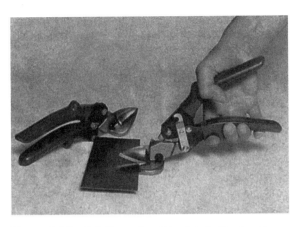

Figure 3.29 Aviation snips (*Frost Auto Restoration Techniques Ltd*)

to 12 in (152–305 mm). *Standard pattern curved* or *bent snips* are designed for cutting circular outlines and inside circles or curved shapes, as the actual cutting blades themselves are curved. (See Figure 3.11.)

The cutting tools known as *aviation snips* were originally developed for the aircraft industry (Figure 3.29). In contrast to the single pivot of ordinary tin snips, aviation snips feature compound leverage for greater shearing power, being able to cut up to 18 gauge (1.2 mm) steel and 16 gauge (1.6 mm) aluminium. The blades are made from heat-treated forged chrome steel and are radiused in order to maintain a constant cutting angle from the back of the jaw to the front, allowing good cutting pressure for the user.

The terms 'left hand' and 'right hand' used for snips in this country have the continental meaning. *Left hand* means that the lower blade, which should be on the underside of the sheet of metal being cut, is on the left-hand side. *Right hand* means that the lower blade is on the right-hand side of the metal being cut. The US reference is exactly opposite to that used in this country.

3.15.2 Monodex cutters

This tool is designed for cutting thin sheet metal. Unlike snips, it has a flat face with one central serrated blade which in use cuts a slot equal in width to the thickness of the blade, about $\frac{1}{8}$ in (3.1 mm). As the material is cut, it coils up in front of the cutters. The advantage of this tool is that it can cut sheet metal internally, and work along either a straight or curved line by drilling a small access hole in the metal to allow the blade to penetrate through the metal to commence cutting. Allowance must be made for the wastage of $\frac{1}{8}$ in (3.1 mm) strip when cutting to a final size.

3.15.3 Cold chisels

These are often used by the body repair worker for cutting out components which have been damaged, chipping off rusted bolts which cannot be removed by normal methods, and chiselling sections which cannot be cut be snips or hacksaws. Chisels are made from high-quality steel. The cutting edge is specially hardened and tempered, while the remainder of the chisel is left soft to enable it to withstand hammer blows.

The chisel is subdivided into cutting edge, shank and head. The cutting end has a wide, flat taper and is ground on both sides to form the cutting edge. Metal working chisels should have sufficient length to allow the shank to be gripped properly, as short chisels are difficult to handle and may lead to accidents through hammer blows if the head projects insufficiently beyond the gripping hand. On the other hand excessively long chisels cannot be guided so well and consequently may vibrate and break easily. Only the cutting edge of the chisel is hardened. When the chisel has been in use for some time a ridge forms on its head which is called a burr; this needs to be removed because it may lead to accidents during use. Chisel joint angles vary between about 30° and 60°, the sharper angles being used for cutting softer metals. The metal worker's chisel is flat, and is used for working on flat or curved surfaces of thin sheet. It is also used for cutting solid bolts and rivets.

3.15.4 Hacksaws

The hacksaw is used by the body repair worker to advantage where a clean, neat cut is required, usually where two pieces have to be cut to form a welded joint, and is ideal for cutting irregular-shaped panels which could not be cut with snips or chisel. There are many different types of hacksaws, but in general they consist of a fixed or adjustable frame fitted with a renewable blade of hardened and tempered high-carbon or alloy steel. The hacksaw blades are classified by length and the number of teeth to the inch. Coarse blades have 14–16 teeth to 1 in (25.4 mm) and are used to cut soft metals; finer blades have 18–24 teeth to 1 in (25.4 mm) and are used to cut harder metals. The blades must be fitted with the teeth pointing away from the operator to the front of the frame, and the blade should be slightly tensioned in the frame.

3.15.5 Mini sheet metal cutter

This is a lightweight metal shear (Figure 3.30) which will produce curved and straight cuts in mild steel up to 1.6 mm (16 gauge). A lever and ratchet action operates the roller cutters, which are adjustable for metal thickness, and the cutting action can be reversed for cutting internal angles. The cutter can be bench mounted or used in a vice. This tool is ideally suited for the small body shop where limited cutting is required.

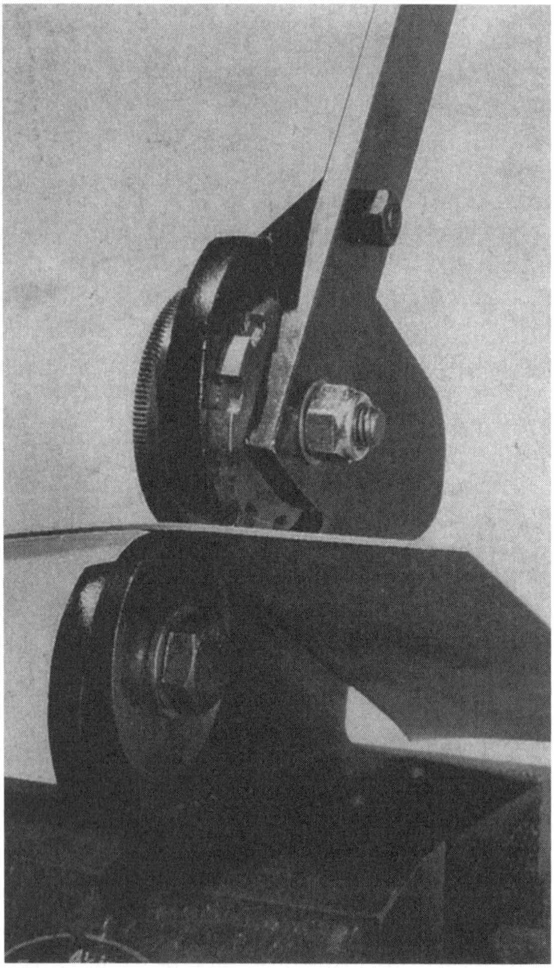

3.15.6 Drill cutting attachment

This is a drill attachment for cutting sheet metal up to 18 gauge (1.2 mm) mild steel and 16 gauge (1.6 mm) aluminium (Figure 3.31). This tool provides the user with clear visibility of the cutting line, allowing an accurate cut. When cutting it removes a continuous 3 mm strip from a panel without distortion. It will fit any drill chuck, air or electric.

Figure 3.31 Drill cutting attachment (*Frost Auto Restoration Techniques Ltd*)

3.16 Bending and swaging tools

3.16.1 Sheet metal folder

This is a small, compact folder constructed and designed for mounting in a vice (Figure 3.32). It can be used to fold mild steel sheet up to 1.2 mm (18 gauge) with a bend length of 24 in (610 mm). The tool is capable of producing bends of up to 100° and will produce both channel and box sections. It is a useful tool for the fabrication of small metal sections needed in vehicle repair.

Figure 3.30 Mini sheet metal cutter (*Sykes-Pickavant Ltd*)

Figure 3.32 Sheet metal folder (*Skyes-Pickavant Ltd*)

Figure 3.33 Swaging tool or edge setter (*Sykes-Pickavant Ltd*)

3.16.2 Swaging tool or edge setter

This tool is designed for bench mounting or to be used in a vice (Figure 3.33). The fixed rollers produce a smooth, uniform stepped panel edge on a straight or curved panel. The bottom roller is adjustable to accept varying gauges of material up to 1.2 mm (18 gauge) mild steel. The swaged or joggled edge can be used to strengthen panel edges or to create a flush-fitted lap joint when working on panel replacement.

3.16.3 Shrinker and stretcher machine

This is a versatile sheet metal working tool, and will quickly and easily pull curved shapes into steel up to 18 gauge (1.2 mm) or aluminium up to 14 gauge (2.0 mm) (Figure 3.34). It shrinks or stretches the metal into smooth even curvature with a radius as small as 3 in (75 mm). It is used to fabricate and repair panels where curves are needed in angle sections, especially around headlights, edges of window frames, door edges and wheel arches.

Figure 3.34 Shrinker and stretcher machine (*Frost Auto Restoration Techniques Ltd*)

3.17 General-purpose assembly and dismantling tools

Having the right tool for the job is a sign of a prepared body repair worker. Knowing how to use the tool is a mark of an experienced repair worker. Knowledge and experience come with study and time, but without the right tools, even the best body worker cannot carry out quality repair work.

3.17.1 Spanners or wrenches

A complete collection of spanners or wrenches is indispensable for the body repair worker. A variety of vehicle body parts, accessories and related parts all utilize common bolt and nut fasteners as well as a range of special fasteners.

The word 'wrench' means 'twist'. A wrench is a tool for twisting or holding bolt heads and nuts. The width of the jaw opening determines the size of the spanner, so that it fits around a nut and a bolt head of equal size. The larger the spanner size, the longer the spanner: the extra length provides the user with more leverage to turn the larger nut or bolt. Spanners are drop forged and made from chrome-vanadium steels.

Open-ended spanner

This is the most commonly used spanner. However, it has the disadvantage that it fits the nut or bolt on two sides only; consequently there is a greater tendency for the open-ended spanner to slip off the bolt or nut, resulting in rounded nuts and injured hands. Open-ended spanners can be single-ended, or double-ended to fit nuts of consecutive sizes. The head of the spanner is usually set at an angle for use in constricted spaces: when the spanner is turned over, the nut can be approached from a different angle. The open-ended spanner fits both square head (four-cornered) or hexagonal (hex) head (six-cornered) nuts. These spanners come in sets ranging in AF from $\frac{1}{4}$ in to $1\frac{5}{8}$ in and in metric from 3.2 mm to 42 mm (Figure 3.35).

Ring spanner

The ring spanner is made in a variety of sizes, points and offsets, and may be flat or cranked. It fits the nut all round by means of teeth cut in the inner surface of the ring. This is a very popular type of spanner as it enables the nut to be tightened quickly and can be used where movement is restricted. The ring spanner is much safer to use as more force can be applied without slipping and rounding the corners of the nut. The handle of the

Figure 3.35 A set of open-ended metric spanners (*Facom Tools Ltd*)

ring spanner can be straight or offset to provide hand clearance. Each end is usually a different size. The ring spanner does have limitations, for there must be sufficient clearance for the jaws to fit over and around the head of the nut or bolt. Ring spanners are available in 6, 8 or 12 points. The advantage of the 12-point ring spanner is that it can hold the nut in twelve different positions, which is ideal for working in confined spaces. Ring spanner sets are available in AF $\frac{1}{4}$ in to $1\frac{1}{4}$ in and metric 6 mm to 50 mm (Figure 3.36).

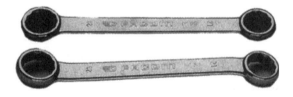

Figure 3.36 Straight and offset ring spanners (*Facom Tools Ltd*)

Combination spanner

The combination spanner has an open-ended jaw at one end and a ring on the other end. Both ends are the same size. The combination is probably the best choice of second set, for it complements either open-ended spanner or ring spanner sets. Combination spanner sets are available in AF $\frac{1}{4}$ in to $1\frac{1}{4}$ in and metric 3.2 mm to 32 mm (Figure 3.37).

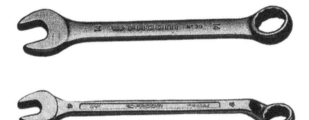

Figure 3.37   Combination and offset combination spanners (*Facom Tools Ltd*)

Adjustable wrench

An adjustable wrench has one fixed jaw and one movable jaw. The movable jaw is adjusted by means of a worm wheel fitted in the handle, which is meshed with teeth in the jaw. These wrenches are useful because they are manufactured in a range of sizes and each one can be adjusted to fit many sizes of nuts (Figure 3.38).

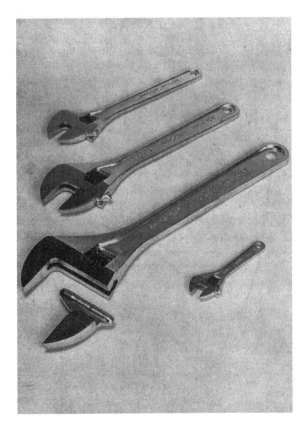

Figure 3.38   A range of adjustable wrenches (*Facom Tools Ltd*)

3.17.2 Socket sets

In many situations a socket ratchet is much faster and easier to use than an open-ended or ring spanner because of its versatility in use, and in some body repair applications it is absolutely essential. A basic socket ratchet set consists of a ratchet, bar extensions, a universal joint and a set of sockets (Figure 3.39).

Figure 3.39   Socket set (*Facom Tools Ltd*)

The barrel-shaped socket fits over and around a given size of nut; inside it is shaped like the ring spanner, having a set of teeth to grip the points of the nut. Sockets are available with 6 pointed teeth and 12 pointed teeth. A 6-point socket gives a tight hold on a hex nut, minimizing slippage and rounding of the nut's points. The 12-point socket does not have the holding power of the 6-point socket, but its numerous positions maximize the possible turning radius (Figure 3.40a).

The closed end of the socket has a square hole into which the drive shaft of the ratchet fits (Figure 3.40b). Socket sets can be purchased in $\frac{1}{4}$, $\frac{3}{8}$, $\frac{1}{2}$ and $\frac{3}{4}$ in drive sizes. The smaller drive sizes are used for turning small fasteners, badges and trim where little torque is required. The larger drive sizes with the corresponding longer handles are used where greater torque is needed. A body repairer will need sets of $\frac{1}{4}$, $\frac{3}{8}$ and $\frac{1}{2}$ in drive sockets.

The size of the individual sockets in a set depends on the drive size of the set as well as the number of sockets in the set. The socket size is

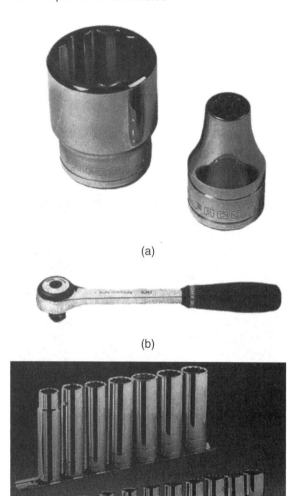

(a)

(b)

(c)

Figure 3.40   (a) Standard sockets (b) ratchet with $\frac{1}{2}$ in drive (c) standard and deep well sockets (*Facom Tools Ltd*)

the same as the face-to-face dimension of the bolt it fits. A $\frac{1}{4}$ in set has sockets ranging from $\frac{3}{16}$ to $\frac{1}{2}$ in (5 to 18 mm), whereas a $\frac{1}{2}$ in socket set has sockets ranging from $\frac{7}{16}$ to $1\frac{1}{4}$ in (13–40 mm).

Sockets are available not only in standard face-to-face diameters, but also in various lengths or bore depths. Normally the larger the socket size, the deeper the well. Deep well sockets are made extra long for reaching nuts or bolts in limited access areas (Figure 3.40c).

Socket set accessories
Socket set accessories multiply the usefulness of the socket set. A good socket set has a variety of accessories such as:

Ratchet
Ratchet attachment
Extension bars
Speed brace
Long-hinged handle with universal joint
Sliding T-bar
Universal joint
Flexible spinner
Flexible extension
Coupler.

Screwdriver attachments
Screwdriver attachments are also available for use with socket ratchets. These attachments are very useful when a fastener cannot be loosened with a regular screwdriver owing to lack of access and space. The leverage that the ratchet handle provides is often all it takes to move a stubborn screw.

3.17.3 Screwdrivers

A variety of threaded fasteners used in the automotive industries are assembled by the use of a screwdriver. Each fastener requires a specific kind of screwdriver, and a body repairer should have several sizes of each type.

All screwdrivers, regardless of type, are designed to have several things in common. The size of the screwdriver is determined by the width or diameter of the blade and the length of the shank. It is important to select the blade tip accurately by matching it to the size of the slot or the type of fastener being used. The larger the handle, the better the grip and the more torque it will generate when turned.

Standard screwdriver (for slotted heads)
A slotted screw accepts a screwdriver with a standard tip. The standard tip screwdriver is probably the most common type used. The blade should match the slot: an oversize screwdriver will not

reach the bottom of the slot and an undersize screwdriver will float in the slot, and either condition could result in damage to the tool and the fastener.

Phillips screwdriver (for cross-heads)

Phillips screws have a four-pronged funnel-shaped depression in the screw head, and are known as cross-heads. The tip of the Phillips screwdriver has four prongs that fit the four slots in the screw head. The four surfaces enclose the screwdriver tip so that there is less likelihood that the screwdriver will slip off the fastener. The most useful set of Phillips screwdrivers ranges from number 0 to number 4.

Pozidriv screwdriver

This screwdriver is like a Phillips in that the head is in the form of a cross, but the tip is flatter and blunter. The square tip grips the screw head and slips less than a Phillips screwdriver. Although Pozidriv and Phillips look alike, they are not interchangeable. The most useful set of Pozidriv screwdrivers ranges from number 1 to number 4.

Torx screwdriver

The Torx fastener is used quite a lot on vehicle bodies. The screwdriver has a six-pronged star profile which provides greater turning power and less slippage, and also ensures optimum pressure to prevent the tool from slipping out of the fastener's head.

Speciality screwdrivers

These are normally in sets of screwdriver tips and assorted drivers which are all interchangeable and prove most useful in vehicle body repair work.

3.17.4 Toolchests

A cabinet-type toolchest as shown in Figure 3.41 is standard equipment in most body shops. A portable toolbox on top of the chest holds large hand and power tools. A chest of drawers holds spanners, sockets and all the necessary assembly tools. The toolchest is on castors so that it can be conveniently located in any working area of the body shop.

Figure 3.42 shows a complete set of all assembly tools.

Figure 3.41   Portable toolchest (*Facom Tools Ltd*)

3.18 Power tools used in body repair work

Choosing between pneumatic or electric power tools has always been difficult because, while they are designed to perform basically the same function, the particular advantages and disadvantages of each are so different.

Electric tools are simple to set up. They require only a source of electric current, and so are suitable for immediate use anywhere in the workshop. The disadvantage is that because they have their own power supply (the motor) incorporated, they tend to be heavy and awkward for some jobs. Not only that, but over-heavy use can wear the motor, causing it to burn out.

Pneumatic tools, by contrast, are more complex. Among the advantages they offer are the regulation of air flow to suit the job, and the fact that they are

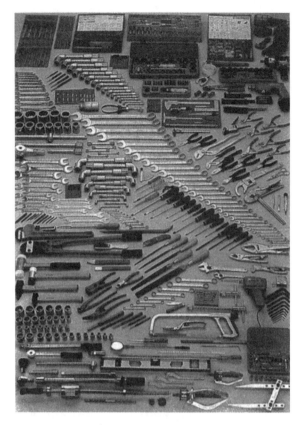

Figure 3.42 A full set of general and assembly tools (*Facom Tools Ltd*)

usually lighter, smaller and easier to handle. They also require less maintenance than a comparable electric tool. These factors can only be achieved, however, if a high-performance air compressor system is available. Moreover, it has long been a disadvantage that, despite filters and regulators, peneumatic tools can still blow oil, rust, dust and other impurities into the atmosphere and on to the work being treated.

Power tools play an important part in the modern techniques used by the panel beater in the repairing of crash damage to motor vehicles. Their use reduces man hours to a minimum and they also enable the operators to achieve better results than those possible using hand tools. A wide range of equipment is available to suit all types of crash damage.

Section 3.19 to 3.25 describe the typical range of power tools.

3.19 Air power chisel

This tool (Figure 3.43) is designed for the removal of damaged panels so that essential stripping can be carried out easily and quickly. It consists of a chisel bit which is retained by a spring through one end of a barrel, and a piston which moves to and fro within the barrel, striking the end of the chisel bit as it does so. The barrel must be kept dirt-free to eliminate the possibility of undue wear by the reciprocating movement of the chisel. A feather-action trigger in the handle allows for variation in the strength of the blow according to the amount of pressure applied. To provide additional safety a beehive or volute spring is fitted over the end of the barrel to prevent the chisel leaving the tool accidentally.

A wide range of special chisel bits is produced for the power tool. Most of these are cutting chisels of various sizes, used for general dismantling, cutting rivets, nuts, bolts and the removal of spot-welded sections. There is also a special thin-gauge metal cutting chisel for use on panel work, and a recessing tool which can be fitted on the end of the chisel. The recessing tool can be used on the edges of panels to form a lip or joggled section so that one panel passes over the face of the other, forming a flush finish which is ideal for spot-weld connections and joints in panels.

3.20 Metal cutting shears and nibblers

3.20.1 Metal shears

These are usually electrically operated (Figure 3.44), but in some cases can be driven by compressed air (Figure 3.45). They are designed to be used in the hand with an on/off trigger control built in the handle. The blade holding shoe is designed in a spiral so that it parts the metal as it cuts it. The machine is based on the principle of shearing metal, and this is done by a pair of very narrow blades, one of which is usually fixed while the other moves to and fro from the fixed blade at high speed. The blades have a very pronounced angle or set to permit the blade to pierce the sheet for internal cutting. The tool can be easily turned when cutting owing to the narrow width of the blades, and so can cut curved as well as straight shapes. The shears are capable of cutting sheet metal up to a maximum of 2 mm

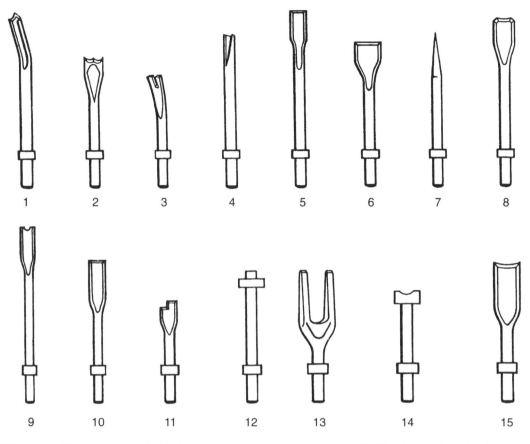

1 Silencer cutting tool
2 Sheet metal tool
3 Panel cutting tool
4 Bush removing tool
5 Rivet cutting tool

6 Undercoat scraper
7 Tapered punch
8 Bush remover chisel
9 Spot weld breaker
10 175 mm nut and bolt cutting tool

11 Shock absorber chisel
12 Bush installing tool
13 Fork chisel
14 Universal joint and track rod tool
15 Exhaust pipe cutting tool

Figure 3.43 Air power chisel and accessories (*Black and Decker Ltd*)

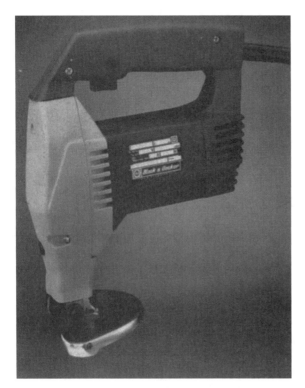

Figure 3.44   Metal cutting shears (*Black and Decker Ltd*)

Figure 3.45   Air shears (*Desoutter Automotive Ltd*)

(14 SWG) or 3 mm (10 SWG) aluminium, which more than covers the automobile requirements. The shear is used extensively in the cutting of light materials for the construction of new vehicles, but also plays an important role in the removal of damaged sections in crash repairs, as it leaves a nice neat cut and there is no risk of flame damage as with the more traditional methods of oxy-acetylene cutting.

3.20.2 Pneumatic metal shear

This tool uses the same action as snips, having two fixed blades and one blade moving vertically between them (Figure 3.46). This central movable blade cuts out a thin ribbon of metal which coils and is discarded. Cutting can be started from the edge of the panel, or from any point on the panel surface using a predrilled hole large enough to insert the central cutting blade. Radiuses up to $7\frac{3}{4}$ in (200 mm) can be cut on material up to 1.2 mm (18 gauge). Care must be taken not to damage the panel surface by causing buckling during the cutting action.

Figure 3.46   Pneumatic metal shear (*Desoutter Automotive Ltd*)

3.20.3 Metal cutting nibbler

This is a portable nibbler which uses a punch and die (Figure 3.47). The actual cutting is done by the edge of the punch which is reciprocated at high speed, leaving a slot or channel in the metal. Cutting can be commenced from the panel edge, or from a hole predrilled in the panel surface to allow entry of the punch. Allowances must be made for the width of the cut and, when accurate positioning of the cut is essential, templates can be used as a guide.

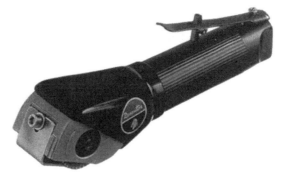

Figure 3.49   Circular metal cutting saw (*Desoutter Automotive Ltd*)

3.22 Power drills

3.22.1 Standard drills

Figure 3.47   Metal cutting nibbler (*Desoutter Automotive Ltd*)

3.21 Power saw

This is a high-speed pneumatic tool using interchangeable, reciprocating hacksaw blades (Figure 3.48). It is able to cut from straight lines to tight curves. Cutting can be started anywhere on a panel without the need for pilot holes. Single- or double-skinned panel assemblies can be easily cut owing to the short blade stroke. The tool can be used to cut a wide variety of materials including light-gauge steel, aluminium, reinforced glass fibre, wood and hardboard. Its fine cut is ideally suited for panel replacement where joints have to be cut and welded.

Another power saw is the rotary or circular saw (Figure 3.49).

There are two types of power drill used in the bodyshop: the majority are usually electrically driven and require a 13 A electrical power point (Figure 3.50), while the others are air powered and require a compressor and air points (Figure 3.51). The most popular sizes in common use are the $\frac{1}{4}$ in (6.3 mm) pistol drill, and the $\frac{3}{8}$ in and $\frac{1}{2}$ in (9.5 mm and 12.6 mm) heavy-duty machines. Drills are used mostly for drilling out spot welds along damaged panels which need removing, and also for drilling out broken rivets and bolts. As well as the standard size drills there is a special drill called an angle drill, which is extremely useful in bodywork as it can be used in confined spaces and enables many otherwise impossible drilling tasks to be carried out (Figure 3.52).

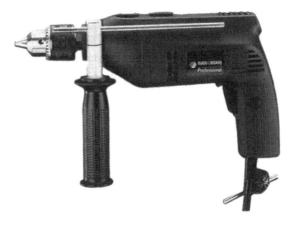

Figure 3.48   Power saw (*Desoutter Automotive Ltd*)

Figure 3.50   General duty drill (*Black and Decker Ltd*)

Figure 3.51 Air drill (*Desoutter Automotive Ltd*)

Figure 3.52 Angle drill (*Desoutter Automotive Ltd*)

Figure 3.53 Cordless screwdriver/drill (*Black and Decker Ltd*)

3.22.2 Cordless drill

This uses an energy pack to supply the power, with a fast charge facility for recharging from a mains supply (Figure 3.53). It can be single speed or two speed, and can drill up to 3 mm steel. When fitted with a reversible adjustable clutch this tool can be used as a power screwdriver, thus making it a tremendously versatile piece of equipment, especially in the body building industry.

3.22.3 Specialized air drill (spot-weld remover)

This is a hand-held clamp-type drilling tool designed for cutting out spot welds (Figure 3.54). It is powered by compressed air with controllable speeds, and has a removable cutting drill supplied in two sizes, 6 mm and 8 mm; the depth of cut is also adjustable. When there is a flange at the rear of the spot weld it is supported by the clamping action of the C-shaped arm, while the drill is cutting out the spot weld to a preset depth, allowing the panels to separate. Owing to the shape and clamping action of the tool it is restricted to removing spot welds from an open panel edge only. Welds on panel edges which have no access

from the rear would entail the cutting of the panel to be replaced, to obtain access for the tool clamp. This tool could be a worthy addition to any body shop because of its time saving over conventional methods and its neatness of finish.

3.23 Sanding machines

Whether they are disc, belt, orbital, flat or finishing, sanders are the most frequently used power tools in the body repair shop.

3.23.1 Disc sander

This is the most popular sander. It is a powerful machine with a hard rubber flexible pad on to which the actual sanding disc is fastened by means of a centre nut. This nut is threaded in such a way that it tightens as the machine rotates, thus avoiding the risk of a disc coming loose. Disc sanders are driven either electrically (Figure 3.55) or by compressed air (Figure 3.56).

The main uses of the disc sander in the repair shop are sanding sections on panels which are either rusted or have been repaired by planishing and require sanding to a smooth finish in readiness for painting; cleaning sections that need to be filled with

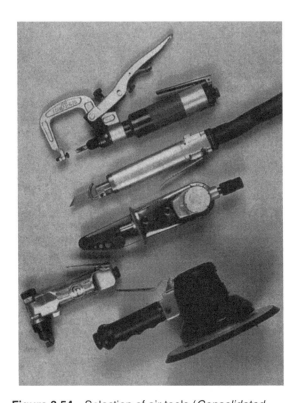

Figure 3.56 Air driven sander (*Desoutter Automotive Ltd*)

minor paint blemishes; and blending in body solder or plastic filler in sections that have been filled.

3.23.2 Belt sander

This can be used for repair to bodywork, and is extremely good on convex or concave surfaces where access with a disc sander would be difficult. This type of machine uses an endless sanding belt, which is 3 in (16.1 mm) wide, instead of the normal round disc. Some models are fitted with a vacuum unit so that the dust is kept to an absolute minimum in the workshop.

3.23.3 Pneumatic sander/filer

This is a dual-purpose tool for either sanding or filing (Figure 3.57). For sanding, an abrasive paper sheet is clipped to the tool. It has a reciprocating straight-line action which speeds up feather edging or filing materials and results in a smooth surface finish. A standard file blade can be fitted to the tool's base to convert it into a power-operated body file with a speed of up to 3000 strokes per minute.

3.23.4 Dual-action sander

Figure 3.54 Selection of air tools (*Consolidated Pneumatic Tool Co. Ltd*).
From top:
Spotle specialized air drill for removing spot welds
Supa-sander (miniature belt sander)
Metal nibbler
Dual-action orbital sander

Figure 3.55 Disc sander (*Black and Decker Ltd*)

solder prior to tinning; stripping paint prior to planishing or dressing out; dressing off surplus weld metal after welding new sections; feather edging

The dual action (DA) on this sander makes the pad oscillate counter-clockwise while the head revolves clockwise, resulting in a non-repeating surface pattern that eliminates scratches on the surface of the panel (Figure 3.54).

Figure 3.57 Straight line sander (*Desoutter Automotive Ltd*)

3.23.5 Geared orbital sander/polisher

This is a multi-orbital action machine which eliminates swirl marks and allows coarse grit paper to be used for high stock removal rates (Figure 3.58). On paint and fibreglass it leaves a perfect key for primers. The multi-orbital action reduces work surface heat generation, which is particularly important when sanding fillers and paint. The tool can also be used for polishing because of its rotary gearing. It can also be connected to a centralized or free-standing dust collection system.

Figure 3.58 Geared orbital sander/polisher (*Desoutter Automotive Ltd*)

3.23.6 Small angle grinder

The small angle grinder (Figure 3.59) is sometimes known as a mini-grinder or a grinderette. It is a versatile one-handed machine, and will cut away damaged metal, remove rust, grind down spot, MIG and gas welds, smooth infills and generally grind down prepared surfaces. Owing to its small size and light weight it is ideal for grinding in confined spaces and tight corners. These machines normally use a hard disc for grinding.

Figure 3.59 Small (178 mm) angle grinder (*Desoutter Automotive Ltd*)

3.23.7 Oscillating sander

This special sander (Figure 3.60) is an oscillating power tool, only available in this form from the company Fein. Oscillating means that the grinding plate – here a triangle rather than a disc shape – does not rotate, but moves to and fro through only 2 minutes of angle at a frequency of 20 000 oscillations per minute. This ensures high grinding performance, with multiple contact between the grinding particles and virtually all materials: wood, paint, plastic, filler, metal, non-ferrous metals and so on. With the aid of the triangular grinding plate you can reach into any corner or inside the narrowest of openings, for ideal sanding of edges and profiles. Good-quality transitions can even be obtained when spot grinding. This sander is the ideal complement to orbital sanders, eccentric

Figure 3.60   Oscillating sander (*Fein/George Marshall Ltd*)

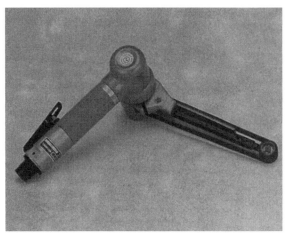

Figure 3.62   Miniature belt sander (*Desoutter Automotive Ltd*)

sanders and other portable grinding equipment used in a body repair workshop.

3.23.8 Rotex sander

This tool (Figure 3.61) can be used as a fine sanding random orbital machine or as a rough sanding machine; in the latter mode it can also be used as a polisher. As part of the process of repairing vehicle bodywork, old layers of paint can be removed with this sander. Because of the efficient dust extraction the sandpaper does not become loaded, and so the

Figure 3.61   Rotex sander with full dust extraction (*Minden Industrial Ltd*)

transition to undamaged paintwork remains smooth and free of scratches, thus considerably reducing repair work.

3.23.9 Supa-sander (miniature belt sander)

This machine (Figures 3.54, 3.62) has a 20 mm wide belt and is designed to handle small and accurate convex and concave contours. It has an adjustable head feature for flexibility of operation.

It can be efficiently operated in one hand owing to its small size and light weight, and thus may be used to complement conventional angle grinders or disc sanders. It can be used for the removal of paint to reveal production spot welds, and for dressing small plug welds or tack welds in awkward locations such as wing and inner valances, floor pans and inner sills.

3.24 Other power tools

3.24.1 Impact air screwdriver

The impact air screwdriver has a quick-change bit holder for inserting different screwdriver bit heads: slot, Phillips and Pozidriv (Figure 3.63). It is ideal for removal of rusted-in screws and other fasteners in locations such as door locks and door hinges.

3.24.2 Ratchet wrench

Ratchet wrenches can have $\frac{3}{8}$ in or $\frac{1}{2}$ in drives (Figure 3.64). They are ideally suited to a wide variety of nut turning operations, either fastening or unfastening. Initial hand ratcheting enables a tight or seized fastener to be slackened off prior to rapid runoff using the air power. Also final tightening can be controlled by hand ratcheting after fast rundown using the air drive.

Figure 3.63 Impact air screwdriver (*Desoutter Automotive Ltd*)

Figure 3.64 Ratchet wrench (*Desoutter Automotive Ltd*)

3.24.3 Bonded windscreen electric cutter

This cutter (Figure 3.65) greatly simplifies what was formerly a time-consuming task for two people. With this tool, windows bonded in with polyurethane adhesive can be taken out without difficulty. There is no environmental hazard, and no fumes which might constitute a health risk. Residual adhesive can be removed without damaging the body.

The special cutter uses an AC/DC motor. An electric circuit permits the blade oscillating frequency to be varied to suit the required cutting speed. A wide selection of special-purpose steel cutting blades is now available for all current car models. To simplify operation at various points on the body, the cutter blade has a twelve-sided mount to vary its position in relation to the tool.

This precision tool is used to cut through seals, joints and hard-bonded connections with exceptional speed and ease. Bonded-in car windscreens, for example, can be removed without damaging either the glass or the frame.

Figure 3.65 Bonded windscreen electric cutter (*Fein/George Marshall Ltd*)

3.25 Dust extraction for power tools

Built-in dust extraction should be a must in any new equipment programme, for it is essential that airborne dust levels in body shops are reduced to a minimum for the safety of the body shop employees. Sanding, grinding and cutting operations in the body shop usually generate quantities of airborne dust; this is a major hazard not only to health but also for refinishers wanting a quality paint finish. Systems therefore have been specifically designed to overcome problems of dust, noise and oil pollution.

The equipment consists of a high-performance dust extractor unit and a range of plug-in air tools which includes random orbital sanders, orbital sanders, saws and grinders (Figures 3.66, 3.67). Dust created during the sanding or sawing process is immediately extracted into a handy, easily emptied vacuum pump. In the case of the sanders, a series of holes in the pad itself is used

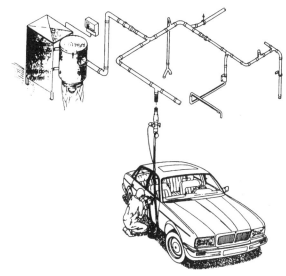

Figure 3.66 Centralized dust extraction system for use with FESTO tools (*Minden Industrial Ltd*)

Figure 3.67 Dust extraction system in a workshop (*Minden Industrial Ltd*)

to remove the dust and debris which, with conventional systems, form a cloud and obscure the surface to be worked on as well as necessitating breathing apparatus for safety's sake. A single quick-fix hose connects the hand tool to the power source. The hose has three separate compartments for compressed air supply, exhaust and dust extraction (Figures 3.68, 3.69, 3.70). Most sanding and flatting tools have adjustable speed control. They have housings made of high-impact GRP, which makes them light in weight and suitable for prolonged usage without operator fatigue.

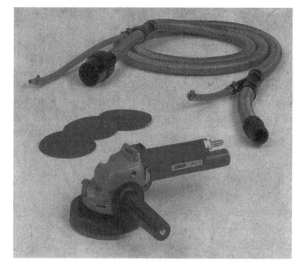

Figure 3.69 Air tool with hose attachment for centralized system (*Minden Industrial Ltd*)

Figure 3.68 Two extraction points with self-closing flaps (*Minden Industrial Ltd*)

Figure 3.70 Dust-free hand sanding tools: Vacbloc and Vacfile (*Minden Industrial Ltd*)

Questions

1 Sketch and describe each of the following hand tools, giving an example of the type of work it would be used for: (a) cross-pein and finishing hammer (b) body spoon (c) planishing hammer.

2 List the hand tools which you would expect to find in a body repairer's toolkit.

3 Describe a suitable application for the use of a flexible body file.

4 Describe a suitable application for (a) a toe dolly (b) an anvil dolly.

5 Explain a repair situation where a body spoon may be used instead of a panel dolly.

6 Sketch and describe four essential hand tools used by the body repairer.

7 Name and explain the use of a hand tool which would have the effect of spreading a blow, thus reducing the stretching of the metal.

8 For what purpose would a sandbag and a wooden hollowing block be used?

9 State a typical use for a pear-shaped mallet.

10 Describe a typical repair application in which a pick hammer would be used.

11 Describe a repair situation in which each of the following special hand tools would be used: (a) impact driver (b) panel puller (c) edge setting tool (d) door skinner.

12 Describe how the following four power tools could be used by the body repair worker: (a) power saw (b) spot-weld remover (c) bonded windscreen cutter (d) miniature belt sander.

13 Describe the movement of a dual-action (DA) sander.

14 In what circumstances would a grid dolly be used?

15 Compare the advantages of a beating file with those of a planishing hammer.

16 When referring to sheet metal snips, what do the terms 'right hand' and 'left hand' mean?

17 Describe three types of sheet metal bench stakes, and sketch one of them.

18 Explain two safety precautions which must be observed when using a cold chisel.

19 What safety precautions must be observed before using an electric drilling machine?

20 Describe the safety measures which would be necessary when using hand and small power tools in a workshop.

21 What action should be taken to render a panel hammer safe for further use after its head has become loose?

22 Why are the teeth on a body file milled in a curved formation?

23 Some beating files and panel hammers have serrated faces. Explain the reasons for this.

24 Name a pneumatic tool which can be used for the removal of a damaged panel section.

25 Name three types of spanner that could be used by a body repair worker.

26 Describe the advantages of socket sets in removing and replacing damaged panels.

27 Sketch a Torx screwdriver bit that could be used with an impact screwdriver.

28 State the function of an air transformer and its importance when used with power tools.

29 Explain the two different types of air hose couplings used with pneumatic tools.

30 Explain the reason why the face of a planishing hammer should be maintained in perfect condition.

31 State the dangers of carrying sharp or pointed tools in pockets or protective clothing.

32 When a fault develops on an electrical power tool, what action must be taken?

33 Explain the reasons why some tools have a built-in dust extraction system while others are connected to a centralized dust extraction system.

34 Identify and sketch two types of mole grips used in body repair applications.

35 With the aid of a sketch, identify the BS symbol used on power tools to indicate double insulation.

Metals and non-metals used in vehicle bodies

4.1 Manufacture of steel coil and sheet for the automobile industry

In the manufacture of steel coils, the raw material iron ore is fed into a blast furnace, together with limestone and coke; the coke is used as a source of heat, while the limestone acts as a flux and separates impurities from the ore. The ore is quickly reduced to molten iron, known as pig iron, which contains approximately 3–4 per cent carbon. In the next stage of manufacture, the iron is changed into steel by reheating it in a steel-making furnace and blowing oxygen either into the surface of the iron or through the liquid iron, which causes oxidation of the molten metal. This process burns out impurities and reduces the carbon content from 4 per cent to between 0.08 and 0.20 per cent.

4.1.1 Casting

The steel is cast into ingots; these are either heated in a furnace and rolled down to a slab, or more commonly continuously cast into a slab. Slabs by either casting process are typically 8–10 in (200–250 mm) thick, ready for further rolling. These slabs are reheated prior to rolling in a computer-controlled continuous hot strip mill to a strip around twice the thickness required for body panels. The strip is closely wound into coil ready for further processing.

4.1.2 Pickling

Before cold rolling, the surfaces of the coils must be cleaned of oxide or black scale formed during the hot rolling process and which would otherwise ruin the surface texture. This is done by pickling the coils in either dilute hydrochloric acid or dilute sulphuric acid and then washing them in hot water to remove the acid. The acid removes both the oxide scale and any dirt or grit which might also be sticking to the surface of the coil.

4.1.3 Cold rolling

In the cold rolling process the coil is rolled either in a single-stand reversing mill (narrow mills using either narrow hot mill product or slit wide mill product) or in a multiple-stand tandem mill to the required thickness. Most mills are computer controlled to ensure close thickness control, and employ specially prepared work rolls to ensure that the right surface standard is achieved on the rolled strip. The cold rolling process hardens the metal, because mild steel quickly work hardens. The cold rolled coils are suitable for applications such as panelling where no bending or very little deformation is needed. At this stage the coil is still not suitable for the manufacture of the all-steel body shell and it must undergo a further process to soften it; this is known as annealing.

4.1.4 Annealing

Coils used for the manufacture of a car body must not only have a bright smooth surface but must also be soft enough for bending, rolling, shaping and pressing operations, and so the hardness of cold rolled coils to be used for car bodies must be reduced by annealing. If annealing were carried out in an open furnace this would destroy the bright surface of the coil and therefore oxygen must be excluded or prevented from attacking the metal during the period of heating the coils.

The normal method of annealing the coils is box annealing. The coils are stacked on a furnace base, covered by an inner hood and sealed. The atmosphere is purged with nitrogen and hydrogen to eliminate oxygen. A furnace is then placed over the stack and fired to heat the steel coils to a temperature of about 650 °C for around 24 hours, depending upon charge size and steel grade.

4.1.5 Temper rolling

During the process of annealing the heat causes a certain amount of buckling and distortion, and a further operation is necessary to produce flat coils. The annealed coil is decoiled and passed through a single-stand temper mill using a specially textured work roll surface, where it is given a light skin pass, typically of 0.75–1.25 per cent extension. This is necessary to remove buckles formed during annealing, to impart the appropriate surface texture to the strip, and to control the metallurgical properties of the strip. The strip is then rewound ready for despatch or finishing as appropriate.

4.1.6 Finishing

The temper rolled coils can be slit to narrow coil, cut to sheet, reinspected for surface critical applications, or flattened for flatness critical applications as appropriate. Material can be supplied with a protective coating oil, and packed to prevent damage or rusting during transit and storage.

4.2 Specifications of steels used in the automobile industry

The motor body industry uses many different types of steel. Low-carbon steel is used for general constructional members. High-tensile steels are used for bolts and nuts which will be subjected to a heavy load. Specially produced deep-drawn steel including micro-alloyed steel is used for large body panels which require complex forming. Zinc-coated steel sheets are increasingly being specified for automobile production, both for body and chassis parts, as improved corrosion protection is sought. Stainless steel is used for its non-rusting, hard wearing and decorative qualities. The many different types of

springs used in the various body fittings are produced from spring steel, while specially hardened steels make the tools of production. Drills, chisels, saws, hacksaws and guillotine blades are all produced from special alloy steels, which are made from an appropriate mixture of metals and elements.

Steel varies from iron chiefly in carbon content; iron contains 3–4 per cent carbon while carbon steels may contain from 0.08 per cent to 1.00 per cent carbon. The chemical composition and mechanical properties of these carbon steels, especially when alloyed with other elements such as nickel, chromium and tungsten, have been gradually standardized over the years, and now the different types of steels used are produced to specifications laid down by the British Standards Institution. A British Standard specification defines the chemical composition and mechanical properties of the steel, and also the method and apparatus to be used when testing samples to prove that the mechanical properties are correct. The tensile strength, and in the case of sheet and strip steel the bend test, are the properties of most interest, but the British Standard specification also defines the elongation, the yield point and the hardness of the steel.

The steels used in the motor trade may be grouped as follows:

1   Cold forming steels
2   Carbon steels
3   Alloy steels
4   Free cutting steels
5   Spring steels
6   Rust-resisting and stainless steels.

As each group may contain many different specifications, some idea of the variety of steels may be gained. However, in the motor body industry the specifications which apply are those pertaining to *cold forming steels*, namely BS 1449: Part 1: 1983.

The greatest percentage of steel used in motor bodies is in the form of coil, strip, sheet or plate. Sheet steel is a rolled product produced from a wide rolling mill (600 mm or wider); to come under the heading of sheet steel, the steel must be less than 3 mm thick. Steel 3–16 mm thick comes under the heading of plate.

Tables 4.1, 4.2 and 4.3 are the specifications for steel sheet strip and coil for the manufacture of motor body shells in the automobile industry.

Table 4.1 Symbols for material conditions: BS 1449: Part 1: 1983

Condition	Symbol	Description
Rimmed steel	R	Low-carbon steel in which deoxidation has been controlled to produce an ingot having a rim or skin almost free from carbon and impurities, within which is a core where the impurities are concentrated
Balanced steel	B	A steel in which processing has been controlled to produce an ingot with a structure between that of a rimmed and a killed steel. It is sometimes referred to as semi-killed steel
Killed steel	K	Steel that has been fully deoxidized
Hot rolled on wide mills	HR	Material produced by hot rolling. This will have an oxide scale
narrow mills	HS	coating, unless an alternative finish is specified (see Table 4.2)
Cold rolled on wide mills	CR	Material produced by cold rolling to the final thickness
narrow mills	CS	
Normalized	N	Material that has been normalized as a separate operation
Annealed	A	Material in the annealed last condition (i.e. which has not been subjected to final light cold rolling)
Skin passed	SP	Material that has been subjected to a final light cold rolling
Temper rolled		Material rolled to the specified temper and qualified as follows:
	H1	Eighth hard
	H2	Quarter hard
	H3	Half hard
	H4	Three-quarters hard
	H5	Hard
	H6	Extra hard
Hardened and tempered	HT	Material that has been continuously hardened and tempered in order to give the specified mechanical properties

4.3 Carbon steel

Carbon steels can be classified as follows (Table 4.3):

Low-carbon steel
Carbon-manganese steel
Micro-alloyed steel
Medium-carbon steel
High-carbon steel.

The properties of plain carbon steel are determined principally by carbon content and microstructure, but it may be modified by residual elements other than carbon, silicon, manganese, sulphur and phosphorus, which are already present. As the carbon content increases so does the strength and hardness, but at the expense of ductility and malleability.

4.3.1 Low-carbon steel

For many years low-carbon steel (sometimes referred to as mild steel) has been the predominant autobody material. Low-carbon coil, strip and sheet steel have been used in the manufacture of car bodies and chassis members. This material has proved an excellent general-purpose steel offering an acceptable combination of strength with good forming and welding properties. It is ideally suited for cold pressings of thin steel sheet and is used for wire drawing and tube manufacture because of its ductile properties.

Low-carbon steel is soft and ductile and cannot be hardened by heating and quenching, but can be case hardened and work hardened. It is used extensively for body panels, where its high ductility and malleability allows easy forming without the danger of cracking. In general low-carbon steel is used for all parts not requiring great strength or resistance to wear and not subject to high temperature or exposed to corrosion.

However, new factors such as a worldwide requirement for fuel conservation for lighter-weight

Table 4.2 Symbols for surface finishes and surface inspection: BS 1449: Part 1: 1983

Finish	Symbol	Description
Pickled	P	A hot rolled surface from which the oxide has been removed by chemical means
Mechanically descaled	D	A hot rolled surface from which the oxide has been removed by mechanical means
Full finish	FF	A cold rolled skin passed material having one surface free from blemishes liable to impair the appearance of a high-class paint finish
General-purpose finish	GP	A cold rolled material free from gross defects, but of a lower standard than FF
Matt finish	M	A surface finish obtained when material is cold rolled on specially prepared rolls as a last operation
Bright finish	BR	A surface finish obtained when material is cold rolled on rolls having a moderately high finish. It is suitable for most requirements, but is not recommended for decorative electroplating
Plating finish	PL	A surface finish obtained when material is cold rolled on specially prepared rolls to give one surface which is superior to a BR finish and is particularly suitable for decorative electroplating
Mirror finish	MF	A surface finish having a high lustre and reflectivity. Usually available only in narrow widths in cold rolled material
Unpolished finish	UP	A blue/black oxide finish; applicable to hardened and tempered strip
Polished finish	PF	A bright finish having the appearance of a surface obtained by fine grinding or abrasive brushing; applicable to hardened and tempered strip
Polished and coloured blue	PB	A polished finish oxidized to a controlled blue colour by further heat treatment; applicable to hardened and tempered strip
Polished and coloured yellow	PY	A polished finish oxidized to a controlled yellow colour by further heat treatment; applicable to hardened and tempered strip
Vitreous enamel	VE	A surface finish for vitreous enamelling of material of specially selected chemical composition
Special finish	SF	Other finishes by agreement between the manufacturer and the purchaser

body structures, and safety legislation requiring greater protection of occupants through improved impact resistance, are bringing about a change in materials and production technology. This has resulted in the range of micro-alloyed steels known as high-strength steels (HSSs) or high-strength low-alloy steels (HSLAs).

4.3.2 Micro-alloyed steel

This steel is basically a carbon-manganese steel having a low carbon content, but with the addition of micro-alloying elements such as niobium and titanium. Therefore it is classed as a low-alloy high-strength steel within the carbon range. As a result of its strength, toughness, formability and weldability,

the car body manufacturers are using this material to produce stronger, lighter-weight body structures.

A typical composition utilized for a micro-alloyed high-strength steel (HSS) is as follows:

	Percentage
Carbon (C)	0.05–0.08
Manganese (Mn)	0.80–1.00
Niobium (Nb)	0.015–0.065

The percentage of niobium used depends on the minimum strength required.

Formable HSSs were developed to allow the automotive industry to design weight out of the car in support of fuel economy targets. A range of high-strength formable steels with good welding and

Table 4.3 Summary of material grades, chemical compositions and types of steel available: BS 1449: Part 1: 1983

Material grade	Rolled condition (see Table 4.1)	Chemical composition								
		C		Si		Mn		S	P	
		min.	max.	min.	max.	min.	max.	max.	max.	
		%	%	%	%	%	%	%	%	
Materials having specific requirements based on formability										
1	HR, HS, –, –,	–	0.08	–	–	–	0.45	0.030	0.025	Extra deep drawing aluminium-killed steel
1	–, –, CR, CS	–	0.08	–	–	–	0.45	0.030	0.025	Extra deep drawing aluminium-killed stabilized steel
2	HR, HS, CR, CS	–	0.08	–	–	–	0.45	0.035	0.030	Extra deep drawing
3	HR, HS, CR, CS	–	0.10	–	–	–	0.50	0.040	0.040	Deep drawing
4	HR, HS, CR, CS	–	0.12	–	–	–	0.60	0.050	0.050	Drawing or forming
14	HR, HS, –, –,	–	0.15	–	–	–	0.60	0.050	0.050	Flanging
15	HR, HS, –, –,	–	0.20	–	–	–	0.90	0.050	0.060	Commercial
Materials having specific requirements based on minimum strengths										
Carbon-manganese steels										
34/20	HR, HS, CR, CS	–	0.15	–	–	–	1.20	0.050	0.050	Available as rimmed (R), balanced (B) or killed (K) steels
37/23	HR, HS, CR, CS	–	0.20	–	–	–	1.20	0.050	0.050	
43/25	HR, HS, –, –,	–	0.25	–	–	–	1.20	0.050	0.050	
50/35	HR, HS, –, –,	–	0.20	–	–	–	1.50	0.050	0.050	Grain-refined balanced (B) or killed (K) steel
Micro-alloyed steels										
40/30	HR, HS, –, CS	–	0.15	–	–	–	1.20	0.040	0.040	Grain-refined niobium- or titanium-treated fully killed steels having high yield strength and good formability
43/35	HR, HS, –, CS	–	0.15	–	–	–	1.20	0.040	0.040	
46/40	HR, HS, –, CS	–	0.15	–	–	–	1.20	0.040	0.040	
50/45	HR, HS, –, CS	–	0.20	–	–	–	1.50	0.040	0.040	
60/55	–, HS, –, CS	–	0.20	–	–	–	1.50	0.040	0.040	

Grade	Condition									Notes
40F30	HR, HS, –, CS	–	0.12	–	–	–	1.20	0.035	0.030	The steels including F in their designations in place of the oblique line offer superior formability for the same strength levels
43F35	HR, HS, –, CS	–	0.12	–	–	–	1.20	0.035	0.030	
46F40	HR, HS, –, CS	–	0.12	–	–	–	1.20	0.035	0.030	
50F45	HR, HS, –, CS	–	0.12	–	–	–	1.20	0.035	0.030	
60F55	–, HS, –, CS	–	0.12	–	–	–	1.20	0.035	0.030	
68F62	–, HS, –, –	–	0.12	–	–	–	1.50	0.035	0.030	
75F70	–, HS, –, –	–	0.12	–	–	–	1.50	0.035	0.030	

Narrow strip supplied in a range of conditions for heat treatment and general engineering purposes

Grade	Condition									Notes
4	–, HS, –, CS	–	0.12	–	–	–	0.60	0.050	0.050	Low-carbon steel available hot rolled, annealed, skin passed or cold rolled to controlled hardness ranges H1 to H6 inclusive
10	–, HS, –, CS	0.08	0.15	0.10	0.35	0.60	0.90	0.045	0.045	For case hardening
12	–, HS, –, CS	0.10	0.15	–	–	0.40	0.60	0.050	0.050	A range of carbon steels available in the hot rolled or annealed condition
17	–, HS, –, CS	0.15	0.20	–	–	0.40	0.60	0.050	0.050	
20	–, HS, –, CS	0.15	0.25	0.05	0.35	1.30	1.70	0.045	0.045	
22	–, HS, –, CS	0.20	0.25	–	–	0.40	0.60	0.050	0.050	
30	–, HS, –, CS	0.25	0.35	0.05	0.35	0.50	0.90	0.045	0.045	
40	–, HS, –, CS	0.35	0.45	0.05	0.35	0.50	0.90	0.045	0.045	A range of carbon steels for use in the hot rolled, normalized, annealed and (except for grade 95) in the temper rolled (half hard) conditions. Grades 40 and 50 may be induction or flame hardened and grades 60, 70, 80 and 95 may be supplied in the hardened and tempered condition
50	–, HS, –, CS	0.45	0.55	0.05	0.35	0.50	0.90	0.045	0.045	
60	–, HS, –, CS	0.55	0.65	0.05	0.35	0.50	0.90	0.045	0.045	
70	–, HS, –, CS	0.65	0.75	0.05	0.35	0.50	0.90	0.045	0.045	
80	–, HS, –, CS	0.75	0.85	0.05	0.35	0.50	0.90	0.045	0.045	
95	–, HS, –, CS	0.90	1.00	0.05	0.35	0.30	0.60	0.040	0.040	

painting characteristics have been developed. The steels are hot rolled for chassis and structural components, and cold reduced for body panels. Through carefully controlled composition and processing conditions, these steels achieve high strength in combination with good ductility to allow thinner gauges to be used: a reduction from 0.90 mm down to 0.70 mm is a general requirement.

There was a danger that the new HSS, thinner but just as strong, would lack the ductility which allows it to be press formed into shape. This problem was overcome by the use of microalloying. In a metal crystal the atoms are in layers; when the crystals are stretched (as in forming), one layer of atoms slides over another. The layers of atoms slide like playing cards in a pack, and in doing so are changing shape in a ductile way. The sliding can be controlled by adding elements to the steel such as niobium or titanium. The element reacts with the carbon to produce fine particles which spread through the steel. The element controls the ductility and strengthens the steel, thereby improving the properties of the material.

These steels are low carbon, employing solution strengthening (cold reduced rephosphorized) or precipitation hardening (hot rolled and cold reduced micro-alloyed) elements to produce fine-grained steels which are suitable for welding by spot and MIG processes only.

4.3.3 Medium-carbon steel

This can be hardened by quenching, and the amount it can be hardened increases with its carbon content. This type of steel can be used for moving parts such as connecting rods, gear shafts and transmission shafts, which require a combination of toughness and strength, but it is being replaced in the car industry by high-alloy steels.

4.3.4 High-carbon steel

This can be hardened to give a very fine cutting edge, but with some loss of its ductile and malleable properties. It is used for metal and wood cutting tools, turning tools, taps and dies, and forging and press dies because of its hardness and toughness, but is seldom used now for motor vehicle parts because of the introduction of high-alloy steels.

4.3.5 Zinc coated steels

The automotive industry, in seeking to provide extended warranties, is turning increasingly to the use of zinc coated steels. Modern automobiles must be not only of high quality but also durable and economical, as perceived by their purchasers. These vehicles are expected to exceed an average of seven years without structural or cosmetic deterioration due to corrosion. Increasingly aggressive environmental influences tend to shorten the life of the car, whereas ever more specialized steel sheets are being incorporated into vehicle construction in the battle against corrosion.

The use of zinc coated steels has dramatically increased to meet these challenges. Different areas of a vehicle require different zinc coatings and coating weights to meet appearance and performance criteria. These are available in both hot dipped (BS 2989: 1982) and electrolytically deposited (BS 6687: 1986) versions in a range of coating weights or thickness. Both types offer barrier and sacrificial corrosion protection, and the choice of product depends on the particular application and requirements. The hot dip product (available as plain zinc, or iron-zinc alloy) is generally used for underbody parts. The electrolytic product is used for exposed body panels, where a full-finish surface quality is available to ensure that a showroom paint finish is achieved. The electrolytic product is available in single-sided and double-sided coating. (See Figure 4.1.)

Single-sided zinc coated steel   Free zinc is applied to one side of a steel sheet by either the hot dip or the electrolytic process for this material. Its uncoated side provides a good surface for paint appearance, so it is used mainly for outer body panels. Since free zinc is towards the inside of the car, it protects against perforation corrosion.

One-and-a-half-sided zinc coated steel   In this case, one side of the sheet is coated with free zinc, and a thin layer of zinc-iron alloy is formed on the other side. This product is produced mainly by the hot dip process. It is primarily used for exposed panels, where the zinc-iron layer is on the outside for cosmetic protection and the free zinc side provides perforation protection.

Double-sided zinc coated steel   This product is manufactured by applying free zinc to both sides of the sheet with equal or differential

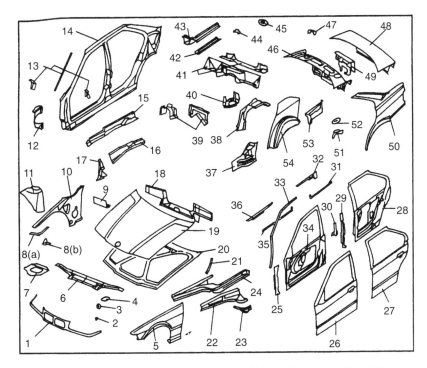

Figure 4.1 Body shell panels showing galvanized protection (*Motor Insurance Repair Research Centre*)

1	Front grille panel	B	25	Front door frame stiffener RH and LH	A
2	Reinforcement	B	26	Front door skin RH and LH	A
3	Front bumper mounting reinforcement RH and LH	B	27	Rear door skin RH and LH	A
4	Bonnet lock reinforcement RH and LH	B	28	Rear door frame RH and LH	B
5	Front wing RH and LH	B	29	Rear door frame stiffener RH and LH	A
6	Bonnet lock panel	B	30	Rear door frame gusset RH and LH	B
7	Front suspension turret stiffener RH and LH	A	31	Rear door skin stiffener RH and LH	A
8a/8b	Front wheel arch gussets RH and LH	B(a)/C(b)	32	Rear door lock stiffener RH and LH	A
9	Front bulkhead stiffener RH and LH	A	33	Front window frame stiffener RH and LH	A
10	Front inner wing RH and LH	B	34	Front door frame RH and LH	B
11	Front suspension turret RH and LH	A	35	Front door skin stiffener RH and LH	A
12	A-post reinforcement RH and LH	B	36	Front door lock stiffener RH and LH	A
13	Body side gussets RH and LH	B	37	Sill rear gusset RH and LH	A
14	Body side RH and LH	A	38	Rear chassis leg RH and LH	A
15	Inner sill RH and LH	A	39	Floor/heelboard gussets	A
16	Inner sill reinforcement RH and LH	A	40	Rear chassis leg gusset RH and LH	B
17	B-post gusset RH and LH	A	41	Boot floor cross members	A
18	Upper dash panel	B	42	RH boot floor brace closing panel	A
19	Bonnet skin	B	43	Boot floor brace RH and LH	A
20	Bonnet frame	B	44	Boot floor bracket RH and LH	A
21	Dash stiffener RH and LH	B	45	Rear bumper mounting gusset RH and LH	B
22	Front chassis leg closing panel RH and LH	A	46	Boot lid frame	B
23	Front chassis leg gusset RH and LH	B	47	Exhaust rear hanger stiffener	B
24	Front chassis leg RH and LH	A	48	Boot lid skin	A
			49	Boot lid lock mounting stiffener	B
			50	Rear wing RH and LH	A
			51	Rear suspension turret RH and LH	A
			52	Suspension turret capping RH and LH	B
			53	Boot floor side extension RH and LH	A
			54	Outer rear wheel arch RH and LH	A

A = Galvanized one side only
B = Galvanized both sides
C = Galvanized layer applied individually

coating weight. All types are readily paintable and weldable. However, care should be taken to ensure that welding conditions are comparable with the material used: for example, higher weld current ratings may be necessary on the heavier coatings.

4.4 Alloy steels

Alloy steel is a general name for steels that owe their distinctive properties to elements other than carbon. They are generally classified into two major categories:

Low-alloy steel   possesses similar microstructures to and requires similar heat treatments to plain carbon steels (see Section 4.3.2 on micro-alloyed steel).

High-alloy steel   may be defined as a steel having enhanced properties owing to the presence of one or more special elements or a larger proportion of element than is normally present in carbon steel. This section is concerned primarily with high-alloy steels.

Alloy steels usually take the name of the element or elements, in varying percentages, having the greatest influence on the characteristics of the alloy.

Chromium   Increased hardness and resistance to corrosion.

Cobalt   Increased hardness, especially at high temperatures.

Manganese   High tensile strength, toughness and resistance to wear.

Molybdenum   Increased hardness and strength at high temperatures.

Nickel   Increased tensile strength, toughness, hardness and resistance to fatigue.

Niobium   Strong carbide forming effect; increases tensile strength and improves ductility.

Silicon   Used as a deoxidizing agent, and has the slight effect of improving hardness.

Titanium   Strong carbide forming element.

Tungsten   Greater hardness, especially at high temperatures; improved tensile strength and resistance to wear.

Vanadium   Increased toughness and resistance to fatigue.

Correct heat treatment is essential to develop the properties provided by alloying elements.

There are many alloy steels containing different combinations and percentages of alloying elements, of which some of the most popular are as follows:

High-tensile steel   Used whenever there is an essential need for an exceptionally strong and tough steel capable of withstanding high stresses. The main alloying metals used in its manufacture are nickel, chromium and molybdenum and such steels are often referred to as nickel-chrome steels. The exact percentage of these metals used varies according to the hardening processes to be used and the properties desired. Such steels are used for gear shafts, engine parts and all other parts subject to high stress.

High-speed steels   These are mostly used for cutting tools because they will withstand intense heat generated by friction and still retain their hardness at high temperatures. It has been found that by adding tungsten to carbon steel, an alloy steel is formed which will retain a hard cutting edge at high temperatures. High-speed steels are based on tungsten or molybdenum or both as the primary heat-resisting alloying element; chromium gives deep hardening and strength, and vanadium adds hardness and improves the cutting edge.

Manganese steel   An addition of manganese to steel produces an alloy steel which is extremely tough and resistant to wear. It is used extensively in the manufacture of chains, couplings and hooks.

Chrome-vanadium steel   This contains a small amount of vanadium which has the effect of intensifying the action of the chromium and the manganese in the steel. It also aids in the formation of carbides, hardening the alloy and increasing its ductility. These steels are valuable where a combination of strength and ductility are desired. They are often used for axle half-shafts, connecting rods, springs, torsion bars, and in some cases hand tools.

Silicon-manganese steel   This is a spring steel using the two elements of manganese and silicon. These steels have a high strength and impact resistance and are used for road springs and valve springs.

4.4.1 High-strength steel

High-strength steel have been introduced into automotive production slowly only because of the need for specialised press tools to form body panels from this stronger material. The die tools need to be harder than for normal low carbon (LC) steel, and the presses need to be stronger and more accurate. HSS came about because of the need to make vehicles lighter following the 1970 fuel crisis. Lighter car means better fuel economy. This lead to American car makers forming the Ultra Light Steel Auto Body (ULSB) group and the Ultra Light Steel Body – Advanced Vehicle Concepts (ULSB – ACV) group. This further research has led to the concept of advanced high-strength steel (AHSS) as the materials have been developed and understood.

Cost
Steel costs about one-fifth of the price of aluminium when bought in the quantities needed by a car maker. Also the iron and steel industry has hundreds of years of practical experience in shaping and forming steel compared to the other materials which could be used to make vehicle bodies.

Properties of HSS
High-strength steel has a yield strength ranging from 300 to 1200 MPa compared to LC steel which has a range of 140–180 MPa. However, although the metal is stronger, it is not necessarily stiffer. That is, the body parts can not necessarily be made of thinner metal as they are likely to sag. If you look at the swage lines on the latest vehicles, you will see that many panels are stiffened by the use of swaging. The current modern shapes are to allow the usage of thinner sheet steel which is lighter and of course cheaper. Oddly however, the new vehicles are not lighter in weight; this is because of the addition of electrical body controls such as electric windows and seats. HSS is not as easy to form as LC steel; also some types of HSS can be drawn better than other. Generally the extra strength of HSS is brought about by changes in the steel microstructure during the steel processing. The following paragraphs discuss the different types of HSS and AHSS steels.

HSS are also known as re-phosphorized – added phosphorous; isotropic – added silicone and bake hardened – strain age hardened. The two most common types used in vehicle body construction are MSLA and HSLA.

Medium-Strength Low Alloy steel has a yield strength of between 180 and 300 MPa. This steel is made by dissolving more phosphorous or manganese alloy into the molten steel during manufacture.

High-Strength Low Alloy steel has a yield strength of between 250 and 500 MPa. This is made by adding small amounts of titanium or niobium to the molten steel which produces a fine dispersion of carbide particles.

Advanced high-strength steel types are aimed at producing steel with suitable mechanical properties for the forming of vehicle body parts, usually through the hydro forming process – using water pressure to mould the metal over the die.

Dual phase (DP) steel has a yield strength of between 500 and 1000 MPa. It is made by adding carbon to enable the formation of (hard) martensite in a more ductile ferrite matrix. Manganese, chromium, vanadium or nickel may also be added. The DP steel may have its strength triggered by either bake hardening or work hardening when it is stressed under the stamping or other forming process.

Transformation induced plasticity (TRIP) steel has a yield strength of between 500 and 800 MPa with greater figures attainable in some cases. TRIP steels may be alloyed with higher quantities of carbon and silicone and aluminium. The strength is triggered by work hardening by the stress induced during the stamping or forming process. That is, the retained austenite is transformed into martensite by the increasing strain during the stamping or other forming process.

Complex phase (CP) steel has a yield strength of 800–1200 MPa. CP steel has a very fine microstructure using the same alloying elements as in DP or TRIP steel with the possible additions of niobium, titanium and/or vanadium. Again the high strength is triggered by applied strain.

Applications of AHSS
TRIP and CP steel is ideal for use in crash zones. It is excellent for absorbing energy during impact. CP steel is often used for 'A' and 'B' posts and bumper attachments. Increasingly AHSS steel is used for strengthening members to which other steel panels are welded, in other words a steel composite structure.

Repair of HSS and AHSS panels
One of the problems is that it is not possible to recognize HSS and AHSS panels by sight. Therefore it is essential to follow the guidelines offered by the vehicle manufacturer on recommended repair methods. As a general guide, look out for parts such as 'A', 'B'

and 'C' posts, screen pillars, cant rails and strengthening cross members on cars made after about 1990. On new cars look out for panels with large swage lines, remember that the pressing process triggers the hardening of the metal and so makes the panel stiff both in shape and in microstructure. Damaged AHSS and HSS panels are not readily repairable as the impact changes the microstructure, making the metal harder. So, panel beating is not an option, replacement panels must be fitted in most cases.

The normal method of joining HSS and AHSS panels is by spot welding. The heat and pressure involved in the welding process changes the microstructure of the metal, so great care is needed in this process. Again, follow manufacturer's instruction on these repairs. It is sensible to do tests before spot welding the new panels. You can use the undamaged sections of the panels which you have removed for test welds, changing weld time and current, then cutting through the welded area to check for penetration and adhesion.

Any form of applied heat to HSS and AHSS panels should be avoided, this includes trying to anneal or soften the panel for the purposes of straightening, heat shrinking or oxyacetylene cutting. Neither MIG plugging nor cold working are recommended. Remember that the nature of these processes will affect a change to the properties of the steel, and that the energy of the impact will have had the same effect on the panel.

Guide to spotting AHSS and HSS panels

- Look for reinforced areas such as door pillars and cross bracing areas, and where two or more panel parts over lap each other.
- Look for body panels with pronounced sharp swage lines.
- Feel for very thin panels.
- Listen for panels which when tapped gently give a crisp metallic ring.

4.4.2 Boron steel

A particular type of HSS is referred to as Boron steel, containing a very small amount of boron, typically 0.001–0.004 percent. This steel is used by many of the major manufacturers such as Vauxhall and requires special repair techniques. To join boron steel panels the method used is often MIG brazing. See 12.13.3.

4.5 Stainless steel

The discovery of stainless steel was made in 1913 by Harry Brearley of Sheffield, while he was experimenting with alloy steels. Among the samples which he threw aside as unsuitable was one containing about 14 per cent chromium. Some months later he saw the pile of scrap test pieces and noticed that most of the steels had rusted but the chromium steel was still bright. This led to the development of stainless steels. The classic Rolls-Royce radiator was one of the first examples of the use of stainless steel.

The designer, engineer or fabricator of a particular component may think that stainless steel is going to be both difficult to work and expensive. This is quite wrong, and perhaps stems from the fact that many people tend to fall into the trap of the generic term 'stainless steels'. In fact, this is the title for a wide range of alloys. Therefore if such materials are to be used effectively and maximum advantage is to be taken of the many benefits they have to offer, there should be very close collaboration and consultation over which grade of stainless steel is best for the particular job in hand.

There are over 25 standard grades of stainless steel specified by BS 1449: Part 2. Each provides a particular combination of properties, some being designed for corrosion resistance, some for heat resistance and others for high-temperature creep resistance. Many, of course, are multipurpose alloys and can be considered for more than one of these functions. In terms of composition, there is one element common to all the different grades of stainless steel. This is chromium, which is present to at least 10 per cent. It is this element which provides the basis of the resistance to corrosion by forming what is known as a 'passive film' on the surface of the metal. This film is thin, tenacious and invisible and is essentially a layer of chromium oxide formed by the chromium in the steel combining with the oxygen in the atmosphere. The strength of the passive film, in terms of resistance to corrosion, increases within limits with the chromium content and with the addition of other elements such as nitrogen and molybdenum. The formation of the passive film, therefore, is a natural characteristic of this family of steels and requires no artificial aid. Consequently, if stainless steels are scratched or cut or drilled, the passive film is automatically and instantaneously repaired by the oxygen in the atmosphere.

Table 4.4   Typical stainless steels used in vehicles: BS 1449: Part 2

Steel grade	Typical alloying elements (%)	Characteristics	Typical applications
Austenitic			
301 S21	17Cr 7Ni	Good corrosion resistance	Riveted body panels, wheel covers, hubcaps, rocker panel mouldings
304 S16	18Cr 10Ni 0.06C	Good corrosion resistance	Mild corrosive tankers
305 S19	18Cr 11Ni 0.10C	Low work hardening rate	Rivets
316 S31	17Cr 11Ni 2.25Mo 0.07C	Highest corrosion resistance of the commercial grades	Road tankers for widest cargo flexibility
Ferritic			
409 S19	11Cr 0.08C + Ti	Reasonable combination of weldability, formability and corrosion resistance	Exhaust systems and catalytic converter components and freight container cladding
430 S17	17Cr	Good corrosion resistance. Can be drawn and formed	Tanker jackets, interior trim, body mouldings, windshield wiper arms
Martensitic			
410 S21	13Cr 0.12C	Corrosion resistant. Heat-treatable composition capable of high hardness	Titanium modified for silencer components

Stainless steels can be conveniently divided into the following three main groups:

Austenitic   Generally containing 16.5–26 per cent chromium and 4–22 per cent nickel.
Ferritic   Usually containing 12–18 per cent chromium.
Austenitic/ferritic duplex   Usually containing 22 per cent chromium, 5.5 per cent nickel, 3 per cent molybdenum, and 0.15 per cent nitrogen.
Martensitic   Based on a chromium content of 11–14 per cent, although some grades may have a small amount of nickel.

Of the above groups, the austenitic steels are by far the most widely used because of the excellent combination of forming, welding and corrosion-resisting properties that they offer. Providing that the correct grade is selected as appropriate to the service environment, and that the design and production engineering aspects are understood and intelligently applied, long lives with low maintenance costs can be achieved with these steels.

HyResist 22/5 duplex is a highly alloyed austenitic/ferritic stainless steel. It has more than twice the proof strength of normal austenitic stainless steels whilst providing improved resistance to stress corrosion cracking and to pitting attacks. It

possesses good weldability and can be welded by conventional methods for stainless steel. The high joint integrity achievable combined with good strength and toughness permit fabrications to be made to a high standard. It is being increasingly used in offshore and energy applications.

Table 4.4 shows typical stainless steels used in motor vehicles.

4.6 Aluminium

In the present-day search for greater economy in the running of motor vehicles, whether private, public or commercial, the tendency is for manufacturers to produce bodies which, whilst still maintaining their size and strength, are lighter in weight. Aluminium is approximately one-third of the weight of steel, and aluminium alloys can be produced which have an ultimate tensile strength of 340–620 MN/m^2. In the early 1920s the pioneers of aluminium construction were developing its use for both private and commercial bodies; indeed, the 1922 40 hp Lanchester limousine body had an aluminium alloy construction for the bulkhead and bottom frame, and aluminium was used for all the body panels. Before the Second World War aluminium was used mainly for body panels, but since the war aluminium alloys have been and are now being used for body structures. Although aluminium is more expensive than steel, it is easy to work and manipulate and cleaner to handle. It also has the advantage of not rusting, and, provided that the right treatment is adopted for welding, corrosion is almost non-existent. In recent years the use of aluminium and aluminium alloys for motor bodies, especially in the commercial field, has developed enormously.

In the modern motor body the saving of weight is its most important advantage, and although on average the panel thickness used is approximately double that of steel, a considerable weight saving can be achieved. One square metre of 1.6 mm thick aluminium weights 4.35 kg while one square metre of 1.00 mm thick steel weighs 7.35 kg; the use of aluminium results in a saving in weight of just under 40 per cent.

The non-rusting qualities of the aluminium group are well known and are another reason for their use in bodywork. An extremely thin film of oxide forms on all surfaces exposed to the atmosphere, and even if this film is broken by a scratch or chip it will reform, providing complete protection for the metal. The oxide film, which is only 0.0002 cm thick, is transparent, but certain impurities in the atmosphere will turn it to various shades of grey.

4.6.1 Production

The metal itself has only been known about 130 years, and the industrial history of aluminium did not begin until 1886 when Paul Heroult in France discovered the basis of the present-day method of producing aluminium. Aluminium is now produced in such quantities that in terms of volume it ranks second to steel among the industrial metals. Aluminium of commercial purity contains at least 99 per cent aluminium, while higher grades contain 99.5–99.8 per cent of pure metal.

In the production of aluminium, the ore bauxite is crushed and screened, then washed and pumped under pressure into tanks and filtered into rotating drums, which are then heated. This separates the aluminium oxide from the ore. In the next stage the aluminium oxide is reduced to the metal aluminium by means of an electrolytic reduction cell. This cell uses powdered cryolite and a very heavy current of electricity to reduce the aluminium oxide to liquid metal, which passes to the bottom of the cell and is tapped off into pigs of aluminium of about 225 kg each.

4.6.2 Types of sheet

Sheet, strip and circle blanks are sold in hard and soft tempers possessing different degrees of ductility and tensile strength. Sheet is supplied in gauges down to 0.3 mm, but it is generally more economical to order strip for gauges less than 1.6 mm.

4.6.3 Manufacturing process

Sheet products are first cast by the semicontinuous casting process, then scalped to remove surface roughness and preheated in readiness for hot rolling. They are first reduced to the thickness of plate, and then to sheet if this is required. Hot rolling is followed by cold rolling, which imparts finish and temper in bringing the metal to the gauge required. Material is supplied in the annealed (soft condition) and in at least three degrees of hardness, H1, H2 and H3 (in ascending order of hardness).

4.7 Aluminium alloys

From the reduction centre the pigs of aluminium are remelted and cast into ingots of commercial purity. Aluminium alloys are made by adding specified amounts of alloying elements to molten aluminium. Some alloys, such as magnesium and zinc, can be added directly to the melt, but higher-melting-point elements such as copper and manganese have to be introduced in stages. Aluminium and aluminium alloys are produced for industry in two broad groups:

1   Materials suitable for casting
2   Materials for the further mechanical production of plate sheet and strip, bars, tubes and extruded sections.

In addition both cast and wrought materials can be subdivided according to the method by which their mechanical properties are improved:

Non-heat-treatable alloys   Wrought alloys, including pure aluminium, gain in strength by cold working such as rolling, pressing, beating and any similar type of process.
Heat-treatable alloys   These are strengthened by controlled heating and cooling followed by ageing at either room temperature or at 100–200 °C.

The most commonly used elements in aluminium alloys are copper, manganese, silicon, magnesium and zinc. The manufacturers can supply these materials in a variety of conditions. The non-heat-treatable alloys can be supplied either as fabricated (F), annealed (O) or strain hardened (H1, H2, H3). The heat-treatable alloys can be supplied as fabricated (F) or annealed (O), or, depending on the alloy, in variations of the heat treatment processes (T3, T4, T5, T6, T8).

4.7.1 Wrought light aluminium alloys: BS specifications 1470–75

Material designations
Unalloyed aluminium plate, sheet and strip:

1080A	commercial pure aluminium	
	99.8 per cent	
1050A	commercial pure aluminium	
	99.5 per cent	
1200	commercial pure aluminium	
	99.0 per cent	

Non-heat-treatable aluminium alloy plate, sheet and strip:

3103	AlMN
3105	AlMnMg
5005	AlMg
5083	AlMgMn
5154A	AlMg
5251	AlMg
5454	AlMgMn

Heat-treatable aluminium alloy plate, sheet and strip:

2014A	AlCuSiMg
Clad 2014A	AlCuSiMg clad with pure aluminium
2024	AlCuMg
Clad 2024	AlCuMg
Clad 2024	AlCuMg clad with pure aluminium
6082	AlSiMgMn

Abbreviations for basic temper

As fabricated: F The temper designation F applies to the products of shaping processes in which no special control over thermal conditions or strain hardening is employed. For wrought products there are no specified requirements for mechanical properties.
Annealed: O The temper designation O applies to wrought products which are annealed to obtain the lowest strength condition.

Abbreviations for strain hardened materials
The temper designation H for strain hardened products (wrought products only) applies to products subjected to the application of cold work after annealing (hot forming) and partial annealing or stabilizing, in order to achieve the specified mechanical properties. The H is always followed by two or more digits, indicating the final degree of strain hardening. The first digit (1, 2 or 3) indicates the following:

H1 strain hardened only
H2 strain hardened and partially annealed
H3 strain hardened and stabilized

The second digit (2, 4, 6 or 8) indicates the degree of strain hardening, as follows:

HX2	tensile strength approximately midway between O temper and HX4 temper
HX4	tensile strength approximately midway between O temper and HX8 temper
HX6	tensile strength approximately midway between HX4 temper and HX8 temper
HX8	full hard temper

Abbreviations for heat-treated materials
The temper designation T applies to products which are thermally treated, with or without supplementary strain hardening, to produce stable tempers. The T is always followed by one or more digits, indicating the specific sequence of treatments as follows:

T3	solution heat treated, cold worked, and naturally aged to a substantially stable condition
T4	solution heat treated and naturally aged to a substantially stable condition
T5	cooled from an elevated temperature shaping process and then artificially aged
T6	solution heat treated and then artificially aged
T8	solution heat treated, cold worked and then artificially aged

The additional digits for the T tempers TX51 indicate that the products have been stress relieved by controlled stretching.

Tables 4.5, 4.6, 4.7 and 4.8 show the characteristics and properties of aluminium alloys.

4.7.2 Aluminium alloys used in bodywork

The choice of material and the condition in which it is required must depend largely upon design requirements and the manufacturing processes within the factory. The alloys most commonly used in vehicle body work are as shown in Table 4.9.

Table 4.5 General characteristics of all wrought forms of aluminium alloys

Purity or alloy	Temper or condition	General characteristics						
		Cold forming	Machining	Durability	Inert-gas shielded arc (MIG or TIG)	Welding resistance (spot, seam, flash butt or stud)[b,c]	Oxy-acetylene[d]	Metal arc
5251	F	V	G	V	V[e]	E	G[e]	F[e]
	O	V	G					
	H22	G	G					
	H24	G	V					
	H28	G	V					
	H39X	F	V	F	V[e]	E	G[e]	G[e]
5454	F	V	V	V	V[e]	E	F[e]	N
	O	V		V				
	H22	G		G				
	H24	G		G				
5154	F	V	V	V[f]	E	E	F	N
	O	V		V[f]				
	H22	G		G[f]				
	H24	F		G[f]				
	H28							
5083	F	G	E	V[f]	E	E	F	N
	O	G		V[f]				
	H22	F		F[f]				
	H24	G		F[f]				
6082	F	V	G	G	V[e]	V	F[e]	G[e]
	O	E	G	G				
	T4	G	V	V				
	T6	F	E	G				
	T451							
	T651							
1200	F	E	F	V	E	V	V	G
	O	E	F					
	H12	V	F					
	H14	V	F					
	H16	G	G					
	H18	F	G					
3103	F	E	F	V	E	E	V	G
	O	E	F					
	H12	V	F					
	H14	V	F					
	H16	G	G					
	H18	F	G					

(a) Materials are graded thus: E excellent; V very good; F fair; P poor; N not recommended.

(b) The mechanical properties of work hardened or heat-treated materials will be reduced in the vicinity of the weld.

(c) The weld zone is generally discernible after anodic treatment, the degree depending on the material and welding process.

(d) The oxy-acetylene process is normally recommended only for material thinner than 6–4 mm.

(e) Filler or electrode of other than parent metal composition is recommended.

(f) Applicable at temperature of 70 °C and less.

Table 4.6 Chemical composition limits[1] and mechanical properties of unalloyed aluminium plate, sheet and strip (BS 1470)

Material designation	Tolerance category	Silicon (%)	Iron (%)	Copper (%)	Manganese (%)	Magnesium (%)	Chromium (%)	Nickel (%)	Zinc (%)	Gallium (%)	Titanium (%)	Others[2] Each (%)	Others[2] Total (%)	Aluminium min. (%)	Temper[4]	Thickness > (mm)	Thickness ≤ (mm)	Tensile strength Min. (N/mm^2)	Tensile strength Max. (N/mm^2)	Elongation on 50 mm — Materials thicker than 0.5 mm min. (%)	0.8 mm min. (%)	1.3 mm min. (%)	2.6 mm min. (%)	3.0 mm min. (%)	Elongation on $5.65\sqrt{S_o}$, over 12.5 mm thick (min.) (%)
1080A	A	0.15	0.15	0.03	0.02	0.02	–	–	0.06	0.03	0.02	0.02	–	99.80[3]	F	3.0	25.0	–	–	–	–	–	–	–	–
															O	0.2	6.0	–	90	29	29	29	35	35	–
															H14	0.2	12.5	90	125	5	6	7	8	8	–
															H18	0.2	3.0	125	–	3	4	4	5	–	–
1050A	A	0.25	0.40	0.05	0.05	0.05			0.07		0.05	0.03	–	99.50[3]	F	3.0	25.0	–	–	–	–	–	–	–	–
															O	0.2	6.0	55	95	22	25	30	32	32	–
															H12	0.2	6.0	80	115	4	6	8	9	9	–
															H14	0.2	12.5	100	135	4	5	6	6	8	–
															H18	0.2	3.0	135	–	3	3	4	4	–	–
1200	A	1.0 Si + Fe		0.05	0.05				0.10		0.05	0.05	0.15	99.99[3]	F	3.0	25.0	–	–	–	–	–	–	–	–
															O	0.2	6.0	70	105	20	25	30	30	30	–
															H12	0.2	6.0	90	125	4	6	8	9	9	–
															H14	0.2	12.5	105	140	3	4	5	5	6	–
															H16	0.2	6.0	125	160	2	3	4	4	4	–
															H18	0.2	3.0	140	–	2	3	4	4	–	–

1 Composition in per cent (m/m) maximum unless shown as a range or a minimum.

2 Analysis is regularly made only for the elements for which specific limits are shown. If, however, the presence of other elements is suspected to be, or in the case of routine analysis is indicated to be, in excess of the specified limits, further analysis should be made to determine that these other elements are not in excess of the amount specified.

3 The aluminium content for unalloyed aluminium not made by a refining process is the difference between 100.00% and the sum of all other metallic elements in amounts of 0.010% or more each, expressed to the second decimal before determining the sum.

4 An alternative method of production, designated H2, may be used instead of the H1 routes, subject to agreement between supplier and purchaser and providing that the same specified properties are achieved.

Table 4.7 Chemical composition limits[1] and mechanical properties of aluminium alloy plate, sheet and strip (non-heat-treatable) (BS 1470)

Material designation	Tolerance category	Silicon (%)	Iron (%)	Copper (%)	Manganese (%)	Magnesium (%)	Chromium (%)	Nickel (%)	Zinc (%)	Other restrictions (%)	Titanium (%)	Others[2] Each (%)	Others[2] Total (%)	Aluminium (%)	Temper[3]	Thickness > (mm)	Thickness ≤ (mm)	0.2% proof stress min. (N/mm²)	Tensile strength Min. (N/mm²)	Tensile strength Max. (N/mm²)	Elong. 50 mm 0.5 mm min.(%)	0.8 mm min.(%)	1.3 mm min.(%)	2.6 mm min.(%)	3.0 mm min.(%)	Elong. 5.65√S₀ over 12.5 mm thick min.(%)
3103	A	0.50	0.7	0.10	0.9–1.5	0.30	0.10	–	0.20	0.10 Zr + Ti	–	0.05	0.15	Rem.*	F	0.2	25.0	–	90	130	20	23	24	24	25	–
															H12	0.2	6.0	–	120	155	5	6	7	9	9	–
															H14	0.2	6.0	–	140	175	3	4	5	6	7	–
															H16	0.2	12.5	–	160	195	2	3	4	4	4	–
															H18	0.2	6.0	–	175	–	2	3	4	4	4	–
3105	A	0.6	0.7	0.30	0.30–0.8	0.20–0.8	0.20	–	0.40	–	0.10	0.05	0.15	Rem.	O	0.2	3.0	–	110	155	16	18	20	20	–	–
															H12	0.2	3.0	115	130	175	3	3	4	5	–	–
															H14	0.2	3.0	145	160	205	2	2	3	4	–	–
															H16	0.2	3.0	170	185	230	1	1	2	3	–	–
															H18	0.2	3.0	190	215	–	1	1	1	2	–	–
5005	A	0.30	0.7	0.20	0.20	0.50–1.1	0.10	–	0.25	–	–	0.05	0.15	Rem.	O	0.2	3.0	–	95	145	18	20	21	22	–	–
															H12	0.2	3.0	80	125	170	4	5	6	8	–	–
															H14	0.2	3.0	100	145	185	3	3	5	6	–	–
															H18	0.2	3.0	165	185	–	1	2	3	3	–	–
5083	B	0.40	0.40	0.10	0.40–1.0	4.0–4.9	0.05–0.25	–	0.25	–	0.15	0.05	0.15	Rem.	F	3.0	25.0	–	275	–	–	–	–	–	–	14
															O	0.2	80.0	125	275	350	12	14	16	16	16	–
															H22	0.2	6.0	235	310	375	5	6	8	10	10	–
															H24	0.2	6.0	270	345	405	4	5	6	8	8	–
5154A	B	0.50	0.50	0.10	0.50	3.1–3.9	0.25	–	0.20	0.10–0.50 Mn + Cr	0.20	0.05	0.15	Rem.	O	0.2	6.0	85	215	295	12	14	16	18	18	–
															H22	0.2	6.0	165	245	325	5	6	6	8	8	–
															H24	0.2	6.0	225	275	–	4	5	5	6	6	–
5251	A	0.40	0.50	0.15	0.10–0.50	1.7–2.4	0.15	–	0.15	–	0.15	0.05	0.15	Rem.	F	3.0	25.0	–	–	–	–	–	–	–	–	–
															O	0.2	6.0	60	160	200	18	18	18	20	20	–
															H22	0.2	6.0	130	200	240	4	5	6	8	8	–
															H24	0.2	6.0	175	225	275	3	4	5	5	5	–
															H28	0.2	3.0	215	255	285	2	3	3	4	–	–
5454	B	0.25	0.40	0.10	0.50–1.0	2.4–3.0	0.05–0.20	–	0.25	–	0.20	0.05	0.15	Rem.	F	3.0	25.0	–	–	–	–	–	–	–	–	–
															O	0.2	6.0	80	215	285	12	14	16	18	18	–
															H22	0.2	6.0	180	250	305	4	5	7	8	8	–
															H24	0.2	3.0	200	270	325	3	4	5	6	–	–

* Remainder

1 Composition in per cent (m/m) maximum unless shown as a range or a minimum.

2 Analysis is regularly made only for the elements for which specific limits are shown. If, however, the presence of other elements is suspected to be, or in the case of routine analysis is indicated to be, in excess of the specified limits, further analysis should be made to determine that these other elements are not in excess of the amount specified.

3 An alternative method of production, designated H2, may be used instead of the H1 routes, subject to agreement between supplier and purchaser and providing that the same specified properties are achieved.

Table 4.8 Chemical composition limits and mechanical properties of aluminium alloy plate, sheet and strip (heat-treatable) (BS 1470)

Material designation	Tolerance category	Silicon (%)	Iron (%)	Copper (%)	Manganese (%)	Magnesium (%)	Chromium (%)	Nickel (%)	Zinc (%)	Other restrictions (%)	Titanium (%)	Others[2] Each (%)	Others[2] Total (%)	Aluminium (%)	Temper[3]	Thickness > (mm)	Thickness ≤ (mm)	0.2% proof stress min. (N/mm²)	Tensile strength Min. (N/mm²)	Tensile strength Max. (N/mm²)	Elong. 0.5 mm min. (%)	Elong. 0.8 mm min. (%)	Elong. 1.3 mm min. (%)	Elong. 2.6 mm min. (%)	Elong. 3.0 mm min. (%)	Elongation on 5.65√S_o, over 12.5 mm thick min. (%)
2014A	B	0.50–0.9	0.50	3.9–5.0	0.40–1.2	0.20–0.8	0.10	0.10	0.25	0.20 Zr + Ti	0.15	0.05	0.15	Rem.*	O	0.2	6.0	110	–	235	14	14	16	16	16	–
															T4	0.2	6.0	225	400	–	13	14	14	14	14	–
															T6	0.2	6.0	380	440	–	6	6	7	7	8	–
															T451	6.0	25.0	250	400	–	–	–	–	–	14	12
																25.0	40.0	250	400	–	–	–	–	–	–	10
																40.0	80.0	250	395	–	–	–	–	–	–	7
Clad 2014A	B	0.50–0.9	0.50	3.9–5.0	0.40–1.2	0.20–0.8	0.10	0.10	0.25	0.20 Zr + Ti	0.15	0.05	0.15	Rem.	T651	6.0	25.0	410	460	–	–	–	–	–	–	6
																25.0	40.0	400	450	–	–	–	–	–	–	5
																40.0	60.0	390	430	–	–	–	–	–	–	5
																60.0	90.0	390	430	–	–	–	–	–	–	4
																90.0	115.0	370	420	–	–	–	–	–	–	4
																115.0	140.0	350	410	–	–	–	–	–	–	4
2024	B	0.50	0.50	3.8–4.9	0.30–0.9	1.2–1.8	0.10	–	0.25	–	0.15	0.05	0.15	Rem.	O	0.2	6.0	100	–	220	14	14	16	16	16	–
															T4	0.2	1.6	240	385	–	13	14	14	–	–	–
																1.6	6.0	245	395	–	–	–	–	14	14	–
															T6	0.2	6.0	345	420	–	7	7	8	9	9	–
															O	0.2	6.0	110	–	235	–	–	–	–	–	–
															T3	0.2	1.6	290	440	–	12	12	14	–	–	–
																1.6	6.0	290	440	–	11	11	11	12	12	–
															T351	6.0	25.0	280	430	–	–	–	–	–	–	10
																25.0	40.0	280	420	–	–	–	–	–	–	9
																40.0	60.0	270	410	–	–	–	–	–	–	9
																60.0	90.0	270	410	–	–	–	–	–	–	8
																90.0	115.0	270	400	–	–	–	–	–	–	8
																115.0	140.0	260	390	–	–	–	–	–	–	7

*Remainder

Table 4.9 Standard aluminium alloys: availability, physical properties, and applications

Purity or alloy (new nomenclature)	Related BS/GE specification (BS 1470–1475) (old BS alloy designation)	Nominal composition: % alloying elements (remainder aluminium and normal impurities)			Standard forms[a]					Physical properties				Typical road transport applications
		Mg	Si	Mn	Sheet	Plate	Extrusions	Hollow extrusions	Tube	Density g/cm³	lb/in³	Melting range °C (approx.)	Coefficient of linear expansion per °C (20–100°C)	
Non-heat-treatable alloys														
1200	IC	–	–	–	×	×	×	×	×	2.71	0.098	660	0.000 023 5	Vehicle panelling where panel beating is required; mouldings and trim. Tank cladding
3103	N3	–	–	1.2	×	×	×ᵇ	×ᵇ	×ᵇ	2.73	0.099	645–655	0.000 023 5	Flat panelling of vehicles and general sheet metal work. Tank cladding
5251	N4	2.25	–	0.4	×	×	×	–	×	2.69	0.097	595–650	0.000 024	Body panelling. Head boards and drop sides. Truss panels in buses. Cab panelling. Tanker shells and divisions
5154A	N5	3.5	–	0.4	×	–	×	–	×	2.67	0.096	600–640	0.000 024 5	Welded body construction. Tipper body panelling. Truss panels in buses
5454	N51	2.7	–	0.8										Tanker shells and divisions
5083	N8	4.5	–	0.5	×	×	×	–	×	2.65	0.096	580–635	0.000 024 5	Pressurized bulk transport at ambient or low temperature
Heat-treatable alloys														
6082	H30	0.7	1.0	0.5	×	×	×	×	×	2.70	0.098	570–660	0.000 023 5–0.000 024ᶜ	All structural sections in riveted vehicle bodies. Tipper body panelling. Highly stressed underframe gussets, truss panels

(a) Other forms may be available by special arrangement.

(b) Not covered by British Standard (general engineering) specification.

(c) Depending on condition.

Alloy 5154A is suitable for use in car panels which are to be pressed into shape; it is supplied in either annealed condition or H2 condition, which are the most suitable for press work on vehicle bodies. Of the other materials, 1200 is a commercial purity sheet, and is widely used for exterior and interior panelling where no great strength is required. Types 3103, 5251, 5154A and 5056A are non-heat-treatable alloys of the aluminium-magnesium range with a strength of 90–325 N/mm². They come in sheet form, and provide a range of mechanical properties to suit different applications. They are used extensively in panel work, and also for forming, pressing and machining, and can be welded without much difficulty. The plate material 5083 is a medium-strength non-heat-treatable alloy particularly suitable for welding. It can be used for parts carrying fairly high stress loads and is often used in the form of patterned tread plate for floor sections.

For internal structure members which need to be stronger than the outer panels, the heat-treatable alloys usually used are 6063 and 6082, and in odd cases 2014A. Type 6082 is a heat-treatable medium-strength alloy which combines good mechanical properties with high corrosion resistance. Permissible stresses in this alloy can be as high as 200 N/mm² under static loading conditions, although some reduction below this would normally be made for transport applications where there is a considerable element of dynamic loading. The alloy is weldable by the inert gas arc process, but there is a considerable loss of strength near the weld owing to the annealing effect of the welding process. Type 6063 is also a heat-treatable alloy but of somewhat lower strength, and is used mainly in applications requiring good surface finish or where the parts are required to be anodized. Alloy 2014A contains a greater percentage of copper than the others, is more expensive, is more difficult to form and is less resistant to corrosion, but has the advantage of a greater tensile strength.

Fastenings and solid rivets can be of commercial purity material or of aluminium alloy 5154A, and for smaller sizes 6082 is sometimes used. Rivets are also available in 5056A material, but should not be used in cases where high temperatures occur in service. Bolts used in bodywork are normally of the 6082 alloy.

The condition in which heat-treatable alloys are supplied should be related to their application or use in bodywork. For example, if a section is to remain straight and is part of a framework which is to be bolted, riveted or welded in place, it is obvious that the material used should already be fully heat treated so that maximum strength is provided to support the framework or structure of the body. On the other hand, if the section has to be shaped, bent or formed in any way the material should be used in the annealed condition and then heat treated after the shaping operations have all been carried out.

Aluminium alloys are now being accepted by the automobile manufacturers as a standard material for exterior and interior trim, and are used for all normal bright trim applications such as radiator grilles, headlamp bezels, wheel trim, instrument panels, body mouldings and window and windscreen surrounds. Alloys used for trimming can be divided into two groups: high-purity alloys bright finished on one side only, in which the majority of the trim components are made; and super-purity alloys for use when maximum specular reflectivity is an advantage, such as would be required by light units.

4.8 Rubber

The value of rubber lies in the fact that it can be readily moulded or extruded to any desired shape, and its elastic quality makes it capable of filling unavoidable and irregular gaps and clearances. It is an ideal material in door shuts and as the gasket for window glass, and in both instances it provides the means for excluding dust and water, although with windscreens and back-lights additional use has generally to be made of a sealing material. Rubber specifications have been built on the basis of the properties of material which has given satisfaction. One major difficulty has been to ensure and measure resistance to weathering; rubber is subject to oxidation by ozone in the atmosphere, and this results in cracking. In addition to natural rubber, a variety of types of synthetic rubbers are used by the motor industry; these vary in price and characteristics, and all are more expensive than natural rubber. For complete ozone resistance, it is necessary to use either butyl or neoprene rubber; both satisfy atmospheric and ozone ageing tests. Butyl rubber, however, is 'dead' to handle and contains no wax, and so whilst neoprene is costly its use is essential for some parts.

Sponge sealing rubbers can be provided with built-in ozone resistance by giving them a live skin of neoprene, and a further way of providing ozone resistance is to coat the rubber components with Hypalon; more recently, continuously extruded neoprene sponge has been adopted. Apart from weather resistance, the important requirement of door and boot lid seals is that their compression characteristics should ensure that they are capable of accommodating wide variations in clearance, without giving undue resistance to door closing.

Various types of foam rubber have been evolved to suit the different parts of the car seating, and the designer's choice of material is governed by cost, comfort, durability, the type of base, the type of car, and whether it is a cushion or a squab, a rear or front seat. When considering the foams available, it is apparent that the number of permutations is large. The types of foam available today fall into seven broad categories. These are:

Moulded latex foam
Low-grade fabricated polyether
Fabricated polyether
Moulded polyether
Fabricated polyester
Polyvinyl chloride foam
Reconstituted polyether.

Latex foam today utilizes a mixture of natural and synthetic latexes to obtain the best qualities of both. After being stabilized with ammonia, natural latex is shipped in liquid form to this country from Malaysia, Indonesia and other rubber producing countries. Synthetic latex, styrene butadiene rubber, is made as a byproduct of the oil cracker plants. Polyether foam can now be made in different grades, and the physical properties of the best grades approach those of latex foam. As a general guide, service life and physical properties improve as the density increases for any given hardness. As the cost is proportional to weight, it follows that the higher-performance foams are more costly.

Flexible polyurethane foam seats are replacing heavy and complicated padded metal spring structures. Moulded seats simplify assembly, reduce weight and give good long-term performance. A major innovation here has been the cold cure systems. These produce foams of superb quality, particularly in terms of strength, comfort and long-term ageing. The systems are particularly suitable for the newer seat technologies such as dual hardness, where the wings are firm to give lateral support, leaving the seat pad softer and more comfortable.

4.9 Sealers

The history of sealers is longer than that of the motor car. Mastic, bitumen compounds and putties of various kinds have been used since the invention of the horseless carriage. It is likely that early coach builders used putties of some kind – possibly paint fillers – to bridge joints in various applications on motor bodies, but it is generally conceded that the first use of specialized sealers on a large scale was in the early 1920s when, in America, the pressed steel body became popular. In this country it was 1927 before one of the first truly effective sealers was introduced. It was known as Dum Dum and is still in use. It was a modified roof sealer, and proved to have many applications in body production. It was not until the late 1930s, when all-steel bodies and unitary construction became a common feature of mass produced cars, that more thought was given to the points that required the use of sealing compounds, and to the nature of these products. Amongst the first developed was interweld sealing compound, primarily to prevent corrosion. Since then, particularly in the post-war years, there have been remarkable developments, probably accelerated by criticisms from overseas markets that British cars were susceptible to dust and water entry. Companies specializing in the manufacture of mastic compounds have developed a range of materials which are now used not only for welding and for general putty application but also for floor pans, drip rails, body joints, exterior trim and many other points, leading to well sealed car bodies equal to any produced elsewhere in the world (Figure 4.2, Table 4.10).

The term 'sealer' covers a wide variety of materials used in the motor industry for sealing against water and dust, from products which remain virtually mastic throughout their life to others which harden up but still retain some measure of elasticity. They range from mixtures of inert fillers and semi-drying oils to heat curing plastisols which may be applied in a thin paste form as an interweld sealer or as extruded beads. Sealing compounds can be categorized into the following general groups: oil-based compositions, rubber-based compositions and

SEALANT, ADHESIVE AND ANTICORROSION CLASSIFICATION CHART

Item	Required Properties	Applications	Brand Names
Thermosetting sealant	• Hardens when heated • Non-running	Sealant and adhesive	Drying sealant
Body sealant	• Highly solid, no volume shrinkage • Non-running	Sealing of sheetmetal seams (drip rails, floor, body side panels, trunk, etc.)	Body sealant 3M PartNo. 8531 3M PartNo. 8646
Spot sealer	• Electro-conductive, and spotwelding can be done after application • Excellent water resistance	Spotwelding locations	Spot sealer
Structural adhesive	• Two-agent mixture adhesive • Low Viscosity	Gluing places that cannot be spot-welded, such as roofs	Two-agent denatured epoxy adhesive (MZ 100320)
Interior trim adhesive	• Highly solid, non-running • Good heat resistance • No rubber swelling • Non-drying	Sealing for gromments, packing and metals	3M 8513 Grommeted Windshields Sealer (Black)
Quick-drying, high-strength adhesive	• Quick drying (10 seconds to 3 minutes) • Colourless and transparent after drying	Excellent adhesive performance with most materials (except polyethylene, polypropylene, fluorocarbon resins and other substances with highly absorbent surface)	(For Europe) 3M PartNo. 8121 (Except for Europe) 3M PartNo. 8155
Wax injection		Wax injection	Tectyl 506 Tectyl 506T Tectyl ML Dinitrol 3122 Dinitrol 3654-1 Mercasol 831-ML Waxoyl Terotex HV200PLUS HV300
Rocker panel primer		Rocker panel primer	Glasurit FX89–7330 (Polyester basis) Glasurit FX90–7103 (Water basis)
Underbody anticorrosion agent		Underbody anticorrosion agent	Tectyl 506T Tectyl 506 Tectyl S Mercasol NON-DRIP Waxoyl Dinitrol 4954 Terotex WAX
Undercoating	• Non-running, good adhesion • Thick application possible • Good low temperature performance	Undercoating agent	3M PartNo. 8864 3M PartNo. 8877 (For Europe)

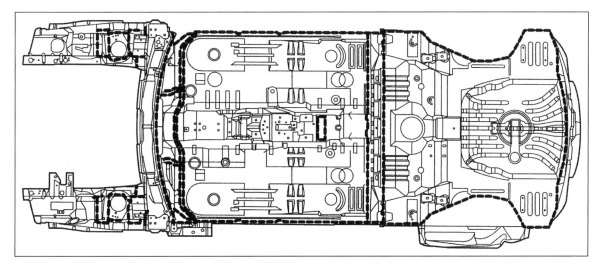

Figure 4.2 Problem areas requiring body sealing

Table 4.10   Types of sealed joints used in vehicle bodywork (*Rover Group Ltd*)

Application		Material description	Application equipment
Bolted joints	Between panels	Performed strip	Hand or palette/putty knife
		Zinc-rich primer	Brush or spray
	Panel edges	Seam sealer light	Applicator gun (hand)
Spot-welded joints	Between panels	Zinc-rich primer	Brush or spray
		Structural adhesive or seam sealer	
	Panel edges	Seam sealer light	Applicator gun (hand)
Bonded joints	Between panels	Metal-to-metal adhesive semi-structural	Caulking gun

synthetic-resin-based compositions. The choice of each of the types will be dependent on the site for application, on the eventual conditions of exposure and often on price. These categories can be subdivided further into the various physical forms in which they can be made available, which include mastic putties for hand application, extruded sections for placing in precise locations, gun grade compositions which have the advantage of speed and economy of application, and pouring and spraying grades.

The properties of sealers will obviously vary according to their type and to their application. Thus preformed strip or putty sealers must adhere to the surfaces to which they are applied, and must not harden or crumble in service. Glazing sealers must be capable of being readily applied from a gun, with the ability to harden off on the surface, but must remain mastic in the assembly so that they are capable of maintaining a leak-tight joint whatever deflection the body undergoes. Heat gelling sealers must be capable of being readily applied by extrusion or possibly by spraying, and then must set up when cured but still retain a degree of flexibility.

As a result of soaring energy costs together with the need for car aerodynamic design, direct glazing of windscreens and fixed body glass was introduced and an adhesive was required to bond glass windscreens to the metal aperture. The material used is polyurethane adhesive, sealant. It possesses a combination of adhesion, sealant and gap filling qualities; it is a one-component adhesive and sealing compound of permanent elasticity. This dual-purpose material is based on a special moisture cured polyurethane with an accelerated setting time. The curing time is dependent upon the humidity levels prevailing, as well as the temperature. For example, at 20 °C with a relative humidity level of 65 per cent, a 6 mm diameter bead will be tack free within 1 hour and fully cured in 24 hours.

Table 4.11 indicates the uses of various sealant materials.

4.10 Sound deadening, thermal insulating and undersealing materials

The type of material used for sound damping or deadening depends on whether or not it is also required to provide undersealing. A material required for sound deadening only will normally be applied to the interior of a vehicle, whereas one required to provide sound deadening and undersealing properties will be applied to the underside of

Table 4.11 Sealers used in vehicle body repair work

Type	Base material	Application
Visible seams	Polyurethane	Extremely adherent sealant used on front and rear aprons, rear panel, engine compartment, bottom of boot, passenger compartment, side panels, wheel arches, vehicle underbody, tank filler caps and wings. Can be painted over with primer and fillers after curing, is non-shrinking, can be brushed and smoothed with a spatula. Cures by means of air moisture
	Synthetic rubber	Particularly suitable for all automotive problem areas where cleaning is difficult. Extremely adherent to raw, degreased, bonded, primed and painted sheet steel. Can be painted over with lacquers after thoroughly drying. Following application, can be passed through drying ovens at a maximum temperature of 90 °C
	Acrylic dispersion (water based)	Particularly suitable for sealing joints, welded seams and butt joints on vehicle bodies. Substrate must be primed and can be readily painted over. When cured, is resistant to water
Structural seams	MS polymer	Applied by means of an air pressure pistol. All structural seams sprayed by the manufacturer can be re-created with this sealant, so that the original finish can be restored after repair. Also can be painted over immediately wet-on-wet
	Nitrile-butadiene rubber	Special brushable sealant used for front and rear aprons, bottom of boot, inside floor, side walls, wheel arches. Has excellent adhesion to raw, primed and painted sheet metal, and can be painted over after drying
Underbody seams	Bitumen rubber	Specially used for the underbody area of the vehicle. Resistant to water, salt spray, alcohol and dilute sulphuric acid
Overlapping joints	Synthetic rubber	Used for sealing bolted wings, headlight units, rear light housings and cable inlet holes
Sealing tape	Synthetic rubber	Suitable for all overlapping and screwable joints on vehicle bodies, metal to metal, metal to wood, metal to plastic, wood to wood or plastic to plastic
Rubber profiled windows	Synthetic resin, synthetic rubber	Particularly suitable for sealing rubber profiled front, rear and side window units between rubber and glass or rubber and the vehicle body

the vehicle. Thus the former need not be fully water resistant, whereas the latter must be water resistant in addition to many other necessary requirements.

The sound deadening properties of a material are related to its ability to damp out panel vibrations, and this in turn is related to some extent (but not solely) to its weight per unit volume. Thus the cheapest sound deadening materials are based on mixtures of sand and bitumen, although these tend to be brittle. A better material is bitumen filled with asbestos; although this is probably less effective as a panel damper than sand-filled bitumen, it is nevertheless more suitable owing to its better ductility. In general, those sound deadeners applied to the interior of the vehicle are water-based bitumen emulsions with fillers, whereas sound deadener/sealers applied externally should be solvent-based materials. A more effective sound deadener than

Table 4.12 Undersealing and protection materials used in vehicle body repair work

Type	Base material	Application
Coatings for underbodies, spray type	Bitumen/rubber	Coating for underbodies, wheel arches, new and repaired parts. Also a corrosion protection for vehicle underbodies against elements such as moisture, road salt, road stone chippings. Good adhesion, and durable at extreme temperatures. Applied with a spray pistol
	Rubber/resin	Suitable for underbodies, wheel arches, front and rear aprons, sills, new parts, repair sheets. Can be painted over and has high abrasion resistance. Applied with a spray pistol
	Wax	Suitable for underbodies, touching up and subsequent treatment of all protective coatings
	Polymer wax	Long-term corrosion protection even when thinly applied. Has good flowing properties
Small-scale repair application (brushable)	Rubber/bitumen	Brushable coating, suitable for underbody and wheel arches. High abrasion resistance and good sound deadening properties
Road stone chip repair material	Synthetic/dispersion	Good protection against stone chips and corrosion. Particularly suitable for front and rear aprons, sills, spoilers, wheel arches

asbestos-filled bitumen is a clay-filled water dispersed polyvinyl acetate (PVA) resin emulsion; this has damping characteristics approximately three times better than the bitumen-based material, but naturally it is more expensive.

Other sound absorbing materials are now used for insulation in the automotive industry. Needle felts are blends of natural and manmade fibres locked together by needle punching. These are used in die-cut flat sheet forms for attachment to moulded carpets, floor boot mats and as anti-rattle pads. Bonded, fully cured felts are similar blends of fibres bonded together with synthetic resins. All the binder is cured during the felt making process. As with needled materials, these are mainly used for flat products, especially where low density is required such as in sound absorption pads and floor mats. In moulded felts and moulded glass wool the binder is only partially cured during the felt making process. The curing sequence is completed under the action of temperature and pressure in matched die compression moulded tools to produce components which have three-dimensional form and a controllable degree of rigidity. Fully cured and needled products can be given various surface treatments including abrasion-resistant and waterproof coatings such as latex, PVC or rubber, or they can be combined with

bitumen to improve the sound insulation properties. Moulded felt can be supplied covered with a range of woven textile covers or with various grades of PVC and heavy-layer bitumen EVA products.

Polyurethane flexible moulded foam can be modified or filled to meet different insulating requirements in the vehicle. The foam can also be moulded directly on to the hard layer, allowing simple tailoring of insulation thickness.

Table 4.12 indicates the use of various undersealing materials.

4.11 Interior furnishings

4.11.1 Carpets and floor coverings

The body engineer has a choice of materials, ranging from carpeting of the Wilton type for prestige vehicles, through polyamide and polypropylene moulded needle felts, to rubber flooring for economy versions. The main requirements of flooring are wear resistance, colour fastness to light and water, and adequate strength to enable the customer to remove the flooring from the car without it suffering damage. The method of manufacture of pile carpets varies: in some cases the pile is bonded to hessian backing; in other cases it is woven simultaneously into the backing and then anchored in position with either a rubber coating on the back surface or a vinyl coating. Quality is normally controlled by characteristics such as number of rows of tufts of pile per unit length, height and weight of the free pile, overall weight of the carpet, strength as determined by a tensile test in both the warp and weft directions, together with adhesion of pile if applicable. Rubber flooring generally has a vinyl coating to provide colour.

4.11.2 Leather (hide)

Large numbers of motor vehicle users all over the world continue to specify hide upholstery when the option is available, and will gladly pay the extra cost involved for a material which defies complete simulation. Great advances have been made in the development of suitable substitutes, and the best of the plasticized materials are to many people quite undistinguishable by eye from leather. The unique character of leather lies in its microstructure, the like of which is not obtained in any manmade material. Under a microscope leather can be seen to consist of the hairy epidermis and under that the corium, or bulk of the hide, this being the basis of the leather as we know it. By virtue of the millions of minute air spaces between the fibres and bundles of fibres, leather is able to 'breathe'. To the motorist this means that leather does not get hot and uncomfortable in warm weather or cold and inflexible in winter, and although permeable to water vapour it offers sufficient resistance if it is exposed to normal liquids. It is also strongly resistant to soiling, and when it does get dirty the dirt can usually be removed fairly easily without special materials. Unlike some plasticized materials, leather does not appear to attract dirt and dust as a result of static electricity. Many people, moreover, regard the distinctive smell of leather as an asset, and this defies imitation by manufacturers of substitute materials. However, natural hide has to go through many complex processes before it attains the form familiar to the upholsterer trimmer or motorist.

4.11.3 Fabrics for interior trim

Vinyl coated fabrics are now well established as trim material. Their vast superiority over the linseed oil coatings, and later the nitrocellulose coatings, of yesteryear are almost forgotten in the march of progress. Vinyl coatings are now sufficiently familiar for their merits to be taken for granted; nevertheless they continue to provide a material which for durability, uniformity and appearance at a reasonable cost so far remains unsurpassed. The resin polyvinyl chloride, the main ingredient of the coating, became available in commercial quantities in the early 1930s, and now a coating based on polyvinyl chloride (PVC) is used for seating material in this country. Although by tradition PVC is produced with a simulated leather appearance, on the Continent and particularly in the USA it is widely used with fancy embosses, and patterns. An extremely wide range of qualities is available, and in recent years there has been an effort to achieve some degree of rationalization. Additional qualities are necessary for tilt covers, headlinings and hoods for convertibles; the material for hooding convertibles must be resistant to mildew, to shrinkage and to wicking, this last term relating to the absorption of water on the inside surface of the cloth from the bottom edge of the hood. The characteristics necessary to provide serviceability over the life of the vehicle are the strength of the material under tension and under tearing conditions; adhesion of the coating to the backing cloth; resistance to flexing, and resistance to cracking at low temperatures; low friction, to enable the owner to slide on the seat; colour fastness, soiling resistance and, of course, wear resistance. In the case of the breathable leathercloth, the air permeability of the fabric has to be controlled.

4.11.4 Modern trends

The interior furnishing of a car is gaining in importance within the automotive trade. In response stylists are endeavouring to upgrade and soften the interior, using fabrics with the appearance and feel of textiles to appeal visually and functionally. This has manifested itself in all areas of the car, including the boot, seating, carpets, door trims and headliner cover fabrics. In all four areas of fabric use in car interiors (seating, door and side panels, bolsters and headliners) fabrics are gaining ground against exposed plastics. Both polyamide and polyester are giving designers new scope for attractive colours and variation in seating upholstery and in panels, while fully meeting light fastness and other performance standards.

4.12 Plastics

4.12.1 Development

Celluloid might well qualify for the honour of being the first plastic, though its inventor, Alexander Parkes, was certainly not aware of that fact. He made it around 1860 and patented his method for making it in 1865. An American, John Hyatt, found a way of solving the technical problems which plagued Parkes, and he set up business in 1870 to sell the same sort of material. He called it Celluloid to indicate its raw material, cellulose.

In 1920 a German chemist, Hermann Staudinger, put forward a theory about the chemical nature of a whole group of substances, natural and synthetic. He called them macromolecules; today we call them polymers. His theory not only explained the nature of plastics, but also indicated the ways in which they could be made. It provided the foundation for the world of plastics as we know it.

4.12.2 Polymerization

The raw materials for plastics production are natural products such as cellulose, coal, oil, natural gas and salt. In every case they are compounds of carbon (C) and hydrogen (H). Oxygen (O), nitrogen (N), chlorine (Cl) and sulphur (S) may also be present. Oil, together with natural gas, is the most important raw material for plastics production.

The term *plastics* in the broadest sense encompasses (a) organic materials which are based on (b) polymers which are produced by (c) the conversion of natural products or by synthesis from primary chemicals coming from oil, natural gas or coal.

The basic building blocks of plastics are *monomers*. These are simple chemicals that can link together to form long chains or *polymers*. The type of monomer used and the way it polymerizes, or links together, give a plastic its individual characteristics. Some monomers form simple linear chains. In polyethylene, for example, a typical chain of 50 000 ethylene links is only about 0.02 mm long. Other monomers form chains with side branches. Under certain circumstances, the individual chains can link up with each other to form a three-dimensional or cross-linked structure with even greater strength and stability. Cross-linking can be caused either chemically or by irradiating the polymer.

To get the advantages of two different plastics, two different monomers can be combined in a *copolymer*. By combining the monomers in different proportions and by different methods, a vast range of different properties can be achieved (see Tables 4.13 and 4.14). The properties of plastics can also be enhanced by mixing in other materials, such as graphite or molybdenum disulphide (for lubrication), glass fibre or carbon fibre (for stiffness), plasticizers (to increase flexibility) and a range of other additives (to make them resistant to heat and light).

4.12.3 Thermoplastics and thermosetting plastics

The simplest way of classifying plastics is by their reaction to heat. This gives a ready subdivision into two basic groups: thermoplastics and thermosetting plastics. *Thermoplastic* materials soften to become plastic when heated, no chemical change taking place during this process. When cooled they again become hard and will assume any shape into which they were moulded when soft. *Thermosetting* materials, as the name implies, will soften only once. During heating a chemical change takes place and the material cures; thereafter the only effect of heating is to char or burn the material.

As far as performance is concerned, these plastics can be divided into three groups:

General-purpose thermoplastics

Polyethylene
Polypropylene
Polystyrene
SAN (styrene/acrylonitrile copolymer)
Impact polystyrene
ABS (acrylonitrile butadiene styrene)
Polyvinyl chloride
Poly(vinylidene chloride)
Poly(methyl methacrylate)
Poly(ethylene terephthalate)

Engineering thermoplastics

Polyesters (thermoplastic)
Polyamides
Polyacetals
Polyphenylene sulphide
Polycarbonates
Polysulphone

Table 4.13   Physical properties of polymers

Material	P or S	Density (kg/m³)	Melting (softening) range (°C)	Specific-heat capacity (J/kg/K × 10³)	Thermal conductivity (W/m/K)	Coefficient of linear expansion (K × 10⁻⁶)
LD polyethylene	P	0.01–0.93	80	2.3	0.13	120–140
HD polyethylene	P	0.04–0.97	90–100	2.1–2.3	0.42–0.45	120
Polypropylene	P	0.90	100–120	1.9	0.09	120
GFR polypropylene	P	1.00–1.16	110–120	3.5	–	55–85
Polyvinylchloride	P	1.16–1.35	56–85	0.8–2.5	0.16–0.27	50–60
Polystyrene	P	1.04–1.11	82–102	1.3–1.45	0.09–0.21	60–80
Polystyrene copolymer (ABS)	P	0.99–1.10	85	1.4–1.5	0.04–0.30	60–130
Nylon 66	P	1.14	250–265	1.67	0.24	80
Nylon 11	P	1.04	185	2.42	0.23	150
PTFE (Teflon)	P	2.14–2.20	260–270	1.05	0.25	100
Acrylic (Perspex)	P	1.10–1.20	70–90	1.45	0.17–0.25	50–90
Polyacetals	P	1.40–1.42	175	1.45	0.81	80
Polycarbonates	P	1.20	215–225	1.25	0.19	65
Phenol formaldehyde	S	1.25–1.30	–	1.5–1.75	0.12–0.25	25–60
Urea formaldehyde	S	1.40–1.50	–	1.65	0.25–0.38	35–45
Melamine formaldehyde	S	1.50	–	1.65	0.25–0.40	35–45
Epoxies	S	1.20	–	1.65	0.17–0.21	50–90
Polyurethanes R	S	3.2–6.0	150–185	1.25	0.02–0.025	20–70
Polyurethanes F	S	4–8	150–185	1.25	0.035	50–70
Polyesters	S	1.10–1.40	–	1.26	0.17–0.19	100–150
Silicones	S	1.15–1.8	200–250	–	0.17	24–30

GFR glass fibre reinforced; P thermoplastic; S thermosetting; R rigid; F flexible; LD low density; HD high density

Table 4.14   Typical mechanical properties of representative plastics

Material	Modulus of elasticity E (MN/m²)	Tensile strength (MN/m²)	Compressive strength	Elongation (%)
LD polyethylene	120–240	7–13	9–10	300–700
HD polyethylene	550–1050	20–30	20–25	300–800
Polypropylene	900–140	32–35	35	20–300
GFR polypropylene	1500+	34–54	40–60	5–20
Flexible PVC	3500–4800	10–25	7–12	200–450
Rigid PVC	2000–2800	40	90	60
Polystyrene	2400–4200	35–62	90–110	1–3
ABS copolymer	1380–3400	17–58	17–85	10–140
Perspex	2700–3500	55–75	80–130	2–3
PTFE	350–620	15–35	10–15	200–400
Nylon 11	1250–1300	52–54	55–56	180–400
GFR nylon 6	7800–800	170–172	200–210	3–4

Modified polyphenylene ether
Polyimides
Cellulosics
RIM/polyurethane
Polyurethane foam

Thermosetting plastics

Phenolic
Epoxy resins
Unsaturated polyesters
Alkyd resins
Diallyl phthalate
Amino resins

4.12.4 Amorphous and crystalline plastics

An alternative classification of plastics is by their shape. They may be crystalline (with shape) or amorphous (shapeless).

Amorphous plastics

Amorphous plastics basically are of three major types:

ABS: acrylonitrile butadiene styrene
ABS/PC blend
PC: polycarbonate.

Amorphous engineering plastics have the following properties:

High stiffness
Good impact strength
Temperature resistance
Excellent dimensional stability
Good surface finish
Electrical properties
Flame retardance (when required)
Excellent transparency (polycarbonate only).

In the automotive industry use is made of the good mechanical properties (even at low temperatures), the thermal resistance and the surface finish. The applications are:

1 Body embellishment
2 Interior cladding
3 Lighting where, apart from existing applications of back lamp clusters, polycarbonate is expected to replace glass for headlamp lenses.

Semi-crystalline plastics

Semi-crystalline plastics are in two basic types:

Polyamide 6 and 66 types
Polybutylene terephthalate (PBT).

Semi-crystalline plastics have the following properties:

High rigidity
Hardness
High heat resistance
Impact resistance
Abrasion, chemical and stress crack resistance.

The semi-crystalline products find major application in the automotive sector, where full use is made of the mechanical and thermal properties, together with abrasion and chemical resistance. Examples include:

1 Underbonnet components
2 Mechanical applications
3 Bumpers, using elastomeric PBT for paint on-line
4 Body embellishment (wheel trims, handles, mirrors)
5 Lighting, headlamp reflectors.

Blended plastics

Blended plastics have been developed to overcome inherent specific disadvantages of individual plastics. For large-area body panels, the automotive industry demands the following properties:

Temperature resistance
Low-temperature impact resistance
Toughness (no splintering)
Petrol resistance
Stiffness.

Neither polycarbonate nor polyester could fulfil totally these requirements. This led to the combination of PC and PBT to form Macroblend PC/PBT, which is used for injection moulded bumpers.

4.12.5 Plastics applications

Plastic products can be decorated by vacuum metallizing and electroplating. They have replaced metals in a lot of automotive applications, such as mirror housings, control knobs and winder handles as well as decorative metallic trim. It is a field which uses their advantages to the full without relying on properties they lack.

Thin parts must be tough and resistant to the occasional impact. They must be impervious to attack by weather, road salts, extremes of temperatures and all the other hazards that reduce older forms of body embellishments to pitted, rusted, dull, crumbling metal. They do not need high tensile strength or flexural strength as they do not have to carry heavy stresses. They must be cheap and capable of being formed into highly individual and complex shapes. All these requirements are satisfied by thermoplastics and thermosetting resins. They can be pressed, stamped, blow moulded, vacuum formed and injection moulded into any decorative shape required.

Apart from their decorative properties, the mechanical properties of acrylic resins are among the highest of the thermoplastics. Typical values are a tensile strength of 35–75 MN/m^2 and a modulus of elasticity of 1550–3250 MN/m^2. These properties apply to relatively short-term loadings, and when long-term service is envisaged tensile stresses in acrylics must be limited to 10 MN/m^2 to avoid surface cracking or crazing. Chemical properties are also good, the acrylics being inert to most common chemicals. A particular advantage to the automotive industry is their complete stability against petroleum products and salts.

Acetal resins are mostly used for mechanical parts such as cams, sprockets and small leaf springs, but also find application for housings, cover plates, knobs and levers. They have the highest fatigue endurance limits of any of the commercial thermoplastics, and these properties, coupled with those of reduced friction and noise, admirably qualify the acetal resins for small gearing applications within the vehicle.

Plastics can be self-coloured so that painting costs are eliminated and accidental scratching remains inconspicuous, and they can be given a simulated metal finish. For large-scale assemblies, such as automobile bodies, painting is necessary to obtain uniformity of colour, especially when different types of plastics are used for different components. Plastics can also be chrome plated, either over a special undercoating which helps to protect and fix the finish, or by metal spraying or by vacuum deposition in which the plastic part is made to attract metal particles in a high-vacuum chamber. The use of a plastic instead of a metal base for chrome plating eliminates the possibility of the base corroding and damaging the finish before the chromium plating itself would have deteriorated. The chrome coating can be made much thinner and yet have a longer effective life, with a consequent saving in cost.

Until fairly recently polymer materials were joined only by means of adhesives. Now the thermoplastic types can be welded by using various forms of equipment, in particular by hot gas welding, hot plate machines which include pipe welding plant, ultrasonic and vibration methods, spin or friction welding machines, and induction, resistance and microwave processes. Lasers have been used experimentally for cutting and welding. The joining of metals to both thermoplastic and thermosetting materials is possible by some welding operations and by using adhesives.

4.12.6 Future of plastics in the automotive industry

The automotive industry has grown to appreciate the potential of plastics as replacements for metal components within their products. The realization that plastics are, in their own right, engineering materials of high merit has led to rapid advancement of material and application technology, with the end result that plastics have gained a firm and increasing footing in the motor vehicle. Many factors have aided the adoption of plastics by the automotive industry, which uses them in the following areas: body, chassis, engine, electrical system, interior and vehicle accessories. Lower costs of plastics parts must, of course, be the major contributing factor in the replacement of existing parts, and this is closely followed by the ease with which modern plastics can be formed by comparatively inexpensive tooling. The inert properties of synthetic materials also contribute greatly; properties like corrosion resistance, low friction coefficients and light weight are of prime importance.

The use of plastics in the automotive industry continues to accelerate at a phenomenal rate as research into plastic technology results in new developments and applications. The future growth of plastics in the automotive industry will be controlled by two factors: the growth of the industry itself, and the greater penetration of plastic per car. A key constituent in world growth, therefore, is the developing nations who are involved in the assembly and production of motor vehicles. They will consequently favour the use of plastics as a first choice, rather than as a replacement for metal.

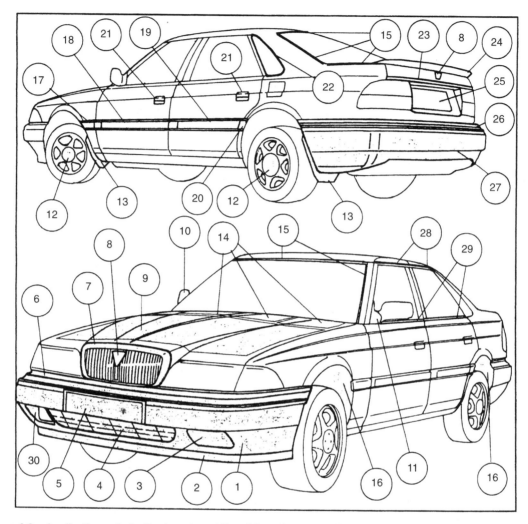

Figure 4.3   Applications of plastics in automobiles *(Motor Insurance Repair Research Centre)*

1 Front bumper (Pocan S7913)
2 Front spoiler (Santoprene grade 123–50 and 121 with aluminium insert)
3 Fog lamp blanking plate (Xenoy EPX500)
4 Lower front grille (Xenoy CL100)
5 Front number plate plinth (Xenoy CL100)
6 Front bumper insert (PVC and EB-type Nylar)
7 Front grille (moulding, ABS; Benzel, MS Chrome)
8 Bonnet/boot lid/tailgate badges (ABS, aluminium and PU skin)
9 Underbonnet felt (moulded felt)
10 Door mirror casing RH and LH (polyamide, 15% glass reinforced)
11 Door mirror mounting RH and LH (polyamide, 15% glass reinforced)
12 Front/rear wheel trims RH and LH (*cap*, Noryl 731; *moulding*, Bayer Duretan BM30X, ICI Maranyl TB570)
13 Front/rear mudflaps RH and LH (front, rubber to BLS.22 RD.27 Ref. 421; rear EPDM mix 4080)
14 Scuttle grille/mouldings (ABS)
15 Front/rear screen upper and side mouldings (PVC with stainless steel co-extrusion)

16 Front/rear wing splashguards RH and LH (PP)
17 Front wing waist moulding RH and LH (Noryl)
18 Front door waist moulding RH and LH (Noryl)
19 Rear door waist moulding RH and LH (Noryl)
20 Rear wing waist moulding RH and LH (Noryl)
21 Front/rear door outer handles RH and LH (*body*, Xenoy; *flap*, Glass-filled nylon)
22 Rear quarterlight moulding RH and LH (*4-door*, PVC/Stainless steel extrusion; *5-door and coupé*, PU with stainless steel moulding)
23 Boot lid moulding (ABS)
24 Rear spoiler (PU core and polyester skin)
25 Rear number plate plinth (ABS)
26 Rear bumper insert (PVC and EB-type Nylar)
27 Rear bumper (Pocan S7913)
28 Front/rear door upper mouldings RH and LH (PVC with stainless steel moulding)
29 Front/rear door outer weatherseals RH and LH (PVC with stainless steel co-extrusion)
30 Fog lamp bezel (PP)

Over the past years, the natural applications for plastics in automobiles (interior fittings, cushioning and upholstery, trim, tail lights, electrical components) have become saturated. The growth for the future can be expected to come from the use of plastic for bodywork and some mechanical components. Already there is a widespread use of plastics for front and rear bumpers. We can expect to see bonnets, boot lids and front wings in plastics.

All the major volume producers of cars are engaged in long-term development work towards the all-plastic car. Whether or not such targets can be realized remains to be seen. Factors such as energy costs and availability of resources may play a greater part in the total picture than simple objects like vehicle weight reduction.

Abbreviations for automotive plastics

4.13 Plastics repair

A new car is made up, by weight, of about 65 per cent steel, 5 per cent non-ferrous metal, 15 per cent plastics material and 15 per cent other non-metallic materials. Plastics materials are very light, so in terms of bulk the percentage is much larger, but nobody appears to have worked out the figures for this yet! What we do know is that the plastics parts are often damaged in even a minor accident and the replacement of these parts costs insurance companies and private owners dearly. That's how we make a profit, you might say; but you could make more profit by repairing the damaged plastics part, this would reduce the cost to the customer and reduce the waste of precious natural resources.

Abbreviations for automotive plastics

Abbreviation	Full name
ABS	Acronitrile butadiene styrene*
PP	Polypropylene*
PE	Polyethylene*
PC	Polycarbonate*
PA	Polyamide*
PBT	Polybutylene tetraphtalate*
PU	Polyurethene (thermoset)
UP	Unsaturated polyester (thermoset)
CS	Chopped strands
SMC	Sheet moulding compound
MF	Milled fibres
WR	Woven roving

* Can be repaired by welding

SPI materials coding system

Material type	Code
Polyethylene terephthalate	1PETE
High density polyethylene	2HDPE
Vinyl	3V
Low density polyethlene	4LDPE
Polypropylene	5PP
Polystyrene	6PS
Other	6OTHER

4.13.1 Types of plastics

The word plastics is being used here because it is technically correct to describe the range of man-made materials. Plastic, without the 's', is used to describe the material state where it can be deformed and it will remain in that state after the force has been removed. In conversation it is normal to say plastic for both cases as it is unlikely that there will be any confusion.

The two main classifications of plastics are: thermoplastics and thermosetting plastics (which are referred to as thermosets). One of the key areas of knowledge needed to repair plastics components is an understanding of these two classifications and being able to identify them in a vehicle component.

A thermoplastic is one which melts when it is heated up. If you get a carrier bag from the supermarket and warm it slightly it will become soft and pliable. It is a thermoplastic. At this point you must remember that plastics are made from petroleum-based chemicals and are therefore easy to set on fire and burn at very high temperatures, so avoid matches and other naked flames when handling them. Conversely, if you put the same carrier bag in a freezer it would go stiff and make a crackling noise when you handle it. Vehicle thermoplastic components are made to operate normally over a wide temperature range, so obviously they need to get very hot before they will melt and very cold before they become brittle. Now if you heat up a thermoplastic to a high enough temperature it is going to melt, this means that you can repair a thermoplastic component by welding. Before you dash out to try to weld that bumper assembly which is sat on the bench, there are a few more things which you need to know. First, that you need a special plastics welder and second, if it is not a thermoplastic bumper you may well damage it beyond repair and set the workshop on fire too. You cannot simply identify thermoplastics just by warming them up; we'll look at ways of identifying plastics later in the article.

A thermoset is one which uses or generates heat during its setting stage. The first thermoset was Bakelite, the heavy dark brown plastics material which was used for distributor caps and ignition coil ends. It does not go soft when you try to heat it up; if you subject it to a flame it will burn and char. It is a brittle material and easily chips. Thermosets cannot be welded; most can be bonded using a suitable bonding agent or glue.

4.13.2 Reinforcement

On their own plastics materials have only a limited amount of strength – interior trim, dashboard panels and lamp lenses are examples of non-reinforced plastics. Apply a strong force from a mechanics hand and these components will break. So, components which are going to take structural loads within the vehicle or be capable of withstanding impact, such as a bumper assembly, need some form of reinforcement. The most common reinforcement material is glass.

You are probably familiar with glass reinforced polyester (GRP) where a piece of glass matting is layed up with a mixture of polyester resin and catalyst (hardener) to effect a body repair, or for the manufacture of kit cars and small boats. Incidentally, GRP is also used to cover all glass reinforced plastics. If you look at a strand of the glass from this matting through a microscope, or with a very strong magnifying glass, you will see that the glass is in fact made from very small diameter round tubes. A round tube gives very high strength, but the length to diameter ratio of these tubes is such that they can bend without collapsing, so that they can be layed up on curved surfaces, then when the resin sets they are firmly held in place like roof beams for maximum strength.

The glass reinforcement used for vehicle components is made to suit the application and can vary between the woven cloth like material which is sandwiched in layers in a bumper assembly and the finely powdered glass particles which are used to strengthen a lamp cluster.

Increasingly materials other than glass are being used to reinforce plastics, although often this is in addition to glass. Carbon fibre is used either in the form of a continuous thread which is wound around the component or as a woven matting similar to glass reinforcement. Where two different materials are used they are referred to as composites.

So, when it comes to identifying a plastics material check to see if it is reinforced, and if it is which type of reinforcement is used. If a glass matting is used you can usually see the woven layers of glass on the underside of the component. Carbon fibre can be recognized by its graphite grey colour.

4.13.3 Identification markings

Many manufacturers now mark their products with a code which will enable the identification of the type of plastics used. The reason for this coded marking is mainly for the identification of genuine parts and recycling purposes. Currently there is no standard system in Europe, nor indeed the UK, for identifying plastics. The British Plastics Federation are trying to encourage all European plastics manufacturers to use the American Society of the Plastics Industry (SPI) material code system. This uses a number and a series of abbreviation letters. The number and abbreviation letters identify a classification of plastics. The code letters used by manufacturers outside the SPI system are often registered trade marks, this creates legal problems as well as identification problems. Again, the SPI classification code is intended mainly for recycling purposes, but it is very useful for general identification information.

4.13.4 Manufacturing processes

There are many different methods of manufacturing plastics components. After a little experience you will be able to work out how components are best manufactured, this is usually a good guide as to the type of material used. Injection moulding (Inj) is used for items such as grille panels, air vents, dashboard panels, wheel covers and lamp units. The material is likely to be a thermoplastic. These items usually have a smooth surface finish and carry markings which show that they have been in a mould, typically the lettering is raised. They also tend to be fairly flexible. Injection moulding can also be carried out with thermosets. In this case the material is much stiffer but still shows the mould lines and has a smooth surface finish with raised lettering. The thermoset injection moulding may use a resin (RIM) or a bulk material (BMC). In some cases the BMC may be a recycled material filler, but this will have low strength. Bumper assemblies, wheel arch extensions and rear lamp holders are typical applications of injection moulded thermosets. Thermosets in the form of GRP using woven glass fibre or woven carbon fibre may be hand layed-up (HLU) or compression moulded (com). To effect a good bond with a carbon fibre material an autoclave is needed to control the finishing process. The texture and colour will allow you to identify GRP and carbon fibre materials.

4.13.5 Safety

Having described the different plastics materials and the manufacturing processes you should be able to start identifying them. As with all things, you will need to practise until you become skilled, and some mistakes are inevitable. When you try repairing a few items you will get a feel for the job, just like tightening nuts and bolts. You will soon become aware of which parts on which vehicles can be repaired. But before you start to work on plastics materials, you need to look at safety. As well as the normal workshop safety procedures, there are a number of specific hazards relating to plastics materials which you must take extreme care with, let's have a look at them before discussing some of the repair procedures.

Plastics materials are made from petroleum-based products, this means that they are highly flammable so you must avoid high levels of heat and naked flames. The most common reinforcing material is glass, but other equally problematic materials may be used. If you start to grind plastics components you will get powdered glass as well as the plastics dust. The powdered glass can cut the blood vessels inside the lungs and stomach. The plastics powder can cause respiratory diseases and the dust from carbon fibre can cause lung and other internal diseases. So ensure that you are using the

correct masks or other breathing apparatus to suit the situation. Heating plastics materials, or using solvents, or bonding agents, can give rise to volatile organic compounds (VOCs); breathing protection is obviously needed in this case. To prevent a buildup of fumes and dust in the workshop the use of an extractor system is advised. Solvents and bonding agents should not come into contact with your skin, gloves and safety goggles are a first line of defence, and you are reminded to consult the COSHH sheet supplied by the manufacturer with all of these products.

4.13.6 Repair procedures

As a general rule thermoplastics can be welded and thermosets bonded. We'll have a look at a few procedures in detail.

Starting with something simple. Often in an accident repair a plastics headlamp binnacle is scrapped because one of the lugs is broken off. If you apply a small amount of acetone to both of the broken surfaces you will often find that the lug can be bonded back into place. The acetone (also used as nail varnish remover and not popular amongst mechanics) actually melts the plastics material, pressing the two parts together causes them to bond and dry.

Dashboards and some flexible bumper assemblies which are made from thermoplastics can be welded. There are two ways of welding thermoplastics. One is to use a hot air welding gun and the other is to use a soldering iron. The hot air welding gun blows out a stream of air which will melt the plastic, the temperature is over 100 °C. The paint should be cleaned off about 20 mm ($^3/_4$ in) on both sides of the joints. The welded joint is made using the blow gun and a plastics filler rod in the same way as you would oxy-acetylene weld steel. If it is a long joint you should tack weld first. The two parts can be held together with strips of masking tape on the reverse whilst you carry out the welding. If a component has cracked, like a bumper assembly, and internal stress might cause the crack to continue during or after the repair, it is a good idea to drill a small hole at each end of the crack. Usually 4 mm ($^3/_{16}$ in) holes at each end of the crack will be sufficient to remove the internal stress. These should then be filled after the welding has been completed. If you are using a filler rod it is a good idea to 'vee' the edges of the joint to accommodate the filler, this can be done using a file.

Thin thermoplastic items which will melt without a great deal of heat being applied can be welded using a soldering iron (without the solder). The procedure is as follows. Remove any paint within about 15 mm of the joint, a P40 disc is usually ideal; drill stress relieving holes at each end if it is a crack; hold the gap closed with tape on the underside; run the soldering iron over the joint so that the material melts and fuses together. When the repair has cooled, remove any excess or unevenness with the P60 and then finish to feather into the existing paint using P600.

If the component, say a bumper, has been holed, it is possible to weld in a piece from a scrap bumper of the same shape. Cut the damaged section out of the bumper, a round or oval shape will prevent further cracking. Then cut a piece out of the scrap bumper which will just fit into the hole. Remove the paint from all the edges and weld in as in the previous examples (blow gun or soldering iron). Finish with P600 production paper. If the joint or the final contour is not satisfactory this can be corrected using body filler in the normal way.

To repair a thermoset component you need to bond or glue on a patch. If the damage is a crack, such as in a bumper assembly, the procedure is to clean up the damaged area on both sides of the crack, drill stress relieving holes at each end, then whilst the crack is held closed bond a patch to the underside of the component. Complete the repair using body filler and P600 paper in the normal way. The patch is preferably the same material as the damaged item; the bonding could be by a number of materials, including superglue. If the bumper has been holed, then cut a patch to fill the hole as you would for a thermoplastic bumper and additionally cut another patch which is larger than the hole. The larger patch is then bonded to the underside so that it attaches to both the original item and the piece which is filling the hole. The job is again completed using body filler and P600 to feather in the paintwork. A small amount of body filler on the underside to blend the patch into the surrounding material will make the job look neat.

Be aware, that not all plastics can be repaired. Those which have a waxy finish will not even let superglue stick to their surface.

4.13.7 Painting

Plastics materials require a suitable keying primer and/or undercoat. You should use the one which is recommended by the vehicle manufacturer. A coat of underseal on the rear of any panel which is open to the elements will give added protection.

4.14 Safety glass

More and more glass is being used on modern cars. Pillars are becoming slimmer and glass areas are increasing as manufacturers approach the ideal of almost complete all-round vision and the virtual elimination of blind spots. Windscreens have become deeper and wider. They may be gently curved, semiwrapped round, or fully wrapped. With few exceptions they are of one-piece construction, sometimes swept back as much as $65°$ from the vertical. Styling trends, together with a growing knowledge of stress design in metal structures, have resulted in a significant increase in the glazed areas of modern car body designs. As a result of this move towards a more open style, the massive increase in the cost of energy has brought growing pressure on vehicle designers to achieve more economic operations, principally in respect to lower fuel consumption through better power/weight ratios. An outcome of these two lines of development has been a situation in which although the area of glass has increased, the total weight of glass has remained constant or even decreased.

Broadly speaking, motor vehicle regulations specify that windscreens must be of safety glass. To quote one section: 'On passenger vehicles and dual-purpose vehicles first registered on or after 1 January 1959, the glass of all outside windows, including the windscreen, must be of safety glass'. The British Standards Institute defines safety glass indirectly as follows: 'All glass, including windscreen glass, shall be such that, in the event of shattering, the danger of personal injury is reduced to a minimum. The glass shall be sufficiently resistant to conditions to be expected normal traffic, and to atmospheric and heat conditions, chemical action and abrasion. Windscreens shall, in addition, be sufficiently transparent, and not cause any confusion between the signalling colours normally used. In the event of the windscreen shattering, the driver shall still be able to see the road clearly so that he can brake and stop his vehicle safely.'

Two types of windscreen fulfil these requirements – those made from heat-treated (or toughened) glass, and those of laminated glass. In addition there are plastic coated laminated or annealed safety glasses. Most windscreens and some rear windows fitted in motor vehicles are of ordinary laminated glass. For the main part, toughened glass is confined to door glass, quarter lights and rear windows where the use of more expensive laminated products has yet to be justified. However, laminated glass is being increasingly used on locations other than windscreens for reasons of vehicle security and also for passenger safety (containment in an accident), especially in estates with seating in the rear. Note that the applicable EEC Directive (see later) has effectively banned the fitment of toughened windscreens from the end of 1992.

Ordinary laminated safety glass is the older of the two types and is the result of a basic process discovered in 1909. Some years earlier a French chemist, Edouard Benedictus, had accidentally knocked down a flask which held a solution of celluloid. Although the flask cracked it did not fall into pieces, and he found that it was held together by a film of celluloid adhering to its inner surface. This accident led to the invention of laminated safety glass, made from two pieces of glass with a celluloid interlayer. An adhesive, usually gelatine, was used to hold them together and the edges had to be sealed to prevent delaminating. However, despite the edge sealing, the celluloid (cellulose nitrate plastic) discoloured and blistered; hence celluloid was replaced by cellulose acetate plastic, but this, although a more stable product than celluloid, still needed edge sealing.

Nowadays a polyvinyl butyral (PVB) self-bonding plastic interlayer is used; no adhesive is necessary and the edges do not need sealing, making it quite practical to cut to size after laminating. When producing glasses to a particular size, however, the glass and vinyl interlayer are usually cut to size first. In the process the vinyl plastic interlayer is placed between two clean, dry pieces of glass and the assembly is heated and passed between rubber-covered rollers to obtain preliminary adhesion. The sandwich of glass and interlayer is then heated under pressure for a specified period in an autoclave. This gives the necessary adhesion and clarity to the interlayer, which is not transparent until bonded to the glass. If a piece of laminated glass is broken, the interlayer will hold the splinters of glass in place and prevent them flying.

Plastic coated laminated safety glass is an ordinary laminated glass which has soft elastic polyurethane films bonded on to the inner surface to provide improved passenger protection if fragmentation occurs. There is some interest in the use of bilayer construction which uses 3 mm or 4 mm annealed glass bonded with a load bearing surface layer of self-healing polyurethane.

Uniformly toughened glass is produced by a completely different process, involving heating of the glass followed by rapid cooling. Although patents were taken out in 1874 covering a method of increasing the strength of flat glass sheet by heating and cooling it in oil, toughened glass was not in common use until the 1930s. Modern toughened glass is produced by heating the glass in a furnace to just below its softening point. At this temperature it is withdrawn from the furnace and chilled by blasts of cold air. The rapid cooling hardens and shrinks the outside of the glass; the inside cools more slowly. This produces compressional strain on the surfaces with a compensating state of tension inside, and has the effect of making the glass far stronger mechanically than ordinary glass. If, however, the glass does fracture in use, it disintegrates into a large number of small and harmless pieces with blunted edges. The size of these particles can be predetermined by an exact temperature control and time cycle in the toughening process, and manufacturers now produce a uniformly toughened safety glass which will, when broken, produce not less than 40 or more than 350 particles within a 50 mm square of glass. This conforms to the British Standard specification.

The main standards for the UK are now:

BS 857
ECE R43 (UN regulation)
EEC Directive (AUE/178) (A common market regulation)
BS 857 glazing is still valid but is seldom used because ECE R43 is accepted throughout Europe, Japan and Australia.

There are other types of safety glass – mostly crossbreeds of the pure toughened glass screen – which are designed to combine vision with safety. These are modified zone-toughened glasses, having three zones with varying fragmentation characteristics. The inner zone is a rectangular area directly in front of the driver, not more than 200 mm high and 500 mm long. This is surrounded by two other zones, the outer one of which is 70 mm wide all round the edge of the windscreen. This type of windscreen has been fitted to various vehicles since 1962. As a result of ECE Regulation 43 this type of windscreen has been superseded by the fully zebra-zoned windscreen. Many countries, including the USA but with the exception of the UK, legislate against toughened windscreens.

Although sheet and plate glass are manufactured satisfactorily for use in doors, rear lights and windscreens, float glass has now largely superseded their use for reasons of economy and improved flatness.

Most laminated windscreens used in the motor vehicle trade are 4.4 mm, 5 mm, 5.8 mm or 6.8 mm in overall thickness, with a 0.76 mm PVB interlayer. However, 4.4 mm is the thinnest laminated glass available, and as this has to be made from two pieces of glass it needs very careful handling during manufacture and is therefore expensive. Windscreens made from float glass should be a maximum of 6.8 mm thick, whether toughened or laminated. However, some large coaches and lorry windscreens are 7.8 mm thick (4 mm glass + 0.76 mm PVB interlayer + 3 mm glass). This gives immense strength and robustness against stone impact. Other body-glasses, because they can be made from sheet glass and also can be toughened safety glass, are usually between 3 mm and 4 mm thick.

From our brief look at the history of glass manufacture it is obvious that the curving of glass presents no problems; in fact the problem has been to produce flat, optically perfect glass. However, to curve safety glass and still retain its optical and safety qualities requires careful control. Glass has no definite melting point, but when it is heated to approximately 600 °C it will soften and can be curved. Curved glasses should be specified as 6.8 mm thick, as it is more difficult to control the curving of 4.4 mm glass. Even 6.8 mm thick glasses will have slight variations of curvature. To accommodate this tolerance, all curved glasses should be glazed in a rubber glazing channel, of which there are many different sections available. Glazed edges of glasses should be finished with a small chamfer known as an arrised edge, while edges of glasses that are visible or which run in a felt channel should be finished with a polished, rounded edge. Should a glass be required for glazing in a frame, a notch will usually be required to clear the plate used to join the

two halves of the frame together. The line of this notch must not have sharp corners because of the possibilities of cracking. Although laminated safety glass can be cut or ground to size after laminating, toughened safety glass must be cut to size and edge finished before the heat treating process.

Nearly all fixed glazing is now glazed using adhesive systems. Shapes are becoming more complex, needing very good angles of entry control to meet bonding requirements. The trend is towards aerodynamical designs involving flush glazing and the removal of sudden changes in vehicle shape; therefore corners must be rounded rather than angular as in older vehicle designs. Glass is often supplied with moulded-on finisher (encapsulation). Consequently bending processes are becoming very sophisticated. Adhesive glazing (polyurethane is the adhesive normally used) has added considerably to the complexity of vehicle glazing in a scientific sense. It has many advantages, however, if carried out correctly: it will reduce water leaks, it suits modern car construction, it results in a load bearing glazing member, and it lends itself to robotic assembly in mass production. As a consequence of adhesive glazing, all the associated glazing is now printed with a ceramic fired-in black band to protect the polyurethane adhesive from ultraviolet degradation, and also for cosmetic reasons so that the adhesive cannot be seen.

By a Ministry of Transport regulation, safety glass was made compulsory in 1937 for windscreens and other front windows. As already indicated, with effect from 1 January 1959 the Road Traffic and Vehicle Order 359 has demanded that for passenger vehicles and dual-purpose vehicles, all glass shall be safety glass. For goods vehicles, windscreens and all windows in front of, or at the side of, the driver's seat shall be safety glass.

Questions

1 What would the following alloy steels be used for:
(a) high-tensile steel (b) manganese steel
(c) chrome-vanadium steel?

2 List the properties of commercially pure aluminium.

3 Explain why, in the construction of a motor vehicle, commercially pure aluminium has a very limited application.

4 Identify the grades of hardness in aluminium sheet and state how the hardness is achieved.

5   Explain how you would identify the following:
(a) low-carbon steel (b) aluminium alloy
(c) stainless steel.

6   Describe the difference between laminated safety glass and toughened safety glass.

7   Give three requirements of a body sealing compound, and describe one type of sealer used in vehicle repair.

8   Suggest reasons why stainless steel is sometimes used for trim and mouldings.

9   Explain the difference between hide and PVC materials.

10   Explain what is meant by micro-alloyed steel or HSS.

11   Give reasons why the car manufacturers are using zinc-coated steels.

12   Name the three main groups of stainless steel.

13   Explain the following terms in relation to plastic:
(a) monomer (b) polymer (c) copolymer.

14   Explain the difference between thermoplastics and thermosetting plastics.

15   Which safety glass, used for vehicle windscreens, shatters into small segments on impact?

16   Describe the basic properties required of a body joint sealing compound.

17   Identify the group of plastics that can be softened or remoulded by the application of heat.

18   Steel panels can be strengthened without adding weight. Name and explain the process.

19   Describe three different ways in which the surface of steel can be protected.

20   State the reasons why certain metals need to be protected from the effects of the atmosphere.

21   Describe how some metals can resist attack by the atmosphere.

22   What is the alloying effect when zinc and copper are added to aluminium?

23   Explain the different properties of heat-treatable and non-heat-treatable aluminium alloys.

24   State the reasons, other than weldability, why low-carbon steel is chosen in preference to aluminium as a vehicle body shell material.

25   Define the term 'HSLA steel'.

26   Define what is meant by the term 'non-ferrous metal'.

27   Explain the difference in properties between low-carbon steel and alloy steel.

28   Describe the two processes which can be used to join plastic.

29   Explain where plastic can be used on a vehicle.

30   State the applications where natural rubber has been replaced by synthetic materials in the automobile industry.

31   Describe two ways of attaching a windscreen to a vehicle body.

32   Describe how to replace a glass in an opening quarter-light frame.

33   Why should you never hit a hammer with another hammer?

34   Which material should not be used for axle stand pins?

35   Why is brass often used for drifting bearings?

Metal forming processes and machines

5.1 Properties of metals

Metals and alloys possess certain properties which make them especially suitable for the processes involved in vehicle body work, particularly the forming and shaping of vehicle body parts either by press or by hand, and some of the jointing processes. These properties are described in the following sections, and some typical values of characteristics are shown in Tables 5.1 and 5.2.

5.1.1 Malleability

A malleable metal may be stretched in all directions without fracture occurring, and this property is essential in the processes of rolling, spinning, wheeling, raising, flanging, stretching and shrinking. In the operation of beating or hammering a metal on a steel block (such as planishing) an action takes place at each blow wherein the metal is squeezed under the blow of the hammer and is forced outwards around the centrepoint of the blow. The thinner the metal can be rolled or hammered into sheet without fracture, the more malleable is the metal.

After cold working, metals tend to lose their malleable properties and are said to be in a work hardened condition. This condition may be desirable for certain purposes, but if further work is to

Table 5.1 Physical properties of metals and alloys

Metal	Melting temperature range (°C)	Density (kg/m³)	Specific heat capacity/ (J/kg/K × 10³)	Thermal conductivity (W/m/K)	Electrical conductivity (% IACS)	Coefficient of linear expansion/ (K × 10⁻⁶)
Aluminium	660	2.69	0.22	218	63	23
Al-3.5 magnesium	550–620	2.66	0.22	125	25	23
Duralumin type	530–610	2.80	0.21	115–140	20–36	23
Copper	1085	8.92	0.39	393	101	17
70/30 brass	920/950	8.53	0.09	120	17	19
95/5 tin bronze	980/990	8.74	0.09	80	12	17
Lead	327	11.34	0.13	35	8	29
Magnesium	650	1.73	1.04	146	35	30
Nickel	1455	8.90	0.51	83	21	13
Monel	1330/1360	8.80	0.43	26	3	10
Tin	232	7.30	0.22	64	13	20
Titanium	1665	4.50	0.58	17	3	8.5
Zinc	419	7.13	0.39	113	26	37
Iron	1535	7.86	0.46	71	7	12
Mild steel	1400	7.86	0.12	45	31	11

Table 5.2   Typical mechanical properties of metals and alloys

Material	Modulus of elasticity E (kN/mm²)	Tensile and compressive strength (N/mm²)	Elongation (%)	Hardness (HV)
Pure aluminium	68–70	62–102	45–7	15.30
Aluminium alloys	68–72	90–500	20–5	20–80
Magnesium	44	170–310	5–8	30–60
Cast irons (grey)	75–145	150–410	0.5–1.0	160–300
SG cast iron	170–172	370–730	17–2	150–450
Copper	122–132	155–345	60–5	40–100
Copper alloys	125–135	200–950	70–5	70–250
Mild steel	190	420–510	22–24	130
Structural steels	190	480–700	20–24	130
Stainless steel	190	420–950	40–20	300–170
Titanium	100–108	300–750+	5–35	55–90
Zinc	90	200–500	25–30	45–50

be carried out the malleability may be restored by annealing. Annealing, or softening, of the metal is usually carried out before or during curvature work such as raising and hollowing, provided the metal is not coated with a low-melting-point material. However, the quality of the modern sheet metal is such that many forming operations, such as deep drawing and pressing, may be carried out without the need for an application of heat.

The following are examples in which the properties of malleability are most evident:

Riveting   Here the metal will be seen to have spread to a marked degree. If splitting occurs the metal is insufficiently malleable or has been overworked (work hardened).

Shaping   The blank for a dome consists of a flat disc which has to be formed by stretching and shrinking into a double-curvature shape. The more malleable and ductile the material of the blank is, the more readily it can be formed; the less malleable and ductile, the more quickly does the metal work hardened thus need more frequent annealing.

The degree of malleability possessed by a metal is measured by the thinness of leaves that can be produced by hammering or rolling. Gold is extremely malleable and may be beaten into very thin leaf. Of the metals used for general work, aluminium and copper are outstanding for their properties. The property of malleability is used to advantage in the manufacture of mild steel sheets, which are rolled to a given size and gauge for the motor industry.

It is also evident in the ability of mild steel and aluminium panels to be formed by mechanical presses into complicated contours for body shells. Malleability and ductility are the two essential properties needed in order to mass produce vehicle body shells by pressing.

The order of malleability of various metals by hammering is as follows: gold, silver, aluminium, copper, tin, lead, zinc, steel.

5.1.2 Ductility

Ductility depends on tenacity or strength in tension and the ease with which a metal is deformed, and is the property which enables a metal to be drawn out along its length, that is drawn into a wire. In wire drawing, metal rods are drawn through a hole in a steel die; the process is carried out with the metal cold, and the metal requires annealing when it becomes work hardened.

Ductile properties are also necessary in metals and alloys used in the following processes:

Pressed components   Special sheets which have extra deep drawing qualities are manufactured especially for press work such as that used in modern motor vehicle body production. These sheets undergo several deformations during the time they are being formed into components, yet because of their outstanding ductile properties they seldom fracture.

Welding electrodes and rods   Ductility is an essential property in the production of electrodes, rods and wires. The wire drawing machines operate at exceptionally high speeds and the finished product conforms to close tolerances of measurement; frequent failure of the material during the various stages of drawing would be very costly.

The order of ductility of various metals is as follows: gold, aluminium, steel, copper, zinc, lead.

5.1.3 Tenacity

A very important property of metals is related to its strength in resistance to deformation; that property is tenacity, which may be defined as the property by which a metal resists the action of a pulling force. The *ultimate tensile strength* of a metal is a measure of the force which ultimately fractures or breaks the metal under a tensile pull. The ultimate tensile strength (UTS) of a material is normally expressed in tons per in^2 or MN/m^2, and may be calculated as follows:

$$UTS = \frac{\text{tensile force in N}}{\text{cross-sectional area in mm}}$$

In this case the load is the maximum required to fracture a specimen of the material under test, and the calculation is based on fracture taking place across the original cross-sectional area. In ductile materials a special allowance must be made for wasting or reduction of original cross-sectional area.

High-carbon steels possess a high degree of tenacity, evidence of which can be seen in the steel cables used to lift heavy loads. The mild steels used in general engineering possess a small amount of tenacity, yet a bar of metal of one inch square (6.5 cm^2) cross-section, made from low-carbon steel, is capable of supporting a load in excess of 20 tonnes.

Methods of increasing tensile strength

It is possible to increase the tensile strength of both sheet steel and pure aluminium sheets by cold rolling, but this has the added effect of reducing their workable qualities. In the manufacture of vessels to contain liquids or gases under pressure it is not always possible to use metals with a high tensile strength; for instance, copper is chosen to make domestic hot water storage cylinders because this metal has a high resistance to corrosion. In this case, the moderate strength of copper is increased by work hardening such as planishing, wheeling and cold rolling. Work hardening has the added effect of decreasing the malleability.

The order of tenacity of various metals in tons per in^2 (MN/m^2) is: steel 32 (494); copper 18 (278); aluminium 8 (124); zinc 3 (46); lead 1.5 (23).

5.1.4 Hardness

When referring to hardness, it should be carefully stated which kind of hardness is meant. For example, it may be correctly said that hardness is that property in a metal which imparts the ability to:

1   Indent, cut or scratch a metal of inferior hardness
2   Resist abrasive wear
3   Resist penetration.

A comparison of hardness can be made with the aid of material testing machines such as those used to carry out the Brinell or Vickers Diamond tests. Hardness may be increased by the following methods:

Planishing   In addition to increasing the tensile strength of a metal, planishing also imparts hardness.
Heat treatment   Medium- and high-carbon steels, such as those used in many body working tools, can be hardened by heating to a fixed temperature and then quenching.

The order of hardness of various metals is as follows: high-carbon steel, white cast iron, cast iron, mild steel, copper, aluminium, zinc, tin, lead.

5.1.5 Toughness

This property imparts to a metal the ability to resist fracture when subjected to impact, twisting or bending. A metal need not necessarily be hard to be tough; many hard metals are extremely brittle, a property which may be regarded as being opposite to toughness.

Toughness is an essential property in rivets. During the forming process the head of the rivet is subject to severe impact, and when in service rivets are frequently required to resist shear, twist and shock loads. Toughness is also a requisite for steel motor car bodies, which must be capable of withstanding heavy impacts and must often suffer severe denting or buckling without fracture occurring. Further, when repairs are to be made to damaged areas it is often necessary to apply force in the direction opposite to that of the original damaging force; and the metal must possess a high degree of toughness to undergo such treatment.

5.1.6 Compressibility

Compressibility may be defined as the property by which a metal resists the action of a compressing force. The *ultimate compressive strength* of a metal is a measure of the force which ultimately causes the metal to fail or yield under compression. Compressibility is related to malleability in so far as the latter refers to the degree to which a metal yields by spreading under the action of a compressing or pushing force, while the former represents the degree to which a metal opposes that action.

5.1.7 Elasticity

All metals possess some degree of elasticity; that is, a metal regains its original shape after a certain amount of distortion by an external force. The elastic limit of a metal is a measure of the maximum amount by which it may be distorted and yet return to its original form on removal of the force. Common metals vary considerably in elasticity. Lead is very soft yet possesses only a small amount of elasticity. Steel, on the other hand, may reveal a considerable degree of elasticity as, for example, in metal springs. The elasticity of mild steel is very useful in both the manufacture of highly curved articles by press work and in the repair of motor car bodies.

5.1.8 Fatigue

Most metals in service suffer from fatigue. Whether the metal ultimately fails by fracture or by breaking depends on a number of factors associated with the type and conditions of service. When metal structures in service are subjected to vibration over long periods, the rapid alterations of push and pull, i.e. compressive and tensile stresses, ultimately cause hardening of the metal with increased liability to fracture. Any weak points in the structure are most affected by the action and become the probable centres of failure, either by fracture or by breaking. Various methods have been devised for testing the capacity of metals to resist fatigue, all of which depend on subjecting the metal specimen to alternating vibratory stresses until failure occurs. The rate and extent of vibration over a specified time form the basis of most fatigue tests.

5.1.9 Weldability

This property is a measure of the ease with which a metal can be welded using one of the orthodox systems of welding. Certain metals are welded very easily by all recognized methods; some metals can only be welded by special welding processes; and other metals and alloys cannot be welded under any circumstances.

5.2 Heat treatment of metals and metal alloys

Heat treatment can be defined as a process in which the metal in the solid state is subjected to one or more temperature cycles, to confer certain desired properties. The heat treatment of metals is of major importance in motor body work. All hand tools used by the body worker are made from a type of steel which is heat treatable so that the tools are strong, hard and have lasting qualities. The mechanical parts of a motor vehicle are also subject to some form of heat treatment, and in the constructional field heat-treatable aluminium alloys are being used extensively in commercial body work.

5.2.1 Work hardening and annealing

Most of the common metals cannot be hardened by heat treatment, but nearly all metals will harden to some extent as a result of hammering, rolling or bending. Annealing is a form of heat treatment for softening a metal which has become work hardened so that further cold working can be carried out. The common metals differ quite a lot in their degree and rate of work hardening. Copper hardens rather quickly under the hammer and, as this also reduces the malleability and ductility of the metal, it needs frequent annealing in order that it may be further processed without risk of fracture. Lead, on the other hand, may be beaten into almost any shape without annealing and without undue risk of fracture. It possesses a degree of softness which allows quite a lot of plastic deformation with very little work hardening. However copper, though less soft than lead, is more malleable. Aluminium will withstand a fair amount of deformation by beating, rolling and wheeling before it becomes necessary to anneal it. The pure metal work hardens much less rapidly than copper, though some of the sheet aluminium alloys are too hard or brittle to allow very much cold working. Commercially pure iron may be cold worked to a fair extent before the metal becomes too hard for further working. Impurities in iron or steel impair the cold working properties to the extent that most steels cannot be worked cold (apart from very special low-carbon mild steel sheets used in the car industry), although nearly all steels may be worked at the red-heat condition.

The exact nature of the annealing process used depends to a large extent on the purpose for which the annealed metal is to be used. There is a vast difference in technique between annealing in a steel works where enormous quantities of sheet steel are produced, and annealing in a workshop where single articles may require treatment. Briefly, cold working causes deformation by crushing or distorting the grain structure within the metal. In annealing, a metal or alloy is heated to a temperature at which recrystallization occurs, and then allowed to cool at a predetermined rate. In other words, crystals or grains within the metal which have been displaced and deformed during cold working are allowed to rearrange themselves into their natural formation during the process of annealing. Iron and low-carbon steels should be heated to about 900 °C and allowed to cool very slowly to ensure maximum softness, as far as possible out of contact with air to prevent oxidation of the surface; this can be done by cooling the metal in warm sand. High-carbon steels require similar treatment except that the temperature to which the steel needs to be heated is somewhat lower and is in the region of 800 °C. Copper should be heated to a temperature of about 550 °C or dull red, and either quenched in water or allowed to cool out slowly. The rate of cooling does not affect the resulting softness of this metal. The advantage of quenching is that the surface of the metal is cleaned of dirt and scale. Aluminium may be annealed by heating to a temperature of 350 °C. This may be done in a suitable oven or salt bath. In the workshop aluminium is annealed by the use of a blowpipe, and a stick or splinter of dry wood is

rubbed on the heated metal; when the wood leaves a charred black mark the metal is annealed. Sometimes a piece of soap is used instead of the wood; when the soap leaves a brown mark the heating should be stopped. The metal may then be quenched in water or allowed to cool out slowly in air. Zinc becomes malleable between 100 and 150 °C, and so may be annealed by immersing it in boiling water. Zinc should be worked while still hot, as it loses much of its malleability when cold.

5.2.2 Heat treatment of carbon steel

Steel is an important engineering material because, although cheap, it can be given a wide range of mechanical properties by heat treatment. Heat treatment can both change the size and shape of the grains, and alter the microconstituents. The shape of the grains can be altered by heating the steel to a temperature above that of recrystallization. The size of the grains can be controlled by the temperature and the duration of the heating, and the speed at which the steel is cooled after the heating; the microconstituents can be altered by heating the steel to a temperature that is sufficiently high to produce the solid solution austenite so that the carbon is dispersed, and then cooling it at a rate which will produce the desired structure. The micrograin structure of carbon steel has the following constituents:

Ferrite   Pure iron.
Cementite   Carbon and iron mixed.
Pearlite   A sandwich layer structure of *ferrite and cementite.*

All low-carbon steels of less than 0.83 per cent carbon content consist of a combination of ferrite and pearlite. Carbon steel containing 0.83 per cent carbon is called *eutectoid* and consists of pure pearlite structure. Steels over 0.83 per cent carbon up to 1.2 per cent carbon are a mixture of cementite and pearlite structures. If a piece of carbon steel is heated steadily its temperature will rise at a uniform rate until it reaches 700 °C. At this point, even though the heating is continued, the temperature of the steel will first remain constant for a short period and then continue to rise at a slower rate until it reaches 775 °C. The pause in the temperature rise and the slowing down of the rate indicate that energy is being absorbed to bring about a chemical and structural change in the steel. The carbon in the steel is changing into a solid solution with the iron and forming what is known as *austenite.* The temperature at which this change in the structure of the steel starts is 700 °C, which is known as the *lower critical point;* the temperature at which the change ends is known as the *upper critical point.* The difference between these points is termed the *critical range.* The lower critical point is the same for all steels, but the upper critical point varies with the carbon content as shown in Figure 5.1. Briefly, steels undergo a chemical and structural change, forming austenite, when heated to a temperature above the upper critical point; if allowed to cool naturally they return to their normal composition.

Steel can be heat treated by normalizing, hardening, tempering and case hardening as well as by annealing, which has already been described in Section 5.2.1.

Normalizing

Normalizing is a process used to refine the grain structure of steel after it has been subjected to prolonged heating above the critical range (as in the case of forging) and to remove internal stresses caused by cold working. The process may appear to differ little from annealing, but as its name suggests the effect of normalizing is to bring the steel back to its normal condition and no attempt is made to soften the steel for further working. Normalizing is effected by slowly heating the steel to just above its upper critical range for just sufficient time to ensure that it is uniformly heated, and then allowing it to cool in still air.

Hardening

It has already been said that if a piece of steel is allowed to cool naturally after heating to above its upper critical point, it will change from austenite back to its original composition. If, however, the temperature of the heated steel is suddenly lowered by quenching it in clean cold water or oil, this change back from austenite does not take place, and instead of pearlite, a new, extremely hard and brittle constituent is formed, called *martensite.* This process makes steels containing 0.3 per cent or more carbon extremely hard, but steels having a carbon content of less than 0.3 per cent cannot be hardened in this way because the small amount of carbon produces too little martensite to have any noticeable hardening effect. The steel to be hardened should be quenched immediately it is uniformly heated to a temperature just above the upper critical point. It is also important not to overheat the steel and to allow it to cool to the quenching temperature. Whether water or oil is used for quenching depends upon the use to which the steel is to be put. Water quenching produces an extremely hard steel but is liable to cause cracks and distortion. Oil quenching is less liable to cause these defects but produces a slightly softer steel. A more rapid and more even rate of cooling can be obtained if the steel is moved about in the cooling liquid, but only that part of the steel which is to be hardened should be moved up and down in the liquid in order to avoid a sharp boundary between the soft and hard portions.

A workshop method of hardening carbon tool steel is to heat the steel, using the forge or oxyacetylene blowtorch, to a dull red colour (see Table 5.3) and then quench it in water or oil. This would harden the article ready for tempering.

Tempering

Hardened steel is too brittle for most purposes, and the process of tempering is carried out to allow the steel to regain some of its normal toughness and ductility. This is done by heating the steel to a temperature below the lower critical point, usually

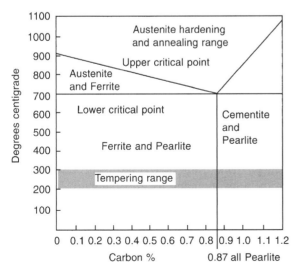

Figure 5.1 Changes in structure of carbon steel with temperature and carbon content

Table 5.3 Temperature colours for steel

Colour	Temperature (°C)
Black	450–550
Very dark red	600–650
Dark red	700–750
Cherry red	800–850
Full red	850–900
Bright red	950–1000
Dark orange	1050–1100
Light orange	1150–1200
Yellow white	1270–1300
White (welding heat)	1400–1550

between 200 and 300 °C, thereby changing some of the martensite back to pearlite. The exact temperature will depend on the purpose for which the steel is intended; the higher the temperature, the softer and less brittle the tempered steel becomes. Methods used for controlling the tempering process depend upon the size and class of the article to be tempered. One method is to heat the hardened steel in a bath of molten lead and tin, the melting points of various combinations of these two metals being used as an indication of the temperature. Another method of tempering small articles is to polish one face or edge and heat it with a flame. This polished surface will be seen to change colour as the heat is absorbed. The colour changes are caused by the formation at different temperatures of thin films of oxide, called tempering colours (Table 5.4). After tempering, the steel is either quenched or allowed to cool naturally.

Table 5.4 Tempering colours for plain carbon steel

Colour	Temperature (°C)	Type of article
Pale straw	220–230	Metal turning tools, scrapers, scribers
Dark straw	240–245	Taps, dies, reamers, drills
Yellow-brown	250–255	Large drills, wood turning tools
Brown	260–265	Wood working tools, chisels, axes
Purple	270–280	Cold chisels, press tools, small springs, punches, knives
Blue	290–300	Springs, screwdrivers, hand saws

Case hardening

Although mild steels having a carbon content of less than 0.3 per cent cannot be hardened, the surface of the mild steel can be changed to a high-carbon steel. Case hardening of mild steel can be divided into three main processes:

1 Carburizing
2 Refining and toughening the core
3 Hardening and tempering the outer case.

A method called pack carburizing is often used in small workshops. After thoroughly cleaning them, the steel parts to be carburized are packed in an iron box so that each part is entirely surrounded by 2.5 cm of carburizing compound. The box is sealed with fireclay, heated in a furnace to 950 °C and kept at that temperature for two to twelve hours, during which time the carbon in the compound is absorbed by the surface of the steel parts. The steel parts are then allowed to cool slowly in the box. At this stage their cores will have a coarse grain structure due to prolonged heating, and the grain is refined by heating the steel parts to 900 °C and quenching them in oil. Next the casing is hardened by reheating the steel to just under 800 °C and then quenching it in water or oil. Tempering of the casing may then be carried out in the normal way.

Small mild steel parts can be given a very thin casing by heating them in a forge to a bright red heat and coating them with carburizing powder. The parts are then returned to the forge and kept at a bright red heat for a short period to allow the carbon in the compound to penetrate the surface of the metal. Finally the parts are quenched and ready for use.

5.2.3 Heat treatment of aluminium alloys

An aluminium alloy is heat treated by heating it for a prescribed period at a prescribed temperature, and then cooling it rapidly, usually by quenching. The particular form of heat treatment which results in the alloy attaining its full strength is known as *solution treatment*. The alloy is raised to a temperature of 490 °C by immersing it in a bath of molten salt. The bath is usually composed of equal parts of sodium nitrate and potassium nitrate contained in an iron tank. This tank is heated by gas burners and, except for its open top, is enclosed with the burners in a firebrick structure which conserves the heat. The temperature of the bath must be carefully regulated, as any deviation either above or below prescribed limits may result in the failure of the metal to reach the required strength. The alloy is soaked at 490 °C for fifteen minutes and then quenched immediately in cold water.

At the moment of quenching the alloy is reasonably soft, but hardening takes place fairly rapidly over the first few hours. Some alloys, chiefly the wrought materials, harden more rapidly and to a greater extent than others. Their full strength is attained gradually over four or five days (longer in cold weather); this process is known as *natural age hardening*. As age hardening reduces ductility, any appreciable cold working must be done while the metal is still soft. Working of the natural ageing aluminium alloys must be completed within two hours of quenching, or for severe forming within thirty minutes. Age hardening may be delayed by storing solution-treated material at low temperatures. Refrigerated storage, usually at 6–10 °C, is used for strip sheet and rivets, and work may be kept for periods up to four days after heat treatment. If refrigerated storage is not used to prevent age hardening it may be necessary to repeat solution treatment of the metal before further work is possible.

Alloys of the hiduminium class may be *artificially age hardened* when the work is finished. Artificial ageing is often called *precipitation treatment;* this refers to the precipitation of the two inter-metallic compounds responsible for the hardening, namely copper and manganese silicon. The process consists of heating the work in an automatically controlled over to a temperature in the region of 170 °C for a period of ten to twenty hours. Artificial ageing at this temperature does not distort the work. The temperature of the oven must be maintained to within a few degrees, and a careful check on the temperature is kept by a recording instrument. In order to ensure uniform distribution of temperature a fan is fitted inside the oven to keep the air in circulation. At the end of a period of treatment the oven is opened to allow the work to cool down. One of the chief advantages of this process is that work of a complicated character may be made and completed before ageing takes place. Moreover, numerous parts may be assembled or riveted together and will not suffer as a result of the ageing treatment.

5.3 How metal is formed to provide strength

It has been established that the strength of a material is governed by its material composition and by the method and direction of loading, i.e. tensile, compressive, torsion, shear and bending. Generally the majority of metals are capable of withstanding greater loads in tension than any other type of stress. One of the properties of steel is that, within certain limits, it is elastic: that is, if it is distorted by a load or force it will change shape, but it will return to its original shape when the force is removed. However, above a certain intensity of load (the elastic limit) the metal will remain distorted when the load is removed. Sheet steel, such as is used in the manufacture of car bodies, has reasonable strength in tension but has little resistance to compressive and/or torsional loads. This lack of resistance is due not so much to poor compressive strength as to lack of rigidity. Low-carbon steels are used extensively in the manufacture of vehicle bodies, and the designer has to ensure that the relatively thin sections of material are capable of withstanding the various types of loading. In addition to the permanent stresses present in the material, the vehicle body as a whole is subject to shock stresses due to road conditions, and these must also be taken into consideration by the designer.

With the development of deep drawing steels and better press equipment, large streamlined panels were designed and formed into contours that were more attractive, gave longer life and greater safety, and at the same time reduced the bulky construction previously required to give similar strength. It is known that the shape of any material is held by the stresses set up in the material itself, such as those given by angles, crowns, channels and flanges. The original shape will be maintained until the material is subject to a force sufficient to overcome the initial stresses. Furthermore, it will tend to return to its original shape providing it has not been distorted beyond the point of elasticity.

5.3.1 Crowned surfaces

The building up of stresses at the bend or peak is also an important consideration in the design and manufacture of the modern car body. The most

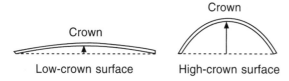

Figure 5.2 Strength in crowned surfaces

common features of the body are the curved surfaces (Figure 5.2); these are called crowns and may be curved either in one direction or in all directions. A crowned surface is stronger than a flat panel, and whilst it will resist any force tending to change its shape, it also has the ability to return to its original shape unless distorted beyond its point of elasticity. These are the features of metal sheet which has been formed in a press into a permanent shape, with die-formed stresses throughout its entire area tending to hold the shape. On the bend or crown, one side of the sheet is longer than the other; and the metal at the surface is more dense than at the centre of the sheet. The final action of the press is to squeeze the surface together, thus setting up stresses and greater strength. The greater the crown or curve of the panel, the greater its strength and rigidity to resist change in its shape. This is illustrated by the fact that a low crown, i.e. a surface with very little curve, such as a door panel, is springy and is not very resistant to change of shape. On the other hand high crowns, that is surfaces with a lot of curve, like wings, edges of roofs and sill panels, are very resistant to change in shape by an outside force.

5.3.2 Angles and flanges

A further method of giving strength to metal is to form angles or flanges along the edges of sheets (Figure 5.3). A right-angled bend greatly increases

the strength of a sheet, as can be demonstrated by forming a right-angled bend in a thin sheet of metal and then trying to bend the metal across the point of the bend. This method is used on inner door panels and at the edges of wings, edges of bonnets and boot lids, and wherever stiffness is required at unsupported edges.

5.3.3 U-channels and box sections

A U-channel, as the name implies, comprises two right-angled bends, one at each of the opposite sides of a piece of metal (Figure 5.4). Much more strength can be gained by making a U-channel instead of a single-angled bend in any reinforcement section. The U-channel is also the most common type of section used in the construction of car frames. Yet another method of increasing strength is known as box construction, which consists of two U-sections welded together to form a square pillar or box. Box sections are used in conjunction with U-channel construction in the manufacture of car chassis frames, underbodies, subframes, cross members and any construction where great strength is necessary.

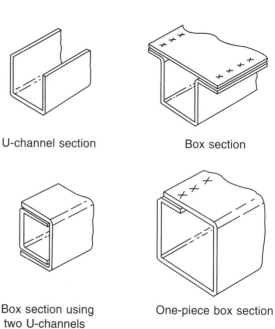

Figure 5.4 Strength in U-channel and box sections

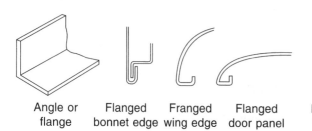

Figure 5.3 Strength in angles and flanges

5.4 Machines for sheet metal fabrication

The mass produced, all-steel bodies of both private cars and light commercial vehicles call for the production of involved panel shapes on a massive scale. This requires heavy power presses employing mating die sets which in effect press the sheet metal and deform the structure of the metal to a degree where it takes on a natural shape and retains that shape. However, since all pressed panels have some degree of tension in the material structure, it is important that this should be taken into account by the body repairer.

It is often necessary to manufacture sheet steel parts by metal forming machines. The most common of these include wheeling machines, guillotines, fly presses, folders and forming presses. In body repair shops the fly press is often hand operated, and is a most useful piece of equipment. Tools for the fly press fit into a central spindle which is moved down in order to pierce, blank and to a limited extent form the metal parts required. Another widely used machine is the folder or bending machine; adjustment of the setting of the blades of this machine enables sharp acute or obtuse angled bends to be produced in long lengths of sheet metal for the making of sections. The swaging machine is used for swaging, wiring, joggling and closing sheet metal work edges. Finally, the body shop is not complete unless it possesses a wheeling machine, which is used to produce double curvature panels from flat sheets by passing the sheets to and fro between rollers or wheels in order to stretch the sheets and to create a curved shape.

5.5 Shearing theory

If a piece of sheet metal is placed on one bottom cutting member or blade, and the top cutting blade is brought to bear on the metal with continuing pressure, after a certain amount of deformation the elastic limit of the metal is exceeded and the top cutting blade penetrates and cuts the surface of the metal (Figure 5.5). Fractures begin to run into the metal thickness from the points of contact of the top and bottom cutting blades, and if these are positioned correctly relative to each other, the fractures meet and the metal is sheared before the top member penetrates and cuts the whole thickness of the metal. The horizontal distance between the two cutting members is called the clearance (Figure 5.6),

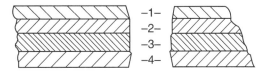

1 The notch made by the upper cutting blade
2 Smoothly cut part
3 Torn or fractured section
4 The notch made by the lower cutting blade

Figure 5.5   The cutting section showing a sheared piece of metal

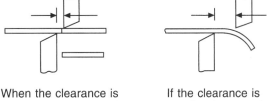

When the clearance is set correctly the metal is cut (sheared) smoothly and cleanly

If the clearance is excessive the material is bent instead of being cut

Figure 5.6   Blade clearance

and it is this distance that determines the quality of the cut and the load on the machine or hand tool being used. There is only one correct clearance allowance for any particular sheet, and this depends mainly upon the thickness of the metal being cut and its physical properties. This clearance allowance is usually 5–10 per cent of the metal thickness.

5.6 Cutting machines

5.6.1 Hand lever shears

There are the simplest sheet metal working machines. For cutting mild steels they are used with blade lengths up to 500 mm for cutting 3 mm material, or with a length of about 200 mm for material up to 6 mm. The cutting members are flat blades and there is generally no provision for adjustment of blade clearance. The bottom blade is fixed to the machine frame, while the top cutting blade is fixed to a moving member which is pivoted in the main frame and operated by a hand lever via a simple link mechanism. An adjustable hold-down is usually fitted to the top member to hold the sheet down during cutting. The body of the shears may be cranked or offset to permit the cut sheets to be forced beyond the blades so that sheets longer than the blades may be cut. The body of the machine is usually fabricated, and this construction is to be preferred.

5.6.2 Universal shearing machines (nibbling machines)

In these machines the cutting is done by a pair of very narrow blades, one of which is fixed while the other moves up and down from the fixed blade at fairly high speeds. Generally the blades have a very steep rake or angle to permit piercing of the sheet for internal cutting. Since the blades are very narrow the sheet can be easily turned during cutting. The term 'universal' is apt, since the machine will cut straight or curved shapes inside or outside. These machines are powered by compressed air or an electric motor, and are held in the hands during the cutting operation. In the machine the top blade is fixed to the moving member or ram and the bottom blade is fitted on an extension which is shaped like a spiral so as to part the material after cutting. The ram and blades are driven at a speed between 1200 and 2100 rev/min, depending on the metal thickness and whether air or electrical power is used. The maximum thickness of metal to be cut is about 4.5 mm, but the machines find their widest application on lighter gauges of metal 1.00 mm and 0.8 mm. The lighter machines have a maximum cutting radius of about 15 mm and the heavier machines one of about 50 mm.

The most important points to remember when using the machine are:

1 See that the blades are set for the correct thickness of metal being cut.
2 Keep the blades sharp (a blunt blade tends to be dangerous).
3 Hold the machine correctly with the blades at right angles to the metal being cut.

5.6.3 Guillotine

A 1.25 m treadle guillotine (Figure 5.7) will meet the requirements of the average panel shop for cutting sheet metal. The guillotine is probably the most widely used of the straight-line cutting

Figure 5.7 Treadle operated guillotine (*Selson Machine Tool Co. Ltd*)

machines. The principle employed in the cutting action is very similar to that of the hand bench shears. A concentrated load is applied across the width of the sheet or plate as cutting proceeds. All have a bottom cutting blade which is flat and is fixed horizontally, and a top blade which is mounted at a rake to the bottom blade and is fixed to a moving beam. The top beam and blade can be brought to meet the bottom blade by hand, foot, mechanical, air or hydraulic power, depending on the type of machine.

The guillotine consists of a bed, a foot treadle, two cutting blades, and a front, back and two side gauges. The face of the bed sometimes has a graduated scale which permits the cutting of sheets to a specified size. The side gauge, which has steel bars bolted on each side of the bed, can also be graduated for measuring purposes. By placing the edge of the sheet against one of these side gauges, a cut can be made at right angles between the side gauge and the bottom blade. The front gauge consists of a rectangular bar which slides in the bed and can be set at any required distance from the blades. For cutting long sheets, extension arms are fastened to the front of the machine. Slots in the extension arms permit the movement of the front gauge away from the bed to accommodate the larger sheets. The back gauge has an angle-shaped bar which slides on two rods fastened to the rear of the machine. This bar is used to cut a number of pieces of the same length by the adjustment that is allowed on the rods. The lower cutting blade is fastened to the bed of the machine, and the upper cutting blade is attached to a beam which is moved by stepping on the foot treadle. This beam is connected to the treadle by heavy springs which return the treadle to its original position after the metal has been cut. Also attached to this beam and blade is a hold-down device which is located in front of the top blade, and as the treadle is pressed down it clamps and holds the metal in place ready to be cut. A guard is used in front of the upper blade to prevent the operator's fingers from coming into contact with the blade or hold-down device. The standard guillotine blades are usually made from plain carbon steel and are suitable for cutting all types of mild steel and non-ferrous metals. For cutting higher-tensile steels such as stainless steel, alloy steel blades are necessary.

Guillotines should never be used to cut metal which exceeds the capacity of the machine. The capacity range for these models is: a 2 m blade, which will cut up to 1.2 mm; a 2.5 m blade, which will cut up to 1.00 mm; and a 1.25 m blade, which will cut up to 1.6 mm. Most foot operated machines will cut up to 1.6 mm metal. Under no circumstances should the guillotine be used to cut wire, rod or bar, as it is intended to cut only flat sheet metal.

When using this machine, make sure that the angle finger guard, which is fitted in a position between the operator and the cutting blades, is set in such a way as to protect the hands from both the crushing action of the hold-down device and the cutting action of the blades. Accurate cutting is accomplished by leaning over and sighting the scribed cutting line on the metal at a point perpendicular to the cutting edge of the bottom blade. Check before using the machine that the clearance between the top and bottom blades is set correctly for the thickness of metal to be cut. Do not cut metal of heavier gauge than the machine is designed to cut. Always hold the metal flat and firmly on the base plate of the guillotine; this will prevent drag (the effect of pulling the metal in towards the blades) which often occurs as the cutting edge loses its sharpness.

The guillotine is used more in the coach building and building of new vehicles than in the body repair side of the industry. It is ideal for cutting a number of sheets to a predetermined size because it guarantees greater accuracy than is possible with hand methods.

The following safety precautions should be observed when working with guillotines:

1 Keep the blades of the machines sharp. Remember that blunt blades drag the metal and can cause the loss of a finger.
2 Hold the metal firmly on the bed plate of the machine to ensure shearing action and also to prevent drag.
3 All guards should be kept in place. They are required to be there by law in the interests of your safety.
4 When using a treadle guillotine make sure that any assistant or onlooker has not got his feet under the treadle.

5.7 Bending theory

Many metals and their alloys may be formed by bending. When considering metals with regard to their bending properties, the following facts are most important.

Behaviour of the metal in bending

In bending sheet metal to an angle, the inner fibres in the metal on the bend are compressed and given a compressive stress, while the outer fibres are stretched and given a tensile stress (Figure 5.8). The boundary line in the metal thickness between the two areas of stress is called the *neutral axis*. The position of the neutral axis may vary with the properties of the metal and its thickness.

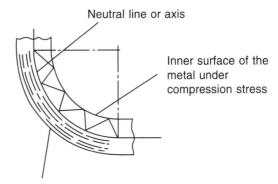

Neutral line or axis

Inner surface of the metal under compression stress

Metal stretched on the outside is subject to tensile stress

Figure 5.8 Bending action

Calculation of the bending length

The formula for calculating the bending allowance is shown in Figure 5.9. An approximate workshop method for calculating the bending allowance is $(r/2) + T$.

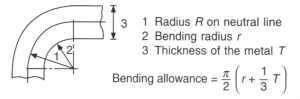

1 Radius R on neutral line
2 Bending radius r
3 Thickness of the metal T

$$\text{Bending allowance} = \frac{\pi}{2}\left(r + \frac{1}{3}T\right)$$

Figure 5.9   Bending allowance

5.8 Bending machines

5.8.1 Angle bender (clamp folder)

The essential factors in producing a clean bend on sheet metal are that the edge or blade over which the metal is bent should be straight, smooth and fairly sharp and that the pressure applied to bend the metal over this edge should be equal throughout the length of the bend. The most generally used machine for bending sheet metal up to 1.6 mm is the angle bender or clamp folder (Figure 5.10). This consists of a

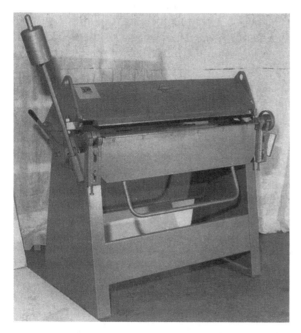

Figure 5.10   Bending machine (*Selson Machine Tool Co. Ltd*)

clamp to which is attached the blade around which the metal is bent, and is operated by a hand lever fastened to the end of the machine. The clamping action applies necessary pressure to hold the work to the bed or base of the machine, in order to prevent movement while the bending is in progress. The actual bending of the metal is done by swinging up the front part of the bed which carries the bottom blade on a hinged centre with the top blade.

The following procedure should be used:

1   Mark the bend lines on the metal.
2   Open the upper blade by pushing the clamping handles backwards.
3   Place the sheet of metal between the upper blade and the bed of the machine with the bend line directly under the edge of the top blade, then clamp the metal by pulling the clamping handle forwards again. Some adjustment to obtain the correct clamping pressure may be necessary.
4   Raise the lower bending blade to a position that will produce the desired angle of bend. Move the blade slightly beyond the required angle to allow for springback of the metal.
5   Drop the bending blade to its normal position and open the clamping handles, which will release the work.

5.8.2 Edge folder

As the name implies, this machine is used for turning the edges on sheets to make hooks for grooving seams or for wire insertion, or for making small flanges, but generally only in mild steel of 1.00 mm or less.

5.8.3 Open-end folder or bending machine

In these machines the end frames in which the clamping beam is fixed have small gaps so that folds can be made in sheets of unlimited length, provided that the depth of the bend is below the width or height of the gap, which is usually about 50 mm. The maximum length of the blades in this type of machine is 2 m, and the maximum gauge which can be bent is about 1.6 mm. The lift of the clamping beam and top blade is about 38–50 mm in height. The clamping beam and top blade are operated by an eccentric shaft running across the front of the machine; this shaft has handles at each end which lift the blade up and down. The bending blade is lifted by a handle at the front of the machine. In larger machines there is provision for counter-balance weights to be fitted to the bending beam. The bending machine is only suitable for single bending operations.

5.8.4 Box and pan bending machine (universal bending machine)

It is possible to form boxes with more than one bend using the clamp folder if there is sufficient clearance between the clamping beam and the top blade to allow the workpiece to be removed. However, most box forming is done using the box bender, which differs from the clamp folder in the design of the upper blade (Figure 5.11); whereas the clamp folder has a single solid bending blade, the box bender has a series of individual blades known as fingers. The advantage of this type of machine is that any number of finger blades can be removed, permitting a great variety of bends to be made. The actual operation of the machine is the same as the clamp folder. Both machines are used to advantage in body work as they are capable of bending to shape many articles suitable for both new and reapir work.

Figure 5.11 Box and pan bending machine (*Selson Machine Tool Co. Ltd*)

5.8.5 Safety measures for working bending machines

1 Before locking the clamping blade down, make sure that your fingers and those of your assistant are clear.

2 Check on the path of the swinging counter-balance weight, warning anyone working close by of your intentions to use the machine.

5.9 Rolling machines

5.9.1 Bending rolls

Bending rolls (Figure 5.12) of 1 m capacity are well suited for panel work. They are used mainly to form curves in panels and to roll complete cylinders, but they can also be used for breaking the grain of coated sheet metal, such as tin plate or galvanized sheet, before it is worked. Machines are made in all sizes: bench rollers for light tin plate work, hand powered rollers for general sheet metal, and motor driven rollers for heavy gauge sheet steel. All rolling machines comprise two front rollers which lightly grip and draw the metal through, and a free roller at the rear to set or bend the metal to the desired radius. They can, however, differ in their mode of operation, and bend rolls fall into two main types: pyramid rolls and pinch rolls.

Figure 5.12 Bending rolls (*Selson Machine Tool Co. Ltd*)

5.9.2 Pinch rolls

The rollers (Figure 5.13) are of the same diameter and are usually geared together to give the same feeding speed and rotation. The two front rollers are adjustable for different thicknesses of metal; the third roller at the rear is also adjustable to provide for various degrees of bend required. A swinging-type end support can be opened, permitting the top roller to be pivoted upwards so that completed cylinders can be removed. Nearly true

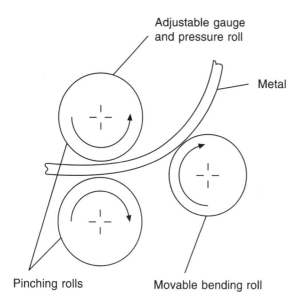

Figure 5.13   Pinch rolls

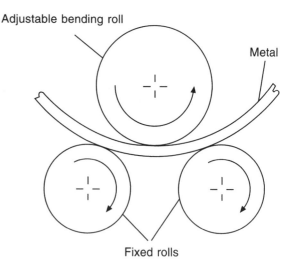

Figure 5.14   Pyramid rolls

cylinders can be formed using pinch rolls; the slight flat which occurs on the leading edge of the plate can be reduced by initially reversing the rollers and feeding the flat plate back into the rollers a little way, which will preform the edge ready for rolling. Pinch rolls grip or pinch the metal between the first two rollers and give good contact during bending, but if these rolls are too tightly adjusted the edges of the metal will be stretched and the result will be an uneven edge. Pinch rolls are used for rolling all types of thin gauge sheet metal up to 2.00 mm.

5.9.3 Pyramid rolls

This machine (Figure 5.14) has two rollers of equal diameter mounted side by side which rotate in the same direction. The third roller, which is situated above the other two, is of larger diameter and is adjustable in the vertical plane in order to control the radius of bend. It pivots at one end to permit easy removal of the finished work. Pyramid rolls tend to leave more flat surfaces than pinch rolls. Again this may be reduced by rocking the plate between the rollers, but it may be necessary to preform the edges of the metal by hand before rolling. This type of machine is best suited for the heavier gauges of metal 3 mm and upwards.

As already stated, common metals used for panel work have a certain degree of elasticity. This means that before a metal sheet can be formed into a curve or cylinder, sufficient force must be applied to deform the structure of the material beyond its elastic limit so that the sheet assumes a permanent curved shape. Some heavier material may require presetting by hammering to provide a lead over the rear roller and also to prevent flats at the ends of the rolled surfaces. However, small flats may occasionally prove to be of assistance in the welding of thin sheet metal. The slight ridge formed by the flats tends to act in a similar manner to a swage and has the effect of stiffening the area to be welded. Excessive distortion is checked by the flats and, after welding, the smooth curve of the cylinder can be restored with a mallet. Bending rolls are sometimes used to repair flat aluminium panels by rolling them, then repairing them in this state, then rolling them flat again.

5.9.4 Cone rolling machine

Cone rolling machines are mostly used for rolling small conical-shaped fabricated articles in light gauge metals. They consist of a pair of conical rolls, geared and mating together and acting as pinch rolls. There are no special sizes. They are usually made to the particular requirements of the company (Figure 5.15).

Figure 5.15 Cone rolling machine (*Frost Auto Restoration Techniques Ltd*)

5.9.5 Safety measures for the operation of rolling machines

1 Loose clothing and long hair can easily be drawn between the rollers.
2 Care should be taken when putting the panel into the rollers that the operator's fingers are kept clear.

5.10 Wheeling machines

These machines (Figures 5.16 and 5.17) can be used to form flat sheets of metal into double-curvature shapes such as are found in automobile bodies. They are also used for removing dents, buckles and creases from sheets or panels previously shaped, for reducing the thickness of welds, and for planishing or smoothing panels which have been preshaped by hand. The main frame of the wheeling machine is of large C form. The size of

Figure 5.16 Wheeling machine (*Frost Auto Restoration Techniques Ltd*)

Figure 5.17 Range of wheeling machines (*Frost Auto Restoration Techniques Ltd*)

the gap in this frame usually determines the machine's specifications, which are from 0.75 m to 1 m in width and about 0.6 m in depth; the wheel diameters are approximately 90 mm. The machine consists simply of an upper wheel which is nearly flat and a second lower wheel which is curved or convex in shape; the two wheels meet at a common centre. The lower wheel runs free on a spindle carried by a vertical arm, which may be raised or lowered by screw movement through a hand wheel which regulates the pressure that can be applied to the work. Some machines have a quick pressure release mechanism attached to the spindle. There are three standard shapes of faces for the bottom wheel – flat, small radius and full radius – and the choice of wheel depends on the required shape of the finished panel. The upper wheel is carried on a horizontal shaft which is allowed to rotate freely in bearings. The top wheel and housing can be swivelled round through 90° to make it more accessible when wheeling certain shaped panels. This also applies to the housing for the bottom wheel, enabling it to move in conjunction with the top wheel.

Up to three times as much lift or stretch is obtainable with aluminium than with steel, and consequently aluminium can be shaped more by wheeling than steel. When wheeling aluminium, care must be taken not to apply too much pressure by the bottom wheel as this could have the effect of overstretching the material.

This machine is one of the few machines that has been used in the panel beating trade since its infancy. The art of wheeling a panel to the correct shape and contour requires a highly skilled craftsman, who is known in the trade as a wheeler.

5.11 Swaging machines

These machines (Figure 5.18) can be used on sheet metal blanks to carry out a large number of different operations such as swaging, wiring, joggling, flanging, beading and many other edge-type treatments. Hence the machine is also called a jennying, swaging, burring or beading machine, despite the fact that the same basic machine is used to perform the different operations. In body work the machine is used in the stiffening and strengthening of panels; for decoration in the form of beading or swaging; for the edge preparation of panels, such as wiring; and for making joints between panels as in lap and joggle joints. It can also be used to give rigidity to large flat panels to prevent 'drumming'.

The basic machine consists of a machine frame carrying two horizontal shafts which are geared

Figure 5.18   Swaging machine (*Selson Machine Tool Co. Ltd*)

together. The shafts are mounted one above the other, one of them in a fixed housing, and the other in a housing which has some vertical adjustment with reference to the fixed shaft. It also has some horizontal adjustment controlled by a gauging device known as the stop. In hand machines the shafts are turned by a handle which is fixed to the end of one shaft; in power driven machines they are turned by an electric motor which is connected through gearing to the shafts. The shafts are usually arranged so that they have a 1:1 rotation. The actual swaging is done by attachments fitted on to the ends of the shafts, and the shape and form of these attachments can be varied as is shown in Figure 5.19. Attached to one end of the shafts are male and female rollers shaped to produce a swage of the desired form. The top section of the frame carrying the upper roller and shaft is hinged at the back and kept in upward tension by a flat spring. The wheels are brought into position vertically by the operation of a small handscrew. A small lever situated near the gear wheel provides horizontal adjustment so that the wheels can be set to match

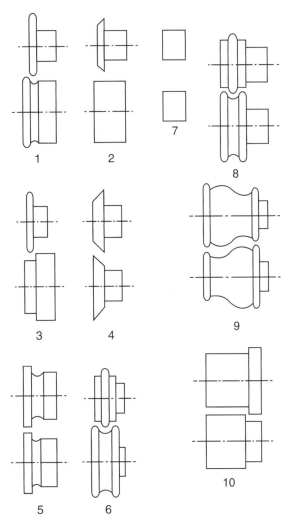

1 Wiring rolls for forming U-seating for wire
2 Closing rolls for closing metal around the wire
3 Jenny rolls (flanging)
4 Panning down rolls
5 Bottom closing rolls
6 Swage rolls
7 Collars or distance pieces
8, 9 Swage rolls
10 Necking rolls used for joggling

Figure 5.19   Swaging attachments

exactly. An adjustable guide is provided to ensure that the swaged impression will be true and parallel with the edge of the panel.

To set up the swaging machine for operation, first select a pair of wheels which will give the required section. Fit these wheels to the shafts and line them up by slackening the locking screw and adjusting the small lever at the rear of the frame until the wheels engage centrally with each other. Next set the stop to the distance required between the centre of the moulding and the edge of the sheet. Adjust the top screw until the wheel forms a depression in the metal and then run the metal through the wheels. Give another half turn on the handscrew and repeat the operation until the wheel 'bottoms', at which stage the swage will be fully formed. The object of forming the swage in gradual stages is to avoid strain on the metal which may result in splitting or distorting the panel.

5.12 Brake presses

Brake presses (Figure 5.20) are devices for bending sheet metal quickly and accurately, and their rapid development in the past twenty years has resulted in a very wide capacity range of up to 1500 tonnes. The bulk of the machines for bending thin gauge metal are of 3 m or less, with capacities starting at 20 tonnes and going up to about 200 tonnes. Brake press capacities are usually given either in tonnage terms or in maximum bending of a certain metal thickness, and are based on V-bending pressures. The tonnage specifications are usually determined by using a V-die opening of eight times the stock thickness, with a corner radius to the bend not less than the metal thickness. There are two important dimensions in brake press specifications: one is the maximum bending length over the bed, and the other is the distance between the housing. The width between the housing must permit the work to pass through the machine, and the brake press length is the overall bending length of the machine.

Brake presses comprise a *frame*, a *ram* or bending beam, and means for moving the beam. The frames are always of all-steel construction with parts welded or bolted together. Obviously these frames should be as rigid as possible to resist deflection – a fact of the highest importance. There are usually three main parts to the frame – a bed and two side frames. The top of the bed has a location slot into which the bending tools slide. The bending beam is usually a steel plate which works by simply sliding in the main frame. The load on the beam acts as near to the centre of the beam as possible to avoid side strain. The bottom of the beam face is made to receive the top bending tool or die, generally by means of a side plate to hold a tongue formed on the tool or die. The general methods of operating the beam are mechanical or hydraulic. The mechanical means comprise a crankshaft rotating in a phosphor bronze bush in the main frame. The working strokes are usually between 50 mm and 150 mm long, and this length can be adjusted to suit varying conditions. The motor drives a flywheel through a multiplate friction clutch and single or double gearing. A brake is fitted to work in conjunction with the clutch and is of the drum-brake shoe-operated type. Its function is mainly to hold the weight of the moving parts when the clutch is disengaged.

The V-type blade and interchangeable die is used extensively for forming light mild steel sheet components, from simple bends to complex multibends in panels or sheet metals (Figure 5.21). Not only can it produce straight, sharp bends, but the tools can be interchanged to give curves or radius bends. Curved sections, ribbing or stiffening sections, notching plates, corrugating sheets, and punching holes in plates can also be formed. The machine is mostly used in sheet metal industries, but it does find a use in the manufacturing side of commercial body work where large and complicated panels are to be formed.

Figure 5.20 Brake press (*Edwards Pearson Ltd*)

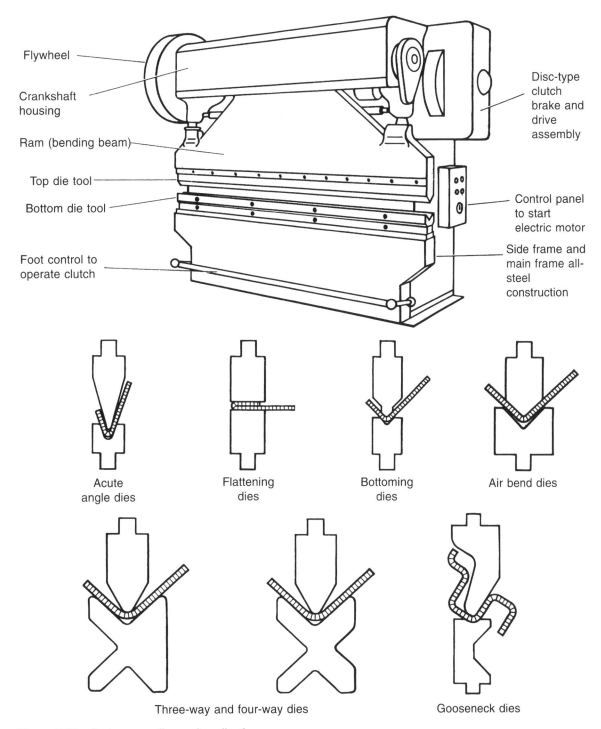

Flywheel

Crankshaft housing

Ram (bending beam)

Top die tool

Bottom die tool

Foot control to operate clutch

Disc-type clutch brake and drive assembly

Control panel to start electric motor

Side frame and main frame all-steel construction

Acute angle dies

Flattening dies

Bottoming dies

Air bend dies

Three-way and four-way dies

Gooseneck dies

Figure 5.21 Brake press dies and applications

5.13 Forming and drawing

During forming, one area of a sheet metal blank is held stationary on a die while a punch forces the other area to assume a new contour or shape. The force applied is sufficient to stress the metal beyond its elastic limit so that the change in shape is permanent. Forming has the distinct characteristic of stressing the metal at localized areas only; in the case of bending, for instance, this localized stress occurs only at the bend radius, resulting in a reduction of the thickness of the metal at the bend. It is only in these localized areas that any structured change occurs within the metal itself. This type of change in shape of the metal, with little structural change, is known as *metal movement* (Figure 5.22).

In drawing, however, total stretching of the metal occurs, with a correspondingly large amount of structural change within the metal itself. This structural change within the metal as a result of applied forces is known as *metal flow* (Figure 5.23). Many irregularly shaped panels are formed by drawing, but the simplest drawing operation, that of cupping, more easily illustrates the theory of drawing. During cupping, metal flows through an opening provided by a clearance between a punch and a die which is in a cup shape. The punch exerts a force on the bottom of the cup so that metal flows away from the bottom. Owing to the compressive forces on the outer edge of the blank, metal tends to flow into this region. These compressive stresses could cause wrinkles at the edge of the blank and cup, and to prevent this a blank holder is added around the punch. Pressure is applied to the blank holder by springs, air, or an outer ram of a press. If the blank holder pressure is too high, metal flow will be restricted and excessive stretching will cause the cup side walls to break. If the pressure is too low at the start of the drawing, wrinkles will occur before the pressure can build up. Thus the blank holder pressure must be low enough to allow the metal to move or flow underneath it and high enough to prevent wrinkling from occurring. This pressure cannot be reduced or wrinkles will occur, and therefore a lubricant must be applied to reduce the friction. Nearly all drawing operations require some lubricant for this reason.

5.13.1 Sheet metal drawing operations

Drawing operations are classified according to the shape of the part drawn, as follows: cupping, box drawing, panel drawing shallow and panel drawing deep.

Cupping

Cupping (Figure 5.24) is the drawing of parts having a circular or cylindrical shape. The cup may be perfectly cylindrical, or be composed of several cylinders of different diameters, or have tapered walls with spherical bottoms or flat bottoms. Changing the flat blank to the cup is called the *drawing* operation. Reducing the cup to a smaller diameter with greater height is called *redrawing*. If the cup is turned inside out during reduction, the operation is called *reverse redrawing*.

Box drawing

The drawing of square or rectangular shapes is called box drawing. Greater blank holder force is required in box drawing than in cupping to prevent wrinkling at the corners of the box. Box drawing

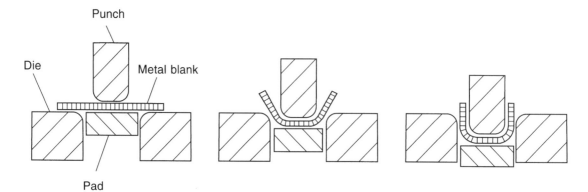

Figure 5.22 Metal movement during forming

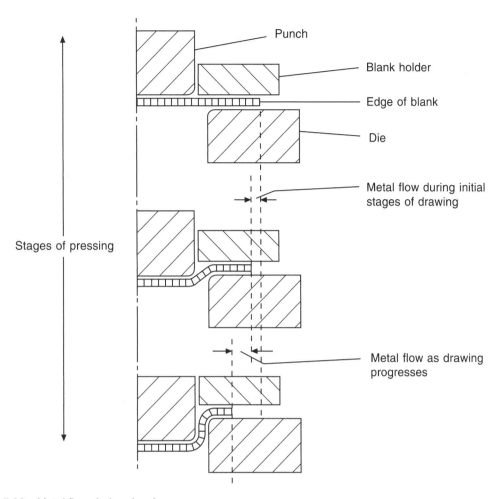

Figure 5.23   Metal flow during drawing

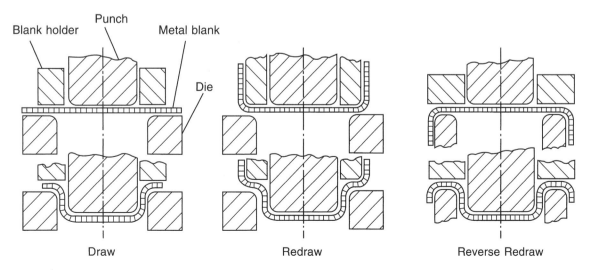

Figure 5.24   Drawing operations

consists basically of forming at the straight sections and drawing at the corners. Some flow of metal does, however, occur at the straight sections.

Panel drawing

Drawing of irregular-shaped parts is called panel drawing. The contour of the parts changes irregularly, causing a wide variation of stresses; for example, wrinkling or excess of metal may occur in one section while the metal tends to cause tears in other sections. It is very difficult to predict these stresses, as wrinkling is not limited to the flange area as in cupping, and tearing or failure is not limited to the bottom of the panel as in cupping. Hence panel drawing dies must go through a try-out period before production is possible. Adjustment in blank holding pressure, surface contour and pressure, draw radii and blank size are made during the try-out period to correct for variations in stresses from the predicted values.

If the panel height is great the operation is called *deep drawing;* low panel heights indicate *shallow drawing* (Figure 5.25). Shallow panels often incorporate a great deal of stretch forming of metal. Deep drawn panels are used for automobile front wings, rear wings, rear quarter panels and instrument panels. Shallow drawn panels are used for automobile roofs, doors, bonnets and boot lids.

5.14 Sheet metal cutting for press work

There are four types of sheet metal cutting for press work: shearing, cut-off, parting and blanking.

In *shearing* the cutting action must be along a straight line. Shearing does not use a press or die and is therefore not a stamping operation. It is used for the following purposes:

1 To cut strip or coiled stock into blanks
2 To cut strip or coiled stock into smaller strips to feed into a blanking or cut-off die
3 To trim large sheets, thereby squaring the edges of the sheets.

A *cut-off* is made by one or more single-line cuts. The line of cutting may be straight, angular, curved or of a broken design. It is important to note that the blanks must interlock perfectly before a cut-off operation is carried out so that no scrap is produced during cutting. The operation is performed on a press by a die and therefore may be classified as a stamping operation.

Parting is used when the blanks will not interlock perfectly. Some scrap is inevitable since the blanks do not fit together, and so parting is not as efficient as shearing or cut-off.

In *blanking* the cutting action cuts out the whole shape of the part in one operation. Excessive scrap is formed during blanking and therefore it is a very inefficient operation. The scrap is in the form of a ribbon commonly called the 'skeleton'. Blanking is performed in a press by a die and is a stamping operation.

Questions

1 Describe the forming and fabricating properties of the steels used in vehicle body manufacture.

2 With the aid of sketches, describe how a low-carbon steel can be shaped into four different forms to provide strength in body construction.

3 Which properties of a low-carbon steel allow it to absorb energy resulting from impact through collision?

Figure 5.25 Drawn parts (*Rover Group Ltd*)

4   Explain the following treatments that are associated with aluminium alloys: (a) solution treatment (b) age hardening (c) precipitation.

5   Explain the following heat treatments to carbon steel: (a) normalizing (b) hardening (c) tempering (d) case hardening.

6   What is meant by the term 'work hardened material'?

7   Explain the process of annealing.

8   State the important conditions to be observed when using a wheeling machine.

9   List three safety measures to be observed when using a treadle operated guillotine.

10   What is meant by the principle of shearing metals?

11   Describe the differences between a bending machine and a rolling machine.

12   Explain the operation of the swaging machine.

13   Explain the three methods by which heat can be transferred from one body to another.

14   Define the importance of the property 'ductility' in a material.

15   Explain the changes which affect metal when heated and then cooled.

16   Describe the condition of a metal panel which has been excessively wheeled.

17   After sheet metal has been work hardened, describe which process will allow further working of the sheet metal.

18   What is the minimum carbon content required for steel which will be hardened by heat treatment?

19   Describe a simple workshop method which can be used to estimate the temperature when tempering carbon steels.

20   When working with carbon steel, what is the difference between normalizing it and annealing it?

21   List the properties contained in commercially pure aluminium.

22   What practical factors limit the width of sheet metal which may be formed on a wheeling machine?

23   By means of a sketch, draw one type of swage which could be produced on a flat piece of sheet metal.

24   What is meant by the upper critical point of steel?

25   Explain the blanking operation.

Measuring and marking-out instruments

6.1 Marking out

A body repair worker must have a good general knowledge of engineering drawing and surface developments in order to mark out replacement parts on sheet metal, as this requires accurate calculation of sizes and angles. While the majority of body workers who work in the repair side of the industry do not have to perform a lot of difficult marking out from elaborate drawings on to sheet metals for the making of the articles, those employed in the manufacturing side of the industry (e.g. coachworks) have to mark out all units and assemblies from drawings, whether for a one-off item or for 'mass' production. Consequently a thorough knowledge is essential of measurement and marking tools, and also of how to check these tools so that a standard of accuracy can be achieved and maintained throughout.

Blueprints, as working or production drawings are generally termed, are photographic reproductions of an original drawing. They take their name from the colour of the finished print when the drawing is reproduced on ferroprussiate paper. This paper produces copies of the drawing in white lines on a blue background. Several other types of coloured prints are also produced by industrial contact photo equipment, but in spite of their colour these are still generally known as blueprints. The practical body worker is rarely called upon to produce a finished drawing, but he will find it necessary in many cases to understand and work from both general and detailed drawings as set out by a skilled draughtsman. The main object of reading a drawing is to obtain a clear mental picture of what another person has represented on paper by means of a conventional arrangement of lines and symbols. To this end the value of hand sketching cannot be too fully emphasized, as well as the making of simple models, for a much clearer impression of form can be obtained when movement and touch are combined with sight. The skilled craftsman, particularly in the small workshop or garage, is generally responsible for seeing his work through all stages of manufacture, commencing with careful study and accurate interpretation of the drawing. Before actual manufacture can begin, it is necessary to set out the desired shape by marking the outline on the surface of the sheet of material.

Automobile work demands a thorough knowledge of the methods of sheet metal work, and although the machine has displaced the man for certain operations there is an increasing demand for craftsmen who are capable of developing the large variety of work met in the industry. Much of the work done deals with double-curvature shapes which are in some cases fabricated by hand in sections. This type of work needs templates made to the exact curvature required so that the work can be checked at each stage of its progression. Also, very accurate alignment jigs have to be manufactured to hold these parts, after they are formed to the template size and shapes, so that they can be welded together into the finished component. Marked-out templates, which are usually in thin metal sheet, cut and then filed to a very accurate size, are used when one is working to rolling radii so that the curve can be checked as it is rolled. Angle templates are useful when bending in the bending machine to check the angle of bends so that they are constant.

6.1.1 Manufacture of templates

These can be made first of a paper or hard card known as template paper, then using materials such as thin-gauge aluminium, tin plate, and in some cases thin-gauge mild steel. These metals are used for their ease of cutting. The next stage in the manufacture of the template, if it is to be marked straight on to metal, is to take the sizes from the appropriate drawings and, using such instruments as scribes, dividers, straight edges, trammels and chalk lines where long sizes are involved, accurately to mark them directly on to the metal. When the template is fully marked out the profile is centre punched very carefully on the scribed lines at close intervals, and then cut, either by machine or by hand using tin snips, to the centre-punched marks. The metal is next filed very carefully to obtain the final finish and checked from the drawing for its dimensional accuracy. The template is now ready to be used either for checking or for marking out predetermined shapes directly on to sheets. Metal templates have the advantage that they can be kept and used again if necessary.

6.2 Basic marking-out and measuring instruments

6.2.1 Rules

A length may be expressed as the distance between two lines (line measurement) or as the distance between two faces (end measurement). The most common example of line measurement is the rule. Checking by way of comparing with a rule is called measuring. Measuring rules can be adjustable or non-adjustable. The adjustable type is usually of the folding variety, but there are also thin spring steel rules which are referred to as steel tapes. Rules are made of hardened and tempered high-grade steel and are usually graduated in the English and/or metric systems of measurement. Rules should not be misused as feeler gauges, scrapers, screwdrivers or levers; the end of the rule particularly should be preserved from wear because it usually forms the basis of one end of the dimension. In use, a rule should be held so that the graduation lines are as close as possible, preferably touching, the faces being measured. The eye which is observing the reading should be vertically above the mark being read; this avoids what is known as

'parallax', which results in an inaccurate reading due to the graduation lines being viewed from a very oblique angle. Rusting of the rule should be avoided by oiling or an occasional rub with metal polish. Chalk is good for cleaning the rule.

6.2.2 Scribers

A scriber is used when marking out work, and will leave a fine line on the metal surface (Figure 6.1). This will more easily be seen if the surface of the metal is first prepared by lightly rubbing it with white chalk or a special colouring agent. Scribers are made of hardened and tempered high-grade steel, with a knurled body which facilitates handling. They are ground at one or both ends to a fine point which should be kept sharp and protected when not in use (see Figure 6.2). Some of the double-ended scribers have points which can be

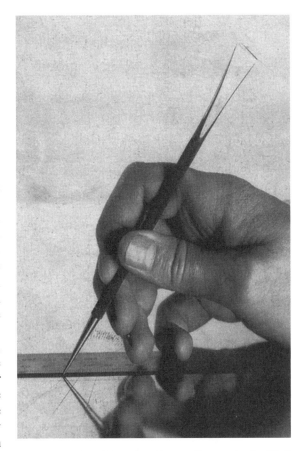

Figure 6.1 Marking out with a scriber and rule (*Neill Tools Ltd (Eclipse)*)

Figure 6.2 Engineer's scriber (double-ended) (*Neill Tools Ltd (Eclipse)*)

6.2.3 Straight edges

These are used to test that surfaces are level or that parts are in line one with another, and are ideal for lining up and jig work. There are two types of straight edge. One (Figure 6.3), made from good-quality hardened and tempered steel about 3.175 mm thick with one bevelled face, is made in lengths up to 2 m. The other type, made from cast iron, is ribbed along its length, and has a flat face. This type can be more accurate than the steel straight edge because its weight and robust construction reduce the possibility of flexing. It is usually made in lengths of straight edge up to 2.5 m. To test the surface contour of a metal panel or sheet, the straight edge is held against the panel and then moved over the surface, and any distortion or unevenness is shown as gaps between the edge of the straight edge and the edge of the article. Straight edges can also be used in sets of three for sighting parts into line. A straight edge should never be used for any purpose other than that for which it is intended.

Figure 6.3 Straight edges (with square or bevelled edges) (*Neill Tools Ltd (Eclipse)*)

replaced if damaged or worn. A handy pocket scriber is also available. When using the scriber, it is inadvisable to mark any lines other than cutting lines on coated metals, thus avoiding the destruction of the protective coating, as this may result in corroding of the parent metal. It is also inadvisable when working on soft metals, such as zinc, aluminium and copper, to scribe along the full length of fold or bend lines, because this could result in cracking during the bending of the metal. Short marks at each end are sufficient for bend lines.

6.2.4 Punches

Centre punch

This cast steel tool (Figure 6.4a) is driven into the metal with a hammer blow. For marking out work, locating centres for curves and defining profiles, the centre punch is struck only lightly. Sometimes lighter pattern punches, called prick or dot punches, are used for marking out. A centre 'pop' should be made as deeply as possible where a hole is to be drilled. It is as well to mark hole positions lightly at first, check them for accuracy, and then increase their depth by heavier hammer blows, holding the work on a firm foundation such as an anvil. Punches should be ground so that the grinding scratches lead to the point. The tapering point of the marking punch is usually ground to an angle of 40°, whereas the centre or pop punch has a more obtuse angle of 60° which facilitates the application of a drill.

Automatic centre punch

This is hand operated, and a hammer should not be used. It delivers a punch when hand pressure is applied to the knurled head. The depth of impression can be adjusted by turning the knurled head.

Nail punch

This is designed to drive nail heads below the surface of wood (Figure 6.4b). The square head prevents rolling and provides a larger face for striking with the hammer, while the cupped and bevelled point limits skidding off the nail head.

Parallel pin punch

This is used to punch out pins and dowels on mechanical assemblies (Figure 6.4c).

6.2.5 Try square (engineer's square)

This is the most common tool for testing squareness, and is used for internal and external testing to check whether work is square (Figure 6.5). It can also be used for setting out lines at right angles to an edge or surface. A try square consists of a stock and a blade, and these may be made separately and joined together, or the whole square may be formed from a single piece of metal. Squares are of cast steel, hardened and tempered and ground to great accuracy. The size of the square is the inside length of the blade. Engineers' squares are made with a groove cut in the stock where the blade enters the stock; this allows for any burr on the edge of the metal being tested.

Before a square can be used, one surface of the workpiece must be made level and true. This surface is termed the 'master' or 'face' side. To use a square to scribe a line at right angles to the face side, the blade of the square is laid flat on the workpiece with the inside of the stock pressed firmly against this face side; the line can then be scribed along the outer edge of the blade at the position required. To check that an adjacent surface is at right angles to this face side, the square is held upright on the workpiece on the corner and the inside edge of the stock is pressed firmly against the face side; any discrepancy in the angle will be indicated by a gap or gaps between the inside edge of the blade and the surface being tested. Engineer's squares are available in three grades of accuracy: B (workshop), A (inspection) and AA (reference).

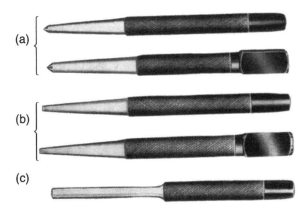

(a)

(b)

(c)

Figure 6.4 (a) Centre punches (b) nail punches (c) parallel pin punch (*Neill Tools Ltd (Eclipse)*)

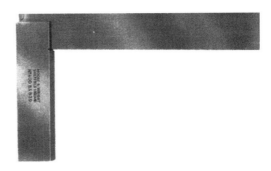

Figure 6.5 Try square (*Neill Tools Ltd (Moore and Wright)*)

6.2.6 Bevel gauge

When two surfaces are at any angle other than 90°, the angle between them can be tested with an instrument called a bevel. Bevels consist of a stock with one or two blades which are adjustable and can be locked at any angle by means of a thumbscrew. There are three main types. First is the plain engineers' bevel, which has an offset slot for testing angles which are not possible with a straight slot. Second is the universal bevel, which has offset blades and can be set at any angle. The thumbscrew on the bevel is recessed into the stock, allowing it, if necessary, to lie flat on the work. Third is the combination bevel, which consists of stock, split blade and auxiliary blade. The split blade swings to any angle and, with the auxiliary blade attached, can be transferred from one surface to another, work can be laid out or any desired angle checked and measured.

6.2.7 Bevel protractors

The alignment of planes or surfaces designed to form a definite angle may be checked by the universal bevel protractor. This is a graduated disc with a fixed adjustable blade and a stock with a straight-edge extension. The disc with the blade is graduated from 0° to 90° each way and rotates, together with its blade, through an entire circle in relation to the stock. For very accurate testing a vernier bevelled protractor must be used.

6.2.8 Adjustable protractors

When the edge of the workpiece forms an angle which deviates from the usual angles of 90° and 45°, this angle may be determined with the aid of an adjustable protractor. The protractor consists of a semicircular segment covering 180°. The movable blade which indicates the angle reading is secured by a locknut which can be set and tightened at any angle, thus allowing angles to be read directly from the workpiece. Such a protractor may also be used for marking out when the marking lines do not form one of the usual angles.

6.2.9 Combination square

This is used to test that work is true, to measure and check angles, and to locate the centres of round bar or centre lines of tubes. It consists of a blade (Figure 6.6) which is usually marked in English

Figure 6.6   Combination square (*Neill Tools Ltd (Moore and Wright)*)

and/or metric scales, and which may be used in conjunction with any one of three heads. The various heads enable the tool to be used as a square, a protractor or a gauge. A central groove along the length of the blade enables each of the heads to slide along this blade and to be locked in any position on the blade. The three heads used on this square are:

1   The square head, which has two surfaces, one at 90° and the other at 45° to the base, so that it can be used as a normal or a metre square, and also as a depth or height gauge. This head is usually fitted with a spirit level and a scriber, which is contained in the head.
2   The protractor head, which is used to measure or check any angle up to 360° by simply reading-off the scale.
3   The centre head, which is V-shaped with two internal surfaces. These form an angle of 90°, and very quickly find the centres of round objects.

The complete set forms a very useful workshop accessory and fulfils many needs, but when a high degree of accuracy is required a normal solid try square should be used instead of the square head of the combination set.

6.2.10 Centre square

This is a flat square made of hardened and tempered tool steel. It has two blades of different thicknesses which are designed for the accurate location of the centres of faces of round bar or discs. The centre square tool is used in conjunction with a scriber. It is held against the circumference of the round bar so that one blade touches the bar on its edge and the other blade lies across the surface, giving a diameter line which is then scribed. When two or more of these lines have been scribed the point of intersection is the centre of the bar.

6.2.11 Dividers

These are metal workers' compasses, made of steel with hardened points. They are used for scribing circles and arcs, and for marking off lengths. The most common type are spring dividers (Figure 6.7). These comprise a circular spring bow which joins the legs at the top and acts as a spring and pivot. The distance between the points is set by means of a fine adjustment screw mounted across the legs. In use, the centre point of the circle to be scribed is lightly marked and is used for the static position of the pivot leg, while the other leg is rotated to scribe the circle. The points of the divider must be kept finely ground and without making one leg shorter

than the other, and the points must be protected against damage when not in use.

6.2.12 Trammels (beam compasses)

Trammels (Figure 6.8) are used to describe arcs and circles of large radii and mark off lengths which are beyond the range of dividers. They comprise two heads fitted with scribers which slide along a bar. The bar can be any length. To use a trammel, one leg is placed at the centre point of the required arc or circle. The other leg is then set and locked at the correct distance by the aid of a locking screw fixed in the head. Light pressure is kept on the pivot leg with one hand while the other hand describes the arc with the other leg. Additional legs can be supplied with curved tips and can replace the scriber legs for use as calipers.

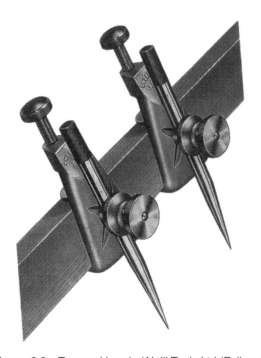

Figure 6.8   Trammel heads (*Neill Tools Ltd (Eclipse)*)

6.2.13 Calipers

These are used to measure or compare distances or size (Figure 6.9). There are two designs: inside calipers, which have straight legs than turn outwards towards the points and are used for measuring bores and internal spaces; and outside calipers, which have curved legs that turn inwards towards

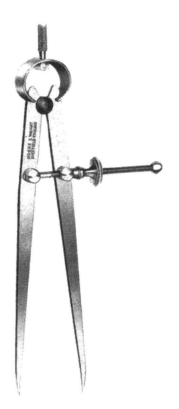

Figure 6.7   Spring dividers (*Neill Tools Ltd (Moore and Wright)*)

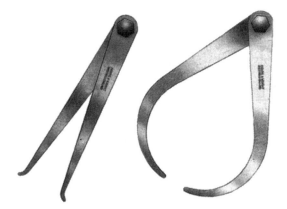

Figure 6.9 Calipers (*Neill Tools Ltd (Moore and Wright)*)

the points, and which are used for external diameters and other outside measurements. In their simplest form calipers are usually termed 'firm joint' calipers and consist of two legs riveted together at the top. Also available are the 'spring joint' type where the legs are joined at the top by a circular spring band and a fine adjustment screw is fitted across the legs. In use, the legs of the calipers are first opened out to the approximate size and held square with the object being measured. The points are next screwed towards each other so that they just touch both sides of the part to be measured. The calipers are then carefully withdrawn from the workpiece and the distance between the two points is measured.

6.2.14 Oddleg calipers

This measuring instrument is half caliper and half divider (Figure 6.10). It consists of two legs riveted together at the top, one of which is ground to a point while the other is turned inwards or outwards at its tip. In use, the curved tip is held against the edge of the workpiece and the line is scribed with the point of the other leg. These instruments are used for marking-out operations such as scribing parallel lines, following perimeters, and drawing intersecting arcs on cylindrical objects.

6.2.15 Gauges

Whereas measuring is the establishing of the actual size of a component, gauging is the positive comparison of the size of a component with a standard. Gauges are normally made from a hard

Figure 6.10 Oddleg calipers (*Neill Tools Ltd (Moore and Wright)*)

wear-resisting metal, and find extensive use in the mass production of car bodies where many components are made to standard sizes which require frequent checking to ascertain their accuracy. Handy 'limit gauges' are available for the routine testing of parts specified to be within certain limits, and come in various sizes and shapes such as plugs, rings, slotted plates and tapered plugs.

6.2.16 Depth gauges with reversible base or protractor

A depth gauge is designed to measure the depth of holes or recessed surfaces. It consists of a reversible head or a protractor base which sits on the top surface of the workpiece in use, and a graduated steel rule which can move vertically through the centre of the head. The rule can be locked in any desired position by a thumbscrew. When the head is placed on the workpiece the rule is lowered to the bottom of the recess, then locked in this position when the size or depth of the recess is indicated. Figure 6.11a and b illustrate the two types.

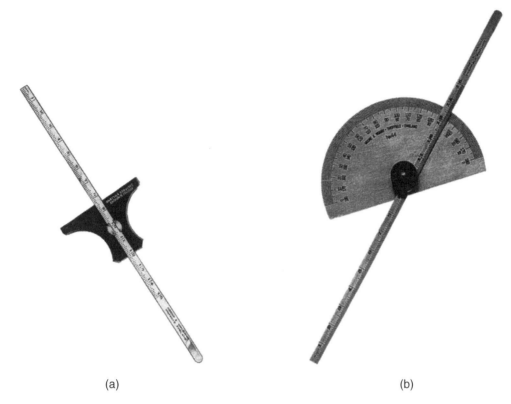

(a)   (b)

Figure 6.11   (a) Reversible depth gauge (b) protractor depth gauge (*Neill Tools Ltd (Moore and Wright)*)

6.2.17 Feeler gauges

A feeler gauge (Figure 6.12) is used to measure small clearances between two objects, and in some cases to establish the amount of wear or distortion of a component part. Each tempered steel blade has a number indicating its thickness in millimetres. The thicknesses are carefully graded in order that various clearances ranging from 0.01 mm to 1.0 mm can be measured. The gauge illustrated in Figure 6.12 is imperial but this tool is now mostly made in metric sizes.

6.2.18 Screw pitch gauge

A screw pitch gauge (Figure 6.13) consists of a metal case with two sets of flat pivoting blades, each having teeth cut on its edge to correspond to a screw thread form. In use, the blades are placed in turn on to the profile of the thread to be checked until an

Figure 6.12   Feeler gauges (*Neill Tools Ltd (Moore and Wright)*)

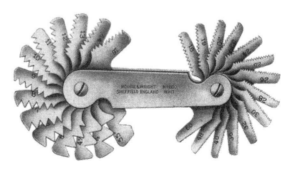

Figure 6.13   Screw pitch gauge (*Neill Tools Ltd (Moore and Wright)*)

exact match is found in the gauge. This matching thread form will give the correct name, size and pitch of the thread in question. The gauge illustrated in Figure 6.13 is imperial, but screw threads in common use are: Système International (SI) metric, International Organization for Standardization (ISO) metric, American National, Whitworth, and Unified.

6.2.19 Radius gauges

These are used to measure small internal and external radii. They are supplied as a set having a number of pivoting flat blades each of hardened and tempered steel, cut to a specific radius size. They are used as a template to check radii sizes on the workpiece (Figure 6.14).

Figure 6.14 Radius gauges (external and internal radii) (*Neill Tools Ltd (Moore and Wright)*)

6.2.20 Drill gauge

The drill gauge (Figure 6.15) is a flat piece of metal having a selection of holes of accurate diameters.

The size of a drill is found by inserting the drill in the holes until a hole is found which fits the drill exactly. These gauges are made in either fractional or decimal sizes, number sizes, or lettered sizes.

6.2.21 Imperial standard wire gauges

A wire gauge (Figure 6.16) can either be oblong or circular in shape. Slots cut into its outside edge give accurate wire gauge sizes, usually numbers 1 to 36 SWG or 0.2 to 10.0 mm. It is still used by the body worker and sheet metal worker, as metal sheet thickness may be given in SWG size. The gauge is used by slipping it over the edge of a metal sheet until the corresponding slot is found, and the gauge size is then read off.

6.3 Precision marking-out and measuring instruments

The average body worker will not normally come into contact with precision marking and measuring instruments. However, measurement is the basis of engineering, and an understanding of the more common precision instruments is essential if he is to realize that greater accuracy is possible than that which is achieved by the rule and scriber.

6.3.1 Marking-off table

This is made of close-grained cast iron, rigidly constructed and supported on very substantial legs. The top surface is machined level and finally scraped to a degree of flatness which is determined by the use to which the table is to be put.

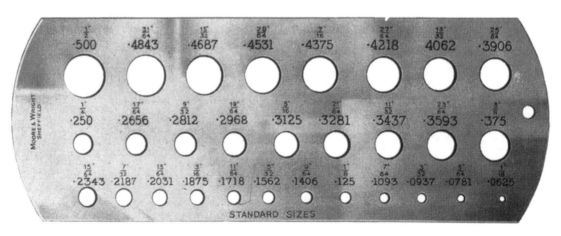

Figure 6.15 Drill gauge (*Neill Tools Ltd (Moore and Wright)*)

Figure 6.16 Wire gauges (metric) (*Neill Tools Ltd (Moore and Wright)*)

The poorest grade is used for marking out material appertaining to body work; it also supports the work during marking out, and is often used as a datum from which all measurements can be taken. Marking-off tables must be used for no other purpose except marking out and measuring. When not in use they should be lightly oiled and covered against accidental damage and corrosion.

6.3.2 Surface plate

A flat surface is one of the fundamentals of engineering, and the flatness of a surface can be verified by testing it against the flatness of a standard surface, i.e. the surface plate. Surface plates are similar in construction and grading to the standard of finish of a marking table, but are smaller in size and have two carrying handles. The back is strongly ribbed to give rigidity to the surface. In use, the surface plate is covered with a special marking compound and then lightly rubbed on the workpiece to be tested. The marking dye is transferred to the parts of the workpiece which actually come into contact with the surface of the plate, and if the surface of the workpiece is uneven, only the high spots will be marked. Any high spots must be scraped and the workpiece must be retested until the whole surface comes in contact with the surface plate. Surface plates should be kept covered at all times when not in use.

6.3.3 Surface gauge or scribing block

There are various types of surface gauges (Figure 6.17), but in the main they consist of a very accurately machined base, an adjustable pillar

Figure 6.17 Universal surface gauges (*Neill Tools Ltd (Eclipse)*)

fastened to the base, and an adjustable clamp attached to the pillar for holding the scriber. The surface gauge is used in conjunction with a marking-off table or surface plate to mark out very accurate, parallel lines to a true surface. The machined base enables it to slide over the surface of the table while the scriber retains its accurately set position. The scriber can be set at various heights and angles in order to mark out different jobs or parts of jobs with horizontal lines at a fixed height above the surface of the marking-off table. Scribed lines show up more readily on a work surface that is first covered with a solution of copper sulphate or marking compound.

6.3.4 Vee blocks

These blocks are made in matching pairs, and have to be used in conjunction with a marking-off table. They are made of cast iron with vees machined accurately to 90° into the top of the block to hold the work (Figure 6.18). Along each side of the block runs a groove into which a clamp can be fitted in order to secure round-shaped work in place during marking off with scriber, try square or scribing block. In order to make sure that you have a matching pair of vee blocks, before using them always check that each of the blocks is stamped with the same number.

6.3.5 Engineer's level

This is used for setting surfaces level and parallel to a marking-off table. The base is machined so that it can slide across the surface of the work-piece, and if the surface is level then a bubble in a tube across the top of the level lies at the centre of its scale. The base of the level has a concave groove along its length which allows it to lie on curved work. Some levels are shown in Figure 6.19.

Figure 6.18   Vee blocks (*Neill Tools Ltd (Eclipse)*)

6.3.6 Vernier caliper gauge

These calipers are made in a variety of types, some being used for external measurements only and others for both external and internal measurements. Vernier calipers consist of a fixed jaw attached to a main beam which is graduated, and another jaw which slides along the main beam. This sliding jaw is marked with the vernier scale which coincides with the main scale when the jaw is moved. The vernier scale is based on the difference between measurement made on two scales which normally have one division difference. By careful use of this adjustment a reading can be taken to 0.02 mm or 0.001 in (Figure 6.20).

Figure 6.19 Engineer's levels (*Neill Tools Ltd (Moore and Wright)*)

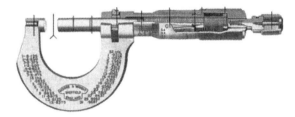

Figure 6.21 Micrometer (*Neill Tools Ltd (Moore and Wright)*)

1 Spindle and anvil faces
2 Spindle
3 Locknut
4 Sleeve
5 Main nut
6 Screw adjusting nut
7 Thimble adjusting nut
8 Ratchet
9 Thimble
10 Steel frame
11 Anvil end

Figure 6.20 Vernier caliper gauge (*Neill Tools Ltd (Moore and Wright)*)

6.3.7 Micrometer

The micrometer (Figures 6.21 and 6.22) is used to measure and record accurately to 0.01 mm, or when using a vernier micrometer to 0.001 mm. The size

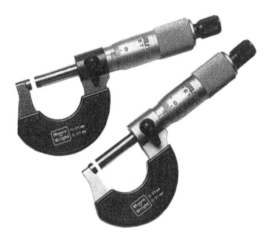

Figure 6.22 External micrometers (*Neill Tools Ltd (Moore and Wright)*)

of an object is measured by the contact principle, which requires it to be positioned between the anvil and the spindle head of the micrometer. The micrometer has a C-shaped steel frame. An anvil with a flat face is attached to one end of the frame and the other end accommodates the spindle and the barrel, which is graduated in linear measure. The bore of the barrel is screw threaded and engages the part of the spindle which is externally cut to the same thread. The outer end of the spindle has a ratchet stud and a thimble which is graduated around its bevelled circumference. When this thimble is turned it moves the spindle forward on to the object to be measured, and in so doing moves along the linear scale on the barrel. A locking ring is usually provided in the frame at the point where the spindle emerges from the barrel, so that the spindle can be locked in position after a measurement has been taken. An object to be measured is held squarely in position between the anvil and the spindle, then the spindle is closed on the object by turning the ratchet stud clockwise until it starts to slip, indicating that there is a predetermined pressure on the object. If there is no ratchet stud the thimble is turned just until resistance is felt to the turning; great care must be taken to avoid over-tightening. Next the spindle is locked in position by turning the locking ring and the reading is taken. The working principle of the micrometer is based on the distance a screw moves forward for each turn made. The screw thread has a lead of 0.5 mm while the thimble and barrel are graduated into divisions of 0.5 mm for the barrel and 0.01 mm for the thimble. Therefore, one revolution of the thimble moves it along the barrel for 0.5 mm. The thimble has 50 equal divisions and as one revolution of the thimble is 0.5 mm, then one of the thimble divisions will equal:

$$\frac{10.5}{50} = 0.01\,\text{mm}$$

There are a number of different types of micrometers made to serve a variety of uses in the field of precision engineering: internal micrometer, tube micrometer, screw-thread micrometer, adjustable micrometer, depth-gauge micrometer, tubular inside micrometer, bench micrometer and gear-teeth micrometer.

Questions

1 When would a body repairer need to make templates?

2 Describe the making of a template and the types of materials which could be used for this purpose.

3 Make a list of measuring and marking-out instruments which would be used in vehicle body work.

4 Explain the type of steel that would be suitable to manufacture a scriber.

5 Suggest an appropriate repair situation which would require the use of a long straight edge.

6 Explain the difference between a try square and a centre square.

7 Explain the uses of a set of trammels.

8 Illustrate the two types of calipers which are used in the workshop.

9 State the reasons why gauges are a necessary piece of equipment in engineering.

10 Make a sketch, and state the operation of, a micrometer.

11 Explain the term 'datums'. Why are they used when marking out?

12 With the aid of a sketch, illustrate the principal parts of an engineer's combination square.

13 Explain the use of the vee block when marking out.

14 What is the purpose of a marking-off table and a surface gauge?

15 Explain the operation of a vernier caliper gauge.

16 State the effects of parallax error when measuring with a steel rule.

17 Explain the purpose of a metre square when marking out.

18 Describe a method by which a surface can be checked to ensure that it is vertical.

19 With the aid of a sketch, show how to locate the centre between two points using dividers.

20 Explain the difference between a radius gauge and a drill gauge.

Methods of joining

7.1 Development of joining methods

In the manufacture of motor vehicles, the development of new and more effective fastenings is almost a branch of engineering in itself. However, the traditional nut and bolt is still in use, and for many applications it is difficult to imagine how it can be replaced. Indeed, there are instances of a return to nuts and bolts from newer methods of fastening; some British and many continental cars now have bolt-on wings or body panels where previously welding was employed or the structure was integral. The bolt has not changed much in design since it was first developed, apart from the use of special materials for certain applications and also greater standardization of threads. Nuts, however, are produced in a variety of special types; for example, high-tensile steel bolts having greater strength than hitherto use special shake-proof nuts which form a very efficient fastening.

Solid rivets have also been used since the early days of motor vehicle manufacture to form a permanent joint, but nowadays riveted joints are being replaced by welding.

In the field of mechanical fasteners, the modern motor vehicle uses special types of spring-clip fasteners in ever-increasing quantities. The most usual form of such fastenings is the simple spring clip made of tempered strip or wire; polythene, nylon and other plastics fastenings have been used. Another method is that of fastening panels together using adhesives; this is proving to be very successful in certain types of construction.

7.2 Solid rivets

The strength of a riveted joint depends on several factors: the material in which the rivets and plates to be riveted are made; whether the holes for the rivets are punched or drilled; and the workmanship involved in the riveting of the joint. In light sheet metal, rivets are usually applied in the cold state so that there is virtually no possibility of distortion. However, in heavier metal plate riveted joints often have to be comparable with the surrounding metal, and so the rivets are inserted in the hot state as this increases the strength of the joint.

Rivets used in hot riveting are made from steel and/or iron, while for cold working rivets made from copper, brass, aluminium and aluminium alloys are used. In body work, steel rivets are used on commercial chassis frames and also on fabricated sections such as subframes and structural load-bearing members. In addition aluminium rivets are used on alloy frame construction in the body building industry.

Rivets are classified according to the shape of the head, and the diameter and the length of the rivet. The shape of the rivet head is selected according to the intended use of the workpiece to be riveted. The diameter is selected according to the required strength and thickness of the component to be riveted. Most important, the length of the rivet must correspond to the total thickness of the components to be joined. The various types of rivets are shown in Figure 7.1, together with their British Standard proportions.

The snap or round head rivet is used where high strength is required. The pan head, similar to the snap head, is also used where strength is required. The mushroom is used in thin sheet metal which would be weakened by the use of countersunk rivets. Flat head rivets are used for riveting flat bar and angle sections to thin metal sheet as the head is fairly flush and does not obstruct. The countersunk head rivet is used when the surface of the component must remain smooth.

7.2.1 Rivet holes

Rivet holes can be punched or drilled. Punched holes are generally slightly conical, and for an efficient joint the work has to be arranged so that

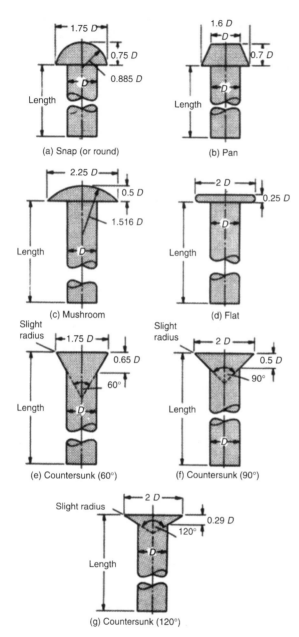

(a) Snap (or round)

(b) Pan

(c) Mushroom

(d) Flat

(e) Countersunk (60°)

(f) Countersunk (90°)

(g) Countersunk (120°)

Figure 7.1   Small rivet heads (BS 641)

the holes are brought together with their smaller ends adjacent. Punched holes also have ragged edges which must be smoothed, otherwise they will decrease the shearing resistance of the rivets. Drilling produces holes with smooth edges, and has the further advantage that it can be carried out with the plates in position so that there is no damage due to badly aligned holes. A joint formed with drilled holes is about 8 per cent stronger than a joint made with punched holes. Table 7.1 gives rivet and hole diameters.

7.2.2 Riveting dimensions

The *diameter* of a rivet is usually determined by $D = 1.2\sqrt{t}$, where D is the diameter of the rivet and t is the thickness of the plate.

The *allowance for riveting* is the allowance for the head of the rivet, and is the amount of rivet showing above the plate before riveting. The allowances are: snap head = 1.5D; countersunk head = 0.15D; flat head = 0.5D.

The *pitch* of a rivet is the spacing between the rivet centres. It can be calculated from the principle that the part of the plate between each pair of holes should be the same strength as one rivet. Therefore if the pitch is less than 3D (Figure 7.3) the plate between the rivets will be too weak. The pitch can then be extended to 8D in special cases, but above that the plate will tend to buckle.

The *width of lap* is the distance from the centre of the rivet to the edge of the plate. It is generally taken as 1.5D, so that a single-riveted lap joint would have an overlap of 3D (Figure 7.2) and a double lap joint an overlap of 5D.

7.2.3 Types of riveted joint

The simplest form of riveted joint is the single lap joint (Figure 7.2), in which the plates overlap a short distance with a single row of rivets along the

Table 7.1   Rivet sizes and hole diameters for solid rivets

Rivet diameter	(mm)	2.38	3.17	3.96	4.76	6.35	7.93	9.52	12.70
	(in)	0.094	0.125	0.156	0.187	0.25	0.312	0.375	0.5
Hole diameter	(mm)	2.43	3.25	4.03	4.85	6.52	8.02	9.80	13.10
	(in)	0.096	0.128	0.159	0.191	0.257	0.316	0.386	0.516

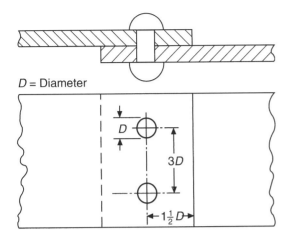

D = Diameter

Figure 7.2   Single-riveted lap joint

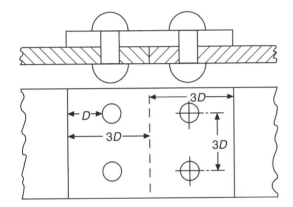

Figure 7.3   Butt joint

centre of the lap. The double-riveted lap joint has two rows of rivets arranged either in line or staggered alternately. The single-strap butt joint consists of two plates butted edge to edge with a strap covering the centre of the butt and riveted down each side (Figure 7.3). The double-strap butt joint is the same as the single-strap butt joint with the addition of a second strap on the opposite side.

A single-riveted joint has an efficiency of only about 55 per cent. In a double-riveted joint this is raised to about 70 per cent, whilst in a treble-riveted joint 80 per cent or greater efficiency may be attained.

7.2.4 Riveting procedure

Plates to be riveted should be clamped together with their rivet holes in alignment. If hot rivets are being used, they should be at forging temperature and the operation should be completed before they become 'black' hot. For snap head rivets, a punch or tool (snap) having a half-spherical cavity similar to the rivet head is used to support the rivet head, while the other-end of the rivet is riveted over by holding a similar punch to the rivet end and striking it with a hammer, thus forming a second cup head on the other end of the rivet. The aim in riveting is to swell the body of the rivet until it completely fills the holes, and to complete the process as quickly as possible while the rivet is still very hot so that there is maximum contraction of the rivets after riveting to pull the plates together. Pneumatic or air operated hammers are used extensively for closing over rivet heads. They are designed to deliver the right weight of blow at the correct speed to form the rivet head speedily and accurately. For cold riveting the closure of the rivet is similar to that of a hot rivet except that the metal is not as plastic and greater difficulty will be encountered in swelling the rivet shank to fill the hole. Cold riveting plates cannot be as tight as with hot riveted because there is no contraction of the rivets to pull them together.

Care should be taken that the rivet head is spread evenly in all directions and not bent over in one direction only. This is often helped by giving the rivet a few preliminary blows with the ball end of the hammer, thus spreading the rivet a little before using the riveting tool (Figure 7.4). Countersunk rivets should be supported with a flat-headed punch. The initial spreading is done with the ball end of the hammer and the head is finished with the flat end of the hammer. When riveted joints are large it is advisable not to start at one end of the workpiece but to rivet the extremities first and next the centre, then the rest of the joint. This eliminates creeping of the plates and misalignment of the holes.

7.3 Bifurcated, tubular and semitubular rivets

These rivets have the outstanding advantage that there is no swelling of the solid portion of the shank during the closure operation. They are used extensively for the joining of soft materials such as plastic, rubber, leather and/or brake and clutch linings to metal. Only a small percentage is used in the body building industry.

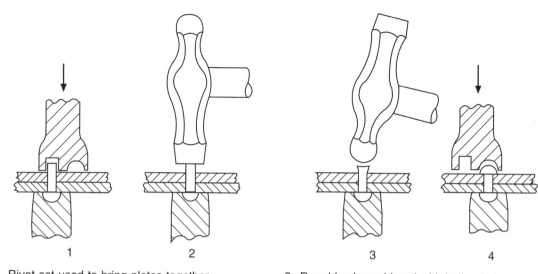

1 Rivet set used to bring plates together
2 Flat pein hammer used to spread rivet and fill up
3 Roughly shaped head with ball-pein hammer
4 Finish head to shape with rivet snap

Figure 7.4 Riveting procedure

7.4 Blind rivets

The blind rivet was originally developed in the 1930s for use in the aircraft industry, but since then it has been adopted in the light engineering industries. It is now employed as a standard sheet metal fastener in cars, buses and all forms of commercial vehicles. Blind rivets, as their name implies are rivets which can be set when access is limited to only one side of a structure. They are invaluable both in car construction work, where many of the panel assemblies are of the double-skinned type so that accessibility would be impossible for conventional riveting methods, and also in body repair work as they eliminate any unnecessary stripping of interiors. Blind rivets are manufactured by several companies under brand-names, the most famous of which is POP. POP is the registered trademark of Tucker Fasteners Limited who were the pioneers of blind riveting.

7.4.1 Blind rivet types

POP rivets

The POP rivet comprises a hollow rivet assembled to a headed steel mandrel or stem (Figure 7.5). It is inserted into a predrilled hole of the correct size in the workpiece, and a special tool containing a gripping device is applied to the mandrel or stem of the rivet. When the tool is operated, either manually or automatically, the mandrel head is drawn into the hollow rivet, expanding the end of the rivet which is on the blind side of the structure and at the same time pulling the material together. When and only when, a tight joint has been formed, the mandrel breaks at a predetermined position, so that the mandrel head is left as a plug in the bore of the rivet (Figure 7.6). The spent portion of the mandrel is then ejected from the tool. POP rivets are manufactured from aluminium alloys, Monel, mild steel, copper and stainless steel. They are supplied with domed or countersunk (90°, 100° or 120°) heads in diameters of 2.4 mm, 3.2 mm, 4.0 mm, 4.8 mm and 6.4 mm, for riveting thicknesses up to 12.7 mm. Shear strengths range from 400 N for 2.4 mm diameter aluminium alloy rivets, to 5400 N for 6.4 mm diameter Monel rivets. Owing to the high clenching action of the rivet, tensile strengths are in excess of these figures.

Seven types of POP blind rivet are shown in Figure 7.7. These and other types are described as follows:

Standard open rivet is subdivided into the break-head mandrel type, which leaves the mandrel head

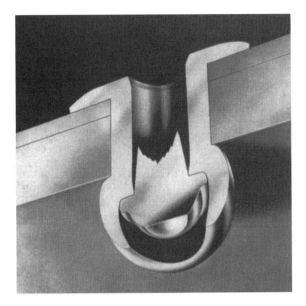

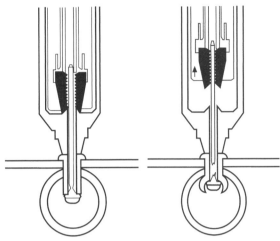

Figure 7.6 POP blind rivet, standard open type: setting sequence (*Tucker Fasteners Ltd*)

Figure 7.5 POP blind rivet: standard open type (*Tucker Fasteners Ltd*)

free to fall away from the rivet (Figure 7.6), and the break-stem mandrel type, which retains the broken portion of the mandrel within the set rivet, thus sealing the rivet to a certain extent. The latter type is intended for use in all normal blind riveting situations where the materials to be fastened do not present structural problems. This is a hollow rivet, preassembled on to a headed pin or mandrel. The mandrel is designed to fracture at a predetermined point during the setting operation, when the materials to be fastened have been drawn closely together and the joint is tight. *Sealed rivet* consists of a tubular rivet with a sealed end containing a steel or stainless steel mandrel. The riveting sequence is similar to that of a POP rivet, but this

type has the advantage that in setting the rivet it is both compressed beyond its elastic limit and expands radially, thus ensuring a joint which is airtight and watertight up to 34 bar. The rivet is available with two alternative mandrels: one, called 'short break' (Figures 7.8 and 7.9), fractures immediately under the mandrel head which is retained in the blind end of the set rivet; and the second, known as the 'long break' type, fractures at a point outside the rivet and can be finished off to result in a flush finish. The sealed rivet is a fastener of high shear and tensile strength and vibration resistance. Owing to its high rate of expansion in setting, it cannot be recommended for use in very soft or brittle materials. It is designed for use where the fastening to be used has to be pressure or water tight.

LSR rivet This range of aluminium rivets is designed to offer two particular advantages in joining soft friable or brittle materials. Its controlled setting gives outstanding reliability and it produces a neat uniform appearance. The LSR is suited to plastics, wood, GRP laminates and thin gauge materials. It has an important role to play in such applications as the assembly of caravans and trailers.

MGR rivet has been designed for use in situations where hole sizes are inconsistent. It also offers a multi-grip facility. The rivet is available in aluminium 2.5 per cent magnesium alloy with a carbon steel mandrel. MGR is valuable where components are supplied to the user with holes already punched or drilled, or where the sheets to

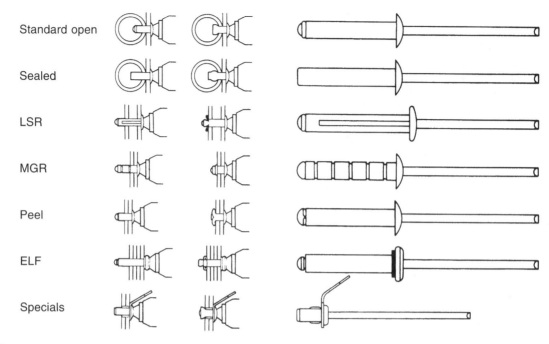

Standard open

Sealed

LSR

MGR

Peel

ELF

Specials

Figure 7.7 Types of POP blind rivet (*Tucker Fasteners Ltd*)

be joined are folded or curved, making it difficult to match up two holes of the correct size.

Grooved rivet has been developed for use in thick sections of soft or brittle materials such as hard board, plywood, glass fibre, asbestos board, con-crete and brick. The POP grooved rivets gain their name from the series of grooves around the shank which engage into the workpiece on setting and set inside the material rather than against the rear face. Construction and setting action is similar to the standard open type. The body is of aluminium alloy with a carbon steel mandrel. When set, this rivet is capable of withstanding high pullout loads (see Figures 7.10 and 7.11).

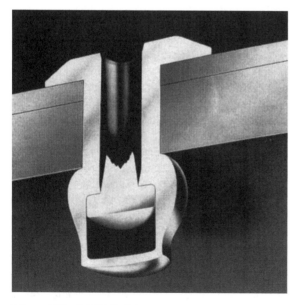

Figure 7.8 POP blind rivet: sealed type (*Tucker Fasteners Ltd*)

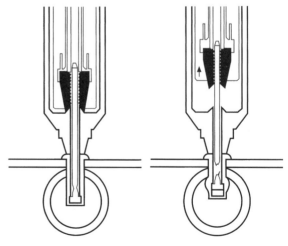

Figure 7.9 POP blind rivet, sealed type: setting sequence (*Tucker Fasteners Ltd*)

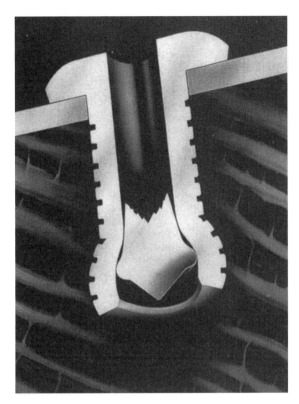

Figure 7.10 POP blind rivet: grooved type *(Tucker Fasteners Ltd)*

Peel rivets are specifically developed for fastening soft or friable materials. They will secure blow-moulded or glass-reinforced plastic, rubber and ply-wood to metal panels or sections up to 13.5 mm thick. The rivet has an aluminium alloy body and a special carbon steel mandrel. On setting, the rivet body is split into four petals by the action of the mandrel head, producing a large blind-side bearing area capable of withstanding high pullout load (see Figures 7.12 and 7.13).

ELF rivet is an aluminium 3.5 per cent magnesium alloy with an aluminium alloy mandrel. The body of this rivet splits and folds on setting to form three neat leaves which spread clamp-up load over a wide area. It is a totally sealed rivet designed for use on roofing but is suitable for materials like composite board, GRP, hard rubber and laminates, and is used in commercial vehicle construction and caravans.

Special POP rivets are shown in Figure 7.14. Earth terminal rivets are designed to provide an effective

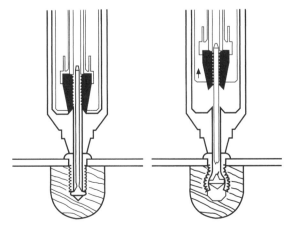

Figure 7.11 POP blind rivet, grooved type: setting sequence (*Tucker Fasteners Ltd*)

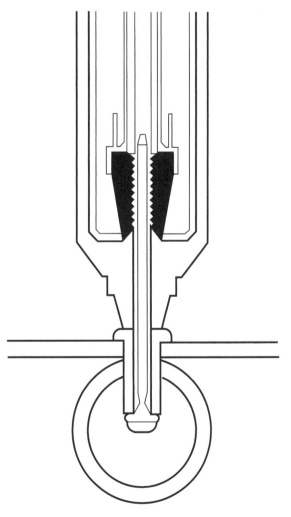

Figure 7.12 POP blind rivet, peel type: before setting (*Tucker Fasteners Ltd*)

Figure 7.13 POP blind rivet, peel type: after setting (*Tucker Fasteners Ltd*)

earth connection to pre-painted sheets without damaging the finish. Tab rivets are designed to perform two or more functions, such as fastening the

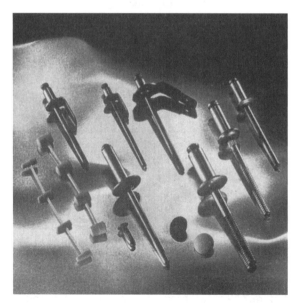

Figure 7.14 Special POP rivets (*Tucker Fasteners Ltd*)

tops of rechargeable batteries and at the same time providing a tab to serve as a connector. Tamper-proof rivets are set in the workpiece with a standard tool; a stainless steel pin is then tapped into the bore of the rivet and its domed head locates in a recess in the rivet head, making the fastening virtually impossible to remove without obvious traces. T-rivets are designed for exceptional strength and versatility; they can be reliable and vibration-proof in holes as much as 0.8 mm oversize.

Blind rivet nut is a threaded insert system, and is a remarkably simple fastening system for use in all kinds of assemblies. It provides a straight, reliable threaded insert to which other components can be attached. It can be used as a blind rivet for permanently fastening one or more panels or sections together, and provides an anchorage with the holding strength of at least six full thread turns. It can be installed at any stage without damaging the work-piece finish, even after painting. It can be used in materials with thicknesses varying between 0.25 and 7.5 mm. There are a wide range of sizes available in steel, aluminium or brass with thread sizes from M4 to M10. (M is the metric screw thread designation, and when followed by a number, e.g. M4 is a metric thread of 4 mm diameter.) Sealed or open types are available with flat or countersunk heads. They can be used in drilled or punched holes with normal tolerances (Figure 7.15).

Well nut This demountable blind screw anchor is designed to provide a vibration-resistant fixing for engineering and automotive structures. It consists of a captive brass nut in a resilient neoprene bush which can be installed from one side of the workpiece using only a screw and a screwdriver (Figure 7.16). It is suitable for materials from light gauge metal panel, too thin for self-tapping or set screws, to plastic sheet up to 16.5 mm thick. The flanged ends prevent electrolytic corrosion at metal-to-metal assembly points.

Repetition rivets

Repetition riveting systems are widely used on vehicle body construction, especially commercial vehicles. Some examples are as follows:

Chobert riveting system operates on an entirely different principle from other blind riveting systems (Figure 7.17). Instead of a break-stem type of mandrel, there is a mandrel with a tapered hardened

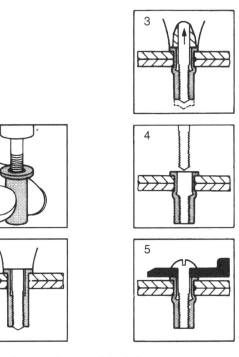

Figure 7.15   Blind rivet nut, installation sequence

1   POP nut screwed on to tool
2   POP nut inserted into hole
3   Tool operated: mandrel retracts into tool, and
    unearthed part of nut expands on blind side of
    workpiece
4   Tool mandrel unscrewed
5   Assembly completed

steel head larger in diameter than the tapered bore of the rivet. The mandrel is drawn through the rivet towards the head, and is thus part of the placing tool rather than the component. It is possible to extend the mandrel length to accommodate a charge of up to 100 rivets and to achieve a rate of placing up to 2000 per hour. The Chobert rivet is made in a wide range of materials and sizes. Head forms currently available include snap, mushroom, flat and countersunk, and sizes range from 2.38 mm to 6.35 mm diameter. Materials include a wide range of aluminium alloys and zinc-plated steel. They are supplied in tube loaded form, each 317.5 mm long and containing up to 100 carefully aligned rivets. The tubes can be used in special tools which can be either manually or pneumatically operated. If additional strength or water tightness is required the rivets can be pinned, giving in effect a solid rivet.

Figure 7.16   Well nut (*Tucker Fasteners Ltd*)

Chobert Grovit system   has been developed so that the widest range of materials (such as wood, plastic, aluminium) may be joined. The Grovit has grooves on its shank which bite firmly into the material when

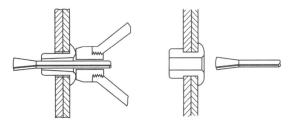

Figure 7.17   The Chobert hollow rivet (*Avdel Ltd*)

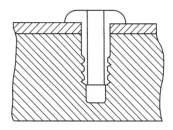

Figure 7.18 The Chobert grooved rivet (*Avdel Ltd*)

the rivet is expanded (Figure 7.18). The rivets can also be used in the high-speed Chobert system. These types of riveting systems are ideal for mass production and are invaluable in the body building industry where speed and economy are essential (Figure 7.19).

Briv system is a high-speed, high-clench riveting system for use with a wide range of assembly materials. Briv is installed from one side of the workpiece with the use of a mandrel loaded into the Avdel power tool. The action of the mandrel draws through the rivet, expanding the shank to fill the hole. The applications range from body component assembly to fixing electronic control panels (see Figure 7.20).

7.4.2 Blind riveting tools

The majority of blind riveting tools have been designed on the assumption that the operator will be moving around a stationary structure, as in body building. However, some stand tools are available.

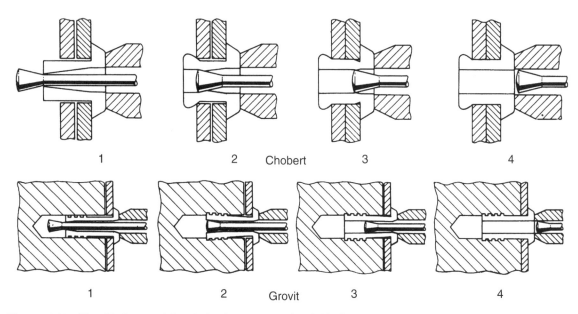

Figure 7.19 The Chobert and Grovit riveting systems (*Avdel Ltd*)

1 Place Chobert rivet or Grovit in prepared hole
2 Draw steel mandrel, which has opposite taper to rivet, through from tail end of rivet, expanding rivet tail around rear side of hole, to form shoulder. (For the Grovit, annual convolutions are forced into materials.)
3 Continue to pull the mandrel through rivet, which symmetrically expands shank to fill hole
4 The rivet has good bearing in hole and a parallel bore is left in the rivet

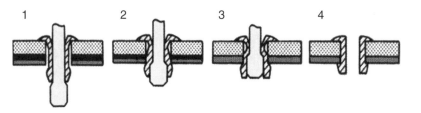

Figure 7.20 Briv placing sequence (*Avdel Ltd*)

1 Magazine loaded rivet is placed into hole
2 Tool draws mandrel through rivet, compressing rivet
3 Mandrel expands shank of rivet to fill hole and clamp materials
4 Completed rivet. Tool reloads automatically

The range of tools has been designed to cater for all riveting conditions, whether they are used on high-speed assembly production lines or for occasional use. The basic types of tools consist of:

1 Manually operated plier types
2 Manually operated lazy tongs
3 Portable pneumatically operated guns
4 Portable pneumatic hydraulic guns
5 Pneumatically operated stand tools.

In addition various heads are available for attaching to either manual or power operated tools which can gain access to restricted places. In order to change the rivet diameter or type of rivet being used, it is necessary merely to change the nose-piece or the gripping jaws which hold the mandrels. It must be remembered that certain special rivets and fasteners can only be used by their appropriate setting tool.

Some examples of riveting tools are given in Figures 7.21–7.24.

Figure 7.22 Hand operated riveting tool, lever type (*Tucker Fasteners Ltd*)

Figure 7.23 Hand operated riveting tool, lazy tongs type (*Tucker Fasteners Ltd*)

Figure 7.21 Hand operated riveting tool, plier type (*Tucker Fasteners Ltd*)

7.4.3 Rivet selection

The following factors must be considered when selecting the type of rivet to be used on a particular job:

Rivet diameter In some cases this may be decided by the size of an existing hole. When designing new work the main factor is usually the shear or tensile strength required from the riveted joint. In load-bearing joints the diameter of the rivet should be at least equal to the thickness of the thickest sheet but should not be greater than three times the thickness of the sheet.

Hole size Drill sizes for each diameter of rivet are usually specified by the manufacturers. These are usually designed to give a clearance of 0.05 mm to 0.13 mm between the rivet and the hole, and in the case of rivets capable of only limited radial expansion these recommendations must be followed if maximum efficiency is to be achieved.

Rivet length Manufacturers also specify the correct rivet length to be used on a given thickness of material. It is often advantageous to use rivets which are longer than those recommended as they will set satisfactorily on thinner materials, and the longer length, when expanded, increases the grip of the rivet, but the higher cost for longer rivets should be taken into account before doing so.

Pitch of rivets In load-bearing joints the distance between rivets in the same row should not exceed six times the rivet diameter. Even when the joint is not load bearing, the rivet pitch should not exceed

Figure 7.24 Automatic riveting system (*Tucker Fasteners Ltd*)

twenty-four times the thickness of the thinnest sheet in the joint.

Edge distance In lap or butt joints which are likely to be subjected to shear or tensile loads rivet holes should not be drilled within a distance equal to two rivet diameters from the edge of the sheet, but should not exceed twenty-four rivet diameters.

Rivet material The choice of rivet material will normally be related to the strength required from the riveted joint. Usual rivet materials are steel, aluminium alloy, copper and Monel. Other factors affecting the choice are weight, high temperatures and corrosion resistance, especially electrolytic corrosion which can occur when different metals are joined together.

Mandrel type The break-stem mandrel is usually chosen where the rivet is to act as a waterproof plug, and also where it would be inconvenient to eject the mandrel head on the blind side of the enclosed structure. The break-head mandrel is used where weight is a factor, and where a comparatively clear hole is required through the rivet set.

7.5 Structural fasteners

Figure 7.25 shows a variety of structural fasteners used on a vehicle body.

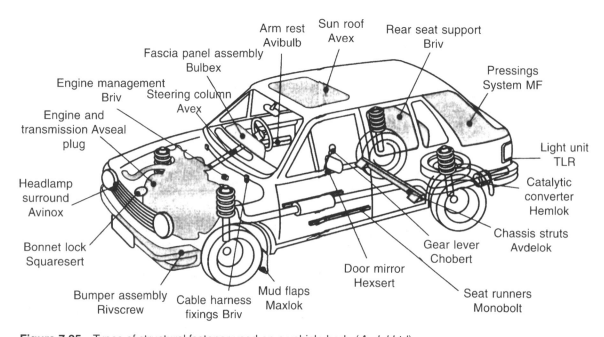

Figure 7.25 Types of structural fastener used on a vehicle body (*Avdel Ltd*)

7.5.1 Avdelok system

This is not a blind fastener, but has a number of advantages over the conventional rivet or nut and bolt which it replaces. The Avdelok is a two-piece high-strength fastener consisting of a bolt and collar made from carbon steel, stainless steel and aluminium alloy (Figure 7.26). It gives a high clench action and a positively locked nut or collar which is proof against vibrations. It can be placed very simply by manual or pneumatic tools. The Avdelok bolt is placed through the prepared hole and the collar is slipped on to the other side of the joint (Figure 7.27). The nose of the tool is then pushed over the tail of the bolt and the trigger is pulled. This draws the bolt tight, clamping the sheets together. The pull continues until the nose of the tool is drawn over the collar, compressing the collar until it is swaged into the grooves of the pin. The tail of the bolt breaks off at the deep breaker groove, flush with the collar. No finishing is needed. This system is employed in such applications as chassis cross-bearer brackets on commercial vehicle bodies.

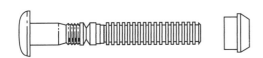

Figure 7.26   The Avdelok fastener (*Avdel Ltd*)

7.5.2 Hemlok system

The Hemlok is a blind fastener which comprises a steel stem and shell. Both components are zinc plated and gold passivated to ensure good corrosion resistance. The fastener is supplied in 6.4 mm diameter in various lengths. It is installed from one side of the workpiece, thus making it a blind fastener. It is designed with a chamfer at both ends of the stem; this facilitates easy insertion into the placing tip of the installation equipment, and into the hole in the workpiece. The installed fastener breaks flush with or below the surface of the low-profile Hemlok head, leaving a clean finish and appearance. Hemlok provides strength and good clamp-up, and the well formed fastener tail achieves this without deforming thin sheet material. This fastener is designed to be used on thin-gauge materials for high-strength joints suited to automotive work (see Figures 7.28 and 7.29).

7.5.3 Threaded inserts

Steel threaded inserts (such as Nutsert) of high strength (Figure 7.30) can be placed by any of three specially designed tools. The Nutsert is placed on the threaded drive screw of the placing tool and inserted into a prepared hole in the workpiece. The drive screw turns clockwise drawing the tapered nose portion of the Nutsert back into its outer shell and thus causing the shell to expand into the preformed hole. The drive screw is then withdrawn, leaving the Nutsert permanently and tightly placed in the hole and ready to receive either a screw or a bolt. It can be used on all types of metals and also plastic and timber, and is made in a wide range of thread sizes, both imperial and metric (Figure 7.31).

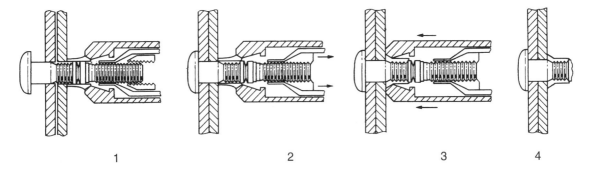

Figure 7.27   The Avdelok system (*Avdel Ltd*)

1   Insert Avdelok pin through predrilled hole and slip collar over pin tail
2   Place tool nose over pin tail and press trigger
3   Swage collar into locking grooves of pin
4   Tail of pin breaks flush with collar and is ejected

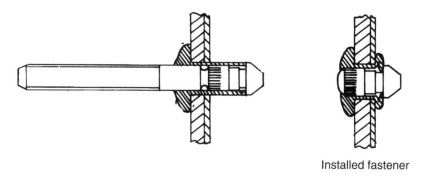

Installed fastener

Figure 7.28   Hemlok fastener (*Avdel Ltd*)

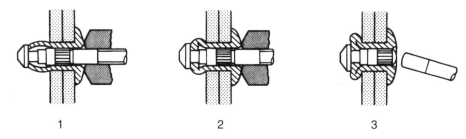

1   2   3

Figure 7.29   Hemlok placing sequence (*Avdel Ltd*)

1   Hemlok inserted into power tool and workpiece
2   Tool operated: Hemlok tail deforms over material

3   Tail fully formed: fastener stem breaks off

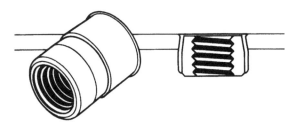

Figure 7.30   Standard Nutsert (*Avdel Ltd*)

7.5.4 Monobolt system

Monobolt is a precision engineered two-part component. It is a high-performance flush-break fastener designed to build in strength, security and quality. It consists of a body and a stem and is supplied as a one-piece assembly. A sealing on the Monobolt stem provides exceptional resistance to moisture, an essential requirement for many applications in vehicle body work. The fasteners are available in aluminium alloy, carbon steel and stainless steel, in diameters of 4.8 mm and 6.35 mm. Fast and simple to install, Monobolt is a blind fastener, ideal for original structural assembly and to replace conventional fastening methods, particularly when access to the workpiece is difficult. A major feature of the Monobolt construction is the visible locking element which allows easy, visible inspection after placement (see Figure 7.32).

7.5.5 Avtainer system

This system has been designed for the joining of composite panels of plywood and glass-reinforced plastic to metal framing. This fastener has the ability to join firmly, but without cracking or pulling the bolt right through the composite. It is a two-piece fastener consisting of a zinc-plated carbon steel pin with a nylon seal and a collar which is pulled tight from the other side of the workpiece, thus causing no damage to the material during installation. One of its main applications is in the building of commercial vehicles (see Figure 7.33).

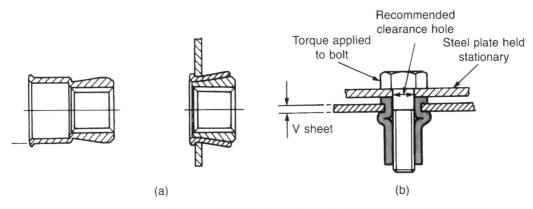

Figure 7.31   (a) Standard Nutsert placement (b) Thin sheet Nutsert with bolt in place (*Avdel Ltd*)

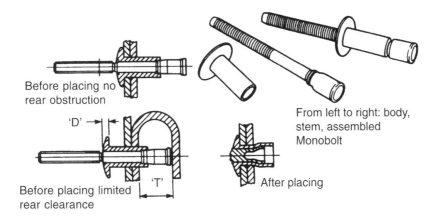

Figure 7.32   Monobolt and placing sequence (*Avdel Ltd*)

7.6 Screws and bolts

7.6.1 Wood screws

Joining by wood screws is an essential part of the construction of vehicle bodies which incorporate timber. Screws are available in steel, brass and bronze, with a variety of alternative finishes. Sizes range from no. 2 (1.6 mm $\frac{1}{16}$ in) to no. 8 (7.9 mm, $\frac{5}{16}$ in), and lengths vary from 6.35 mm, $\frac{1}{4}$ in, up to 101.6 mm, 4 in and longer in special cases. The head types include the standard countersunk (used in most timber fixings), round head (sometimes used when fixing metal fittings to wood), and raised head (used when fixing mouldings where the raised head gives a decorative effect) (Figure 7.34). A new development is the Supadriv recess head, specially engineered to give a 'cling-fit' between driver and screw; it is increasingly used in robotic assemblies. In body shops wood screws are applied by hand or with special power operated screwdrivers which are either electrically or pneumatically operated.

7.6.2 Metal screws (self-tapping screws)

Often there is a need for a fastening which will fix two or more parts together securely for indefinite periods, yet can be dismantled if necessary. Since the Second World War, the self-tapping screw has in many cases taken the place of more traditional fastenings and has justified itself completely on all types of motor vehicles. It is a very hard steel screw designed for joining metal to metal and is used for joining metal pressings and exterior fittings. The screw is inserted into a predrilled hole and it cuts its own thread in the metal, thus making a very secure joint. When joining thin sheets, improved characteristics can be obtained if the hole is plunged or extruded (Figure 7.35). This in effect thickens the material and gives greater

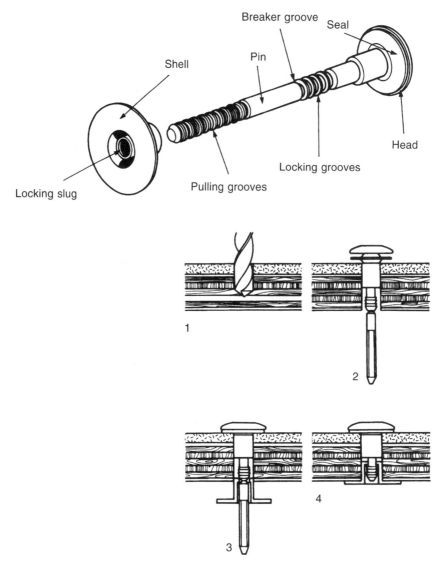

Figure 7.33 Avtainer and placing sequence (*Avdel Ltd*)

1 Drilled hole   3 Shell placed over pin tail
2 Avtainer in position   4 Tool applied: fastener completed. Seal protects against ingress of moisture

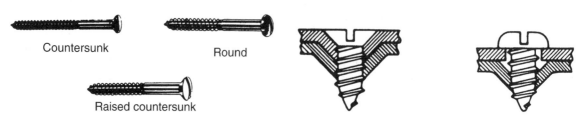

Figure 7.34 Unhardened wood screws
(*European Industrial Services Ltd*)

Figure 7.35 Fixing self-tapping screws
(*European Industrial Services Ltd*)

thread engagement; thus the best conditions are obtained if both sheets are extruded together. If this device is used, the total sheet thickness should be used in assessing the required hole size. Pilot holes may be punched, drilled, rim extruded or moulded when using selftapping screws.

Screws are available in stainless steel and in steel with various finishes, and gauges range from no. 4 to no. 14 and lengths from 6.35 mm, $\frac{1}{4}$ in, to 63.5 mm, $2\frac{1}{2}$ in. The types of heads available are counter-sunk, round head, raised countersunk, pan head, mushroom head and hexagon head. Most heads have a single slot for insertion of a screwdriver; some, known as Supadriv heads, have a star-type slot and must be placed with a special type of screwdriver (Figure 7.36).

Self-tapping screws are particularly useful to the body engineer and trimmer since no access is needed to the rear side of the work, it being necessary simply to drill the requisite size of hole and drive the screw home. The screw will stay secure until removal is necessary, and can be removed and replaced many times without becoming slack if reasonable care is taken. Typical applications are the mounting of door-pillar trim pads, carpeting or rubber matting, parcel shelves, glove boxes, window cappings, ashtrays and kick plates.

A more recent development is the Taptite high performance thread forming screw which has a thread like that of a conventional machine screw (Figures 7.37, and 7.38). The thread on the Taptite screw has a trilobal cross-section instead of the customary circular form. As the screw is rotated the lobe profile rolls the material out of its path to form

Figure 7.37 Supadriv Taptite thread forming screw (*European Industrial Services Ltd*)

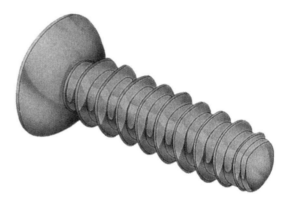

Figure 7.38 Plastite screw (*European Industrial Services Ltd*)

the thread. Not only does this eliminate the need for an additional tapping operation, but it also produces inherent resistance to vibrational loosening, thereby obviating the need for separate locking components. This is achieved by an increase in prevailing torque which comes from the radial pressure interference produced at and around the major diameter and not present on conventional fastenings because of the need for a clearance fit. These factors mean that in the assembly shop the operators (and drivers) can insert more screws in less time and with less fatigue. In maintenance applications, a standard machine screw can be substituted if necessary.

The AB screw combines the benefits of Taptite in a heavy-duty thread forming fastener and is available in sizes M6 and above (see Figure 7.39). Another version of the Taptite screw is CA Taptite (see Figure 7.40), designed in response to the increasing use of thinner-gauge materials. Fastening thin sheet poses problems with conventional self-tapping screws, and Taptite offers two solutions. Either a standard Taptite screw can be driven into holes which have been rim extruded to provide more

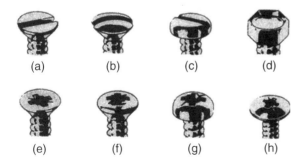

Figure 7.36 Self-tapping screw head types: slotted and (a) countersunk (b) raised countersunk (c) pan; (d) hexagon; Supadriv and (e) countersunk (f) raised countersunk (g) pan (h) flange (*European Industrial Services Ltd*)

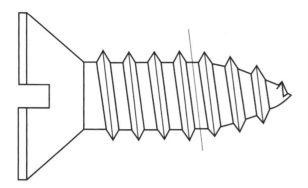

Figure 7.39   AB type self-tapping screw (*European Industrial Services Ltd*)

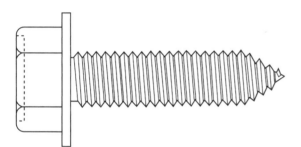

Figure 7.40   CA Taptite screw (*European Industrial Services Ltd*)

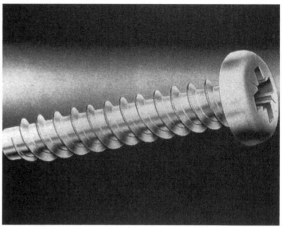

Figure 7.41   Polymate screw (*European Industrial Services Ltd*)

Figure 7.42   Hammer driven screw (*European Industrial Services Ltd*)

metal. Or, where this is not acceptable, the CA Taptite can be used; this has a gimlet point capable of self-extruding in thin sheet sections.

As plastic materials are increasingly used to replace aluminium, zinc die castings and other metal parts, so new fasteners have been designed capable of joining plastic components together. A screw can be driven directly into thermoplastics, making it a very cost effective way of making positive joints. The screw can be removed and reinserted into the same hole many times and will always pick up the same thread, thus maintaining a strong joint. Polymate is a round bodied, twin-threaded self-tapping screw with a closely controlled critical helix angle and thread pitch. The torque required to drive the screw is low, while the clamping load achieved is very high (see Figure 7.41).

7.6.3 Steel hammer drive screws

These sheet metal fasteners (Figure 7.42) are used only where permanent fixing is required, as they are difficult to extract once placed. They are made in a very hard steel, and the head shapes available are round head and countersunk. The screws are inserted in predetermined holes of the correct size and hammered so that the spiral thread is forced to cut into the material. The cutting action of the thread reduces the size of the hole after the pilot point has passed through.

7.6.4 Screw nails

These are made of hardened steel and are used for fastening thin sheet metal on to wood structures (Figure 7.43). They are hammer driven through light-gauge sheet metal, taking care not to bend or break them, so that the hardened spiral thread cuts into the burr in the sheet metal and then worms its way into the wood. This makes a secure metal-to-wood joint which is ideal for panelling in the

Figure 7.43 Screw nail (*European Industrial Services Ltd*)

construction of vehicle bodies. The screws are obtainable with countersunk, flat and round heads.

7.6.5 Coach screws

These are designed for securing heavy-gauge metal and fittings to timber, and are obtainable with either hexagon or square heads suitable for using a spanner or socket wrench. Sizes range from 4.763 mm, $\frac{3}{16}$ in diameter up to 12.7 mm, $\frac{1}{2}$ in and in lengths from 19 mm to 254 mm, $\frac{3}{4}$ in to 10 in (Figure 7.44).

Square head

Figure 7.44 Coach screw, square head (*European Industrial Services Ltd*)

7.6.6 Bolting

Bolts are used extensively in the manufacture of motor vehicles, as components which are built in sections and then bolted together can easily be dismantled for repair or replacement. One example of this type of construction is the car body, which has certain panels which are made separately and later bolted in position so that they can be replaced as individual units if damaged. Another application of bolting is in cases where fabrications are too large for workshop assembly; then the article is made in sections which are assembled on site.

A bolt is a cylindrical rod having a head on one end and a thread cut along part of its length from the other end. Bolts are available in mild steel, alloy steel, stainless steel and brass, and can be made of other metals and alloys for special applications. A nut and bolt is the most common means of joining. The unthreaded portion of the shank of the bolt is intended for fitting into clearance holes, and a washer can be used to minimize the effect of

friction between the head of the bolt and the parent metal. A spring washer placed under an ordinary nut assists in preventing the slackening of the nut when the joint is subject to vibrations.

The basic differences between bolts, set screws and carriage bolts are illustrated in Figure 7.45. The identification of bolts is outlined in Figure 7.46.

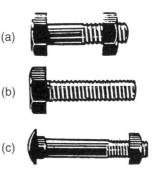

Figure 7.45 (a) Bolt (b) set screw (c) carriage bolt (*European Industrial Services Ltd*)

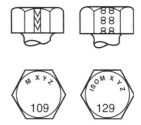

Figure 7.46 Bolt identification. An ISO metric bolt or screw made of steel and larger than 6 mm in diameter can be identified by either of the symbols ISO M or M embossed or indented on top of the head. In addition to marks to identify the manufacturer, the head is also marked with symbols to indicate the strength grade, e.g. 8.8; 10.9; 12.9; 14.9. As an alternative, some bolts and screws have the M and strength grade symbol on the flats of the hexagon (*Rover Group Ltd*)

The following types of head are available onbolts:

Hexagon head This is the most frequently used shaped, and it is suitable for all spanners.
Square head The square headed bolt is used mainly for heavy engineering and structural work.
Countersunk head A bolt of this type has a head shaped to fit into conical recesses in the work surface so that it lies level and flush. A slot is provided in the head for insertion of a screwdriver.

7.6.7 Set screws (or machine screws)

A set screw is similar to a bolt but has the whole of its cylindrical shank threaded (Figure 7.47). Set screws are used either to fix two pieces of metal together or to adjust the distance of one piece relative to the other. They are placed by passing each through a clearance hole in the top piece of metal and screwing it into a hole tapped out with a thread in the lower piece of metal. Countersunk, round, cheese and pan heads are available, with either slotted or star recesses (Figure 7.48).

Figure 7.47 Set screw (*European Industrial Services Ltd*)

7.6.8 Nuts

A nut is a shaped block of metal internally cut to form a thread. It is usually made from mild steel, although other metals and alloys can be used for special purposes. Common nuts are:

Hexagonal plain nut Used in all classes of engineering.
Square nut Used in heavy engineering and structural work and also in coach building.
Castellated nut and slotted nut Used where the nut must always remain tight, and slotted to take split pins which act as a locking device.
Lock nuts These are thinner than the normal nut and are fitted beneath the main nut to act as a locking device.
The identification of nuts is outlined in Figure 7.49.

7.6.9 Clinch nut

This is a captive nut (Figure 7.50) used for blind hole fixing in thin sheet metal. It is so called because a threaded nut is held securely in position by an annular rivet, the head of which is hammered flat around the hole to form a secure anchorage (Figure 7.51).

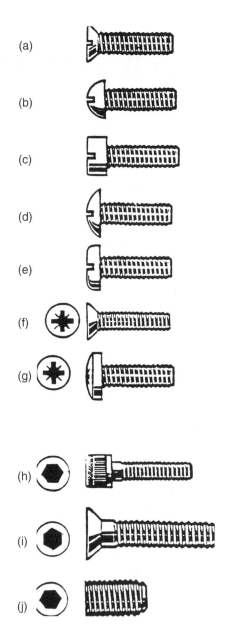

Figure 7.48 Machine screw head types: slotted and (a) countersunk (b) round (c) cheese (d) mushroom (e) pan; Pozidriv and (f) countersunk (g) pan; socket and (h) cap (i) countersunk (j) cup point (*European Industrial Services Ltd*)

7.6.10 Aerotight nut

This nut is designed for use in conditions of severe vibration. On the top of the nut there are two threaded cantilever arms which are deflected downwards and inwards. When the bolt is passed

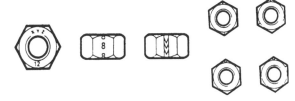

Figure 7.49 Nut identification. A nut with an ISO metric thread is marked on one face or on one of the flats of the hexagon with the strength grade symbol 8, 12 or 14. Some nuts with a strength grade 4, 5 or 6 are also marked, and some have the metric symbol M on the flat opposite the strength grade marking. A clock face system is used as an alternative method of indicating the strength grade. The external chamfer of a face of the nut is marked in a position relative to the appropriate hour mark on a clock face to indicate the strength grade. A dot is used to locate the 12 o'clock position and a dash to indicate the strength grade. If the grade is above 12, two dots identify the 12 o'clock position (*Rover Group Ltd*)

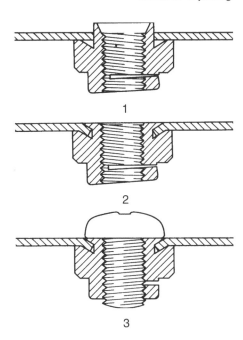

Figure 7.51 Clinch nut fixing procedure (*European Industrial Services Ltd*)

1 Insert shank of nut into predrilled hole
2 Rivet by closing spigot down with hammer. A convex punch may be used for large sizes to spread the spigot but a flat tool must be used for the riveting operation
3 An ordinary screw or bolt is then used for making an attachment to the clinch nut

Figure 7.50 Clinch nut (*European Industrial Services Ltd*)

through the nut these arms are forced out towards their original position, thus causing them to grip the thread very tightly.

7.6.11 Nyloc self-locking nut

This is a self-locking nut with a built-in moulded nylon insert made of smaller diameter than the nut's normal thread (Figure 7.52a). When the nut is screwed on to the bolt it runs freely until the end of the bolt meets the nylon insert. Further tightening of the bolt forces a thread on to the nylon insert, making the nut grip the bolt more tightly and thus creating a shock- and vibration-proof nut.

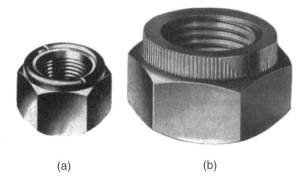

(a)   (b)

Figure 7.52 (a) Nyloc (b) Cleveloc nuts (*Forest Fasteners*)

7.6.12 Cleveloc self-locking nut and flange nut

The locking collar is an integral part of the Cleveloc nut (see Figure 7.52b), but in the flange nut the locking element is integral within the nut and has no collar. They are precision formed to a geometric ellipse and provide two locking elements of uniform shape and thread contour. These locking elements utilize most of the threads that come within the depth of the collar (on the Cleveloc nut) or the element (on the flange nut) and distribute locking pressure over wide areas. This ensures high fatigue life, with the flexibility necessary to give consistent performance and dependability in service.

There are no thread interruptions or pitch errors, and there is no deformation of the nut body or its hexagon form. The nut runs freely on the bolt threads until contact is made with the locking elements, each of which has a predetermined area of contact. These locking elements are designed to create gradually increasing areas of friction on the bolt threads.

Further tightening forces the locking elements to engage more fully with the bolt threads, and this increasing resistance to the entry of the bolt brings the full length of the nut threads into close contact with the working faces of the bolt threads. Further friction is created and both forces combine to give a smooth, progressive and increasingly self-locking action.

7.7 Fastening devices

Spring steel fasteners directly contribute to an ever-increasing degree of automation, and have probably saved the motor industry millions of pounds by enabling massive economies of labour and time in assembly. Industry data points to the fact that over 50 per cent of the total cost of a motor body is in the area of assembly. A large portion of this arises from the time and labour expended in picking up and putting together a number of parts, some of them relatively small, and this is where spring steel fasteners achieve marked savings compared with earlier methods. Another ancillary industry affected by the advent of spring fasteners is the repair industry; here economies are less dramatic but spring fasteners still make valuable savings in time, labour and handling. All of these types of clips are made from a specially treated spring steel which retains its original spring locking power so that the clips are reusable, and hence economical. Other advantages of these clips are that they are easy to apply, they eliminate the use of washers and nuts, and glass, plastic and other materials can be fastened together without fear of damage.

One of the firms who manufacture these clips is Forest Fasteners, who have developed the Spire speed nut (see below) which is used considerably in the car industry. A selection of fasteners is shown in Figure 7.53.

7.7.1 Spire speed nut

This nut has a double-locking action which operates by means of an arched base and arched prongs. Since the introduction of this nut, hundreds of different fastenings have been introduced over the years, and in some private cars over 200 spring steel fastenings are used.

The flat Spire speed nut (Figure 7.54) has replaced the threaded nut plus lock or plain washer. It is available in a wide variety of sizes to suit machine screws and sheet metal self-tapping screws. First a hole is drilled in the appropriate panels, then the self-tapping screw is placed through the panels and pushed into the spire nut or clip, making sure that the prongs point outwards. When the nut is tightened it is locked both by the self-energizing spring lock of the base and by the compensating thread lock as the arched prongs engage the thread. These free-acting prongs compensate for tolerance variations, and the combined forces of the thread and spring locks are claimed to eliminate any risk of loosening by vibration.

7.7.2 Captive nut, U type

This nut (Figure 7.55) is widely used for fastening blind assemblies in the motor industry. It can be assembled to the panels by hand and no welding or riveting is required. The nut allows a certain degree of float, which facilitates speedy assembly. It remains captive to the panel, anchored by means of a sheared tongue on the lower leg which drops into the mounting hole and holds the nut in the screw receiving position. The range available covers many panel thicknesses and screw sizes.

As the nut is pushed on the panel edge and over a predrilled hole, the locking tongue seats into the hole, thus holding the nut in position. The second panel is aligned and the screw is driven through this panel into the captive nut, which holds it in place. This type of fastener has the advantage that it can be fitted before or after panels have been

Figure 7.53 Selection of fasteners used in the automotive industry (*Forest Fasteners*)

painted, because there is no danger of clogging during any spraying operation.

7.7.3 Captive nut, J type

This is similar in concept to the U-nut but with a shorter leg designed to snap into a clearance hole (Figure 7.56). The J-nut is easily started over the edge of a panel and pressed into position with the thumb. A typical application for J-nuts in the motor industry is the replacement of reinforcing rings and blind, bushed-on headlamp assemblies. The short leg on the front side of the nut is clipped into screw receiving positions on the wing aperture, and ensures a good seal between gasket and wing, thus precluding mud leakage.

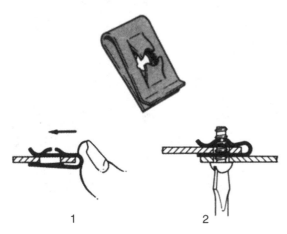

Arched prongs   Compensating thread lock

Arched base

Self-energizing spring lock

1   2

Figure 7.54   Spire speed nut (*Forest Fasteners*)

1   Pre-locked position
The two arched prongs move inwards to engage and lock against the flanks of the screw thread. They compensate for tolerance variations
2   Double locked position
A self-energizing spring lock is created by the compression of the arch in both the prongs and base as the screw is tightened, and vibration loosening is eliminated

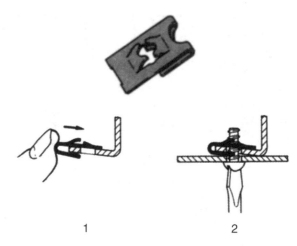

1   2

Figure 7.56   'J' type captive nut (*Forest Fasteners*)

1   Snap nut into position
2   Drive screw into nut

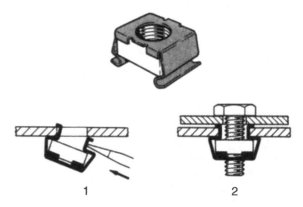

1   2

Figure 7.57   Nut grip (*Forest Fasteners*)

1   Insert one spring steel leg of nut into hole and snap in other leg using a suitable tool
2   Complete assembly with bolt

1   2

Figure 7.55   'U' type captive nut (*Forest Fasteners*)

1   Nut pressed into position: locating tongue snaps into predrilled hole and locks nut into position
2   Assembly completed with screw

7.7.4 Nut grip

The nut grip (Figure 7.57) is used in a square hole and replaces costly welded cage nuts and similar fastenings. It is installed into the panel by hand,

where it remains captive. The fully threaded nut in the cage is permitted to float slightly to overcome the problem of misaligned holes, and is fitted after the finishing process at any convenient point on the production line. Three sizes of cage, covering threads from M3 to M10 are available.

7.7.5 Cable clip

This is a special clip used for carrying cables. It is fixed by clipping into predrilled holes, thus eliminating the need for screw fixings (Figure 7.58).

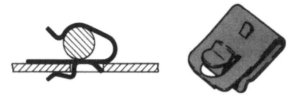

Figure 7.58 Cable clip used for latching: sectional view shows clip in position in a panel (*Forest Fasteners*)

7.7.6 Push-on clips

In addition to spring steel fasteners that are associated with threaded members, there is a wide variety of push-on clips (Figure 7.59). However, these clips, unlike the nuts, are not pitched to follow the helix of a thread, and the two sheared arms are of equal height. As the push-on clip is forced over a

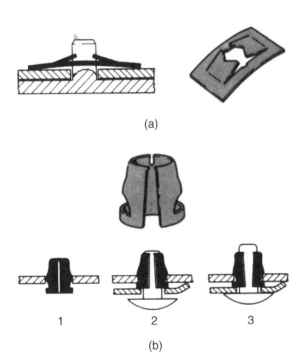

(a)

1   2   3

(b)

Figure 7.59 (a) Simple push-on clip showing clip in position (b) tubular clip, showing fitting sequence (*Forest Fasteners*)

1 Clip enters hole: cut-outs register with panel, allowing clip to expand and hold
2 Second part of assembly is lined up and rivet and integral stub inserted
3 Rivet pressed home, engaging turned-in end of clip which expands to bite on rivet

plain stud the fixing legs bite into the surface and, on finally depressing the arched base, which of course reacts as soon as the pressure is released, the fixing legs are given a strong upward and inward pressure which firmly holds the fastener in position. This has the effect of drawing the assembly together and removing any possibility of rattle.

When using the push-on clip, it is important to ensure that the tolerance on the stud diameter is held to within reasonable limits; to get the best results from this type of fastener a tolerance of 0.05–0.08 mm is recommended.

Push-on clips take many forms, from the simple sheared type which is still widely used, to the more modern blanked types, the multi-pronged clips for rectangular studs and now the plastic-capped type. To fix this type of fastener it is, of course, necessary to have access to the back of the panel, but if assembly is possible from one side only a tubular type can be used which consists of a small spring steel split tube. This is pushed into a hole in the panel, where it remains captive and ready to receive the studs of the badge or nameplate. Such tubular clips permit the assembly of a badge to the grille as a final operation. A removable version of this clip is also made for applications which have to be taken apart from time to time.

7.7.7 Non-metal clips

The use of plastics has given rise to the development of new types of fastenings. The common fault with the strip or wire clip is that they become rusty and break, either on removal or on replacement, and plastics have the advantage of being rust-proof.

Tough rubber has great possibilities for certain types of fastenings. Some time ago a tough rubber strap was offered as an alternative to the traditional metal clip for retaining loose cables and wires. It is still widely used in this way, and rubber with a synthetic content is sometimes used for door check straps; this demonstrates the strength of the material, which clearly lends itself to many applications as yet unexploited.

7.7.8 Plastic captive nuts and trim panel fasteners

Plastic captive nuts are available in a large range of shapes and sizes for vehicle body applications. They can be used in high-corrosion areas, and are designed to snap into a suitably punched hole in a metal panel, where they will remain captive until final assembly. The shallow head gives highly desirable minimum clearance between panels. The combination of design and material permits the recommended screw to form its own thread by displacement, resulting in high torque load and good anti-vibration characteristics. Screws may be removed and the nuts reused, providing the screws are the same size and thread form. This makes them ideal for applications such as inspection covers and service access. They are manufactured from glass filled nylon 66, which permits higher torque loadings. All other parts are manufactured from polypropylene (see Figure 7.60).

Push-in panel fasteners (W-buttons) are designed to hold board, plastic, rubber or other soft materials to secondary panels. They are equally suitable for assembling components to panels. No special tools are required, as they assemble on the principle of a push-in rivet. Simply push the W-button through the aligned panel holes so that it contracts on entry

and then relaxes as it snaps through the total panel thickness, securing the panels under tension. Many applications exist for W-buttons where a positive, attractive, light-weight non-corrosive fastening solution is required: to assemble painted or PVC covered trim boards, to fasten instruction badges, or to blank off holes. In addition, a special feature which makes the buttons particularly suitable for the automotive industry is the various styles of textured and colour-moulded heads to match adjacent panels (see Figure 7.61a).

Christmas-tree buttons are a variation of push-in fasteners, and when they are pushed into the hole they take up great variations in panel thickness. These parts have a good tolerance of poor-quality hole conditions. Although it is possible to remove the buttons, they are not normally reusable (see Figure 7.61b).

Trim panel fasteners are used principally in the automotive industry to attach door and tailgate panels to inner door panels. They are blind fixes, ensuring a neat, attractive assembly appearance.

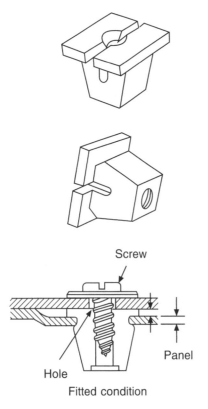

Figure 7.60 Plastic capture nut (*TRW United-Carr Ltd*)

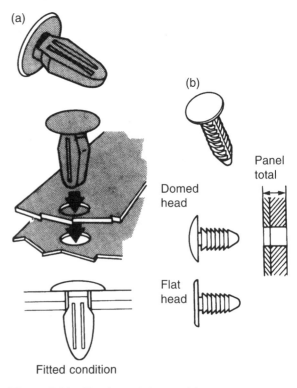

Figure 7.61 Plastic push-in panel fasteners: (a) W button (b) Christmas tree button (*TRW United-Carr Ltd*)

They allow panel removal for repair, service or inspection purposes. Easily fitted without the use of special tools, they are positioned in a keyhole slot in the trim panel, then firmly pushed home into the inner door panel mounting hole. Various types and sizes are available to suit different conditions, and most have a flexible skirt which seals the mounting hole against entry of dirt or water (Figure 7.62).

Retainers – where, for a technical or styling reason, a keyhole slot is not feasible, assembly can be simply achieved by using a separate retainer. With its own integral keyhole slot feature, it can be attached to the trim panel either with a suitable adhesive or by heat staking. This will provide a secure, attractive attachment with no evidence of fastener location.

Snapsacs provide increased retention and sealing for trim panel fasteners, and their smooth surface eases assembly and removal.

Quick-release fasteners (quarter turn) consist of a stud, cam and an optional retaining washer. A selection of lengths and head forms is available to suit a variety of assembly conditions. The washer allows the stud to be positively and securely located prior to and away from the final assembly point. Where the assembly is subject to routine servicing, the stud, washer and cam stay securely mounted, avoiding any risk of component loss (see Figure 7.63).

Edge fasteners of the D type are lightweight fasteners which are ideal for assembling leather-cloth, soft plastics and fabrics to metal, rigid plastic or fibreboard panels. Fitted without the use of special tools, they are pushed or tapped on to a convenient panel edge, using a light hammer. The barbs retain the assembly securely in position on one side, and the flat D side provides a neat flush appearance on the other. The friction grip version offers good retention without paint damage where corrosion resistance is paramount. They are manufactured from austempered carbon steel, and normally supplied in phosphate and black finish (see Figure 7.64).

7.8 Adhesives

Adhesives are to be found almost everywhere in the modern world. There are natural as well as manmade adhesives: for example, spiders use adhesive to spin and stick their webs and to catch their prey, and limpets and shell fish use adhesive to anchor themselves to rocks.

A bond between two surfaces may be regarded as a chain of three links, the strength of the bond

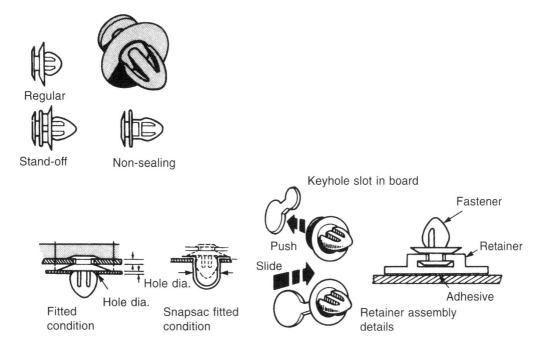

Figure 7.62 Trim panel fasteners (*TRW United-Carr Ltd*)

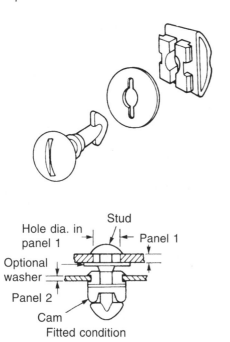

Figure 7.63 Quick-release fasteners, quarter turn (*TRW United-Carr Ltd*)

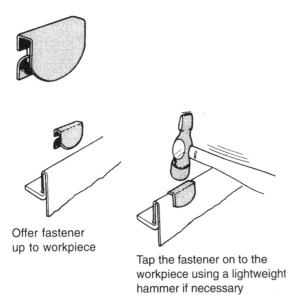

Figure 7.64 Edge fasteners, D type (*TRW United-Carr Ltd*)

being that of the weakest link. The central link is the adhesive film between the surfaces, and the outer links are formed where the adhesive film meets the bonded surface. To form a strong surface

linkage, the adhesive must thoroughly wet the surface whilst liquid, and when dry must adhere by penetration into the pores or fibres (Figure 7.65). In the case of smooth, non-porous surfaces such as glass or metal, a strong film of adhesive is best formed between the surfaces if these are joined at the appropriate stage in the drying of the applied adhesive. In this process the molecules of the adhesive and substrate are brought close together and a variety of forces operate; the most important are atomic forces, which can be likened to magnetic attraction. Many variations on this simple theme are possible and practicable, but the character of the bond in a particular case will always be governed by how closely the technique used approximates to this ideal. Almost everything that is manufactured or made will use adhesives. A good example is the motor car. Paint has to adhere to the metal on which it is placed. Inside the car, the upholstery will generally be stuck together with adhesives. All the brake shoes on the car are bonded together. Finally, some panel assemblies are bonded with adhesives.

7.8.1 Adhesive types

To facilitate the selection of adhesives, it is convenient to classify them into the following major groups (see Tables 7.2, 7.3 and 7.4):

Anaerobics Often known as sealants or locking compounds. Acrylic based, they normally set in the presence of metal and absence of air (or to be exact, atmospheric oxygen). They are generally used to lock, seal and retain all manner of turned, threaded and fitted parts, and are often used to seal flanges. *Cyanoacrylates* Also based on acrylic resins. Unlike anaerobics, they require surface moisture as the vital catalyst in hardening. Generally, they harden in seconds. Often used in car trim applications to bond rubber trim to metal, rubber to rubber and also rubber to plastic.

Epoxies Based on an epoxy resin, which is mixed with a hardener. This allows great variety in formulation. Their strength is often employed in bonding larger components. Single-part epoxies (ESP) are a development in epoxide technology. Resin and catalyst are premixed, so they give high performance without mixing by the user. In the motor vehicle they see a great deal of use as a supplementary

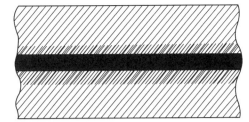

Bulk of adherend 1

Surface layer of adherend 1
Adhesive
Surface layer of adherend 2

Bulk of adherend 2

Figure 7.65   Simplified cross-sectional representation of a typical joint (*Permabond Adhesives Ltd*)

bond to welding: the process is known as weld-bonding.

Hot melts   A refined form of the earliest adhesive, hot wax. They are convenient for assembling small, lightly loaded items for use in less severe environments. They are too viscous for use on the smallest parts, but are often used for the assembly of motor vehicle trim.

Phenolics   One of the earliest types of structural adhesive. Their use often involves specialized equipment and complex procedures. Nevertheless, they perform well in severe environments. The phenolics are hardly ever seen in motor vehicles, but they do have a significant residual use in the assembly of brake shoes.

Plastisols   Based on PVC dispersions. They cure only at elevated temperatures, and are generally used only on large-scale work or where there is access to a heat source, intended for another purpose.

Polyurethanes   Like epoxies, they offer variety in formulation. However, polyurethanes are difficult

Table 7.2   Adhesives: their general nature and uses (*Permabond Adhesives Ltd*)

Adhesive	Toxicity	Capital cost	Material cost	Process complexity	Process temperature	Solvent resistance	Heat resistance	General durability	General comment
Anaerobic	1	1	2–3	1	E	3	2–3	3	Assembly of machined, coaxial components
Cyanoacrylate	1	1	3	1	C	2	1	1–2	Almost every type of small plastics, metal and rubber assembly
Epoxide	1–3	1–2	2	2	E	3	2–3	3	Usually used on larger objects where good performance required
Toughened acrylic	1–2	1	2–3	1–2	E	2–3	2–3	2–3	Excellent general performance, particularly structures
Toughened epoxide (heat cured)	1	1	2–3	1	H	3	3	3	Superb structural adhesives
Phenolic	1–2	3	2	3	E	2–3	2–3	2–3	Usually used in large stressed and critical structures
Polyurethane	2	2	1	2	E	2–3	2–3	1–2	Used when rapid assembly needed in large structures

1 low; 2 medium; 3 high
H hot; C cold; E either

Table 7.3 Compatibility of the principal structural adhesives with a variety of composite and associated materials (*Permabond Adhesives Ltd*)

Material to be bonded	Acrylic		Epoxy		PU Two part
	Pseudo one part	*Two part*	*One part (heat cured)*	*Two part*	
Metal (also see paint):					
Aluminium	1	1	1	1	4
Steel	1	1	1	1	4
Zinc	2	2	2	2	4
Thermoplastic:					
Polyamide (nylon)	2	2	3	2	2
Polyphenylene oxide	3	3	3	2	1
Polypropylene	2	2	4	3	1
Thermoset:					
Epoxy	2	2	1	1	2
Phenolic	2	2	1	1	2
Polyester:	1	1	2	1	1
hand lay					
VARI	1	1	2	1	1
SMC	1	1	2	1	1
cold press	1	1	2	1	1
Polyurethane, RIM	3	3	4	2	1
Paint					
Cataphoretic	1	1	1	1	1

Scale: 1 excellent, 2 good, 3 good but possible problems, 4 unsuitable.

Note: this categorization is given in relation to the types of application usually seen in association with the materials nominated. Therefore, each line should be considered to be unique.

to handle and usually require specialized mixing equipment. Generally used for load-bearing applications in dry conditions, as they are prone to moisture attack. They can be successfully used for bonding painted metal surfaces and polyester components together.

Solvent-borne rubber adhesives Based on a rubber solution, where the solvent evaporates to effect bonding. Not suitable for loaded joints or harsh environments.

Tapes Adequate for bonding small components, but cannot support heavy loads. Some will withstand quite harsh environments.

Toughened adhesives Toughened variants are hailed as a breakthrough in adhesive technology. They incorporate low-molecular-weight rubbers that build in exceptional resistance to peel and impact forces. Acrylic-based, epoxy, and single-part epoxy adhesives can be toughened in this way. Toughening reinforces the best features of these adhesives with the unique shock absorption and strength of the rub-

ber matrix. Toughened adhesives are used in much the same way in motor vehicles, trucks and vans but they may be used on unpainted metal. They are extremely durable in poor operating environments.

7.8.2 Selection of adhesives

Joint types involved
There are three basic joint types:

Co-axial joints, where one part fits into another, usually require an anaerobic. Other adhesive types may be too viscous, or lack the appropriate grades of strength.

Plain lap joints, where the adhesive is primarily loaded in either shear or compression. As a rule, cyanoacrylates are better for unloaded and toughened variants for loaded lap joints.

Butt joints, where the adhesive is in tension. These joints are very susceptible to peel and cleavage forces, which toughened adhesives withstand exceptionally well.

Table 7.4 Summary of the characteristics of the principal structural adhesives for composite bonding (*Permabond Adhesives Ltd*)

	Main characteristics	*Principal advantages*	*Principal disadvantages*
PU one part	Low modulus Very low strength	Very simple to use. Hot melt variants very convenient on suitably sized components. Fills large gaps. No mixing	Sensitive to moisture. Not true structural adhesives. Slow curing. Must be applied to non-metal surface for long-term durability
PU two part	Very low to medium modulus Very low to medium strength	Fast curing possible. Very good application characteristics. Fills large gaps	Sensitive to moisture. Often requires heating to achieve acceptable production times. Must be applied on a non-metallic surface for long-term durability. Some versions cannot be considered to be structural adhesives. Must be mixed
Acrylic pseudo one part	Medium modulus Medium strength	Very fast. Very easy to apply. Extremely durable. Bonds metals particularly well. Copes with contamination well. A true structural adhesive. No mixing	Need good fit and narrow gaps to function effectively. Best below 1 mm
Acrylic two part (VOX)	Medium modulus Medium strength	Fast. Easy to apply. Benefit of delayed action cure (DAC). Extremely durable. Fills large gaps. Copes well with light contamination. A true structural adhesive	Must be mixed
Epoxy one part	High modulus Very high strength	Fast curing. Easy to apply. Extremely durable, with robust all round performance. No mixing	Needs to be heat cured
Epoxy two part	Medium to high modulus Medium to high strength	Easy to apply. Durable. Can be speeded by warming/heating. True structural adhesive	Must be mixed. Slow curing

Joint function

Where a coaxial joint needs dismantling for maintenance, the weaker grades of anaerobic should be used.

Service conditions

It is possible to indicate conditions using a three-point scale:

Benign Room temperature, where components are not expected to suffer high loads or impacts.
Normal Room temperature with occasional excursions to about 80 °C. Loads could be heavy, and components may experience occasional shocks.
Severe High humidity and/or temperature. Components heavily loaded, with heavy impacts during their use or assembly.

Curing times

This has important consequences for later production processes. Cyanoacrylates set in seconds and reach full strength in minutes. Anaerobics, epoxies and the toughened variants take longer to harden. However, all three can be made to cure rapidly using a variety of techniques.

Selection by computer

One firm, Permabond, has a computer program called the Permabond Adhesive Locator and Sealant Guide (PAL II). PAL links the common engineering materials, generic adhesive types and Permabond's own products to assess the relationship between materials, design, production and use. It can assess both mechanical and structural joints and is specifically intended to cover such individual issues as: the assembly of lap, sandwich, honeycomb, and butt joints; the fitting of bushes, bearings, splines, gears, shafts and gaskets; and the retention and sealing of all types of threaded fittings and pipes, together with the use of adhesive related sealing compounds. The program selects the adhesives that are most suitable for the job from among different types in order to join two of the numerous possible surfaces. It uses a variable question-and-answer sequence to determine the exact nature of the bond required and the conditions under which it must operate. The program then offers a selection of first-and second-choice adhesives, together with notes indicating why a particular choice has been made and which factors have excluded which adhesives.

7.8.3 Achieving good results with adhesives

Many applications require no preparation. For example, anaerobics needs no pretreatment unless contamination is excessive. This is because the locking or jamming mechanism of the coaxial joint plus adhesion is sufficient to hold components together. Similarly, cyanoacrylates are almost always used without surface preparation. On rubber and plastic they are capable of penetrating surface debris. The toughened acrylic adhesives are even more hardy, and some will actually bond through a film of oil (so too will the heat cured epoxies). Therefore, surface preparation can safely be omitted, or restricted to simple degreasing, on a wide range of jobs. However, where maximum performance is required, or conditions are particularly severe, it is worth spending time on surface preparation.

Preparation involves degreasing (to remove oil grease and contaminants) and abrading or etching (to increase surface grip).

Degreasing Best done with a chlorinated solvent, except for sensitive materials.

Abrasion Metals are best abraded by grit blasting. However, all the following give satisfactory results: abrasive discs, belts, and cloths; medium-grit emery paper; and wire brushes. Plastics, when bonded with cyanoacrylates, generally need no abrading. When using epoxy-based systems, abrade lightly with abrasive discs, belts or cloths, or use medium-grit emery paper.

Chemical etching Some materials require chemical treatment to ensure optimum performance. After chemical etching, wear clean gloves to handle materials; even a fingerprint can contaminate the etched surface, and so weaken the bond.

Table 7.5 shows surface preparation for engineering materials, which needs to be carried out in order to maximize the performance of any adhesive.

The principles of good adhesive joint design are shown in Figure 7.66. Note the following in particular:

1 Do use an adequate overlap, as this gives a stronger joint.
2 Do choose a rigid adherence where loads are carried.
3 Do form joints from thick, rigid sections where possible.
4 Do avoid butt joints.
5 Do refer to the manufacturer's instructions for specific adhesives.

7.8.4 Adhesives and the automotive industry

Non-structural adhesives

The primary function of non-structural adhesives is to attach one material to another without carrying functional loads on the components. Loads are usually light, particularly for those used inside vehicles. Some interior applications include fabric on trim and door panels, headlinings, carpets, and sound deadening panels. Exterior applications include moulding, wood grain, striping, and vinyl roof. The materials used for these bonds include pressure sensitive tapes, hot melts, epoxy spray, vinyl plastisol, elastomer solvent, silicon, solvent-based rubber (used for large gap sealing) and butyl rubber (formerly used to mount windscreens).

Hot melts

The advantages of hot melt adhesives and sealants are their ability to bond to both pervious and impervious substrates and the speed with which the ultimate bond strength is attained, the latter being particularly valuable in high-rate production. There are, however, several other significant advantages in these solvent-free substances. With other types of adhesive, full bond strength is not achieved until all the solvent or liquid carrier has evaporated. A hot

Table 7.5   Surface preparation materials (*Permabond Adhesives Ltd*)

Material	Cleaning	Abrasion or chemical treatment	Procedure
ABS acrylonitrile butadiene styrene) plastics	Degrease with detergent solution, except for cyanoacrylates when cleaning and other preparations are probably unnecessary	Etch in solution of (parts by weight) Water 30 Conc. sulphuric acid (SG 1.84) 10 Potassium dichromate or sodium dichromate 1 Add acid to 60% of the water, stir in sodium dichromate and add remaining water Add acid to water, never vice versa	Immerse for up to 15 min at room temperature Wash with clean, cold water, followed by clean, hot water Dry with hot air
Aluminium and alloys	Degrease with solvent	Etch in a dichromate solution Prepare as for ABS	Heat solution to 68 °C ± 3 °C Immerse for 10 min Rinse thoroughly in cold, running distilled (or deionized) water Air dry, oven dry or use infrared lamps at not over 66 °C for about 10 min Treated aluminium should be bonded as soon as possible and never be exposed to the atmosphere of a plating shop. Even brief exposure reduces bonding strength Care should be taken in handling as the surfaces are easily damaged Bonding surfaces should not be touched (even with gloves) or wiped with cloths or paper
Cellulose plastics	Degrease with methyl alcohol or isopropyl alcohol	Roughen surface with fine grit emery paper	Repeat degreasing If using epoxies, heat plastics for 1 h at 95 °C and apply adhesive while still warm NB Follow manufacturer's instructions to avoid premature curing of epoxy adhesives
Ceramics, porcelain and glazed china	Degrease in vapour bath, or dip in solvent	Use emery paper or sandblast to remove ceramic glaze	Repeat degreasing Let solvent evaporate completely before applying adhesive

Table 7.5 (*Continued*)

Material	Cleaning	Abrasion or chemical treatment	Procedure
Diallylphthalate plastics	Degrease with solvent, unless using cyanoacrylates	Abrade with medium grit emery paper	Repeat degreasing
Epoxy plastics	Degrease with solvent	Abrade with medium grit paper	Repeat degreasing
Expanded plastic (foams, etc.)	Do not use solvent	Roughen surface with emery paper	Remove all dust and contaminants
Furane plastics	Degrease with solvent	Abrade with medium grit emery paper	Repeat degreasing
Glass and quartz (non-optical)	Degrease with solvent	Etch in solution of (parts by weight) Distilled water 4 Chromium trioxide 1 Or, use a silane primer in accordance with the manufacturer's instructions	Immerse for 10–15 min at 23 °C ± 1 °C Rinse thoroughly with distilled water Dry for 30 min at 98 °C ± 1 °C Apply adhesive while glass or quartz is hot
Glass reinforced polyesters (GRP)	Degrease with solvent	Abrade with medium grit emery paper	Repeat degreasing
Graphite	Degrease with solvent	Abrade with fine grit emery paper	Repeat degreasing Allow graphite to stand to ensure complete evaporation of solvent
Iron (cast iron)	Degrease with solvent	Grit blast or abrade with emery paper	Repeat degreasing
Melamine and melamine faced laminates including Formica, Warite, etc.	Degrease with solvent	Abrade with medium grit emery paper	Repeat degreasing
Nickel	Degrease with solvent	Immerse for 5 s in conc. nitric acid solution (SG 1.41) at 25 °C	Rinse thoroughly in cold running distilled (or deionized) water Dry with hot air
Nylon	Degrease with solvent	Roughen the surface with medium grit emery paper	Repeat degreasing
Paper laminates	Degrease with solvent	Abrade with fine grit emery paper	Repeat degreasing
Paper (unwaxed)	Do not use solvent	Requires no treatment before bonding	–
Phenolic, polyester and polyurethane resins	Degrease with solvent	Abrade with medium grit emery paper	Repeat degreasing
Polyacetals	Degrease with detergent solution	Etch in solution of (parts by weight) Water 33.0 Conc. sulphuric acid (SG 1.84) 184.0 Potassium dichromate or sodium dichromate 1.43	Immerse for 5 min at room temperature Wash with clean cold water followed by clean hot water Dry with hot air

Table 7.5   (*Continued*)

Material	Cleaning	Abrasion or chemical treatment	Procedure
Polycarbonate, polymethylmethacrylate (acrylic) and polystyrene	Degrease with methyl alcohol or isopropyl	Add acid to water, never vice versa Abrade with medium grit emery paper alcohol	Repeat degreasing
Polyester plastics	Degrease with solvent except when using sensitive materials which require detergent	Roughen with emery cloth cloth or etch in sodium hydroxide solution (20% by weight) for 2–10 min. at 70–95 °C	After abrasion, repeat degreasing After etching wash thoroughly in cold running distilled (or deionized) water

melt adhesive, on the other hand, relies on mechanical keying to the surface roughness of the substrate surfaces, and the adhesive's chemical affinity with the substrate. A good hot melt adhesive or sealant must exhibit six properties: high physical strength; a degree of flexibility to cater for differential movements in joints and vibrational stresses; good specific adhesion; a melt temperature well above maximum

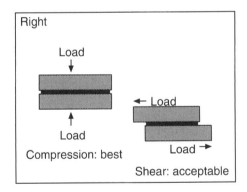

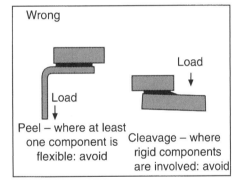

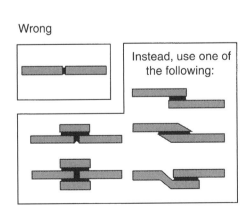

Figure 7.66   Adhesive joint design (*Permabond Adhesives Ltd*)

service temperature; a low viscosity at the application temperature; and a good substrate wetting.

Apart from the polyamide and EVA-based products, there are a number of other hot melt products that are useful to the car industry. The thermoplastic rubber-based materials have many of the properties of vulcanized rubber systems, yet can be applied as hot melts; they are especially useful in trim applications as derived pressure-sensitive adhesives. Where sealing is most important, such as to prevent moisture penetration into spot-welded seams, fairly soft butyl-based materials may be used; these do not liberate corrosive hydrochloric acid as a decomposition product when welded through. Another development is the hot melt systems that cure after application, either by reaction with moisture or by the application of heat.

For assembling trim components at the end of an assembly line, hand or robot guns can be employed. For continuous rapid assembly, wheel or roller applicators dipping into reservoirs of the heated adhesive can be most effective. Machines with one or more applicator heads, and developed for high rates of production, are now available. Hot melt adhesives may even be sprayed. There is also available a hot melt adhesive in the form of a strip about 0.5 mm thick, impregnated in a synthetic fabric scrim made up into a composite tape backed with aluminium foil. This has been used to join pieces of carpet butted together. The heat is generated by passing an electric current through the foil. Light rolling ensures that wetting is uniform over the whole area.

Toughened structural adhesives

These materials span both the acrylic and epoxy technologies, where a specific technique is used to prevent catastrophic crack propagation when joints are overloaded. Improved performance of the adhesive film is brought about by the introduction of a rubbery distortable phase within the load-bearing matrix of the body of the adhesive. It is this physically separate but chemically linked zone which absorbs fracture energy and prevents crack propagation. In this manner the resistance of the adhesive to catastrophic failure is enhanced considerably, and such adhesive compositions show a marked resistance to peel, cleavage and impact forces (see Figure 7.67).

Since the introduction of high-performance toughened structural adhesives, many designers and

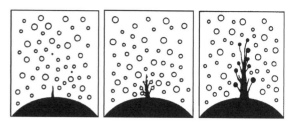

Figure 7.67 Toughened adhesive: when overloaded, crack propagation is stopped by the dispersed rubbery phase (*Permabond Adhesives Ltd*)

manufacturers are turning to bonded structures in order to reap the economic and technical benefits given by this technology. Materials used for vehicle manufacture are often chosen for strength, rigidity or light weight, and in some cases surfaces will be prefinished to enhance appearance. Bonding will not only overcome some of the assembly problems which such materials often present, but will give a stiffer and stronger structure than could be obtained from either riveting or welding. In the correct circumstances, satisfactory bonds can be made on many metallic and non-metallic materials.

Adhesives may be used alone; or in a secondary role to supplement welding, brazing or riveting; or in a primary role, complemented by welds or rivets. Weld bonding and rivet bonding are already standard procedures in many types of vehicle construction. Intermittent spot welding or riveting is a useful technique for pinning components during periods when the adhesive is uncured or temporarily softened from passage through a stoving oven.

A number of car manufacturers are already researching the possibility of a totally bonded car, especially those who want to use plastic or alloy bodies. If motor engineers are to build extremely lightweight vehicles, they will have to consider the application of adhesive engineering. Motor vehicle manufacturers are now extensively using high-performance epoxides (often toughened) to supplement welded joints in their vehicles. This has resulted in a major reduction in the number of welds required and the gauge of metal needed, and at the same time has increased body stiffness and reduced corrosion.

Typical applications of structural adhesives include the following:

Toughened acrylic   Aluminium or steel fitments bonded to GRP roof sections; patch repairs to metal or plastic panels; internal steel fixtures bonded to steel.

Toughened epoxy   Aluminium floor sections bonded to wood; inner and outer door skins bonded together.

Cyanoacrylates   Rubber seals and weather strips.

Adhesives used in vehicle body repair

Weatherstrip adhesive   Bonding rubber weatherstripping to door shuts, boot lid.

Fast tack adhesive   Repairs to trim fabrics, headlinings and carpets.

Disc pad adhesive   Bonding paper discs to backup pads.

Tape adhesive   Trim applications.

Adhesive/sealant   (polyurethane-based compound) Structural bonding and sealing of replacement windscreen.

Questions

1   Sketch four types of solid rivet.

2   What is meant by riveting allowance and rivet pitch?

3   Name three types of metals used in making solid rivets, and state a property which they all have in common.

4   Sketch and name the type of joint that would be used in solid riveting.

5   Name, and sketch, the type of rivet required when fastening a panel to a frame; where the riveting is only possible from one side.

6   Describe four types of blind rivets and their placing procedure.

7   Explain the types of materials which are used in the manufacture of blind rivets.

8   Explain the use of a blind rivet nut.

9   Explain the advantages of the Hemlok system of structural fasteners.

10   Describe the advantages of the Monobolt system of structural fasteners.

11   What is meant by a permanent fastening method?

12   Describe the difference between a set screw and a bolt.

13   Explain what is meant by a self-tapping screw.

14   Explain the advantages of the Taptite screw system.

15   Describe the use of a screw nail in coachwork.

16   What purpose do nuts and bolts serve in vehicle body work?

17   List, and describe, four types of self-locking nut.

18   Explain the principle of the Spire speed nut and state where you would expect to find them on a vehicle body.

19   Explain the use of captive nuts in vehicle assembly.

20   What is the function of a spring washer when used as a securing device?

21   Name three types of plastic trim panel fasteners.

22   Sketch a pull-on panel fastener used in trim work.

23   Name the classified groups of adhesives.

24   Explain how an adhesive can be selected for use.

25   Explain what is meant by a non-structural adhesive.

26   Describe the importance of toughened structural adhesives.

27   List the applications where toughened adhesives would be used.

28   Explain how crack propagation is achieved in structural adhesives.

29   List the types of adhesive that could be used by the body repairer.

30   What type of adhesive will resist heat, water and acid, and is used to join metal?

31   What purpose do nuts and bolts serve in vehicle body repair work?

32   List and describe four types of self-locking nuts.

33   What is the purpose of a spring washer when used with a nut and bolt?

34   With the aid of a sketch, illustrate and name a type of self-securing joint.

35   Give the advantages of blind riveting when compared with solid riveting.

36   With the aid of a sketch, name a type of joint that would be used when solid riveting.

37 Explain where quick-release fasteners would be used on a vehicle body.

38 Explain the principle of the Nyloc and Cleveloc nuts.

39 Name three types of blind riveting tools that could be used in assembly work.

40 Explain the purpose of using coach screws in vehicle body assembly.

41 Describe two ways of locating a bonnet stay.

42 How are two bars safely attached to a vehicle body?

43 How can you ensure that road wheels are fitted securely?

44 Why are cross-head screws popular?

45 State one advantage of self-tapping screws.

Soft and hard soldering methods

8.1 Comparison of fusion and non-fusion jointing processes

The jointing of metals by processes employing fusion of some kind may be classified as follows:

Total fusion
Temperature range: 1130–1550 °C approximately. Processes: oxy-acetylene welding, manual metal arc welding, inert gas metal arc welding.

Skin fusion
Temperature range: 620–950 °C approximately. Processes: flame brazing, silver soldering, aluminium brazing, bronze welding.

Surface fusion
Temperature range: 183–310 °C approximately. Process: soft soldering.

In *total fusion* the parent metal and, if used, the filler metal are completely melted during the jointing. Thin sheet metal edges can be fused together without additional filler metal being added. Oxy-acetylene welding and manual metal arc welding were the first processes to employ total fusion. In recent years they have been supplemented by methods such as inert-gas arc welding, metal inert-gas (MIG/MAG) and tungsten inert-gas (TIG) welding, carbon dioxide welding and atomic hydrogen welding. Welding is normally carried out at high temperature ranges, the actual temperature being the melting point of the particular metal which is being joined. The parent metal is totally melted throughout its thickness, and in some cases molten filler metal of the correct composition is added by means of rods or consumable electrodes of convenient size. A neat reinforcement weld bead is usually left protruding above the surface of the parent metal, as this gives good mechanical properties in the completed weld. Most metals and alloys can be welded effectively, but there are certain exceptions which, because of their physical properties, are best joined by alternative methods.

In *skin fusion* the skin or surface grain structure only of the parent metal is fused to allow the molten filler metal to form an alloy with the parent metal. Hard solders are used in this process, and, as these have greater shear strength than tensile strength, the tensile strength of the joint must be increased by increasing the total surface area between the metals. The simplest method of achieving this is by using a lapped joint in which the molten metal flows between the adjoining surfaces. The strength of the joint will be dependent upon the wetted area between the parts to be joined. Skin fusion has several advantages. First, since the filler metals used in these processes have melting points lower than the parent metal to which they are being applied, a lower level of heat is needed than in total fusion and in consequence distortion is reduced. Second, dissimilar metals can be joined by applying the correct amount of heat to each parent metal, when the skins of both will form an alloy with the molten hard solder. Third, since only the skin of the parent metal is fused, a capillary gap is formed in the lap joint and the molten filler metal is drawn into the space between the surfaces of the metals.

In *surface fusion* the depth of penetration of the molten solder into the surfaces to be joined is so shallow that it forms an intermetallic layer which bonds the surfaces together. The process employs soft solders, which are composed mainly of lead and tin. As these also have a low resistance to a tensile pulling force, the joint design must be similar to that of the skin fusion process, i.e. a lapped joint.

This chapter covers skin and surface fusion methods; Chapters 9–12 deal with total fusion.

8.2 Soft and hard solders

In spite of the growing use of welding, the techniques of soldering remain one of the most familiar in the fabrication of sheet metal articles, and the art of soldering still continues to occupy an important place in the workshop. While soldering is comparatively simple, it requires care and skill and can only be learnt by actual experience.

Soldering and brazing are methods of joining components by lapping them together and using a low-melting point alloy so that the parent material is not melted. Soldering as a means of joining metal sheets has the advantage of simplicity in apparatus and manipulation, and with suitable modifications it can be applied to practically all commercial metals.

8.3 Soft soldering

The mechanical strength of soft soldered sheet metal joints is usually in the order of 15–30 MN/m^2, and depends largely upon the nature of the solder used; the temperature at which the soldering is done; the depth of penetration of the solder, which in turn depends on capillary attraction, i.e. on the power of the heated surface to draw liquid metal through itself (Figure 8.1); the proper use of correctly designed soldering tools; the use of suitable fluxes; the speed of soldering; and, especially, workmanship.

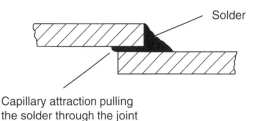

Capillary attraction pulling
the solder through the joint

Figure 8.1 Capillary attraction through a soldered lapped joint

8.3.1 Solders

Soft solder is an alloy of lead and tin, and is used with the aid of a soldering flux. It is made from two base metals, tin and lead. Tin has a melting point of 232 °C and lead 327 °C, but the alloy has a lower melting point than either of the two base metals and its lowest melting point is 183 °C; this melting point may be raised by varying the percentage of lead or tin in the alloy (see Table 8.1). A small quantity of antimony is sometimes used in soft solder with a view to increasing its tenacity and improving its appearance by brightening the colour. The small percentage of antimony both improves the chemical properties of the solder and increases its tensile strength, without appreciably affecting its melting point or working properties.

There is a great variety of solders, e.g. aluminium, bismuth, cadmium, silver, gold, pewterer's, plumber's, tinman's; solders are usually named according to the purpose for which they are intended. The following solders are the most popular in use today:

95–100 per cent tin solder, is used for high-quality electrical work where maximum electrical conductivity is required, since the conductivity of pure tin is 20–40 per cent higher than that of the most commonly used solders.

60/39.5/0.5 (tin/lead/antimony) solder, the eutectic composition, has the lowest melting point of all tin–lead solders, and is quick setting. It also has the maximum bulk strength of all tin–lead solders, and is used for fine electrical and tinsmith's work.

50/47.5/2.5 solder, called tinman's fine, contains more lead and is therefore cheaper than the 60/40 grade. Its properties in terms of low melting range and quick setting are still adequate, and hence it is used for general applications.

45/52.5/2.5 solder, known as tinman's soft, is cheaper because of the higher lead content, but has poorer wetting and mechanical properties. This solder is widely used for can soldering, where maximum economy is required, and for any material which has already been tin plated so that the inferior wetting properties of the solder are not critical.

30/68.5/1.5 solder, known as plumber's solder, is also extensively used by the car body repairer. Because the material has a wide liquidus–solidus range (about 80 °C), it remains in a pasty form for an appreciable time during cooling, and while in this condition it can be shaped or 'wiped' to form a lead pipe joint, or to the shape required for filling dents in a car body. Because of its high lead content, its wetting properties are very inferior and the surfaces usually have to be tinned with a solder of higher tin content first.

8.3.2 Fluxes

The function of a flux is to remove oxides and tarnish from the metal to be joined so that the solder will flow, penetrate and bond to the metal surface, forming a good strong soldered joint. The hotter the metal, the more rapidly the oxide film forms. Without the chemical action of the flux on the metal the solder would not tin the surface, and the joint would be weak and unreliable. As well as cleaning the metal, flux also ensures that no further oxidation from the atmosphere which could be harmful to the joint takes place during soldering, as this would restrict the flow of soldering.

Generally, soft soldering fluxes (see Table 8.1) are divided into two main classes: corrosive fluxes and non-corrosive fluxes.

Corrosive fluxes

These are usually based on an acid preparation, which gives the fluxes their corrosive effect. They are very effective in joining most metals. If the flux is not completely removed after use, corrosion is set up in and around the joint, and the risk of this happening prevents the use of these fluxes in electrical trades and food industries.

The following substances are corrosive fluxes:

Zinc chloride (killed spirits) This is made by dissolving pure zinc in hydrochloric acid until no more zinc will dissolve in the acid. This changes the acid into zinc chloride – hence the name killed spirits. As an all-round flux for soft soldering zinc chloride is without equal, but it has one disadvantage for some purposes – its corrosive action if the joint is not afterwards cleaned thoroughly with water.

Hydrochloric acid Although hydrochloric acid is not a good substitute for zinc chloride, it is nevertheless used with excellent results on zinc and galvanized iron. It can be used neat, but it is better to dilute it with at least 50 per cent zinc chloride.

Ammonium chloride (salammoniac) This may be used as a solution in water in much the same way as zinc chloride, but is not quite so effective for cleaning the metal.

Phosphoric acid This is effective as a flux for stainless steel, copper and brass, and does not have the corrosive effect of other acid types of flux.

Non-corrosive fluxes

These prevent oxidation on a clean or bright metallic surface during soldering. In general non-corrosive fluxes are not so active in cleaning the metal and

Table 8.1 Soft solders, fluxes, and their method of application for different sheet-metals

| Sheet metal | Flux | Solder constituents (%) | | | Used with |
		Tin	Lead	Other	
Aluminium	Stearin	85	–	Zn8, Al7	Soldering bit or blowpipe
Brass	Zinc chloride or resin	65	34	Sbl	Soldering bit
Copper	Zine chloride, ammonium chloride or resin	65	34	Sbl	Soldering bit
Galvanized steel	Dilute hydrochloric acid	50	50	–	Soldering bit
Lead	Tallow or resin	40	60	–	Soldering bit or blowpipe
Monel	Zinc chloride	66	34	–	Soldering bit or blowpipe
Nickel	Zinc chloride	67	33	–	Soldering bit
Pewter	Olive oil, resin or tallow	25	25	Bi5o	Soldering bit or blowpipe
Silver	Zinc chloride	70	30	–	Soldering bit
Stainless steel	Phosphoric acid + zinc chloride	66	34	–	Soldering bit
Terne steel	Zinc chloride	50	50	–	Soldering bit
Tin plate	Zinc chloride	60	40	–	Soldering bit
Tinned steel	Zinc chloride	60	40	–	Soldering bit
Zinc	Zinc chloride or hydrochloric acid	50	50	–	Soldering bit
Iron	Zinc chloride or ammonium chloride	50	50	–	Soldering bit
Steel	Zinc chloride or ammonium chloride	60	40	–	Soldering bit

serve chiefly as a measure of protection against further oxidation, when the material is hot. Therefore these fluxes should only be used when the metal has been cleaned prior to soldering.

The following substances are non-corrosive fluxes:

Resin and linseed oil Resin, finely powdered and dissolved in linseed oil, forms a good flux where non-corrosion is important. It is necessary that the parts to be soldered should be quite clean before applying oil and resin as a flux.

Tallow and palm oil Tallow and palm oil is often used as a flux for soldering lead. The surface of the lead must be first scraped clean before the flux is applied and the joint soldered. Tallow is a popular flux with plumbers for the purpose of wiping joints in lead pipes.

Olive oil A thin oil such as olive oil is generally used as a flux for soldering pewter. The soldering temperature for pewter is rather lower than for most soldering operations, hence the use of a thinner oil as a flux. As a flux, olive oil is sometimes called Gallipoli oil.

8.3.3 Soldering tools

Soldering irons are made in different sizes and shapes. The head is nearly always made of copper, although for soldering aluminium a nickel bit is necessary. Electrolytic copper is the best as it gives longer life and holds the solder well on the working faces, and forged soldering bits are far better heat retainers than cast copper bits as they are less liable to crack at the tips. A small soldering tool is not suitable for soldering any heavy or comparatively large sheet metal articles, because the heat loss by conduction is too fast to allow an even temperature to result which will allow the solder to flow freely and sweat into the seams or joints. Soldering tools heated by gas, electricity and oxyacetylene are now available to speed up the process of soldering in mass production work.

Figure 8.2 shows a selection of soldering bits.

8.3.4 Soft soldering process

Soldering is a process of joining two lapped metal surfaces together by fusing another metal or alloy of a lower melting point in between them in such a way that the melted metal bonds firmly to the other

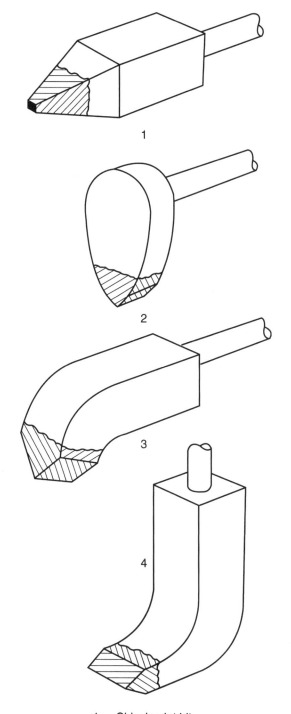

1. Chisel point bit
2,3,4. Hatchet soldering bits used for soldering the bottoms of deep containers

Figure 8.2 Soldering bits

two. A soldering iron or copper bit is usually used to apply the solder, although sometimes a blowpipe is used to sweat the solder into the joint. A liquid flux such as zinc chloride or resin or linseed oil is generally used to assist the solder to flow and run smoothly into the joint (Figure 8.3).

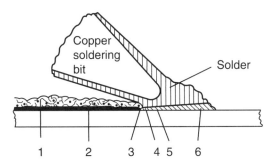

1 Flux solution lying on the surface
2 Boiling flux removing oxides
3 Bare metal oxides removed
4 Liquid solder replacing flux
5 Tin reacting with metal to form alloy compound
6 Solder solidi

Figure 8.3 Soldering process

The soldering process comprises the following steps:

Choose the right materials The choice of the soldering iron and its shape is governed by the size and accessibility of the material to be soldered; in general always use as large a soldering iron as is practicable. Some means of heating the soldering iron is necessary, together with a quantity of selected solder, file, emery cloth and a tin of flux, either corrosive or non-corrosive to suit the work in hand.

Clean the soldering iron Solder will not adhere to or bond to a dirty or greasy soldering iron, and whether the iron is new or old it must be clean and bright on its working surfaces, i.e. approximately 20 mm up from the point on each face. An emery cloth could be used for this purpose, but generally a file is preferred. It is important to see that just sufficient copper is removed to get rid of the pitting or scale and leave a clean bright surface.

Heat the soldering iron A clean flame such as gas is best for this purpose. Care must be taken not to allow the bit to become red hot, as overheating of the bit causes heavy scaling of the surface due to oxidation; this will mean refiling and

recleaning the bit, and hence unnecessary wear and a shorter service life. When the flame turns green around the soldering bit it is at the correct temperature; if the colour is allowed to change into a bright green the copper bit will begin to become red hot. With an electric soldering iron the heat is automatically controlled at the right temperature because the tool has a built-in thermostat for controlling the temperature.

Tin the soldering iron Dip the heated copper bit into flux to obtain a complete coating of flux on the surface faces of the bit, then rub the fluxed portion of the bit on a piece of solder to obtain a film of solder over the copper surface or working faces of the soldering iron. This operation is referred to as tinning the bit, and makes it easier for the bit to pick up solder and then discharge it on the workpiece. Tinning also protects the bit against further oxidation, thus increasing its life.

Clean the surface of the workpiece All metals have a covering of oxide on their surface, although it may not be visible to the naked eye, and this oxide film will prevent the solder from bonding to the metal to be soldered and therefore create a weak joint. First clean the surface to a bright finish with coarse emery paper or steel wool, then immediately apply the flux. The flux helps to clean the surface chemically so that the molten solder can flow and penetrate into the metal, forming a strong joint. Also it prevents oxides reforming on the surface of the work as the soldering is carried out.

Reheat the bit until a green flame forms around it, again taking care not to overheat it and destroy the tinning on its surface. Dip the bit into flux, then hold the tinned face of the bit against the solder stick until all the face that is tinned is covered with molten solder. The soldering iron is now ready for use.

Apply the soldering iron loaded with solder to the face of the workpiece or joint which has previously been smeared with flux. The metal surrounding the iron is heated to the melting point of the solder by conduction of heat from the soldering iron, and the solder will start to flow. Draw the iron slowly along the face of the joint, allowing solder from the bit to flow into the joint. A good joint has only a very thin film of solder, as too much solder weakens the joint. The length of joint that can be soldered before the bit needs recharging with solder depends upon the size of the bit, its temperature and the size of the job to be soldered.

The most important points in soft soldering are:

1 A perfectly cleaned joint.
2 A soldering iron that has been tinned and heated to the correct temperature.
3 The correct flux for the particular job in hand.
4 A good quality tin–lead solder.
5 Allow the heat from the soldering iron to penetrate into the metal before moving the iron along the joint; this will give the solder a chance to flow into the joint.
6 The correct type of joint must be used. Soft soldering can only be used on metals that are lapped one over the other to form the joint. This allows for capillary attraction of the solder (Figure 8.4).

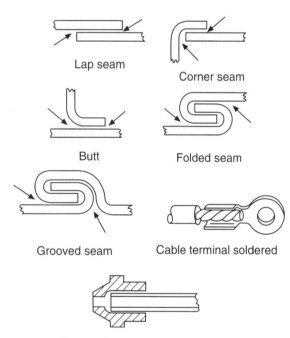

Lap seam

Corner seam

Butt

Folded seam

Grooved seam

Cable terminal soldered

Brass union joint sweated to tube

Figure 8.4 Types of soft soldered joints

The strength of a soft soldered joint is not governed by the amount of solder between the plates. The more solder, the weaker the joint; therefore a good joint has only a very thin film of solder between the metal plates, which forms a surface alloy by using the tin in the solder.

8.4 Hard soldering

8.4.1 Brazing

Brazing is used extensively throughout the panel beating trade as a quick means of joining sheet metal panels and other automobile parts. Although a brazed joint is not as strong as a fusion weld, it has many advantages which make it useful to the panel beater. Brazing is not classed as a fusion process, and therefore cannot be called welding, because the parent metals are not melted to form the joint but rely on a filler material of a different metal of low melting point which is drawn through the joint. The parent metals can be similar or dissimilar as long as the alloy rod has a lower melting point than either of them. The most commonly used alloy is of copper and zinc, which is, of course, brass. Brazing is accomplished by heating the pieces to be joined to a temperature higher than the melting point of the brazing alloy (brass). With the aid of flux, the melted alloy flows between the parts to be joined due to capillary attraction, and actually diffuses into the surface of the metal, so that a strong joint is produced when the alloy cools. Brazing, or hard soldering to give it its proper name, is in fact part fusion and is classed as a skin fusion process.

Brazing is carried out at a much higher temperature than that required for the soft soldering process. A borax type of powder flux is used, which fuses to allow brazing to take place between 750 and 900 °C. There are a wide variety of alloys in use as brazing rods; the most popular compositions contain copper in the ranges 46–50 and 58.5–61.5 per cent, the remaining percentage being zinc (Table 8.2).

The brazing process comprises the following steps:

1 Thoroughly clean the metal to be joined.
2 Using a welding torch, heat the metals to a temperature below their own critical or melting temperature. In the case of steel the metal is heated to a dull cherry red.
3 Apply borax flux either to the rod or to the work as the brazing proceeds, to reduce oxidation and to float the oxides to the surface.
4 Use the oxy-acetylene torch with a neutral flame, as this will give good results under normal conditions. An oxidizing flame used for

Table 8.2 Copper–phosphorus brazing alloys

BS 1845 ref.	Nominal composition (wt%)											Melting range (°C)	Tensile strength (N/mm²)	Elongation (%)	Hardness (HV)	Notes
	Cu	Zn	Mn	Ni	P		Bi	Si	Sn	Others					
CP1	Balance	–	–	Ag 14–15	4.3–5.0		–	–	–	For details of impurities see British Standards	645–800	670–700	10	187	Fluxless brazing High strength, good ductility
CP2	Balance	–	–	1.8–2.2	6.1–6.9		–	–	–		645–825	490–560	5	195	Good strength
CP3	Balance	–	–	–	7.0–7.8		–	–	–		710–810	490–550	7	192	Good strength
CP4	Balance	–	–	4.5–5.5	5.7–6.3		–	–	–		645–815	490–530	7	192	Good strength
CP5	Balance	–	–	–	5.6–6.4		–	–	–		690–825				
CP6	Balance	–	–	–	5.9–6.5		–	–	–		710–890				

materials having a high percentage of brass content will produce a rough-looking brazed joint, which nevertheless is slightly stronger than if brazed with a neutral flame.

5 Use only a small amount of brazing rod; if too much is used this weakens the joint.

6 The two pieces of material to be brazed must be either lapped or carefully butted after edge preparation and must fit tightly together during the brazing operation (Figure 8.5). Iron, steel, copper and brass are readily brazed, and metals of a dissimilar nature can also be joined. Typical examples are as follows:

Copper to brass
Copper to steel
Brass to steel
Cast iron to mild steel
Stainless steel to mild steel.

Also, coated materials like zinc-plated mild steel can be better brazed than welded.

7 Carefully select the types of metal to be joined. Although dissimilar metals can be joined by hard or soft soldering, corrosion may occur due to the electrolytic action between the two dissimilar metals in the presence of moisture. This action is an electrochemical action similar to that of an electric cell, and results in one or other of the two metals being corroded away.

The main advantages of brazing are:

1 The relatively low temperature (750–900 °C) necessary for a successful brazing job reduces the risk of distortion.

2 The joint can be made quickly and neatly, requiring very little cleaning up.

3 Brazing makes possible the joining of two dissimilar metals; for example, brass can be joined to steel.

4 It can be used to repair parts that have to be rechromed. For instance, a chromed trim moulding which has been deeply scratched can be readily filled with brazing and then filed up ready for chroming.

5 Brazing is very useful for joining steels which have a high carbon content, or broken cast iron castings where the correct filler rod is not available.

8.4.2 Silver soldering

Silver solder probably originated in the manufacture and repair of silverware and jewellery for the purpose of securing adequate strength and the desired colour of the joint, but the technique of joining sheet metal products and components with silver solder has come into wide usage in the automobile industry. The term 'soft soldering' has been widely adopted when referring to the older process to avoid confusion with the newer hard soldering process, known generally as either silver soldering or silver brazing. The use of silver solder on metals and alloys other than silver has grown largely because of the perfection by manufacturers of these solders which makes them easily applicable to many metals and alloys by means of the oxy-acetylene welding torch. This process is employed

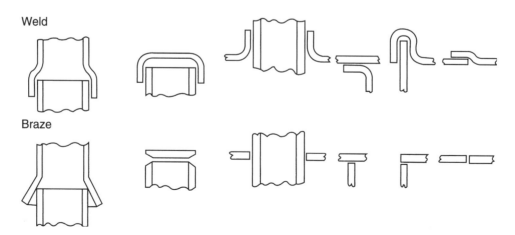

Figure 8.5 Joint design for brazing, showing the brazing equivalents to welding

for joining metal parts when greater strength is required than can be obtained by soft soldering, when the parts have to withstand a temperature that would cause soft solder to melt, and in cases where the high temperature developed by welding would seriously distort the metal parts. Vehicle parts which are manufactured from light-gauge sheet brass, stainless steel, very thin mild steel, sheet products or components fabricated from nickel, bronze or copper, can be very effectively joined by silver soldering.

Solders and fluxes

Silver solders are more malleable and ductile than brazing rods, and hence joints made with silver solder have a greater resistance to bending stresses, shocks and vibration than those made with ordinary brazing alloys. Silver solders are made in strip, wire (rod) or granular form and in a number of different grades of fusibility, the melting points varying between 630 and 800 °C according to the percentages of silver, copper, zinc and cadmium they contain (Table 8.3).

As in all non-fusion processes the important factor is that the joint to be soldered must be perfectly clean. Hence special care must be taken in preparing the metal surfaces to be joined with silver solder. Although fluxes will dissolve films of oxide during the soldering operation, sheet metal that is clean is much more likely to make a stronger, sounder joint than when impurities are present. The joints should fit closely and the parts must be held together firmly while being silver soldered, because silver solders in the molten state are remarkably fluid and can penetrate into minute spaces between the metals to be joined. In order to protect the metal surface against oxidation and to increase the flowing properties of the solder, a suitable flux such as borax or boric acid is used.

Silver soldering process

In silver soldering the size of the welding tip used and the adjustment of the flame are very important to avoid overheating, as prolonged heating promotes oxide films which weaken both the base metal and the joint material. This should be guarded against by keeping the luminous cone of the flame well back from the point being heated. When the joint has been heated just above the temperature at which the silver solder flows, the flame

should be moved away and the solder applied to the joint, usually in rod form. The flame should then be played over the joint so that the solder and flux flow freely through the joint by capillary attraction. The finished silver soldered joint should be smooth, regular in shape and require no dressing up apart from the removal of the flux by washing in water.

When making a silver solder joint between dissimilar metals, concentrate the application of heat on the metal which has the higher heat capacity. This depends on the thickness and the thermal conductivity of the metals. The aim is to heat both members of the joint evenly so that they reach the soldering temperature at the same time.

The most important points during silver soldering are:

1   Cleanness of the joint surfaces
2   Use of the correct flux
3   The avoidance of overheating.

8.4.3 Aluminium brazing

There is a distinction between the brazing of aluminium and the brazing of other metals. For aluminium, the brazing alloy is one of the aluminium alloys having a melting point below that of the parent metal. For other metals, the brazing alloys are often based on copper-zinc alloys (brasses – hence the term brazing) and are necessarily dissimilar in composition to the parent metal.

Wetting and fluxing

When a surface is wetted by a liquid, a continuous film of the liquid remains on the surface after draining. This condition, essential for brazing, arises when there is mutual attraction between the liquid flux and solid metal due to a form of chemical affinity. Having accomplished its primary duty of removing the oxide film, the cleansing action of the flux restores the free affinities at the surface of the joint faces, promoting wetting by reducing the contact angle developed between the molten brazing alloy and parent metal. This action assists spreading and the feeding of brazing alloy to the capillary spaces, leading to the production of well filled joints. An important feature of the brazing process is that the brazing alloy is drawn into the joint area by capillary attraction: the smaller the gap is between the two metal faces to be joined, the deeper is the capillary penetration.

Table 8.3 Silver solders

BS 1845 ref.	Nominal composition (wt%)							Melting range (°C)	Tensile strength (N/mm²)	Notes
	Ag	Cu	Zn	Mn	Ni	Cd	Sn			
Cadmium-containing alloys										
AG1	49–51	14–16	14–18	–	–	18–20	–	620–640	470	Low melting point, very fluid, high strength, general purpose
AG2	41–43	16–18	14–18	–	–	24–26	–	610–620	470	Lowest melting point, very fluid. General purposes, especially small components
AG3	37–39	19–21	20–24	–	–	19–21	–	605–650	–	Very fluid and strong
AG9	49–50	14.5–16.5	13.5–17.5	–	2.5–3.5	15–17	–	635–655	480	Limited flow, useful for fillet joints
AG11	33–35	24–26	18–22	–		20–22	–	610–670	–	
AG12	29–31	27–29	19–23	–		20–22	–	600–690	485	
Cadmium-free alloys										
AG5	42–44	36–38	18–22	–	–	0.025	–	690–770	400	Cadmium-free for food equipment, etc.
AG7	71–73	27–29		–	–	0.025	–	780	–	Fluxless brazing of copper
AG7V	71–73	27–29		–	–	0.025	–	780	–	High-purity alloy for vacuum assemblies
AG8	99.99	–		–	–	0.25	–	960	–	Pure silver
AG13	59–61	25–27	12–16	–	–	0.025	–	730–695	–	–
AG14	54–56	20–22	21–23	–	–	0.025	1.7–2.3	660–630	450	–
AG18	48–50	15–17	21–25	6.5–8.5	4–5	0.025	–	705–680	360	For carbide brazing
AG19	84–86			14–16	–	0.025	–	960–970	–	–
AG20	39–41	29–31	–	–	–	0.025	1.7–2.3	710–650	–	–
AG21	29–31	35–37	–	–	–	0.025	1.7–2.3	755–665	–	–

The various grades of pure aluminium and certain alloys are amenable to brazing. Aluminium-magnesium alloys containing more than 2 per cent magnesium are difficult to braze, as the oxide film is tenacious and hard to remove with ordinary brazing fluxes. Other alloys cannot be brazed because they start to melt at temperatures below that of any available brazing alloy. Aluminium-silicon alloys of nominal 5 per cent, 7.5 per cent or 10 per cent silicon content (Table 8.4) are used for brazing aluminium and the alloy of aluminium and 1.5 per cent manganese.

The properties required for an effective flux for brazing aluminium and its alloys are as follows:

1   The flux must remove the oxide coating present on the surfaces to be joined. It is always important that the flux be suitable for the parent metal, but especially so in the joining of aluminium-magnesium alloys.
2   It must thoroughly wet the surfaces to be joined so that the filler metal may spread evenly and continuously.
3   It must flow freely at a temperature just below the melting point of the filler metal.
4   Its density, when molten, must be lower than that of the brazing alloy.
5   It must not attack the parent surfaces dangerously in the time between its application and removal.
6   It must be easy to remove from the brazed assembly.

Many types of proprietary fluxes are available for brazing aluminium. These are generally of the alkali halide type, which are basically mixtures of the alkali metal chlorides and fluorides. Fluxes and their residues are highly corrosive and therefore must be completely removed after brazing by washing with hot water.

Brazing method

When the cleaned parts have been assembled, brazing flux is applied evenly over the joint surface of both parts to be brazed and the filler rod (brazing alloy). The flame is then played uniformly over the joint until the flux has dried and become first powdery, then molten and transparent. (At the powdery stage care is needed to avoid dislodging the flux, and it is often preferable to apply flux with the filler rod.) When the flux is molten the brazing alloy is applied, preferably from above, so that gravity assists in the flow of metal. In good practice the brazing alloy is melted by the heat of the assembly rather than directly by the torch flame. Periodically the filler rod is lifted and the flame is used to sweep the liquid metal along the joint; but if the metal is run too quickly in this way it may begin to solidify before it properly diffuses into the mating surfaces. Trial will show whether more than one feed point for the brazing alloy is necessary, but with proper fluxing, giving an unbroken path of flux over the whole joint width, a single feed is usually sufficient.

8.4.4 Bronze welding

Bronze welding is carried out much as in fusion welding except that the base metal is not melted. The base metal is simply brought up to tinning temperature (dull red colour) and a bead is deposited over the seam with a bronze filler rod. Although the base metal is never actually melted, the unique characteristics of the bond formed by the bronze rod are such that the results are often

Table 8.4   Aluminium filler alloys for brazing

BS 1845 ref.	Nominal composition (wt%) (balance aluminium)						Melting range (°C)
	Si	Cu (max.)	Mn (max.)	Zn (max.)	Ti (max.)	Mg (max.)	
4004	9.0/10.5	0.25	0.10	0.20	–	1.0/2.0	555–590
4043A	4.5/6.0	0.30	0.15	0.10	0.15	0.20	575–630
4045	9.0/11.0	0.30	0.05	0.10	0.20	0.05	575–590
4047A	11.0/13.0	0.30	0.15	0.20	0.15	0.10	575–585
4104	9.0/10.5	0.25	0.10	0.20	–	1.0/2.0	555–590
4145A	9.0/11.0	3.0/5.0	0.15	0.20	0.15	0.10	520–585
4343	6.8/8.2	0.25	0.10	0.20	–	–	575–615

comparable with those secured through fusion welding. Bronze welding resembles brazing, but only up to a point. The application of brazing is generally limited to joints where a close fit or mechanical fastening serves to consolidate the assembly and the joint is merely strengthened or protected by the brazing material. In bronze welding the filler metal alone provides the joint strength, and it is applied by the manipulation of a heating flame in the same manner as in gas fusion welding. The heating flame is made to serve the dual purpose of melting off the bronze rod and simultaneously heating the surface to be joined. The operator in this manner controls the work: hence the term 'bronze welding'.

Welding rods and fluxes

Almost any copper-zinc alloy or copper-tin alloy or copper-phosphorus alloy (see Table 8.2) can be used as a medium for such welding, but the consideration of costs, flowing qualities, strength and ductility of the deposit have led to the adoption of one general purpose 60–40 copper-zinc alloy with minor constituents incorporated to prevent zinc oxide forming and to improve fluidity and strength. Silicon is the most important of these minor constituents, and its usefulness is apparent in three directions. First, in the manner with which it readily unites with oxygen to form silica, silicon provides a covering for the molten metal which prevents zinc volatilization and serves to maintain the balance of the constituents of the alloy; this permits the original high strength of the alloy to be carried through to the deposit. Second, this coating of silica combines with the flux used in bronze welding to form a very fusible slag, and this materially assists the tinning operation, which is an essential feature of any bronze welding process. Third, by its capacity for retaining gases in solution during solidification of the alloy, silicon prevents the formation of gas holes and porosity in the deposited metal, which would naturally reflect unfavourably upon its strength as a weld.

It is essential to use an efficient and correct flux. The objects of a flux are: first, to remove oxide from the edges of the metal, giving a chemically clean surface on to which the bronze will flow, and to protect the heated edges from the oxygen in the atmosphere; second, to float oxide and impurities introduced into the molten pool to the surface, where they can do no harm. Although general-purpose fluxes are available, it is always desirable to use the fluxes recommended by the manufacturer of the particular rod being employed.

Bronze welding procedure

1 An essential factor for bronze welding is a clean metal surface. If the bronze is to provide a strong bond, it must flow smoothly and evenly over the entire weld area. Clean the surfaces thoroughly with a stiff wire brush. Remove all scale, dirt or grease, otherwise the bronze will not adhere. If a surface has oil or grease on it, remove these substances by heating the area to a bright red colour and thus burning them off.

2 On thick sections, especially in repairing castings, bevel the edges to form a 90° V-groove. This can be done by chipping, machining, filing or grinding.

3 Adjust the torch to obtain a slightly oxidizing flame. Then heat the surfaces of the weld area.

4 Heat the bronzing rod and dip it in the flux. (This step is not necessary if the rods have been prefluxed.) In heating the rod, do not apply the inner cone of the flame directly to the rod.

5 Concentrate the flame on the starting end until the metal begins to turn red. Melt a little bronze rod on to the surface and allow it to spread along the entire seam. The flow of this thin film of bronze is known as the tinning operation. Unless the surfaces are tinned properly the bronzing procedure to follow cannot be carried out successfully. If the base metal is too hot, the bronze will tend to bubble or run around like drops of water on a warm stove. If the bronze forms into balls which tend to roll off, just as water would if placed on a greasy surface, then the base metal is not hot enough. When the metal is at the proper temperature the bronze spreads out evenly over the metal.

6 Once the base metal is tinned sufficiently, start depositing the proper size beads over the seam. Use a slightly circular torch motion and run the beads as in regular fusion welding with a filler rod. Keep dipping the rod in the flux as the weld progresses forward. Be sure that the base metal is never permitted to get too hot.

7   If the pieces to be welded are grooved, use several passes to fill the V. On the first pass make certain that the tinning action takes place along the entire bottom surface of the V and about half-way up on each side. The number of passes to be made will depend on the depth of the V. When depositing several layers of beads, be sure that each layer is fused into the previous one.

Questions

1   Describe the differences between the following classifications: (a) total fusion (b) skin fusion (c) surface fusion.

2   Which three conditions are essential to achieve a good soldered joint?

3   What are the main functions of a flux in soft soldering?

4   Name two types of flux used in soft soldering, and state a typical application for each.

5   Explain the reason why corrosive and non-corrosive fluxes are used in soft soldering.

6   List the step-by-step procedure for the process of soft soldering.

7   Explain the sequence of operations required to tin a soldering iron.

8   What is the difference between bronze welding and brazing?

9   State the advantages of brazing.

10   Explain the reason why certain joints are brazed in preference to being soldered or welded.

11   Name the flux and detail the composition of the rods used in the brazing process.

12   Name one non-fusion welding process, and give a detailed description of the process.

13   Name the constituents of a brazing rod suitable for brazing low-carbon steel.

14   Define the term 'capillary attraction'.

15   Explain the type of application in which the process of silver soldering would be used.

16   Describe the differences between the processes of aluminium welding and aluminium brazing.

Gas welding, gas cutting and plasma arc cutting

9.1 Development of gas welding

The first successful oxy-acetylene welding equipment was developed in France at the turn of the century (1901–3). The equipment was introduced to the commercial industries a year later and it immediately proved successful. Acetylene was first discovered in 1836, but was not produced successfully on a commercial basis until many years after this when the use of calcium carbide was first discovered. Oxygen was discovered many centuries ago, but its preparation on a commercial basis was not started until 1892.

Gas welding processes are so called because the welding heat is provided by a flame produced by the combustion of a mixture of gases. A variety of gases are used commercially, but an oxy-acetylene mixture is the most common because of its high flame temperature and because the gases are convenient to handle.

Oxygen forms 20 per cent of the atmosphere, the rest being nitrogen and a small percentage of rare gases such as helium, neon and argon. To obtain the oxygen in a state that makes it usable for welding, it is necessary to separate it from the other gases. Oxygen is produced commercially either by the electrolytic process or by a separation method known as the liquid-air process. Unlike oxygen, acetylene is not a natural but a manmade gas. It is a hydrocarbon gas produced by the action of water on calcium carbide. Acetylene is a colourless gas with a very distinctive nauseating odour. It is highly combustible when mixed with air or oxygen. It burns in ordinary air with a luminous smoky flame, but in pure oxygen it burns with an intensely hot non-luminous flame. Although it is stable under low pressures, it becomes very unstable and dangerous if compressed to any great pressure. Hence acetylene for welding purposes is dissolved in a liquid chemical known as acetone, which is capable of absorbing up to 25 times its own volume without changing the nature of the gas itself, and then stored in steel cylinders.

When acetylene and oxygen are mixed in the correct proportions and ignited, the resulting flame reaches a temperature of 3200 °C. This is intense enough to melt all commercial metals so completely that metals to be joined actually flow together to form a complete bond without mechanical pressure or hammering. Except on very thin material, extra metal in the form of a wire rod is usually added to the molten metal in order to strengthen the seam slightly. If the weld is performed correctly, the section where the bond is made will be as strong as the base metal itself. Metals which can be welded with oxy-acetylene flame include iron, steel, cast iron, copper, brass, aluminium and bronze, and many alloys may be welded; it is possible also to join some dissimilar metals, e.g. steel and cast iron, brass and steel, copper and iron, brass and cast iron. The oxy-acetylene flame is also employed for cutting metal, case hardening and annealing (see Section 5.2).

The welding industry grew very rapidly, and by the year 1907 the use of oxy-acetylene equipment was very popular in all parts of Europe. Today this type of welding is no longer used in the construction but only in the repair of motor vehicles. A casual inspection of a modern car reveals little evidence of welding, yet no modern car could be assembled without it. In body repair work welding techniques are used for frame straightening, reinforcing frame members, and welding in new or patching old panel assemblies. It is often necessary to repair panels where the metal has partially disintegrated through rust, or has been torn or otherwise broken. At one time this type of repair was effected by either riveting a patch over the rusted or torn parts, or renewing the part concerned. With the development of small portable gas welding equipment, however, the edges of the material can be heated locally until they melt, and then fused or welded together.

9.2 Systems of gas welding

There are two systems of gas welding. The low-pressure system is no longer widely used, and not at all in body repair work. The high-pressure system is used in all fields of engineering, and especially in the body repair trade. The difference between them lies in the method of manufacture of the acetylene. Acetylene for the high-pressure system is manufactured and stored ready for use in a dissolved state and at a high pressure in steel cylinders. In the low-pressure system the acetylene gas is manufactured as and when required by an acetylene generator in the workshop; in this case the gas is produced at a

very low pressure. Oxygen for use in both systems is supplied in steel cylinders at high pressure.

In the low-pressure system (Figure 9.1) acetylene gas, produced by a special generator situated in an isolated part of the workshop, is piped to convenient positions round the workshop at a pressure between $1.7\ kN/m^2$ and $62\ kN/m^2$. There are two types of generator: water to carbide, and carbide to water.

For the high-pressure system (Figure 9.2) the dissolved acetylene (DA) is supplied in cylinders at a pressure of 17.7 bar maximum at 15 °C. The cylinders generally contain between $0.57\ m^3$ and $8.73\ m^3$ of acetylene, from which a maximum supply of up to 20 per cent of the contents can be obtained during any one hour.

The high-pressure system has the following advantages:

1 Portable equipment is used for welding and cutting metals, so that the operator can have his equipment close up to the work.

2 The gases in the cylinders are under perfect control and do not need any maintenance either in or out of use.

3 The acetylene from the cylinder is always of the highest degree of purity, which enables the operator to make welds of the best quality.

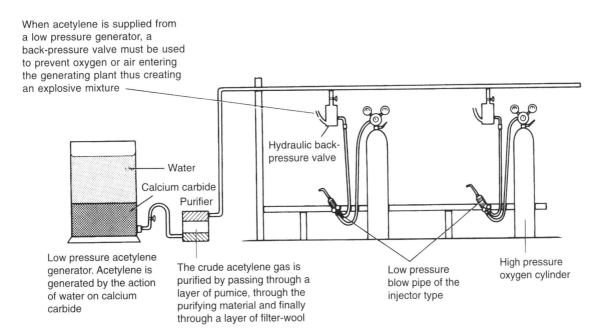

When acetylene is supplied from a low pressure generator, a back-pressure valve must be used to prevent oxygen or air entering the generating plant thus creating an explosive mixture

Water

Calcium carbide

Purifier

Hydraulic back-pressure valve

Low pressure acetylene generator. Acetylene is generated by the action of water on calcium carbide

The crude acetylene gas is purified by passing through a layer of pumice, through the purifying material and finally through a layer of filter-wool

Low pressure blow pipe of the injector type

High pressure oxygen cylinder

Figure 9.1 Low-pressure welding equipment

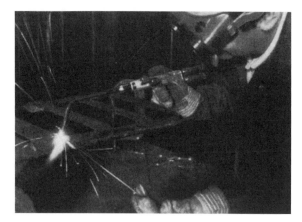

Figure 9.2 High-pressure gas welding equipment in use (*Murex Welding Products Limited*)

9.3 Oxy-acetylene welding equipment

High-pressure welding equipment (Figure 9.3) comprises:

1 Cylinder of dissolved acetylene
2 Cylinder of oxygen
3 High-pressure welding torch with separate oxygen control and acetylene control, and various-sized tips for different thicknesses
4 Acetylene regulator with gauges
5 Oxygen regulator with gauges
6 Two 5 m lengths of 10 mm high-pressure rubber canvas hose with clips and fitted with hose check valves (red for acetylene and blue for oxygen)
7 Acetylene flashback arrester and oxygen flashback arrester
8 Keys to fit cylinders and spanners for gland nuts, and for fixing the regulators to the cylinders
9 Pair of welding goggles
10 Cylinder trolley.

9.3.1 Acetylene supply

In the high-pressure system the acetylene is supplied in steel cylinders painted maroon and containing between $0.57\,\mathrm{m}^3$ and $8.73\,\mathrm{m}^3$ of gas when fully charged. These cylinders contain a porous filling which acts as an absorbent for the acetone used to dissolve the acetylene. Acetone has the remarkable property of being able to dissolve large volumes of acetylene gas. At normal charging pressures (about 17.7 bar) the acetone will dissolve approximately 25 times its own volume. To prevent the interchange

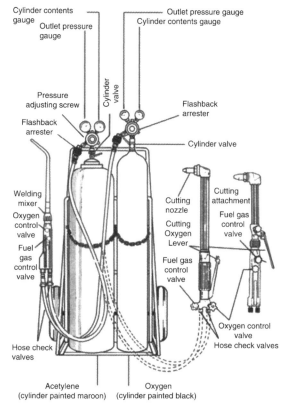

Figure 9.3 A typical oxy-fuel gas welding and cutting system (*Murex Welding Products Limited*)

of fittings between cylinders, the acetylene cylinder valve into which the regulator is fitted has a left-hand thread, while the oxygen cylinder valve has a right-hand thread (see Figure 9.4). The actual cylinder valves are all opened anti-clockwise and closed clockwise in the usual way.

Most acetylene cylinders supplied by the manufacturers are protected by at least one safety device. The majority of cylinders have a bursting

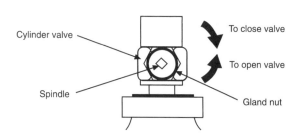

Figure 9.4 Cylinder valve (*BOC Ltd*)

disc on the back of the cylinder valve opposite the spindle. The disc operates at a pressure of about 90 bar. Others have either two fusible plugs fitted on the shoulder near the cylinder valve, or a bursting disc. These plugs melt at about 100 °C and relieve any excess pressure if the cylinder is subjected to undue heat or catches fire. Care is necessary in handling the cylinders as any leak is liable to lead to fire, especially in confined spaces.

9.3.2 Oxygen supply

The oxygen supply in the high pressure system comes in seamless steel cylinders containing between 0.68 m^3 and 9.66 m^3 of oxygen and painted black. As this is a non-combustible gas, the cylinder has a right-hand threaded valve outlet. Oxygen is supplied at a maximum pressure of 200/230 bar at 15 °C.

Cylinder labelling
To comply with the Classification, Packaging and Labelling of Dangerous Substances Regulations 1984, BOC cylinders now appear with a label

designed to give more safety information. The format is shown in Figure 9.5 and the various sections are identified. Never use gas from a cylinder which has no label; return the cylinder to the supplier. If a cylinder is involved in an accident, withdraw it from service and set it aside clearly marked; contact the supplier.

Gas cylinders are designed and constructed in accordance with Home Office specifications and/or British Standards. These standards define the material of which the cylinder is made, the method of construction, its test pressure, the maximum permissible filled pressure and the method of regular testing (Figure 9.6).

All cylinders containing gas at high pressure are fitted with a cylinder valve which ***must not*** be removed or tampered with at any time except to tighten the gland nut when necessary. Some cylinders have a security cap over the cylinder valve indicating that they have been filled and checked. This cap is removed by rotating the hexagon nut in either direction using a regulator spanner: this will cause the cap to split for easy removal. To open the cylinder valve, rotate the

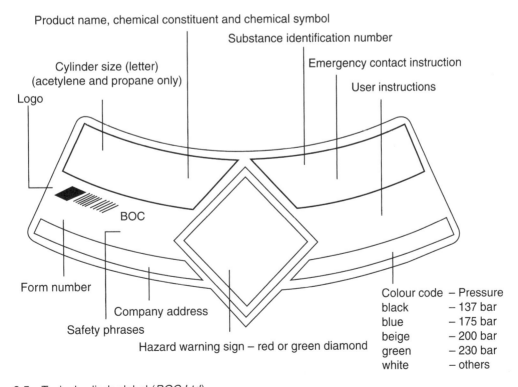

Figure 9.5 Typical cylinder label (*BOC Ltd*)

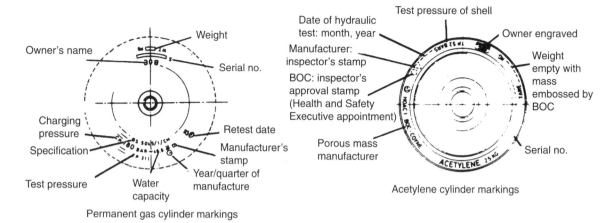

Permanent gas cylinder markings

Acetylene cylinder markings

Figure 9.6 Identification markings on gas cylinder necks (*BOC Ltd*)

spindle anticlockwise using the special spindle key. Some cylinders are fitted with hand wheels, obviating the need to use a spindle key. These cylinders, together with others which are fitted with valves, normally have valve guards or valve protection caps. Valve guards should not be removed. Valve protection caps should always be replaced after use.

Only the owner of a cylinder can authorize its scrapping, and before this is done the cylinder must first be destroyed as a pressure vessel.

9.3.3 Welding torches

The welding torch comprises two needle valves (one for regulating the oxygen and the other the acetylene), a mixing chamber and an interchangeable welding nozzle, all incorporated in a shank. The torch, sometimes known as the blowpipe, mixes acetylene and oxygen in the correct proportion and permits the mixture to flow to the end of the welding tip or nozzle, where it is burned. Blowpipes vary to some extent in design, but basically they are all made to provide complete control of the flame during the welding operation. The equal-pressure torch is designed to operate in the high-pressure system (Figure 9.7), and will operate when acetylene is supplied from cylinders. The acetylene and oxygen are fed independently to the mixing chamber, and then flow through the tip. On the rear end of the torch there are two female unions for connecting the relevant hoses; to

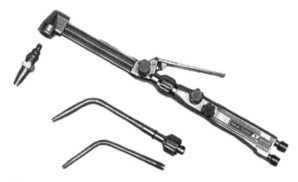

Figure 9.7 Welding and cutting blowpipe, high pressure (*Murex Welding Products Limited*)

eliminate any danger of interchanging the hoses, the oxygen union has a right-hand thread and the acetylene union a left-hand thread.

Interchangeable welding nozzles
To make possible the welding of different thicknesses of metal, welding torches are equipped with an assortment of different size heads or nozzles. The size of the nozzle is governed by the diameter of the opening at the welding end. The most common sizing system uses numbers which are marked on the nozzle, and the larger the number, the greater is the diameter of the opening in the nozzle; nos 1–35 are for use with thicknesses up to 10 mm and nos 35–90 are for 10–25 mm (see Table 9.1).

Table 9.1 Welding data (*BOC Ltd*) Swaged nozzle, 6.3 mm × 10 m hose, resettable flashback arresters

Mild steel thickness			Nozzle size	Operating pressure				Gas consumption			
				Acetylene		Oxygen		Acetylene		Oxygen	
mm	in	SWG		bar	lbf/in²	bar	lbf/in²	l/min	ft³/h	l/min	ft³/h
0.9	–	20	1	0.14	2	0.14	2	0.47	1	0.47	1
1.2	–	18	2	0.14	2	0.14	2	0.94	2	0.94	2
2.0	–	14	3	0.14	2	0.14	2	1.42	3	1.65	3.5
2.6	–	12	5	0.21	3	0.21	3	2.36	5	2.83	6
3.2	$\frac{1}{8}$	10	7	0.21	3	0.21	3	3.30	7	3.77	8
4.0	$\frac{3}{32}$	8	10	0.28	4	0.28	4	4.7	10	5.2	11
5.0	$\frac{3}{16}$	6	13	0.28	4	0.28	4	6.6	14	7.1	15
6.5	$\frac{1}{4}$	3	18	0.35	5	0.35	5	8.5	18	9.4	20
8.2	$\frac{5}{16}$	0	25	0.4	6	0.48	7	11.8	25	12.7	27
10.0	$\frac{3}{8}$	4/0	35	0.66	9.5	0.66	9.5	16.5	35	17.9	38
13.0	$\frac{1}{2}$	7/0	45	0.4	6	0.4	6	21.2	45	22.6	48
25+	1+	–	90	0.62	9	0.62	9	42.5	90	44.8	95

9.3.4 Regulators

There are two types of regulator, the two-stage and the single-stage. The oxygen and acetylene pressure regulators perform two functions (see Figures 9.8 and 9.9): they reduce the cylinder pressure to the required working pressure, and they produce a steady flow of gas under varying cylinder pressures. With the two-stage regulator the gas flows from the cylinder into a first chamber where a predetermined high pressure is maintained by means of a spring and diaphragm. From the high-pressure chamber the gas passes into a second, reducing chamber where the pressure is governed by an adjusting screw. The principal advantage of the two-stage regulator is that the working pressure remains constant until the gas in the cylinder is exhausted, and little adjustment of gas flow is needed. In the single-stage regulator there is only one, low-pressure chamber; the gas from the cylinder flows into the regulator and is controlled entirely by the adjusting screw. The disadvantage of this regulator is that as the cylinder pressure drops, the actual working pressure also falls and thus needs continual readjustment.

Both types of regulator have two gauges, one indicating the pressure in the cylinder and the other showing the actual working pressure or line pressure

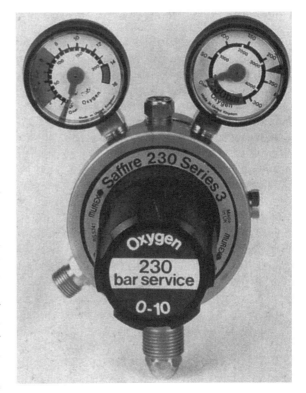

Figure 9.8 Oxygen regulator (*Murex Welding Products Ltd*)

Figure 9.9   Acetylene regulator (*Murex Welding Products Ltd*)

Figure 9.10   Flashback arresters for oxygen and acetylene regulators (*Murex Welding Products Limited*)

at the torch. The content gauge on the oxygen regulator is calibrated up to 300 bar, whereas the working pressure gauge is calibrated from 0 to 6 bar (Figure 9.8). The content gauge on the acetylene regulator is calibrated up to 40 bar, whereas the working pressure gauge is calibrated from 0 to 2.5 bar (Figure 9.9). It is most important to remember to release (turn out) the adjusting screw on the regulator before opening the cylinder valve, otherwise the tremendous pressure of the gas in the cylinder is forced into the working pressure gauge and this could result in gauge damage. Before a regulator is attached to a cylinder, the valve of the latter should be 'cracked' by quickly opening and then closing it to blow out any particles of dirt that may be present; if the gas is acetylene, care must be taken to guard against any possibility of it becoming ignited. As previously mentioned, oxygen and acetylene regulators also have female unions, which have right and left threads respectively to prevent the risk of hose interchange. To distinguish the connections still further, the oxygen regulator gauges when new are painted black and those of acetylene regulators are red. An acetylene regulator may be used on a propane cylinder, but not on a hydrogen or coal gas cylinder as the initial high pressure would damage or burst the cylinder contents gauge.

9.3.5 Automatic flashback arrester

This is an automatic safety device (Figure 9.10) for use in the high-pressure system with either oxygen or acetylene cylinders. An arrester is fitted to the pressure regulator outlet of a cylinder with its opposite end connected to the hose leading to the welding torch. If excess pressure builds up, the arrester acts by completely sealing off the gas supply. Only when the pressure has subsided to normal working level via the built-in escape valve will the arrester allow the main supply of gas to continue flowing. The increase in safety which this piece of equipment affords makes it an essential part of the high-pressure system of welding.

9.3.6 Gas hose

Gas hoses are of a special non-porous rubber, reinforced with canvas. They are designed to withstand the working pressures which transmit gas from the regulator to the welding torch. Hoses are available with internal diameters of 5–20 mm and are coloured red for acetylene and blue for oxygen to prevent the risk of hose interchange. The nut on the acetylene hose is distinguished from the nut on the oxygen hose by a groove that runs around its centre, indicating that it has a left-hand thread. A ferrule is used to squeeze the hose around the union shank to prevent it from working loose. The life of the hose may be prolonged by reasonable care in use, e.g. by keeping it away from heat, sparks, oil and

Figure 9.11 Hose fittings: nut, nipple, and O-clips (*Murex Welding Products Ltd*)

grease and by preventing it from being crushed. Hose fittings are shown in Figure 9.11.

9.3.7 Hose check valve

This is a safety device in the form of a non-returnable valve which is fitted behind the welding torch and connected to the hose (Figure 9.12). There are check valves for both oxygen and acetylene connections. If the pressure or gas flow tries to reverse, the hose check valve immediately seals off the torch from the main supply, thus eliminating the possibility of flashback and back-firing of the welding torch through the hose to the regulator and cylinder.

Figure 9.12 Hose check valves (*Murex Welding Products Ltd*)

9.3.8 Cylinder key

This key is an essential part of the equipment, and is used for turning the valves on the gas cylinders on and off. It should be always attached to the welding equipment or plant so that it is readily available in case of emergency.

9.3.9 Welding goggles

Goggles fitted with tinted glass of an approved type complying with the requirements set out in British Standard specification 679 should be worn to protect the eyes from the intense light of the flame as well as from sparks and small metal particles thrown up from the weld. As the goggles become pitted in time and obscure the work, it is usual to protect the tinted glass with a plain one which can be replaced at low cost. Typical welding goggles are shown in Figure 9.13.

Figure 9.13 Types of welding goggles (*Murex Welding Products Limited*)

9.3.10 Cylinder trolley

It is advisable to have all welding equipment portable and mobile when in use in a body repair shop, and for this purpose a cylinder trolley can be used. Both cylinders must be securely fastened but at the same time easy to replace. The trolley itself should be of sturdy construction and should be as narrow as possible so that it can pass through restricted spaces.

9.3.11 Assembly of high-pressure welding equipment

The cylinders must be kept upright and all metal jointing surfaces should be free from oil or grease. Before fitting the regulators, blow out the cylinder valves to remove any dirt or obstruction. To ensure gas-tight connections between cylinder valves and regulators, first screw down the hexagon or wing nuts by hand, then give the regulators a twist to bed them down on their seats, and finally tighten the nuts properly, but without the use of excessive force. New hoses in use for the first time should be blown through to remove any grit that may be present. Attach the appropriate hoses to the regulators and the welding torch, and fit the latter with a suitable welding tip. Next slacken off the pressure regulating screws, open the regulator outlet valves and turn on the gases with the cylinder key; this must be done slowly to avoid damage to the regulators. The cylinder valves must be opened at least two full turns to ensure that the flow of gases to the regulators is unrestricted. Next set the regulators to the required pressure, open the acetylene control valve on the welding torch, check working pressures when pure acetylene is coming out of the end of the nozzle, and then light the gas. Reduce or increase the acetylene supply by the welding torch valve until the flame just ceases to smoke, then turn on the oxygen by the welding torch control valve until the white inner cone is sharply defined, and check working pressures again. The welding torch is now ready and is burning with a neutral flame, which is used on most welding operations.

On completion of the welding operation the following procedure must be carried out to render the equipment safe. Turn off the acetylene first by the welding torch control valve, then the oxygen. Close the cylinder valves. Open the welding torch valves one at a time to release the pressure in the hose; open the oxygen valve and shut it; and open the acetylene valve and shut it. Unscrew the pressure regulating screws on the oxygen and acetylene regulators.

9.4 Definitions of welding terms

The material of the parts to be welded is described as the *parent metal*. Any material that it may be necessary to add to complete the weld is known as *filler metal* which, if in rod or wire form, is obtained from a *welding rod*. The surface to be welded is called the *fusion face*. That part of the weld where the parent metal has been melted, if filler is used and interdiffusion has taken place, is called the *fusion zone*, the depth of which is termed the *weld penetration*. Bordering on the fusion zone is the *zone of thermal disturbance*, consisting of that portion of the parent metal which, although not melted by the flame, has been heated sufficiently to disturb the grain structure; where the fusion zone and zone of thermal disturbance meet is known as the *junction* (Figure 9.14). A *bead* is a single longitudinal deposit of weld metal laid on a surface by fusion welding; a local deposit laid on a surface is termed a *pad*, and is usually formed by a series of overlapping beads. *Tack welds* are local welds used to hold parts in their correct relative positions ready for welding.

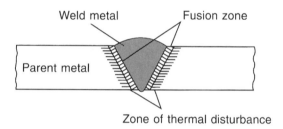

Figure 9.14 Section through a welded joint

Most welded assemblies are made by *butt joints* in which the ends or edges directly face each other. In some cases the joining weld is made between these edges. Additionally or alternatively *fillet welds* may be used, which are of approximately triangular cross-section and lie externally to the parts joined. These fillet welds may consist of *T-joints* or *lap joints*; in the former the parts are usually set at right angles to one another, and in the latter the weld is made in the angle formed by the overlap. An alternative to the butt joint is the *edge joint*, in which the two metals are put together to form a corner.

Various terms are used to indicate the different parts of a weld. The *weld face* is the exposed surface of any weld, a *leg* is the fusion face of a fillet weld, and the *toe* is a border line where the weld face adjoins a welded part; along this line *undercut* or wastage of the parent metal in the form of a grooving may occur. The *root* is the zone at the bottom of a space provided for or occupied by a fusion weld, and the *throat* is the minimum depth of the weld measured along a line passing through the root. The condition which arises when the filler metal flows on to heated but unfused joint surfaces, and the interfusion of the filler and parent metals does not take place, is known as *adhesion*.

9.5 Welding rods and fluxes

9.5.1 Filler rods

Filler rods for use in oxy-acetylene welding are available in the following metals: mild steel, wrought irons, high-carbon steel, alloy steel, stainless steel, cast iron, copper, copper alloys, aluminium, aluminium alloys, hard facing alloys, zinc-based die cast alloys.

The metal from the rod has considerable influence on the quality of the finished weld. Good welding rods are designed to give deposited metal of the correct composition, and have allowance in their chemistry for changes which take place in the welding process. Rods are obtainable in sizes ranging from 1.6 mm to 5.0 mm diameter. Some have a copper coating to keep their surfaces free from oxides or rust, but uncoated rods are equally efficient provided that they are clean. Sound welds, comparable in strength with the material welded, can be produced with satisfactory filler rods, but similar results cannot be expected with dirty or rusty filler rods or with any odd piece of wire that may come to hand.

9.5.2 Welding fluxes

It is impracticable to incorporate in welding rods all the elements necessary to overcome oxidation. Therefore it is necessary to use with them certain chemical compounds to act as deoxidizing agents or fluxes, which must be of the correct composition to ensure perfect welds. A flux *must* be used with the following metals: cast iron, high-carbon steel, stainless steel, copper, copper alloys, aluminium, aluminium alloys, magnesium alloys. In the majority of cases it is essential that the flux residues should be removed from the surface of the metal after the welding operation has been completed.

9.6 Flame control and types of flame

9.6.1 Temperature of flame

When acetylene is burned with an equal volume of oxygen, a maximum flame temperature of over 3200 °C (about 2.5 times hotter than the melting point of cast iron and steel) is obtained just beyond the inner luminous cone (Figure 9.15). Variations of the proportions of oxygen and acetylene can be made at the welding torch, and three distinct types of flame are obtained (Figure 9.16) as follows.

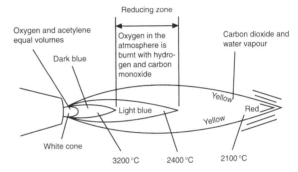

Figure 9.15 Flame temperature

Carbonizing (carburizing) flame (excess acetylene)

Neutral flame (equal quantities of oxygen and acetylene)

Oxidizing flame (excess oxygen)

Figure 9.16 Regulation of welding flame

Carbonizing, carburizing or reducing flame

For this flame the gases are adjusted so that all the acetylene gas is not completely burned and there is an excess of acetylene or an insufficiency of oxygen. When the welding torch is lit, the acetylene is turned on first and ignited, giving a very smoky yellow flame of abnormal size which shows two cones of flame in addition to an outer envelope; this is an exaggerated form of the carbonizing flame, but gives out comparatively little heat and is useless for welding. The oxygen is turned on and the supply is gradually increased until the flame, though still of abnormal size, contracts towards the welding torch tip, where an inner white cone of great luminosity commences to make its appearance. The increase in the supply of oxygen is stopped before the cone becomes clearly defined and while it is still an inch or so long; this results in a carburizing flame, which is characterized by a dullish white feather surrounding a brilliant, clearly defined white cone. The flame can be used to advantage in the welding of high-carbon steel, aluminium, Inconel and Monel metal in the technique of stelliting, for stainless steel, and wherever excess of oxygen on metals would cause detrimental oxidation. However, in some cases the flame is detrimental owing to the fact that it deposits carbon; for example, if it is used on mild steel the weld deposit will be higher in carbon and therefore becomes hardenable material, and cracks may result.

Neutral flame

As the supply of oxygen to the welding torch is increased beyond the point at which the carbonizing flame is formed, the flame contracts and the white cone assumes a definite rounded shape. At this stage approximately equal quantities of acetylene and oxygen are being used and combustion is complete, all the carbon supplied by the acetylene being consumed and the maximum heat given out. This flame is known as the neutral flame, and is the one most extensively used by the welder. It does not oxidize or carburize the material, and it is used on mild steel, copper and magnesium.

Oxidizing flame

A further increase in the oxygen supply will produce an oxidizing flame in which there is more oxygen than is required for complete combustion. The inner cone will become shorter and sharper and the flame will turn a deeper purple colour and emit a characteristic slight hiss. With this, the oxidizing flame, the molten metal will be less fluid and tranquil during welding and excessive sparking will occur. An oxidizing flame is only used for special applications, and should always be avoided when welding steel, because it causes rapid oxidation of the metal. It is, however, used in the welding of brasses, bronzes, copper-silicon alloys and in the process of bronze welding, the degree of oxidation required being determined experimentally.

9.6.2 Power of flame

The power or heat value of the flame is governed by the size of the orifice in the tip used. The power required depends on the thickness, mass, melting point and heat conductivity of the metal to be welded. Manufacturers of welding torches provide information regarding the size and rated gas consumption of tips for different thicknesses of metals and alloys. The power of a welding torch is measured by the number of litres or cubic feet of acetylene which is consumed in an hour with the flame perfectly regulated, and this figure is sometimes marked on the welding torch itself. The heat value of the flame must be adjusted by changing the tip and not by unduly increasing or decreasing the pressure and volume of gases used. The pressure of gases used for any size of tip determines the length of the inner cone, which should be 3 to 3.5 times as long as its breadth at the base. Too low a pressure gives too short a cone, which may cause lack of penetration and fusion; it also causes frequent backfiring and will be deflected by particles of metal and slag thrown up from the weld. On the other hand, too high a gas pressure giving too long a cone causes overheating and lack of control of the molten metal, resulting in adhesion and over-penetration.

9.6.3 Flame adjustment and gas pressure

The adjustment of a good welding flame is affected by the pressure of the gas entering the welding torch. If the pressure from either the acetylene or the oxygen regulator is excessive, the result will be a fierce or harsh flame which is very difficult to manipulate, especially when welding panel steel with the neutral flame. Excessive pressure also increases the tendency to blow the edges of the steel away as they are melted. On the other hand, if the pressure of either gas is too low a backfiring of the welding torch will result.

9.7 Methods of welding

Leftward or *forward* welding is the technique or method of welding in which the welding torch flame is directed towards the uncompleted joint. When the flame is directed towards the completed weld, this is termed *rightward, backward* or *backhand* welding. The positions in which welding is performed are termed respectively *flat, downhand* or *underhand* when the weld is made on the upper side of a horizontal or flat surface, *overhead* when on the underside, and *vertical* when on an upright or vertical surface.

Leftward and rightward welding are shown in Figure 9.17.

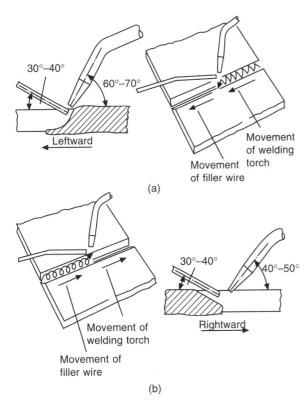

(a)

(b)

Figure 9.17 Methods of welding (a) the leftward method and (b) the rightward method

9.7.1 Leftward welding

This is used on steel for flanged edge welds, for bevelled steel plates up to 3.2 mm, and for bevelled plates up to 4.8 mm. It is also the method usually adopted for cast iron and non-ferrous metals. The weld is started on the right-hand end of the joint

and welding proceeds towards the left. The welding torch is given a forward motion, with a slight sideways movement to maintain both edges melting at the desired rate, and the welding wire is moved progressively along the weld seam. The sideways motion of the welding torch should be restricted to a minimum. The advantages of the leftward welding technique are that it is faster because the flame, in preceding the weld, has a preheating effect; the molten metal is easily controlled; and penetration (complete fusion of the edges) is easily obtained over the range of application of the technique. One of the important features of a good weld is complete penetration. A disadvantage of the leftward method of welding is that if incorrect manipulation of the flame occurs, molten metal will flow ahead of the molten pool at the bottom of the weld and adhere to the comparatively cold plate, thus causing poor penetration.

The characteristics of the leftward welding technique are as follows:

Position Flat.
Direction of welding Flame points away from the finished weld.
Angle of welding torch 60°–70°.
Movement of welding torch Straight along seam, with side to side movement reduced to a minimum.
Gas consumption of the welding torch 86 l/h (litres per hour) and 3 ft³/h (feet cubed per hour) for 1.6 mm thickness.
Type of flame Neutral.
Position of filler rod Precedes the welding torch.
Angle of filler rod 30°–40°.
Movement of filler rod Slight movement along seam in and out of the molten pool.
Size of filler rod Approximately equal to the plate thickness up to 2.4 mm, and then 3.2 mm for thickness above 2.4 mm.
Plate preparation Up to 1.00 mm: edges flanged. From 1.00 mm to 3.2 mm: square edge preparation. From 3.2 mm to 4.8 mm: 80 °V preparation.

9.7.2 Rightward and all-position rightward welding

Rightward welding is recommended for steel plates over 4.8 mm thick. Plate edges from 4.8 to 7.9 mm need not be bevelled. Plate over 7.9 mm should be bevelled to 30° to give an included angle of 60° for the welding V. The weld is started at the left-hand end of the joint and the welding torch is moved towards the right, the welding torch preceding the filler rod in the direction of travel. The wire is given a circular forward action and the welding torch is moved steadily along the weld seam. It is quicker than leftward welding and consumes less gas. The V-angle is smaller, less welding rod is required and there is less distortion. The advantages of the rightward technique are as follows: the direction of the flame holds back the molten pool, preventing any tendency to adhesion; the flame, which points directly towards the root of the weld, causes a hole to form as the edges melt, thus ensuring complete penetration; no preparation is necessary on thicknesses up to 7.9 mm.

The characteristics of the rightward technique are as follows:

Position   Flat.
Direction of welding   Flame points towards finished weld.
Angle of welding torch   40°–50°.
Movement of welding torch   Straight along seam, with sufficient side swing to ensure complete fusion of the edges.
Gas consumption of the welding torch   140 l/h which is 5 ft³/h for 2.6 mm thickness.
Type of flame   Neutral.
Position of filler rod   Follows the welding torch.
Angle of filler rod   30°–40°.
Movement of filler rod   Circular forward action.
Size of filler rod   Half the plate thickness up to a maximum of 6.4 mm.
Plate preparation   From 4.8 mm to 7.9 mm: square edge, with gap of half thickness. From 7.9 mm to 15.9 mm: 60° bevel, gap of 3.2 mm, and weld made in one pass. From 15.9 mm upwards: as above but step weld, using multipass technique.

In the rightward and leftward techniques it is seen that edge preparation becomes necessary at 7.9 mm and 3.2 mm respectively, in order to obtain complete fusion without fear of adhesion. This involves extra cost and additional filler material. If therefore the bevel edge preparation can be obviated, the cost of the welding is reduced.

9.8 Edge preparation and types of joint

In order to produce a satisfactory butt weld it is essential for the plate edges to be joined throughout their entire thickness; this necessitates complete melting of the edges. With thin materials this result can be achieved with square edged plates. However, the same procedure applied to thicker metal is unlikely to result in complete fusion, and the fusion faces must be bevelled to enable the torch flame to be directed into the root of the weld in order to obtain complete fusion.

Square edge preparation is used for material up to 3.2 mm thickness. The edges are left square and separated by a distance equal to one-half the thickness of the sheet used. Such joins are termed *open butt joint*.

Single V preparation consists of bevelling the edges of each plate so that a V is formed when they are brought together. For thicknesses between 3.2 mm and 4.8 mm an 80° bevel is used, leaving a small gap at the bottom edges. For material over 7.9 mm thickness the angle of V should be 60°. It is not necessary to bevel materials up to 7.9 mm thick if the rightward method of welding is used.

Double V preparation is used for material thicker than 15.9 mm, which must be welded from both sides of the plate; for this reason a V must be provided on each side. The top V should be 60° and the bottom V 80°, and the edges separated by a gap of 3.2 to 4.0 mm.

Figure 9.18 gives details of edge preparation.

9.8.1 Control of distortion

A proper understanding of the problems of distortion or buckling associated with the welding of sheet metal is of the utmost importance to the panel beater working on thin sheet metal. This distortion results from the shrinkage which occurs when molten iron passes from the liquid to the solid state. In consequence the welded seam tends to shorten in length, causing the parent metal to buckle. If the welded seam is so placed that it may be planished then the hammering along the weld will correct the distortion. As iron shrinks approximately 10 mm per metre when solidifying from the molten condition, the tendency to produce distortion is considerable. Where the work to be welded is of thick steel plate, the plate itself may be sufficiently strong to counteract or minimize some of the distortion, but where the work is of thin mild steel the distorting stresses take full effect in buckling the sheet. In the process of

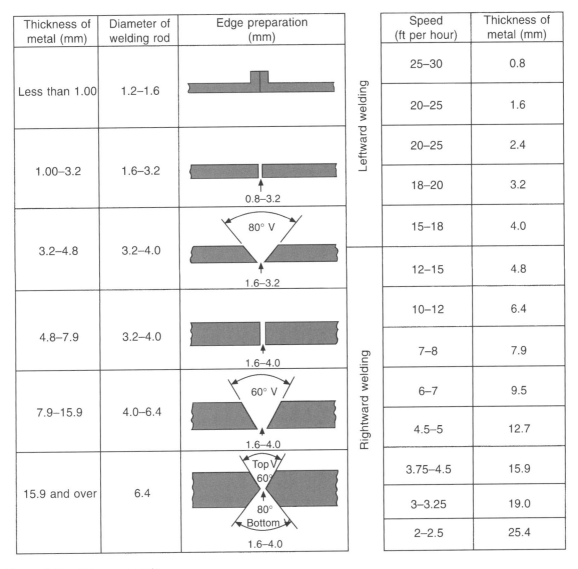

Thickness of metal (mm)	Diameter of welding rod	Edge preparation (mm)		Speed (ft per hour)	Thickness of metal (mm)
			Leftward welding	25–30	0.8
Less than 1.00	1.2–1.6			20–25	1.6
				20–25	2.4
1.00–3.2	1.6–3.2	0.8–3.2		18–20	3.2
		80° V		15–18	4.0
3.2–4.8	3.2–4.0	1.6–3.2		12–15	4.8
				10–12	6.4
4.8–7.9	3.2–4.0	1.6–4.0	Rightward welding	7–8	7.9
		60° V		6–7	9.5
7.9–15.9	4.0–6.4	1.6–4.0		4.5–5	12.7
		Top V 60°		3.75–4.5	15.9
15.9 and over	6.4	80° Bottom		3–3.25	19.0
		1.6–4.0		2–2.5	25.4

Figure 9.18 Edge preparation

welding it is necessary to consider and make allowances for this distortion which, unless controlled, may buckle the work to such an extent that it is useless. When the joint edges are heated they expand, and as welding proceeds, contraction of the deposited weld metal takes place owing to the loss of heat by radiation and condition. The rate at which cooling takes place depends on various factors such as the size of the work; the amount of weld metal and the speed at which it is deposited; the thermal conductivity of the parent metal; and the melting point and specific heat of the weld metal.

Some methods of controlling distortion are as follows (Figure 9.19):

1 Efficient tacking or clamping, which maintains the edge positions of the plates
2 Tapered spacing of plates so that they pull together the process of welding
3 Use of intermittent weld technique and back-step weld technique

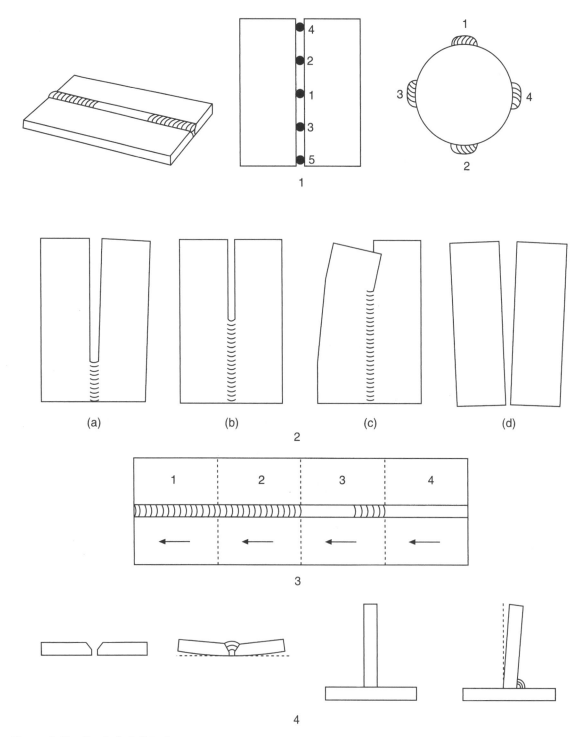

Figure 9.19 Control of distortion

4   The offsetting and presetting of plates so that they are pulled correct by the contraction of the welds
5   The use of chilling bars and chemical foam barriers
6   The use of planishing when the weld is in the cold state
7   The use of jigs and fixtures.

9.9 Welding technique: butt joint in mild steel

To master the skill of welding with an oxy-acetylene torch, you will have to practise a series of operations in a definite order. The torch may be held in either one of two ways, depending on which is the more comfortable for you. When welding light-gauge metal, most operators prefer to grasp the handle of the welding torch with the hose over the outside of the wrist, which is the way in which a pencil is usually held. In the other grip, the torch is held like a hammer, with the fingers lightly curled underneath. In either case the torch should balance easily in the hand to avoid fatigue. Hold the torch in the direction in which you are going to weld and at an angle of about 65° with the completed part of the weld. If you are right-handed, start the weld at the right edge of the metal and bring the inner cone of the neutral flame to within 3 mm of the surface of the plate. If you are left-handed, reverse this direction.

Hold the torch still until a pool of molten metal forms, then move the puddle across the plate. As the puddle travels forward, rotate the torch to form a series of overlapping ovals. Do not move the torch ahead of the puddle, but slowly work forward, giving the metal a chance to melt. If the flame is moved forward too rapidly, the heat fails to penetrate far enough and the metal does not melt properly, but if the torch is kept in one position too long the flame will burn a hole through the metal.

On some types of joints it is possible to weld the two pieces of metal without adding a filler rod, but in most instances the use of a filler rod is advisable because it builds up the weld, adding strength to the joint. The use of a filler rod requires coordination of the two hands. One hand must manipulate the torch to carry a puddle across the plate, while the other hand must add the correct amount of filler rod. Hold the rod at approximately half the angle of the torch, but slant it away from the torch.

It is advisable to bend the end of the rod at right angles, since this permits holding the rod so that it is not in a direct line with the heat of the flame.

Melt a small pool of the base metal and then insert the tip of the rod in this pool. Remember that the correct diameter rod is an important factor in securing perfect fusion. If the rod is too large the heat of the pool will be insufficient to melt the rod, and if the rod is too small the heat cannot be absorbed by the rod, with the result that a hole is burned in the plate. As the rod melts in the pool, advance the torch forward. Concentrate the flame on the base metal and not on the rod. Do not hold the rod above the pool, as if you do the molten metal will fall into the puddle, combining with oxygen in the air as it falls so that part of it burns up and will cause a weak, porous weld. Always dip the rod in the centre of the pool.

Rotate the torch to form overlapping ovals and keep raising and lowering the rod as the molten puddle is moved forward. An alternative torch movement is the semicircular motion. When the rod is not in the puddle, keep the tip just inside the outer envelope of the flame. Work the torch slowly to give the heat a chance to penetrate the joint, and add sufficient filler rod to build up the weld about 2 mm above the surface. Be sure that the puddle is large enough and that the metal is flowing freely before you dip in the rod. Watch the course of the flame closely to make sure that its travel along both edges of the plate is the same, maintaining a molten puddle approximately 6–10 mm in width. Advance this puddle about 2 mm with each complete motion of the torch. Unless the molten puddle is kept active and flowing forward, correct fusion will not be achieved. Keep the motion of the torch as uniform as possible, as this will produce smooth, even ripples and so complete the weld.

9.10 Welding various metals

Mild steel Select the appropriate method and always use a neutral flame. Fluxes are not necessary.
Carbon steel A flux must be used, and the flame kept in a neutral condition. After welding is completed the metal should be cooled slowly to avoid the weld metal becoming brittle.
Alloy steels It is very important that the correct welding rod is used for the appropriate alloy. The

metal should be first preheated and then welded. It is essential to cool the metal slowly after welding.

Stainless steel A welding rod having the same composition as stainless steel should be used. A flux should be used, and the welding flame kept in the neutral position. Do not interrupt the welding sequence, and carry out the weld as quickly as possible. On completion of the weld allow it to cool slowly. Remove all oxide and scale from the weld when it is finished.

Cast iron A silicon cast iron welding rod should be used, together with a cast iron welding flux. The metal must be preheated to dull red before the welding commences. It is very important that the cast iron cools very slowly, or the metal will crack.

Aluminium For pure aluminium sheet, either paint both sides of the metal with flux or dip the hot rod into the flux and allow the flux to coat the rod like varnish. Before welding, remove all traces of oil or grease and brush the edges to be welded with a wire brush. Tack the weld at frequent intervals using a neutral or slightly carburizing flame. The welding technique is leftward and should be carried out more quickly than when welding mild steel. On completion of the weld, wash it thoroughly to remove the remaining flux as this could be harmful.

Aluminium castings Use cast aluminium alloy rods with silicon and flux. Preheat before welding. Melt the rod well into the weld, using the welding rod to puddle the molten metal. On completion cool slowly.

9.11 Gas cutting

The oxy-acetylene process is widely used to cut metal, especially in the body building industry where heavy sections have to be cut to special shapes for construction, and also in the repair side of the industry where special nozzles have been developed for cutting away damaged parts of sheet metal body sections. Cutting may be done by means of a simple hand cutting torch, or by a more complicated, automatically controlled cutting machine.

9.11.1 Flame cutting

Flame cutting, known also as oxygen cutting, is made possible by the fact that oxygen has a marked affinity for ferrous metals which have been previously heated to their ignition temperature.

Thus the cutting of iron and steel merely involves the direction of a closely regulated jet or stream of pure oxygen on to an area that has been previously heated to ignition temperature (1600 °C, or when the metal has reached a bright cherry-red colour). As the iron is oxidized, the oxygen jet is moved at a uniform speed so that a narrow cut is formed. Since only the metal within the direct path of the oxygen jet is acted upon, very accurate results can be obtained if close control is exercised; when hand cutting 150 mm thick steel, the working tolerance is about 1.5 mm (Figure 9.20).

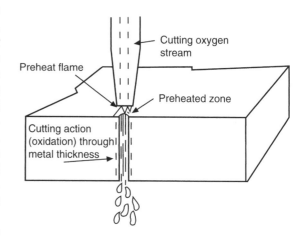

Figure 9.20 Flame cutting process (*BOC Ltd*)

9.11.2 Cutting torch

This differs from the regular welding torch in that it has an additional lever for the control of the oxygen used to burn the metal (Figures 9.21 and 9.22). It is possible to convert a welding torch into a cutting torch by replacing the mixing head with a cutting attachment (Figure 9.7). The torch has conventional

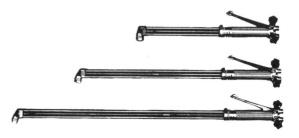

Figure 9.21 Cutting torches (*Murex Welding Products Limited*)

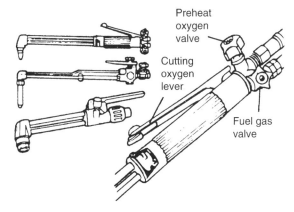

Figure 9.22 Types of cutting torches (*BOC Ltd*)

oxygen and acetylene valves, and these are used to control the passage of oxygen and acetylene when heating the metal. The cutting tip has an orifice in the centre for the oxygen flow, surrounded by several smaller holes for a preheating flame which generally uses acetylene, propane or hydrogen. The preheating flame has two purposes.

1 To provide sufficient heat to raise a small area of the steel surface to the ignition temperature
2 To transmit sufficient heat to the top surface of the steel to offset the thermal conductivity of the metal.

A wide variety of cutting torches are available, but in essentials they are all similar. Oxygen and a fuel gas for the preheat flame, for example acetylene, enter the torch separately and are mixed in the body of the torch or in the nozzle. Their respective flow rates are adjusted by two hand valves on the torch. The cutting oxygen stream is bled off inside the torch through a lever operated valve (Figure 9.22).

Two basic methods of gas mixing are employed. In the first type, the two gases enter the torch at approximately the same pressure and are mixed in a separately designed chamber either in the body of the torch or in the nozzle. This is termed the equal-pressure or high-pressure cutting torch. Alternatively, oxygen enters the torch at a very much higher pressure than the fuel gas and sucks the fuel gas in through an injector which also mixes the two gases. These are termed injector cutting torches (Figure 9.23). Both designs are equally good for flame cutting.

9.11.3 Nozzles

A number of different designs of nozzles are available to suit the various combinations of fuel gases and torch design. In principle, however, nozzles are the same: they have a central orifice for the cutting oxygen stream, surrounded by a ring of orifices for the preheat flame.

Cutting torches in which oxygen and the fuel gas are mixed in the body of the torch (injector type) have two tubes leading to the nozzle, one for cutting oxygen and one for mixing oxygen

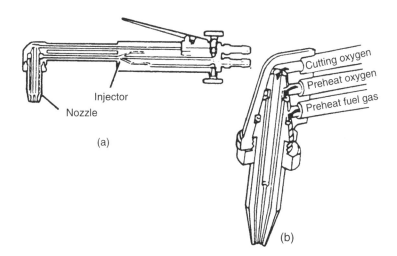

Figure 9.23   (a) Injector cutting torch and two-seat nozzle (b) three-seat nozzle (*BOC Ltd*)

and fuel gas. A so-called two-seat nozzle fits this design (Figure 9.23a). Some torches are designed for mixing oxygen and fuel gas in the nozzle itself. These torches have three tubes leading to the nozzle, and require special three-seat nozzles (Figure 9.23b).

A typical oxy-acetylene nozzle has circular pre-heat holes surrounding a central circular cutting oxygen orifice, and is frequently of a one-piece construction, made of copper and often chrome plated (Figure 9.24a). On the other hand a liquid propane gas (LPG) or natural gas nozzle is mainly of two-piece construction: it has fluted ports for the preheat flame, and the central part of the nozzle is recessed: (Figure 9.24b). Nozzles should not be interchanged between different fuel gases.

The nozzle is one of the main keys to good quality efficient cutting. There are different sizes of nozzle for different metal thicknesses, and the nozzle manufacturers will state the correct size on their data sheets (Table 9.2).

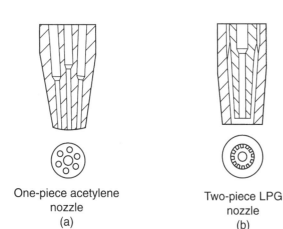

One-piece acetylene
nozzle
(a)

Two-piece LPG
nozzle
(b)

Figure 9.24 (a) One-piece acetylene nozzle (b) two-piece LPG nozzle (*BOC Ltd*)

9.11.4 Hand cutting procedure

First remove any oxide or scale from the line of the cut. Set the preheating flame to neutral and hold the torch with two hands, one to act as a steady and the other to control the oxygen flow, and position the cutting torch so that the white cone is 6 mm from the work surface (Figure 9.25). When the metal reaches a bright red colour, switch on the oxygen cutting supply, and move the cutter steadily and at a speed which produces a smooth cut. Keep the white cone just clear of the work surface, and ensure that the cut is penetrating the surface. Whenever possible the operator should draw the cutter towards him.

9.11.5 Machine cutting

Much oxygen cutting is done with machines, particularly if the cuts are long. Machine cutting has many advantages over hand cutting and results in greater accuracy and better edge finish, particularly when used for making single and double V edge preparation. The principles of machine cutting are similar to those for hand work. A wide variety of types of machines are available, including stationary, general-purpose or universal models, multi-burner machines, straight-line and circle cutting and joint cutting machines. Some incorporate pantograph or electronic devices, enabling profiles to be accurately copied from templates or direct from drawings.

Table 9.2 gives material thickness, nozzle size and pressures for use with Saffire equipment.

9.12 Gases: characteristics and colour coding

The following is a summary of gas characteristics and cylinder colour codes.

Oxygen
Cylinder colour: black.
Characteristics: no smell. Generally considered non-toxic at atmospheric pressure. Will not burn but supports and accelerates combustion. Materials not normally considered combustible may be ignited by sparks in oxygen-rich atmospheres.

Nitrogen
Cylinder colour: grey with black shoulder.
Characteristics: no smell. Does not burn. Inert, so will cause asphyxiation in high concentrations.

Argon
Cylinder colour: blue.
Characteristics: no smell. Heavier than air. Does not burn. Inert. Will cause asphyxiation in absence of sufficient oxygen to support life. Will readily

Table 9.2   Flame cutting data (*BOC Ltd*)

(a) ANM/ANM1E nozzle, 6.3 mm × 10 m fitted hose, resettable flashback arresters

Mild steel plate Thickness			Operating pressure				Gas consumption					
			Oxygen		Fuel gas		Cutting oxygen		Heat oxygen		Fuel	
mm	in	Nozzle size	bar	lbf/in²	bar	lbf/in²	l/min	ft³/h	l/min	ft³/h	l/min	ft³/h
6	$\frac{1}{4}$	$\frac{1}{32}$	1.4	20	0.3	4	14.15	30	8.5	18	8	17
13	$\frac{1}{2}$	$\frac{3}{64}$	2.1	30	0.35	5	30.7	65	10.4	22	9.4	20
25	1	$\frac{1}{16}$	2.8	40	0.4	6	67.5	143	13.2	28	11.8	25
50	2	$\frac{1}{16}$	3.1	45	0.4	6	78.3	166	13.2	28	11.8	25
75	3	$\frac{1}{16}$	3.5	50	0.4	6	88.7	188	13.2	28	11.8	25
100	4	$\frac{5}{64}$	3.1	45	0.31	4.5	121	256	14.6	31	13.2	28
150	6	$\frac{3}{32}$	3.1	45	0.4	6	175	370	20	43	18.4	39
200	8	$\frac{1}{8}$	4.1	60	0.45	6.5	283	600	26	55	23.5	50
250	10	$\frac{1}{8}$	4.8	70	0.45	6.5	377	800	26	55	23.5	50
300	12	$\frac{1}{8}$	6.2	90	0.45	6.5	434	920	26	55	23.5	50
Sheet		A-SNM	1.4	20	0.14	2	14.15	30	2.4	5	2.4	5

(b) AFN nozzle, Saffire Lite cutting attachment (valved version), 6.3 mm × 10 m fitted hose, resettable flashback arresters

Mild steel plate Thickness			Operating pressure				Gas consumption					
			Oxygen		Fuel gas		Cutting oxygen		Heat oxygen		Fuel	
mm	in	Nozzle size	bar	lbf/in²	bar	lbf/in²	l/min	ft³/h	l/min	ft³/h	l/min	ft³/h
6	$\frac{1}{4}$	$\frac{1}{32}$	2	30	0.14	2	11.8	25	4.2	9	3.8	8
13	$\frac{1}{2}$	$\frac{3}{64}$	2	30	0.2	3	23.5	50	4.2	9	3.8	8
25	1	$\frac{1}{16}$	3	45	0.28	4	56.6	120	4.2	9	3.8	8
50	2	$\frac{1}{16}$	3.8	55	0.35	5	75.5	160	5.2	11	4.7	10
Sheet		A-SFNM	1.7	25	0.4	6	14.2	30	2.1	4.5	1.9	4

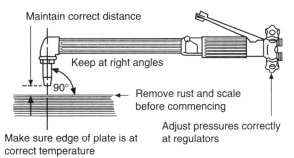

Maintain correct distance

Keep at right angles

90°

Remove rust and scale before commencing

Adjust pressures correctly at regulators

Make sure edge of plate is at correct temperature

Figure 9.25   Cutting technique (*BOC Ltd*)

collect in the bottom of a confined area. At high concentrations, almost instant unconsciousness may occur followed by death. The prime danger is that there will be no warning signs before unconsciousness occurs.

Propane

Cylinder colour: bright red and bearing the words 'Propane' and 'Highly flammable'.

Characteristics: distinctive fish-like offensive smell. Will ignite and burn instantly from a spark or piece of hot metal. It is heavier than air and will collect in ducts, drains or confined areas. Fire and explosion hazard.

Acetylene

Cylinder colour: maroon.
Characteristics: distinctive garlic smell. Fire and explosion hazard. Will ignite and burn instantly from a spark or piece of hot metal. It is lighter than air and less likely than propane to collect in confined areas. Requires minimum energy to ignite in air or oxygen. Never use copper or alloys containing more than 70 per cent copper or 43 per cent silver with acetylene.

Hydrogen

Cylinder colour: bright red.
Characteristics: no smell. Non-toxic. Much lighter than air. Will collect at the highest point in any enclosed space unless ventilated there. Fire and explosion hazard. Very low ignition energy.

Carbon dioxide

Cylinder colour: black, or black with two vertical white lines for liquid withdrawal.
Characteristics: no smell but can cause the nose to sting. Harmful. Will cause asphyxiation. Much heavier than air. Will collect in confined areas.

Argoshield

Cylinder colour: blue with green central band and green shoulder.
Characteristics: no smell. Heavier than air. Does not burn. Will cause asphyxiation in absence of sufficient oxygen to support life. Will readily collect at the bottom of confined areas.

9.13 Safety measures

9.13.1 General gas storage procedures

1 Any person in charge of storage of compressed gas cylinders should know the regulations covering highly flammable liquids and compressed gas cylinders as well as the characteristics and hazards associated with individual gases.
2 It is best to store full or empty compressed gas cylinders in the open, in a securely fenced compound, but with some weather protection.

3 Within the storage area oxygen should be stored at least 3 m from fuel gas supply.
4 Full cylinders should be stored separately from the empties, and cylinders of different gases, whether full or empty, should be segregated from each other.
5 Other products must not be stored in a gas store, particularly oils or corrosive liquids.
6 It is best to store all cylinders upright, taking steps, particularly with round bottomed cylinders, to see that they are secured to prevent them from falling. Acetylene and propane must *never* be stacked horizontally in storage or in use.
7 Storage arrangements should ensure adequate rotation of stock.

9.13.2 Acetylene cylinders

1 The gas is stored together with a solvent (acetone) in maroon painted cylinders, at a pressure of 17.7 bar maximum at 15 °C. The cylinder valve outlet is screwed left-handed.
2 The hourly rate of withdrawal from the cylinder must not exceed 20 per cent of its content.
3 Pressure gauges should be calibrated up to 40.0 bar.
4 As the gas is highly flammable, all joints must be checked for leaks using soapy water.
5 Acetylene cylinders must be stored and used in an upright position and protected from excessive heat and coldness.
6 Acetylene can form explosive compounds in contact with certain metals and alloys, especially those of copper and silver. Joint fittings made of copper should not be used under any circumstances.
7 The colour of cylinders, valve threads, or markings must not be altered or tampered with in any way.

9.13.3 Oxygen cylinders

1 This gas is stored in black painted cylinders at a pressure of 200/230 bar maximum at 15 °C.
2 Never under any circumstances allow oil or greases to come into contact with oxygen fittings because spontaneous ignition may take place.
3 Oxygen must not be used in place of compressed air.
4 Oxygen escaping from a leaking hose will form an explosive mixture with oil or grease.

5   Do not allow cylinders to come into contact with electricity.
6   Do not use cylinders as rollers or supports.
7   Cylinders must not be handled roughly, knocked or allowed to fall to the ground.

9.13.4 Safe usage and handling of gas cylinders

Leaks

As a matter of routine, check for leaks. Test with a solution of 1 per cent Teepol in water; apply with a brush. Never use a naked flame to trace leaks. In hoses, leakages, cuts or local surface damage may be repaired by cutting out faulty sections and inserting an approved coupling. Worn ends should be cut back and refitted with the appropriate hose and connections and clips. Discard hoses that show signs of general deterioration.

If an acetylene valve shows a minor leak and it cannot be stopped by closing the valve or tightening the gland nut, eliminate all sources of ignition, move the cylinder outside to a safe area and contact your supplier.

If a cylinder is leaking and on fire, and you suspect the valve is damaged, do *not* attempt to extinguish the fire yourself. Call the fire brigade and evacuate the area. Use a fire extinguisher to put out any fire caused.

Fire: action to be taken

Gas cylinders involved in a fire may explode. Therefore the key actions to be taken are as follows:

1   Evacuate the surrounding area to a minimum distance of 100 metres.
2   Call the fire brigade immediately.
3   Advise persons between 100 and 300 metres from the cylinder to take cover.
4   Any attempt to fight the fire should be done from a protected position using copious quantities of water.
5   When the fire brigade arrives, inform them of the location, the number of gas cylinders directly involved in the fire, and the names of the gases they contain.
6   Cylinders which are not directly involved in the fire and which have not become heated should be moved as quickly as possible to a safe place, provided this can be done without undue risk to personnel. Make sure the valves of these cylinders are fully closed.
7   As soon as possible, inform your gas supplier of the incident.
8   Be aware of the fact that, even after the fire has been extinguished, some cylinders which have been heated can *explode*, particularly acetylene cylinders. Therefore do not approach any cylinder until the key actions given below have been taken.
9   When the cylinder content are unknown, treat as acetylene cylinders.

Acetylene cylinders in a fire: key actions

Refer to Figure 9.26.

1   Never approach or attempt to move cylinders.
2   From a safe position, drench the entire surface of all cylinders with water for at least one hour after the fire has been extinguished. Do not use a jet of water of such strength that it would knock over a free-standing cylinder.
3   Check visually from a safe position. If steam is seen to be coming from the surface of the cylinder when water spray is interrupted, continue spraying with water. Then check at hourly intervals until it is seen that steaming has ceased.
4   Once steaming has stopped, observe from a safe distance whether the surface of the cylinder remains wetted. If patches dry quickly, continue to cool with water and observe again after half an hour. Repeat this operation until all surfaces remain wetted after water spray is stopped. Pay particular attention to the centre cylinders of manifold cylinder pallets (MCPs) and to any cylinder where some difficulty has been experienced in maintaining a good supply of cooling water.
5   Once all cylinder surfaces remain wetted after the water spray is discontinued, check using the bare hand that the cylinder remains cold for 30 minutes. Wait a further 30 minutes and check again: if any part of the cylinder feels warm to the touch, reapply the cooling water for 30 minutes and repeat the procedure until cylinders remain cold for one hour.
6   When you are satisfied that the entire surface has remained cold for one hour, submerge the cylinder in water carefully, avoiding shocks and bumps. Normally after 12 hours immersion the cylinder will be safe to be disposed of by the gas supplier.

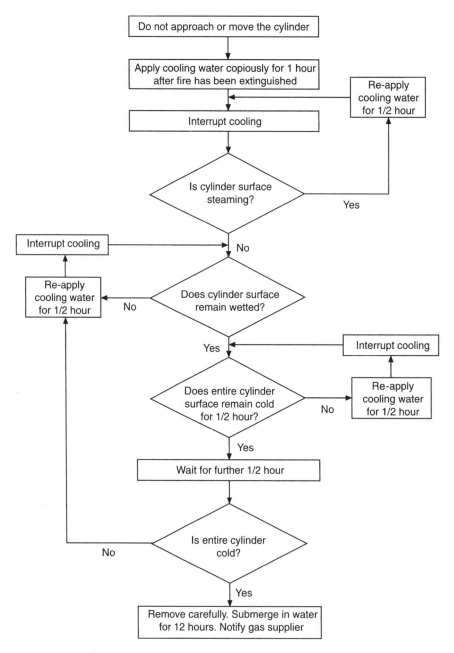

Figure 9.26 Acetylene cylinders in fires (*BOC Ltd*)

Ventilation and oxygen-enriched areas

Whenever cylinder gases are used, constant and thorough ventilation should be maintained. This is particularly important when they are used in confined spaces.

The normal content of air is 21 per cent. If this becomes enriched to 25 per cent there is an increase in the speed at which materials will burn. At 30 per cent the typical characteristics of an oxygen-fed fire are apparent. The fire is in two phases: an initial flash fire, followed by local burning at a number of points. Fires in oxygen-enriched atmospheres are very difficult to extinguish, and can spread rapidly across combustible materials from a single point source such as a spark from a cigarette.

Nitrogen, argon and carbon dioxide, if allowed to replace the oxygen in the atmosphere, can cause asphyxiation. The dangers occur typically when gas is released in a confined space. It is especially important to beware of argon and carbon dioxide; both are heavier than air, and will displace air in confined spaces or spaces with no ventilation at floor level.

9.13.5 General equipment safety

All equipment should be subjected to regular periodic examination and overhaul. Failure to do so may allow equipment to be used in a faulty state, and may be dangerous.

Rubber hose Use only hose in good condition, fitted with the special hose connections attached by permanent ferrules. Do not expose the hose to heat, traffic, slag, sparks from welding operations, or oil or grease. Renew the hose as soon as it shows any sign of damage.

Pressure regulators Always treat a regulator carefully. Do not expose it to knocks, jars or sudden pressure caused by rapid opening of the cylinder valve. When shutting down, release the pressure on the control spring after the pressure in the hoses has been released. Never use a regulator on any gas except that for which it was designed, or for higher working pressures. Do not use regulators with broken gauges.

Welding torch When lighting up and extinguishing the welding torch, the manufacturer's instructions should always be followed. To clean the nozzle use special nozzle cleaners, never a steel wire.

Fluxes Always use welding fluxes in a well ventilated area.

Goggles These should be worn at all times during welding, cutting or merely observing.

Protection Leather or fire-resistant clothing should be worn for all heavy welding or cutting. The feet should be protected from sparks, slag or cut material falling on them.

9.13.6 Backfiring and flashbacks of welding torch

A welding torch is said to *backfire* when it goes out with a loud pop, and then relights itself immediately, providing the heat of the job is sufficient to ignite the acetylene. Backfires result from defective equipment, incorrect pressures or incorrect lighting up; or careless use of the welding torch, permitting the nozzle to touch the work, overheating the nozzle tip, or working with a loose nozzle. Usually the backfire is arrested at the mixer or injector of the welding torch. If prompt action is taken, turning off first the oxygen then the acetylene valve, no damage occurs and the welding torch may be relit provided that the cause of the trouble has been eliminated.

Sometimes a backfire may pass beyond the injector and travel back into either the oxygen or fuel gas hoses. This is termed *flashback,* and its effect is more serious as immediate damage to hoses and regulators may result; there is also a risk of injury to the operator. A flashback should be suspected if there is a squealing or hissing noise, sparks coming from the nozzle, heavy black smoke, or if the welding torch gets hot. If the flame burns back far enough it may burst through the hose.

With oxy-fuel gas equipment, flashbacks can and do occur because the recommended pressures and procedures have not been observed, and because of nozzle blockage, faulty equipment or leaking equipment. One of the main causes of flashback is due to backfeeding, which occurs when higher-pressure gas feeds back up a lower-pressure stream. The presence of hose check valves will prevent the oxygen and fuel gas mixing in the hose and subsequently causing fire, injury and damage.

A flashback arrester is a device designed to quench the flashback. When it incorporates a cut-off valve, this will automatically shut the gas flow. These multifunction devices afford an additional safeguard, particularly in locations where a fire following flashback could not be tolerated, such as garages, body shops and other workplaces with flammable or hazardous materials.

9.14 Plasma arc cutting

9.14.1 Plasma cutting process and equipment

The plasma cutting process relies on the fact that if a gas or mixture of gases, such as air, is subjected to a very high temperature it becomes ionized: negative electrons are separated from the atom, which is then positively charged. This ionized state of the gas is called 'plasma', and in this state the gas is electrically conductive.

The high temperature necessary to create the plasma is achieved, in the case of plasma cutting, by a standing electric arc. This is constricted by forcing the plasma through a small nozzle which increases the temperature of the arc to over 24 000 °C and concentrates it on to a very small area. When this plasma is directed at a conductive material the arc is transferred through the plasma (transferred arc operation) to the material. The high energy of the arc melts the material which, as long as it is within the cutting range of the equipment, will be displaced by the gas flow.

In order to initiate the standing arc it is necessary to produce an ionized path in the gases. This is achieved either by applying a very high voltage at a high frequency between the electrode and the tip and work, causing a high-frequency spark, or by momentarily touching the electrode and nozzle together and then quickly breaking the contact, causing an arc between the electrode and the nozzle. As soon as the gas between the tip and the nozzle is ionized, the main arc will ignite.

Plasma cutting and welding torches are designed with a small orifice which constricts the arc. Gas flows under pressure through the arc, and is subsequently heated to an exceptionally high temperature. The heated gas cannot expand owing to the constriction of the nozzle, and it is forced out to form a supersonic jet hotter than any known flame or a conventional electric arc. The jet melts down and patially vaporizes the base material, and the force of the jet blasts away the molten metal. The plasma jet is controlled by adjusting current, gas velocity and type of gas.

The distance between electrode and workpiece is made electrically conductive (ionization) by an auxiliary arc (pilot arc) between electrode and plasma nozzle. As the intensity of the high-voltage impulses of the pulse generator is not sufficient, part of the cutting current (limited by a resistor) is admitted for the sufficient ionization of the pilot arc distance. When the ionized gas jet contacts the workpiece the main circuit is closed and the cutting process is introduced (see Figure 9.27).

Clean, dry workshop air can be used when severing thin sheet steel, but it is recommended to use an argon/hydrogen mix or nitrogen when cutting

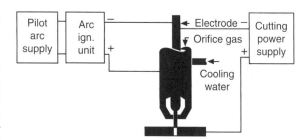

Figure 9.27   Simplified diagram of plasma arc cutter (*Motor Insurance Repair Research Centre*)

gauges in excess of 5 mm, and for stainless steels, as these gases produce a higher thermal conductivity and will transfer more heat to the material to be cut. Reference to the manufacturers' recommendations are essential, as the correct cutting speed must be maintained to ensure the quality of the cut.

The equipment can be grouped as follows:

High-output units (approximately 50 amperes) operating on input voltages of 380/415 V three phase, fitted with water-cooled torches and operating with workshop air or special cutting gases (see Figure 9.28).

Figure 9.28   Plasma cutting unit: three phase, 40 amperes (*Olympus Welding Supplies Ltd*)

Medium-output units (approximately 30 amperes) operating on input voltages of 220/240 V single phase. These cutters are normally available with air-cooled torches using clean workshop or bottled air (see Figures 9.29 and 9.30).

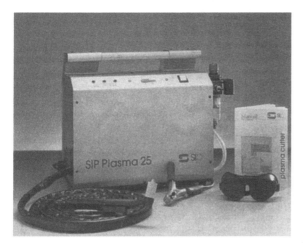

Figure 9.29 Plasma cutting unit: single phase, 30 amperes (*SIP (Industrial Products) Ltd*)

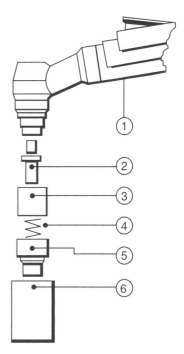

Figure 9.30 An air-cooled plasma cutting torch (*SIP (Industrial Products) Ltd*)

1 Handle
2 Electrode
3 Insulator bush
4 Spring
5 Plasma nozzle
6 Protection nozzle

There are two types of electrode available for the plasma cutting torches. The first is a tungsten electrode intended for cutting with nitrogen or an argon/hydrogen mix. The second is an electrode intended for cutting with air (see Figure 9.31).

9.14.2 Setting up the equipment for use

1 Select the electrode type according to the cutting gas used. Check that the gas nozzle and electrode of the torch are not damaged, and set the electrode distance as required for the selected electrode type using the nozzle tool. For the best cutting results, the electrode distance should be regularly checked and adjusted.
2 Before performing any work in the torch, *always* disconnect the main voltage.
3 Check that the correct pressure (approximately 5 bar or 75 psi) is set on the regulator, and that any valves ahead of the regulator are open.
4 Connect the earth lead, ensuring as good a contact as possible to the workpiece (if necessary, clean off any surface coating rust). The welding clamp must not be moved during cutting; the machine should be disconnected.

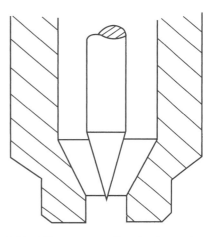

Figure 9.31 Plasma arc cutting torch nozzle, showing position of electrode (*Motor Insurance Repair Research Centre*)

5   Set the main switch of the machine. The cooling unit and the fan will start and the indicator 'on' lamp will light.

6   Check for correct gas flow by activating the switch on the torch handle. The gas valve will open and the gas will start to flow from the nozzle (3.5–4.1 bar). After this gas test, reset the switch to its initial position.

7   Press the trigger button once only. Air should flow from the torch; if it does not, check the warning lamps on the machine. The red lamps should not be illuminated. The left-hand one indicates low air pressure, and the right-hand one that thermal cut-outs have operated. If either is lit then appropriate action should be taken. The green lamp should be illuminated, as it indicates that the machine is on. The green light is switched off when a red light goes on.

8   Whilst the air is flowing, check that the pressure is 3.5 bar (50 psi). If not, adjust the regulator to give 3.5 bar.

9   Set the machine to either maximum or minimum according to the thickness to be cut.
    Caution: do not attempt to cut material beyond the range specified, as this will damage the torch.

10 Position the torch on the workpiece, ensuring that the tip has a good contact.

11 Press the trigger button once and release it. Allow the air to flow for a few seconds. Press the button again and then, within one second, press and hold the button. The arc will strike and a hole will be cut in the metal.
    Caution: if the arc does not penetrate the metal then stop, as the machine is either on the wrong setting or not sufficiently powerful for the thickness being cut.

12 Move the torch along so that the metal is cut right through. If the metal is not cut, reduce the speed of travel until it is. The torch head should be held at right angles to the work (Figure 9.32).

13 When the cut is complete, release the trigger button.
    Caution: do not switch off the machine until the air has stopped flowing, otherwise the torch will be damaged, as the air is required to cool the torch.

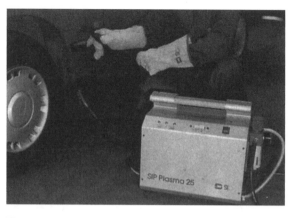

Figure 9.32   Plasma arc cutting machine in use (*SIP (Industrial Products) Ltd*)

14 The operator must be constantly aware of the risk of fire and fumes, and therefore proper health and safety precautions must be adhered to at all times.

9.14.3 Cutting methods

The equipment can be used for contact or distance cutting. When cutting is started, the techniques differ between nitrogen and argon/hydrogen plasma cutting and air plasma cutting.

Contact cutting
Contact cutting is used for materials up to 5 mm in thickness. In this method the torch nozzle is in contact with the workpiece. The torch is tilted at an angle with the work surface to obtain plasma flow, and gradually straightened up until it lies flush with the workpiece when cutting.

Distance cutting
This is used for materials over 5 mm thick. With this method the nozzle is not in contact with the workpiece, and an even distance must be maintained between them when cutting.

9.14.4 Process advantages

Plasma arc cutting is one of the most effective processes for high-speed cutting of many kinds of ferrous and non-ferrous metals. The quality of the cut is very high, with little or no dross and reduced distortion. Better economy is also a feature of plasma-arc cutting as compared with oxy-fuel methods.

With plasma arc equipment a saving in time and energy can be achieved in certain areas of repair, particularly in the removal of damaged panels and structural members. Plasma arc cutting will prove of benefit in a modern body repair shop as an alternative cutting technique to speed up damaged panel removal and to complement traditional mechanical methods.

9.14.5 Safety

General

1 Do not operate the machine with any of the panels removed.
2 Ensure that the machine is connected to the correct voltage supply, that the correct fuse is used and that the equipment is earthed.
3 Under no circumstances must the plasma nozzle be removed or any other work be carried out on the torch with the machine switched on. Ignoring this precaution could lead to contact with high DC voltage.
4 The torch should be kept free of slag at all times to ensure a free passage of air.
5 The tip and electrode will require changing when they have become pitted and the arc will not strike. The life of these components depends upon the current used and any misuse of the torch.

Fire

1 All inflammable materials must be removed from the area.
2 Have a suitable fire extinguisher available close by at all times.
3 Do not cut containers which have held inflammable materials or gases.

Glare and burns

1 The electric plasma arc should not be observed with the naked eye. Always wear goggles of the type used for oxy-acetylene welding.
2 Gloves should be worn to protect the hands from burns.
3 Non-synthetic overalls with buttons at neck and wrist, or similar clothing, should be worn.
4 Greasy overalls should never be worn.
5 Do not wear inadequate footwear.

Compressed air

Compressed air is potentially dangerous. Always refer to the relevant safety standards for compressed gases (see Chapter 2).

Ventilation, fumes, vapours

Ventilation must be adequate to remove the smoke and fumes during cutting. Toxic gases may be given off when cutting, especially if zinc or cadmium coated steels are involved. Cutting should be carried out in a well ventilated area, and the operator should always be alert to fume build-up. In small or confined areas, fume extractors must be used.

Vapours of chlorinated solvents can form the toxic gas phosgene when exposed to UV radiation from an electric arc. All solvents and degreasers are potential sources of these vapours and must be removed from the area being cut.

Questions

1 Describe the general precautions which should be taken in the storage and handling of oxygen and acetylene cylinders.

2 State the safety precautions to be taken when assembling oxy-acetylene equipment.

3 State how gas cylinders are visibly identifiable.

4 State the correct pressures for full cylinders of oxygen and acetylene.

5 Compile a list of safety measures which should be applied in the preparation and use of gas welding equipment.

6 What method should be used to find the location of a leak in an acetylene connection?

7 What safety precautions should be taken if an acetylene cylinder was to take fire internally?

8 State the reason why copper or high-copper-content alloy tubing should not be used for an acetylene connection.

9 Why is it dangerous to allow grease or oil to come into contact with the oxygen cylinder valve or fittings?

10 What is meant by the following terms: (a) high-pressure welding (b) low-pressure welding system?

11 What are the essential components of a high-pressure welding system?

12 Describe the reasons for both leftward and rightward welding methods, and state any benefits derived from these methods.

13 What is meant by an oxidizing flame?

14 What is meant by a carburizing flame?

15 Suggest two reasons for a welding torch backfiring.

16 Explain the meaning of the term 'penetration' in a welded joint.

17 Sketch and describe the best condition for a welding torch flame for welding low-carbon steel sheet.

18 When making gas-welded butt joints in sheet steel, which common faults should be avoided?

19 Sketch and name the type of flame suitable for welding each of the following: (a) aluminium (b) brass.

20 Describe the technique which would be most suitable for welding low-carbon steel up to 5 mm.

21 With the aid of a sketch, give a detailed explanation of the working and function of the hose check valve on oxy-acetylene welding hoses.

22 List the procedure sequence necessary to shut down a gas welding plant high-pressure system.

23 Explain the difference between line pressure and contents pressure as shown on the regulator.

24 Describe the process of gas welding pure aluminium sheet.

25 Explain the precautions to be taken when welding aluminium.

26 State two disadvantages which limit the use of oxy-acetylene welding in vehicle body repair.

27 What principle makes possible the cutting of metal by means of oxy-acetylene?

28 Explain the basic principle of plasma arc cutting.

29 Explain the type of gases used in plasma arc cutting.

30 Describe how the plasma arc process can be used in vehicle body repair.

31 Name the oxy-acetylene flame that would burn equal quantities of the gases.

32 State the purpose of the two pressures that are shown on the gauge of an acetylene gas cylinder regulator.

33 Explain the safety device used on oxy-acetylene welding equipment to stop the risk of the flame travelling back down the torch supply hose.

34 With the aid of a sketch, outline two methods of limiting distortion when welding thin sheet metal.

35 Explain the difference in welding low-carbon steel and aluminium of equal thickness.

36 List the consumables used in plasma arc cutting.

37 State the welding process that is not recommended when welding HSLA steels.

38 What is meant by 'contact cutting' when using plasma arc?

39 What is the electrode made from in a plasma arc cutting torch?

40 What are the two gases that can be used in the flame cutting process?

Electric resistance welding

10.1 Resistance welding in car body manufacture

Resistance welding is a joining process belonging to the pressure welding sector. With its locally applied heat and pressure it has an obvious relationship with the forge welding technique practised by blacksmiths when joining metal. The resistance welding process was invented in 1877 by Professor E. Thomson of Philadelphia, USA, when an accidental short circuit gave him the idea for what was originally termed short circuit welding. From the beginning of the twentieth century it was used on a small scale in industry, but it was only after the Second World War that resistance spot welding had its real beginning in the automobile industry. It has since grown to be the most important method of welding used in the construction and mass production of vehicle bodies.

Resistance welding is extensively used for the mass production assembly of the all-steel body and its component sheet metal parts (see Figure 10.1). Its wide adoption has been brought about by its technical advantages and the reductions in cost. Most mass produced car bodies are assembled entirely by welding steel pressings together to produce an integral rigid chassis and body structure. Low-carbon steel thicknesses used in this unitary construction range from 0.8–1 mm for skin or floor panels to 3 mm for major structural pressings such as suspension brackets. Intermediate gauges such as 1.2 mm are used for hinge reinforcements, 1.6 mm for chassis structural members, and 1.8 mm to 2.5 mm for suspension and steering members. With the introduction of high-strength steels (HSLA steels), car manufacturers are producing body panels as thin as 0.55 mm, and structural members with gauges of between 1.2 and 2 mm. This reduction in thickness can be made without loss of strength.

There are a number of resistance welding processes. Resistance spot welding is the most widely used welding process in car body construction; there are approximately 4500–6000 spot welds per body, which accounts for approximately 80 per cent of the welding used. A further 10 per cent is composed of other resistance welding processes: seam, projection, flash and butt welding. The remaining 10 per cent is divided between MIG/MAG welding and gas welding. Of the many welding techniques used in the mass production of car bodies, resistance welding dominates the field.

The fundamental principle upon which all resistance welding is based lies in the fact that the weld is produced by the heat obtained from the resistance to flow of electric current through two or more pieces of metal held together under pressure by electrodes which are made from copper or copper alloys. The engineering definition of heat (heat being the essence of all welding) is (energy) × time. This indicates a balance between

Figure 10.1 Automatic spot welding using robots on body framing assembly line (*Vauxhall Motors Ltd*)

energy[2] input and weld time; therefore the faster
the welding, the greater the clamping force. However, the engineering definition of resistance is such
that the higher the clamping force, the greater the
current needed to produce a constant heat. Heat is
generated by the resistance of the parts to be joined
to the passage of a heavy electrical current. This
heat at the junction of the two parts changes the
metal to a plastic state. When the correct amount
of pressure is then applied fusion takes place.

There is a close similarity in the construction
of all resistance welding machines, irrespective of
design and cost. The main difference is in the type
of jaws or electrodes which hold the object to
be welded. A standard resistance welder has four
principal elements:

Frame   The main body of the machine, which
differs in size and shape for robot, stationary and
portable types (Figures 10.2, 10.3, 10.4).

Electrical circuit   Consists of a step-down transformer, which reduces the voltage and proportionally increases the amperage to provide the
necessary heat at the point of weld.

Electrodes   Include the mechanism for making
and holding contact at the weld area.

Timing control   Represents the switches which
regulate the volume of current, length of current time
and the contact period. Many now include adaptive
in-process control units (pulsing weld timer).

The principal forms of resistance welding are
classified as: resistance spot welding; resistance
projection welding; resistance seam welding; resistance flash welding; and resistance butt welding.

Figure 10.3   Stationary pedestal spot welding
machine (*SIP (Industrial Products) Ltd*)

Figure 10.2   Robot spot welder on an assembly line
(*Vauxhall Motors Ltd*)

Figure 10.4   Portable spot welding machine in use
(*SIP (Industrial Products) Ltd*)

10.2 Resistance spot welding

Resistance spot welding (Figure 10.5) is basically confined to making welds approximately 6 mm diameter between two or more overlapping sheet metal panels. This type of welding is probably the most commonly used type of resistance welding. The art of production planning for spot welding is to simplify the presentation of panels in the region of mutual panel overlap. The limitation of spot welding is that the electrode assemblies have to withstand applied forces ranging from 2200 N to 4450 N for the range of sheet steel thicknesses used in vehicle construction and repair. Product

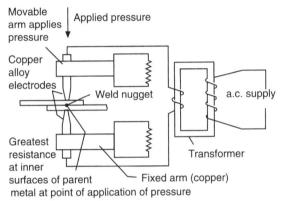

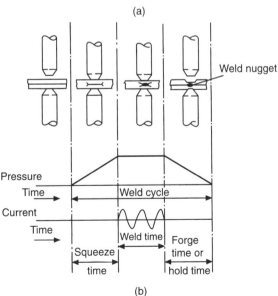

Figure 10.5 (a) Resistance spot welding system (b) relationship between weld formation, current and pressure in welding

design of planning must account, therefore, for the requirement that the electrodes, which are constructed from relatively weak copper alloys, need normal access to both sides of the overlapping panels to withstand such electrode forces.

The material to be joined is placed between two electrodes, pressure is applied, and the electric current passes from one electrode through the material to the other electrode. There are three stages in producing a spot weld. First, the electrodes are brought together against the metal and pressure is applied before the current is turned on. Next the current is turned on momentarily. This is followed by the third hold time, in which the current is turned off but the pressure continued. The hold time forges the metal while it is cooling.

Regular spot welding usually leaves slight depressions on the metal which are often undesirable on the show side of the finished product. These depressions are minimized by the use of larger-sized electrode tips on the show side. Resistance spot welding can weld dissimilar metal thickness combinations by using a larger electrode contact tip area against the thicker sheet. This can be done for mild steel having a dissimilar thickness ratio of 3:1.

There are three kinds of distortion caused by resistance spot welding which are relevant to car body manufacture and repair. The first is the local electrode indentation due to the electrode sinking into the steel surface. This is a mechanical distortion – a byproduct of the spot welding process. Second, there is a small amount of thermal distortion which is troublesome when attempting to make spot welds on show surfaces such as skin panels without any discernible metal distortion. Last, there is the gross distortion caused when badly fitting panels are forced together at local spots. This is a mechanical distortion totally unconnected with the spot welding process; the same type of distortion would occur with rivets or with any similar localized joining method. A combination of all these distortions contributes to the general spot-weld appearance, which is virtually unacceptable on a consumer product. The technique on car bodies is to arrange for spot-welded flanges to be either covered with trims (door apertures), or with a sealing weather strip (window and screen surrounds). The coach joint is one of the features that distinguish a cheap, mass produced body from an expensive hand built one.

Spot welders are made for both DC and AC. The amount of current used is very important. Too little current produces only a light tack which gives insufficient penetration. Too much current causes burned welds. Spot welds may be made one at a time or several welds may be completed at one time, depending on the number of electrodes used. One of the most significant advantages of resistance spot welding is its high welding production rate with the minimum of operator participation. Typical resistance spot-welding rates are 100 spots per minute. To dissipate the heat at the weld as quickly as possible, the electrodes are sometimes water cooled. Although many spot welders are of the stationary design, there is an increased demand for the more manoeuverable, portable type. The electrodes which conduct the current and apply the pressure are made of low-resistance copper alloy and are usually hollow to facilitate water cooling. These electrodes must be kept clean and shaped correctly to produce good results. Spot welders are used extensively for welding steel, and when equipped with an electronic timer they can be used for metal such as aluminium, copper, stainless and galvanized steels.

10.3 Resistance projection welding

Projection welding (Figure 10.6) involves the joining of parts by a resistance welding process which closely resembles spot welding. This type of welding is widely used in attaching fasteners to structural members. The point where the welding is to be done has one or more projections which have been formed by embossing, stamping or machining. The distortion on the embossed metal face is low and is negligible on heavy metal thicknesses, although in the case of dissimilar thicknesses it is preferable to emboss the projection on the thicker of the two sheets. The projections serve to concentrate the welding heat at these areas and permit fusion without the necessity of employing a large current. The welding process consists of placing the projections in contact with the mating fixtures and aligning them between the electrodes. The electrodes on the machine are not in this case bluntly pointed as in spot welding, but are usually relatively large flat surfaces which are brought to bear on the joint, pressing the projections together.

The machine can weld either a single projection or a multitude of projections simultaneously. The many variables involved in projection welding, such as thickness, kind of material, and number of projections, make it impossible to predetermine the correct current. Not all metals can be projection welded. Brass and copper do not lend themselves to this method because the projections usually collapse under pressure.

10.4 Resistance seam welding

Seam welding (Figure 10.7) is like spot welding except that the spots overlap each other, making a continuous weld seam. In this process the metal pieces pass between roller-type electrodes. As the electrodes revolve, the current supply to each one is automatically turned on and off at intervals corresponding to the speed at which the parts are set to move. With proper control it is possible

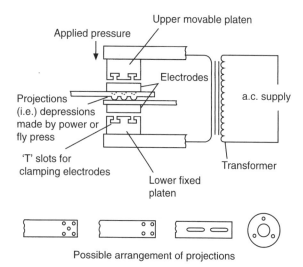

Possible arrangement of projections

Figure 10.6 Projection welding system

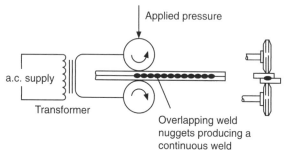

Figure 10.7 Continuous resistance seam welding

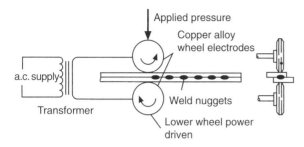

Figure 10.8 Seam welding (stitch welding)

to obtain air-tight and water-tight seams suitable for such parts as fuel tanks and roof panels. When spots are not overlapped long enough to produce a continuous weld, the process is sometimes referred to as roller spot or stitch welding (Figure 10.8). In this way the work travels the distance between the electrodes required for each succeeding weld cycle. The work stops during the time required to make each individual weld and then automatically moves the proper distance for the next weld cycle. Intermittent current is usually necessary for most seam welding operations. Each individual weld contains a number of overlapping spot welds, resulting in a joint which has good mechanical strength but which is not pressure tight.

Mild steel, brass, nickel, stainless steel and many other alloys may be welded by this method, although its use is largely limited to operations on mild and stainless steel. A maximum economy is obtained if the combined thickness of sheets to be joined does not exceed 3.2 mm. Before welding, the surfaces must be clean and free from scale, and this may be done by sand blasting, grinding or pickling.

10.5 Resistance flash welding

In the flash welding process (Figure 10.9) the two pieces of metal to be joined are clamped by copper alloy dies which are shaped to fit each piece and which conduct the electric current to the work. The ends of the two metal pieces are moved together until an arc is established. The flashing action across the gap melts the metal, and as the two molten ends are forced together fusion takes place. The current is cut off immediately this action is completed.

Flash welding is used to butt or mitre sheet, bar, rod, tube and extruded sections. It has unlimited

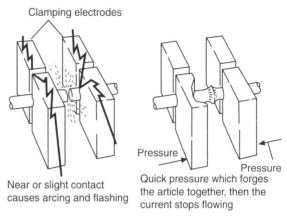

Figure 10.9 Flash welding

application for both ferrous and non-ferrous metals. For some operations the dies are water cooled to dissipate the heat from the welded area. The most important factor to be considered in flash welding is the precision alignment of the parts. The only problem encountered in flash welding is the resultant bulge or increased size left at the point of weld. If the finish area of the weld is important, then it becomes necessary to grind or machine the joint to the proper size.

10.6 Resistance butt welding

In butt welding (Figure 10.10) the metals to be welded are brought into contact under pressure, an electric current is passed through them, and the edges are softened and fused together. This process differs from flash welding in that constant pressure

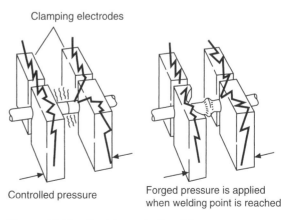

Figure 10.10 Resistance butt welding

is applied during the heating process, which eliminates flashing. The heat generated at the point of contact results entirely from resistance. Although the operation and control of the butt welding process is almost identical to flash welding, the basic difference is that it uses less current, has a constant pressure and allows more time for the weld to be completed.

10.7 Resistance welding in body repair work

The object of car body repair is to put damaged vehicles back into a pre-accident condition. Today's chassisless bodies hold engine, suspension and steering in the right places and are designed to absorb the impact of crashes by crumpling, thus shielding the passenger compartment (and its passengers) from shock and deformation. From the viewpoint of safety as well as mechanical efficiency, proper welding is vital in this kind of repair.

Car body design demands careful choice of the sheet metal, which was, until recent years, all mild steel. Tensile strength and ductility, which are good in mild steel, are vital to 'crumplability' (the ability to absorb the impact energy), and this is why resistance spot welds are used. The average body shell is joined together by approximately 4500–6000 such spot welds. These remain ductile because the welding process does not alter the original specifications of the steel. Lighter body weight reduces the load on the car engine and therefore has a direct influence on petrol consumption.

For weight and fuel saving reasons, high strength steels have been introduced for some panel assemblies on body shells. As these steels have a different character from low-carbon steel (mild steel), which still accounts for 60 per cent of the body shell (see Figure 10.11), they cause repair welding problems. Higher-strength steels have been made specifically for motor car manufacturers to produce body shells from thinner but stronger steel sheet. These steels are less ductile and are harder. Above all, they do not tolerate excess heat from bad welding, which makes them brittle or soft or can cause panel distortion. Low-carbon steel tolerates excess heat well. While older all-mild-steel bodies essentially needed only to have the welding machine set for the metal thickness, current new bodies can contain steel of up to four different strengths, hardnesses

and ductilities, some coated on one side, some coated on both sides, and some uncoated. Zinc coated sheet materials require the use of heavier welding equipment capable of producing a higher current to penetrate the zinc coating, and the electrodes must be constantly maintained by cleaning to avoid zinc pick-up when welding.

The repair of bodies incorporating low-carbon steels and HSLA steels therefore demands very different welding routines from those for low-carbon steel alone. Low-carbon steel bodies can be resistance spot welded or gas (TIG) welded or arc (MIG) welded; but higher-strength steels should not undergo the last two processes, because they involve nearly three times the heat of resistance spot welding. The temperatures generated are over 3300 °C for gas or arc welding but only 1350 °C for resistance spot welds, for joints of similar strength. Higher-strength steels, however, with their higher tensile strength, limited ductility and greater hardness, are particularly vulnerable to heat, and are apt to lose strength and change ductility when overheated.

Developments in welding equipment combined with the use of electronic controls have opened the way to new body repair welding techniques that help to simplify the practical problems posed by bodies made from a mixture of low-carbon and high-strength low-alloy steels. The traditional resistance welding method has had to be improved to join higher-strength steels. Because the weld current flow is hindered by the steel's coating, these may require higher temperatures to break them down before a weld can be formed. For producing consistently good welds it is necessary to use two or three stages for welding, the duration of each stage being adapted to the nature of the steel and its coating.

10.8 Resistance spot welding of high-strength steels

The rigidity of the body and its ability to withstand high torsional and other stresses depend on the assembly method used to bring the various body panels together. Spot welding is used throughout the industry (see Figure 10.12) for two reasons: first, because it is the strongest and most reliable method of joining two pieces of metal; and second, because of the total absence of panel distortion through the welding. In order to effect a satisfactory

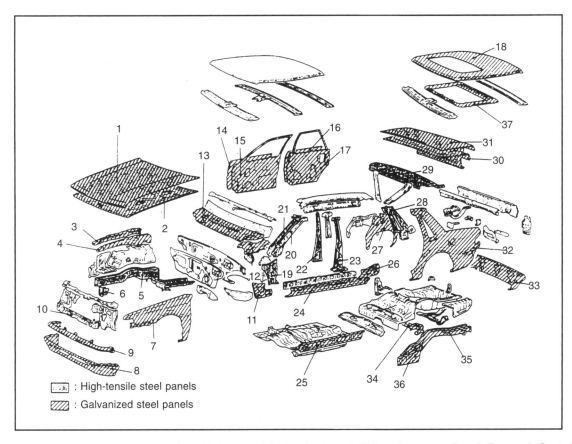

: High-tensile steel panels

: Galvanized steel panels

Figure 10.11 Body panels made from high-strength/galvanized steels (*Motor Insurance Repair Research Centre*)

		Material			*Material*
1	Hood panel, outer	SGAC35R	20	Front pillar, outer, upper	SPRC35
2	Hood panel, inner	SGACC	21	Front Pillar, inner, upper	SPRC35
3	Upper frame, outer	SGACC	22	Centre pillar, inner	SPRC35
4	Upper frame, inner	SGACC	23	Centre pillar, outer	SPRC35
5	Front side member	SPRC35	24	Front floor, side sill, outer	SGACC
6	Front end gusset	SPRC35	25	Front floor, side sill, inner	SGACC
7	Front fender	SENCE	26	Side sill, outer rear extension	SGACC
8	Front skirt panel	SGACC	27	Quarter panel, inner	SENCE
9	Grille filler panel	SGACC	28	Rear pillar, inner	SPRC35
10	Front end cross member	SPRC35	29	Rear shelf panel	SPRC35
11	Side sill, inner front	SGACC	30	Trunk lid inner panel	SGACE
12	Front upper frame extension	SGACC	31	Trunk lid outer panel	SGAC35R
13	Cowl top outer panel	SGACC	32	Quarter panel, outer	SENCE
14	Front door outer panel	SGACC	33	Rear skirt panel	SGACC
15	Front door inner panel	SGACE	34	Rear floor cross member	SGACC
16	Rear door outer panel	SGACC	35	Rear floor side member	SGACC
17	Rear door inner panel	SGACE	36	Rear floor side sill	SGACC
18	Roof panel	SGACC	37	Roof reinforcement	SGAHC
19	Front pillar, outer, lower	SPRC35			

SPRC: phosphorus added

SGACC, E ⎤
SGAHC ⎦ Galvanized steel plate

SGAC35R: phosphorus added (also galvanized)
SENCE: SPCE with electrogalvanized zinc-nickel coating
The numbers in the material codes indicate the tensile strength (kg/mm^2)

Figure 10.12 Weldmaster 2022 in use (*ARO Machinery Co. Ltd*)

repair, vehicle welds having the same characteristics as the original are essential.

A comparison between an electrical spot weld and a forged weld shows that in both these processes a union is formed by an amalgamation of the metal molecules. These have been consolidated in each of the two pieces, despite the difference in the two processes employed. In the case of forge welding the pieces are heated in the forge furnace and then hammered until a homogeneous material results. In the case of the spot-welding process the gun must have a pressure device which can be operated by the user and transmits the pressure to the electrodes, and a transformer to enable current at high intensity to be fed to the electrodes.

Spot welding is the primary body production welding method. Most production welding is carried out on metal gauges of less than 2.5 mm, although greater thicknesses can be spot welded. With the introduction of high-strength steels, car manufacturers are producing body panels as thin as 0.55 mm and using gauges of between 1.2 and 1.5 mm for structural members. By using the correct equipment, two-sided spot welding can be successfully accomplished on these steels.

Weld quality

Careful control of three factors is necessary to make a good-quality spot weld:

Squeeze time
Weld time (duration of weld flow)
Hold time.

The main disadvantage associated with welding in general is that any visual inspection of the weld gives no indication of the weld quality. It is therefore essential, that spot-welding equipment used in motor body repair takes outside circumstances automatically into account, such as rust, scale, voltage drop in the electric main supply, and current fluctuations.

The quality of a spot weld depends on:

1 The strength of the weld must be equal to the parent metal. This is a problem of tensile strength and homogeneity of the weld nugget itself.
2 The welding heat must not in any way alter the inherent qualities of the parent metal.
3 The welding pressure must prevent parent metal separation so as not to cause panel distortion and stresses in the assembly.

Heat application

With resistance welding the welding time is controlled by either electronic or electro (electrical) devices. Hand-held pincer-type weld guns, as supplied to the car body repair trade, are usually controlled electronically with the minimum of presetting. Weld time can range from a fraction of a second for very thin gauges of sheet steel, to one second for thicker sheet steel. Weld time is an important factor, because the strength of a weld nugget depends upon the correct depth of fusion. Sufficient weld time is required to produce this depth of fusion by allowing the current time to develop enough heat to bring a small volume of metal to the correct temperature to ensure proper fusion of the two metals. If the temperature reached is too high, metal will be forced from the weld zone and may induce cavities and weld cracks. In some grades of high-strength steels, cracking within the vicinity of the heat affected zone has appeared after the welding operation.

Heat balance

In the resistance welding of panels of the same thickness and composition, the heat produced will be evenly balanced in both panels, and a characteristic oval cross-sectional weld nugget will result (see Figure 10.13). However, in the welding of panels produced from steels of a dissimilar composition, i.e. conventional low-carbon steel to high-strength low-alloy steels, an unbalanced heat rate will occur. This problem will also arise when spot welding two different thicknesses of steel; more heat will usually be lost by the thicker gauges, resulting in unsatisfactory welds. To compensate, the welder must either select an electrode made from materials that will alter the thermal resistance factor, or vary the geometry of the electrode tip, or use pulse equipment.

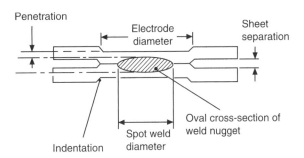

Figure 10.13 Welding together panels of the same thickness (*Motor Insurance Repair Research Centre*)

Electrodes

The functions of the electrode (chromium/copper) are to:

1 Conduct the current to the weld zone
2 Produce the necessary clamping force for the weldment
3 Aid heat dissipation from the weld zone.

The profile and diameter of the electrode face is dependent upon the material to be welded (Figure 10.14). The diameter of the face will directly influence the size of the nugget. The proper maintenance of the electrode tip face is therefore vital to ensure that effective current flow is produced. For example, if a tip diameter of 5 mm is allowed through wear to increase to 8 mm, the contact area is virtually doubled; this will result in low current density and weak welds. Misalignment of the electrodes and incorrect pressure will also produce defective welds.

Electrode tip dressing is best carried out with a fine-grade emery cloth. Coated steels will often cause particles of the coating to become embedded in the surface of the electrode tip face (pick-up). This necessitates frequent tip dressing; when the faces

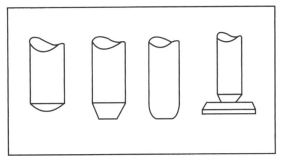

Figure 10.14 Types of electrode profile (*Motor Insurance Repair Research Centre*)

Table 10.1 Recommended tip pressures according to the gauge of steel

Gauge of panel (mm)	Pressure (kg)	Length of arm (mm)
0.55 + 0.55	50	250
0.80 + 0.30	60	250
0.95 + 0.95	70	250
1.00 + 1.00	80	250
1.25 + 1.25	100	250
1.25 + 2.00	120	250

These pressures relate to HS steels: a reduction in pressure may be necessary for conventional steels

have become distorted, the profile or contour must be reshaped. Cutters are available to suit most profiles and tip diameters. Domed tips are recommended; they adapt themselves best to panel shapes and last longest. The adjustment of electrode clamping force is critical for any type of steel, particularly when long-arm arrangements are used and complex configurations of arms are used for welding around wheel arches. It should also be noted that the tip force is difficult to maintain with long arms fitted to hand-held pincer-type guns (see Table 10.1).

10.8.1 Factors affecting weld quality

The following factors can have a large influence on the final outcome of the spot weld:

1 The type of metal being welded (for example, coated metals)
2 Joint configuration
3 Current and timing settings
4 The type of welding equipment being used
5 Electrode maintenance.

Clearance between welding surfaces
Any clearance between the surfaces to be welded causes poor current flow. Even if welding can be made without removing such gaps, the welded area would become smaller, resulting in insufficient strength. Flatten the two surfaces to remove the gaps, and clamp them tightly with a clamp before welding.

Metal surface to be welded
Paint film, rust, dust, or any other contamination on the metal surface to be welded causes insufficient current flow and poor results. Remove such foreign matter from the surface before commencing welding.

Corrosion prevention on metal surfaces
Coat the surface to be welded with anti-corrosion agent that has a high conductivity (e.g. zinc primers). It is important to apply the zinc primer uniformly between the panel surfaces to be welded.

Minimum welding pitch
The strength of individual spot welds is determined by the spot weld pitch (the distance between spot welds). The bond between the panels becomes stronger as the pitch is shortened, but if the pitch becomes too small and the spot welds are too close together this can allow current shunting, which in effect drains away the weld current towards the previous spot weld (Figure 10.15).

Positioning of welding spot from the edge and end of the panel
The edge distance is also determined by the position of the welding tip. Even if the spot welds are normal, the welds will not have sufficient strength if the edge distance is insufficient.

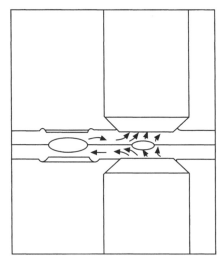

Figure 10.15 Current shunting in resistance welds positioned in close proximity

Inspection of spot welds
Spot welds are inspected either by outward appearance (visual inspection) or by destructive testing. A visual inspection is used to judge the quality of the outward appearance (tip bruises, pin holes, spatter, number of spots, spot positions), while destructive testing is used to measure the strength of a weld. Most destructive tests require the use of sophisticated

equipment which most body shops do not possess. Consequently a simpler method called a peel test has been developed by general use in the workshop.

Destructive testing: peel test

After setting up the equipment a test weld should be made, using sheet metal of the same thickness and condition as the job to be done. When the test piece is completed, it should be torn apart. If the settings are correct, the joint will 'unbutton', that is

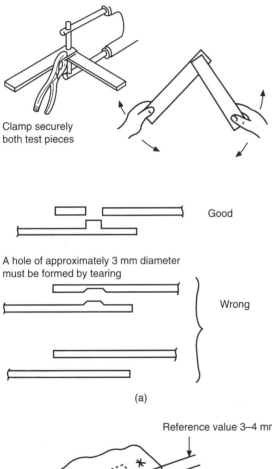

Clamp securely
both test pieces

Good

A hole of approximately 3 mm diameter must be formed by tearing

Wrong

(a)

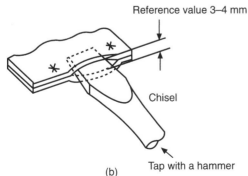

Reference value 3–4 mm

Chisel

Tap with a hammer

(b)

Figure 10.16   (a) Peel test (b) destructive test (*Motor Insurance Repair Research Centre*)

the weld nuggets will be on one sheet and the holes on the other (Figure 10.16(a)).

Weld tests for shear and peel for HSLA steels

When shear testing spot welds produced when welding together two HSLA steels, partial spot-weld failure may occur. If the partial failure does not exceed 20 per cent of the total nugget area, the strength of the weld should not be impaired (see Figure 10.17).

10.9 ARO Spotrite Pulsa resistance welding system

For resistance spot welding in vehicle body repair there is now a British invention called the Spotrite (made by ARO Machinery Co. Ltd). This is an adaptive (self-setting timing) in-process control unit which operates in several stages. Stage 1 is the monitoring stage for recognizing the material (low-carbon or HSLA steel) and for sensing through any

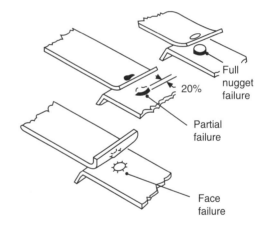

Full
nugget
failure

20%

Partial
failure

Face
failure

Figure 10.17   Modes of spot weld failure (*Motor Insurance Repair Research Centre*)

protective coatings, whilst judging how closely the clamped materials are fitting between the electrodes which will form the joint. Stage 2 is automatically linked and self-setting to provide a burst of current to produce the weld nugget. Stage 3 (and 4 if needed) will provide further precise timed pulses to bring the weld to the desired size, taking into account the number of sheets of metal and their thickness and the distance of the weld from the edge of the workpiece. The adaptive part of the control decides the number of stages needed and their duration. A weld cycle which includes all these stages added together, will take less than 0.75 s.

With this equipment much of the technology is transferred from the operator to the controls, only because the operator has no means of knowing which type of material he is welding. The control monitors what the welding gun is doing, and uses the information to take the decision by computing variables such as electrode tip wear, spot-weld pitch, voltage fluctuation, changes in metal thickness, and the presence of scale, dirt and rust. Consequently quality control is no longer the operator's responsibility.

Audiovisual signals warn the operator if there is overheating, if the electrode tip needs reshaping or cleaning, and if the weld arms are open too early. For difficult or unusual work, the experienced body repairer can switch to manual setting to control the unit's weld heat and time tolerance controls (see Figure 10.18).

A pincer-type welding gun with a powerful weld transformer (to maintain the very short weld times) is fitted with a strong pressure system to deal with the majority of work, which can be approached from two sides in pincer-type fashion. A range of interchangeable weld arms and electrodes of different shapes and lengths can be used to convert the standard gun instantly into the special tool that a job may require (see Figure 10.19). Special

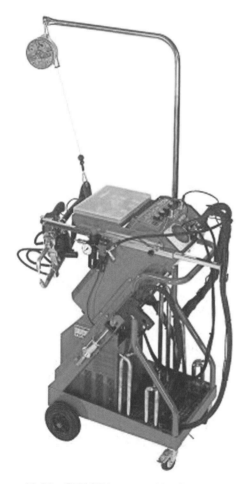

Figure 10.18   CEBORA spot welder (Mig Tig Arc)

Figure 10.19   Pincer welding gun used in different locations in repair work (*Motor Insurance Repair Research Centre*)

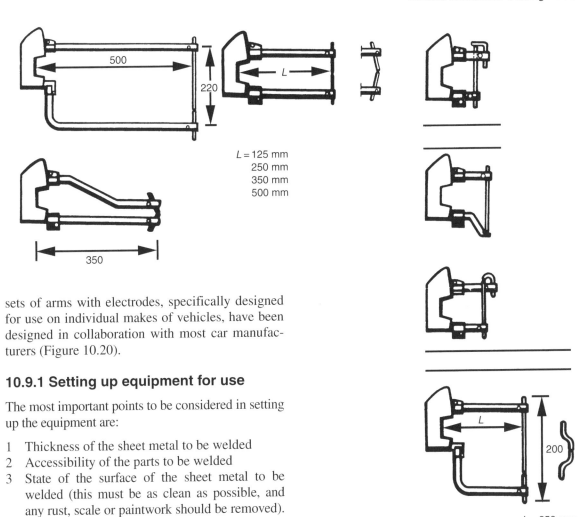

$L = 125$ mm
250 mm
350 mm
500 mm

$L = 250$ mm
125 mm

Figure 10.20 Electrode arm sets for different makes of vehicles (*SIP (Industrial Products) Ltd*)

sets of arms with electrodes, specifically designed for use on individual makes of vehicles, have been designed in collaboration with most car manufacturers (Figure 10.20).

10.9.1 Setting up equipment for use

The most important points to be considered in setting up the equipment are:

1 Thickness of the sheet metal to be welded
2 Accessibility of the parts to be welded
3 State of the surface of the sheet metal to be welded (this must be as clean as possible, and any rust, scale or paintwork should be removed).

The setting-up procedure is as follows:

1 With the equipment switched off, check that the electrode tips are aligned and correctly shaped. If domed tips are used (these generally give the best results), the tip radii should be 51 mm and 77 mm. When truncated cone tips are used, check that the diameter is correct for the gauge of material to be used: the diameter should be $5\sqrt{t}$, where t is the thickness of single sheet in millimetres. Check that the tip force is enough without bending the electrode arms or sliding the tip one upon another.
2 Switch on mains supply and check that the mains light (red) is on.
3 Switch function selection switch to resistance spot welding.
4 Switch on Spotrite control and check that the ready light is on.

5 Test earth leakage unit by pressing test button.
6 Set spot size to zero.
7 Make splashless welds by selecting the lowest heat and the longest time setting to make the desired size of weld nugget.
8 Make two welds approximately 51 mm apart. Peel test the second weld, i.e. tear the test piece apart and check whether the nugget is the size approximately to the weld strength needed. The nugget size should be $4.5\sqrt{t}$.
9 Increase the spot size setting from zero in single steps, and make five trial welds in each setting.

If the reject alarm is heard (a high pitched pulse note: red light illuminates) for more than three welds, reduce the spot size setting until the reject alarm is only activated two or three times.

10 Increase the heat step by step to reduce the actual weld time until splashing occurs. Then set back heat one step to stop splashing. The equipment is now fully set.

10.9.2 Welding procedure

1 Prepare the panel surface by removing any paint, primer and in general any insulating material covering the surface to be welded. This preparation is a key factor to good-quality spot welding. Paint must be removed by either paint remover or sanding. If rust is present on the panel surface, sanding is a better method of preparation as it will leave the surface with a bright metallic finish which facilitates the flow of electric current.

2 Obtain the correct adjustment of welding gun electrodes and arms.

3 Determine a suitable weld pitch for the panel assembly to be welded to obtain maximum strength.

4 Make sure that the correct distance is set from the edge of the sheet metal panel to the nearest spot weld.

10.10 Single-sided spot welding

As a result of the operational limitations of conventional double-sided spot welding, single-sided spot welding equipment has become more widely used in the vehicle body repair industry. This system offers the benefits of conventional spot welding without the inherent problems of accessibility with double-skinned panel sections. Single-sided spot welding is therefore an alternative to conventional double-sided spot welding, or can be used in conjunction with it.

In the single-sided spot welding process the operator manually forces the single electrode against the panel, with the electrical circuit being completed by an earth clamp and the cable back to the transformer. This allows welds to be made in positions where access is possible only from one side.

The manufacturers are now using different types of steel to construct their vehicle bodies, the main three being low-carbon steel, galvanized steel and high-strength steel. This has led to confusion and difficulty in identification and welding. The main

problem with high-strength steel is that it is heat sensitive and difficult to identify on vehicles without data (MIRRC Thatcham *Methods Manual*). In double-sided welding this problem is solved by using pulse welding, which keeps the heat in the weld as low as possible. Some single-sided equipment, especially that which operates at high current, does not require pulse control welding. Instead these machines use a massive burst of power at 8000 A DC in a very short interval, which keeps the temperature of the spot weld below the recommended temperature. The other advantage of this system is that it does not need to identify whether the steel is high strength or low carbon. The problem with galvanized steel is that it has a very high contact resistance, making it difficult to weld; therefore the high DC machines use a special preheat function for this type of steel. In the first stage of the weld the machine lowers the resistance of the steel by melting the coating, causing it to flow from the weld. Then, once the resistance is low enough, it will automatically carry out the second stage weld at the correct setting.

10.10.1 Single-sided spot welding equipment

This equipment (Figure 10.21) ranges from 2500 to 9200 amperes as follows:

1 2500 A using two phases at 415 V
2 8000 A using three phases at 415 V
3 9200 A using three phases at 415 V.

The equipment can be used to carry out the following operations:

Single-sided spot welding Ideal where access is difficult from both sides of the material. It is suitable for welding on wings, front panels and rear quarter panels (Figure 10.22).

Two-electrode single-sided spot welding Suitable when unable to attach the earth clamp satisfactorily. It also allows two spot welds to be made simultaneously and is suitable for welding sill sections in place.

Pulse control roller spot welding This gives a continuous spot weld along an overlapped edge of metal. It is ideal for roof gutters and for welding patches to vehicle panels when dealing with corrosion repair without creating distortion (Figure 10.23).

Figure 10.21   Single-sided and double-sided spot welding equipment (*Stanners Ltd*)

Figure 10.22   Single-sided spot welding used in panel repair (*Stanners Ltd*)

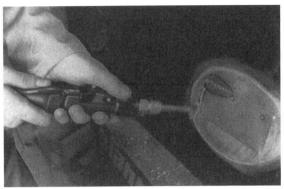

Figure 10.23   Pulse controlled roller spot welding (*Stanners Ltd*)

Welding copper ring washers for pulling   This allows rows of washers to be welded to the panel surface. Ideal for pulling out large dints by using a slide hammer with a special hook attachment which fits through the rings, which are later broken off by twisting.

Rapid puller   This is designed for pulling out small dints quickly and effectively by welding the puller to the panel, pulling out the dint, then twisting the tool to release it from the panel (Figure 10.24).

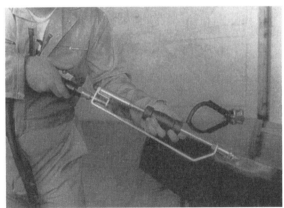

Figure 10.24   Rapid spot puller (*Stanners Ltd*)

Copper shrinking of high spots   This uses a copper tool for shrinking stretched panels which have been overworked by hammering (Figure 10.25).

Carbon shrinking for over stretched panels   This uses a carbon pencil rod and is used for retensioning a panel surface that is only slightly stretched (Figure 10.26).

Figure 10.25   Copper shrinking attachment (*Stanners Ltd*)

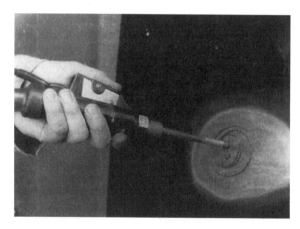

Figure 10.26   Carbon pencil shrinking attachment (*Stanners Ltd*)

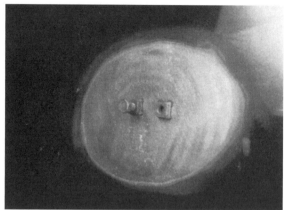

Figure 10.27   Captive nuts welded on to panel (*Stanners Ltd*)

Welding captive nuts to vehicle This piece of equipment can be used to weld captive nuts, trim studs and threaded bolts to panel surfaces when they need to be replaced (Figure 10.27).

Single-sided spot welding equipment has a low current intensity, which ensures safety for the operator. The equipment uses direct current at a maximum of 12 volts. All electrodes can be quenched in water, without danger to the operator, when they become overheated, and this extends their life. Single-sided welding with one electrode and one earth clamp results in current loss in the metal panel, and also the metal panel is a bad conductor of electricity. To overcome this problem, Stanners

Ltd use a system which gives good results by using two earth clamps positioned correctly on the panel being welded. With this system 20 per cent more current is passed through the weld, resulting in better and stronger welds.

10.10.2 Setting up equipment for single-sided spot welding

1   Set the weld timer dial.
2   Set the welding mode switch.
3   Set the welding power dial.
4   Connect the earth clamp plate to the negative side of the welding output cable holder, and the single-sided electrode to the positive side of the cable holder.
5   Connect two earth clamp plates as close as possible to the area to be welded.
6   Ensure that both pieces of metal are clean and making good contact. Also make sure that the lower piece of metal is well supported to allow pressure to be applied to the electrode by the operator.
7   Press the electrode against the workpiece to be welded. Apply pressure to both pieces of metal and press the switch to carry out the weld.
8   It is important that the electrode is quenched in water after prolonged welding to prevent it becoming hot or overheating.
9   The electrode tip should maintain its original profile if good spot welds are to be achieved. Therefore, when necessary, the electrode should be redressed back to its correct shape.

Questions

1   What is meant by the term 'resistance welding'?

2   Name and describe three methods of resistance welding.

3   Give an example where resistance welding is used in vehicle manufacture.

4   Resistance welding uses various electrodes: select and sketch one type.

5   How is a workshop test carried out on a spot weld, to test its strength?

6   Suggest one application of projection welding in the assembly of a vehicle body.

7   Illustrate types of joints that would be suitable for spot-welding applications.

8   Describe the three stages in the production of a spot weld.

9   Which important points should be considered when reshaping a resistance welding electrode?

10   Explain the importance of resistance welding in the repair of vehicle bodies.

11   Which metals are used for making electrodes in resistance welding?

12   What is the difference between butt welding and flash welding?

13   Name the principal parts of resistance welding equipment that would be used in the body repair shop.

14   State the qualities required of a good spot weld.

15   Explain the term 'robotic welding'.

16   What are the important factors to be considered when spot welding high-strength steel.

17   Explain the importance of the adaptive self-setting timing control unit when welding high-strength steels.

18   Explain the working procedure of a pincer-type welding gun.

19   Describe, with the aid of a sketch, the formation of a weld nugget during the spot welding cycle.

20   Describe a method of resistance welding that would be used on the assembly line to weld a roof panel into position.

21   What is the purpose of 'hold time' when carrying out the process of spot welding?

22   Describe typical repair situations that would use a pincer welding gun and a single-sided welding gun.

23   State where resistance seam welding could be found on a vehicle body.

24   Show, with the aid of a sketch, the principle of twin-spot welding.

25   Describe, with the aid of a sketch, the resistance projection welding process.

26   List the advantages of the resistance spot welding technique used in the repair of vehicle bodies.

27   Sketch one type of electrode arm sets used in vehicle repair.

28   With the aid of a sketch, explain how heat is generated to form the spot weld.

29   State the treatment that should be carried out before replacing a spot welded panel.

30   State why water-cooled electrode tips are used in the manufacture of vehicle body shells.

Manual metal arc welding

In the sphere of welding the electric arc has become an efficient and reliable means of welding sheet and metal plate. It is useful for welding the heavier-gauge plates used for commercial vehicle body building and also for the type of metal plate processes in which the metal ranges in thickness from 3 mm to 75 mm.

The use of arc welding depended naturally upon the development of electricity, and dynamos or generators were not developed until 1880. The first actual arc welding, meaning the melting of metal by means of electrodes and thus fusing them together, was developed by Bernardoz in 1885; he created a mechanism using a carbon electrode which produced an arc between the carbon and metal, melting the edges and thus performing a weld. The arc form of welding, using the metallic electrodes, was discovered by Slavinoff in 1892, but had very little success because of the use of bare metal electrodes. However when Kjellberg, a Swedish inventor, developed the flux electrode in 1907, the success of the metallic electrode was assured. Progress accelerated as a result of the First World War, when productivity and speed of welding was of prime importance. However, it was not until the 1930s that good-quality joints could be reliably and consistently produced by the arc welding process. This was achieved by the development of coatings which gave adequate protection and improved arc stability whilst transferring metal between the electrode and the parent metal. From that time arc welding gradually displaced gas welding techniques, especially when joining heavy-gauge metal, although the major development in arc welding was due to the production of portable and automatic welding machines.

Although metal arc welding is only used a small amount in private car construction for heavy-gauge assemblies in cars which have chassis, it is still used in the commercial vehicle body building industry for the assembly of the fully welded, trailer-type bodies in mild steel, stainless steel and aluminium.

11.1 Principles of manual metal arc welding

Manual metal arc welding (MMAW) is used extensively in modern practice. It is based on the principle whereby intense heat is obtained from an electric current which creates an arc between a metal electrode and the plates which are to be welded (Figure 11.1a). The heat produced is sufficient to fuse the edges of the plates at the joint, forming a small pool of molten metal. Additional molten metal from the tip of the electrode is deposited into the molten pool, and when solidified it results in a strong welded joint. With this process the electrode from which the arc is struck is made of the same metal as the parent metal which is being welded; metal from the electrode is transferred to the weld, partly as drops under the influence of gravity and partly as high-velocity particles. The maintenance of a stable arc between a bare metal electrode and the workpiece is an unreliable procedure, and for this reason (among others) various coatings are applied to the electrode wire. Not only do these coatings help stabilize the arc, but they also perform the important functions of stopping oxidation of the heated electrode tip and molten workpiece. By their slagging properties they dissolve and segregate oxides and other impurities whose presence might otherwise adversely affect the quality of the welded joint. When the weld has cooled the slag forms a brittle coating which can be fairly easily removed by chipping and brushing (Figure 11.1b). The electric supply for this type of welding may be

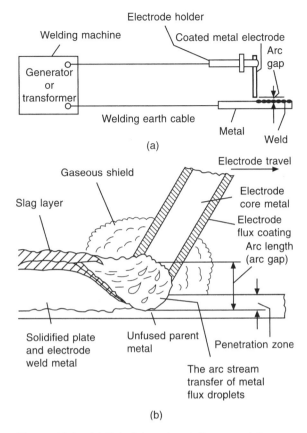

(a)

(b)

Figure 11.1 (a) Principles of metallic arc welding (b) cross-section of a coated electrode in the process of welding

either direct current (DC) or alternating current (AC); both systems possess certain advantages, depending on the purpose for which they are employed. Many different types and sizes of electric welding machines are now available, and two of the most generally used are the DC generator and the AC transformer.

11.2 Electrical terms used in arc welding

Circuit A circuit is the path along which electricity flows. It starts from the negative (−) terminal of the generator where the current is produced, moves along the wire or cable to the load or working source and then returns to the positive terminal (+).
Amperes Amperes or amperage refers to the amount or rate of current that flows through a circuit.
Voltage The force (electromotive force or EMF) that causes electrons to flow in a circuit is known as

voltage. This force is similar to the pressure used to make water flow in pipes. In the water system the pump provides the pressure, whereas in an electrical circuit the generator or transformer produces the force that pushes the current through the wires.
Direct and alternating current There are two kinds of current used in arc welding: DC and AC. In DC the current flows constantly in one direction. In AC the current reverses its direction in the circuit a certain number of times per second. The rate of change is referred to as frequency and is indicated as 25, 40, 50, 60, cycles per second (hertz).
Voltage drop Just as the pressure in a water system drops as the distance increases from the water pump, so does the voltage lessen as the distance increases from the generator. This fact is important to remember in using a welding machine, because if the cable is too long, there will be too great a voltage drop. When there is too much drop, the welding machine cannot supply enough current for welding.
Open-circuit voltage and arc voltage Open-circuit voltage is the voltage produced by the welder when the machine is running and no welding is being done. After the arc is struck, the voltage drops to what is known as the arc or working voltage. An adjustment is provided to vary the open-circuit voltage so that welding can be done in different positions.

11.3 Metal arc welding equipment

11.3.1 DC generator

With a DC welding machine the electric current is produced by means of a generator, which is driven by a petrol or diesel engine (Figure 11.2) or alternatively by an AC or DC electric motor. The motor driven type of generator set is chiefly used for the type of welding work performed inside a workshop and is therefore often permanently mounted on the floor, but types are available for site work. The electric motor provides a good constant-speed drive for the generator and is not affected by the load imposed upon it. The generator is specially designed for welding purposes so that it generates about 60 volts on open circuit; this drops to about 20 volts immediately the arc is struck.

Generators are built in various current ratings ranging from about 100 to 600 amperes, and most

(a)

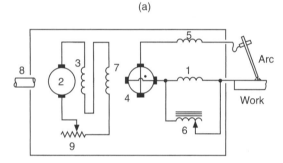

1 Differential series field coils
2 Exciter armature commutator and brush gear
3 Exciter field coils
4 Main armature, commutator and brush gear
5 Spark-prevention interpole coils
6 Current regulator wound on laminated iron core
7 Main shunt-field coils
8 Prime mover: electric motor, petrol or diesel engine
9 Voltage control rheostat

(b)

Figure 11.2 Diesel engined DC generator and circuit

modern types incorporate means for automatically adjusting the voltage to meet variations in the demand of the arc. Although the arrangement of the control unit may vary for machines made by different manufacturers, most consist of a wheel or lever control which selects the correct current for the right size of electrode and thickness of plate being welded. DC generators usually have a polarity switch which enables the welder to reverse the polarity, as is occasionally required when welding with special electrodes.

11.3.2 AC transformers

The AC welding machine employs a transformer instead of a generator to provide the required welding current. The AC transformer, as its name implies, is an instrument which transforms or steps down the voltage of the normal mains electrical supply to a voltage suitable for welding between 60 and 100 volts (Figure 11.3). In its simplest form an AC transformer consists of a primary and a secondary coil with an adjustment to regulate the current output. The primary coil receives the alternating current from the source of supply and creates a magnetic field which constantly changes in direction and strength. The secondary coil has no electrical connection to the power source but is affected by the changing lines of force in the magnetic field and, through induction, delivers a transformed current at a higher value to the welding arc. The output current is controlled either by an

(a)

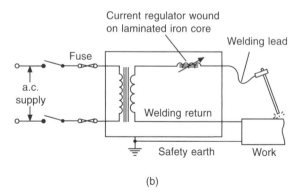

(b)

Figure 11.3 AC transformers and circuit (*Murex Welding Processes Ltd*)

adjustable reactor in the output circuit of the transformer or by the adjustment of the internal reactance of the transformer. The current adjuster is operated by turning a handwheel or crank. As the control handle is moved, a calibrated dial shows the current setting in amperes. Unlike the DC generator, the AC transformer has no moving parts and for this reason is usually referred to as a static plant. AC welding equipment has many advantages, namely:

1 Low initial cost
2 No moving parts and therefore negligible maintenance
3 Higher electrical efficiency
4 Ease of transport
5 AC can be converted to DC by means of a motor generator or rectifier, thus making both types available.

The disadvantages of the AC system are:

1 Coated electrodes must always be used
2 Voltage higher than in DC system and therefore risk of shock greater
3 Welding of non-ferrous metals more difficult than in DC system.

11.3.3 Choice of system

As far as the quality of the weld is concerned there is little to choose between AC and DC systems. An AC system gives a smoother arc when using very high current, although a DC system is essential when welding certain non-ferrous metals. The availability of the mains supply is obviously a criterion.

11.3.4 Welding accessories

In addition to the actual welding generator or transformer, the following accessories should form a part of every welder's equipment (Figure 11.4):

Electrode holder fitted with a length of flexible cable for connection to the plant. It should be of sufficient capacity to hold the largest electrode to be used, light in weight to reduce fatigue, well balanced, and able to locate and eject the electrode easily. It should also be able to carry the maximum welding current without overheating. It can be a partially or fully insulated type, the latter being by far the safer.
Welding earth A length of cable which is flexible and connects the work to the plant. The diameters of these cables are governed by the voltage and

Figure 11.4 Welding accessories (*Murex Welding Products Ltd*)

distance to be carried from the machine. From the safety point of view it is essential that the welding circuit is efficiently earthed. The earth clamp (Figure 11.5) which is fixed on the end of the earth

Figure 11.5 Welding accessories showing electrode holders and earthing clamps (*Murex Welding Products Ltd*)

cable should be as near to the work as possible. Its function is to keep the work at earth potential to safeguard personnel against any breakdown of the welding set. If the work is earthed and the return current lead is broken, no welding current can flow. *Head shield* or *face screen* fitted with special coloured lenses is an absolute necessity, because an electrode arc produces a brilliant light and gives off invisible ultraviolet and infrared rays which are very dangerous to the eyes and skin. Never attempt to look at the arc with the naked eye. The helmet type of head shield fits over the head and leaves both hands free (Figure 11.6). The face screen provides adequate protection but needs holding by hand. The coloured lenses are classified according to the current used.

Figure 11.6 Welder wearing headshield and protective clothing (*Murex Welding Products Ltd*)

Gloves (Figure 11.6) are another important item for arc welding. These must be worn to protect the hands from the ultraviolet rays and from hot metal. Gloves, irrespective of the type used, should be flexible enough to permit proper hand movement, yet not so thin as to allow any heat penetration.

Leather aprons are often used by beginners in order to protect their clothes (11.6), although most experienced welders seldom wear an apron on the job except in situations where there may be an excessive amount of metal spatter resulting from awkward welding positions.

Goggles should be worn when chipping slag from a weld; this is a thin crust which forms on the deposited bead during the welding process. Whilst removing slag, tiny particles are often deflected upwards, and without proper eye protection these particles may cause a serious eye injury.

Cleaning tools (Figure 11.4) To produce a strong welded joint, the surface of the metal must be free of all foreign matter such as rust, oil and paint. A wire brush is used for cleaning purposes. After a bead is deposited on the metal, the slag which covers the weld is removed with a chipping hammer. The chipping operation is followed by additional wire brushing. Complete removal of slag is especially important when several passes must be made over a joint. Otherwise, gas holes will form in the bead, resulting in porosity which weakens the weld.

Welding booth for use in production welding is designed to protect all other personnel from the arc glare. It comprises a steel bench, insulated from the booth, and a wooden duckboard to safeguard the welder from damp floors.

11.4 Electrodes used in welding: BS coding

Most modern electrodes are coated or covered and consist of a metal core wire surrounded by a thick coating applied by extrusion or other processes. The coating is usually a mixture of metallic oxide with silica which, under the heat of the electric arc, unites to form a slag which floats on the top of the molten weld metal. To a large degree the success of the welding operation depends on the composition of the coating, which is varied to suit different conditions and metals. The principal functions of the coating are to:

1 Stabilize the arc and enable the use of arc welding
2 Flux away any impurities present on the surface being welded

3 Speed up the welding operation by increasing the rate of melting

4 From a slag over the weld in order to protect the molten metal from oxidation by the atmosphere, slow the rate of cooling of the weld and so reduce the chances of brittleness, and provide a smoother surface.

In the past each manufacturer employed different means of identifying the properties of his electrodes. To overcome the confusion arising from this the British Standards Institution drew up a scheme for classifying all electrodes from a code number which enables the user to know the main features of an electrode irrespective of the source of supply.

The BS classification of an electrode is indicated by the following coding, in the order stated.

Strength, toughness and covering (STC) code

1 The letter E indicates a covered electrode for manual metal arc welding.

2 Two digits indicate the strength (tensile, yield and elongation properties) of the weld metal (see Table 11.1).

3 A digit indicates the temperature for a minimum average impact value of 28 J (see Table 11.2).

4 A digit indicates the temperature for a minimum average impact value of 47 J (see Table 11.3).

5 A letter or letters indicate the type of covering:

E — — — — B	basic
E — — — — BB	basic, high efficiency
E — — — — C	cellulosic
E — — — — R	rutile
E — — — — RR	rutile, heavy coated
E — — — — S	other types.

The covering should be free from defects, such as cracks and abnormalities which would affect the operation of the electrode. The gripped end of each electrode should be free from covering for a minimum distance of 15 mm and a maximum of 40 mm.

Additional coding

The following additional coding must follow the STC coding:

1 When appropriate, three digits indicate the nominal electrode efficiency, which is the ratio of the mass of weld metal to the mass of nominal diameter core wire consumed for a given electrode.

2 A digit indicates the recommended welding positions for electrodes:

1 all positions
2 all positions except vertical/down
3 flat and, for fillet welds, horizontal/vertical
4 flat
5 flat, vertical/down and, for fillet welds, horizontal/vertical
6 any position or combination of positions not classified above

3 A digit indicates the power supply requirement (see Table 11.4).

4 Where appropriate, a letter H indicates a hydrogen-controlled electrode.

Coding example

A typical classification for an electrode is therefore:

E51 54 BB [160 3 0 H]

the STC code is identified as follows:

E51 strength (510–650 N/mm^2)
5 temperature for minimum average impact strength of 28 J ($-40\,°C$)

Table 11.1 Designation for tensile properties (*British Standards Institution*)

Electrode designation digit	Tensile strength (N/mm^2)	Minimum yield stress (N/mm^2)	Minimum elongation (%)		
			When digit of Table 11.2 is 0 or 1	When digit of Table 11.2 is 2	When digit of Table 11.2 is 3, 4 or 5
E43— — —	430–550	330	20	22	24
E51— — —	510–650	360	18	18	20

Table 11.2 First digit for an impact value (*British Standards Institution*)

Digit	Temperature (°C) for minimum average impact value of 28 J using 4 mm diameter electrodes only
E— — 0 — —	Not specified
E— — 1 — —	+20
E— — 2 — —	0
E— — 3 — —	−20
E— — 4 — —	−30
E— — 5 — —	−40

Table 11.3 Second digit for an impact value (*British Standards Institution*)

Digit	Temperature (°C) for minimum average impact value of 47 J using 4 mm diameter and largest diameter electrodes submitted for classification
E— — — 0 —	Not specified
E— — — 1 —	+20
E— — — 2 —	0
E— — — 3 —	−20
E— — — 4 —	−30
E— — — 5 —	−40
E— — — 6 —	−50
E— — — 7 —	−60
E— — — 8 —	−70

Table 11.4 Welding current and voltage conditions (*British Standards Institution*)

Digit	Direct current: recommended electrode polarity	Alternating current: minimum open-circuit voltage (V)
0	See manufacturer	Not suitable for use on AC
1	+ or −	50
2	−	50
3	+	50
4	+ or −	70
5	−	70
6	+	70
7	+ or −	80
8	−	80
9	+	80

4 temperature for minimum average impact strength of 47 J (−30 °C)

BB covering (basic, high efficiency)

The additional code, is as follows:

160 efficiency
3 welding positions
0 welding current and voltage conditions
H hydrogen controlled

Electrode dimensions and tolerances

The *size* of an electrode should be the specified nominal diameter of the core wire. The *length* of an electrode should be within ±2 mm of the nominal value given. Table 11.5 gives the nominal lengths of electrodes.

Table 11.5 Nominal lengths of electrodes (*British Standards Institution*)

Diameter (mm)	Nominal length (mm)
1.6	200
	250
2	200
	250
	300
	350
2.5	250
	300
	350
3.2, 4, 5	350
	450
6 to 8	350
	450
	500
	600
	700
	900

Electrode and bundle identification

The STC code should be marked on the covering of each electrode as near as possible to the gripped end, except on electrodes of 1.6 mm and 2 mm diameter where it is not practicable.

Each bundle or package of electrodes should be clearly marked with the following information:

1 The number and date of the British Standard Classification (STC code)
2 Name of manufacturer

3 Trade designation of electrodes
4 Size and quantity of electrodes
5 Batch number
6 Recommended current range, polarity and power supply
7 Recommendations for special storage conditions
8 Any other significant information on electrode characteristics or limitations on use
9 Health warnings.

11.5 Arc welding positions

Arc welding operations can be carried out with work in almost any position, but the degree of skill required of the welding operator will vary considerably depending on the position of the joint, which may be flat or downhand, horizontal, vertical or overhead (Figure 11.7).

11.5.1 Flat position (downhand)

Although welding can be done in any position the operation is simplified if the joint is flat. The speed of welding is then increased because the molten metal has less tendency to run, better penetration can be secured, and the work is less fatiguing for the welder in this position. Some jobs may at first glance appear to require horizontal, vertical or overhead welding, but upon examination it may be possible to change them to the easier and more efficient flat position, and for this reason positioning jigs are used in mass production. When welding in this position the electrode should be at an angle of 60–70° to the plate surface or horizontal.

11.5.2 Horizontal position

Occasionally the welding operation must be done while the work is in a horizontal position, which means that the welder must use a slightly shorter arc than for flat position welding. The shorter arc will minimize the tendency of the molten pool to sag or run down and cause overlap. An overlap occurs when the pool runs down to the lower side of the weld and solidifies on the surface without actually penetrating the metal. For horizontal welding, hold the electrode so that it points 5–10° and slants approximately 20° away from the depositing weld. When welding, use a narrow weaving motion which will still further reduce the tendency of the molten weld pool to sag.

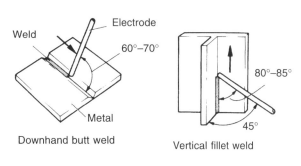
Downhand butt weld Vertical fillet weld

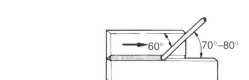
Horizontal fillet weld

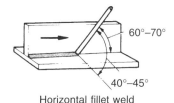
Vertical butt weld

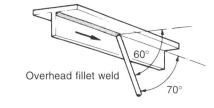
Overhead fillet weld

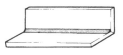
Continuous weld

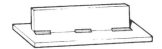
Intermittent or tack weld

Normal weaves on weld bead

Figure 11.7 Arc welding positions

11.5.3 Vertical position

Vertical welding is done by depositing a weld in an upward or downward direction. Downward welding is very good for welding light-gauge metals because penetration is shallow, therefore forming an adequate weld without burning through the metal. Downward welding can also be performed more rapidly, which is important in production work. On heavier plates of 6 mm or more, vertical upward welding is more practical since deep penetration can be obtained. Welding upwards also makes it possible to create a shelf upon which successive layers of weld can be placed. For vertical downward welding the electrode should be held at 10–15° tilted from the horizontal, and starting at the top of the seam, moving downwards with little or no weaving motion. For vertical upward welding start with the electrode at right angles to the plate, then lower the rear of the electrode 10–15° with the horizontal.

11.5.4 Overhead position

Welding in an overhead position is probably the most difficult of operations. It is difficult because the welder must assume an awkward stance, and at the same time work against gravity, which means that the molten pool has a tendency to drop, making it more difficult to secure a uniform weld and correct penetration. Heavily coated rods must not be used because of the continual dropping of the slag, and for this reason medium coated rods are generally used. The most important points about overhead welding are to obtain the correct control of the current and keep a very short arc. Correct current control gives a pool that is sufficiently molten to ensure good penetration. Hold the electrode at right angles to the seam, then tilt the rear of the electrode until it forms an angle of 10–15° to the horizontal.

11.6 Essential factors of arc welding

11.6.1 Correct choice of electrode

In selection, consideration must be given to position of weld, type of metal, size of electrode, type of joint, and current setting. The main group of electrodes include mild steel, high-carbon steel, special alloy steel, cast iron and non-ferrous aluminium, classified as bare and flux-covered. The bare electrodes have a very light coating which affords some protection against oxidation of the surface. The flux-covered or shielded electrode has a very heavy coating of several chemical substances which protect the molten metal from the oxidation and help to keep a steady arc. When this type of electrode melts, the coating produces a shield of gas around the molten metal, safeguarding it against the atmosphere. Some of the coating also forms a slag over the molten part, which also serves as a protection against oxidation.

11.6.2 Correct arc length

This will depend upon the type of electrode used and the nature of the welding operation. If the arc is too long, a wide irregular bead is produced with insufficient fusion between the plate and weld metal. When the arc is too short, insufficient heat is generated to melt the plate properly and the electrode will stick frequently, resulting in a very uneven bead (Figure 11.8).

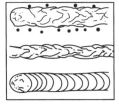

1   This is how the beads of weld appear when the arc is too long
2   In this case the arc is too short
3   Here the arc is the correct length

Figure 11.8   Correct arc length

11.6.3 Correct speed of travel

This must be slow and even to ensure sufficient penetration without excessive build-up of bead. When the speed is too fast, the molten pool solidifies rapidly and impurities are retained in the weld. If the speed is too slow, excessive metal produces a high and wide bead (Figure 11.9).

11.6.4 Correct current

The correct current setting in accordance with the size of electrode used is most important (Table 11.6). When the current is too great the electrode melts fast, causing a large pool of metal and excessive spatter. If the current is too low the heat generated will be insufficient to melt the plate and the molten pool will be small, resulting in lack of fusion (Figure 11.9).

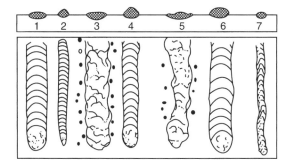

1 Current, voltage and speed normal
2 Current too low
3 Current too high
4 Voltage too low
5 Voltage too high
6 Speed too slow
7 Speed too fast

Figure 11.9 Correct speed and current setting

Table 11.6 Current setting for electrode

Sheet thickness (mm)	Size of electrode (mm)	Current (A)
1.0	1.6	30
1.2	1.6	35
1.6	2.0	50
2.0	2.5	80
2.5	3.2	110
3.2	4.0	120–160
3.2–9.5	5.0–6.3	250–400
9.5 and over	8 mm	400–600

11.6.5 Striking the arc

The successful step towards electric arc welding is learning to strike and maintain the arc and run a straight bead of weld metal. First set the control unit to the correct setting specified for the size of electrode being used. Then bring the electrode into contact with the plate by one of the following two methods.

11.6.6 Methods of welding

Tapping method

The tapping motion method is as shown in Figure 11.10. The electrode is brought straight down on the plate and instantly withdrawn a distance of

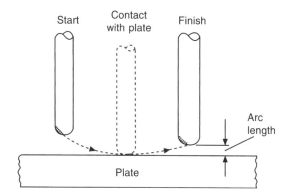

Figure 11.10 Tapping method of striking the arc

3–5 mm, this distance being equal to the core diameter of the electrode.

Scratching method

The scratching method (Figure 11.11) is where the electrode is tilted at an angle and is then given a slight circular movement similar to that of striking a match. As in the previous method the electrode is promptly raised a distance equal to its diameter, otherwise it will stick to the plate.

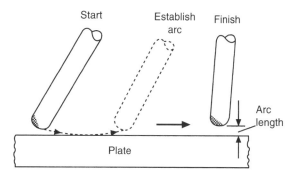

Figure 11.11 Scratching method of striking the arc

11.6.7 Welding currents

Welding currents may vary from 20 to 600 A. For striking a DC arc an open circuit of 55–60 volts is required, whilst an AC set requires 80–100 volts. Once the arc is struck the arc voltage will drop to 20–25 volts. Before striking the arc the operator should have his head shield or face screen in position and observe the arc through the glass filter.

11.6.8 Weld defects

Undercutting

This is a condition that results when the welding current is too high. The excessive current leaves a groove in the base metal along both sides of the bead, which greatly reduces the strength of a weld (Figure 11.12).

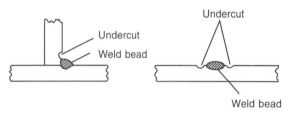

Figure 11.12   Undercutting

Overlapping

This occurs when the current is set too low. In this instance the molten metal falls from the electrode without actually fusing with the base metal, resulting in a defective weld (Figure 11.13).

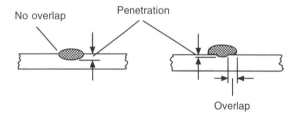

Figure 11.13   Overlapping

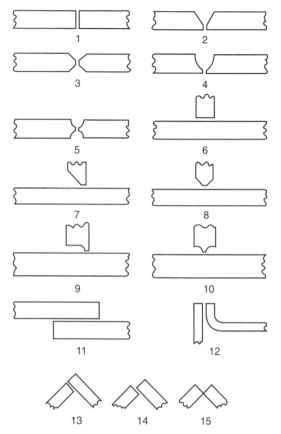

1	Square butt joint	9	Single-U-T joint
2	Single-V butt joint	10	Double-U-T joint
3	Double-V butt joint	11	Lap joint
4	Single-U butt joint	12	Edge joint
5	Double-U butt joint	13	Flush corner joint
6	Square-T joint	14	Half open corner joint
7	Single bevel-T joint	15	Full open corner joint
8	Double bevel-T joint		

Figure 11.14   Types of arc welded joint

11.6.9 Types of joint used in manual metal arc welding

The selection of the type of joint to be used for a particular application is governed by the following factors (Figure 11.14):

1   The load and its characteristics
2   The manner in which the load is applied
3   The cost of joint preparation and welding.

Butt joints are of several types, each having a number of variations. However, they are generally classified as plain, single-V, double-V, single-U and double-U. T-joints also have several variations which can be applied in appropriate places. The other joints commonly used in manual metal arc welding are the lap joint and corner joints.

11.6.10 Welding joint design

Joint configuration

The three basic joint types are usually classified as fillet, butt or groove, and lap (Figure 11.15). These may be further subdivided depending on the joint detail and the joining process being used.

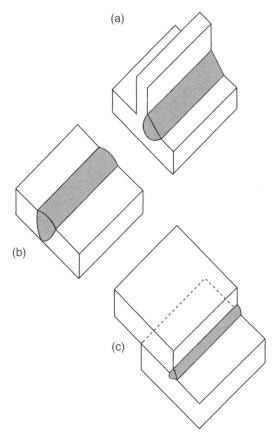

Figure 11.15 Basic joint types: (a) fillet (b) butt (c) Lap (*BOC Ltd*)

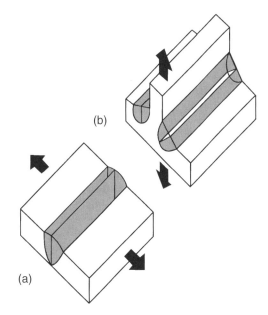

Figure 11.16 Static loading conditions: (a) butt weld (b) fillet weld, load carrying area shaded (*BOC Ltd*)

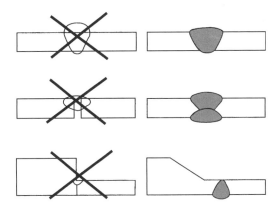

Figure 11.17 Acceptable and undesirable joint detail (*BOC Ltd*)

Strength and loading conditions

The achievement of the required static strength is a primary criterion in the selection of joint design. For fusion welding processes which employ a filler material of composition matching that of the base material, it is normally only necessary to ensure that the cross-sectional area of the loaded joint is adequate (Figure 11.16). This is relatively easy to achieve where simple tensile or compressive forces are involved, but the effect of shear loading must also be considered in the case of lap joints. In dynamic loading situations the joint profile is more significant and it is important to avoid discontinuities which may act as stress raisers. This is particularly important in joints subjected to fatigue (Figure 11.17).

Economic considerations

The cost of any weld connection will be influenced by the joint design. In general it is desirable to minimize the time taken to prepare and complete the joint to the required quality. This is usually associated with a reduction in the amount of material added to the joint. Narrow gap butt welds, for example, give a reduction in weld time and, depending on the process, may reduce the complexity of plate preparation. Overwelding, that is the use of more weld metal than is needed to meet the strength requirements, may involve a significant higher cost.

Practical weld constraints

Some of the practical considerations which may affect joint design are:

1   Access
2   Welding position
3   Process/consumable/equipment availability
4   Dissimilar thicknesses.

If access is limited to one side of a butt joint it may be necessary to employ integral backing or complex plate preparation to ensure that adequate quality is achieved in the weld root (Figure 11.18).

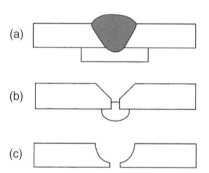

Figure 11.18   Plate preparation for limited access butt joints: (a) permanent backing strip (b) fusible insert (c) U preparation for root access (*BOC Ltd*)

In positional applications, non-symmetrical preparations may be necessary to improve joint profile and avoid excess weld metal. The joint preparation may be significantly affected by the welding process, consumables and equipment which are available. Dissimilar thicknesses may be joined using lap and fillet configurations, but it may be necessary to equalize the cross-section at the joint to perform satisfactory edge or butt welds. In some cases it may be possible for the joint to be self-jigging. This is particularly useful for site assembly, saving time and improving repeatability.

A range of approved joint designs is provided in international standards covering welding construction BS 5135.

11.7 Technique of welding

11.7.1 Welding mild steel plate in the flat position (downhand)

Prior to setting up the job it is necessary to clean the area to be welded. Where the material is heavier than 3 mm, V-ing of the welding edges is necessary to ensure penetration. Next make sure that the earth lead is securely fastened to the work. Painted or rusted surfaces provide a poor earth and should be scraped before connecting the earth clamp. Adjust the current setting of the transformer or generator in accordance with the electrode maker's specifications. Place the electrode in the holder, making sure that good contact is gained between the jaws of the holder and the bare end of the electrode. Care should be taken that the flux coating of the rod is clear of the electrode holder jaws. Never grip this holder in a tense manner; a firm but relaxed grip will produce a steady hand which is essential while striking and maintaining the arc. The angle at which the electrode meets the plate being welded should be about 70°. The speed of travel should be such that a continuous penetrating bead of electrode is deposited. Welds on heavy plates are best made by making multiple runs rather than by excessive weaving of the electrode from side to side to build up the weld.

11.7.2 Welding light-gauge steel

The manual metal arc process offers a number of advantages for the welding of thin-gauge mild steels. The intensely localized heat input of the arc ensures that the minimum of disturbance occurs to the surrounding metal plate, therefore resulting in very little distortion. At present 1 mm is the thinnest gauge which can be readily welded by the MMAW process, and both butt and outside corner welds in this thickness can be produced with complete satisfaction. Inside corner fillet welds in this gauge are difficult to make because of the low heat input, which restricts the free flowing of the slag away from the depositing metal so that fusion is obstructed and the resulting weld is unsatisfactory.

Successful welding of light-gauge steel depends on three important conditions:

1   The close fitting of the edges to be welded
2   The right type and gauge of electrode
3   The accurate current adjustment.

Either DC or AC is suitable providing the supply is steady and capable of fine adjustment. The open-circuit voltage should not be less than 60 volts in the case of DC and 70 volts in the case of AC. Unless care is taken to see that edges or surfaces to be joined are in close contact along the seam, the heat conductivity of the weld area and the local heat will cause burning through wherever a gap occurs. Joint contact is maintained by tacking at 75–150 mm apart. The welding operation, which takes a little time to practise, consists of striking the arc and passing the electrode tip with the shortest possible arc rapidly along the seam to be welded at a uniform rate, without any sideways movement. If this is done correctly it will produce a satisfactory weld.

11.7.3 Welding stainless steels (austenitic group only)

The technique of welding stainless steel does not differ greatly from that of welding mild steel, but as the material being handled is very expensive, extra precautions and attention to detail at all stages of welding are desirable. In principle all stainless steel for high-class work should be welded with a short arc using a DC supply with the positive polarity applied to the electrode. For less important work an AC supply can be used with an open-circuit voltage of 100 volts. The current ranges recommended for different thicknesses will depend on the nature of the work and the size of the plate. Usually the lowest convenient current should be used; weaving should not be wider than twice the diameter of the electrode, and electrodes of like composition to the parent metal must be used. The edges of the welding should be free from scale. Clamps and jigs are advisable when welding thin sheet. Tack welds, particularly on thin sheets, should be placed much closer together than is the usual practice for welding mild steel.

11.7.4 Welding aluminium and its alloys

The equipment used is standard DC welding equipment. The operating voltage is about 25 volts, depending upon arc length and type of electrode, and over a current range of 40–450 amperes. The electrode coating on the rods usually consists of a mixture of chloride and fluorides though other salts may be present. For manual metal arc welding, mixtures generally similar to those used for gas welding are supplied as the coating, but in this case the flux may include binders which prevent the coating from chipping and also help to stabilize the arc. The composition of the core wire is important and in general falls into three types: pure aluminium, aluminium silicon or aluminium manganese. If a wrought alloy is to be welded containing less than 2 per cent alloying element, then silicon-type electrodes are used. Pure aluminium should be welded with a pure aluminium electrode.

Vertical welding is possible with materials thicker than 5 mm, although downhand welding is preferable with the electrode moving in a straight line along the seam without weaving. The speed of welding depends on the current and the operator's skill, but is usually about three times that of mild steel. It increases as the electrode size is increased and it should also be increased as the weld progresses in order to allow for the rise in temperature of the parent metal. Too rapid welding on low currents does not give the required penetration, while welding that is slow, or a current that is too high, gives an excessive bead which may result in burning through the parent metal. When the arc is broken the coating tends to cover the tip of the electrode, thus obstructing the re-establishing of the arc. It is usually sufficient to tap the end of the electrode on the work to crack off this coating, although it may be necessary to cut the tip off. The slag on MMAW should flake off readily, especially if the weld is allowed to cool before the slag is removed. With all electrodes all traces of slag should finally be removed by scrubbing with a wire brush and washing.

11.8 Safety precautions for the welder

1 Never look at a welding arc without a shield.
2 Always replace the clear cover glasses when they become pitted and encrusted due to metal spatter.
3 Examine the closed lenses in the helmet. If they are cracked, replace them immediately.
4 Wear goggles when chipping slag off a weld.
5 Always wear gloves and an apron when welding.
6 Use a holder that is completely insulated. Never lay it on the bench while the machine is running.
7 Welding processes should be carried out only in the areas where there is adequate ventilation.

8 When welding outside a permanent welding booth, be sure to have screens around so the arc will not harm persons nearby.

9 Prevent welding cables from coming in contact with hot metal, water, oil or grease.

10 Make sure that you have a good earth connection.

11 Keep the cables in an orderly manner to prevent them from becoming a stumbling hazard. Avoid dragging the cables over or around sharp corners. Whenever possible, fasten the cables overhead to permit free passage.

12 Do not weld near inflammable materials.

13 Be sure tanks, drums or pipelines are completely cleaned of inflammable liquids before welding.

14 Always turn off the welding machine when not in use.

Questions

1 What is the difference between alternating and direct currents?

2 Describe, with the aid of a line diagram, the principle of manual metal arc welding.

3 Name four factors which control the quality of manual metal arc welding.

4 Describe the process of manual metal arc welding and explain how the necessary heat is produced.

5 What is the function of the coating of a manual metal arc welding electrode?

6 What is meant by the term 'arc length'?

7 Give four essential factors in making a good arc weld.

8 In what way does a current affect a weld?

9 State the current in amperes and the size of electrode needed, for welding over 5 mm thick steel plate.

10 Explain the function of an AC metal arc welding transformer.

11 What would be the resulting effect if the amperage is set too high in manual metal arc welding?

12 Why is it important not to look at an electric arc without proper eye protection?

13 Compare the use of a head shield to that of a hand shield for continuous welding operations.

14 What causes the defect known as undercutting, and how can this be avoided?

15 How is the weld affected when the arc is too short?

16 What happens when the welding arc is too long?

17 What determines the number of runs that should be made on a weld?

18 Why must a welder take into account the expansion and contraction of metal?

19 How are welding electrodes identified?

20 Why is it important to have adequate air extraction when welding in confined situations?

Gas shielded arc welding

12.1 Development of gas shielded arc welding

Originally the process was evolved in America in 1940 for welding in the aircraft industry. It developed into the tungsten inert-gas shielded arc process which in turn led to shielded inert-gas metal arc welding. The latter became established in this country in 1952.

In the gas shielded arc process, heat is produced by the fusion of an electric arc maintained between the end of a metal electrode, either consumable or non-consumable, and the part to be welded, with a shield of protective gas surrounding the arc and the weld region. There are at present in use three different types of gas shielded arc welding:

Tungsten inert gas (TIG)   The arc is struck by a non-consumable tungsten electrode and the metal to be welded, and filler metal is added by feeding a rod by hand into the molten pool (Figure 12.1).
Metal inert gas (MIG)   This process employs a continuous feed electrode which is melted in the intense arc heat and deposited as weld metal: hence the term consumable electrode. This process uses only inert gases, such as argon and helium, to create the shielding around the arc (Figure 12.2).
Metal active gas (MAG)   This is the same as MIG except that the gases have an active effect upon the arc and are not simply an inert envelope. The gases used are carbon dioxide or argon/carbon-dioxide mixtures.

The following should also be noted:

Gas tungsten arc welding (GTAW)   This is the terminology used in America and many parts of Europe for the TIG welding process, and it is becoming increasingly accepted as the standard terminology for this process.

Figure 12.1   TIG welding equipment AC/DC (*Murex Welding Products Ltd*)

Gas metal arc welding (GMAW)   This is the terminology used in America and many parts of Europe for the MIG/MAG welding processes, and it is becoming increasingly accepted as the standard terminology for these processes.

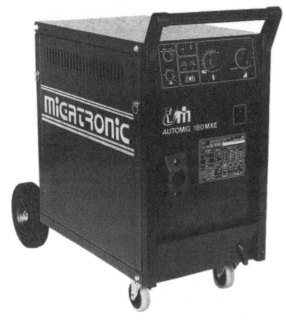

Figure 12.2   MIG welding equipment (*Migatronic Welding Equipment Ltd*)

12.2 Gases used for shielded arc processes

The shielding gases used in the MIG/MAG and TIG welding processes perform several important functions:

1   Protection from atmospheric contamination
2   Arc support and stabilization
3   Control of weld bead geometry
4   Control of weld metal properties.

It is necessary to prevent contamination of the weld pool by atmospheric gases which cause deterioration of the weld bead quality, by surface oxidation, porosity or embrittlement. In the consumable electrode process MIG/MAG, it is also necessary to consider the potential loss of alloying elements in the filler wire owing to preliminary oxidation in the arc atmosphere. In TIG welding, oxidation of the non-consumable tungsten electrode must be prevented. For these reasons most welding shielding gases are based on the inert gases argon (Ar) and helium (He). Active gases such as carbon dioxide (CO_2), oxygen (O_2) and hydrogen (H_2) may be added to the shielding gas to control one or other of the functions stated, but the gas chosen must be compatible with the material being welded (Table 12.1).

Table 12.1   Chemical and physical compatibility of welding shielding gases and materials

Material	Gas	Compatibility
Plain carbon steel and low alloy steel	Argon, helium	No reaction
	CO_2, oxygen	Slight oxidation of alloying elements
	Hydrogen	Porosity and HICC risk
	Nitrogen	Porosity and loss of toughness
Austenitic stainless steel	Argon, helium	No reaction
	H_2	Reduces oxide
	O_2	Oxidizing
	CO_2	May cause carbon pick-up
Aluminium and alloys	Argon, helium	No reaction
	H_2	Gross porosity
	O_2	Oxidizing
Copper	Argon, helium, N_2	No reaction
	H_2	Porosity
Nickel	Argon, helium	No reaction
	N_2	Porosity
Titanium	Argon, helium	No reaction
	O_2, N_2, H_2	Embrittlement

The successful exclusion of atmospheric contamination depends on the ability to provide a physical barrier to prevent entrainment in the arc area, and in the case of some reactive metals such as titanium it may be necessary to extend this cover to protect the solidified weld metal whilst it is cooling. The gas cover depends on the shielding efficiency of the torch and the physical properties of the gas. The higher the density of the gas the more resistant it will be to disturbance, and gases which are heavier than air may offer advantages in the downhand welding position (Table 12.2).

Table 12.2   Density of common welding shielding gases

Gas	Density (kg/m^3)
Argon	1.784
Helium	0.178
Hydrogen	0.083
Nitrogen	1.161
Oxygen	1.326
Carbon dioxide	1.977

The shielding gas used in MIG/MAG processes displaces the air in the arc area. The arc is struck within this blanket of shielding gas, producing a weld pool free from atmospheric contamination. The type of gas used determines the heat input, arc stability and mode of transfer, as well as providing protection for the solidifying weld metal. With any gas shielded arc process the type of shielding gas used greatly influences the quality of the weld deposit, weld penetration and weld appearance. The heat affected zone can also be influenced by the composition of the gas.

One of the important functions of the shielding gas is to protect the weld zone from the surrounding atmosphere and from the deleterious effects of oxygen, nitrogen and hydrogen upon the chemical composition and properties of the resulting weld. In this capacity the gas fulfils the major function of the fluxes used as electrode coverings or deposited as an enveloping layer during welding with other processes. The obvious advantages derived from the use of gas shielding are that the weld area is fully visible; that little, if any, slag is produced; and that the absence of abrasive flux increases the life of jigs and machine tools. In the MIG welding process, gas shielding enables a high degree of mechanism of welding to be achieved. Few gases possess the required shielding properties, however, and those that do with certainty – the inert gases, notably argon – are relatively expensive.

12.2.1 Argon

Argon is one of the rare gases occurring in the atmosphere and is obtained from liquefied air in the course of the manufacture of oxygen. Argon is an inert gas. It does not burn, support combustion, or does not take part in any ordinary chemical reaction. On account of its strongly inactive properties it can prevent oxidation or any other chemical reaction from taking place in the molten metal during the welding operation.

The argon is supplied in steel cylinders coloured blue, a full cylinder having approximately 200 bar pressure which is reduced by a regulator to approximately 15 litres per min when drawing for welding. Normally the cylinder should not be allowed to empty below 2 bar. This prevents air from entering the cylinder and helps to preserve the purity of the contents. Gas pressures are shown as bar: 1 bar $\simeq$ 14.505 lbf/in^2; 10 lbf/in^2 $\simeq$ 0.689 bar. Gas consumptions are in litres per hour (litre/h): 1 ft^3/h $\simeq$ 28.316 litre/h.

Argon has been more extensively used than helium because of a number of advantages:

1   Smoother, quieter arc action
2   Lower arc voltage at any given current value and arc length
3   Greater, cleaner action in the welding of such materials as aluminium with AC
4   Lower cost and greater availability
5   Lower flow rates for good shielding
6   Easier arc starting.

The lower arc voltage characteristics of argon are used to advantage in the welding of thin metals, because the tendency towards burn-through is lessened. Pure argon can be used for welding aluminium and its alloys, copper, nickel, stainless steel and also for MIG brazing.

12.2.2 Helium

Helium is a colourless, odourless inert gas almost as light as hydrogen. It is found in the United States in a natural gas and is therefore more widely used in America than anywhere else. Helium has a high specific heat, so that a given quantity requires much more heat to raise its temperature by 1 °C than does air; therefore the weld temperature is reduced and so distortion is minimized. A disadvantage of helium is that, because of its lightness, two and a half times the gas flow is required for a given performance than would be needed with argon. Helium is more favourable than argon where high arc voltage and power are desirable for welding thick material and metal with high heat conductivity such as copper. Helium is more often used in welding heavy than light materials.

12.2.3 Carbon dioxide

This gas is used as a shielding gas in the MAG welding process. It is not an inert gas: when it passes through the welding arc there is some breakdown into carbon monoxide and oxygen. To ensure that the oxygen is not added to the weld, deoxidants such as silicon, aluminium and titanium are included in the welding wire, which is specially made for carbon dioxide MAG welding. The deoxidant combines with the released oxygen to form a sparse slag on the surface of the completed weld. A gas heater with an electrically heated element is used to prevent freezing of the gas regulator after prolonged use with carbon dioxide gas. Since the development of the carbon dioxide process it has become widely used for the welding of plain carbon steels.

12.2.4 Argon mixtures

The gas mixtures that are suitable for vehicle body repair work consist of 95 per cent argon and 5 per cent carbon dioxide, or 80 per cent argon and 20 per cent carbon dioxide. These mixtures give smoother results with a better, cleaner and more attractive spatter-free weld. They also improve metal transfer and weld finish. These gases can be used to weld low-carbon mild steel, high-strength steel (HSS) and very low-carbon rephosphorized steels.

12.2.5 Helium mixes

These are specially formulated for MIG welding of stainless steels. They are a mixture of high-purity helium and argon, with small controlled additions of carbon dioxide. They contain no hydrogen, and are suitable for welding all grades of stainless steel including weldable martensitic grades, extra-low-carbon grades and duplex stainless steel.

12.2.6 Choice of shielding gases

Shielding gases must be carefully chosen to suit their application (Table 12.3). The selection will depend on:

1 The compatibility of the gas with the material being welded
2 Physical properties of the material
3 The welding process and mode of operation
4 Joint type and thickness.

12.3 TIG welding

12.3.1 Principles of operation

The necessary heat for this process (Figures 12.3 and 12.4) is produced by an electric arc maintained between a non-consumable electrode and the surface of the metal to be welded. The electrode used for carrying the current is usually a tungsten or tungsten alloy rod. The heated weld zone, the molten metal and the electrode are shielded from the atmosphere by a blanket of inert gas (argon or helium), fed through the electrode holder which is in the tip of the welding torch. A weld is made by applying the arc heat so that the edges of the metal are melted and joined together as the weld metal solidifies. In some cases a filler rod may be used to reinforce the weld metal.

Before commencing to weld it is essential to clean the surfaces that are to be welded. All oil, grease, paint, rust and dirt should be removed by either mechanical cleaning or chemical cleaners.

Striking the arc may be accomplished in one of the following ways:

1 Using an apparatus which will cause a spark to jump from the electrode to the work (arc stabilizer AC equipment); or
2 By means of an apparatus that starts and maintains a small pilot arc which provides a path for the main arc.

Once the arc is struck, the welding torch is held with the electrode positioned at an angle of about 75° to the surface of the metal to be welded. To start welding, the arc is usually moved in a small circle until a pool of molten metal of suitable size is obtained. Once adequate fusion is achieved, a weld is made by gradually moving the electrode along the parts to be welded so as progressively to melt the adjoining edges. To stop welding, the welding torch is quickly withdrawn from the work and the current is automatically shut off.

The material thickness and joint design will determine whether or not filler metal in the form of welding rod needs to be added to the joint. When filler metal is added it is applied by feeding the filler rod from the side into the pool of molten metal in the arc region in the same manner as used in oxy-acetylene welding. The filler rod is usually held at an angle of about 15° to the surface of the work and slowly fed into the weld pool.

The joints which may be welded by this process include all the standard types such as square

Table 12.3 Gas mixtures available and their applications

Gas	Applications	Features
Argon	TIG: all metals. MIG: spray pulse, aluminium, nickel, copper alloys	Stable arc performance. Poor wetting characteristics in MIG Efficient shielding. Low cost
Helium	TIG: all metals, especially copper and aluminium. MIG: high-current spray, aluminium	High heat input. Increased arc voltage
Argon + 25 to 80% He	TIG: Aluminium, copper, stainless steel. MIG: aluminium and copper	Compromise between pure Ar and pure He. Lower He contents normally used for TIG
Argon + 0.5 to 15% H_2	TIG: austenitic stainless steel, some copper nickel alloys	Improved heat input, edge wetting and weld bead profile
CO_2	MAG: plain carbon and low-alloy steels	Low-cost gas. Good fusion characteristics and shielding efficiency, but stability and spatter levels poor. Normally used for dip transfer only
Argon + 1 to 7% CO_2 + up to 3% CO_2	MIG/MAG: plain carbon and low-alloy steels. Spray transfer	Low heat input, stable arc. Finger penetration. Spray transfer and dip on thin sections. Low CO_2 levels may be used on stainless steels but carbon pick-up may be a problem
Argon + 8 to 15% CO_2 + up to 3% CO_2	MIG/MAG: plain carbon and low-alloy steels. General purpose	Good arc stability for dip and spray pulse Satisfactory fusion and bead profile
Argon + 16 to 25% CO_2	MIG/MAG: plain carbon and low-alloy steels. Dip transfer	Improved fusion characteristics for dip
Argon + 1 to 8% O_2	MIG/MAG: dip, spray and pulse, plain carbon and stainless steel	Low O_2 mixtures suitable for spray and pulse, but surface oxidation and poor weld profile often occur with stainless steel No carbon pick-up
Helium + 10 to 20% argon + oxygen + CO_2	MIG: dip transfer, stainless steel	Good fusion characteristics, high short-circuit frequency Not suitable for spray pulse transfer
Argon + 30 to 40% He + CO_2 + O_2	MIG: dip, spray and pulse welding of stainless steels	Improved performance in spray and pulse transfer. Good bead profile. Restrict CO_2 level for minimum pick-up
Argon + 30 to 40% He + up to 1% O_2	MIG: dip, spray and pulse welding of stainless steels	General-purpose mixture with low surface oxidation and carbon pick-up. (It has been reported that these low-oxygen mixtures may promote improved fusion and excellent weld integrity for thick-section aluminium alloys)

groove, V groove, as well as T and lap joints (Figure 12.5). It is not necessary to bevel the edges of material that is 3.2 mm or less thickness.

12.3.2 Modes of operation

The TIG process may be operated in one of the following modes:

DC electrode negative In this mode the electrode remains relatively cool whilst the workpiece is effectively heated. This is the most common mode of operation for ferrous materials, copper, nickel and titanium alloys.

DC electrode positive With DC electrode positive there is a tendency for the electrode to overheat and fusion of the workpiece is poor. The advantage of this mode of operation is the cathodic cleaning effect which removes the tenacious oxide film from the surface of aluminium alloys.

AC alternating current This offers a good compromise between plate heating and the cathodic cleaning effect and is used with aluminium and with manganese alloys.

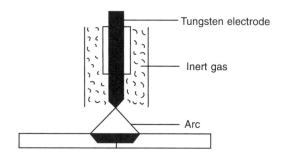

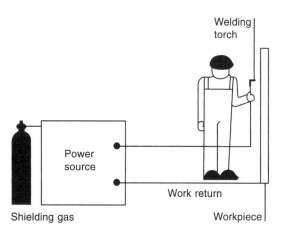

Figure 12.3 Basic principles of the TIG welding process (*BOC Ltd*)

12.4 TIG spot welding

TIG spot welding is an adaptation of the main process. This method utilizes the heat from the tungsten arc to fuse the base material in much the same way as ordinary TIG welding. The tungsten electrode is set inside the argon shield to ensure that fusion takes place in a completely shrouded atmosphere. The sheets or parts to be joined are held in close contact by the manual pressure of the gun, and fusion is made from one side of the joint only (Figure 12.6). The equipment comprises a water-cooled torch with associated cables for argon, water and power, a standard AC/DC welding set and a timer and contact unit. The arc is struck by pressing a switch on the gun activating the timer which carries the welding current to the tungsten electrode. The arc is struck automatically and fusion takes place between the top and bottom components of the joint to be made. The depth of penetration through the component part is controlled by the current and time cycle.

The process is used for joining mild steel and stainless steel not exceeding 1.6 mm, but is not suitable for aluminium and magnesium base alloys. The results of the spot welds differ according to the type and quality of the materials used, but providing the

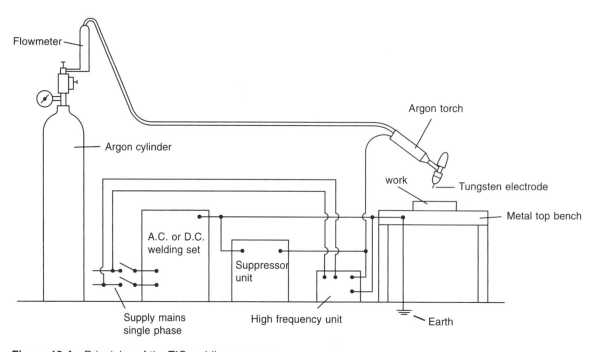

Figure 12.4 Principles of the TIG welding process

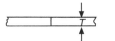

1 Square edge close butt

T should not exceed 3 mm for manual welding or 5 mm for machine welding without filler wire. With filler wire addition, machine welds up to approximately 6 mm thick material are possible with this edge preparation

2 Flanged edge close butt

Used on material thickness up to 2 mm

3 Square edge open butt

This form of preparation is only used when filler wire is to be added. Gap should not normally exceed 3 mm and then only used when plate is to be welded from both sides. *T* should not exceed 10 mm

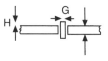

4 Open butt inserted strip

This method of preparation avoids the use of a separate filler wire and is suitable mainly for machine welding applications. Thickness should not normally exceed 8 mm and the thickness of the filler strips should not exceed approximately 3 mm. The edges of the plate must butt closely to the filler strip throughout the whole length of the weld seam. *H,* the height of the strip above the plate, should not exceed 3 mm for the whole range of metal thickness involved

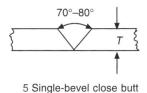

5 Single-bevel close butt

Single-pass welds up to 10 mm *T* are possible with both hand and machine applications, but multipass welds are recommended where *T* exceeds 8 mm. Single-pass machine welds with filler wire addition and heavy currents are possible on thicknesses up to 3 mm but the preparation shown in 6 is preferred where *T* exceeds 10 mm

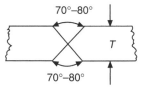

6 Double-bevel close butt

This preparation is recommended for *T* in excess of 10 mm, assuming both sides of the joints are accessible

Figure 12.5 Recommended edge preparation

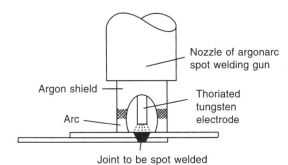

Figure 12.6 Principles of the TIG spot-welding process

surfaces of the joint are clean and free from rust, scale, greases or dirt, satisfactory spot welds can generally be made.

12.5 Equipment used in TIG welding

The equipment required for TIG welding (Figure 12.1) consists of a welding torch equipped with a gas passage and nozzle for directing the shielding gas around the arc, a non-consumable electrode, a supply of shielding gas (argon), a pressure reducing regulator and flow meter, a power unit and a supply of cooling water (Figure 12.7) if the welding

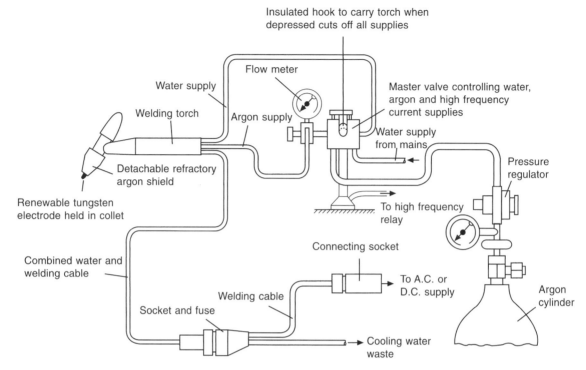

Figure 12.7 Gas, water and power supply for TIG welding

torch is water cooled. The individual components may differ considerably, depending upon power requirements and the type of work to be carried out.

12.5.1 Power unit

Standard AC and DC welding equipment may be used, but in most cases special welding units are used which are capable of producing AC or DC, have automatic control of argon and water flow, and have fine current control switches for the stopping and starting of welding. The choice of welding current is determined by the material to be welded. Metals having a refractor surface oxide film, like magnesium, aluminium and its alloys, are normally welded with AC, while DC is used for carbon steels, stainless steels, copper and titanium. When direct current is used the electrode may be connected either to the positive or to the negative side of the power source, depending on the material to be welded. Usually the majority of general welding requires direct current with negative polarity, as the heat distribution and current loading are used to the best advantage and the tungsten arc electrode can carry at least four times as much current, without

overheating the electrode, as an equivalent positive arc. Practically all metals other than magnesium and aluminium are suitable for this method, which gives deep penetration with a very narrow weld bead, whereas DC positive gives a shallow penetration with a wide bead (Figure 12.8).

12.5.2 Welding torch

Various types of torches are available to suit the different applications and current requirements. Torches may be water or air cooled: for currents below 150 A air-cooled torches are used, while from 150 to 400 A water-cooled torches are used (Figure 12.9). The air-cooled torches are used for welding light materials and the water-cooled torches for welding heavier materials, as the water cooling then prevents cracking of the ceramic shield at the tip of the torch. The torch is fully insulated electrically and has a quick release collet arrangement to facilitate convenient adjustment or changing of the tungsten electrodes. Tungsten having a melting point of 3400 °C, is used as the electrode material owing to its refractory nature. It is almost non-consumable when used under ideal welding conditions. A

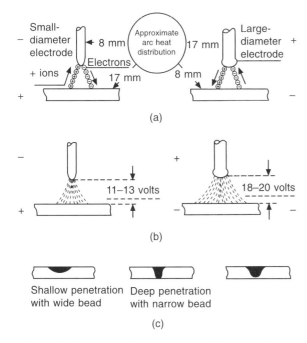

(a)

(b)

Shallow penetration with wide bead

Deep penetration with narrow bead

(c)

Figure 12.8 Alternative methods of DC connections (a) Theoretical distribution of heat in the argon shielded arc with the alternative methods (b) Average differences in arc voltages with equal arc lengths, using (left) negative polarity at the electrode and (right) positive (c) Relative depths of penetration obtainable with (left to right) DC positive, DC negative and AC

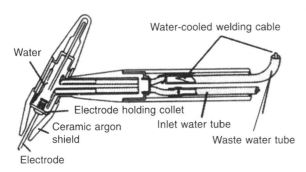

Figure 12.9 A water-cooled TIG welding torch

ceramic shield, which is interchangeable, directs the gas so as to form a shroud around the arc and weld metal. The argon and the electric current are supplied to the torch through a combined cable and gas hose. In the water-cooled models a third cable is added to carry the water to and from the torch. A water flow switch can be provided to give complete protection to equipment and operator by shutting off the welding current if the water supply should fail.

To avoid contamination of either the electrode tip or the work, which would occur if the normal method of arc striking were employed, a high-frequency spark unit or an arc stabilizing device is used to stop the operator from having to touch the electrode on the surface of the work.

12.5.3 Gas supply

The inert gas, argon or helium, is supplied to the welding torch from the storage cylinder via a gas pressure regulator and a gas economizer valve (Figure 12.7), which may be a dual-purpose valve when cooling water is used, and a special flow meter calibrated in cubic feet per hour or litres per minute of gas flow. The gas flow required varies with current setting, shroud size, material being welded, and type of weld joint.

12.5.4 Filler metals

Filler materials for joining a wide variety of metals and alloys are available for use with the gas tungsten arc welding process. Among them are filler rods for welding various grades of carbon and alloy steels, stainless steels, nickel and nickel alloys, copper and copper alloys, aluminium and most of its alloys, magnesium, titanium, and high-temperature alloys of various types. There are also filler materials for hard surfacing. Wherever a joint is to be reinforced, a filler rod is added to the molten puddle. In general, the diameter of the filler rod should be about the same as the thickness of the metal to be welded. For sound welds, it is important that the physical properties of the rod be similar to the base metal.

12.5.5 TIG electrodes

Normally pure tungsten electrodes are used, but to improve are striking and stability an addition of either thorium oxide or zirconium oxide is added to the tungsten. For alternating current welding zirconiated electrodes are used, while for direct current welding thoriated electrodes are used. The chosen electrode must be of a diameter which suits the current.

Improved performance can be obtained by alloying the electrodes lanthanum, yttrium and cerium, particularly in automatic TIG welding where consistency of operation is important (Table 12.4). The electrode diameter is determined by the current and

Table 12.4   Electrodes for TIG welding

Electrode type	Use
1–2% thoriated tungsten	DC electrode negative Ferrous metals, copper, nickel, titanium
Ceriated tungsten	As above Improved restriking and shape retention
Zirconiated	AC Aluminium and magnesium alloys

polarity: recommended diameters are given in Table 12.5. The angle to which the electrode is ground depends on the application. The included angle or vertex angle (Figure 12.10) is usually smaller for

Table 12.5   Recommended diameters and current ratings (BS 3019: Part 1) for TIG electrodes

Diameter (mm)	Maximum current (A) DC Electrode (−) thoriated	AC electrode (+) zirconiated
0.8	45	–
1.2	70	40
1.6	145	55
2.4	240	90
3.2	380	150
4.0	440	210
4.8	500	275
5.6	–	320
6.4	–	370

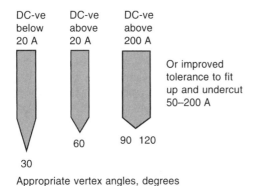

Appropriate vertex angles, degrees

Figure 12.10   Appropriate vertex angles of electrodes (*BOC Ltd*)

low-current DC applications. In order to obtain consistent performance on a particular joint it is important that the same vertex angle is used.

12.6 TIG welding techniques

The normal angles of the torch are 80–90° and of the filler rod 10–20° from the surface of the horizontal plate respectively (Figure 12.11). The arc length, defined as the distance between the tip of the electrode and the surface of the weld crater, varies between 3 mm and 6 mm, depending on the type of material and the current used. The filler rod is fed into the leading edge of the molten pool and not directly in the arc core, and should be added with a

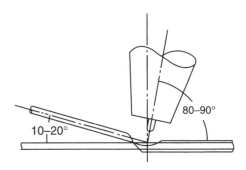

Figure 12.11   Recommended angles for torch and filler rod in TIG welding

slightly transverse scraping motion, with the tip of the filler rod actually making contact with the weld metal. This ensures that the rod is at the same electrical potential as the plate during transfer of metal to the weld and avoids any tendency of the rod to spatter. The heated end of the filler rod should always be kept within the influence of the shrouding argon gas in order to prevent its oxidation.

Butt welds in thin-gauge materials are carried out with a progressive forward motion without weaving, but a slightly different technique is required when tackling medium- and heavy-section plate. As the filler rod diameter increases with increasing thickness of plate, there is a tendency for the end of the filler rod to foul the tungsten electrode. Contamination of the hot tungsten by particles of molten metal causes immediate spattering of the electrode and particles of tungsten may, according to the degree of contamination, become embedded in the weld pool. Loss of tungsten in this manner will cause the arc to become erratic and unstable, and the electrode will certainly require to be replaced before further welding is attempted. To avoid repetition of this occurrence, the arc length must be increased slightly to accommodate the insertion of the larger filler rod. This procedure cannot be taken too far, however, because there is a maximum arc length beyond which good welding becomes impossible. The upper limit is usually about 6 mm thick for aluminium, which does not allow for the free access of a 6 mm diameter filler rod. For heavier sections, therefore, a forward and backward swinging motion of the torch is employed. The weld area is melted under the arc; the torch is withdrawn backwards for a short distance from 6 mm to 3 mm along the line of the seam and the filler rod is inserted in the molten pool (Figure 12.12). The torch is moved forward and the filler rod is withdrawn from the pool simultaneously. A rhythmical motion of both torch and filler rod backwards and forwards in a progressive forward motion melts down filler rod and plate without the filler rod entering the core of the arc, and is recommended when welding plate in excess of 6 mm thick.

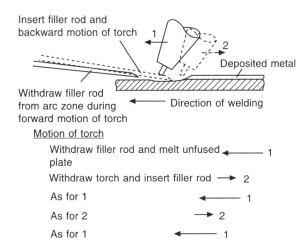

Figure 12.12 Motion of torch and filler rod for TIG welding heavy sections

12.7 Application of TIG welding

This process has found an application in the body building side of the industry, where it is used to ideal advantage for fabricating components by welding in materials such as aluminium and stainless steel. The advantage of the method is that materials which would normally require flux can be welded without it; therefore cleaning of the weld is minimized and the effects of distortion are greatly reduced. This process is especially adapted for welding light-gauge work requiring the utmost in quality or finish because of the precise heat control possible and the ability to weld with or without filler metal. It is one of the few processes which permit the rapid, satisfactory welding of tiny or light-walled objects.

Among the materials which are readily weldable by this process are most grades of carbon, alloy or stainless steels, aluminium and most of its alloys, magnesium and its alloys, copper, copper-nickel, phosphor-bronze, tin bronzes of various types, brasses, nickel, nickel-copper (Monel alloy), nickel-chromium-iron (Inconel alloy), high-temperature alloys of various types, virtually all of the hard surfacing alloys, titanium, gold, silver and many others.

12.8 MIG/MAG welding

The development of the MIG/MAG processes is in some ways a logical progression from the manual metal arc and TIG processes. The effort throughout has been to obtain and maintain maximum versatility, weld quality, speed of deposition, simplification of the welding operation and lower operating costs.

MIG/MAG welding has been adapted for many industrial applications, and over the past years has become widely used for car body repairs. This method of welding is an electric arc process using DC current and a continuous consumable wire electrode without the addition of flux. On account of the absence of flux, gas is used to shield the arc and weld pool from atmospheric contamination. The process can utilize argon, argon/carbon dioxide mixture or carbon dioxide as the shielding gas, the choice being dependent upon the type of material being welded and the economics associated with the selected gas. If non-ferrous metals or stainless steels are to be welded, argon is the usual choice for shielding gas on grounds of compatibility. However when mild steel, low-alloy steels or high-strength steels are to be welded, argon/carbon dioxide mixture or carbon dioxide is generally used for reasons of overall efficiency, weld quality and economy.

Many types and grades of metal can be welded using this method: aluminium, aluminium alloys, carbon steel, low-carbon and alloy steels, microalloy steel, nickel, copper and magnesium. The success of this welding method is due to its capability of giving a consistently high-quality weld while also being very easy to learn. In addition it has the advantage of spreading very little heat beyond the actual welding point, and this helps to avoid distortion and shrinking stresses which are a disadvantage in the oxy-acetylene process.

12.8.1 Principles of operation

Metal inert-gas or active-gas shielded arc welding (consumable) is accomplished by means of a gas shielded arc (Figures 12.13 and 12.14) maintained between the workpiece and a consumable (bare wire) electrode from which metal is transferred to the workpiece. The transfer of metal through the protected arc column provides greater efficiency of heat input than that obtained in the TIG welding process. The resultant high-intensity heat source permits very rapid welding. In this process a con-

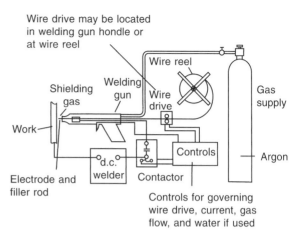

Figure 12.14 Principles of the MIG/MAG welding process: argon, argon/CO_2 or CO_2

tinuously fed electrode passes through a gun, during which it passes through a contact area which impresses the preselected welding current upon the wire. The current causes the wire to melt at the set rate at which it is fed. The shielding gas issuing from the nozzle protects the weld metal deposit and the electrode from contamination by atmospheric conditions which might affect the welding process. The arc may be started by depressing the trigger of the welding gun and scratching the electrode wire end on the work.

The equipment is designed so that the wire automatically feeds into the weld area as soon as the arc is established. Most MIG/MAG welding sets that are manufactured for the automobile repair trade are semi-automatic, the operator only being concerned with the torch-to-work distance, torch manipulation and welding speed. Wire feed speeds, power settings and gas flow are all preset prior to commencement of welding.

12.9 MIG/MAG spot/plug welding

An advantage of MIG/MAG welding is the ability of the process to be adopted for single-side spot welding applications, either semi or fully automatically. By extending the welding gun nozzle to contact the workpiece, one-sided spot welds may be performed using dip transfer conditions. Predetermined weld duration times may be employed, the gun trigger being coupled to a suitable timer and, if desired, fully mechanized. Unlike resistance spot welding, no pressure is

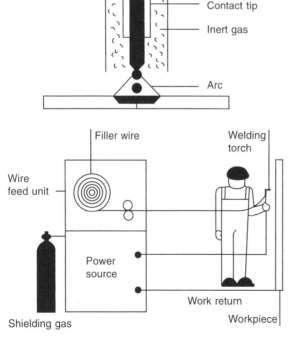

Figure 12.13 Basic principles of operation of MIG/MAG welding (*BOC Ltd*)

required on the workpiece with MIG/MAG spot welding, and neither is a backing block. Mismatch of the sheets is permissible with a maximum gap equivalent to half the sheet thickness, the extra metal being provided by the electrode wire. Up to 30 spots per minute may be welded, which compares reasonably well with the 100 spots per minute from resistance welding techniques. The deep penetration characteristics of this welding process enable spot welding of widely differing metal thicknesses to be performed successfully, together with multisheet thicknesses.

MIG/MAG plug welding differs from MIG/MAG spot welding in that the outer metal panel has a predrilled or punched hole which is filled up with the weld metal to form the 'plug'. The hole sizes used are 5–6 mm with currents of 50–60 A for panel thicknesses of 0.75–1.2 mm. Care should be taken when MIG/MAG plug welding to avoid an excess build-up of weld metal, to reduce the necessity of dressing the finished weld. This method is used to advantage in the fitting of new panel sections, where the original was resistance spot welded and access is now difficult to both sides.

Higher-strength steels are used for selected panel sections, but as they are vulnerable to heat they are not as easily welded as mild steel. MIG seam and butt welding produce hard weld joints; they will tear from the sheet metal on impact. However, MIG spot welds and particularly puddle (or plug) welded MIG spots can be made ductile. The weld time has to be short to reduce the heat-affected zone around the weld.

Modern car body designs are constructed with deformation zones both front and rear, and the shear impact properties of the original welding have been carefully calculated to ensure that the energy of an impact is fully absorbed and contained within the zone. Changing the manufacturers' original welding specification could impair the safety of the vehicle. The crumplability (impact energy absorbing) design of car bodies makes new demands on welds. For its success the body design relies on the sheet metal to crumple or fold rather than tear in a collision. This protective design depends on the ductility of the metal, and the welds too have to be ductile. If they are soft they will break without forcing the assembly to crumple, but if they are hard the welds will unbutton and the assembly will fly apart instead of folding

up slowly. Welds therefore must be ductile as well as large and strong.

12.10 Equipment used in MIG/MAG welding

The basic equipment required comprises a power source, a wire feed unit and a torch. The power sources commonly used have constant-voltage characteristics, and controls are provided for voltage and inductance adjustment. This type of power source is used in conjunction with a wire feed unit which takes the wire from a spool and feeds it through a torch to the arc. A control on the wire feeder enables the speed of the wire to be set to a constant level which will in turn determine the arc current.

The welding torch should be reasonably light and easy to handle, with provision for gas shield shrouding, control switch, easy wire changes and good insulation. The torch is connected to a wire feed and control unit by means of:

1 A flexible armoured tube carrying the welding wire
2 A plastic tube carrying the shielding gas
3 A pair of plastic tubes carrying cooling water, the return tube often carrying the welding supply to cool the welding cable (light-duty torches are air cooled)
4 The control wires for the switchgear.

The whole feed unit is bound together by a plastic sleeve, and is 1.5–3 m long. The wire feed unit houses the drive unit for the wire feed, which can be varied in speed to suit current/voltage conditions. The wire reel is also carried in the unit and is loaded with some 15 kg of welding wire.

12.10.1 Power unit

DC power units are used as either rotary generators or rectifying units which are specially designed to give full versatility of arc control. The equipment is either single phase (130–240 A) or three phase (250–500 A). The principal components of most machines are welding transformer, rectifier, choke coil, wire feed unit, gas solenoid valve and electronic control box.

The welding transformer is dimensioned so as to achieve optimal welding properties. The transformer is manufactured from materials able to withstand a working temperature of up to 180 °C. By way of a further safeguard against overloading, there is a built-in thermal fuse which cuts out the machine at 120 °C. The thermal fuse is automatically reconnected once the transformer has cooled down.

The rectifier is constructed from a fan, thyristors, diodes and a capacitor battery. During welding, or when the machine is hotter than 70 °C, the fan cools the rectifier and the transformer. The rectifier is electronically protected against overloading in the event of any short-circuit between welding positive and negative, with the machine cutting out approximately 1 second after the onset of short-circuiting.

The transformer converts the high mains voltage into a low alternating voltage, which is rectified and equalized into DC voltage in the rectifier module. A choke coil reduces the peaks in the welding current and thus eliminates the cause of welding spatter. When the switch on the torch is depressed, the thyristors come on. The rectifier emits a DC voltage, which is determined by the remote control in the welding trigger and the gas/wire matching switch on the box. Simultaneously, shielding gas is turned on and the wire feed motor is started up at a speed also determined by remote regulation from the torch.

When the trigger is released the motor decelerates, and after a short time lag the gas flow and welding voltage are interrupted. This time delay is called burn-back and causes the welding wire to burn a little way back from the molten pool, thus preventing it from sticking to the workpiece.

Depending on the type of equipment selected, the following functions are available: seam, spot, stitch and latching (four-cycle).

Seam Welding starts when the switch on the welding trigger is actuated and stops when the switch is released. For use in short-duration welding and tacking.

Spot Welding starts when the switch is actuated and is subsequently controlled by the welding timer for a time between 0.2 and 2.5 s. It makes no difference when the switch is released. This function ensures uniform spot welds, providing the correct setting has been found.

Stitch The wire feed motor starts and stops at intervals which are set on the welding timer and pause timer. When welding is interspersed with pauses in this way, the average amount of added heat is reduced, which prevents any melting through on difficult welding jobs.

Latching Welding starts when the switch is actuated; the switch can then be released and welding continues. By reactuating the switch, welding stops when the switch is released. For use on long seams.

A typical welding control panel (Figure 12.15) has the following features:

1 *Selection switch* This selects between the functions seam, spot, stitch and latch as described above.
2 *Power light* This lights up when the machine has been turned on.

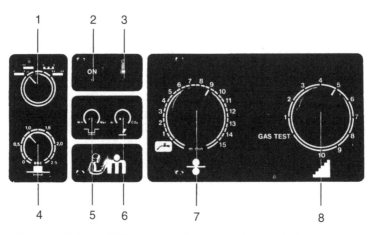

Figure 12.15 MIG welding control panel (*Migatronic Welding Equipment Ltd*)

3 *Overheating warning light* If this light comes on, the welding equipment is automatically switched off owing to overheating of the transformer. When the temperature is back to normal, the welding can be continued.

4 *Welding timer switch* With this switch the welding time is chosen, when the selection switch is in the stitch or spot position.

5 *Adjustable pause time button* With this button the pause time is chosen, when the selection switch is in the stitch position.

6 *Burn-back button* Pre-adjustment of the burn-back delay button indicates the time for stopping the wire feed until the arc is switched off. This varies between 0.05 and 0.5 s.

7 *Adjustment of wire speed switch* This gives the adjustment of the wire feed from 0.5 to 14 m/min.

8 *Welding voltage switch* This sets the welding voltage of the transformer. When set at gas test, the gas flows by pressing the switch on the torch handle.

The characteristics built into the welding power supply are such as to provide automatic self-adjustment of arc conditions as the weld proceeds. Depending on the relevant current and voltage used, metal transfer between electrode and work takes place in the following distinct forms each of which has certain operational advantages (Figure 12.16).

Dip transfer (*short arc*)

This condition requires comparatively low current and voltage values. Metal is transferred by a rapidly repeated series of short circuits when the electrode is fed forward into the weld pool.

Metal dip transfer is the most suitable mode of metal transfer for welding on car repairs, as it offers good bead control and low heat input, thus cutting down distortion when welding in panel sections. This type of transfer will also permit the welding of thinner gauges of sheet steel, and is practical for welding in all positions.

The principle of the method (short-circuiting transfer) is briefly as follows. The molten wire is transferred to the weld in droplets, and as each drop touches the weld the circuit is shorted and the arc is momentarily extinguished. The wire is fed at a rate which is just greater than the burn-off rate for the particular arc voltage being used; as a result

the wire touches the weld pool and short-circuits the power supply. The filler wire then acts as a fuse, and when it ruptures a free burning arc is created. After the occurrence of this short-circuit, the arc re-ignites. This making and breaking, or arc interruption, takes place from 20 to 200 times per second according to the setting of the controls. The result is a relatively small and cool weld pool, limiting burn-through.

Free flight transfer

In this metal transfer a continuous arc is maintained between the electrode and the workpiece and the metal is transferred to the weld pool as droplets. There are three subdivisions of the system:

Spray transfer In this type the mode of transfer consists of a spray of very small molten metal droplets formed along the arc column, which are projected towards the workpiece by electrical forces within the arc and collected in the weld pool. No short-circuiting takes place; the welds are hotter, and the depositing of weld metal is faster. It is ideally suited for the rapid welding of thick sections in the downhand position and the positional welding of aluminium and its alloys.

Globular transfer This occurs at currents above those which produce dip transfer, but below the current level required for spray transfer. The droplet size is large relative to the wire diameter and transfer is irregular. This mode of transfer occurs with steel wires at high currents in carbon dioxide, but is generally regarded as unusable unless high spatter levels can be tolerated. The use of corded wires gives a controlled form of globular transfer which is acceptable.

Pulse arc transfer In this mode the droplets are transferred by a high current which is periodically applied to the arc. Ideally, one drop is transferred with each pulse and is fired across the arc by the pulse. Typical operating frequencies are 50–100 droplets per second. A background current is maintained between pulses to sustain the arc but avoid metal transfer. Maximum control is obtained with this type of metal transfer, which utilizes a power supply to provide a pulsed welding condition in which the metal transfer takes place at pulse peaks. This leads to extreme control over the weld penetration and weld appearance.

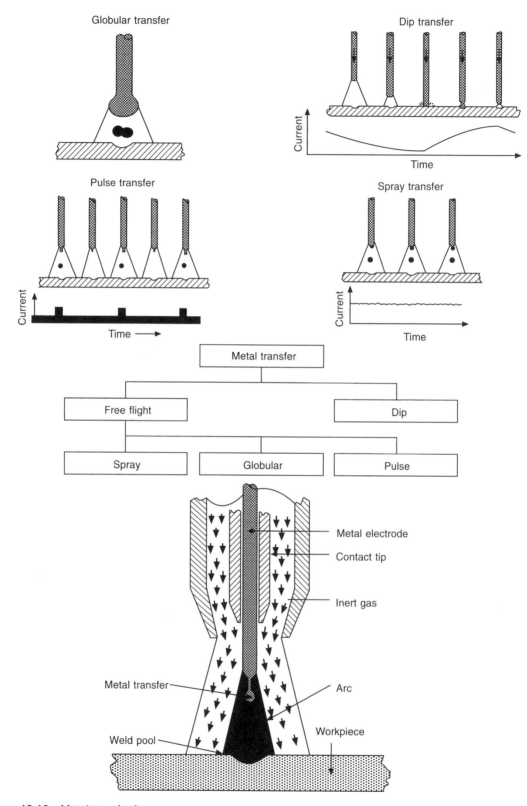

Figure 12.16 Metal transfer forms

12.10.2 Welding torch

Air cooled torches are available for the various welding applications ranging from 180 A to 400 A. The design of the torch is in the form of a pistol or is curved similar to the shape of an oxy-acetylene torch (Figure 12.17), and has wire fed through the barrel or handle. In some versions where the most efficient cooling is needed, water is directed through passages in the torch to cool the wire contact means as well as the normally cooled metal shielding gas nozzle (Figure 12.18). The curved torch carries the current contact tip at the front end through which the feed wire, shielding gas and cooling water are also brought. This type of torch is designed for small diameter feed wires, is extremely flexible and manoeuvrable and is particularly suitable for welding in confined areas. The service lines consisting of the power cable, water hose and gas hose on most equipment enter at the handle or rear barrel section of the torch. The torch is also equipped with a switch for energizing the power supply and controls associated with the process. Some welding torches also have a current control knob located in the torch so that the welding current can be altered during welding. This ensures that the welding voltage is altered at the same time as any current alteration, so allowing the welder to respond immediately to variations in weld gaps and misaligned joints. Welding characteristics are excellent. Fingertip control increases both productivity and weld quality (Figure 12.19).

12.10.3 Feed unit

The consumable electrode for welding ferrous and non-ferrous materials is supplied as a continuous length of wire on a spool or reel, and the wire varies from 0.6 to 0.8 mm diameter. The feeding of the wire is achieved by the unit feed mechanism housing the necessary drive motor, gear box and feed rolls which draws wire from the adjacent reel, or by an integrally built motor drive connected to the torch which pulls the wire in the desired direction. The feed wire unit also houses the controls which govern the feeding of the wire at the required constant

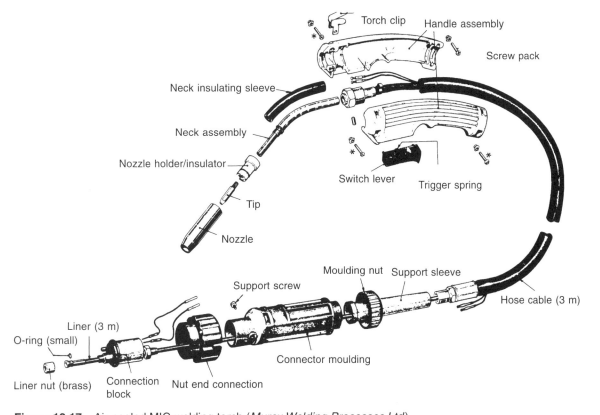

Figure 12.17 Air-cooled MIG welding torch (*Murex Welding Processes Ltd*)

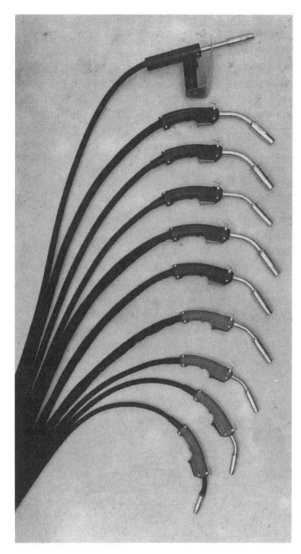

Figure 12.18   Range of MIG welding torches, air and water cooled, 180–600 A (*Murex Welding Products Ltd*)

Figure 12.19   Migatronic Dialog torch with current control in the torch handle (*Migatronic Welding Equipment Ltd*)

speed, plus the automatic control of current and gas flow. The wire from the feed rolls is fed along a carefully designed conduit system which also carries the welding current and shielding gas. The outer end of this conduit system is connected directly to the welding torch (Figure 12.20).

Figure 12.20   Wire feed unit (*Migatronic Welding Equipment Ltd*)

12.10.4 Gas supply

The primary purpose of the shielding gas used in MIG/MAG welding is the protection of the molten weld metal from contamination and damage by the oxygen, nitrogen, water vapour and other gases in the surrounding atmosphere. However, in use there are a number of other factors which influence the desirability of a gas for arc shielding. Some of these are the influence of the shielding gas on the arcing and metal transfer characteristics during welding, on the weld penetration, width of fusion and surface shape patterns, and on the speed of welding. All of these influence the finished welds and overall results. Normal shielding gases in use are argon, carbon dioxide, and mixtures of argon and carbon dioxide, although other mixtures such as argon and oxygen can be used.

12.10.5 Welding wire

The welding wires used for MIG/MAG welding are quite small in diameter compared with those used for other types of welding. Welding wire is 0.6–1.0 mm in diameter, and is supplied in reels of 5–15 kg. Owing to the small size of the wire and comparatively high currents used for welding, the wire melts off very rapidly. Therefore the wires used must always be as long as possible, with suitable temper for being fed smoothly and continuously by motor driven mechanical means through the welding apparatus. For this reason the wire is normally provided on convenient size coils or spools.

In composition the wires are usually quite similar to those used by most other bare wire processes. As a rule the composition as nearly matches that of the base material as is practicable, while also ensuring good weld properties and welding characteristics. Steel wires are often given a light coating of copper to prevent them from rusting and to improve the electrical contact and current pick-up when passing through the welding apparatus. Welding wires are available for joining most ferrous and non-ferrous metals:

Solid wire   The composition of solid filler wire is usually chosen to match the parent metal being welded. Solid wire is the most common and widely used filler material for MIG/MAG welding car body replacement panels.

Cord wires   Metal and flux corded wires are available. These materials are, however, more suitable for welding thicknesses in excess of 6 mm. Metal corded wires can be obtained with additions of alloying elements to meet special requirements (HS low-alloy steels).

MIG brazing wires   There are a number of compositions of filler wire that can be used in MIG brazing, but usually they are of aluminium-bronze composition. When using these filler wires, pure argon must be used as the shielding gas.

12.11 MIG/MAG welding techniques

In the operation of a MIG/MAG welding machine there are two parameter settings which must match each other: the wire speed and the welding voltage. It is important therefore that the wire speed and welding voltage are adjusted to conform with each other. The wire speed gives the welding current, which must match the component panel section being melted. An increase in wire feed rate increases the welding current, at the same time reducing the length of arc, resulting in a shorter arc. When the wire speed is reduced, the current intensity is diminished, at the same time increasing the length of the arc. Raising the open-circuit voltage increases the

arc voltage, though the current intensity remains almost constant. A reduced open-circuit voltage will produce a shorter arc length. A change in welding wire diameter will produce a change in current and voltage, since thinner wires require higher voltage and higher wire speeds in order to produce the same current. If the limit values are exceeded a satisfactory weld cannot be obtained.

When the wire speed setting is too low, this will cause a long, drawn-out arc and spattering and a very unsatisfactory weld condition (Figure 12.21). When the wire speed becomes too great in relation to the voltage, knocking or stubbing will be felt in the welding torch, because the wire travels to the bottom of the molten pool (Figure 12.21). Welding under these conditions will normally produce adhesion defects owing to the lack of fusion. If the voltage becomes too great in relation to the wire speed, large drops will form on the end of the wire. These will be reluctant to leave the end of the wire and will often settle as spatter beside the weld (see Figure 12.22). When the ratio between the voltage and the wire speed is correct, a highly characteristic hiss or hum will be heard from the arc.

The mutual position of the welding torch and workpiece in relation to each other is of importance to the appearance and quality of welding. In practice many different combinations of inclination, torch

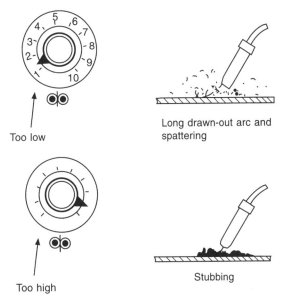

Figure 12.21   Incorrect settings for wire speed (*Stanners Ltd*)

FAULT		CAUSE
Porosity	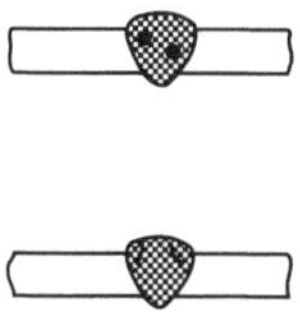	Insufficient Si, Mn in wire Insufficient CO_2 shielding because of $\begin{cases} \text{flow rate} \\ \text{frozen value} \\ \text{clogged nozzle} \\ \text{draughts} \end{cases}$
Cracking	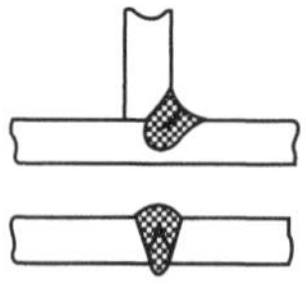	Dirty work – grease, paint, scale, rust (i) Weld bead too small (ii) Weld too deep, greater than 1.2:1 (iii) High sulphur, low manganese, slow cooling rate
Undercutting	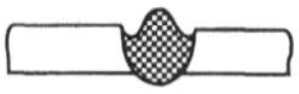	Travel speed too high Backing bar groove too deep Current too low for speed Torch angle too low

Lack of penetration	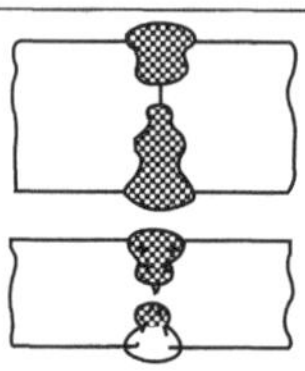	Current too low – setting wrong
		    Wire feed fluctuating
		    Electrode extension too great
		Joint preparation too narrow
		    Angle too small
		    Gap too small

Lack of fusion	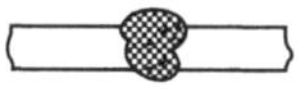	Uneven torch manipulation Insufficient inductance (short circuiting arc) Voltage too low
Slag inclusions		Technique – Too wide a weave Current too low Irregular weld shape

Spatter – on work
          on nozzle
          in weld
	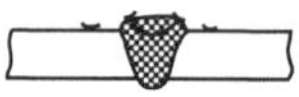	Voltage too high
		Insufficient inductance
		Insufficient nozzle cleaning

Irregular weld shape	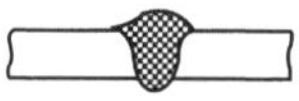	Excessive electrode extension Wire temper excessive, no straightening rolls Cutent too high for voltage Travel speed too slow

Figure 12.22   Weld defects and their causes

handling and welding position will be experienced by the body repairer (see Figure 12.23).

12.11.1 MIG/MAG welding positions

Welding positions can be as follows: flat, horizontal, vertical, overhead and, owing to the complexities of panel shapes, a combination of these (Figure 12.24). In collision repair, the welding position is usually dictated by the location of the weld in the structure or panel assembly of the vehicle. Both heat and wire speed parameters can be affected by the welding positions.

Flat welding is generally easier and faster and allows for better penetration. When welding a panel section that is off the vehicle, try to place it so that it can be welded in a flat position.

When welding a horizontal joint, angle the welding torch upwards to hold the weld puddle in place against the pull of gravity. When welding a vertical

	Thrusting weld	Drawing weld
Width of seam	wider	narrower
Upper bead	smaller	larger
Penetration	decrease	increase
Tendency to lack of fusion	greater	lesser

Drawing weld: torch sloped in direction of weld
Thrusting weld: torch sloped away from direction of weld

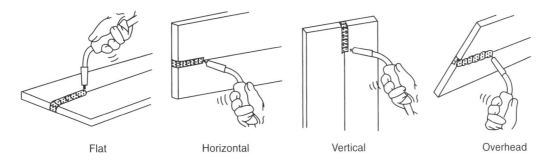

Pushing weld   Pulling weld

Pushing weld sloping   Pulling weld sloping   Pulling weld sloping
downwards   downwards   upwards

Figure 12.23   Welding position (*Migatronic Welding Equipment Ltd*)

Flat   Horizontal   Vertical   Overhead

Figure 12.24   Typical MIG/MAG welding positions

joint, the best procedure is usually to start the weld at the top of the joint and pull downwards with a steady drag.

Overhead welding is the most difficult. In this position the danger of having too large a weld puddle is that some of the molten weld metal can fall down, either into the nozzle where it can create problems, or on to the operator. Therefore always carry out overhead welding on a low voltage setting, which will cut down spatter while keeping the arc as short as possible and the weld puddle as small as possible. Press the nozzle of the torch against the work to ensure that the wire is now moved away from the puddle. It is

better to pull the torch along the joint with a steady drag.

12.11.2 MIG/MAG basic welding processes

There are six basic welding processes used in collision repair work on vehicle bodies.

Tack weld

A tack is a relatively small temporary MIG/MAG weld that is used instead of a clamp or a self-tapping screw, to tack and hold the panel in place while proceeding to make a permanent weld (Figure 12.25).

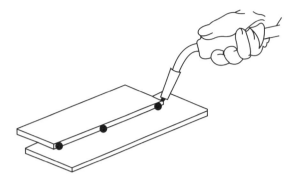

Figure 12.25 Tack weld

Like the clamp or self-tapping screw, the tack weld is always and only a temporary device. The length of the tack weld is determined by the thickness of the metal panel to be welded and is approximately a length of 15–30 times the thickness of the metal panel. Tack welds must be done accurately, as they are very important in maintaining proper panel alignment.

Continuous weld

A continuous weld is an uninterrupted seam or weld bead which is carried out as one steady progressive movement (Figure 12.26). While supporting the torch securely, use a forward movement to move the torch continuously at a constant speed, checking the weld bead frequently as you progress. The welding torch should be inclined between 10 and 15 degrees to obtain the best weld bead shape, weld

line and gas shielding effect. It is essential to maintain both the distance between the tip base and the metal surface, and the correct torch angle, otherwise a problem can arise owing to the wire length being too long. When this occurs, penetration of the metal will not be adequate; therefore to improve penetration and produce a better weld, bring the torch closer to the base metal. When the torch handling is smooth and even, the weld bead will be of consistent height and width with a uniform closely spaced ripple.

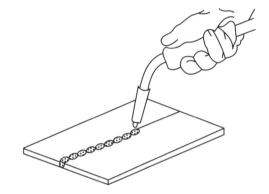

Figure 12.26 Continuous weld

Plug weld

A plug weld is used to join two metal panels together (Figure 12.27). The outer metal panel is first predrilled or punched. Next the arc is directed through the hole to penetrate the inner metal panel. The hole is then filled with molten metal to form the plug.

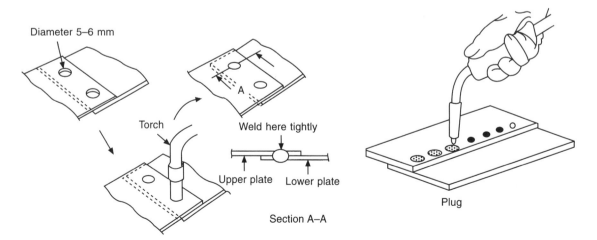

Figure 12.27 Making a plug weld

Spot weld

In MIG/MAG spot welding the arc is directed to penetrate both metal panel pieces through the joint position while triggering a timed impulse of wire feed (Figure 12.28).

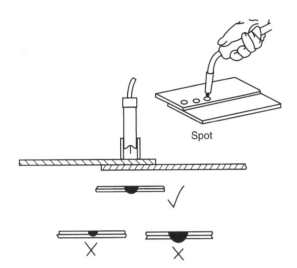

Spot

Figure 12.28   MIG/MAG spot welding

Lap spot weld

In the MIG/MAG lap spot weld technique, the arc is directed to penetrate equally between the edge of the upper metal panel and the top surface of the lower metal panel of the lap joint, and the puddle is allowed to flow on to the edges of the top piece. This is achieved by using the correct angle of the welding torch to the joint (Figure 12.29).

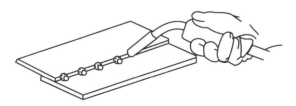

Figure 12.29   Lap spot weld

Stitch weld

A MIG/MAG stitch weld is a series of intermittent seam welds carried out at intervals along a lapped metal panel joint, where a continuous weld is not necessary or would cause distortion due to heat input (Figure 12.30).

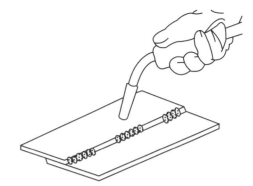

Figure 12.30   Stitch weld

12.12 Applications of MIG/MAG welding in vehicle body construction

The continuous search for increased efficiency in production techniques in the automotive industry has led to all major manufacturers adopting the use of MIG/MAG welding on a very wide scale. This welding process is ideally suited to flow line production processes owing to its inherent automatic characteristics. MIG/MAG welding has numerous advantages over the majority of processes likely to be encountered in automotive production. The most obvious is its capability of being used either semi or fully automatically, thus giving it virtually unrivalled versatility. In automobile construction it has almost totally replaced oxy-acetylene and manual metal arc welding by virtue of its high metal deposition rates, electrode efficiency and high duty cycles. Further advantages of the process are good penetration, even without edge preparation, the ability to handle all metal thicknesses and high weld strength ($618 \ MN/m^2$ being easily achieved).

MIG/MAG dip transfer welding is a low heat input process owing to the relatively low amperages required, and it is ideal for welding fine sheet thicknesses because of the absence of distortion with minimal jigging. With the absence of flux further advantages accrue, the most obvious ones being the elimination of deslagging and chipping back when second weld starts are required. The absence of flux also means that the welder has a clear view of arc operation and weld bead appearance and is not subjected to excessive fumes. Spatter is minimal and weld bead appearances enhanced owing to the improved wetting of the parent material by the molten metal on account of the various additives present in the wire composition.

The MIG/MAG welding process can in many cases replace existing welding techniques, usually manual metal arc and oxy-acetylene, to give superior welding conditions and economies. Semi-automatic and more particularly fully automatic MIG/MAG welding processes with a wide range of technical advantages have opened up new areas of production technique. Numerous companies in the United Kingdom have already adopted MIG/MAG welding to their own particular requirements. On a modern mass produced passenger car the body shell is a relatively low-stressed unit, high stresses imposed by road conditions being carried by subframe assemblies. However, where continuous welds are required or specific components carrying high stresses are to be fitted MIG/MAG welding is employed using semi-automatic equipment. Thus components such as safety harness attachments, door hinges, jacking points, spring shackles, steering column brackets and seat frame guides are attached using the dip transfer technique, which enables welds of this type to be made on thin shell panels with no distortion of the panel and deep penetration into the thicker workpiece. Where MIG/MAG welding has been introduced into automotive production lines, its value as a production process is well established.

12.13 Applications of MIG/MAG welding in body repair work

The semi-automatic MIG/MAG welding process which is used in the construction of vehicle bodies is also used for the repair of these bodies. Equipment is now specially produced for thin-gauge welding in the repair of vehicle bodywork.

Welding in the repair of car bodies was mainly carried out using oxy-acetylene and resistance spot-welding techniques. Both of these techniques have certain disadvantages. In the oxy-acetylene method the weld metal and surrounding panels are liable to distort owing to heat input. In spot welding if the repair is visible the joints, which in most cases are lapped, must be solder filled to achieve an acceptable finish.

The advantages of MIG/MAG welding equipment in repair are first that it is a dual machine being capable of welding in any position, and it

can weld material from 0.5 mm to 4.4 mm in thickness. Second, it can be used for spot welding from one side of the panel only, which is an advantage when welding inaccessible panel assemblies. The equipment itself is very portable, consisting of a transformer, rectifier, built-in feed wire system, lightweight welding torch with interchangeable nozzle heads for conventional welding and spot welding, earthing cable and clamp, and gas supply in steel cylinders, all of which are incorporated on a manoeuvrable trolley which lends itself to body work application in the workshop (see Figure 12.31).

Figure 12.31 CEBORA synergic pulse (Mig Tig Arc)

Most equipment used in motor body repair usually has the following controls and functions: a weld timer which can be set from 0.2 to (usually) 2.5 seconds; some form of wire speed control which ranges from 2 to 12 m/min; and a voltage control usually ranging from 0 to 10, the 0 setting giving the lowest welding current. On some machines the settings for wire speed and welding voltage are coupled together into one control, which can either be on the machine or positioned on the welding torch for instant adjustment. The following function controls are the most widely used: seam for continuous welding; spot, which makes use of a time setting to ensure uniform spot welds; and stitch, which is used through the weld timer and pause timer and provides intermittent welding so that the welding is interspersed with pauses, therefore reducing heat input and preventing melt-through on thin materials and gapped joints.

12.13.1 Preparing the equipment for welding

1 Select the correct dimensions of wire: 0.6 mm is best for welding thin materials and panels; 0.8 mm is used for general purposes; and 1.0 mm is used for aluminium only (using aluminium wire).

2 Select the tip for the type and dimension of wire being used and fitted to the equipment. Also select the torch nozzle for seam or spot welding application.

3 Mount the welding wire reel on the reel holder, roll off the tip of the wire and insert it into the inlet between the feed rollers of the wire feed unit.

4 Connect the machine to the main supply, set the wire speed selector, pull the welding torch trigger and feed the wire through the hose to the welding torch till it appears at the end of the contact tip. (Some machines have a wire inching switch which is used when a new reel of wire is put into the machine and fed to the hose, and the wire is fed at a gentle speed to the torch.)

5 Connect the hose for the shielding gas to be used and set the gas flow regulator between 5 and 15 litres/min.

6 Set the wire speed and voltage control and select the mode of welding operation.

7 Connect the earth return clamp as close as possible to the point of welding, making sure of a secure connection to facilitate a good welding circuit.

8 Before commencing to weld, a visor-type (see Figure 12.32) or hand-held welding screen must be used for the protection of the welder. Then hold the welding torch down to the component to be welded and an arc will be struck between the wire electrode and the workpiece, thus commencing the welding.

9 Before welding on a car it is essential that the electrical connections to the battery, starter motor and generator/alternator are disconnected in order to avoid damage to the electrical equipment of the car.

Figure 12.32   Visor-type welding screen in use (*Motor Insurance Repair Research Centre*)

For the best results for seam, spot or stitch welding, the surface of the panel should be cleaned to bare metal condition as paint or grease cause bad welds. In body work applications the heat input is limited on each side of the weld owing to the shielding gas; this has the effect of reducing possible distortion to a minimum so that adjacent parts do not get so hot as to damage trim, screen rubbers and windscreen glass. It can also be used to great advantage in repairing by sectioning or patching, as two panels can be butt welded and then the weld completely dressed off on one side with a sanding machine and finished off, using normal panel beating hand tools, to a perfect finish. As this system of welding gives a perfectly controlled penetrating weld of great strength, it will be found ideal for replacing new panel sections such as subframe (see Figure 12.33) and underframe members, where the welding is highly critical because of the stress factor present in load-bearing assemblies. The speed at which this welding can be carried out in the repair of the modern mono constructed body shell is proving itself to be an economical asset.

Figure 12.33 Application of MIG welding in body repair work (*Migatronic Welding Equipment Ltd*)

12.13.2 MIG/MAG welding of low-carbon steel and HSLA steels

Until the mid 1970s most if not all vehicle bodies were constructed from plain low-carbon steel, but since then there has been an increase in the worldwide demand for corrosion resistance in body structures. This has led to the introduction of single-sided and double-sided coated galvanized steel, high-strength steel and aluminium. The Japanese, in particular, and the European industries have moved to the use of HSLA steels in further attempts to reduce body weight. Body skin panels have been reduced from around 0.9 mm to 0.7 mm, with structural components reduced from 3 mm to between 1.2 mm and 2.0 mm.

The use of these different materials can lead to problems for the welder. Welding galvanized steels (zinc coated) can reduce rust resistant properties around the weld areas if they are not correctly treated.

Cosmetic body panels are the thinnest panels on the vehicle, and are generally repaired by the use of butt, fillet, lapped edge or corner joints. It is therefore necessary to use as low a current as possible when using the MIG/MAG process on these thin panels.

The choice of shielding gas for the process is important as it affects not only the quality of the weld, but also the depth of penetration of the weld into the material and the amount of weld spatter that is formed around the finished weld. It also controls the temperature, which affects the suitability of the process for the thickness of materials to be welded. The choice of shielding gases commonly used can be seen in Table 12.6.

Under the right conditions, any of the gases in the table can be used for the range of thicknesses encountered in the body shop. However, the ideal for welding thin body panels is the colder mixture of argon with 2 per cent oxygen and 5 per cent carbon dioxide. When welding structural components and chassis members the material tends to be thicker, up to 3 mm or even more, and so it is necessary to use a hot arc in order to penetrate the material. For this reason a mixture of argon with 2 per cent oxygen and 12 per cent carbon dioxide is used to give better heat flow, greater tolerance to variation in setting, and low spatter generation. It is also a common practice to increase the wire diameter from 0.6 mm to 0.8 mm when welding these thicker panel assemblies.

12.13.3 MIG brazing

As well as being used in vehicle repair, this process is also used in vehicle production on some assembly lines as it offers increased speed and a smooth flat bead requiring little dressing after brazing. The MIG brazing process can be used with the traditional brazing filler materials of aluminium bronze, phosphor bronze and silicon bronze in the form of continuous wire on a reel. With this process the choice of filler material and shielding gas is important if the necessary flat bead is to be achieved, otherwise excessive amounts of dressing would be required. The shielding gas used is normally pure argon. However, gases that are a mixture of helium and argon produce a more fluid weld pool, which again leads to a flatter bead than with argon. A 70 per cent argon and 30 per cent helium mix is generally preferred for thinner cosmetic panels, since the richer 80 per cent argon and 20 per cent helium mix can produce excessive penetration with an increased risk of distortion, although this can be used to advantage on thicker panel structures.

Table 12.6 MIG/MAG recommended operating conditions

Application	Metal thickness (mm)	Shielding gas	Wire dia. (mm)	Current (A)	Voltage (V)	Wire feed speed (m/min)	Welding speed (mm/min)	Gas flow (l/min)
Horizontal/vertical fillet	1.0	Argoshield 5	0.8	66	14	3.4	900	12
	1.6	Argoshield 5	0.8	130	18	9.7	600	14
		Argoshield 5	1.0	160	17	5.9	563	14
	3.0	Argoshield TC	1.0	180	20	6.4	391	15
		Argoshield TC	1.2	180	17	3.8	430	15
	6.0	Argoshield TC	1.2	260	27	7.0	480	16
	10.0	Argoshield TC	1.2	290	27	7.4	430	16
Butt weld, flat position	1.0	Argoshield 5	0.8	66	14	3.0	750	12
	1.6	Argoshield 5	0.8	84	18	4.0	800	14
		Argoshield 5	1.0	125	16	4.0	640	14
	3.0	Argoshield TC	1.0	140	19	4.0	350	15
	6.0	Argoshield TC	1.2	300	28	7.6	650	16
	10.0	Argoshield TC	1.2	300	28	7.6	680	16
				320	28	7.6	580	16
Butt weld, overhead	1.0	Argoshield 5	0.8	50	15	4.0	360	12
	1.6	Argoshield 5	1.0	150	17	5.9	600	14
	3.0	Argoshield TC	1.2	180	19	4.3	520	15
Butt weld, vertical up	1.0	Argoshield 5	0.8	45	15	3.5	350	12
	1.6	Argoshield 5	1.0	130	17	4.7	410	14
	3.0	Argoshield TC	1.2	135	17	3.1	275	14

12.13.4 MIG welding of aluminium

While most welding equipment is supplied primarily for the welding of ferrous metals, some can also be used for the welding of aluminium. Equipment with low amperages is really not suitable, although it can be used for short periods of welding. The larger-amperage machines (180 A and over) are better equipped to handle aluminium. The wire sizes used are 1.0 mm and 1.2 mm for the larger machines for welding thicker aluminium. The torch contact tip must be of the correct size for the wire to be used. When welding with aluminium wire a Teflon liner must be used in order to prevent the aluminium from sticking and damage occurring to the wire itself. Also, pure argon must be used as the shielding gas owing to its total inert characteristics, and not argon mixes or carbon dioxide.

12.14 Welding stress and distortion in the MIG/MAG process

The MIG/MAG welding process and its effectiveness may be measured by its ability to produce joints of acceptable quality. Unacceptable joint quality may arise from a failure in the specific design or from the incident of weld defect. These problems, however, are preventable if their causes are understood and the correct operating procedures are followed.

The most common welding problems are:

1 Residual stress and distortion
2 Incorrect bead geometry and appearance
3 Defects
4 Loss of properties.

12.14.1 Residual stress and distortion

The solidification and contraction of the weld bead will induce strain and consequently stresses in the joint. In the case of the contraction of the solid metal the stress will be equal to the change in temperature as the metal cools from its melting point to its ambient temperature. Under normal circumstances movement of the weld bead is restricted by the adjacent body panels in the vehicle structure, and the stress which could be generated is given by the Young's modulus of the material. This stress level often exceeds the elastic limit of the material and plastic deformation takes place. The stresses locked in the material, which may reach levels up to the elastic limit of the material, are called residual stresses, and the deformation of the material is known as distortion.

Residual stresses

In a butt weld the weld bead will tend to contract longitudinally and transversely, and this will induce tensile stresses in the weld and also balancing compressive stresses in the sheet material (Figure 12.34). In the joining processes which rely on heating or fusion, it is difficult to prevent the formation of residual stresses. If the heat affected zone is ductile and defect free (as in thin-sheet panel steel) the presence of some residual stresses may be acceptable. The most common technique used to relieve residual stress in thicker materials is post-weld heat treatment. This consists of uniform heating of the joint to a temperature at which the yield stress of the material is lowered and the residual stresses are relieved by plastic deformation.

(Figure 12.35). Distortion may result in unacceptable appearance (buckled body panels), prevention of subassembly fabrication or the inability of the structure to perform its intended function (alignment of body panels after welding).

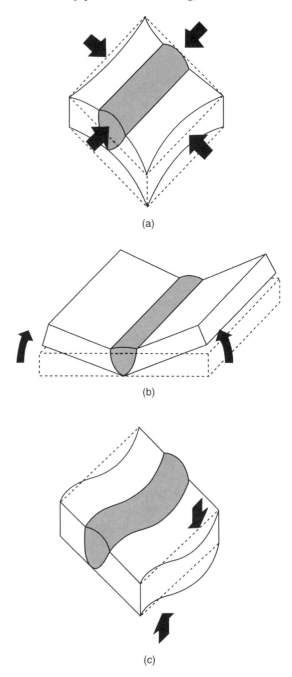

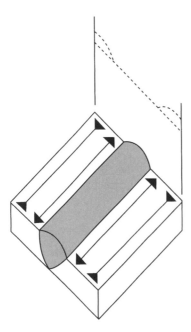

Figure 12.34 Residual stress in a simple butt weld (*BOC Ltd*)

Distortion

Distortion may take the form of a change in dimensions of the joint, transverse or longitudinal shrinkage, angular movement, or out-of-plane buckling

Figure 12.35 Types of distortion: (a) longitudinal and transverse shrinkage (b) angular distortion (c) out-of-plane buckling distortion (*BOC Ltd*)

The following steps may be taken to minimize distortion:

1 Use the minimum amount of weld metal. Over-welding and excessive reinforcement should be avoided in fillet welds and flat butt welds. Intermittent or stitch welding may be used.
2 Use square edge on narrow gap procedures to reduce angular distortion when welding.
3 Use high travel speed and low heat input to limit heat build-up in the panels to be welded.
4 Use the backstep weld sequence or preset the joint.

12.14.2 Incorrect weld geometry and appearance, defects, loss of properties

Overfill

Overfill or excessive reinforcement may be described as the presence of weld metal which exceeds that required for the joint (see Figure 12.36). This creates an unacceptable appearance or surface finish and would require weld dressing.

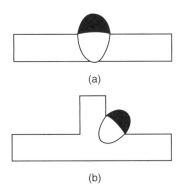

(a)

(b)

Figure 12.36 Excessive reinforcement in (a) butt weld and (b) fillet weld (*BOC Ltd*)

Root convexity

This results from excessive penetration on the underside of a full penetration weld. The cause of this defect is incorrect welding parameter selection (shielding gas or operating technique) (Figure 12.37).

Figure 12.37 Root convexity: excessive penetration (*BOC Ltd*)

Root concavity or suck-back

This may occur if the rear of the weld pool is too hot or large, and the combined effect of contraction and surface tension results in a root surface which is concave (Figure 12.38). The surface profile will often be smooth and less likely to produce stress than over-penetration or lack of penetration; however, it may result in unacceptable loss of cross-section. The cause of this defect is incorrect parameter selection, in particular the use of high current or slow travel speed.

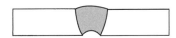

Figure 12.38 Root concavity: suck-back (*BOC Ltd*)

Underfill and undercut

If insufficient weld metal is deposited in the joint, the parent material may remain unfused and the joint may be underfilled (Figure 12.39a). Alternatively, the weld area may be melted but the combined effect of the arc force and the flow of the molten metal may prevent complete joint filling or produce a depression in the surface at the weld boundaries: this is known as undercut (Figure 12.39b). Underfill is usually avoided by careful placement and operating technique. Undercut normally occurs at high speeds and high current, and may be avoided by reducing both current and speed.

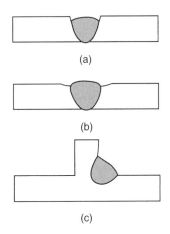

(a)

(b)

(c)

Figure 12.39 (a) Underfill and (b), (c) undercut (*BOC Ltd*)

Lack of penetration

This is classed as partial penetration and is not acceptable, particularly with respect to fatigue, loading and corrosion resistance on the joint, where full penetration is required. The defect results from poor joint preparation and unsuitable welding operating parameters, especially current settings (Figure 12.40).

Figure 12.40 Lack of penetration (*BOC Ltd*)

Lack of fusion

Lack of fusion between successive runs or between the weld bead and the parent metal can produce serious crack-like defects. These defects reduce the load-carrying area of the joint and produce stresses, and are therefore unacceptable in welds which are subjected to fatigue loading or static loading. The cause of the defect is usually incorrect operating parameters, high deposition rate, or low heat input (Figure 12.41).

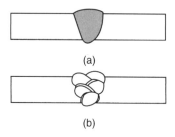

(a)

(b)

Figure 12.41 Lack of fusion (a) at sidewall and (b) inter-run (*BOC Ltd*)

Spatter

Spatter takes the form of small particles of metal which are rejected from the arc or the weld pool and adhere to the surface of the metal being joined. Spatter may be unacceptable for purely cosmetic reasons: it may also inhibit the application of surface coatings (as in vehicle painting) and may initiate corrosion. Spatter usually results from either excessive voltage or insufficient induction in dip transfer or argon enriched gas mixtures.

Porosity

Porosity results when bubbles of gas are nucleated in the weld pool and trapped during solidification.

The source of the gas may be a chemical reaction such as the formation of carbon monoxide from carbon and oxygen, or the expulsion of gases which have been dissolved by the liquid metal but are relatively insoluble in the solid phase. The problem is best controlled by eliminating the source of the pore generating gas. This may be achieved by effective shielding and the prevention of leaks, the ingress of air or moisture to gas lines, and joint surface contamination from oil, grease, moisture or paint.

Inclusions

Non-metallic inclusions may be trapped in the joint owing to inadequate cleaning. Inclusions tend to be more angular than gas pores and can result in higher stress levels as well as forming a plane of weakness in the weld. The cause of this defect is usually insufficient inter-run cleaning, incorrect welding parameters or incorrect gas shielding.

Solidification cracks (hot cracking)

Solidification cracking occurs when the solidifying weld pool is exposed to transverse shrinkage (Figure 12.42). The last part of the weld pool to solidify is usually rich in low-melting-point impurity elements, and liquid films are often present at the centre line of the weld; as a result the ductility of the weld metal is lowered. The transverse contraction strain can be significant and, if the low-ductility area along the centre line of the weld is unable to accommodate this strain, cracking takes place. The problem is usually avoided by careful selection of the weld metal composition.

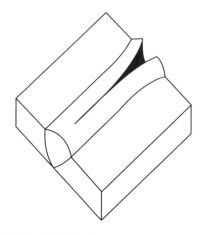

Figure 12.42 Solidification cracks (*BOC Ltd*)

12.15 Weld testing and inspection

Weld testing and inspection can be divided into two types. In non-destructive testing, the test samples are not destroyed in the process. In destructive testing, the test samples are destroyed in the process.

12.15.1 Visible examination

This is normally carried out during and after welding and prior to any other non-destructive or destructive test being used. This visual check will usually determine the following:

Weld size
Profile or weld face shape
Surface defects in weld face
Undercut and overlap
Root defects
Weld penetration
Surface slag inclusions.

12.15.2 Non-destructive testing

Non-destructive testing of weld samples is normally carried out using the following methods.

Macro examination
This method uses a low-power magnification to examine weld specimens which have been levelled, polished and etched to detect the following: lack of fusion, lack of penetration, porosity, oxide inclusions, internal cracks.

Crack detection
By using dye penetrant, surface defects in both ferrous and non-ferrous metals may be detected. A solution of coloured dye is sprayed on to the weld and parent metal and allowed to soak. The dye is then washed off and the surface dried. A liquid developer is then sprayed on to the weld to give a uniform dry powder coating which is white in colour. The coloured dye oozes out of any crack in the weld into the white coating and can be seen in normal lighting conditions.

Magnetic particle method
Surface defects in mild steel and low-alloy steels may be revealed by this method. The test specimen is connected to a special power source. The magnetic test liquid is sprayed along the weld. The magnetic particles then collect along a line of crack when the current flows through the weld. Better test results are obtained if the surface of the weld has first been ground smooth.

As these penetrants and magnetic test liquids give off harmful vapours, the work must be carried out under well ventilated conditions.

Ultrasonic inspection
This method uses sound waves which are passed through the weld. They are transmitted as pulses by a probe connected to an ultrasonic test set. Defects reflect these pulses back to the probe through a flexible cable which is attached to the ultrasonic test set, where it is displayed on an oscilloscope screen as deflections of a trace. These deflections measure the location and size of the defect. Grease is used between the probe and the work to improve sound transmission. Also light grinding of the weld or parent metal surface to remove spatter may be required before testing.

Radiography
This method uses penetrating X-rays or gamma-rays which are passed through the weld and recorded on photographic film held in a light-proof container. When the film has been processed, any defect in the weld will show up as shadows in the photographic image. Only trained and radiation classified personnel may operate radiography equipment, as the equipment is extremely dangerous.

12.15.3 Destructive testing

Mechanical testing is a destructive procedure and therefore cannot be carried out on any component required for use. Representative test samples produced under similar conditions to the in-service components, for example welding procedure tests, are normally used, and accurate comparisons made.

The tests most frequently used to assess the properties of welded joints are as follows.

Tensile test
This test can be used to assess the yield point, ultimate tensile strength and elongation percentage of the weld specimen. Failure usually occurs in the parent material; therefore exact measurements are not usually obtained for the weld itself, although in this case the ultimate tensile strength (UTS) of the

weld is higher than that of the parent metal. Test specimens are cut from the designated area of the weld assembly, the edges are smoothed and the corners are radiused. If an elongation value is required, then two centre punch marks often 50 mm apart are applied one on either side of the weld; this is called a gauge length. The test equipment can vary, but the basic principles are that two sets of vice jaws are used to clamp the specimen, hydraulic power is applied to force the jaws apart, and a dial calibrated in tonnes or newtons records the load. As the load increases, the dial registers the amount of applied load until fracture occurs.

Bend test

Bend tests are carried out on butt welds and are used to determine the soundness of the weld zone. The test piece is the full thickness of the material, with the weld bead reinforcement dressed flush on both sides. It is common practice to take two specimens from a test piece: one can be bent against the face and the other bent against the root. A bend test machine consists of a roller or former connected to a ram and operated hydraulically. Unless otherwise stated, the diameter of the former should be four times the weld specimen thickness. The specimen is placed on support rollers and the centre point of the former is brought into contact with the weld face or root. Pressure is then applied to bend the weld specimen through either 90 or 180 degrees.

Nick-break test

In this test procedure the butt joint weld must first be ground flush. Then two nicks (saw cuts) are made, one on either end of the joint in the weld specimen. The weld specimen may be broken during bending, or alternatively the specimen may be placed in the jaws of the tensile test machine and struck with a hammer when tensioned.

Impact test

There are two main types of impact test, the Charpy V notch and the Izod test. Both tests employ a swinging pendulum which breaks accurately prepared specimens to which a notch has been applied. Both the tests have the same principle, which is to determine the energy measured in joules absorbed by the notched test piece at a specified temperature as it is broken from a single swinging pendulum. During fracture, energy from the pendulum will be absorbed by the specimen, more energy being absorbed by strong materials than by brittle materials. The distance the pendulum swings after fracture is measured by a pointer on a dial calibrated in joules. The lower the value indicated, the more brittle the specimen; conversely, the higher the reading, the greater the toughness.

12.15.4 Vehicle repair weld testing quality control BS EN ISO 9001–2000

Those vehicle body repair companies wishing to obtain British Standards will have to conduct weld test procedures. These procedures will have to become an integral part of the company's policy on quality in the repair of vehicles while maintaining safe working practices. In addition regular testing will demonstrate the skill and ability of body technicians to maintain a high quality of weld standard in all types of crash repair work carried out by their company.

12.16 Equipment maintenance and safety

For the best results the gas and contact nozzle of the welding torch must be cleaned of welding spatter at regular intervals. Unless this is done, there is a risk of the welding current arcing between the contact nozzle and the gas nozzle to the workpiece, which could cause severe damage to the welding torch. When cleaning the gun of spatter and deposits the condition of the gas distributer should also be checked, and if it appears to be burnt and worn by spatter then it should be replaced. This is because in addition to distributing the gas it also acts as an insulator, and is of vital importance to the service life of the swan neck of the torch. It is recommended that the machine should be cleaned with compressed air once yearly, or more frequently if it is used in dusty conditions. The feed mechanism and rollers require more frequent cleaning. Regular maintenance to the following parts is important: gas nozzle, gas distributor, contact nozzle and wire guide rollers. It is never advisable to weld with a worn-out contact nozzle and deformed gas nozzle.

Any exposure to a naked welding arc, even for a fraction of a second, is sufficient to damage the cornea of the eye (welding flash). Consequently a welding screen should always be used, preferably of the helmet and visor type. By using this type both hands are free for control over the welding equipment. The hand-held type of welding screen is of course fully adequate from a safety point of view but only leaves one hand free. This could result in a shakier movement of the gun and a loss of precision in following the joint.

Wear welding gloves and suitably fitting protective clothing, especially when welding overhead so that the welding spatter does not run down the neck or sleeves of the overall worn.

Another hazard is the build-up of fumes given off by the welding process, which could be dangerous to health when welding in a confined space. When welding in a normal workshop there is no problem, but if welding in a tightly enclosed area, such as inside the boot of a car, make sure that the area is well ventilated.

Precaution must be taken when welding vehicle panels in case there are flammable materials attached to the other side of the panel which could ignite and burst into flames. Therefore always have adequate fire-fighting appliances available.

Questions

1 Explain the term 'MIG/MAG welding'.

2 Explain the term 'TIG welding'.

3 What is an inert gas? Explain its function in welding.

4 What are the advantages of gas shielded arc welding?

5 Name three types of gases that could be used in MIG/MAG welding.

6 What is the difference between water-cooled and air-cooled torches in TIG welding?

7 What determines the size of the ceramic nozzle to be used for TIG welding?

8 Which material are the electrodes made from in the TIG welding process?

9 State the advantages of MIG/MAG welding when compared with oxy-acetylene welding, for the repair of vehicle panels.

10 When MIG welding aluminium and its alloys, which type of gas would be most suitable for use?

11 Describe the difference between plug welding and spot welding.

12 Explain the mode of metal transfer used for welding thin-gauge materials.

13 Describe the three functions that most MIG/MAG welding equipment is capable of performing.

14 Explain the term 'spray transfer'.

15 Describe the purpose of a contact tip in a MIG/MAG welding torch.

16 Show, with the aid of a sketch, a repair situation where resistance spot welding can be substituted by a MIG/MAG spot weld.

17 With the aid of diagrams, describe the following defects made by MIG/MAG welding: (a) porosity (b) incomplete fusion (c) spatter on workpiece (d) poor penetration.

18 What colours are used to identify argon and argon-mix cylinders?

19 Why is it important to disconnect a battery on a vehicle before commencing any MIG/MAG welding operation?

20 List the safety precautions to be observed before and during welding.

21 Explain the form in which the electrode is manufactured for MIG/MAG welding.

22 With the aid of a sketch, show a repair application which involves MIG/MAG spot welding.

23 Describe the major features of a TIG welding torch.

24 Which mode of metal transfer uses the short-arc process in MIG/MAG welding?

25 With the aid of a sketch, illustrate one method of metal transfer used in MIG/MAG welding.

26 With the aid of a sketch, explain how plug welding is carried out in repair work.

27 Describe the differences between the processes of TIG welding and MIG/MAG welding.

28 By using a sketch, illustrate a typical weld fault in MIG/MAG welding.

29 State the function of the controls of standard MIG/MAG welding equipment.

30 Explain the setting to be used when welding a large gap with MIG/MAG welding.

Craft techniques and minor accident damage

13.1 Panel beating: forming panels by hand

Essentially, panel beating is a hand method of producing hollow or double-curvature shapes by means of hammering. Nevertheless, the panel beater's craft still retains its place in body work and as yet is irreplaceable by more modern methods; in spite of the tremendous developments in recent years of mechanical methods of forming, panel beating remains an essential means of fabrication of special parts. Some metal shapes cannot be produced at all by mechanical methods and others only with great difficulty, and in such cases panel beating is used to finish the shape that has been roughed out by power processes. Often, too, the prototype of a component ultimately to be made in quantity by stamping or pressing is hand made to allow minor modifications to be studied before mechanical production begins; the part produced by panel beating is used as the pattern for press or stamp tools. Panel beating may also be used where a small number of components only are required and where the cost of press or stamp tools would be uneconomic. In body repair work, panel beating is used to advantage where sections which are either unobtainable or uneconomical to replace completely can be fabricated by hand either in part or as a whole. In many cases corroded areas can be repaired by fabricating new sections for replacement purposes. In the body building trade, panel beating is still used to a large extent where new vehicles are being built either in aluminium or mild steel. Many of the components for these vehicles are still made using the traditional hand shaping methods. Also a lot of the aluminium moulds used in fibreglass construction, where highly developed double-curvature shapes are needed, are made by hand, welded and dressed and planished or wheeled to a final finish.

Panel beating is essentially a hammering process, involving different kinds of blows that can be struck on sheet metal. It should be borne in mind that most metals used in body work possess high malleability and may be overstretched even with a wood tool. The types of blow that can be struck on sheet metal are three:

Solid blow, where the work is struck solidly over a steel stake.
Elastic blow, where either the head or the tool or both is made of a resilient material such as wood.
Floating blow, where the stake is not directly under the hammer.

Each type of blow has its uses for particular purposes. A solid blow will stretch the sheet, and may be necessary when forming a panel, bending a curved strip or angle, removing a loose or tight place in a sheet, or throwing an edge over when thickness is not a consideration. An elastic blow will form metal without undue stretching; indeed, metal can be thickened if desired, as in working out a tuck or pucker. The floating blow is given to the metal when it is held over a suitable head and hit 'off the solid', so forming 'dents' at the points of impact.

13.2 Shaping metal by hand

Thin-gauge mild steel and light aluminium sheet material up to 1.2 mm, and in some cases even 1.6 mm, can be satisfactorily hand beaten by wood mallets into double curvature forms (Figure 13.1). The usual practice is to beat the metal in a suitable recess in a wood block or upon a sandbag in what is known as the *hollowing* or blocking process. Alternatively, the metal can be hammered into a wood block hollowed out to the shape of the job.

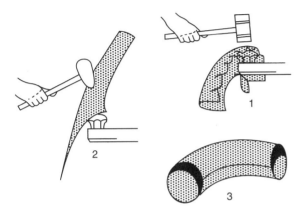

Figure 13.1   Shaping metal by hand

Another method is that of *raising* the metal by means of floating blows over steel stakes or wood formers. This raising process can be speeded up considerably by 'taking in' or making tucks at the edges of the metal. This method is generally called *puckering*, and after being made each pucker must be eliminated by careful blows to drive the metal into itself, so thickening the work at the edges. Much greater advantage may be taken of puckering in aluminium alloy sheet than in most metals because of its malleability and ductility. The exact degree of shaping of the metal which is permissible depends upon the particular alloy being worked, its malleability and the ability of the craftsman.

Another method of shaping metal which can be used in conjunction with hollowing and raising is that of *wheeling*. A panel can either be wheeled from a flat sheet to the desired double-curvature shape, or it can be preshaped using either hollowing or raising techniques and its final shaping and smoothing completed by wheeling. The technique of wheeling is not entirely done by hand, as a wheeling machine is used. The finished result depends on the skill of the craftsman as he manipulates the sheet by hand in the wheeling machine.

An experienced panel beater uses the beating method suitable to the job in hand, complex shapes

often being beaten up by using both hollowing and raising or wheeling methods. For light-gauge material too much hollowing or blocking is not to be recommended, as it tends unduly to thin the metal. The skilled panel beater is a craftsman who, to a great extent, relies on a good eye for line and form; this can only be cultivated by years of experience which, combined with dexterity in the use of hand tools, is the secret of his craftsmanship. Wood formers or jigs, upon which the beaten shape can be tried in order to obtain uniformity of shape for each workpiece, are used for many jobs. The shape of the job is retained in the vision of the panel beater, who can, by beating on the sandbag and raising over a suitable stake, obtain a very good approximation to the desired form. This only needs hammering lightly on the wood former to obtain the finished shape. Until the metal fits the former it is often necessary to check repeatedly to find any high or low areas which may prevent the workpiece from fitting the former correctly. Instead of using formers or jigs many panel beaters use templates only, which are cut to the shape of the various cross-sections of the workpiece. These templates are used towards the end of the beating process to check the finished shape of the job. After shaping, the surface of the finished panel has to be smoothed by the technique of *planishing*, using a steel bench stake, or by wheeling.

13.2.1 Hollowing

One of the methods of shaping metal by hand is that of hollowing (Figure 13.2). In this method the metals used for the purpose of shaping panels for body work applications are usually aluminium and its alloys and mild steel; aluminium is by far the easier of the two to shape owing to its higher ductility and malleability properties. This is a process of shaping a sheet metal blank into a double-curvature shape by beating the blank into a sandbag or hard wood block with the aid of a pear-shaped boxwood mallet, or for thin metal a rubber mallet, or for steel a hollowing hammer with a steel head.

First the blank which is going to be shaped is cut to size and the edges trimmed for any rag by filing. If this rag is not removed as the blank is shaped, it may tend to split from the edge of the rag. The next step is to place the metal blank, which should be held tilted, on to a sandbag, and to give it a series

Figure 13.2 The technique of hollowing

Figure 13.3 A panel being shaped by hollowing (*Frost Auto Restoration Techniques Ltd*)

of blows around its outside edge, working towards the centre by means of a pear-shaped mallet. This hammering has the effect of sinking the metal into the sandbag, which is of course resilient to the blows. After each blow the disc is turned and the next blow struck near the first one, and so on until a series of overlapping blows is made round the circumference of the blank. The tendency will be for creases or wrinkles to appear on the edges of the metal and these must be gently malleted flat to avoid overlapping of the metal and eventual cracking. The hammering of each course has to be done with steady and even blows to bring up a regular curved shape. On completion of the first course of blows, a second course is begun further in from the edge of the metal than the first set. This process of hammering in courses and rotating the blank is continued until eventually the centre of the blank is reached; by this time it will have taken on a double-curvature shape, but should greater depth be required the whole operation must be started again, working gradually from the edge towards the centre of the work until it is completely to the desired shape and size (Figure 13.3).

At this stage templates cut to the correct size and shape can be used to check that the panel being shaped is the correct size and curvature. Alternatively it is possible to use a jig constructed to the correct size and shape, usually in wood. It may be necessary during the beating-up process to anneal the work-piece to restore its malleability, because the hammering tends to harden the metal by work hardening. In some cases, instead of using a sandbag for the shaping to be carried out on, a hollowed-out recess in a wood block can be used. When the panel has reached the desired shape by hollowing it can then be smothed to a final finish by hand planishing using a hand dolly or over a stake, or wheeling to obtain the final smoothness.

13.2.2 Raising

Raising is another method of shaping metal by hand into a double-curvature shape (Figure 13.4). The method of raising is carried out by drawing the metal in courses over a suitably shaped steel stake or wood block, using floating blows which are struck slightly off the stake with a boxwood pear-shaped mallet. A series of blows is made in the metal starting at the centre, the blows being struck slightly off the stake. This has the effect of shrinking or reducing the circumference of the blank, thus forcing it down and around the stake. The disc or blank is rotated after each blow as in hollowing, but working from the centre in courses outwards towards the edges of the blank. The same process is repeated with frequent annealing of the metal until the final degree of raising is reached

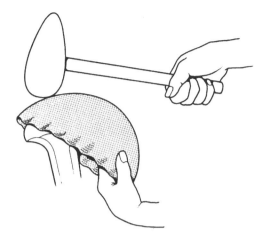

Figure 13.4 Raising

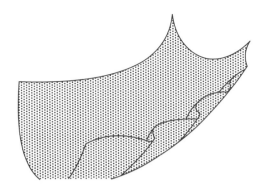

Figure 13.6 A pucker or tuck made at the edge of the panel to quicken the raising process

and the desired shape obtained (Figure 13.5). In the course of the raising, the edges of the metal around the circumference will be continually subjected to creasing, and care and skill is needed to avoid allowing these creases to become too sharp. If they are not worked out the edge will crack or fracture as the shape proceeds.

Figure 13.5 A panel being shaped by raising (*Frost Auto Restoration Techniques Ltd*)

When the article is partly beaten up over the stake, a series of tucks can be made in the outer edges of the work in order to quicken up the beating or raising process by taking in or shrinking surplus metal. This method is to make a pucker or tuck at any point on the edge of the blank by bending the metal on a stake into an ear-shaped form (Figure 13.6). After the tuck is formed, the disc is placed on a steel stake and each side of the

tuck is lightly malleted to stiffen the metal and hold the tuck in position. This is done in three or four places on the circumference of the blank, thus decreasing its diameter. Then working from the base or point of the tuck, this surplus metal is malleted out over a stake towards the edge of the disc. This has to be done very carefully with the minimum of blows, and overlapping must not take place. As each tuck is formed and worked out, the blank deepens towards its final shape.

The processes of hollowing and raising in sheet metal are often applied together in the making of articles in the form of double-curvature shapes, bowl-like shapes, etc. When the work is only slightly domed, the process of hollowing alone may be sufficient to complete the work. There are limitations to the depth which can be obtained by hollowing. This is governed by the diameter of the finished article and access for hand tools. Where the diameter is going to be small and the article deep, the raising method will have to be used to shape the work. Again as in the hollowing process, the final finish can be obtained by planishing the preshaped article to a smooth surface finish.

13.2.3 Wheeling

The craft of wheeling has been used for many years in the production of curved panel assemblies that are used to make up the modern vehicle body. Wheeling was a very skilled art when vehicles were coach built and hand methods were employed to make the component panels (Figure 13.7). With the advent of mass production and the development of the motor car, speed of production became an essential factor. Consequently hand-made panels,

Figure 13.7 Wheeling a panel to restore a vintage vehicle (*Frost Auto Restoration Techniques Ltd*)

Figure 13.8 Sports car panel being shaped by wheeling (*Autokraft Ltd*)

which were usually made by wheeling, were replaced by pressed panels made on power presses with mating dies, which very speedily produced accurate panels having a good surface finish. Today wheeling is still used to produce panels for prototype vehicles; one-off assemblies which are specially built to order, and small-volume production, where the panels are prepressed to somewhere near the finished shape and then finished off by wheeling (Figures 13.8 and 13.9). The vehicle building industry still uses wheeling in producing panels for commercial and private coaches, road transport vehicles and any assembly requiring a double-curvature shape in its production.

The art of wheeling lies in the operator's ability to handle the panel successfully in the wheeling machine (Figure 13.10). Wheeling is simply the stretching of metal between two steel rolls known as wheels. The upper wheel has a flat face and revolves freely on its own bearings. The lower wheel also revolves but has a convex curved face and is made interchangeable so that different shaped lower wheels, from nearly flat to full curved, can be used. The two wheels meet at a common centre

at which the stretching takes place when the wheels are tensioned by applying pressure through the bottom wheel against the panel which is pressed on to the top wheel. As the metal panel is pushed through the wheels a stretched area, the length of

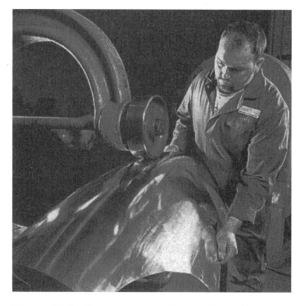

Figure 13.9 Sports car panel being shaped by wheeling (*Autokraft Ltd*)

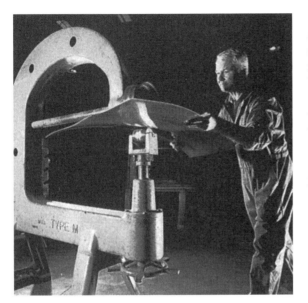

Figure 13.10 The technique of wheeling
(*Autokraft Ltd*)

the panel, is produced, and is known as the wheel track. By carefully allowing these wheel tracks to overlap on the entire surface of the flat panel being wheeled, a curve in one direction only develops. To create double curvature on the same panel, stretching must take place in two opposing directions; therefore a second set of wheel tracks must cross the first set transversely, and this is achieved by turning the panel through approximately 90° before making the second tracks. As the panel is pushed through the wheels the pressure can be increased gradually until the desired curvature is obtained. Therefore the skill of wheeling lies in the use of correct wheel pressures and careful manipulation of the panel through the wheels; this can only be learnt by experience. An alternative method of achieving double curvature on a flat panel is to form the first curve by using a rolling machine, and then to wheel it in one direction only, allowing the tracks to cover the panel in the opposite direction to the first curve, thus stretching the metal in such a way to form the double curvature shape. Wheeling can therefore be used, first, for shaping a flat metal blank to a finished double-curved panel; second, to finish a preshaped panel which has been hollowed or raised to its final shape; and third, to smooth or planish a preshaped panel to its final finish.

The materials best suited for wheeling are aluminium, some of the aluminium alloys and mild steel, all of which possess the properties of malleability and ductility to a certain degree. When using the wheeling machine for wheeling aluminium and its alloys, care should be taken not to put too much pressure on the work of raising the bottom wheel. Up to three times as much lift or stretch is obtainable with aluminium than with steel, and much more shaping by wheeling is possible in the case of aluminium than with harder metals like mild steel. Thus too much pressure could have the effect of overstretching the particular panel or workpiece. The wheeling machine may be used simply to planish, producing a smooth surface by the friction and rolling action derived from passing the sheet backwards and forwards between the wheels when these have just the right amount of pressure.

Where components of moderate curvature are to be produced by wheeling alone, the sheet or blank is placed between the two wheels at one edge or in the case of a round blank in the centre; pressure of the exact amount suitable for both the thickness and the type of material is applied and the sheet is wheeled. In the case of aluminium the wheeling lines are difficult to see, but by smearing the surface of the panel or sheet with mineral turpentine or very light oil these tracks become more easily seen and are therefore more easily lapped. Care should be taken at this stage when passing the metal through the wheels not to twist or jerk the panel, as this could result in ridged sections and an uneven surface. Movement of the sheet is varied until the desired shape is obtained, those parts of the panel which require to be only slightly curved receiving less wheeling than other parts which must be more curved.

To wheel a panel that has been preshaped by blocking in a sandbag, it is necessary to smooth its surface without altering its shape, and for this reason the panel should be pulled right through the wheels at every stroke. This edge-to-edge wheeling will result in an evenly stretched panel surface. When carrying out this smoothing work, or planishing as it could be termed, the pressure exerted on the work by the wheels should be very slight. Another common task that can be successfully carried out with the wheeling machine is to tighten up the loose wavy edges which sometimes occur when shaping a panel by hand. The method of overcoming this trouble is to wheel directly adjacent to the stretched edge. This stretches the area being wheeled and so tightens up the loose edge. The reverse sort of problem, that of a panel with a fullness or stretched area just in from the edge of the sheet, can also be overcome by wheeling. The method is to start wheeling from the centre of the fullness on a track parallel to the edge and to work right out to the edge of the panel.

With most machines three standard wheels are supplied which are generally referred to as flat, medium and full curved. The widths of the wheel tracks are flat 25 mm (1 in), medium 10 mm ($\frac{3}{8}$ in), and full curve 5 mm ($\frac{3}{16}$ in). The best rule to follow when selecting a wheel is to use the flattest wheel possible for the job you are doing. This not only speeds up the wheeling but prevents to a large extent marking the surface of the panel with wheel tracks, which are really the evidence of an over-stretched panel (Figure 13.11).

Figure 13.11 Standard wheel sets (*Frost Auto Restoration Techniques Ltd*)

The most important points when using the machine are:

1 The wheels must be kept clean, free from dirt and in perfect condition.
2 The pressure exerted by the bottom wheel must be correct for both the thickness and type of metal being wheeled.
3 The right-shaped wheel must be used to suit the required shape of panel.

4 When wheeling, the panel should be held without tension, allowing it to move freely without twisting or jerking it.
5 The wheel tracks must be carefully overlapped to achieve a smooth curved surface.

13.2.4 Split-and-weld system of shaping metal

The introduction of welding into the panel beater's craft has led to the development of split-and-weld panel beating, which is at once less laborious and much quicker than the older methods of hollowing and raising. The system consists of making a pattern on a panel jig with pattern paper. The paper is held off the jig by tension at its edge. To allow the paper to drop on to the jig, the paper is slit at suitable points, the edges then opening out to let the pattern fall into position (Figure 13.12). It is obvious, then, that additional material is required at the slits. This may be obtained in the panel either by stretching the metal at these points until enough is obtained to meet the requirements, or by welding in V-shaped pieces of metal. The final shape is then achieved by wheeling or planishing.

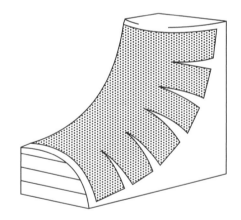

Figure 13.12 Split and weld pattern

13.2.5 Planishing

The technique of planishing is a very old and established craft in the history of hand-fabricated metal articles. Basically planishing takes over from hollowing and raising, which shape the article, to smooth its entire surface and finalize its shape.

Planishing can be performed in three different ways. First, there is the technique which is used mostly by the panel beater in planishing new work. In this case a planishing hammer is used in conjunction with a steel stake, both having highly polished faces. The steel stake is mounted on a bench and is of suitable curved shaped for the article being planished. The work is taken to the stake and planished over it to achieve the final finish (Figure 13.13). Second, there is the technique used by the body repair worker, where the planishing hammer is used in conjunction with a dolly block which is in fact a miniature stake or anvil, again with polished faces. The dolly block is held in place under the panel by hand, while the blows are directed on to the panel surface and transmitted through to the dolly block by the force of the blow being in direct contact with hammer face, work surface and surfaces of dolly block. In this method the tools may be taken to the job and the work carried out on the spot. This fact makes planishing ideal for the repair of vehicle body panels (Figure 13.14). Third, there is the technique of planishing using the wheeling machine as a means of smoothing the work surface (Figure 13.15). This is accomplished by the friction and roll action of the workpiece as it passes between the two steel rollers. This method is normally used by panel beaters in smoothing and finalizing new work; it can also be used by body repair workers, but the difficulty arises that the panel has to be dismantled and removed from the body shell, and is therefore an uneconomical proposition.

Consequently planishing using hand dolly and hammer is accepted universally as the best technique in the repair of panel surfaces by planishing.

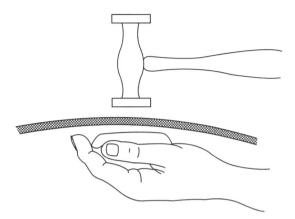

Figure 13.14 Planishing using a hammer and dolly block

In some cases the three techniques can be used together; for instance a panel can be planished using a stake and then finished off by wheeling, or a panel can be wheeled, fitted to the job, and then minor rectification carried out using hand dolly and block. All three techniques have one common feature: when planishing the metal surface is slightly stretched because of the metal-to-metal contact between the working faces of the tools and the work face of the panel. The skill in this process lies in the fact that the craftsman has to merge, by careful hammer blows or wheel action, all these stretched blows into one to create a continuous smooth surface.

Where planishing hammers are employed, the process is carried out over a metal stake or hand dolly. The planishing is carried out over the whole surface of the workpiece; the blows are light and

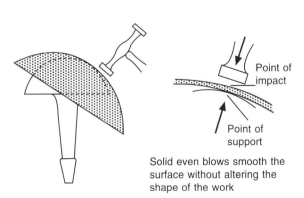

Point of impact

Point of support

Solid even blows smooth the surface without altering the shape of the work

Figure 13.13 Planishing using a steel stake

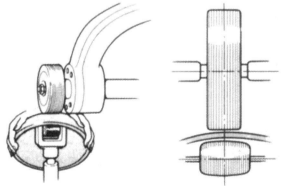

Figure 13.15 Planishing using the wheeling machine

given squarely, otherwise they will produce crescent marks difficult to eliminate. Each hammer blow produces a flat spot, and the blows are so directed that the spots merge imperceptibly into one another over the whole surface. Any low places or 'valleys' on the surface of the workpiece can be eliminated by careful hammering on the head, which slightly stretches the metal, causing it to rise to the correct contours. Both the hammer face and the steel stake must be kept scrupulously clean and perfectly smooth, otherwise it will be impossible to avoid marking the sheet. Planishing should leave the metal with a dead smooth surface. If this is not attained, small hammer marks can be removed by smoothing off with emery cloth glued to a piece of wood and used like a file.

13.3 General guide to the fabrication of hand-made panels

When a panel is required to be made by hand, the first essential is that a jig or former should be made to resemble the exact line and contour. This jig or former can be made in either wood or metal, but preferably in wood as this is easily shaped to double-curvature shapes. Moreover, if the panel has to be fastened to the jig this can be done by putting small panel pins through the metal into the wood jig. The holes left by these tacks can be later filled in by welding.

Once the former has been built, the next stage is the making of the blank template, which is the developed flat form for the total shape of the panel being made. Some shapes can be developed using geometric drafting methods, but where there is no recognized method only the trial and error basis can be used. This pattern can be made in strong brown paper or special template paper. At this point it is necessary to decide whether the panel is to be made in a one-piece construction or made in several pieces which can then be fabricated by welding. On very complicated shapes it is sometimes necessary to use joints; therefore the exact location on the panel should be considered very carefully, taking into consideration facts such as the length of the joint, and its position in relation to accessibility for planishing with hand tools when assembled. Once the pattern draft has been developed it can then be marked from the pattern paper on to the surface of the metal from which the

panel is to be made, the two most popular metals being aluminium and mild steel. The golden rule of hand-made panels is that an allowance should be made all the way round the developed size; in other words, the developed blank should be bigger than the pattern as it is easier to cut off surplus metal on the finished component after shaping than to have to weld pieces to it.

The metal blank can now be shaped by any hand methods. Care must be taken to check continually the shaping against the jig or former to see that no part is overshaped. In some cases where opposing curves meet or combine, small metal templates cut to the correct curvature are useful to check the relative position of individual curves as the panel is being shaped. As each piece is completed it is tried on the jig and made to fit it exactly by planishing.

When all the shaping has been carried out, the next stage is to join these pieces by welding. The only accurate method of doing this is to fasten the appropriate pieces to the jig, making sure that the joints are butted and not overlapped together, leaving no gaps which may need extra welding filler rod which leads to later difficulties in planishing. Once the panels are secured using either clamps or nails, the appropriate sections should be tack welded together at intervals no larger than 19 mm. After each tack the assembly should be cooled for two reasons; first, because the former or jig is usually made from wood, and second, the smaller the heat input the greater will be the accuracy of the alignment of the job. The work is then carefully removed from the jig, when welding of the joints can commence. The utmost care must be taken when welding to ensure that there is adequate penetration, but not so much as to leave unwanted surplus weld metal on the underside, while reinforcement on the face side should be slightly above the surface of the panel to allow the weld to be completely filed off without losing its strength. It is best to weld small sections at a time, planishing with a hammer and dolly whilst the weld is still hot; this allows the weld to be flattened easily and it gives the weld inherent strength.

Once the weld is completely finished and flattened in this manner the final finish can be obtained by further planishing and filing which, if done correctly, should make the weld indistinguishable from the parent metal. The whole assembly is now tried back on the jig and any small rectification required is carried out by further planishing methods (Figure 13.16). If this is then satisfactory it is trimmed to the correct size. Any flanges, wired edges or safe edges can then be formed either by using hand tools or flanging jigs or a combination of hand tools and swaging machine. The final finish can be achieved by using different grades of emery paper or a sanding machine with a very fine grit sanding disc.

Figure 13.16 Planishing a hand-made panel (*Autokraft Ltd*)

Aston Martin Lagonda are one of the few companies who still specialize in producing a car which is hand built or using traditional fabrication methods together with press work. It was Lionel Martin, with Robert Bamford, who began the Aston Martin story in 1913, and he achieved a great reputation for the standard of finish of his cars and for his infinite attention to detail. This high standard has constantly been maintained from those early years.

The vehicle starts with sheets of steel and various sections of rectangular or round tubing from which is constructed the basic chassis frame and body structure. The choice of the thickness of sheet steel will vary according to the load for that particular panel. With the aid of a number of high-precision jigs, which have to be made by specialist tool and jig makers, construction of this first stage of the car takes approximately three weeks. A number of checks on dimensional accuracy are made during the course of building, before the chassis frame can be ready for the panel shop. The last of these checks takes place on a special surface table where random checks can be made on each part of each chassis to enable the accuracy to be maintained (Figure 13.17).

Figure 13.17 Random chassis checking on surface table (*Aston Martin Lagonda Ltd*)

Next comes the anti-rust treatment, which involves a thorough cleaning of the chassis frame with a solvent, followed by hand spraying a heavy coat of zinc-phosphate paint. The whole of the lower part of the structure is then sprayed over with a coat of chip-proof underseal. Finally, into all the tube sections, or any other closed areas that have been fabricated, a wax-based preparation is injected through predrilled holes to give protection to all these inner surfaces. The chassis frame is now mounted on a wheeled subframe which makes it mobile, and it is then ready for the body shell to be fitted.

By tradition all Aston Martins have had aluminium bodywork because aluminium is a medium ideally suited to small production runs and for hand building; it is also light in weight. Up to the 1960s the whole of the body shell was built by the company from a number of different panels, all made with the help of a stretch-press. The angular shape of the succeeding DBS in 1967 and the DBSV8s in 1969 largely made this impossible, and for the first time a number of basic rubber-pressed panels had to be secured from an outside concern. This practice continues today. The bought-in panels include those for front and rear wings (each in two parts) and the roof, leaving the boot lids, bonnets and doors to be made on site as before.

Each part of the front and rear wing will be laid on its respective hammer former. This is an accurate jig made of Kersite, which is a very dense and durable material capable of withstanding a large amount of hammering and beating into shape by the panel beater using highly individual hand tools (Figure 13.18) and his special skills. Following the initial shaping, the two parts of each wing would be welded together, 'dressed' or reworked to the extent that the joint will effectively disappear ready to be finished off, and then fitted to the now fully treated chassis frame (Figure 13.19). This is done by the simple means of riveting them together, taking care to place a layer of insulation material between the two to prevent the electrolytic reaction that will occur between these dissimilar metals. The rivets are similarly treated by dipping them into an anti-corrosion compound before use. The roof, another pre-formed panel, is similarly shaped and is the first to be fitted.

Figure 13.18 Panel beater's hand tools (*Aston Martin Lagonda Ltd*)

The doors, bonnets and boot lids have always been hand made on site, the door skins (panels) because they have a relatively simple shape which can be made more economically than a rubber-pressed panel which will always have a certain amount of wastage. The company has always made bonnets and boot lids but these are far from simple, and demonstrate the full range of the panel beater's

Figure 13.19 Fitting hand-made panels (front wings and roof) to chassis (*Aston Martin Lagonda Ltd*)

skill and art, which has changed little in all the years it has been practised. The bonnet is made by taking three pieces of flat aluminium which are cut to shape, wheeled up, then welded into one piece and finished off by wheeling; this operation takes almost 30 hours and is very skilled indeed. The boot lid is made in a similar fashion. The wing panels are fitted and the doors, boot and bonnet are individually fitted on to their particular chassis frame. After any rectification work, an inspection will allow the vehicle to go forward into the paint shop. The bonnets and boot lids will need to be removed for the later assembly work, so these and the doors are numbered with their particular vehicle.

The vehicle's progression through to the paint process is carried out in two distinct operations, which are longer in total than any other part of the vehicle build. The vehicle starts with the new body shell being cleaned down with a mild acid preparation called deoxidine. When dried off, it is sprayed over with a heavy epoxy-type primer surfacer and hand rubbed before a second similar coat is applied (Figure 13.20). Rubbed again, it is sprayed with a base coat and then various apertures, door shuts, screen areas and rear ends will be sprayed with a coat of the car's final colour. These are areas where components will be fitted, such as front and rear screens, door handle locks and rear lamp assemblies, which will remain in place when the main part of the paintwork is carried out later. The interior of the car is sprayed over in black and is

Figure 13.20 Panelled body shell sprayed in primer surfacer (*Aston Martin Lagonda Ltd*)

now ready to move into the assembly area. This begins with the laying in of various different sound insulation panels, followed by the wiring harness and the air conditioning unit.

The vehicle now moves forward to the area where the suspension is fitted, for which purpose the subframes will now have to be removed. Front and rear suspension components are made by specialist concerns to the company's own design. With all these parts assembled, and with the final drive unit and the power steering unit in place, a set of temporary road wheels can be fitted which make the car mobile. Next, the preassembled engine (which takes 60 hours to build) is fitted by hand to the vehicle. At the end of the production line the bonnet will rejoin the car, the underpanels and stone guards will be fitted, and the water system, oil and fuel will be checked.

The vehicle is then passed over to the road test engineer who will run it for approximately 100 miles. On the satisfactory completion of the road test, the vehicle can be made ready for its second entry into the paint shop, but not before its body panelling has been thoroughly inspected for small marks or minor imperfections.

On the car's re-entry into the paint shop, the bodywork is masked up and then flatted by hand, as with every part of this operation. Final colour coating is now begun, alternatively spraying and rubbing each coat of paint until the finishing overlay lacquer can be applied (around 12–14 coats of paint). This is followed by the final fit of the

vehicle's interior trim and seats, provided in a special colour at the request of the owner, detailed fittings, body items and carpets (Figure 13.21). Then after a short final road test by the same tester the vehicle receives a final inspection (Figure 13.22) and is passed over to the sales section.

The annual build figure for this truly hand-made vehicle is between 250 and 300.

Figure 13.21 Fitting interior trim (*Aston Martin Lagonda Ltd*)

Figure 13.22 Final vehicle inspection (*Aston Martin Lagonda Ltd*)

13.4 Edge stiffening of sheet metal

Edge treatment is a general term used to cover the many methods of forming the edges of sheet metal, panels and components. The body worker frequently has to increase the strength and rigidity of the edges of large unsupported metal panels, to stop movement and vibration when the vehicle is in motion and to create resistance against buckling and twisting. This can be provided on the sheet itself or by adding stiffening agents. The various types of edge treatment are normally classified as follows:

Formed In this case the edge stiffening is formed from the metal panel itself.

Applied Here the edge stiffening is made up as a separate piece and then fixed to the panel edge.

Safe edges, flanges, wiring and swaging can all be classed as formed edge treatments, while the attachment of strips, half-round beads, mouldings angle sections and false wire edges are classified as stiffening agents. Panel edges are treated in these ways for the following reasons:

1 To stiffen and strengthen the panel or component at its extreme edge
2 To act as a safe edge, as it is important that panels which are to be handled frequently should be effectively treated to avoid the risk of injury due to exposed raw edges
3 To ornament and decorate the panel.

Often edge treatment is used for more than one reason; an example of this may be found on the modern motor vehicle wing, where the edge treatment provides a stiffening effect, gives a safe edge, and also has the effect of being pleasing to the eye.

13.4.1 Folding

Folding is the simplest form of edge treatment (Figure 13.23) and is satisfactory when neatness and speed are the main factors. Folds may be creased to give a flushed effect on one side of the panel, or doubled over twice to increase the strength. These edges can be formed either with hand tools or with the aid of a folding machine.

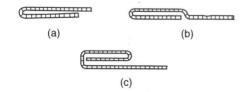

(a)   (b)

(c)

Figure 13.23 Folded edges (a) single fold (b) folded and creased edge (c) double folded edge

13.4.2 Flanging

In flanging, the edge is formed at right angles inside the panel (Figure 13.24). In addition to imparting rigidity it can be readily cleared of road dirt, thus reducing the possibility of corrosion from moisture-retaining matter. One disadvantage of this type of edge treatment is that when the edge suffers a severe blow, the metal tends to crease badly and the edge may crack. When repairing such a fracture, care should be taken to avoid rigidity at one point, which often results in further cracks appearing some distance from the repaired portion. In the case where the edge is formed outside, a plastic moulding is clipped over the protruding edge. Stiffening is provided by the formed edge, the plastic moulding being used for decoration and also to render the edge safe.

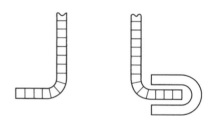

Figure 13.24 Flanged edges

Flanging is the process of hammering the edge of a piece of sheet metal in such a manner that the required width of metal is worked into a position usually at right angles to its original form. This has the effect of strengthening the metal in the area of the bend. It is used to advantage on panel edges to stiffen, to make the edge safer and to improve its appearance, and in many cases to perform the function of a flanged joint where one metal panel is joined to another. If the flange is on a straight piece of metal the flanging technique can be carried out either by using a stake and mallet and hammering the edge over the stake to form a right angle, or by using a bending machine which gives a more consistent edge.

Difficulty may be experienced when forming flanges on curved sections of panels depending on whether the flange is on an external or an internal curve, as these require different techniques of working. This type of flanging can be carried out by using a swaging machine fitted with flanging rolls, which turns the metal edge at right angles as it is fed through the rolls. Owing to the awkward shape of some panels they cannot be flanged in the swaging machine and so must be processed by hand, using either a stake and mallet or a hand dolly and mallet and finishing off with a planishing hammer. If the panel is of a complicated curved nature and requires flanges, it is sometimes necessary to make a flanging jig of two identically shaped pieces of wood. The panel is inserted between the two pieces and clamped, then the metal edge protruding is carefully hammered over. In some cases annealing has to be carried out where the corners are very sharp so as not to split the metal. When taken from the jig the panel retains the curved shape with the flange following its contour.

To form a flange around the edge of a cylinder, it is placed against the edge of a stake so that the width of the metal to be flanged lies on the stake. A stretching hammer is then used to stretch the flange metal, working the cylinder steadily round and keeping the width of the flange constantly on the face of the stake. The maximum amount of stretching must take place on the outside of the flange, gradually diminishing to nothing at the inside. The metal must be kept flat on the top of the steel stake, so that the hammer strikes the metal hard on the stake top. In the case of a flange raised around the edge of a flat disc, a mallet is used instead of a hammer and the metal is gradually drawn inwards or upwards by careful working round the edge over a curved end of a steel bench stake. By allowing the outer edge of the flange to be drawn in slowly a good deal of creasing is avoided; moreover, the creases which do occur around the edge are carefully worked out as they appear. In the working of a deep flange on a disc the metal may be annealed at intervals except, of course, in the case of coated metals such as tinned or galvanized steel.

13.4.3 Swaging

A swage is a moulding or indentation raised upon the surface of sheet metal by means of male and female rollers, the rollers being made in a variety of contours. The machine to which the rollers are attached may be driven either by hand or electric motor, the choice of machine often being influenced by the type or amount of work to be undertaken. Although swaging has many similar functions to that of wired edges, it is not confined to edge treatment but may be used some distance

from the edge within the limits of the throat of the machine. In addition, as no extra allowance of metal is required, a saving in material with consequent reduction in weight is achieved with the use of swaging (Figure 5.19). Special composite swages are also available for use on panels (Figure 13.25); the rollers are made in sections so that their width may be adjusted. If it is found that the throat of the normal swaging machine will not accept curved panels or wings, the rollers may be fitted on to a wheeling machine which is virtually throatless.

The projecting shape of the swage above the surface imparts considerable strength to sheet metal articles. Panels which would otherwise be slack and lacking in rigidity can be stiffened by the use of swaging. Motor vehicle body panels are subject

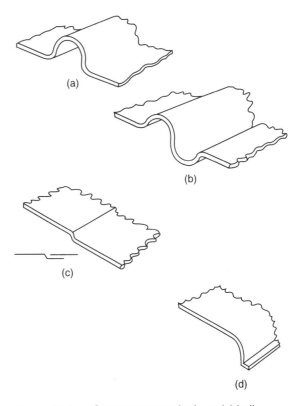

Figure 13.25 Common swaged edges: (a) ball swage – may be produced in various sizes according to the size of the wheels used; is used as a stiffener on sheet metal (b) return curve swage – may be produced in various sizes; is used as a stiffener in body work (c) joggle swage – used to produce a creased lap joint in sheet metal (d) radius swage – used to produce radius corner joints; the small flange may be trimmed off thus forming a butt joint

to vibrations and fluctuating stresses, and for this reason swages are used to obtain a suitably rigid body shell. The return curve swage is frequently used to strengthen the centre portions of cylindrical containers because of its high resistance to externally or internally applied forces. In addition to strengthening purposes, swaging is often used to relieve plain surfaces; this kind of decorative effect is a common feature on motor vehicle bodies.

13.4.4 Wiring

Wiring is the process of forming a sheet metal fold round thin wire to give extra strength to the edges of a panel and also to improve its appearance (Figure 13.26). The wired edge, although not so popular as in the past, is still used where impact strength is the most important requirement of the edge. Mudguards on earth-moving equipment meet conditions in service to which the wired edge lends itself most admirably, and repairs can be readily made, if necessary, with the aid of welding equipment. The wired edge is still found to be a useful form of edge treatment on the wings of public service vehicles, particularly where the wings are not integral with the body.

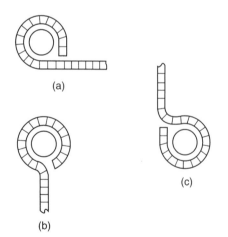

Figure 13.26 Wired edges: (a) plain wired edge and (b), (c) creased wired edges

To carry out the technique of wiring, the edge of the metal should be bent up in the folding machine or by hand, using a mallet and stake, at the corresponding bending line. The bend should not be too sharp, as the metal has to be worked round the wire. Next a length of wire is cut to size and placed

in position, and the metal is beaten over by a mallet to hold the wire in position. The edges can be finally closed by using a wiring hammer or by passing through a swaging machine fitted with wiring rolls (Figure 13.27).

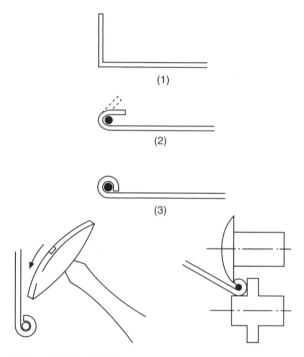

Figure 13.27 Wiring process:

1 Allowance marked off and sheet folded
2 Metal beaten over wire with mallet
3 Edge closed using wiring hammer or wiring rolls

13.5 Techniques of damage rectification

Before a systematic approach to body repairs is possible, it is necessary to understand the characteristics of sheet metal as used for body panels. When a flat sheet of metal is bent to a wide arc or radius it will regain its former shape when released; that is, it is elastic or possesses elasticity. However, if this sheet is bent to a short arc or radius it exceeds the limits of elasticity or flexibility; the metal in the bend becomes stiff and will take on a permanent set and retain the curvature. This is the result of the stresses which have been set up at the bend, making the material work hardened. Before the sheet is formed in the press the grain structure is constant and the thickness uniform throughout (see Figure 13.28). When the metal is formed to make the body panel it is bent beyond its elastic limit. The outer surface stretches or lengthens while the inner surface shrinks or shortens (Figure 13.28). The pressure exerted on the metal by the press also changes the grain structure to work harden the surface layers. This build-up of stresses in the bend or curve is an essential factor in the design of vehicle body panels which together form the body shell. A common feature in the design and manufacture of a motor vehicle is the many curved surfaces which are normally referred to as *crowns*. Vehicle body panels consist of flat or slightly curved areas, sometimes quite

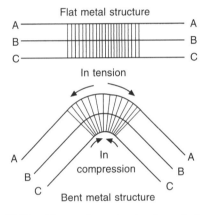

Figure 13.28 Change in grain structure during pressing (*Sykes-Pickavant Ltd*)

large and elastic in nature (low crowns), such as door panels; these are held in position by stiffened, rigid sharp bends and swages which are non-elastic in nature (high crowns), such as the cant of roof panels (see Figure 13.29).

If a panel is damaged in an accident the buckled area, being sharply bent, will create additional stiffness in the panel, whether in an elastic or non-elastic area. The slope of the buckles surrounding the sharp creases will be fairly elastic, but a greater amount of effort will be needed to reshape the sections of the panel which are made rigid either in manufacture or through accidental damage. When a panel becomes damaged due to impact, the resulting force on the metal causes buckling in the form of creases or ridges which are created because the panel has gone beyond its elastic limits to become non-elastic, therefore establishing unwanted rigid sections within the damaged area on the panel. The characteristic stiffness of the ridges prevents the panel returning to its original shape unless additional force is applied to release the stress in the ridges in the damaged area. When these stresses in the unwanted rigid areas are released, the elastic areas will also be allowed to return to their original shape. It is important that these corrections be made in the right sequence on the individual panel, otherwise additional damage will be caused to the panels. Repairs must be

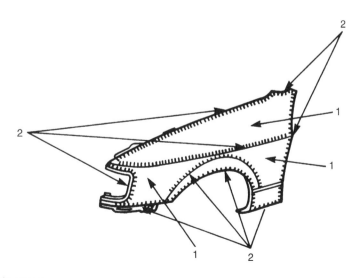

1 Low-crown, elastic areas
2 High-crown, rigid areas

Figure 13.29 Prepressed wing panel

performed using the reverse order and an opposing force to that of the original force which caused the damage. Consequently the correct sequence should be first to remove the last ridge which was formed, and then to work towards the first point of impact of the damaged area.

Rectification of vehicle bodies, following a true assessment of the damage, can be divided into two stages: roughing out or straightening of the reinforced sections and panels to approximately their original shape, and the finishing or preparing of the surface to a smooth appearance for repainting. Both stages are of prime importance, and many man hours can be saved if the job is processed correctly. In cases of damage where the body is distorted, the temptation is to use rough-and-ready methods depending on brute force to restore some resemblance of shape. Whilst this may speed up the first stage of a repair, it will be found that such methods result in additional marking of panels; considerably more time will be spent on the final stage of finishing than would be required if more thought had been given to the job in the first instance, and better methods had been used to rectify distortion. Damaged panels should be restored by relieving the stresses which have been set up by the force of impact. The skill of all body repair techniques lies in the correct handling of the basic hand tools, in a variety of combinations best suited for the job in hand.

13.6 Hammering techniques

Unlike most other trades, where the hammer is used with a follow-through action from a combination of wrist, elbow and shoulder, in the skilled hands of a body repair worker the planishing hammer swing is a rhythmic action involving finger and wrist movement, producing a ringing blow (Figure 13.30). The hammer should not be held tensely, but during the complete cycle of movement it should be held loosely in the hand. This will achieve a higher degree of accuracy and at the same time help to reduce fatigue. This loose holding of the hammer applies equally to dolly blocks, as it permits them to bounce back naturally and to assume the correct position for striking the next blow. With practice the wrist becomes strengthened, and consequently working in restricted places becomes easier where an even

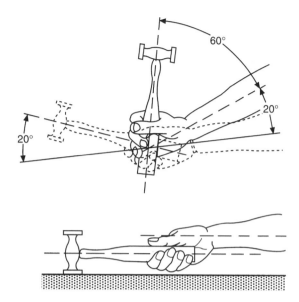

Figure 13.30   Swing of the planishing hammer

wrist action is impossible. The dolly should be allowed to lie naturally in the hand with the face to be used uppermost and, as with the hammer, should be held firmly but not tightly. Tap lightly at the dolly to obtain the feel of metal to metal, and check for control of force of blow; each blow will give a metallic ring which should be the same for each stroke of the hammer. When no metallic ring is heard the hammer is not hitting the metal in alignment with the dolly.

13.6.1 Roughing out damage

In minor repair work which can be carried out using hand tools, the first major operation is to reshape the damaged area back to its original contour. This is done by a technique known as roughing out, which must be carried out prior to any finishing process such as direct hammering or planishing. Roughing out is the reshaping of the area by hand with the aid of a heavy dolly, which forces back the damaged section to its original shape (Figure 13.31).

When repairing collision work, the normal method of correction is to reverse the process which caused the original damage. In a case of minor repair the point of impact is now the lowest part of the damage. To reverse the process this point on the underside of the panel should be struck using the same force as was originally directed against it.

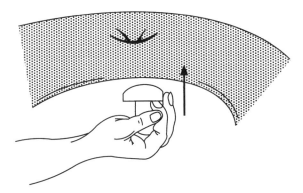

Figure 13.31 The technique of roughing out damage

If this spot is hit accurately with a roughing-out dolly using the same force, the panel will spring back almost to the contour it had prior to the damage. In some cases you will be able to correct the panel damage with a single blow which will spring the panel back to its original shape. In other cases, where the repair is larger, it will be found that several blows are necessary. Hold the roughing-out dolly lightly in the hand and strike the hardest blow at the centre of what appears to be the lowest point of the damaged area, then direct the blows around the first one and gradually work outwards, decreasing the force of the blows until all the damaged area has been roughed out (Figure 13.32). However, in most cases the damage will not be completely restored to its original contour, although it will be roughed out and can be straightened to its correct contour by direct hammering or by combination of direct hammering and indirect hammering.

The use of a heavy hammer for roughing out is not advisable, for this permits heavy blows which are concentrated in small areas and invariably results in stretching or otherwise distorting the metal, whereas a well directed blow with a dolly that matches the original contour of the repair spreads the blow over a larger area, resulting in very little distortion of the metal. In some cases body repair workers use a boxwood mallet for roughing out, because there is less chance of stretching the metal. The technique is similar to that of using a dolly, as the mallet is used on the inside of the panel to hammer the damaged section back to its original shape; then the work is finished off by direct hammering using a panel hammer and dolly. A disadvantage in using a mallet is that on modern panel assemblies there is insufficient space to use a mallet for roughing out; therefore most body workers find a dolly more useful.

13.6.2 Direct hammering

Direct hammering is in fact the process of planishing, and the body repair worker uses it as a finishing process after the work has been preshaped and roughed out (Figure 13.33). It is the essential practice to master, and develops as a result of continuous experience.

Before using the hammer and dolly together, it will be necessary to clean the underside of the portion of the wing or panel on which you will be working. Body panels and wings are covered with a sound deadening material which must be removed before starting the work. If you fail to clean the surface of this material it will not only stick to your dolly but will to a large degree destroy its effectiveness.

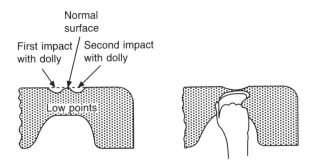

Figure 13.32 Positioning of blows in the roughing-out technique

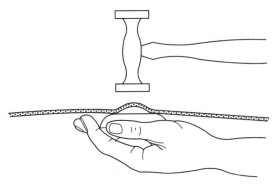

Figure 13.33 The technique of direct hammering

It is most important to choose the correct dolly block for the job, because they differ in shape, curvature and weight. In repairing the high crowned radius of a wing you will have to use a dolly block with a high-crowned radius (Figure 13.34). In repairing large body panels and door panels which are fairly flat it is necessary to use a dolly block with a low-crowned radius (Figure 13.35). In direct hammering, by having a dolly which matches the original contour under the damaged area and striking it with a hammer, you are pushing the uneven displaced metal surface back to its original contour to give a smooth and level finish. The dolly provides support and prevents the undamaged areas that have been previously roughed out from being pushed out of place. If you do not strike squarely over the dolly, you will be hitting an unsupported area of the repair and will displace the metal, creating further damage that must be rectified later.

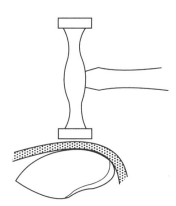

Figure 13.34 Using a high-crowned dolly

Direct hammering requires skill in directing the hammer blows and close observation of what you are doing so as not to hit the metal too hard, thereby displacing it. Perfect coordination between your two hands is necessary to enable you to move the dolly around under the damaged area and still continue to hit squarely over it with the hammer. Start hammering by using light blows; these will not do the job, but will show you whether or not you are hitting squarely over the dolly. Do not forget to let the dolly just lie in your hand and to grip the hammer loosely. A true ring will be heard if you are directly over the dolly otherwise the sound will be dull. Increase the force of the blow gradually until you have found just the right force

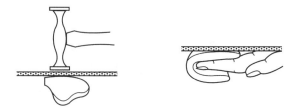

Figure 13.35 Using a low-crowned dolly

to push the raised points of the roughed-out section back without flattening the surrounding metal. The hammer should bounce back of its own accord so that it is ready for the next stroke. Likewise the dolly will spring away from the surface, and the normal resilience of your arms will bring it back, striking a blow on the metal from underneath. These things will occur normally only if you hold both hammer and dolly loosely.

13.6.3 Indirect hammering (off the dolly block)

Indirect hammering is another technique which uses hammer and dolly to level a panel surface. A low area can be raised by hammering round the outer edges in such a manner that the rebound action of the block tends to push the low area upwards to its original contour. This in fact is achieved by the sequence of hammering just off, or at the side of, the dolly block; hence the name of indirect hammering (Figure 13.36). This technique is used in conjunction with direct hammering or planishing to achieve a final finish on the panel surface. Metal that has not been excessively hammered, displaced or stretched will have a tendency

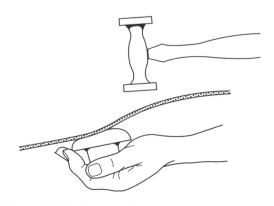

Figure 13.36 The technique of indirect hammering

to return to its original contour of its own accord. This is due to the internal strain imparted to the metal by the forming dies in manufacture. If the metal is prevented from springing back by other strains imparted to it by additional bends or creases that have been formed by accident, the metal can then be restored to its normal contour by relieving whatever new strain is holding it out of position.

In direct hammering a dolly block having the correct contour to match the original shape of the panel is held under the low spot, and a series of light blows are aimed around the outer edge of this low spot, and slightly off the dolly block. The light blows will not displace the surrounding undamaged area, but the force of the downward blow will be transferred to the dolly block. As a result of receiving the hammer blows indirectly, the dolly will rebound and the hand holding the block will automatically bring it back in place so that it imparts a light push upwards on the area. The centre of the damaged area will slowly rise until the original contour is restored.

13.6.4 Spring hammering

This is another technique of using hand tools to smooth and level a panel surface. In this case only a hammer is used, and it is not supported with a dolly block. The technique is used to reduce high spots which sometimes form as a panel is planished. In some cases these high spots can be reduced by careful, controlled hammering which spreads the force of the blow over the area of the metal, thus reducing the high spot. When a crown or curved surface is formed in a metal panel, it becomes strong in that it resists any change to its shape. The strength of this crowned surface can be used to support the surface being hammered without the use of a dolly. This type of hammering is called spring hammering, and can be used to correct high spots on metal panel surfaces (Figure 13.37). To take advantage of a great amount of the natural support provided by the crown of the metal, the force of the hammer blow is spread over a larger area. Once the metal is back to its original contour, additional hammering will cause the surface to sink below its original contour line, and it may not be possible to raise it readily. Always start with light blows, and as the repair nears completion, inspect the work after

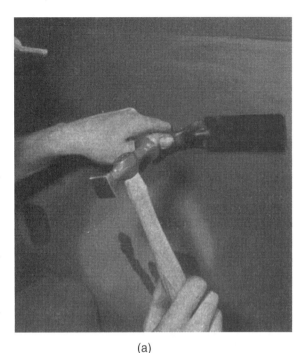

(a)

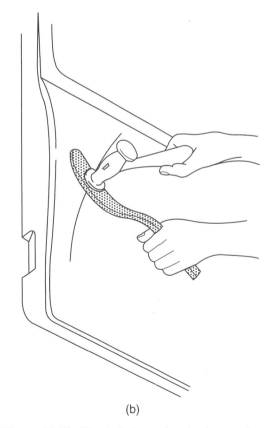

(b)

Figure 13.37 The technique of spring hammering (*Sykes-Pickavant Ltd*)

each blow. This will reduce the possibility of sinking the surface too low. Keep the surface of the hammer face clean and highly polished. Any marks on the surface of the hammer will be transferred to the surface of the metal and create additional work.

13.6.5 Removing low spots

Low spots can be removed in several ways, the two most common being the use of a pick hammer or a dolly block. When using the dolly block, start by holding it so that it can strike the underneath of the low spot on the panel with one of its rounded corners. It must be noted that if the operator does not hit exactly in the centre of the low spot, he will raise metal in some unwanted place. Accuracy is therefore essential, and can be achieved by holding a finger in the low spot and lightly tapping the underside of the panel with the rounded corner of the dolly until you feel that it is exactly beneath your finger, then strike a sharp blow and raise the metal at this point. After each low spot has been raised in this manner, these points can be filed to check that they are level with the surrounding panel surface.

The second common method of raising low spots is by pick hammering. Bringing up low spots with a pick hammer is more difficult than by the use of a rounded corner of a dolly block. With a pick hammer more accurate placing of the blow is required. Likewise greater control over the force of the blow is necessary. Start using the pick hammer in a manner similar to the dolly block. Hold the end of your finger in the low spot. Tap the under surface of the panel until the pick is directly below your finger. Then strike a light blow from beneath the panel, of sufficient strength to form a pimple in the low spot (Figure 13.38). Care must be taken to avoid overstretching the metal by using too hard a blow. These pimples represent stretched metal, but in being formed also raise the surrounding metal. When all the low spots have been raised with a pick hammer in this manner, the pimples can then be lightly hammered level by direct hammering, and finished by filing.

13.7 Filing

Filing is one of the most important aspects of finishing a body panel. It is carried out using an adjustable file holder, fitted with flexible blades which can be adjusted concave or convex to suit most contours on the average vehicle body. Initially the file was used for smoothing off panels prior to sanding and locating high and low spots. With the introduction of body solder and later metal and plastic fillers, filing took on an even greater importance in the finishing of repairs on body panels. Filing indicates any irregularities in the repaired surface of a panel, and is carried out as the panel is planished. First of all fasten the correct file blade to the file holder with the cutting edges of the teeth facing away from the handle or operator. Adjust the contour of the file holder so that it is almost, but not quite, matching the contour of the surface on which you intend to work.

One hand is used to hold the file handle, while the other grasps the knob at the opposite end. The file should be applied with long, straight strokes, pushing it away from you along the length of the panel (Figure 13.39). Short, jabbing strokes should never be used, as these will only scratch the panel and will not indicate low spots. If the file digs in, too much pressure is being applied and hence a need for reduction is essential. At the end of the first stroke, raise the file and, without dragging it over the metal, bring it back to the starting position and make a second stroke. Repeat this procedure until the area has been covered, making the file marks parallel to one another. This

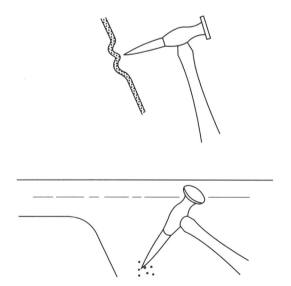

Figure 13.38 Pick hammering used to remove low spots

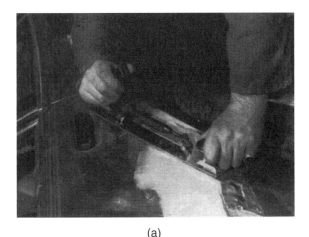

(a)

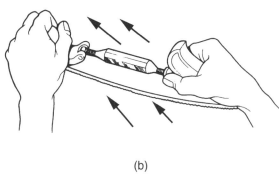

(b)

Figure 13.39 The technique of filing
(*Sykes-Pickavant Ltd*)

is termed *line filing*, and indicates the levelness of the panel in the direction in which it has been filed. At this point both the high and low areas will show up. The high spots can be corrected by spring hammering and the low spots by direct hammering, pick hammering, or in some cases by using the corner of a dolly block. Line filing indicates curvature in one direction only, and as most panels are double curved the panel surface must be cross filed to give an accurate contour check. *Cross filing* means a change in the direction of the file strokes so that the file is moved at an angle between 45° and 90° over the previous file strokes, thus checking the accuracy of the curvature in that direction.

After filing, and prior to refinishing the panel, the damaged area is sanded using a fine-grit sanding disc which leaves a smooth, even surface ideally suited for painting.

13.8 Sanding or disc grinding

13.8.1 Use of grinder

Several general rules govern the use of the disc grinder. If these are observed they will enable the operator to become proficient very quickly in the use of the grinder. The rules are considered good shop practice and are directed towards the safety of the operator. In the first instance, if the device is electrically operated see that it is properly connected and earthed. Shop floors are usually of cement; they are generally moist and, therefore, relatively good conductors of electricity. If the grinder is not properly earthed it is possible to receive a fatal electric shock when the machine is in use. Always wear goggles to protect the eyes from flying particles of metal and from small abrasive particles that come loose from the grinding disc (Figure 13.40). Always replace worn discs as soon as a tear is noticed; torn discs may catch in the work and twist the grinder out of your hands.

Figure 13.40 Ear and eye protection when sanding
(*Motor Insurance Repair Research Centre*)

Always maintain a balanced position when using the grinder. This position not only permits perfect control over the machine at all times, but it will also produce less fatigue over longer periods. When operating the grinder, hold it as flat as possible without permitting the centre connecting bolt to come in contact with the surface being ground. Hold the grinder so that only 40–65 mm of the outer edge of the disc is in contact with the surface being ground. The grinder must never be tilted so that only the edge of the disc contacts the surface. Failure to observe this will cause gouges or deep scratches in the metal which will be hard to remove. Move the grinder from left to right, overlapping the previous stroke with each new stroke. Make the cutting lines as clean and straight as possible. Move the grinder in the same manner whether using it for the removal of paint, rough or finish grinding. For most grinding operations, finish grind in the longest direction possible on the repaired surface.

13.8.2 Sanding discs

The coated abrasive disc is the part of the sander that does the actual sanding, and selection of the right grit and coating for each job is important. There are five different minerals which are commonly used for manufacturing abrasives. These are garnet, flint, emery, aluminium oxide and silicon carbide. Aluminium oxide is the most important of these. Because of its toughness and durability it is used in the motor body repair trade, where it is chiefly used on metal. The abrasive is fixed on a backing which is either of paper, cloth or a combination of the two. For dry grinding or sanding, high-quality hide glues are used for anchoring the abrasive grains to the backing. For wet sanding, resins are used as the bonding agent.

Coated abrasives fall into two additional classifications based on how widely the minerals are spaced. If the minerals are close together the abrasive is close coated. When they are widely spaced it is open coated. In close-coated abrasive discs, the abrasive is applied in such quantity as to entirely cover the backing. In open-coated abrasive discs, the backing is from 50 to 70 per cent covered. This leaves wider spaces between the abrasive grains. The open coating provides increased pliability and good cutting speeds under light pressures. Open-coated discs are used where the surface being ground is of such a nature that closely spaced abrasive materials would rapidly fill up and become useless. When grinding a body panel, always use an open-coated disc up to the time that the area being sanded is completely free of paint. Then use a close-coated abrasive to grind the metal to the point where the surface needs no further correction. The final finish is accomplished with a fine grit to get the surface smooth enough for re-finishing by the painting department.

13.9 Hot shrinking

One of the most important skills in the repair of damaged panels is that of hot shrinking. It is important because in most cases of collision causing the damage of body panels, stretching of the metal takes place. The actual process is carried out by gathering the stretched metal into a common centre or area and then by heating this section. The panel steel is then at its best condition to be hammered down, thus reducing the surface area and so making shrinking possible (Figure 13.41).

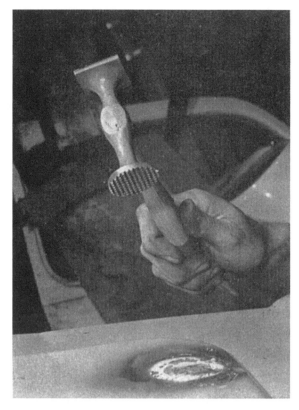

Figure 13.41 Hot shrinking panel (*Sykes-Pickavant*)

The oxy-acetylene flame is used as a means of heating the panel, and care must be taken to reduce the spread of heat to the surrounding area of the panel. This can be done by cooling the panel with water after every shrink. The advisable welding nozzle size when using for a 1.00 mm panel is a number 2 nozzle.

13.9.1 Tools required for hot shrinking

Before commencing to heat up the stretched section, it is essential that all tools and material needed to carry out the shrink should be conveniently placed so that they can be brought into use quickly. The tools required are as follows:

Wire brush The wire brush and scraper are used in preparing the panel prior to shrinking. With the scraper any anti-drum or underseal compound must be removed before applying the heat.

Mallet or shrinking hammer The mallet face has a soft surface much larger than that of a planishing hammer. If used together with a suitable dolly block it will bring the metal down to a level surface whilst hot, and also avoid stretching the panel as the mallet is made of wood and not metal like the panel hammer. Another tool that can be used is the shrinking hammer; this is similar to a planishing hammer but the faces have cross-milled serrations, which reduce the tendency for the hammer to stretch the metal because of the very small contact area between the points on the serrations and the panel.

Planishing hammer This should be used to complete the shrink after malleting, as the hammer is better for levelling out the surface of the panel than the mallet.

Dolly block This should not be fuller in shape than the actual panel being repaired. Also it should be of a rather light weight so that it forms a relatively weak backing for the malleting, thus reducing the stretching during the levelling operation. A grid dolly has a serrated face to reduce the possibility of stretching in the panel.

Damp cloth This is helpful in checking the spread of heat, thus reducing the risk of panel distortion.

13.9.2 Hot shrinking process

Arrange your tools so that they are within easy reach, as it is necessary to change quickly from one tool to another when performing a shrinking operation (Figure 13.42). Locate the highest point in the stretched section of the panel with which you are working. Light the torch and heat the spot to approximately 10 mm diameter in the centre of the high spot or stretched area to a cherry red, using a circular motion when heating the spot.

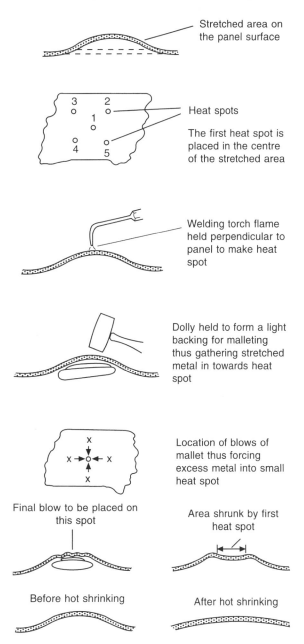

Figure 13.42   The technique of hot shrinking

Be careful not to burn through the metal by overheating. As soon as the spot is cherry red place the shrinking dolly (grid dolly), which is usually fairly flat, under the spot and strike several sharp blows with a mallet around the heat spot; this will force the surplus stretched metal into the heat spot. Then the blows are directed on the centre of the heat spot, which pushes the metal down while it is still red hot. At all times the dolly block must be held loosely against the underside of the panel, exactly centred under the heat spot. After four or five mallet blows, the last being on the centre of the heat spot, the heat spot will turn black, and should be quenched immediately with a water filled sponge. Repeat this operation taking the next highest spot in the stretched section of the panel, until the bulge or fullness is finally shrunk down below the level of the surrounding surface. It can then be brought up to its correct level by planishing with a hammer and dolly and then finishing with a panel file.

The following points should be noted:

1 Never quench a red hot heat spot; wait until the metal has turned black.
2 Never heat an area greater than that which can be hammered with mallet and dolly.
3 Never use anything but an oxy-acetylene welding torch for heating a stretched section.
4 Never attempt to shrink a panel until it has been roughed out.
5 Always hammer the stretched section outwards before applying heat.
6 If the stretched part of the surface is small, make a smaller heat spot.
7 It is possible to shrink metal without quenching each spot. However, the shrinking operation is much faster when each spot is quenched with water, as fewer heat spots are required if the heat expansion is drawn out by quenching than by additional spots.
8 In some panels it is possible to use a spoon for the backing tool for the hammering operation, especially on door panels or over inner constructions.
9 When performing a shrinking operation, take care to avoid overshrinking the panel. This will cause the metal to warp and buckle both in and out of the stretched area, owing to overheating of the stretched section. If this does arise, heat a small spot in the area where the panel is buckling, apply a dolly block or spoon with enough pressure to hold the buckling section up, then allow the metal to cool. Do not use the mallet or water in this case. In extreme cases like this it may be necessary to repeat this operation in several different places in the buckled section.

13.9.3 Shrinking aluminium

Sheet aluminium can, with the application of heat, be subjected to shrinking. The process is similar to that used for sheet steel except that the work should be carried out faster than when shrinking steel. This speed is essential because aluminium is a good heat conductor and the spread of heat must be prevented from distorting adjoining panels. Unlike steel, aluminium does not change colour when heated and, because of the melting temperature differences, great care must be taken not to melt holes in the panel while it is being heated.

13.9.4 Shrinking equipment

Shrinking can also be carried out using shrinking equipment, which can be either a specialized piece of equipment or an attachment to a MIG welder. Basically the equipment consists of a power source to which two cables are attached: one is to the shrinking torch fitted with a tungsten or carbon electrode (which is interchangeable), and the other is to the earth return clamp for completing the circuit.

To use the equipment, first position the earth clamp, making sure of a good connection. Switch on the power source, which in some cases has a built-in timer and then apply the tip of the electrode to the highest point on the surface of the stretched panel to be shrunk. The resulting arc will produce an extreme concentration of heat to the small area on the panel surface. Heat until bright red or to the preset time, then immediately cool with water. This high concentration of heat in one area together with rapid cooling is extremely effective in reducing high spots by heat shrinking in vehicle repair.

13.10 Cold shrinking

Cold shrinking is another method of repairing stretched sections on vehicle body panels. In this case a hammer and special dolly block are used in conjunction with one another. This dolly block is a shrinking dolly and is shaped like the toe dolly but has a groove running along the full length of its top face. The dolly block is placed under the high spot formed by the stretched metal of the damaged panel and the panel is hammered down into the groove with hammer or mallet to form a valley, care being taken not to make the valley any longer or deeper than is necessary to draw the stretched metal back to its original contour. The valley formed is then filled using body solder, which is filed to give a good finish. From the reverse side of the panel it will be seen that a rib has been formed, which will in effect give strength to the damaged area should this be required. This method is also most useful in reducing welds: the weld on the face of the panel is depressed into the channel of the dolly to form a valley below the line of the face of the panel, and the valley is then filled by the body soldering method.

13.11 Body soldering

Body soldering (or loading, filling or wiping, as it is sometimes termed) has become a general practice where owing to the structural design of the all-steel body, the use of normal panel beating methods of repair using hand tools is not possible. It is ideally used to hide a lapped joint in the construction of a body shell, and is very useful to the panel beater repairing a windscreen or door pillars, or when a dent or crease is backed by a bracket or is double skinned, which prevents the placing of a dolly behind the panel and beating the section out. Body solder was first used in the motor trade in the mid 1930s and was introduced into the repair industry by the car manufacturers who were using it on the mass production of car bodies. It was during this period that the all-steel body was developed, along with rapid advances in streamlining. To finish a panel in the area of the production welds, which in most cases were lapped, spot-welded joints, body solder was used and found most useful. This method has been continued, and today solder filling plays a major part in the finish and repair of the modern car body.

13.11.1 Composition and requirements of body solder

Body solder consists of 68.5 per cent lead, 1.5 per cent antimony and 30 per cent tin. The solder has to possess the following characteristics:

1 It must remain plastic over a large temperature range, so that it stays workable.
2 It must wipe with the solder stick and not crumble, thus providing a clean surface finish.

3 The lead and tin must not separate, as it is worked in vertical or overhead positions.
4 It must be capable of being reheated and reworked without forming hard spots.

13.11.2 Process of body soldering

During the repairs of body damage, some means of reproducing the normal contour of the damaged area is necessary when it cannot be restored by normal panel beating methods. This is a case where body solder can be used to advantage so the area can be filled and then dressed down to a smooth and perfect finish.

Body soldering requirements are soldering blocks, which are usually hardwood blocks shaped to suit the panel being soldered; a suitable grease or light oil which is rubbed on to the surface of the blocks to stop them sticking to the solder, resulting in a smoother surface finish (tallow is the optimum substance for this purpose); a sanding machine, files and emery cloth for cleaning purposes prior to tinning the metal for soldering; a welding torch or similar heating appliance which produces a low-temperature flame (if a welding torch is used the flame should be feathered, which means slightly carbonizing, resulting in a soft flame): a quantity of body solder and a suitable tinning paste complete with a tinning brush for ease of application; and a clean rag for rubbing off the tinning paste. The operator must also use a self-contained air-fed mask together with a fume extractor for his own protection and to conform with health and safety regulations. Before commencing work make sure that the appropriate tools and other materials are close to the job and within easy reach, thus avoiding a delay during working operations.

The first step in body soldering is to clean the surface to be soldered to a bright metallic finish. This can be done by using a sanding machine with the right type of sanding disc until the area to be soldered is cleaned to the bare metal. Any small particles of paint which the sander will not remove can be cleaned off by using emery paper and a file; this precleaning is very important, as the tinning paste will not tin the surface unless it is perfectly clean and free from paint. Apply the tinning compound by brushing it over the cleaned section so that an area slightly larger than that to be soldered is covered with tinning paste. Using the welding torch with the flame set in the carbonizing condition, which is slightly feathered and gives a very soft flame, heat the tinning compound until it becomes fluid. Then with a clean rag wipe the tinned area to spread the tinning over the cleaned surface, making sure that every part to be soldered is perfectly clean and completely tinned. An important point at this stage is not to overheat the tinning paste when tinning or the surface will turn discoloured, usually blue, and the tinning will be burnt; solder will not then adhere to this surface unless it is recleaned and tinned again.

Hold the welding torch in one hand and a stick of body solder in the other and play the flame over the tinned section, heating it just sufficiently to cause the tinning to begin to flow. Whilst doing this, hold the stick of body solder near the work and apply the flame over both the tinned area and the solder stick so that the melting stage of each coincides. When the end of the solder stick begins to melt, press it against the tinned section, thus causing a quantity of body solder to adhere to the tinned surface of the work. After sufficient solder has been deposited on the surface of the work, select a suitably shaped solder block, which should have previously been greased or dipped in oil or tallow, and commence to push the body solder over the damaged section of the panel. From time to time the flame should be played over the solder to keep it in a plastic or movable state; then, using the solder block, the solder should be moulded to the general contour of the panel. The solder blocks must be kept continually coated with oil or grease to allow them to slide over the surface of the solder without sticking and picking up pieces of solder whilst it is in the plastic state. This coating also produces a very smooth surface on the face of the area being soldered. It is very important to make sure that the tinned surface is heated to the melting point as the solder is smeared across the area being built up; unless this is done a poor bonding between the solder and the panel will be the result and the solder may fall out.

After the required shape has been formed and the solder built up to a level slightly above the existing panel, the final finish is gained by filing the body solder with a flexible panel file, being careful to ensure that the level does not fall below that of the surrounding area. After filing, the solder can then be rubbed down with emery paper to give a finer surface finish for painting. The sander should never be used for dressing down the solder except as a last resort, because it is too severe and tends to cut deeply and unevenly into the solder. The dust given off from the sander when using body solder is also injurious. When the final shaping and smoothing of the loaded area is complete it is essential to remove all traces of soldering flux, oils and grease which may have been used during the loading operation. If these were not removed they would have a harmful effect when the section is finally spray painted.

Figure 13.43a–f illustrates the stages of body soldering.

Safety points

1 Avoid skin contact with fluxes.
2 Use applicators of some form when necessary.
3 Do not inhale fumes from heat/flux application.
4 The use of sanding machine in finishing produces injurious lead dust.
5 Health and safety regulations demand protection for the operator in the form of a self-contained air-fed mask.
6 A fume extractor reduces the risk of adjacent working areas being contaminated and protects the operator.

13.12 Chemically hardening fillers (plastic fillers)

Although body soldering still provides the best quality of filling for a repair, there are cases where an alternative method is required. Much research on this matter has led to the development of plastic fillers which would come up to the standard of body solder. These fillers are used in the body repair trade as an alternative method to body soldering. They are based on the polyester group of thermo-setting resins, and require a catalyst or activator to cure them. Therefore the fillers obtainable are of a duo-pack type containing paste and hardener. When they are mixed together a catalytic action takes place, resulting in the filler hardening very quickly. As this plastic filler does not require heat during its application it has certain clear advantages. Its use eliminates fire risk, especially when filling next to petrol tanks or any inflammable material. It also eliminates the problem of heat

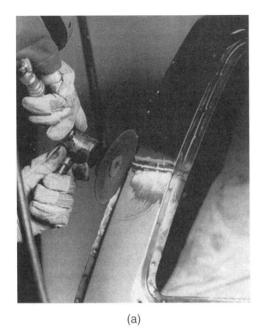

(a)

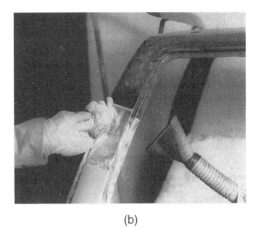

(b)

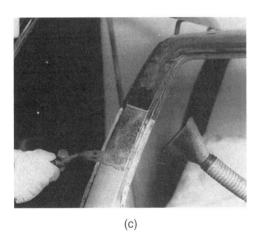

(c)

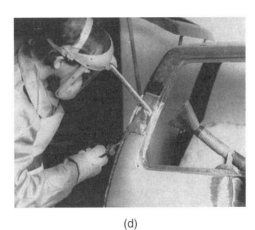

(d)

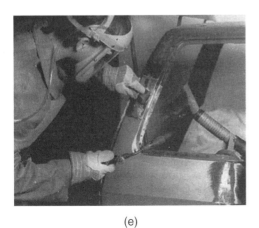

(e)

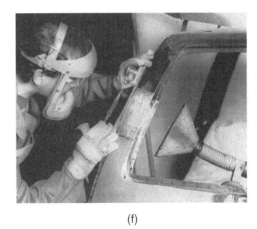

(f)

Figure 13.43 Body soldering: (a) panel preparation (b) applying the tinning paste (c) heating the tinning paste ready for soldering (d) applying the solder (e) forming the solder to shape (f) dressing and filing the finished solder (*Motor Insurance Repair Research Centre*)

distortion which can occur when solder filling flat body panels. It is cheaper than body solder and much easier to apply. The first of these fillers developed were released some years ago and many disappointments were associated with their use; problems encountered were poor bonding to panels, too much delay in hardening and too hard to file when dry. Several fillers now available have been vastly improved; they will harden in as short a period as twelve minutes provided the quantities are correct and the two chemicals thoroughly mixed. When hard these fillers have excellent bonding qualities and will feather out to a fine smooth edge. Filler materials can be dispensed either from wall mounted, air operated filler dispensers or from portable dispensers. These dispensers take a 10 kg tin of polyester filler together with a matching cartridge of benzoyl peroxide hardener (Figure 13.44).

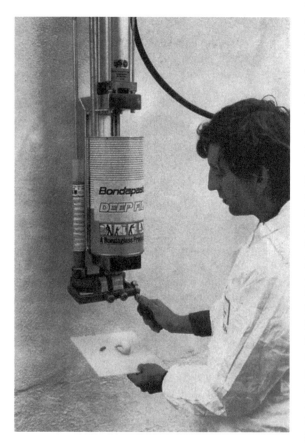

Figure 13.44 Wall mounted, air operated, filler dispenser (*Bondaglass-Voss Ltd*)

Portable dispensers offer the user the advantage of being able to take the filler from repair to repair or from workshop to paint shop. In some bodyshops it is the normal practice to use a general-purpose filler for the main filling operation; then once shaped and flatted, it is surfaced with a fine filler (30 per cent unsaturated polyester resin plus 70 per cent inert filling material) to bring the repair up to prepaint condition. Alternatively a body shop may require different types of filler such as a general-purpose filler, a glass fibre reinforced filler (with added fibreglass strands, which provide great strength for bridging holes and strengthening weakened areas in metal and GRP), a fine surfacing filler (a filler with very fine surface finish to fill fine pinholes, shallow scratches, file marks or sanding marks) and, where repairs are being made to galvanized panel surfaces, a filler suitable for zinc (a filler possessing high adhesion to coated steel surfaces). Aluminium metallic fillers are body fillers containing aluminium. The aluminium content makes them particularly easy to mix and apply. Unless the body shop is very large it would not be cost effective to have many wall mounted dispensers, and the portable dispenser offers a viable alternative.

The portable dispenser is operated by pushing the handle down; the filler is then extruded together with the correct amount of hardener for that quantity of filler (Figure 13.45). With most fillers this is normally 2 per cent by weight. Using a dispenser ensures that the correct proportion is added, making it impossible to over-catalyse the filler, and thus eliminating or very much reducing the possibility of spoilt paintwork. With over-catalysed filler there is a risk, particularly with metallic paints, of the repaired area showing through owing to the action of the peroxide in the hardener bleaching the paint. Another major advantage of using filler dispensers is that the filler is kept clean and uncontaminated from the beginning to the end of the tin because there is no need to replace the lid every time the filler is used. This is a feat which is commonly accepted as impossible in most body shops owing to the rim of the tin becoming encrusted with filler as the applicator is cleaned off, or simply owing to forgetfulness. In any event, even with the lid being replaced every time the filler is used, there is exposure to air, and therefore a tendency for the filler to lose its styrene content and thus to become stiffer and less spreadable.

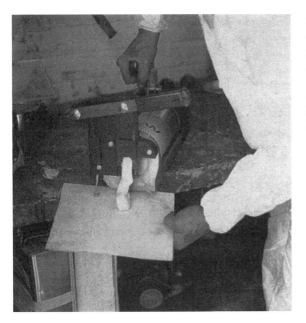

area has been patched, by hammering the patch down below the original level of the panel. The area is then ground using a sander to remove all rust and paint present. A coarse-grained sanding disc should be used for this operation as this will provide better adhesion for the plastic filler. Next mix the paste with its hardener on a flat surface using a stopping knife or flexible spatula, making sure that the materials are mixed in accordance with the maker's instructions (Figure 13.46). The filler can then be applied to the damaged area after making sure that the surface is absolutely clean and free from any trace of oil (Figure 13.47). If the area to be filled up is deep, make several applications, allowing each layer to dry before adding more filler. After the filler is applied and allowed to set, it can be shaped to the contour of the panel using a plastic filler file with an abrasive paper backing, or sanded using a sander (Figure 13.48).

Figure 13.45 A portable dispenser extruding correct amounts of filler to hardener (*Bondaglass-Voss Ltd*)

The technique of repair using this plastic filler is carried out by first roughing out the damaged area to as near the original shape as possible, or, if the

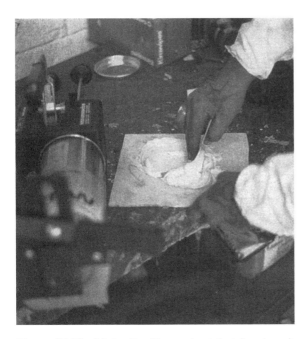

Figure 13.46 Mixing the filler and catalyst (hardener) on a non-porous surface (*Bondaglass-Voss Ltd*)

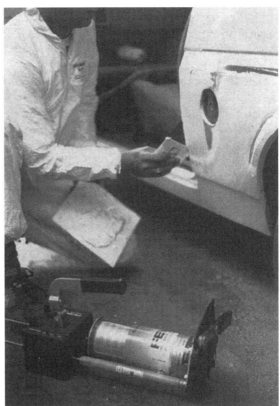

Figure 13.47 Applying filler to damaged area (*Bondaglass-Voss Ltd*)

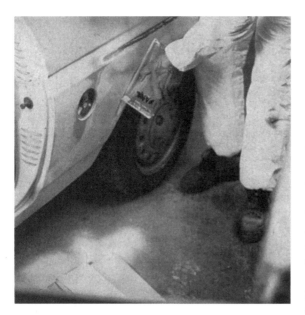

Figure 13.48 Shaping panel contour with abrasive paper and backing block (*Bondaglass-Voss Ltd*)

Plastic filler should never be used over rusted areas without patching the area with new metal so that the filler is applied to a solid base. It should not be used in areas on a panel or body surface which are continually in direct contact with water. Never drill holes in a panel to give the filler a better grip, as the holes will allow water or dampness to work in between the filler and the metal surface, leading eventually to the separation of filler from the metal surface. The filler should not be applied in any great thickness, especially where excessive vibration occurs, or the surface will crack and fall out. A good standard of finish can be achieved using these plastic fillers if the correct working procedure is adopted.

13.13 Body jack (hydraulic)

13.13.1 Development and principles

The evolution in motor car bodywork design and construction has called for many changed methods of handling repairs. This in turn has demanded an increasing amount of repair equipment to augment the traditional hand tools and equipment of the body builder, panel beater and sheet metal worker. Modern equipment has been made necessary by the chassisless construction of modern mass produced cars and vans, which require careful alignment of the complete structure following any serious impact in addition to panel damage rectification. An essential piece of equipment for repair work is the hydraulic body jack, which is used to push and pull body shells and component body parts into alignment following an accident. Blackhawk Automotive Ltd markets a hydraulic body jack under the name of Porto Power. A range of kits of various sizes is available to suit the needs and capital expenditure of various repair shops. Kits are made up in carrying cases (Figure 13.49), on wall-board storage, and on trolleys incorporating a small press or in the form of a bench rack (Figure 13.50). Attachments and fittings have been developed to use with the body jack equipment.

The use of hydraulic body jacking equipment is not new, nor was it invented suddenly. It has developed to its present sophisticated state over many years. At one time the only type of jack available was the ratchet or screw-type body jack. The use of hydraulic body jacking equipment developed from the use of an ordinary hydraulic jack for this purpose. The hydraulic hand jack had all the advantages of providing tons of closely controlled torque-free power for the minimum of effort by the operator. It soon became apparent that the hydraulic jack was ideally suited to repair work because it could be operated in any plane and controlled from outside the car. In the essential repair of collision work, a large percentage of the work will require the use of the body jack to push or pull large areas or sections back to, approximately, their original positions. Hydraulic body jacks can be extended to any desired length by incorporating a number of attachments which are available for pushing or pulling.

The outer skin, or panels, of a body is made from light-gauge metal, placed over a framework of heavier, stiffened metal which is reinforced with various types of supports and braces. In addition to damage as a result of a collision to the outer panels of the body, the inner construction which is attached to the outer panels also becomes damaged, which means that the surface of the outer panel is prevented from being restored to its original contour; hence the inner construction must be restored to its original shape and position either before or at the same time as the outer panel metal is corrected. In some instances it will be found that

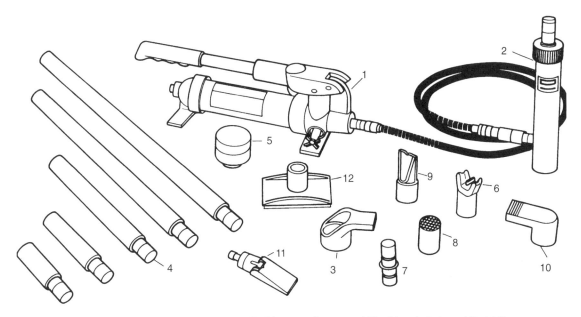

Figure 13.49 Snap 4 tonne standard set supplied for carrying case (*Blackhawk Automobile Ltd*)

1 Manual pump with hose	5 Snap flex head	9 Wedge head
2 4-T Bantam hydraulic hose, 18 m	6 90° V base	10 Plunger toe
3 4-T ram toe	7 Tube connector	11 Spread ram
4 Extension tubes	8 Serrated saddle	12 Flat base

once the inner construction has been restored to its normal position, the outer one will have been corrected at the same time. Before any correction can be made of such damage it is necessary to restore this inner construction. This is generally done by applying pressure to the damaged member or members. Where sharp kinks or creases have been formed at any point in the inner construction, it may be advisable to use heat while the pressure is being applied, but only on low-carbon steel and not on high-strength steel. This permits the metal to return to its original shape with little danger of cracking.

In using the body jack, it is important to understand that pressure is being applied at both ends of the jack simultaneously; therefore there could be a danger of distorting adjacent undamaged panel assemblies during the jacking process if the pushing points are not carefully selected. Such pressure can if necessary be applied either locally or spread over larger areas by the introduction and use of pressure pads, which are usually hard wood blocks. The body jack is also useful for providing support

or pressure at otherwise inaccessible portions of the outer panels, as well as applying controlled pressure in a higher degree than is possible with hand tools on the various panels.

13.13.2 Basic equipment

A body jack consists of three basic units (Figure 13.51): a pump, a flexible hose connecting pump to ram, and the ram unit. The pump comprises reservoir, pump handle and hose, and is controlled by a simple open and close release valve. The handle can be screwed into the pump in two different positions for ease of operation. The hose is connected to the ram by a simple quick-release coupler which needs to be only finger tight. Pressure is applied by closing the pump release valve and operating the pump handle. The pump will build up sufficient pressure only to overcome the external resistance against the ram. The need to apply excessive pressure to the pump handle indicates that the ram has reached the limit of its movement. The ram is designed with a snap-on

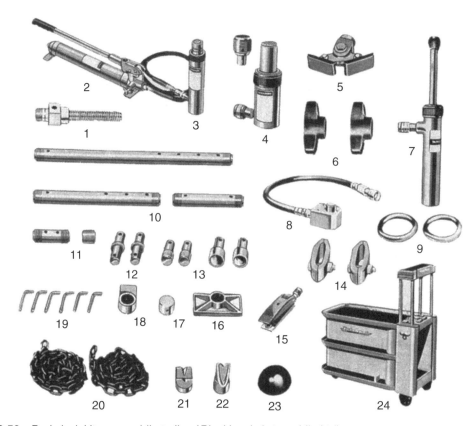

Figure 13.50 Body jack kit on a mobile trolley (*Blackhawk Automobile Ltd*)

1	Slip lock extension	9	Pull rings	17	Serrated saddle
2	Pump	10	Lock-on extension tubes	18	Spreader plunger toe
3	10 tonne ram	11	Male connector	19	Clamp toe lock pins
4	10 tonne short ram	12	Lock-on tube connector	20	Pull chain
5	Wide-angle wedge head	13	Lock-on female and male connectors	21	90° 'V' base
6	Pull plates	14	Pull clamp	22	Wedge head
7	Pull ram	15	Wedgie ram	23	125 mm flex head
8	Midget ram	16	Flat base	24	Trolley and press stand

system to enable extension tubes and attachments to be positioned to harness the hydraulic power for any desired type of application.

All collision damage repair work which makes use of the body jack equipment is carried out by using one or another of a number of simple set-ups or, in the case of more complicated repair, a combination of set-ups (Figure 13.52). The first important step therefore is to understand the set-ups that can be built with this equipment, the attachments required and their application; then it is a question of breaking down a job into its basic set-ups and applying the corrective force in the correct sequence. The corrective force should be applied as near as possible in the direction opposite to the force which caused the damage. The body jack set-up should be applied so as not to push at the deepest point on the damaged section; instead work round the outer edges in ever-decreasing circles which will tend to spring the remaining damage into the final position. If the body jack set-up is applied directly against the lowest part of a damaged section without relieving the strain, then as the pressure is applied the metal surface may become kinked and stretched and require further attention to return it to its correct level.

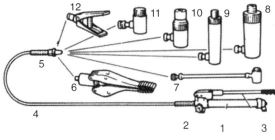

1	Hydraulic pump	2	Release valve	3	Pump handle
4	High-pressure hose	5	Speed-coupler	6	Spread ram
7	4-ton midget ram	8	20-ton ram	9	10-ton ram
10	10-ton hollow ram	11	7-ton shorty ram	12	Wedgie

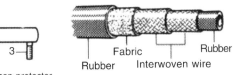

1 Thread cap protector
2 Protector ring
3 Spring protector

A section of the
high-pressure hose

Figure 13.51 Hydraulic body jack (*Blackhawk Automotive Ltd*)

13.14 Application of the body jack

13.14.1 Pushing

Pushing (Figure 13.53) is the simplest operation of all and is achieved by inserting the ram between two points and operating the pump. The plunger extends until it touches the point at which the load is to be applied, and as pumping is continued pressure is built up to overcome the resistance of the metal at the point of application. Movement of the damaged area will take place as long as pumping is continued.

Care must be taken when selecting a ram anchor; for example, if a ram was placed between two chassis members and it was intended to push from the undamaged member to straighten the damaged member, the force applied would not rectify the damage but would distort the undamaged member! The first essential is to ensure that the pushing anchor point is stronger than the point which is receiving the corrective force. This can be done by attaching a base plate to the bottom

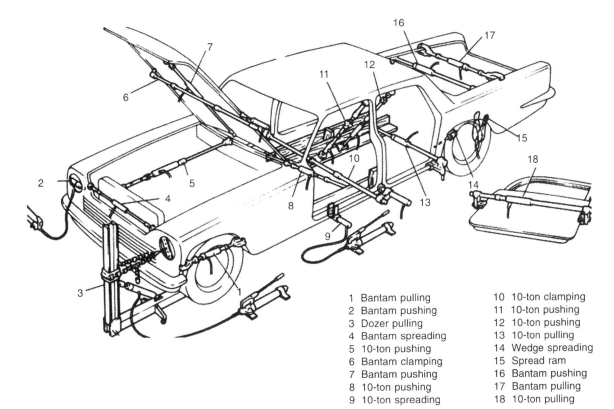

1	Bantam pulling	10	10-ton clamping
2	Bantam pushing	11	10-ton pushing
3	Dozer pulling	12	10-ton pushing
4	Bantam spreading	13	10-ton pulling
5	10-ton pushing	14	Wedge spreading
6	Bantam clamping	15	Spread ram
7	Bantam pushing	16	Bantam pushing
8	10-ton pushing	17	Bantam pulling
9	10-ton spreading	18	10-ton pulling

Figure 13.52 Body jack combinations (*Blackhawk Automotive Ltd*)

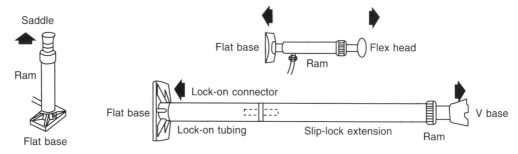

Figure 13.53 Pushing with body jack (*Blackhawk Automotive Ltd*)

end of the ram to spread or distribute the load, which can be spread over an even larger area by putting a piece of solid hardwood timber between the base plate and the pushing point. It is seldom that a pushing application can be achieved using the ram only, because of its limited travel; therefore there are available various combinations of extension tubes, couplings and pusher heads to cater for any repair requiring straightforward push.

13.14.2 Pulling

Pulling (Figure 13.54) is also a simple operation but uses a slightly different type of ram. With the standard equipment a pull converter set is available to enable the one push ram to fulfil both functions, but it gives an off-centre pull and is less convenient to use than the separate pull ram. With the direct set-ups using the pull rams, the tubes are under tension and there is no risk of bending. The thread sections of tubes and couplings are now under tension and it is the threads themselves that carry the load, so it is essential to ensure that they are kept clean, free from dirt and damage and are securely mated. The pulling combination obtainable makes use of the pull rams and direct pulling attachments, though it is also possible to use chain plates and chains for obtaining a pull with a push ram. The latter method is quite convenient in such cases as pulling across the width of a body from an undamaged door pillar which has been bowed outwards. It would be necessary to reinforce the undamaged pillar with timber to prevent distortion and to protect both pillars from marking by the chains with suitable packing material.

13.14.3 Spreading

Spreading (Figure 13.55) is a similar application to pushing except that in the latter case there is sufficient room for access for the ram and extension between the two members to be moved apart to

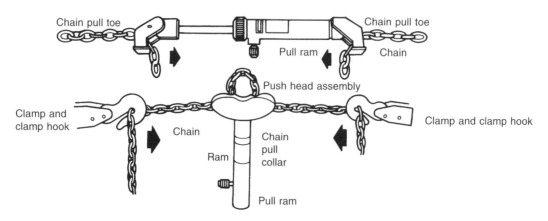

Figure 13.54 Pulling with body jack (*Blackhawk Automotive Ltd*)

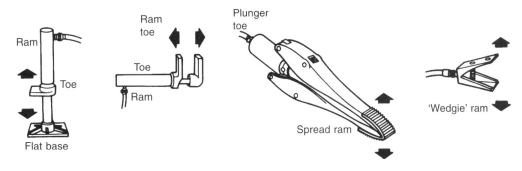

Figure 13.55 Spreading with body jack (*Blackhawk Automotive Ltd*)

permit a direct push. When this is not possible a means must be found of inserting jaws or attachments which are capable of applying an indirect thrust. The most obvious means of spreading is provided by the wedgies and spread rams. This is an off-centre load once more, and even under the most favourable conditions it is not possible to apply a force of more than about 7 tonnes with the 10 tonne ram, or 2.5 tonnes with the 4 tonne ram.

13.14.4 Stretching or tensioning

The technique of stretching or tensioning (Figure 13.56) is another means of obtaining a pull using a push ram. It is different from the type of pulling previously described, which was a method of applying force to pull towards one another, sections which have been forced apart. Here the reverse takes place and an external pull is applied to pull apart or draw outwards areas that have been pushed or drawn in towards each other. The combination used to obtain this external pull employs toes, links and clamps on the end of the push ram and extension tubes. Two types of clamps are used, the pull clamps for attachment to flat edges and the wing clamps which have deep throats or lipped edges. Both clamps have an alligator type of action

and are first tightened down on the centre bolt until the jaws are parallel and in contact with the surface of the panel to be gripped. Pressure is then applied by tightening the rear bolt, which cants the jaws forward and causes them to bite into the surface of the metal. It may well be that the time taken to set up this combination is greater than the time to pull out the damage once it is in position. On flat panels where it is not possible to get at the edges or where there is not an edge to locate a clamp, the same result can be achieved by locating the toes in the bosses of a pair of solder plates which have been sweated on to the panel. A roof panel or car door is a typical example of this. To get at both edges of a car door it is usually necessary to remove the door, but this can be eliminated by using a clamp at the free edge of the door and a solder plate at the hinged edge. If properly used the solder plates will withstand a pull of 13 800 kN/m^2. The plates should be sweated on with a layer of body solder about 3 mm thick, using the minimum of heat, and quenching with a wet rag to prevent distortion. The same technique may be applied to boot lids, front wings, bonnets and rear quarter panels, and can be the means of repairing a panel in a position where direct pressure from inside would not have achieved a satisfactory result.

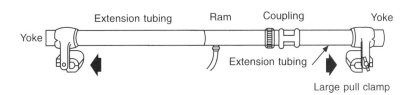

Figure 13.56 Stretching with body jack (*Blackhawk Automotive Ltd*)

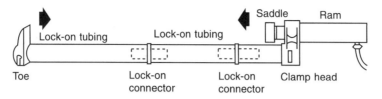

Figure 13.57   Clamping with body jack (*Blackhawk Automotive Ltd*)

Another combination, which is less popular in use, is clamping (Figure 13.57).

13.14.5 Example of use of body jack

Figure 13.58 demonstrates the jack being used diagonally to rectify a door opening. Pressure is applied until the clearance round all sides of the door is equal and it opens and closes freely. Note the use of a pull ram fitted with chains and connected to a swivel clamp at the upper corner and a pull ring and clamp at the lower corner.

Figure 13.59 shows rectification to a door opening using a push ram connected at both ends to push-pull clamps, which are bolted on to the door flange edges. The doors are left suspended on their hinges to act as templates during the operation.

Figure 13.60 shows a twin-linked aperture restraint holding the door opening in shape while pulling or pushing takes place, so that the door aperture does not go out of alignment as the repairs are carried out. The centre screw jack allows preloading to put the restraining unit in tension. The door may be left in its correct position and closed with the restraint in place.

Where the rear end of a car becomes accidentally pushed in, the panels can be returned to their correct alignment in the manner illustrated in Figure 13.61. The jack is placed across the aperture of the boot lid, with a wedge head attached to

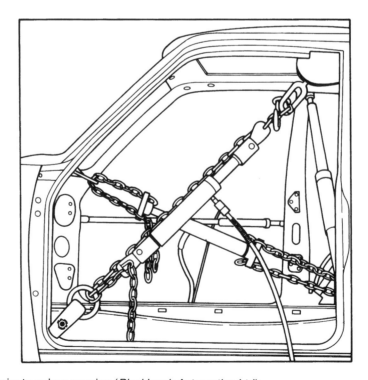

Figure 13.58   Repairs to a door opening (*Blackhawk Automotive Ltd*)

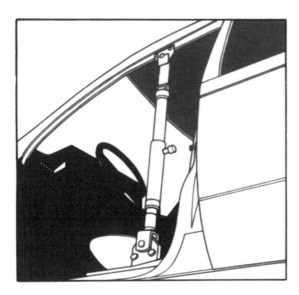

Figure 13.59 Repairs to a door opening using push ram (*Blackhawk Automotive Ltd*)

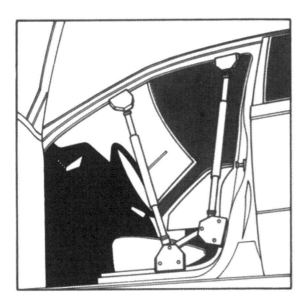

Figure 13.60 Repairs to a door opening using aperture restraint (*Blackhawk Automotive Ltd*)

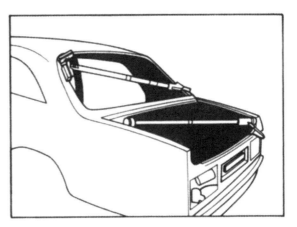

Figure 13.61 Repairs to boot lid aperture and rear windscreen (*Blackhawk Automotive Ltd*)

one end and the rubber flex head to the other. This is to spread the force of the pressure evenly over a larger area. The boot lid, when repaired or replaced, is fitted in position and will act as a guide for alignment when pushing out the surround panel.

A rear windscreen opening can be restored to its original shape by placing the body jack diago-nally across the corners which are out of square, as shown in Figure 13.61. Where the standard rubber flex head does not suit the shape of the body, the wide-angled wedge head must be used, as it will automatically adjust to fit the corner. The rear wind-screen glass may be used as a template, butthis must be handled with great care in case of breakage.

Figure 13.62 shows a rear end collision and the body jack being used to straighten the sub-frame and wheel arch sections of the underbody while also correcting the tension on the D-post, thus allowing the rear door to open.

13.15 Care and maintenance of body jack

As with all hydraulic equipment, little trouble is experienced with the working of the jack provided the unit is kept free from oil leaks. When topping up with oil it is necessary to use the correct type of oil, taking care not to allow any dirt or grit to enter the oil track while adding or checking the oil level. Air sometimes becomes trapped in the oil track, in which case it is necessary to bleed the pump.

13.15.1 Bleeding the body jack

Clamp the pump in a vice. Close the release valve and operate the pump handle until the ram plunger is fully extended. If the plunger will not move by pumping, withdraw it by hand. Remove the filler plug from the end of the pump and release the valve. Place the plunger on the floor and slowly push down until it collapses, expelling all the air.

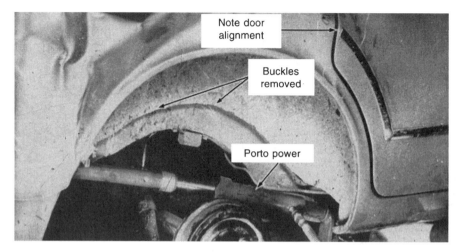

Figure 13.62 Repairs to a wheel arch and subframe (*Blackhawk Automotive Ltd*)

Pump the handle rapidly then close the release valve and replace the filler plug. The unit is now ready for operating.

13.15.2 Care of the hose

On remote control jacks, where the hydraulic pressure is supplied to the ram through a hose, it is well to exercise care so that the hose does not become damaged. The hose is made from oil-proof rubber reinforced by woven steel wire which is covered on the outside by a fabric and rubber combination (Figure 13.51). Do not permit heavy objects to fall or drop on the hose, as a sharp, hard impact may kink the wire strands in the hose. Because of the rubber covering, the kink may not be noticeable, and the application of pressure will eventually cause the strands to break and the hose will leak. In making set-ups with the jack, always be careful to anchor the ram unit so that its pushing force will not tend to bend or break the hose fittings.

13.15.3 Care of threads on ram and attachments

When the ram is not in use, attachments provided for protection of the plunger thread and ram body should be in place. Use all of the threads to make connections, and always turn the attachments until they are tight. Always keep the threads in all attachments clean and free from grease. Whenever

threads become bent or damaged, they should be repaired so that the proper fit can be obtained when connections are made. Most attachments are now snap-on connections and therefore have no threads and need no maintenance.

13.16 Repair of component motor body panels

To repair damaged motor body panels requires great skill in the use of hand tools and repair techniques, and an ability to assess the cause, extent and sequence of the damage sustained. Damage by accident and its subsequent repair covers a wide range of incidents from minor scratches and cosmetic damage to the write-off. The repair will only be approved by an insurance company after considering the car's age, condition and relevant market value.

Vehicle body repair work can be divided basically into two groups: minor accident repair work and major accident repair work. Body repair workshops vary in their opinions as to the dividing line between the two assessments. Generally they can be defined as follows.

Minor accident repair work

To be able to carry out minor repairs, a good level of ability with hand tools and all types of power tools, both electrical and air operated, is essential to the body repair worker. He or she needs a knowledge of body construction, an appreciation of materials and their properties, and a practical expertise in appropriate repair techniques, including the use of all types of welding equipment.

Minor repair work can be classified as: the simple cosmetic repair requiring the use of hand tools only; the cutting out of damaged panel sections and their effective replacement; the replacement of complete panels such as wings, bonnets, boot lids, doors and sill panels; and the finishing of these panels to a point where paint can be applied. A minor repair can also involve adjustment to door locks and window mechanisms, the restoration of body apertures, and the use of hydraulic equipment to realign body panels and apertures to their original shape.

Major accident repair work

Major repair work can be classified as a repair which involves the use of manufacturer's replacement parts, and the reconstruction of the body or parts of the body on an approved body jig. A major repair will normally involve all the features listed for minor repairs. In addition it will include: the rectification of any misalignment of the underbody and body shell using conventional hydraulic pulling equipment, together with either a fixed bracket jig or a universal jig with measuring system; the realignment of the body shell, correcting damage with a combination of pulling and pushing equipment; and the cutting out of panels and reinforcing members which are damaged beyond repair, and the welding in of replacements. Major repairs are dealt with in Chapter 14.

The conclusion of all repairs must restore the vehicle to its original safe, roadworthy condition to the satisfaction of the owner and the insurance company. In order to achieve this the repairer needs to use his acquired skill and knowledge and, whilst each accident repair has its own individual features, he or she must be able to assess the damage and make the choice of appropriate methods of repair and the best equipment to effect a speedy first-class result.

13.16.1 Repairing rust damage

In this work the body repairer relies on his own judgement, skill and experience. Rust is a corrosion, known chemically as iron oxide, which occurs on the surface of iron and most of its alloys when they are exposed to air and moisture. The designs of all-steel

body shells over the past years have provided many pockets which could be termed water or moisture traps, thus creating suitable conditions for the rusting process. In particular, rust attacks wings, sill panels, wheel arches, floors of luggage compartments and bottoms of doors, the worst being the sills because of their close proximity to the road. During the past few years some manufacturers have made improvements which have reduced the rusting of bodies. In some cases this has been achieved by good design and in other cases by treating the metal prior to painting.

When the body repair worker is faced with rust problems, he has a choice of two methods of repair, depending upon the extent of the damage. The best possible results are obtained by cutting away the corroded panel or section and replacing it with a hand-formed section or a factory-pressed panel. The replacement panels can then be fitted by welding them into position and finished by planishing, filing and sanding. In the case of the older vehicle where new panels are not available or where the vehicle's age or condition make it uneconomical to fit new panels, the method of fabricating or patching these areas, by cutting out and replacing the sections from flat sheets fabricated by hand, is adopted. In many cases if the original sections are carefully cut out and not too distorted, they can be used as an excellent guide to the manufacture of the new sections. Templates can also be made up before the old sections are cut out to check alignment and curvature when fitting the new section. Often double-curvature panels such as wheel arch fabrications must be made up in three or four parts for ease of patching on to the original body. The main difficulty experienced in this type of repair is the welding of the new metal on to the old; this is one of the reasons why it is important to cut away all the rusted section so that the new metal can be welded to a rust-free section. In cases where the sections are very badly corroded and welding is difficult, the section can be cleaned of surface rust and sometimes brazed to form the joint.

The second method of rust repair, which is only a temporary measure and not recommended for longevity, is to fill the corroded section with either body solder or a chemically hardening filler. If the corrosion is only very slight, giving a pinhole effect, the area can be sanded down to bare metal, hammered down carefully to below its original level and then filled up with body solder and filed to a finish. Plastic filler should not be used in this case because the moisture would seep through, parting the filler from the panel. Where the corroded area has turned to holes it can only be repaired by placing a patch over the corrosion and welding or brazing it in place. This patch must be tapped down below the panel level, then filled with body solder or plastic filler to obtain the final finish. In some cases where it is difficult to weld a patch, the perforated section can be reinforced by using glass-fibre matting impregnated with resin and bonded to the underside of the repair. The surface can then be filled with plastic filler and finished by filing.

13.16.2 Repair of minor accident damage

Repairs arising from minor accidents are usually of a relatively straightforward nature, as the damage is either a dent, a scratch or a scruff of the outer panel surface of the body and does not always involve structural distortion.

When repairing it is essential to know the nature or properties of the sheet metal used in body panels. When a flat sheet of metal is bent to a wide radius it will regain its former shape when released; in other words, the metal is flexible. If this metal is bent to a small radius it exceeds the limits of its flexibility, and the metal in the bend becomes stiff and retains its curved shape. These are known as rigid sections. Therefore a body shell and panels are a combination of rigid and flexible areas. The rigid areas are such assemblies as door posts, sill members, roof cantrails and flanges around wing edges and bonnets. The flexible areas are large areas of mainly flat or slightly curved sections such as bonnet tops, outer door panels, sections of wings and centres of roof panels. In the main after a collision if the stresses are removed from the rigid areas first, then the flexible areas (except where badly creased in the impact) will regain their former shape. If a panel is damaged in an accident then the buckled area, being sharply bent, will create additional stiffness in the panel, whether in an elastic or a non-elastic area. The slopes of the buckles surrounding the sharp creases will be fairly elastic, but a greater amount of effort will be needed to reshape the sections of the panel which are made rigid either in manufacture or through accidental damage. The theory of repair is that areas which are elastic in manufacture, and remain so after damage, can generally be reshaped

by their characteristics once the rigid areas in the surrounding buckles are forced back into position.

In general the method of repair is to analyse the crash, establish the order in which damage occurred, and reverse the order when correcting the damage. The majority of repairs to minor accident damage requires the skill of the body repair worker in the use of hand tools and general repair techniques. Although every panel assembly has its own individual repair procedure, the basic approach is the same for all minor repairs and is as follows:

1   Carefully inspect the damaged area and analyse the severity of the damage and its ease of accessibility for repair using hand tools.
2   Decide if the repair can be carried out using hand tools only or whether the hydraulic body jack is required.
3   Select the necessary hand tools and, if using the hydraulic body jack, decide which basic set-ups and attachments are best suited to the job.
4   If the repair is carried out using hand tools only, then the back of the damaged section under repair, if accessible, must be cleaned of antidrum compound.
5   If the accessibility for the use of hand tools is difficult owing to the presence of brackets behind the panel or double-skinned sections, rough out the damaged area either by cutting out the back and using spoons to lever the damage out, or by using a panel puller on the front of the damaged section to pull out the damage. The area should then be cleaned, filled with the appropriate material and finished.
6   When using hand tools only, the damaged area should be roughed out to its original shape, although this operation can also be carried out using hydraulic equipment to push the damaged section back to its original shape.
7   Once the damaged section has been reshaped, either by hand or hydraulic means, planish it using a combination of direct and indirect hammering and filing.
8   Then check the low spots, if any, by cross filing and raise them by further planishing or pick hammering. At this stage high spots may develop due to stretching of the metal, and hot shrinking of these points will be necessary.
9   Sand the panel surface over the damaged area using a very fine sanding disc.

13.16.3 Minor repairs to wings

Removal of a small dent from a wing is one of the simplest of repair jobs. The first step is to check the underside of the wing to ensure that it is free from road dirt and anti-drum compounds; then the damaged area is roughed out and the wing reshaped to its original contour. The reshaping is done with a rather heavy hand dolly of a similar shape to that of the contour of the damaged wing section. The blows should be heavy with a follow-through action which will speed up the roughing-out operation. It is very important, while roughing out damage, that full-shaped or very curved dollies and heavy ball pein hammers are not used, as they cause excessive stretching of the already damaged metal. Metal that is returned to its original pressed shape during roughing out will smooth out more quickly and evenly.

When the roughing out is complete, the next step is to planish the roughed area. In the case where the wing has been heavily coated with cellulose, it is advisable to remove this coating using the sander fitted with an open-coated disc suitable for paint, or with paint remover; the latter must be neutralized to stop the burning action on the paint. The first step in smoothing is to try to level out the uneven surface by working the high areas into the low areas. This can be done by holding the dolly block under the low area and tapping down the adjacent high spot with a planishing hammer. In some cases there may be no high spots and no adjacent low areas, in which case the high spots, if any, can be tapped down often with just the planishing hammer and without the dolly block backing. These high and low areas are found by passing the hand over the section under repair. As planishing continues, filing will show up any low and high areas, which will need correcting by tapping down high areas and lifting the low areas by further planishing with hammer and dolly or using a pick hammer. The final finish is achieved by cross filing the original file strokes; this is a double check on curvature and smoothness of the panel surface.

13.16.4 Minor repairs to door panels

When repairing door panels the main difficulties arise through the fact that most door panels are nearly flat sheets of metal, which means that even the slightest stretching due to over-hammering will produce a fullness which will stand out after the repair has been painted. The greatest care must be taken.

The first step is to strip the door of interior trim and window. The amount of stripping necessary depends on the extent and position of the damage. With most door repairs it is quicker to remove the door from the vehicle by releasing the check strap and then taking out the set screws in the door side of the hinges, or, if the hinge is welded, by knocking the hinge pin out. Having prepared the door and the work bench, an inspection can then be made of the door frame to determine the accessibility for the use of hand tools and hydraulic equipment. When carrying out the repairs it is often necessary to cut the back of the door frame and bend it, thus permitting greater freedom of movement for the use of hand tools. The thorough removal of all anti-drum material is essential before any actual planishing is commenced.

The shape is gained by roughing out with the mallet and dolly block, after which the finish is attained with the planishing or pick hammer in conjunction with the file. In selecting a dolly it is advisable to choose one that is rather light and slightly flatter than the panel being planished. This will reduce the risk of stretching the panel. In some cases where the door panel is badly creased and is inaccessible for the use of hand tools, it would be better to rough out the damage with the aid of the body jack or spoon acting as a lever, smooth it over with hammer and dolly, then fill it with body solder and file the area to obtain the final finish.

13.16.5 Reskinning door panels

This process affords considerable saving in time and cost compared with the practice of replacing complete door assemblies. The technique of this process is the same regardless of the shape or type of door to be reskinned:

Removal of a damaged skin panel This is achieved by drilling out the spot welds around the flanged areas of the door using either a Zipcut or cobalt drill or Spotle tool and also, on some types of door, using a power saw to make joints across the pillars (Figure 13.63a). The flange can be then eased up using a flange removing tool (Figure 13.63b), or in some cases the outer edges of the flange are ground through using a sander (Figure 13.63c). If this method is used, care must

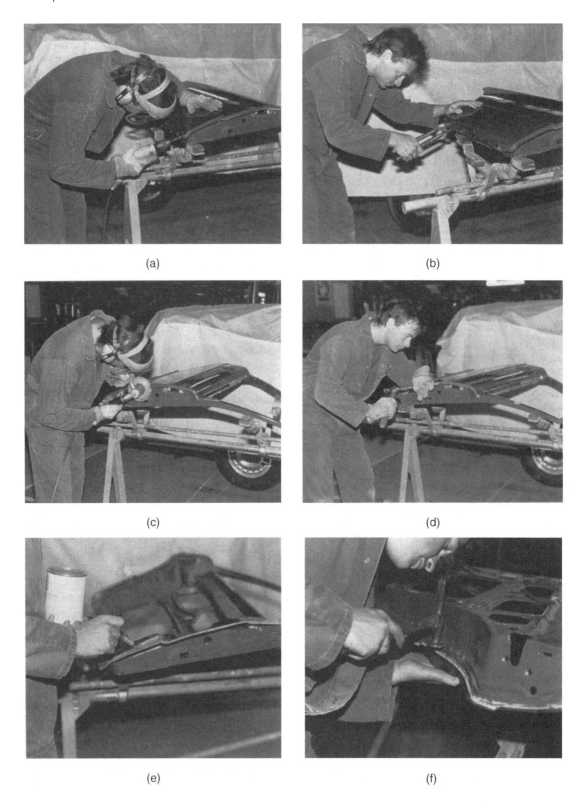

(a)

(b)

(c)

(d)

(e)

(f)

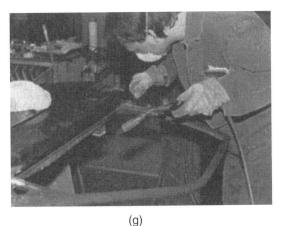

(g)

(h)

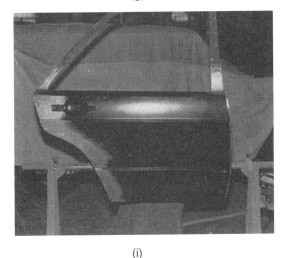

(i)

Figure 13.63 Reskinning door panel: (a) removing spot welds (b) removing flanges (c) grinding through flanged edges (d) removing damaged skin panel (e) priming door frame (f) folding door edge flanges (g) soldering the pillar joints (h) finishing panel edges (i) completed reskinned door (*Motor Insurance Research Repair Centre*)

be taken not to damage the door frame. The damaged panel skin can then be removed using a hammer and chisel and easing up the flange strips that are left (Figure 13.63d).

Preparing door flanges The door flanges should now be straightened up using a hammer and dolly and the face of the flanges sanded to make a smooth surface for the attachment of the new skin panel.

Preparing the new surfaces The door flanges should now be painted with a zinc-rich inter-weld primer to stop any chance of corrosion between the door flange and the new skin (Figure 13.63e).

Fitting the new skin panel The prepainted skin panel is now placed over the frame and aligned using the door lock holes as a guide, and MIG tacked on to the joint pillars if necessary. The edges of the skin are now hammered over the door frame using a hammer and dolly until the flange is folded over completely. During this operation great care is essential if the panel face is not to be marked (Figure 13.63f).

Welding in the new skin panel The spot welds should now be replaced to fix the new skin panel to the frame, taking care to use the correct electrodes, especially on the face side of the panel. Adhesive is used on certain doors to fix the skin panel to the door frame. Where joints have been made in the pillars they must be MIG welded, ground flush and soldered (Figure 13.63g).

Final cleaning and inspection The soldered joints must be filed and sanded with a DA sander. Finally, inspect the skin panel for any defects, which will need correcting by filing and sanding (Figures 13.63h and 13.63i).

13.16.6 Pulling out double-skin panel damage

When inaccessible body panels are damaged, a method of repair for the removal of the damage, without leaving holes or having to remove interior trim, is to use a special pin welding gun. Also pin welding attachments can be fastened to spot welding equipment (Figure 13.64). The welding gun attaches pins, screws, washers or studs to the surface of any damage so it may be pulled out. There are two sizes of pins: small (2 mm) which are normally used for skin panel damage, and larger (2.5 mm) which are used for chassis legs or thicker-gauge metals.

First the damaged area must have its paint removed and be sanded down to a bare metal finish in order to give a good electrical contact between the pin and the welding gun. Insert the pin into the spring-loaded nozzle of the gun. The two points of contact are the tip of the pin and the outer copper ring on the welding gun, which should, when pushed down, make as even a contact as possible with the panel surface under repair. Press the gun to the surface of the metal, press the trigger for 1–2 seconds only and remove the gun from the pin, which now should be welded to the panel. The number of pins which need to be welded on to the panel will vary in accordance with the depth and extent of the damage, starting at the deepest section of the damage and working outward.

The dint is then gradually pulled out using a slide hammer, which tightens on to both sides of the pin by the rotating action of the chuck. The dinted area is then pulled out by using the slide hammer action, on one pin at a time, beginning at the deepest part of the damage and working outwards. The pins are removed by breaking them from the panel surface and sanding down their residue.

When the damage is a crease and very deep, ring washers may be used instead of pins to facilitate a stronger pull. A clamp may be fastened to these washers and attached by chain to pulling equipment which would then pull out the damage. Alternatively a hook is inserted into the slide hammer to pull out each washer individually (Figures 13.65 and 13.66).

When the damage has been completely pulled out the area may be filled using body solder or plastic filler according to the position and extent of the damage. The filled area is then sanded and filed down to an acceptable finish for painting.

13.16.7 Minor repairs to bonnets

The bonnet of a car is firmly fastened when closed. Many models have spring loaded hinges so that the tension of the springs holds the bonnet firmly in place at the hinge side. At the point of locking a sturdy catch holds the bonnet down under spring tension. When open the bonnet is virtually unsupported as it is only held in place by the hinges. This means that only minor repairs can be carried out with the bonnet in place, and if hydraulic power tools are necessary it is advisable to remove the bonnet from the car. Large areas of almost flat sections running into gradual return sweeps are incorporated in the designs of most bonnets. The flat panel and the return sweep are both shapes that need careful repairing to avoid stretching of the metal.

A rather common trouble that arises when repairing bonnets is the development of a loose edge. This looseness usually occurs along the back edge of the bonnet, which in most cases is a raw edge, and even the slightest stretch will cause a buckle. It is essential that this buckle be removed, and the best way of carrying this out is by hot shrinking the edge. Another characteristic of bonnets is that the edge flanges crack, and these should be welded as early as possible to prevent the extension of the crack and to keep the bonnet in its correct shape.

The repair procedure for minor, accessible dents is first to rough out the damaged area using a mallet to minimize stretching, and then to planish the area using a beating file in conjunction with a dolly block which will also reduce the possibility of stretching. If the repair is in such a position as to make the use of hand tools difficult because of the strengthening struts, the area should be roughed out using very thin body spoons and then filled with plastic filler, as the heat produced when soldering might distort the panel. The prominent position of the bonnet requires its finish to be perfect because ripples and low spots can be easily seen when painted.

13.16.8 Minor repairs to sill panels

The sill panel on a car is attached to the door pillar and floor, and is of double-panel box-like construction having both inner and outer panels. Because of its position it is one of the most likely places to receive damage.

Two methods of repair can be used for double-panel assemblies. First, cut away part of the inner panel, or in the case of the sill, along the top. Then bend the panel or top section upwards to make the outer panel accessible for the use of hand tools. Push out the damage using the spread ram which acts as a wedge, or force it out using body spoons as levers, then planish and file. Bend the top section or back panel back into position and weld. Care should be taken here to avoid overheating the repaired area.

The second method, which is less time consuming, can be used if the dent is small and not very deep. The area can be sanded clean to bare metal and filled using body solder. If the damage is very deep it will need pulling out, and this can best be done using a panel puller on the outer panel surface. The holes needed for using this tool should then be welded up and the damaged area should be sanded, filled with body solder and filed to achieve the final finish.

13.16.9 Minor repair of body panels using adhesives

With the advancement of adhesive technology, Permabond have developed an adhesive specially formulated for car repair applications and called Autobond. This is a two-part product which has been developed to meet the specific requirements of the car repairer. It is designed to replace panel welding with a minimum of surface preparation; it will also protect against corrosion. It provides substantial labour saving in avoiding the necessity to remove fuel tanks and soft trim as required before any welding repair. The adhesive not only bonds metal to metal but is also suitable for bonding GRP components.

For ease of use the adhesive is supplied as a layer of resin on top of a layer of hardener inside a plastic tub; the two layers will not react until mixed with the stick provided. The resin and hardener are coloured light grey and black respectively to provide a visual check that thorough mixing is achieved. After mixing is complete, setting begins at a rate depending on the ambient temperature. The best performance is achieved by heating the bond area to 40 °C for 1 hour; however, at 60 °C full hardening is achieved in 30 minutes. Curing at normal ambient temperature will give joints of reduced strength and is not recommended. Figure 13.67a–f shows the application sequence.

13.16.10 Removal and replacement of exterior and interior soft trim and hard trim

Body exterior mouldings are attached by weld stud retaining plastic clips; weld stud retaining plastic clips with attaching screws; an adhesive bond using either tape or sealant; a self-retaining spring clip or clinch-type clip; or special attaching screws.

During repairs, observe the following:

1 Adjacent paint surfaces need protecting with masking tape to help prevent any possible damage.
2 Always use the correct tools for the job, and take great care with mouldings.
3 Water can leak into the interior of the body through any body panel holes made by screws, bolts and clips unless they are securely sealed.

Door trim

The door trim pieces are normally assembled in one-piece or two-piece trim pads. Removal of door trims is carried out in the following order: door lock handle and door locking knob; mouldings; window regulator handle and door latch handle; arm rest assembly; mirror remote control bezel nut; and door trim retaining screws (if any). Then lift the trim panel retaining clips from the door inner panel, and disconnect all wiring so that the trim panel may be removed. Prior to reinstallation of the trim pad, always make certain that the watershield is in position and correctly sealed.

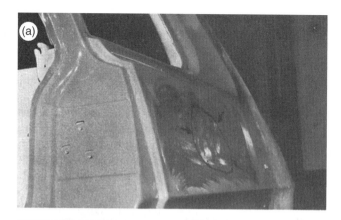

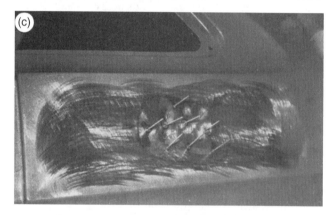

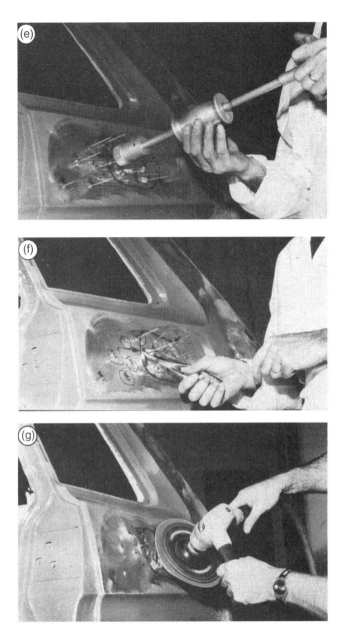

Figure 13.64   Pull pin kit used with a spot welding gun (*ARO Welding Ltd*). (a) Clean the damaged area with an abrasive disc. (b), (c) Set the spot welding gun controls. In case of doubt, perform a trial weld on a section similar to the dented one to help you select the best setting. Weld the pins at the selected positions on the damaged section. (d) Once the pins have been welded to the section, use the inertia weight tool to pull the dented section back into its proper shape. (e) Make sure you do not exercise more pull than is needed to bring the panel back into its correct position, to avoid bulging. (f) Use your normal shearing tool to cut off the projecting portion of the pins. (g) For a smooth finish, go over the area with an abrasive disc in the usual way

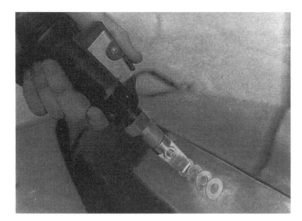

Figure 13.65 Fastening ring washers to the damaged area ready to pull (*Stanners Ltd*)

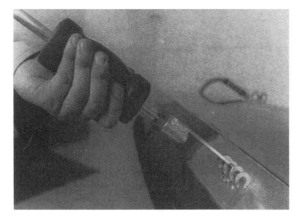

Figure 13.66 Pulling damage out using hook and slide hammer (*Stanners Ltd*)

Quarter trim panel

This is the panel which fits over the interior of the quarter panel. Removal of the quarter trim panel is carried out in the following order: rear seat cushion; rear seat back; rear window handle (if any); and any screws or mouldings. Then unfasten the panel retaining clips and pull the trim panel away from the quarter panel.

Floor carpet

Carpet for car floors is made either in one piece which is moulded to fit the shape of the floor, or in two pieces to fit the front and rear sections of the floor respectively. In order to remove the one-piece carpet it is necessary to remove everything in the way, including seatbelts, seats, centre console, and door sill scuff plates. However, the two-piece carpet can be removed with the seats still in place because they are made with cut-outs to allow for easier removal. In some cases the floor of a car is covered with heavy rubber flooring instead of carpet, and this can be removed and replaced in a similar manner.

Headlinings

Headlinings are of two types. The first is soft and made from cloth or vinyl coated. The second is harder, being formed from moulded hardboard coated with foam or cloth which has a vinyl facing. The replacement of the moulded headlining is a simple, straightforward operation because of its one-piece rigid structure. The soft type is more complicated in its removal owing to the necessity of removing a large amount of trim in addition to the windscreen and rear window in some cases.

13.17 Aluminium panel repair

13.17.1 Differences

The repair of aluminium panels and other components is slightly different to repairing steel ones. Aluminium panel repair is not difficult; it is just different to repairing steel ones. Not greatly different, just that you should follow a set of procedures and remember the characteristics of aluminium. More and more vehicles are using aluminium every year, more correctly we should say aluminium alloy. Let us have a look at the characteristics of the material, particularly its advantage over steel.

Aluminium is light in weight, strength for strength it is about a third of the weight. In an accident, it will absorb double the energy of a similar steel component. Given the correct environment it does not corrode. It is a very abundant material and easy to recycle. Because aluminium is easier to work than steel, and it can be worked in different ways, aluminium bodies can be made for a lesser number of parts than steel ones. The big environmental advantage is that by reducing the weight of the vehicle the fuel efficiency is increased; a 10% weight reduction typically gives an 8% reduction in fuel usage.

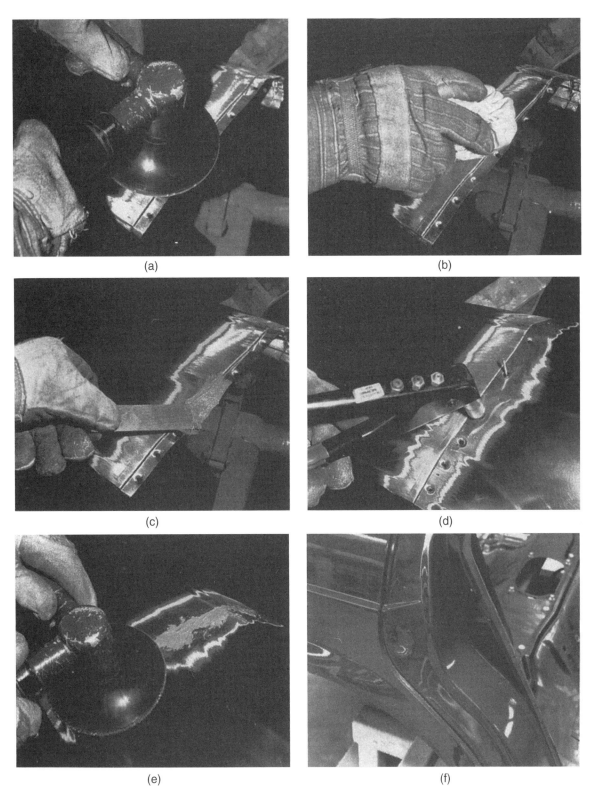

Figure 13.67 Minor panel repair using adhesive; (a) preparing surfaces (b) cleaning with solvent (c) applying adhesive (d) assembling joint with rivets, tack welds or clamps as appropriate (e) removing excess adhesive, heating bond area, buffing joint (f) completed panel (*Permabond Adhesives Ltd*)

13.17.2 Galvanic corrosion

Aluminium is the fifth least noble metal, it is almost at the end of the anode–cathode scale, and it has an anodic index of 0.95. This means that whenever steel and aluminium come into contact there is likely hood that galvanic corrosion is going to take place. If there is dampness of any kind, such as when rubbing down or cleaning, then some form of corrosion is bound to take place. Although this might appear insignificant at first, the smallest pin prick of corrosion can lead to a large rust patch. The smallest corrosion bubble will look unsightly if it is in the centre of the bonnet or other highly visible panel.

Steel particles from tools and abrasives can cause corrosion to aluminium, so careful consideration must be given to every possibility of cross contamination between tools and panels to prevent corrosion taking place.

13.17.3 Contamination

Repairers specializing in working with aluminium bodied vehicles, such as Audi (Figure 13.68), Jaguar (Figure 13.69) and Aston Martin use specialist enclosed work booths for those vehicles. That is, a booth which is fully enclosed and heated and ventilated separately from the rest of the workshop (see Figure 13.70). Also the tools and equipment are dedicated for use only on aluminium bodied cars. The tools used, where possible, have non-ferrous working faces. That is, tools made

from plastics materials, or plastics coated tools, are used where possible.

However, this is not usually practical for the smaller repairer, or the general accident repairer. In which case the following is advised:

- Position the aluminium bodied vehicle in the workshop so that it is quite separate from steel bodied ones, especially when welding, sanding, grinding or other intensive repair procedures are being carried out.
- To avoid cross contamination by minute steel particles from steel bodies to aluminium ones use a set of tools for aluminium use only. That is, you should dedicate a new set of tools for aluminium repair use only. Colour code the tools by painting, or plastic dipping their handles.
- Keep the tools for use with aluminium bodies in an enclosed tool box separate from those used on other vehicles.
- If possible, dedicate sanding and grinding tools for use only on aluminium bodied cars. Use the same colour coding system and store them separately under covers. If this is not possible, then clean the equipment with solvent before use on aluminium bodies and use new discs or abrasives to ensure that there is no cross contamination.
- Always wear a clean set of overalls and gloves when working on aluminium bodies cars.
- Always use new wipers and cloths when carrying out aluminium repairs.

Figure 13.68   Audi A8 – Aluminium body (*Courtesy of POWER-TEC®*)

Figure 13.69   Aluminium bodied Jaguar (*Courtesy of POWER-TEC®*)

Figure 13.70   Enclosed work booth for aluminium repair (*Courtesy of POWER-TEC®*)

13.17.4 Riveting

When drilling out steel rivets from aluminium panels be sure to collect any steel chipping to prevent contamination and possible corrosion. Punches are available to remove old rivets.

When replacing rivets be sure to use the correct type. Often coated rivets are used to reduce the risk of galvanic corrosion. The coating is usually anodizing giving an identifiable colour.

Rivets are best applied with a rivet gun, however copper rivets can sometimes be used and closed using snaps which are hit with a hammer.

13.17.5 Oxidation

Oxidation on aluminium panels is invisible to the naked eye except as there may be a slight dulling of colour. As the oxidation layer will be over the whole panel, the dulling will not be distinguishable. A coating of oxidation is formed within a period of two hours even in a well heated dry workshop. Therefore a primer coat should be applied within two hour of sanding or other preparation.

If the aluminium panel is left for a long period in damp conditions, then the oxidation will become visible as a coating of white powdery aluminium oxide. Aluminium oxide (Al_2O_3) is both infusible, difficult to melt without very high temperatures, and amphoteric – can form a base or acid – which implies it can be dissolved in caustic soda. Be careful when using caustic soda as a cleaner, always wear rubber gloves, goggles and apron as the minimum PPE.

Immediately before welding aluminium panels the areas to be welded must be de-oxidised clean – the best technique is to use a stainless steel wire brush which as been dedicated for use only on aluminium panel work.

13.17.6 Heating aluminium

Most aluminium body panels are hardened and tempered for strength and dent resistance. The aluminium will work harden on impact. Therefore if the panel is dented the damage will be firmly set in place. It will be necessary to heat the panel to soften it to remove the dent and bring the panel back into shape and contour.

Aluminium is a very good conductor of heat, so spot heating is difficult. That is, the whole panel becomes hot very quickly, as does the surrounding area.

Aluminium fuses (melts) at about 650 °C; but the aluminium oxide will not burn off at less than 1250 °C. Therefore it is important to clean off the oxide prior to welding, to be able to keep the aluminium fluid during the welding process at 650 °C.

When heating the panel to carry out a repair the temperature should not exceed 200 °C. This is the threshold temperature for the aluminium. A usual temperature range for the repair of panels is between 110 °C and 160 °C. It is advisable to keep the temperature below 160 °C.

Special attention must be given to adhesively fixed or bonded areas. Whether it is the panel which is being repaired which is bonded or the bonding on an adjacent panel or nearby supporting member. Heat maybe applied to help disassemble an adhesively secured or bonded joint; but be careful not to apply any heat to a joint which should be kept intact. Always pay special attention to these areas both in terms of accidentally applying direct heat or allowing heat to be conducted to reach them. The use of heat soaks in the form of damp material is one way of doing this; but make sure that this is clean and thoroughly dried off afterwards to ensure that no corrosion can occur later.

13.17.7 Heat application

Aluminium panels can be heated in the same way as any other panel; but the working temperatures are very low compared to steel. Heat sources are:

- Oxy/acetylene torch – be careful with the flame to ensure even temperature distribution
- Gas blow torch – use a low to moderate setting (Figure 13.71)
- Hot air gun – probably the best method, especially if it has a temperature control

13.17.8 Temperature measurement

Aluminium does not change colour or otherwise give an indication of its temperature as it gets hot. That is, until it suddenly melts.

There are several ways of measuring the temperature, the choice depends on personal preference and required accuracy. The more accurate the better,

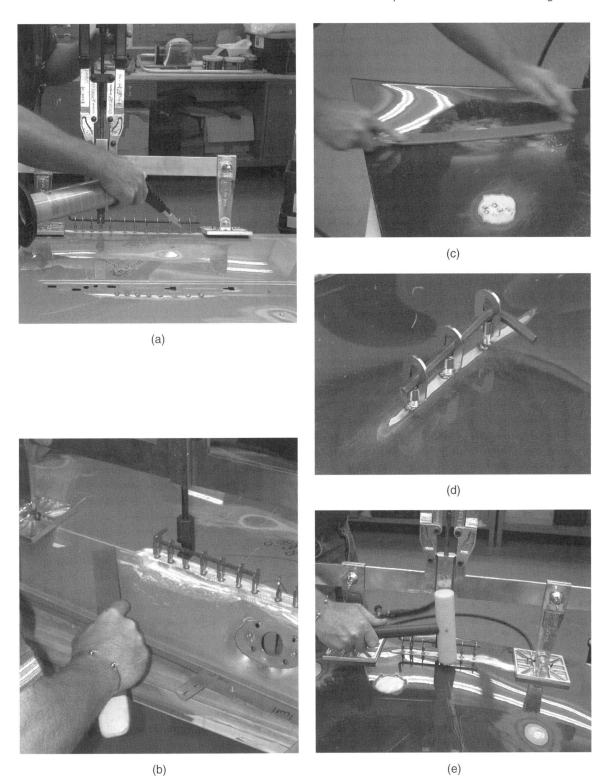

Figure 13.71 (a) Heating the panel with propane gas (b) Conventional bumping file being used to raise dent (c) Using a panel file (d) Threaded studs and loops for pulling repair from outside (e) High density nylon compound panel hammer (*Courtesy of POWER-TEC®*)

especially when dealing with vehicle with large numbers of bonded joints.

- Infrared or laser, 'point and shoot' digital thermometer – These are very accurate, being able to measure to a fraction of a degree. They allow quick measurement not only of the area which is being worked on, but any other areas, for example adjacent bonded joints which must be kept cool. The readings are instantaneous. (Figure 13.72)

Figure 13.72 Laser point and shoot digital thermometer (*Courtesy of POWER-TEC®*)

- Digital thermometer strips – These are placed on the panel, their plastic backing strip holding them to the panel and allowing reading to be observed whilst work is taking place. The readings take several seconds to change, as the material changes colour with the change in temperature to indicate the temperature. Several of these may be needed at any one time, and it may not be possible to secure them to bonded frame areas if there is not a suitable smooth surface. Accuracy is limited to the pre-set limits of the strip.
- Colour crayon – The area to be heated is given an outline mark with a crayon which changes colour at a given temperature. With crayons the correct one must be chosen for the job, the accuracy is limited to pre-determined constituents of the crayon and the speed is that of the chemical colour change. These are limited too in that it is not possible to track heat travel to adjacent panels.

13.17.9 Dent repair

Clean the whole area with soap and water to remove any road dirt then dry off the panel. Remove the paint coating using 80-grit or finer abrasive. Use only a light pressure when sanding, especially on the bare metal.

Wipe the panel with wax and grease remover and dry off with lint free cloth or a wiper. Allow any vapours to evaporate.

Warm the area to be repaired, checking that the temperature does not exceed 160 °C. Also check that adjacent areas, especially where there are bonded joints, are not getting too warm. Use an accurate thermometer for this, don't guess.

Work the damaged metal using conventional straightening techniques, keeping in mind how soft the aluminium panel is compared to steel.

Use the tools which have been dedicated for use only with aluminium bodies. Make sure that any metal tools such as hammers, dollies and spoons have smooth edges to avoid gouging the aluminium panel. Pick hammers are likely to be unsuitable because they are too sharp. The use of wooden bossing hammers and purpose-made high density nylon hammers are recommended.

Remember to re-heat the panel continually throughout the repair process.

13.17.10 Miracle repairs

Lifting and repairing damaged panels from the outside without removing interior trim or on otherwise blind sections can be done using Miracle tools (Figure 13.73). The system uses treaded studs and a range of rings, pulling loops or collars and pulling tools. The stud is welded to the damaged area of the panel after appropriate removal of the paint coating. A suitable pulling loop or collar is screwed onto the stud. The panel is heated up if needed and the pulling tool applied to pull the dent out.

When the dent is out the stud can be snipped off (Figure 13.74), and then its remains filed or ground down level with the surrounding area. Check for low and high spots using a panel file taking care not to remove too much metal. Aluminium clogs files, so keep the file clean using a stainless steel or brass wire brush.

Figure 13.73 'MIRACLE' pulling tool (*Courtesy of POWER-TEC®*)

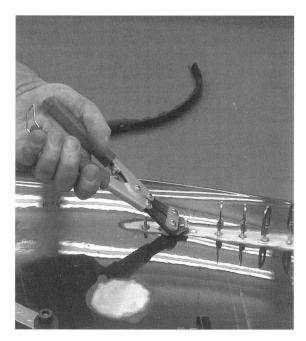

Figure 13.74 Snip-off the studs (*Courtesy of POWER-TEC®*)

Applying a guide coat then using a long sanding block, as is used for filler, high spots can be identified without removing metal.

Heat shrinking

Applying heat to any dent, without any kind of manipulation, may reduce the size of the dent due to the thermal expansion of the aluminium.

Heat shrinking may also be used with hammer and dolly to manipulate the panel and reduce the dent.

Cutting aluminium

Aluminium, though it is soft, requires special care when cutting. The following points should be observed:

- Always ensure that the panel is fully supported on both sides. If it is not, it may sag and bend causing damage.
- Use a reciprocating saw, such as an electric jig saw.
- Use a fairly course blade, Forrest Blades aluminium saws have alternate six and four sided teeth. Cutting fluid must be used with this tooth shape.

#2

WWW.forrestblades.com

- Use the correct cutting speed, up to 5.5 m/s (1000 feet per minute) – much faster than for steel.

Aluminium welding

MIG and TIG are the normal welding processes for use with aluminium. Follow normal procedures using aluminium wire when MIG welding and adjusting the amperage accordingly. Clean the panel immediately before welding with a stainless steel

Figure 13.75 Control box to convert conventional MIG welder to weld aluminium (*Courtesy of POWER-TEC®*)

wire brush. If possible pre-heat the panel before welding to give better weld quality and reduce the risk of distortion. Control boxes are available to convert conventional MIG welders to weld aluminium (Figure 13.75).

13.18 Body electrical and electronic systems

There are very few body panels on current vehicles which do not have some form of electrical or electronic device attached or nearby. This means that the repair of any of the panels will probably require contact with an electrical component or the power supply to a nearby component. This section aims to look at the basics of body electrics as may be tackled by the personnel in a body repair shop. For more complex systems and system fault diagnosis the services of a vehicle electrical/electronics specialist should be sought.

13.18.1 Electrical terms

Amp
The unit of current is the amp and is the quantity of electricity which flows in a circuit. If it were water you would refer to it in gallons or litres. To measure current flow the ammeter must be connected in series with the circuit, the ammeter must be able to handle the expected current or it will be burnt out. To measure large current flow, such as in a starter circuit where the motor may be drawing 200 A, you will need an induction ammeter which clamps over the starter cable without making electrical contact. The most common use of an ammeter is to check the charging rate of a vehicle's alternator; you'll need a meter which goes up to 60 A for this job. A centre zero scale is useful for measuring the current flow when working on vehicles, as it can show the direction of the flow and you do not need to change the connection of the leads. Some ammeters which do not have a centre zero have a button to change the polarity to show charge or discharge.

Volt
This is the pressure of the electricity. The water pressure from a fire hose can knock you over; the electrical pressure from a mains socket can do the same. Any voltage over 50 V is potentially lethal. Vehicle electrical circuits operate at a safe 12 V or 24 V on HGV/PSV applications. The reason for the higher voltage on the large diesel vehicles is to reduce the amperage needed by the starter motor, it also gives better interior lighting in buses. The voltage in a circuit is measured by connecting the voltmeter red to positive and black to negative each side of a component, that is, in parallel with the component.

Watt
This is the unit of electrical power.

$$\text{Watts (W)} = \text{Volts (V)} \times \text{Amps (A)}$$

746 W equals 1 HP.
  Bulbs are usually identified by their wattage, headlamps are typically 60 W.

Resistance
This is measured in ohms. If you have a narrow hosepipe it will restrict the flow of water, in the same way the diameter of the electrical wire will affect the flow of the electric current. As a long hosepipe will reduce the flow of water, the length of the wire will increase the electrical resistance.

$$\text{Resistance (ohms)} = \frac{\text{Volts}}{\text{Amps}}$$

Resistance is measured with an ohmmeter. If there is no resistance the meter needle will move across the scale and read zero. Electrical cables and earth connections should show zero or very small amounts of resistance. An exception is plug leads which have an inbuilt resistance for radio suppression purposes, typically 30 000 ohms per meter of length.

Analogue and digital meters
Analogue meters are ones which have a needle which moves across a scale; digital ones show numbers on an LCD (liquid crystal display) or illuminated display.

Resistor
A resistor is used to control the flow of electricity, it allows electricity to flow, but the flow is at a set rate.

Capacitor
A capacitor (also called a condenser) stores electricity, or absorbs current surges, for a limited period of time. The capacitor in a distributor absorbs the flow of electricity in the LT circuit when the CB points open.

Diode

This is a one-way valve for the flow of electrical current; zener diodes only allow electricity to flow when the voltage exceeds a set figure.

Transistor

This term covers a number of different solid state devices, including diodes. The most common transistor has three connections, base, collector and emitter. By applying a small current to one terminal a larger current can be switched between the other two, remove the small current and the larger current is stopped. A thyristor is such that a small current starts the flow which remains after the small current is stopped. A thermistor is temperature operated, its resistance may increase with temperature or it may be of the negative temperature coefficient (NTC) type where its resistance decreases as it gets warmer.

Series

This refers to a circuit where components are connected in line with each other, that is, positive to negative. As the number of items in the circuit are increased the resistance will increase, and the voltage across each component will decrease.

Parallel

A parallel circuit is one where items are connected electrically alongside each other. The resistance decreases and the voltage remains constant. Vehicle lighting is connected in parallel.

Earth return

This is the name used to describe the use of the vehicle body/chassis as a carrier of electricity. This system is used on most cars and light vans, all the circuits returning to the battery negative via the body/chassis. The battery must have a good earth connection. Tankers and buses use a two-wire system where each component has a feed wire and a return wire to the battery negative.

Short circuit

When a live bare wire or terminal touches the chassis earth or another earth point it produces a short circuit. This may lead to the wire becoming hot and burning, even causing damage to the car.

Open circuit

When a wire breaks or becomes disconnected and the circuit is no longer complete an open circuit is produced. In this case the component involved will not operate.

Fuse

A fuse is a breakable link in a circuit. The fuse will blow instead of starting a fire in the case of a short circuit.

13.18.2 SAE J1930

The Society of Automotive Engineers – SAE International have a task force to produce a list of standardised terms and acronyms for automobile electrical and electronic systems. The table below lists some of the common terms.

Standardised term	Standardised acronym
Accelerator pedal	AP
Air cleaner	ACL
Air conditioning	A/C
Charge air cleaner (intercooler)	CAC
Data link connector	DLC
Diagnostic test module	DTM
Diagnostic trouble code	DTC
Distributor ignition system	DI
Electronic ignition	EI
Engine control module	ECM (Engine ECU)
Engine coolant temperature sensor	ECT Sensor
Engine speed sensor	RPM Sensor
Exhaust gas recirculation	EGR
Fan control	FC
Generator (alternator)	GEN
Ignition control module	ICM (Ignition ECU)
Intake air temperature sensor	IAT Sensor
Knock sensor	KS
Malfunction indicator lamp	MIL
Manifold absolute pressure sensor	MAP Sensor
Mass air flow sensor	MAF Sensor
On-board diagnostic system	OBD System
Open loop	OL
Park/neutral position switch/sensor	PNP Switch/Sensor
Scan tool	ST
Service reminder indicator	SRI
Throttle body	TB
Transmission control module	TCM (Transmission ECU)
Turbocharger	TC
Vehicle speed sensor	VSS

Battery

The battery is the source of power for starting the vehicle. Most vehicle batteries are 12 volts. Before carrying out any dismantling which entails disconnecting any electrical or electronic components or connections it is good practice to disconnect the battery. Always disconnect the chassis earth lead first and re-connect the same lead last. It is good practice to remove the battery completely from the vehicle and put it on charge to ensure that it will re-start the vehicle when the repair is completed. The battery must always be disconnected before any arc/MIG/TIG welding is carried out to prevent damage to the electrical system by the high voltage of the welding apparatus.

Before disconnecting the battery you should ensure that the security codes are available for the radio/cassette/CD player. Also, if appropriate, any other security item codes or personal setting are stored.

Some special vehicles have two batteries. One battery for starting the engine, and one battery for maintaining the supply to the electronic components that operates when the vehicle is not running. The second battery is usually a gel-battery. This is used to maintain a power supply to the clock and the security system, when the vehicle is parked. In this case, both batteries will need to be disconnected.

Alternator

The alternator generates the electrical power to charge the battery when the engine is running. The rectifier inside or, in some cases, near to the alternator contains a number of diodes. These diodes will be damaged if the alternator is not disconnected before carrying out any form of electrical resistance welding. Disconnecting the battery will usually also disconnect the alternator.

Electronic control unit (ECU)

Vehicles are fitted with at least one ECU. The ECU may also be called by other names including ECM or 'brain'. Typically ECUs are used for the engine management system, the ABS, the security system and electronically controlled transmission systems. ECUs are usually attached to the panel with two screws. They are connected to the wiring loom with a 32-pin plug. Pull back the spring clip holding the connector plug into the ECU then pull the 32-pin plug out gently without pulling on any of the wires as this may pull the wired out of the connector.

Before replacing the plug ensure that the plug and ECU are both clean and that each of the 30 pins are aligned before pushing them into the ECU.

Central locking

Central locking may be operated with or without a key. The keyless type uses either an infrared or a wireless signal to a sensor on the vehicle from the key-fob control, or 'zapper'. The small mercury battery in the key-fob control should be replaced every two years. The central locking mechanism may be operated by either a centrally pumped pneumatic system or individual solenoids. Look out for the locking mechanism when working on vehicle doors. Always ensure that the connections are in place before re-fitting the door trim. It is good practice to ensure that the windows are wound down before testing the central locking mechanism.

Electric windows

Most electric windows are operated by a small electric motor and a continuous cable system. The cable is only about 3 mm in diameter and will break if subjected to normal rusting. The cable goes stiff and brittle when rusty. The windows may be operated by switches on either the centre consul or the inside door trim. Some have two switches for each window arranged to work in parallel. You will need to disconnect the electric window power supply leads and the central locking wires from inside the door, after removing the door trim, before attempting to remove the door or repair the door skin panel. Figures 13.76 and 13.77 shows some of the electrical components inside the door of a Mercedes McLaren SLR.

Electric sunroof

Electric sunroofs operate in a similar way to electric windows. The control for these is usually a switch mounted just in front of the interior rear view mirror. Access to the sunroof usually entails removal of the interior head lining (also called roof lining).

Electric seat

Electric seat controls are usually available for height, backwards and forwards position, rake and tilt. Some seats have memory settings available for different drivers, referred to as setting 1, 2 and 3 and so on. The connectors for these are usually underneath the seat cushion. Be careful when

Figure 13.76 Mercedes McLaren SLR tyre pressure sensor

Figure 13.77 Mercedes McLaren SLR door

disconnecting and re-connecting the seat wiring because seats also may incorporate an electrical earth cable to reduce the risk of travel sickness and connections for the pyrotechnic anti-submarine seatbelt pre-tensioner.

Lights
Most light fittings are screwed to the front panel or wing. Removal is usually simply unscrewing and unplugging the cable socket. Re-fitting is the reverse procedure. However, the head lamp beams will need adjusting using a beam setter to ensure that they do not dazzle on-coming traffic and comply with the VOSA regulations.

Supplementary restraint system (SRS)
Restraint systems to protect the driver and passengers in the event of an accident are designed to work in conjunction with the vehicle body crumple zones, the idea being to slow down the vehicle occupants in a controlled manner. That is, if the vehicle hits a wall, then the occupants will decelerate at a gentle rate as possible in the time available. Typically, the amount of time is 120 ms (0.120 sec) – the time it takes to blink. One of the major causes of death in car accidents is damage to the brain caused by deceleration rates which are too high.

The standard three-point seatbelt is defined as the primary restraint system. Other features, namely seatbelt pre-tensioners and airbags, come within the category of SRS.

Airbags
The air bag is designed to protect the occupant's head and upper torso – not to take all the weight of the body. It is inflated by gas generated by the pyrotechnic device. The gas is usually nitrogen which is inert and poses no after-effect problems. The air bag module, comprising the airbag, airbag cover, the pyrotechnic device and its fitting must be treated with extreme care when working on the SRS, or any other related system. For instance, if you are replacing the indicator or ignition switch, you will probably need to disturb the airbag module. The pyrotechnic device is ignited by a low voltage electrical current and small tablets fill the airbag with nitrogen in about 30 ms (0.030 sec).

Always carry, or handle, an airbag module so that it is kept in the upright position. If the module were to accidentally go off, the bag would inflate upwards and the heavy part of the module would go down to the floor. Airbags can not be repaired, and any tampering is probably on a par with taking the pin out of a hand-grenade and hoping that the handle will stay in place.

Control unit
The control unit, sometimes called the diagnostic and control unit, is central to the operation of the SRS and is a form of ECU. The unit has three jobs: it monitors the system for faults, controls the operation of the airbag and seatbelt pre-tensioners, and stores electrical charge for emergency use. The electrical charge is stored in capacitors within the control unit, so if the main vehicle battery is

damaged or disconnected before or during impact, the capacitors can supply an electrical charge to ignite the pyrotechnic materials in the airbag and seatbelt pre-tensioner. As the electrical current from the control unit can operate the SRS independently of the battery, obviously precautions must be taken to avoid accidental operation before working on the SRS. Typically the procedure is to switch off the ignition and completely remove the key, then disconnect the battery – remember always to remove the earth lead first – and then wait about 10 minutes for the capacitors in the control unit to discharge themselves. The capacitors will re-charge when the battery is reconnected, before the ignition is switched on, so it is prudent to ensure that nobody is in the vehicle in case there is a fault which might accidentally operate the SRS when you are first re-fitting the battery terminals.

Release sensors

Most vehicles are fitted with two negative acceleration (deceleration) sensors wired in series. This is a form of logic gate so that the airbag and seatbelt pre-tensioners can be activated only if both sensors agree. The sensors may be fitted in a number of different places. Generally the main crash sensor is located close to, or in front of the vehicle centre of gravity, and is usually bolted to a longitudinal structural member of the body/chassis.

The crash sensor operates the SRS at a deceleration of about 15 g, that is like hitting a wall at 20 mph. The second sensor, called the safing sensor, will trigger at about a tenth of the crash sensor's deceleration, that is at about 1.5 g. If a sensor is faulty, the control unit should find it and illuminate the warning light.

Rotary coupling

The SRS rotary coupling, between the steering wheel and the outside of the steering column, is unlike any other electrical coupling. It is not a friction terminal. The SRS rotary coupling consists of a length of wound wire ribbon which is permanently connected at both ends so that there is no risk of a bad connection. The ribbon coils and uncoils inside the coupling housing as the steering wheel is turned. Before removing the coupling, the steering wheel must be centralised and the road wheels set to the straight ahead position.

Wiring harness

The SRS wiring harness is a special construction; do not attempt to replace any damaged wires or change the plug terminals. The wiring harness can be damaged by heat, so if an airbag or seatbelt pre-tensioner is fired the SRS wiring harness must be replaced too. Similarly, if the vehicle suffers from any fire damage, the SRS wiring harness must be replaced. The damage caused by heat can not always be seen from a visual inspection.

Warning light

The warning light is usually located on the dashboard; but sometimes can be found on the airbag casing in the steering wheel. When the ignition is first switched on, the warning light will illuminate for between three and five seconds to indicate that the system is functional and that the control unit is carrying out a self test of the system.

The amount of time that the lamp is lit indicates the length of the system check. If the system is correct the light will go out. If the light stays on then the system has a fault. On some vehicles the warning lamp will flash to indicate a fault code – this may need to be triggered by a diagnostic tool. Some vehicles have an SRS OFF switch. This is useful when carrying babies or small children in seats which are protected by airbags, or the seatbelts are used as part of a child restraint system. Usually, a small white lamp adjacent to the switch is illuminated when the SRS is in the switched off mode.

13.18.3 Safety check

Do

- Check all procedures in the workshop manual before working on the SRS.
- Ensure that all replacement parts are of the correct type and in good condition.
- Carry out any diagnostic procedures using the correct equipment.
- Carry and store the airbag with the cover facing upwards.
- Keep all SRS components dry.
- Store all SRS components in a locked safety approved area.
- Ensure that nobody is in the vehicle when re-connecting the battery.

Do not

- Drop any SRS components.
- Ever rest anything on an SRS component.
- Place an SRS component near electrical equipment, any source of heat or flames, or anything which gives off electromagnetic radiation.
- Carry SRS components inside the passenger compartment of a vehicle.
- Attach anything to, or otherwise mark components.
- Attempt to tamper, dismantle or repair any components.
- Attach any other wires to the SRS circuit.

Crash sensors

As well as the release sensors for the SRS, many vehicles are fitted with a crash sensor which cuts off the fuel pump in the event of a serious impact. This can usually be reset by pressing a button on the actual sensor. The sensor is usually positioned behind the glove box.

Security and alarm systems

On most vehicles there are three different systems in operation:

- Immobilisation – An electronic system, usually in conjunction with a key, or a key pad, that immobilises the engine unless the correct key is inserted or the correct numbers are entered in the key pad.
- Perimetric – A series of micro-switches on the doors, bonnet and boot which set off the alarm if they are opened when the system is alarmed.
- Volumetric – A sensor inside the car which senses movement or changes in pressure if something moves inside the car or a door is opened.

Air conditioning

Air conditioning heat exchangers (or radiators) are often incorporated in, or mounted adjacent to, the cooling system radiator. Be very careful when dealing with air conditioning radiators and other parts. The fluid in the system is under pressure and should be safely removed before disconnecting any parts. After changing parts the system will need re-charging and re-setting – this requires special equipment. On older cars the air conditioning fluid was R12 – this contains CFCs and must not be allowed to escape as it causes damage to the ozone layer and can poison the operator. An expensive evacuator and re-filler is needed to re-charge the air conditioning system.

Fuses

You will find up to 20 or more individual fuses on vehicles. Do not replace a blown fuse unless you are sure that the circuit fault which caused the fuse to blow as been repaired.

Sensors

When carrying out body repair look out for sensors. Sensors for a wide range of uses are fitted, including outside temperature sensors, rain sensors and on the Mercedes McLaren SLR (Figure 13.76) tyre pressure sensors on the wheel inner arches.

Checklist

After completing any repair it is a good idea to check all the electrical/electronic components. A PDI checklist for the particular vehicle is useful to do this job.

13.18.4 Suggested further reading

Tom Denton, *Automobile Electrical and Electronic Systems,* Elsevier, ISBN 0 7506 6219 0.
Allan Bonnick, *Automotive Computer Controlled Systems,* Elsevier, ISBN 0 7506 5089 3.

Questions

1 Describe, with the aid of sketches, the processes of hollowing and raising. List the necessary tools for these processes.

2 Describe the process of planishing.

3 Compare the differences between planishing and wheeling a panel to produce a smooth surface.

4 Explain the method and name the tools used in fabricating by hand an aluminium component with a high-crowned surface.

5 State two important conditions to be observed when using a wheeling machine.

6 After extensive wheeling on an aluminium body panel, the metal becomes work hardened. Describe a method used to anneal the panel, and

a workshop test which will indicate when the correct temperature is reached.

7 Describe, with the aid of sketches, the technique of split and weld for producing a hand-made panel.

8 What are the purposes of forming angles and flanges along the edges of sheet metal panels?

9 With the aid of sketches, describe three methods of edge stiffening a sheet metal body panel.

10 Explain the terms 'direct hammering', 'indirect hammering' and 'pick hammering', as applied to body repair techniques.

11 Describe the technique of cross filing.

12 Describe, with the aid of sketches, three different methods of using a portable hydraulic body jack in the realigning and straightening of vehicle body damage.

13 Explain each of the following stages of repairing a front wing which has received minor damage: (a) the analysis and assessment of the extent of the damage (b) selection of the appropriate tools needed to carry out the repair (c) the repair techniques required (d) the alignment and final check.

14 Explain the process of body soldering when used in the repair of a damaged body panel, and list the necessary equipment.

15 Describe, with the aid of sketches, the technique of hot shrinking to restore a panel to its original contour.

16 Explain the difference between the hot shrinking and cold shrinking techniques.

17 When is it considered necessary to fill with body solder, in preference to any other method of repair?

18 Describe a method of repairing a sill panel which is showing signs of corrosion.

19 What precautions must be taken, after body soldering a repair, to prevent subsequent paint defects?

20 Explain the importance of tinning a panel before commencing body soldering.

21 State the disadvantages associated with the use of hydraulic push rams when used on the inside of a vehicle.

22 Give one practical application of the body jack when being used for each of the following operations: pushing, pulling and spreading.

23 Describe a suitable method of removing the damaged skin during the repair process of reskinning a door panel.

24 Explain how a new door skin is secured to the door frame, and list the tools required.

25 Indicate the type of welding necessary when securing a new door skin to the door frame.

26 Why is it important to clean the underside of a damaged panel before commencing repairs?

27 Illustrate a tool which could be used in the removal of damage from a double-skinned section.

28 When, and how, would a pin welding gun be used in repair work?

29 What type of attachment would be used on the end of a body jack to spread the force of the load during a repair?

30 Explain the differences between a hydraulic push ram and a hydraulic pull ram.

31 The lower section of an A-post has suffered corrosion damage. Explain a suitable method of repair.

32 Describe a suitable process for the removal of spot welds from a body panel without distorting the flanges.

33 Describe the sequence of repair (not replacement) to a rear quarter panel which has received minor accident damage.

34 With the aid of a sketch, show how a body jack can put a door panel in tension.

35 A small corroded area of a wing needs a patch repair. Explain how this process will be carried out.

36 Describe, with the aid of sketches, how the hydraulic system of a body jack works.

37 With the aid of a sketch, show the techniques involved in the manufacture of a small double-curvature panel by hand.

38 Describe, with the aid of a sketch, a vector pull for correcting vehicle body alignment.

39 Explain why careful study of the accident damage is important before any stripping or repair work is carried out.

40 Explain the importance of correct alignment when fitting new body panels and state how this is achieved.

Major accident damage

14.1 Damage classification and assessment

Damaged bodywork is corrected by first observing the extent of the damage, then deciding how it was caused and the sequence in which it occurred. The resulting damage can be classified into two groups:

Direct or primary damage This results from the impact on the area in actual contact with the object causing the damage. This will result in the largest area of visible damage and is the cause of all other consequent damage. Primary damage is identified by first determining the direction of the primary impact. This knowledge will help in the search for concealed damage.

Indirect or secondary damage This is usually found in the area surrounding the direct damage which causes it, although in certain cases it may be some distance from the actual point of impact. After the impact, internal damage is caused by the forced movement of objects and passengers towards the point of impact, and can be seen in the form of damaged dash panels, broken seat frames and twisted steering wheels.

These two groups can each be subdivided in two further ways:

Visible damage This is damage that can be readily seen in the area of actual contact, such as a vehicle having suffered frontal impact causing damage to the bumper, grille, bonnet and front wings. A detailed examination may discover distortion of the inner wing valances, which would indicate visible indirect damage.

Concealed damage This is indirect damage, but is not easily detected by visual examination unless the vehicle has been partially dismantled to allow a detailed inspection. In most situations measuring equipment in the form of body jigs must be used to detect concealed damage, because complicated monoconstructed vehicles may hide further damage such as misalignment, which could therefore affect the steering and roadworthiness of the vehicle.

Direction of damage

Direction of damage, or line of impact, is particularly important to the body repairer. It is used to identify the sequence and direction in which the damage occurred, and consequently the reverse sequence to be followed for the repair. Direct damage marks are usually scratch marks where the paintwork has been damaged; they are an excellent guide to what happened, and indicate the possible location of any hidden damage.

By careful study of the damage sustained it should be possible to ascertain the direction and strength of the impact force, and this is always the preliminary stage of a detailed assessment.

Methods of describing major damage

Parallel side damage is caused by the impact object moving parallel to the vehicle and causing substantial damage along the full length of the side of the vehicle, e.g. wing, doors, rear quarter panel.

Direct side damage is caused when the vehicle is struck at an angle to its side, causing substantial damage at the point of contact.

Front-end damage is the result of head-on collision, collapsing panels from the bumper to the front bulkhead.

Three-quarter frontal damage is the result of an angled front end collision, sustaining damage to one front wing, the grille and bonnet.

Rear-end damage is the result of an impact direct or slightly angled, to the rear end of a vehicle and causing substantial damage to bumper, rear panel, boot lid, boot floor panel and quarter panels.

Roll-over damage is caused by the vehicle rolling completely over and returning to its wheels. In some cases this type of movement could cause damage to almost every panel.

Total write-off is damage so extensive that the vehicle is either unrepairable or the total cost of repair would be greater than the value of the vehicle.

The proportions of damage sustained in different directions is shown in Figure 14.1.

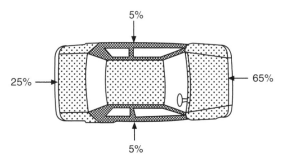

Figure 14.1 Proportions of damage sustained in different directions in UK accidents (*Motor Insurance Repair Research Centre*)

14.2 Pulling, alignment and repair systems used on major accident damage

Since the monocoque construction of car body-work based on engineering principles took the place of coach-built bodies mounted on a separate chassis, the whole concept of repair work has undergone radical changes in both methods and the equipment employed. Repairs to damaged body shells mean a great deal more than the beating out of a few dents, or even the cutting away of a damaged portion and welding in a new pressing.

The majority of motor cars on the road today are of unit or mono construction. Instead of a body built on to a chassis, the whole unit is constructed round a framework of light sheet metal box sections which provide an overall rigidity of exceptional strength and twist resistance. Owing to this construction, collision damage usually is confined to the area of impact; the box sections tend to crumple or concertina, thus absorbing much of the shock and rarely permitting end damage to travel beyond the centre of the car.

In order to deal with the major damage which occurs to the mono constructed vehicle, specialist equipment has been developed. The equipment consists of frame and body straighteners to pull the damaged body shell back into alignment. These are used in conjunction with body repair, alignment and measuring jigs and thus provide accurate location points where exact positioning becomes critical, especially when the panel assemblies carry major mechanical components and suspension units.

The new generation of pulling and pushing equipment began in the 1950s with the introduction of Dozer equipment, which enabled the repairer to use external hydraulic pulling to aid the repair. Then came the idea of repairing a vehicle on an alignment jig using attachable/detachable brackets, so that it was unnecessary to check the repair after every pull. Next came the universal measuring system or bracketless jig, which involved a measuring bridge or in some cases laser measurement. Multipulling repair systems were introduced, which were either static or mobile and which used the vector pulling principle. Later came mobile jig benches and lift operated systems with pulling attachments. Finally, dual repair systems give the repairer the advantage of precision measuring allied to the use of brackets.

The types of pulling, alignment and repair systems described in this chapter include the following:

1 Portable pulling equipment (Dozer system)
2 Stationary pull-post systems (floor anchored)
3 Stationary floor mounted rack (Korek system)
4 Stationary floor mounted rails
5 Floor anchored or anchor pot pulling systems (Mitek)
6 Stationary bench-type jig mounted pulling systems (universal or brackets)
7 Mobile bench-type jig mounted pulling systems (brackets, mechanical measuring, electronic measuring, laser measuring)
8 Mobile bench-type jig mounted pulling system using universal brackets
9 Centre-post hoist jig system with pulling equipment (brackets/measuring)
10 Four-post hoist jig system (brackets/measuring)
11 Scissor-type hoist jig system with pulling equipment (mechanical or electronic measuring)
12 Drive-on bench systems with tower ram pulling (mechanical or laser measuring)
13 Multifunctional repair systems with pulling and measurement systems.

Pulling equipment: general description

Pulling equipment available for major accident damage can range from a single mobile pulling unit to multipulling units; these may be arranged around a bracket jig system or a measuring bridge system; they may be set up on mobile benches, centre-post or four-post lifts; and they may incorporate brackets, or measuring systems or a combination of the two. The choice of systems is varied and wide and depends on the cost involved in the initial outlay, the requirements of the individual workshop and the workload of the shop (based on the number of vehicles repaired each week).

14.2.1 Dozer equipment

Blackhawk Automotive Ltd have a portable pulling frame that can be taken to the car in any part of the body repair shop and then moved on to the next job once straightening has been completed. This frame is called Dozer, and is designed to cater for various classes of repair work. The Dozer (see Figure 14.2) is designed for heavy major frame and body

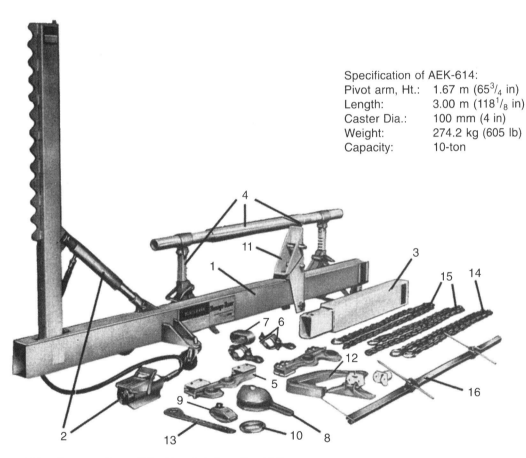

Specification of AEK-614:
Pivot arm, Ht.:   1.67 m ($65^3/_4$ in)
Length: 3.00 m ($118^1/_8$ in)
Caster Dia.:   100 mm (4 in)
Weight: 274.2 kg (605 lb)
Capacity: 10-ton

Figure 14.2   Damage Dozer (*Blackhawk Automotive Ltd*)

1	Basic frame assembly	9	Pull clamp
2	Air hydraulic unit	10	Pull ring
3	Extension beam	11	Multiposition anchor post
4	Support stands and cross tube (set)	12	Multipull dozer hook
5	Underbody clamps (pair)	13	Frame horn puller
6	Cross tube clamps (pair)	14	Chain with two hooks 2.70 m
7	Chain positioning loop	15	Chain with one hook 2.70 m
8	Multi-direction, self tightening clamp	16	Tram track gauge

damage, for light frame and body straightening, and for use in conjunction with a repair bench.

Basically the Dozer consists of a base beam which is the backbone of the unit, an upright beam pivoted to the base beam, a hydraulic ram across the angle and an adjustable rear anchor post. The secret of this equipment's success lies in its anchorage to the car, using body anchor clamps in pairs, either standard for cars with vertical pinch welds (Figure 14.3) or right-angled clamps for those with horizontal pinch welds (Figure 14.4). These clamps each have two sets of jaws so that the thrust of the equipment is spread over four points at the strongest part of the car's underframe, and should be fixed as near to the

Figure 14.3 Standard body anchor clamps (*Blackhawk Automotive Ltd*)

Figure 14.4 Right-angled body anchor clamps (*Blackhawk Automotive Ltd*)

damaged area as possible to restrict the stretching action to this area only. The support cross tube with reinforcing pipe is slid through the bosses in the body anchor clamps, thus providing a rigid thrust point for the rear anchor post (Figure 14.5). The ends of the cross tube provide a lifting point for the adjustable safety stands, enabling the car to be raised to give access under-neath. The stands are adjustable to compensate for uneven flooring. The hydraulic ram may be actuated either by a manual pump or by the air/hydraulic pump, which is quicker and easier to use and leaves the operator with free hands. The pull is made by a chain taken round the pivot post, located by means of the sliding chain guide, and passed through the selected clamp or attachment. The chain then is hooked back on itself, thus providing a double chain. A range of self-tightening clamps and accessories is available (Figure 14.6) designed to clamp on to or be fixed to the wide variety of sections which may require to be pulled.

With this equipment and a little ingenuity it is possible to get a direct pull to the front or rear, or to either side by altering the position of the equipment, an upward or downward pull by varying the height of the chain, and a side pull, so that by varying these set-ups it is possible to pull over a full 360°. This technique also can be applied for panel tensioning to correct indentations in boot lids, bonnets and doors. In the majority of straightening operations it is also necessary to use a hydraulic ram with special attachments in conjunction with the frame and body straightener to keep the metal moving in the correct direction and to remove secondary damage and stresses set up in bent metal. This equipment overcomes the limitations of the body jack in three ways. First, it takes the reaction to more than one of the strongest points on the car underframe, so that the problem of a single point strong enough to push from is overcome. Second, the Dozer can be lined up to pull in any direction, so that the problem of direction of pull does not arise. Finally, all connections are made externally so that there is no necessity to strip interior trim or remove assemblies to provide access (Figure 14.7).

14.2.2 Korek body/frame straightening system

The system commences from a flat base area which is actually a high-tensile steel, box-section modular

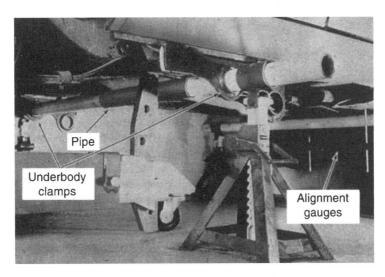

Figure 14.5　Anchor clamps in position on body sills

Figure 14.6　Self-tightening clamp attachments

Figure 14.7　Dozer in position

frame with a slotted upper face. It can be installed in the workshop either by concreting it into a suitable position in the floor or by siting it on a level area of the floor. If the latter method of installation is used, approach ramps are provided for positioning the car.

The equipment uses multiple anchorage and hook-up points which provide tremendous versatility in the direction of pull available (Figure 14.8). Four automatic air/hydraulic pumps with 10 tonne rams are used for the pulling. These are connected to the base frame by special ram foot clamps and are secured by a tapered steel wedge that is hammered into position through the clamp, forcing it tight on to the frame.

To repair a damaged car it must first be positioned over the main frame and raised on safety stands, which are fastened through tubes and pinch-weld clamps to the underside of the vehicle. At this point the vehicle should be anchored to the main frame by the use of chains and clamp attachments to hold it against the corrective pulling forces. On selecting the appropriate pulling positions, clamps are attached to the damaged section and chains are hooked on to the clamps and placed in line with the required direction of pull. The ram units are positioned at the correct angle for pulling by setting the bottom into a ram

Figure 14.8   Korek system with accessories, including P188 Mk2 measuring system (*Blackhawk Automotive Ltd*)

foot clamp which has been secured to the frame, and at the top by the use of a cross-pin through the chain. The loose end of the chain is then fastened to another ram foot anchorage point. This procedure is repeated with each of the four hydraulic units when a multi-pull set-up is required. The foot pumps actuate the ram units which in turn tension the chains and slowly pull out the damaged section to its original position.

The advantage of this system is that cars can be required singly, or several vehicles can be straightened at the same time using the rigid base (see Figure 14.9). Multiple pushes and pulls can be delivered by the vector-type rams, singly or simultaneously, thereby equally distributing the forces applied in repairing damaged components. The system can also be used with various measuring systems, for instance a mobile bench or Blackhawk's P188 Mk2 universal measuring and correction system.

14.2.3 Mitek anchor pot collision repair system

The Mitek is a transportable floor-based hydraulic pulling system using anchor pots set into the workshop floor (see Figure 14.10). These anchor pots are like long plugs with chains attached to the base and a cap on top which leaves a flat floor when the system is not in use. The anchor pot is pretensioned to expand the pot sleeve and give a firm anchorage point. These pots. to which the vehicle and Mitek power unit are secured by the use of tie-down chains, can be fitted into any workshop floor, usually to form a working bay ranging from 12 anchor pots to as many as 28 according to the type of repairs to be carried out. The system uses the vector principle of exerting pressure on chains to pull or push in any direction to carry out repairs on any part of the vehicle.

The vehicle is secured to the anchor pots using tie-down chains and wedge anchors which are fastened to the anchor pots by pulling each chain tightly by hand to remove excess slack and then inserting it into the fork of the wedge anchor. The anchor pot chain is then placed in the slot of the wedge anchor base, driving the wedge into the anchor to take up any slack which remains in the chain. Single or multiple Mitek pulling units are then also secured to the anchor pots using tie-down

Figure 14.9   Korek system straightening two cars at the same time (*Blackhawk Automotive Ltd*)

Figure 14.10   Mitek system with accessories (*Blackhawk Automotive Ltd*)

Figure 14.11   Mitek system in use, with P188 Mk2 measuring system (*Blackhawk Automotive Ltd*)

chains and wedge anchors and siting them in the direction of the pull or push required. Pulling heights can be adjusted by using various attachments to push or pull in all directions up to a height of 2.5 metres. The Mitek can be used with a measuring system so that repairs can be checked as they are pulled (see Figure 14.11).

14.2.4 Celette UK 5000 pulling system

This is powerful pulling system which is equally suitable for cars or light commercial vehicles. It can be used with any Celette mobile bench.

The system 5000 consists of one or more mobile pulling towers, which can be easily moved in any direction on six retractable wheels. The tower can be instantly held in position using chains and self-tightening eccentric anchoring blocks, which fit into slotted channel sections set into a concrete floor in such a way that they form rectangular shapes, thus enabling pulls to be made on four sides of any repair. A special anchoring beam attached to the tower ensures complete safety when the system is in use by locking it into the channel section. The

maximum pulling force on the system is 6 tonnes and it can be used at any height up to 3 metres, which is particularly useful for pulling out damage at roof level on a car or a commercial vehicle. The ram stroke has a length of travel of 500 mm, which has the advantage of speeding up the pulling operation and hence the repair. The angle of pull can be controlled throughout the traction.

The system can also be used for pushing or as a crane using optional accessories. Several simultaneous pulls can be carried out using a number of towers. The equipment can be used with a mobile bench, which can be anchored to the slotted channel sections using chains and anchor blocks. The vehicle itself is then held in place using individual jacking sill clamps attached to the bench (Figure 14.12).

14.2.5 Alignment and repair systems

Jig brackets: general development
Vehicle design and manufacture have progressed considerably over the years. The development of crash repair systems has likewise mirrored these changes. When cars were chassis built, the chassis

Figure 14.12 System 5000 pulling tower and frame (*Celette UK*)

tended to be relatively flat, so that the repair was mainly two dimensional. The change to lighter monocoque construction introduced one other dimension which demanded further consideration. First, the height of the critical points on the vehicle's underside were controlled relative to an imaginary datum line, and so height became of vital importance. Second, accuracy became a far more critical consideration than in the past, as suspension systems became less adjustable and more complex.

Jig brackets were introduced to serve as simple go/no-go gauges, providing a continuous visual indication of the extent of damage. During repairs brackets were found to yield further benefits. They provided progressive anchorage, meaning that once a crititical point had been repaired and locked in place with the jig bracket, it would not be disturbed by subsequent repairs. In addition, during building the jig brackets act as a welding jig to ensure that new parts are correctly installed and that the positions of important steering and suspension pick-up points are not disturbed by welding heat.

Concurrent with these developments, vehicle manufacturers were concentrating on particular aspects of car design. They were attempting to improve body weight by the introduction of high-strength steels and vehicle safety by incorporating strong passenger compartments with front and back and crumple zones to absorb impact on collision. These two developments mean that cars are more susceptible to damage in the crumple zone areas at the front and

rear ends as a direct result of the built-in safety characteristic of deformation of structures. Consequently more cars now need alignment and repair requiring the use of jig equipment. In particular there has been a growth in the number of cars sustaining damage which, while not major, still requires the use of proper diagnostic and repair systems to effect a repair. The repairer must therefore choose the repair system most appropriate to the degree of damage present. Consequently there has been a growth in the development of measuring systems which can be used in a wide variety of repair situations with speed and accuracy. With dual systems, which combine both brackets and measurement, the repairer can carry out fast and accurate repairs whatever the degree of damage. However, with any body alignment system, whether jig brackets or a measuring system, it is vital that the completion of repair should be followed by a full suspension and wheel alignment check to make sure that the car is completely roadworthy.

Since the early 1950s there has been considerable progress surrounding jig brackets. On the technical side, developments mean that the jig brackets are now available for a wide range of vehicles, are lighter, easier to use and more readily available. For ease of use and to prevent possible errors, integral brackets fit to a regular-spaced hole pattern on top of five integral beams which are common in all integral bracket sets. The front integral beams run longitudinally, as do chassis legs, thus allowing first improved access to the engine compartment, and second the production of stronger, simpler brackets. All integral sets incorporate a number of check brackets which are fitted without stripping out mechanical components. Each integral bracket carries a plate which shows a reference number, which side of the jig it fits to and which edge points forward.

14.2.6 Celette UK alignment and repair systems

Celette UK offers a wide range of fixed or mobile jigs, from the mobile bench to the four-post jigs lift.

Celette benches

The Celette bench (Figure 14.13) is mounted on four heavy-duty wheels, two of which are fitted with brakes to allow the bench to be locked in position during use. Being completely mobile, it can be

Figure 14.13 Mobile bench showing pulling system and MZ bracket system (*Celette UK*)

moved to wherever it is most convenient and the space cleared after use. It can also be fitted with four bench legs, which allows it to be used with existing in-ground pulling systems. The vehicle can be fastened in place on the bench using either four sill clamps or the modular cross beams. These cross beams are bolted directly on to the bench and have the MZ bracket system attached above them. Once the tops and towers of the bracket system are fitted they give immediate visual indication of the location and extent of damage, and continue to do so throughout the course of the repair (see Figure 14.14). The brackets also act as a solid welding fixture, clamping new parts in their correct position at the critical points during welding up (see Figure 14.15). The most important feature of the brackets is that they provide progressive anchorage, locking newly repaired areas in place so that they are not disturbed by subsequent pulls to other areas.

A hydraulic push and pull unit can be quickly clamped to the jig bench at any point. Its double adjustments at the base of the vertical arm and on the horizontal beam allows the push or pull to be widely adjusted to any angle or height without the need to relocate the unit. In addition this equipment can be used for vector pulling and pushing and upward or downward pulls. The bench can also be used with the Metro 2000 measuring system.

Celette jig lifts

The jig lifts (Figure 14.16) are of wheel-free design which enables rapid loading and anchoring of damaged vehicles. The wheel-free system allows the vehicle to be moved easily lengthways and sideways for easy bracket or sill clamp location. The design of the decking plate enables the sill clamp outriggers to be left permanently in place which gives the benefit of a drive-on and drive-off pulling facility.

When not in use as a body repair system, the jig makes an ideal general service lift, as the vehicle can be supported either on its own four wheels or on its underside using the wheel-free system. The lifting capacity of the jig lifts ranges from 2.5 tonnes to 3 tonnes, and so large cars and light commercial vehicles can be worked on. The jig bed can be fitted with the full range of Celette brackets, including the MZ repair system and the Metro 2000 measuring system. The hydraulic push and pull unit and adjustable sill clamps can be used for anchorage and pulling, providing a dual repair system (see Figure 14.17).

Celette UK MZ system

This system for crash repairs incorporates all the traditional benefits of jigs and brackets, but, in addition, has been designed to reduce the model-specific content of the system to a minimum while maximizing the number of brackets which can be fitted, without the need to strip out the mechanical parts first. It consists of 22 universal lower brackets called 'base towers', into which upper brackets called 'tops' are fitted that are specific to the model being repaired. These tops are much smaller and lighter than the conventional brackets, and so are easier to insert or remove when the vehicle is on the jig. This system was made possible by the use of computer-aided design (CAD) equipment. Details of vehicles currently in production were inputted on to the computer; it then identified the most common reference points and evaluated a number of alternative arrangements. From this a set of universal towers was developed to suit the wide range of different vehicle models. The CAD system is also being used to produce manufacturing and user drawings for the top sets, resulting in the faster introduction of tops for newly designed vehicle models.

The basis of the MZ system is in effect a two-piece bracket set. A model-specific top slides under a piston action into the universal tower unit, and a locating pin holds the two together at the correct height. The top locates into the bodywork of the vehicle in the same way as a conventional bracket, and likewise the base of the tower is fitted to the jig using four bolts into the modularly

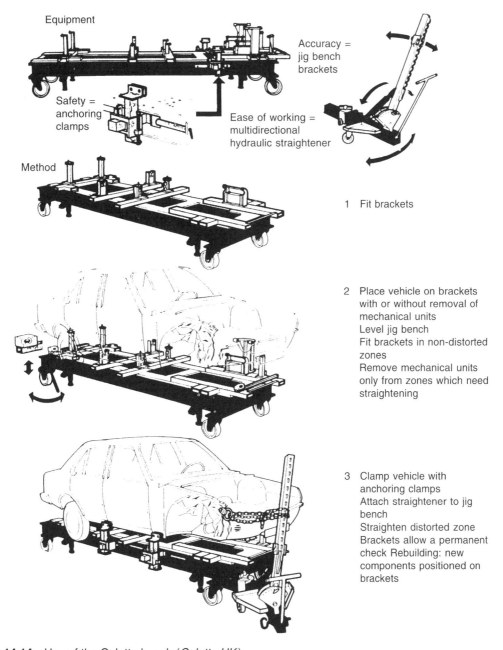

Equipment

Accuracy =
jig bench
brackets

Safety =
anchoring
clamps

Ease of working =
multidirectional
hydraulic straightener

Method

1 Fit brackets

2 Place vehicle on brackets
   with or without removal of
   mechanical units
   Level jig bench
   Fit brackets in non-distorted
   zones
   Remove mechanical units
   only from zones which need
   straightening

3 Clamp vehicle with
   anchoring clamps
   Attach straightener to jig
   bench
   Straighten distorted zone
   Brackets allow a permanent
   check Rebuilding: new
   components positioned on
   brackets

Figure 14.14 Use of the Celette bench (*Celette UK*)

spaced holes. The towers come in five different heights to cater for various requirements. A complete set comprises 22 units, with between 2 and 6 of each height depending on the frequency of use. To cater for variations in width or length, the tower unit is mounted offset on its base. It can therefore be set in one of four mounting points. In so doing the area within the 100 mm modular grid

spacing can be covered; any further variations are covered by additional offset on the model-specific top unit.

The choice of tower, plus its required location and rotation, is clearly marked on the information sheets which accompany all MZ sets, showing both plan and three-quarters view. To simplify identification and rotation, each tower has a coded

Figure 14.15 Car fixed on M8 mobile bench (*Celette UK*)

Figure 14.17 Bodymaster 2000 with Metro 2000 measurement system (*Celette UK*)

Figure 14.16 MUF 61 Bodymaster jig lift with Cobra 3 puller and MZ brackets (*Celette UK*)

arrow which corresponds with arrows on the sheet (see Figure 14.18). When the tower has been fixed, the required top unit can be slid in and held in place by pushing in the locating pin. To cater for mechanicals in requirements two sets of holes are drilled in some towers and tops, the upper set being for mechanicals out and the lower set for mechanicals in.

Celette UK Metro 2000 measuring system

This is a universal diagnostic and repair measuring system based on simultaneous three-dimensional verification of important points of the vehicle. The measuring system is dual purpose as it fits directly to any Celette bench or jig lift, and there is a choice of using brackets or measurement. Thus any degree of damage can be repaired using the most appropriate system. A complete damage diagnosis can be carried out without any stripping of mechanical components, facilitating speedy and effective repairs.

The system is calibrated using points on the vehicle's underbody which are accurately jigged during manufacture. The jacking sill clamps ensure that the vehicle's underbody is brought parallel to the measurement frame; this is vital if height measurements are to be accurate. To ensure accuracies, either the vehicle or the measuring frame must be adjusted so that they are parallel. The calibration procedure is that the vertical scale built into the sill clamps and telescopic pointers set the vehicle at the correct height. The pointer carries slides on transverse rules with their own scales to allow width positions to be set. A tape runs along each side of the frame, and slides within its housing for instant setting. Three or four datum points are selected on any undamaged area, and the system is calibrated to the height, width and length position of these points. The frame can be bolted down after calibration. All eight pointers are

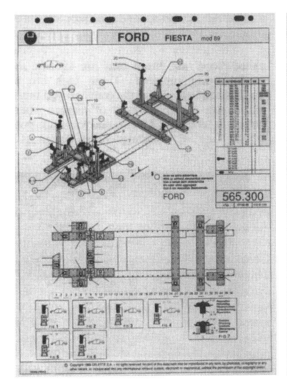 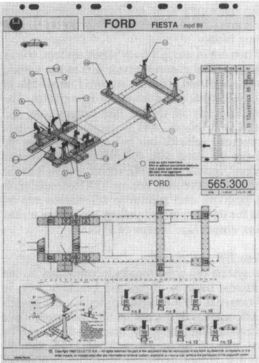

Figure 14.18   Example information sheet for use with the MZ bracket system (*Celette UK*)

available for taking measurements, since none needs to be left in the calibrated position; therefore all eight locations are available for simultaneous checking during repair.

One data sheet is usually required for each vehicle, giving published dimensions from manufacturer's drawings. These data sheets show the initial sill clamp settings, and the critical dimensions for all mechanical units in place and removed. They also give the MacPherson strut data, which enables a check on the position of the front suspension fixing from above without any stripping of suspension components. The checks include the length, width and height positions of the strut fixings and also the angle of the flitch plate; these are all vital to the correct steering geometry of any vehicle under repair (see Figure 14.19).

Celette UK Bodymaster 2000 drive-on repair system

The Bodymaster 2000 drive-on system is easy to operate. Faster repairs are achievable, as vehicles can be loaded and off-loaded very quickly. In order to facilitate the loading and unloading of a damaged

Figure 14.19   Metro 2000 measurement system (*Celette UK*)

vehicle that cannot be driven on to the bench, a winch is provided.

The system has other features which make it attractive to body shop operators because they greatly facilitate the work on crash repairs. These features include a hydraulic adjustable platform and an integral high-capacity twin pulling tower system. The hydraulic platform provides easier and speedier loading and off-loading of a vehicle and allows the operator to set the bench at any suitable working height to a maximum of 700 mm.

Bodymaster 2000 is particularly versatile owing to its 5 metre long tubular steel rigid base. As a result the system is suitable for the repair of passenger vehicles, vans, four-wheel-drive vehicles, and light commercial and chassis-based vehicles. The design of the system enables it to be used with Celette's Metro 2000 measuring system and also with Celette's MZ diagnostic and repair bracket system (Figure 14.17).

Celette UK Metro 90 measurement system

The Metro 90 is a complete crash repair system based on both bracket and measurement methods of repair. Centred around the Celette MT10 mobile bench and Metro 2000 measuring system, Metro 90 is a complete package including a pulling and clamping system, loading kit and full range of accessories (Figure 14.20). Pulling capacity is provided by the Caiman pulling unit, which also provides extra adjustment, making the system easy

to use. An additional wide range of checking accessories also allows more points on the vehicle to be checked, and a telescopic measuring gauge can be used in such areas as engine, upper body compartments, and screen apertures.

The Metro measuring system consists of a rigid alloy frame with eight trolleys that run along the frame. Each trolley is fitted with a half-bridge carrying the vertical pointers and measuring adaptors which allow location on various points of the vehicle's underbody, including horizontal axis locations. Thus asymmetric points can be checked and repaired simultaneously.

The vehicle is supported on four three-way adjustable sill clamps and is calibrated in all three dimensions – height, width and length – on an undamaged area of the vehicle underside. With conventional structural sills being increasingly replaced by aerodynamic skirts, the clamps can use swivel jaws that clamp on to the lower sill panels where conventional clamps cannot be used. The jaws of the clamps rotate through 360 degrees, ensuring that damaged vehicles can be securely held in place, enabling structural alignment to be guaranteed.

The unit is supplied complete with a full set of laminated colour data sheets, and regular updates are available.

Another advantage of the system is that, with the use of beams and towers, it can be used simply as a bracket system, as this mobile bench has also been specifically designed to accept the MZ diagnostic repair system. The Metro 90 can also be used in conjunction with Celette's ECO 2000 mobile lifiting trolley.

Celette UK ECO 2000 mobile lifting trolley

This unit is a mobile lifting trolley which operates using a compressed air supply at a pressure of 7 bar. Capable of lifting weights of up to 2000 kg to a height of 1 m, it provides a simple and effective means of loading damaged vehicles on to a mobile bench, thus creating a comfortable and adjustable working height for various mechanical and servicing work. It is very maneuverable in any workshop area. With its controlled lowering speed and integrated safety system it is ideal for any shop needing a mobile lifting unit, not just the bodyshop but also the general workshop, workbays and mechanical lifting bays (Figure 14.21).

Figure 14.20 Metro 90 measurement system (*Celette UK*)

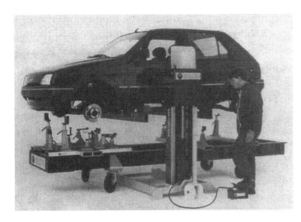

Figure 14.21 The ECO 2000 mobile lifting trolley (*Celette UK*)

14.2.7 Car Bench alignment and repair systems

There is a wide range of body repair equipment available from Car Bench. This equipment is both flexible and versatile, covering a full range of systems designed to suit each individual workshop and all repair situations. It is dual-purpose equipment, suitable both for jig repair using brackets and for repairs using a measurement system. The equipment is described in the following sections.

Mobile bench

The mobile bench (Figure 14.22) is the simplest of all the Car Bench models. It was primarily introduced into the range of equipment for those bodyshops with limited working space and height restrictions. It is mounted on four robust wheels, allowing it to be easily moved around the workshop. This allows the repairer to utilize valuable workshop space. The wheels are completely detachable, allowing the bench to be lowered directly to floor level. This enables a damaged car to be mounted on to the jig frame with the minimum amount of effort. Separate support legs are included for workshops with uneven floors, and they can be adjusted until the jig bed is perfectly levelled. The hydraulic pulling unit can be mounted in any position, so pulls can be made from any angle around the frame at the front, the back or either side. This bench can be used with the autocontrol measuring system, the minibracket system and the traditional bracket system. It can also be converted into a hoist by using the scissor-action lift.

Mobile lift

The lift is an extremely versatile unit which can be easily moved around the workshop (Figure 14.23). It is unique in that the frame may be disengaged from the lifting mechanism, allowing a repaired vehicle to

Figure 14.22 Mobile bench with autocontrol measuring system and swivel pull post (*Car Bench International*)

Figure 14.23 Mobile lift with mini bracket support system (*Car Bench International*)

be wheeled away. It incorporates the same hydraulic pulling equipment and jig/measuring facilities as the mobile bench.

Scissor-action lift

The scissor-action lift is quite unique in that once the frame has been lowered to rest on its wheels, the base unit can be automatically raised from the floor so that it retracts directly into the frame. This gives the necessary clearance needed to enable the mobile bench to be moved freely around the workshop. It is raised by using a mobile control unit with a lifting capacity of 3.5 tonnes, which allows adjustment to height levels to secure the frame in any comfortable working position.

Standard Car Bench four-post lift

The jig itself is a rectangular steel frame incorporated within a four-post 2.5 tonne lift (Figure 14.24). The damaged vehicle can be either driven or pushed on to the bench using a trolley jack. A special trolley is placed under the damaged vehicle, enabling it to be moved easily along the frame to meet the locating points on the jig bed. Once mounted to the vehicle the anchorage brackets are bolted to the predetermined holes in the frame as specified on the jig charts. This ensures that the strongest part of the vehicle is securely anchored to the jig bed and that it remains so throughout realignment and repair. The anchorage bracket incorporates hollow bolts fitted over spring hanger bolts on the front mounting of the rear suspension, enabling the vehicle to be pivoted. This allows further brackets to be fitted to specific points on the vehicle, which may then be used as checking locations. Further brackets, for repair and panel assembly, are provided for fitting to the front and rear suspension and mechanical pick-up points of any type of vehicle. The swivel pull post can be positioned at any point around the frame and locked to it, to allow pulling to be carried out at any angle.

Figure 14.24 Four-post lift with autocontrol measuring system (*Car Bench International*)

Alignment can be precisely checked at any stage of the repair whilst pulling is in progress. This is done by relating the jig brackets to the corresponding holes on both the cross beams and the frame. No further measuring of any kind is required. Therefore the repairer can see how far to push or pull to realign the section under repair. The essential function of this system is to pull and realign simultaneously and rebuild the damaged sections with absolute precision while they are held in place by the brackets. This four-post lift allows work at any comfortable height giving complete freedom and access around and under the bench.

Centre-post bench

This is suitable for workshops with an existing centre piston installed. It has all the characteristics of the standard Car Bench System.

Mini bracket/support system

This system, although greatly reduced in size and weight, still retains the main features of the Car Bench System. The special support arms, to which modified brackets are fixed, are mounted on the cross beams in the same manner as the original bracket system (see Figure 14.23). They can be moved both laterally and longitudinally in order to locate the pick-up points on the underbody of the vehicle. They may also be extended vertically and horizontally to achieve unlimited positions for the attachment of brackets. By doing this, Car Bench has transformed the idea of a traditional bracket, fixed and rigid, to a more functional system which can be adapted to suit each individual repair regardless of the make or model of the vehicle. Mini cross beam plates have consequently been designed to attach to the special support arms and have exactly the same function as the original cross beams, although they can now be moved nearer the car for the attachment of this style of bracket.

Unified support system

The unified bracket system enables Car Bench users to repair any vehicle using a bracket repair technique but without the necessity for hiring brackets. Because of their universal anchorage, this functional system may be used in any repair situation, regardless of the make or model of vehicle being repaired. The unified system may be applied for use with all the models in the Car Bench range of equipment. It consists of four specific anchor points and eight modular base blocks. These modular blocks can be moved both vertically and horizontally to obtain unlimited positions. Once the unified brackets are attached to the base units, they are positioned according to the technical data sheets provided. Because the base blocks, incorporating unified brackets, are fully adjustable, they give total accuracy in all measurement readings. The data sheets provide accurate measurements of vital control points for mechanical parts both in position and dismantled. The system has been designed in such a way that it may be used in conjunction with the Car Bench measuring system and mini bracket system using the base modular blocks as starter packs.

Autocontrol measuring system

This system can be used with any Car Bench model. It gives the repairer the dual advantage of precision measuring allied to the use of brackets. It is designed for use when damage occurs which is not severe enough to necessitate jigging. The system incorporates graduated rails for longitudinal measurements, four sliding cross members for external and internal measurements, and varying lengths of tube with built-in millimetre scales for obtaining height measurements. Four adjustable sill clamps are also supplied to ensure a quick and easy centralization and anchorage of the car to the frame, whilst a special bridging device enables the repairer to check for external damage to the body shell. The system includes a wheeled storage trolley to hold the measuring rods and adaptors, and a catalogue of data sheets.

14.2.8 Globaljig universal bracket system

The Globaljig universal bracket system is manufactured in Italy and is available as a mobile (Figure 14.25), four-post lift or centre-post lift unit (Figure 14.26). It is also available with a side lift, making positioning of the vehicle on the bench very simple; and, because of the mobility of the side lift, the vehicle can be easily moved around the workshop area (Figure 14.27). Conversion sets are also available for other jig benches. It is a universal bracket system, comprising one set of jig brackets that are designed to fit any type of vehicle and light commercials, eliminating costly hire or purchase of specific vehicle model brackets. Each

Figure 14.25 Globaljig mobile universal bracket system (*Globaljig, Tri-Sphere Ltd*)

Figure 14.26 Globaljig centre-post lift unit (*Globaljig, Tri-Sphere Ltd*)

Figure 14.27 Globaljig mobile side lift jig loader (*Globaljig, Tri-Sphere Ltd*)

bracket is designed as a fixing point, providing anchorage throughout the vehicle, with mechanical parts in or out. With the Globaljig there is an option of brackets or sill clamps, providing a precise and progressive anchorage. Precision and strength are the main features of this system: accuracy is guaranteed to within 1 mm. The brackets included in the system are strong and truly universal; each has a single bolt fixing, providing quick and simple assembly. All calibrated scales are easy to read, and are attained by sliding the beams and base blocks to their specific location.

The pulling unit is a multivector system, easily positioned around the jig bed. When placed at the front or the rear of the jig, the puller will pivot through 220 degrees, allowing pulls to be obtained from both sides of the vehicle without removing the unit from the jig bed. This unique design allows the pulling capacity to remain constant at any given height or angle.

The overhead measuring bridge is a standard feature of the Globaljig. It provides a strong, accurate measuring device for suspension mountings and is also used for checking the upper body measurements (Figure 14.28). Once measurements are satisfactorily achieved, the accurate positioning of new panels can be made prior to welding.

Figure 14.28 Globaljig overhead measuring bridge (*Globaljig, Tri-Sphere Ltd*)

Two types of data sheet are supplied. One is for quick diagnostic checks with mechanical parts in position, and the other is for more detailed repairs with mechanical parts removed.

The system can also be used as a building jig for classic cars or tubular framed sports cars, or as control equipment for estimation or diagnosis of damaged or previously repaired vehicles.

In the case of anchor points varying from their engineering specifications owing to manufacturer's tolerances, this system can still carry out the repair because of the total adjustability in all dimensions and the capability to handle tolerances, even allowing for positioning and movement during repair.

14.2.9 Universal Bench car body repair systems

Universal Bench offer a wide range of mobile benches, a four-post jig lift, and a scissor bench. The repair system offers specific anchorage that can guarantee a precise and reliable repair. The UB Modular system gives positive repair with jig brackets. This offers quick and easy mounting on to cross-beams and allows the possibility of checking and anchoring up to 22 points on the underbody of a vehicle (4 anchorage and 18 mini brackets). The pulling equipment can be attached to the bed, giving a pulling facility all round the vehicle. There is a standard four-post bench incorporating lifting facilities, and a four-post lift with a disengageable jig frame. This unit is ideal for busy workshops with fast throughput of work, giving easy mounting and positioning of the vehicle and the possibility of rotating the vehicle on the frame. Disengageable frames allow other frames to be used with the same lift.

Universal Bench have developed a universal repair system that can be mounted to any jig frame by using an adaptor. This is a positive system which allows the repairer to check and hold the most important underbody points on a vehicle. It can also be used for repairs which require the replacement of body panels because of the totally accurate positioning of the panels to be replaced. Data sheets for the system use manufacturers' specifications and measurements. UB Electronics is a measuring system with digital readout. This allows quick and accurate reading of longitudinal, transverse and vertical measurements. The measuring system can be used with all Universal Bench units.

Bracketless universal systems: general description

Bracketless universal systems offer certain advantages over the traditional fixed bracket systems.

The universal jig does not require the attachment of special mounting brackets, and therefore may be used in the repair of any vehicle model. Each system is a self-contained unit, with its own measuring system for checking body distortion and its own pulling and pushing equipment for the repair of any damage, whether minor or major.

14.2.10 Dataliner repair system

Dataliner: general description

The Swedish company Nicator has developed a universal alignment jig combined with a special measuring method. The system is called Dataliner (see Figure 14.29). The straightening bench, around which the system is designed, eliminates model-specific brackets altogether. It allows the repair of collision damage without the extensive dismantling and reassembling of components. Its most outstanding feature is the precise measurement obtainable (± 1 mm) with the use of a laser beam in conjunction with comprehensive data charts suited to different makes and models of vehicles.

Figure 14.29   Complete Dataliner repair system showing laser equipment and pulling equipment (*Dataliner, Geotronics Ltd*)

The bench is made from strong hollow beams, known as the mainframe, to which transport beams are attached in a transverse manner. These transport beams with chassis clamps are first positioned under the vehicle, which is raised from the floor by the use of a high-lift trolley jack. Then the four chassis clamps are positioned and tightened, to the bottom edge of the sill on the vehicle and the top edge of the

transport beams, by using quick-release and snap-on bolts. Once the transport beams have been fitted in place and tightened, their wheels allow the vehicle to be moved to any convenient area in the workshop.

The mainframe of the bench has two wheels which are centrally located. This allows the frame to be pushed under the vehicle. Using the jack support on the mainframe, the frame is lifted and anchored to the transport beams, after which four floor stands are placed under the frame. The floor stands are individually adjustable to raise the vehicle to the most suitable working height for the repair.

Measuring system

The measuring system of the Dataliner is based on the principle that a laser beam can act as an optical square, creating a perfect right angle. By means of transparent scales the plane of measurement is transferred under the vehicle (Figure 14.30). This provides a clear view of all measuring points and allows measurements to be taken and alignment work to be carried out simultaneously. The main components of the measuring system are: a set of measuring rails; a laser unit, which emits a low-power safe laser beam; a beam splitter; and laser guides and transparent scales.

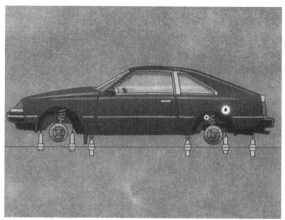

Figure 14.30   Laser beam passing through the transparent scales underneath a vehicle (*Dataliner, Geotronics Ltd*)

When the measuring equipment has been connected to the bench, the long rail is positioned on the least damaged side of the vehicle and the short rail on the front or rear of the vehicle. The transparent scales are hung under the vehicle in accordance with information on the data charts (see Figure 14.31).

Figure 14.31 Transparent scales hung in position for alignment (*Dataliner, Geotronics Ltd*)

The transparent scales transfer the measuring point on the vehicle to a level which can be illuminated with the laser beam. The scales are mounted on special yokes designed for bolt heads and fixture holes. The centre of the scale has a target area through which the laser beam passes when the measuring point is at the correct position. As soon as the scales are in place, calibration can be carried out. This is done by using the laser guide to deflect the light on scales on two undamaged measuring points under the vehicle, vertically adjusting the measuring rail and angling the laser beam using a wheel on the top of the laser guide (see Figure 14.32). The laser guide is equipped with a self-levelling system, which automatically repeats the height angle of the deflected

beam when the unit is moved to other locations on the rail. Calibration is complete when the laser beam passes through the target areas on the scales on three undamaged measuring points. The beam splitter is designed to make it possible for measuring height, width and length at the same time (Figure 14.32). It is sited in front of the laser unit and splits the beam to travel simultaneously along the two rails. It can also be used for the alignment of upper body measurements by deflecting the beam at the appropriate angle. Using the other laser guide on the short rail, the centre line of the vehicle and the width of the underbody can be checked from the front or back of the vehicle (Figure 14.33).

Figure 14.33 Checking the centre line of a vehicle (*Dataliner, Geotronics Ltd*)

Straightening equipment

The straightening equipment is designed on the vector principle using hydraulic rams and chains. Pulling and pushing can be carried out at all possible angles, both lengthwise and crosswise, and at any height. The basic Dataliner kit includes two complete 10 tonne pulling units. A counter-support (blocking arm) prevents the vehicle from moving in the direction of pull, and holds the vehicle so that the pulling forces affect only the damaged zone of the vehicle (Figure 14.34). This blocking arm can be used anywhere on the bench, and with extension tubes it can reach both high and low points of a repair. Downward pulls can be carried out using a chain wheel which can be mounted on the bench frame. As this wheel is pivoted it automatically adjusts itself to the angle of pull (Figure 14.35). On

Figure 14.32 Beam splitter and laser guide (*Dataliner, Geotronics Ltd*)

Figure 14.34 A counter-support (blocking arm) in use (*Dataliner, Geotronics Ltd*)

Figure 14.36 Simultaneous multipulling (*Dataliner, Geotronics Ltd*)

Figure 14.35 A downward pull using a chain wheel (*Dataliner, Geotronics Ltd*)

severe major damage several independent hydraulic units can be used simultaneously at any angle round the bench frame (see Figure 14.36).

A major feature of the Dataliner is that measurement and pulling can be carried out at the same time. This guarantees complete control over the straightening operations from the first pull to the final adjustment of the damaged area. Measuring and alignment also takes place at the same time, which means that a continuous check on the measuring during the whole repair is possible. The position of the laser spot on the transparent scale shows when the correct alignment point has been reached.

Accessories are available to measure the MacPherson suspension strut tower without any dismantling (Figure 14.29). A flex lock arm will fix and hold panels in position during assembly and welding.

Dataliner drive-on rack system

This system has been designed to repair cars and small commercial vehicles. It is a complete workstation that can also be employed for body preparation, mechanical work and as a platform for wheel alignment. The system features versatility as a multifunctional piece of equipment. The construction of the rails permits limitless tie-down points. Chassis clamps for unibody vehicles can accommodate vehicles of all widths. The working height of the track is adjustable, and the work platform makes it easy to reach all points of vehicles being repaired. This equipment is used in conjunction with Dataliner's laser measuring system, and has multipulling facilities (see Figure 14.37).

Dataliner laser check

Dataliner laser check has introduced a system for checking car chassis. This system is faster, more reliable and more accurate than conventional measuring methods. The laser check is based on Dataliner's laser measuring system in combination with a computer program for fast entry and

Figure 14.37 Dataliner drive-on rack system (*Dataliner, Geotronics Ltd*)

checking of measurements. The equipment consists of:

1 Laser guide with electronic calibration and digital measurement display
2 Hand unit with a memory capacity for four different car models
3 Personal computer (PC) for connecting to hand unit for registration of data and specifications
4 Printer for printout of measurements
5 Modern with access to central register and specification for all car models.

With this system, calibration of laser measurements has been simplified considerably. Measurements are made electronically and are accurately displayed digitally on the laser guide. It is not possible to obtain incorrect readings. With the hand unit connected, measurements appear immediately on the unit display. The hand unit has a storage capacity for data and specifications for up to four different car models (Figure 14.38). The test car measurements can be quickly compared with specifications, with deviations displayed in millimetres. When measurements and deviations need to be documented, the hand unit can be connected to a printer for printout.

Figure 14.38 Hand-held unit with memory (*Dataliner, Geotronics Ltd*)

All marketed car models will be found digitally registered in the company's main computer. Measurements are available either on disc or via modem from the main computer. All data and specifications can be stored in the PC and are easily transferred to and from the hand unit via an adaptor. The program is simple to understand and requires no special knowledge of computers.

14.2.11 Blackhawk mobile repair system PB30

The Blackhawk mobile repair system PB30 offers a completely portable repair system that can easily be moved around the workshop. This ensures ease of access to the most difficult repair job even in a limited space. The equipment is available in a single-pull or a two-pull version. Easy working access is ensured by the absence of lift corner posts. It provides a body shop with a versatile and comprehensive means of undertaking crash repairs accurately and efficiently, with minimal disruption to vehicle components.

The complete repair system consists of a base frame and two main elements: a totally independent measuring system allowing complete flexibility of operation, and a pulling system using the Vector principle for precise directional pulling and pushing. The base frame is manufactured from polarized steel. It supports the vehicle, the pulling arm and the measuring system. The vehicle is attached to the frame using sill clamps which are wedged into lateral beams. The vehicle may be loaded on to the sill clamps by using a high-lift trolley jack, a two-post lift or a side loading lift (Figure 14.39).

Figure 14.39 Blackhawk PB 31/8 mobile straightening and measuring system (*Blackhawk Automotive Ltd*)

The measuring system works on the principle that a measuring scale can locate vital alignment points in the length, width and height on every vehicle model. These measurements are taken from the zero plane or datum line. The zero plane is the most basic but accurate of all available references, and is used by the vehicle manufacturers to establish vehicle dimensions. The extension scale not only measures height, but when installed on a telescopic cross slide it also measures width. These cross slides glide backwards and forwards on a longitudinal beam to record length measurements. Data sheets provide the necessary information to identify the car's main measuring points. Pneumatic activators (air bellows) align the system to key datum points on the vehicle using respective adaptors from the range supplied. These adaptors rise to contact the car underbody and are held firmly in position by air pressure. They stay with the vehicle owing to the elasticity of the air bellows, and any pressure changes on the measuring unit are controlled by internal regulators. The measuring unit is then perfectly located and parallel to the zero plane and the manufacturer's datum line. These adaptors are engineered to locate precisely, and in use are selected according to the information given on the vehicle's data charts which show adaptor type, datum points and their coordinates (see Figure 14.40).

The hydraulic repair system includes a multidirectional push-pull facility together with accessories. The measuring system can be used with ground anchorage collision repair systems as well as mobile repair systems.

14.2.12 Car-O-Liner repair and measuring systems

Car-O-Liner have introduced the Mark 5 alignment jig system together with a drive-on collision repair system called BenchRack and an electronic measuring system called Car-O-Tronic which can be used on both jigs.

The Mark 5 collision repair system

This is a universal measuring system developed from the Mark 4 system. It can be either mobile, static or with a built-in lift. The Mark 5 is capable of lifting a car to a height of 1.4 metres using the scissor lift, thus allowing the bodyshop staff to work standing up, and therefore reducing stress and increasing speed and productivity. It provides increased working room between the jig and the vehicle, therefore allowing ease of access for movement and tools. Positioning the vehicle is achieved with the help of rolling ramps. After the ramps have been used, they can be folded up and used as stands

Figure 14.40 Blackhawk PB30 measuring system (*Blackhawk Automotive Ltd*)

or as a work stool, thus helping to keep the work-shop free from additional equipment. The correct pulling angle is essential for successful alignment work. The Draw aligner finds the pulling angles easily with almost continuous adjustment and has a built-in telescopic extension arm which allows for repairs needing high pulls.

The Mark 5 equipment consists of the following: alignment bench, Draw aligner, measuring system (Figure 14.41).

The base frame of the alignment bench is a welded structure built from square cross-section tubing, and can be fitted with an integral scissor lift or a single column lift. The upper surface of the frame is milled flat to serve as a reference plane, and there are milled tracks on all four sides for attachment of bench wheels, bench mountings and jack plates. The bench mountings are located in the side tracks of the base frame and can be adjusted for height. The mounting arms can be turned either inwards or outwards to suit the width of the vehicle. The chassis clamps are bolted to the arms of the bench mountings and are fitted with tooth jaws which grip the sill flanges of the vehicle.

Figure 14.41 Mark 5 collision repair system (*Car-O-Liner (UK) Ltd*)

The dolly sets consist of a number of components which can be used in various combinations and are mounted on the bench frame. Any required point on the vehicle can be fixed in position by the use of these dolly combinations. They can also be used for controlling the pull in certain directions during alignment and as a welding fixture when new parts are to be welded in.

The Draw aligner is used with a 10 tonne hydraulic ram operated by a pneumatic foot pump, and can be placed at any position along the four sides of the base frame, being locked on to the base frame by the locking wedge. The Draw aligner pivots in the sideways direction and is locked in the required position by a peg. Next the arm can be inclined in the sideways direction to give the optimum pulling angle, then locked in position. The arm is fitted with an extension for high pulls. A safety wire is fitted between the arm and the hydraulic ram mounting in order to limit the outward movement of the arm (Figure 14.42).

Figure 14.42 Mark 5 bench showing the scissor lift and Draw aligner (*Car-O-Liner (UK) Ltd*)

The measuring system can be either mechanical or electronic. The mechanical system consists of a measuring bridge which is made of aluminium and is provided on both sides with movable measuring tapes for length measurements. It is designed to be placed on the flat milled surface of the base frame. The measuring slides decide the width measurements and run along the measuring bridge. There is a pull-out slide on each side of these which is provided with three mounting holes for height tubes. Readings are made on scales on the side of the tubes. These height tubes are of various lengths and marked with sliding scales with millimetre graduations for reading height measurements. There are various measuring adaptors such as sockets, cones and angle blocks which are fitted into the top of these height tubes. The combination of height tube and scale adaptor that should be used at the point to be measured is shown in the

data sheet for the particular vehicle (Figure 14.43). Equipment is also available to measure the upper anchor points of MacPherson struts together with the dimensions of the engine compartment and the whole of the upper body (Figure 14.44).

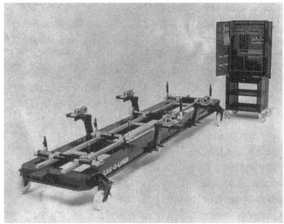

Figure 14.43 Mark 5 measuring system (mechanical) (*Car-O-Liner (UK) Ltd*)

Figure 14.44 MacPherson strut measuring equipment (manual) (*Car-O-Liner (UK) Ltd*)

The data sheet files contain checklists for most car models and show height, length and width dimensions at a number of measuring points on the vehicle. The data sheet also gives information on the correct points for placing the chassis clamps for locating the vehicle on the bench.

Drive-on collision repair system BenchRack

The BenchRack can be used on passenger vehicles and light transport vehicles. The ramp system enables the vehicle to be secured to the bench quickly. The BenchRack is equipped with a built-in hydraulic scissor lift which can lift a weight of 3 tonnes to a height of 1.4 metres, allowing the operator to select quickly a suitable height for any type of repair undertaken. The 12 removable ramp sections are made from lightweight alloy and fitted in the tracks on the side of the bench. The ramps can be removed when necessary to facilitate access when measurements are taken or alignment work is carried out. Adjustable chassis clamp arms may be raised or lowered as needed, or can also be turned in for use on pick-up trucks and other narrow framed vehicles. Chassis clamps hold the vehicle secure while pulling is taking place. Two 9 tonnes Draw aligners provide the pulling force, 360 degrees around the BenchRack. The side support system is used to hold undamaged and already repaired components in place while pulling other areas. The measuring can be carried out using the Mark 5 mechanical system or the Car-O-Tronic electronic system, for fast and accurate measurements of any vehicle damage.

Car-O-Tronic measuring system

The Car-O-Tronic is an electronic computerized measuring system that is universal. It is designed primarily to measure and check the dimensional correctness of vehicle chassis. Car-O-Tronic measures either with reference to the Car-O-Liner data sheet or on an absolute basis. Advanced mathematics and computer technology mean that this system can be used without any other special equipment. All that is needed to make the system operational is a flat smooth surface, a measuring bridge, the electronic measuring arm and a hand terminal (Figure 14.45). The system is easy to operate: the measuring arm is moved along the

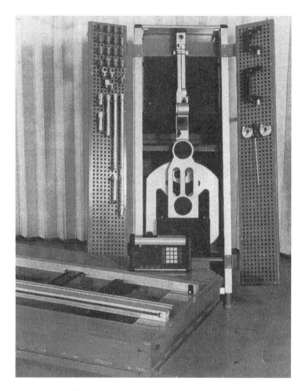

Figure 14.45   Car-O-Tronic electronic computerized measuring system (*Car-O-Liner (UK) Ltd*)

bridge underneath the vehicle and simultaneously measures the length, width and height at the precise data measuring points, displaying this on the hand terminal and then comparing the results with the manufacturer's data stored in its data bank. This immediate, accurate information regarding the extent of the distortion sustained by the vehicle determines whether or not the vehicle will need straightening and subsequently pulling. This information is then fed back to a remote computer terminal upon which it is displayed. A hard copy can be taken via a printer linked to the computer. The measuring system can be left in place while the pulling process is being carried out and can be used to determine the amount of overpull required to bring the vehicle into alignment with its specific manufacturer's data (Figure 14.46).

The Car-O-Tronic can also be linked into the Auto-quote management control computer package. This effectively allows for complete computerization of the entire bodyshop operation.

Figure 14.46 Pulling and measuring using the Car-O-Tronic (*Car-O-Liner (UK) Ltd*)

Figure 14.47 Chief E-Z Liner II system showing the gauge measuring equipment (*Chief Automotive Ltd*)

14.2.13 Chief E-Z Liner II universal alignment and straightening system

The Chief E-Z Liner II system consists of all the equipment necessary for damage analysis and repair to current vehicle models. It uses a precision universal gauge measuring system and special pulling and anchoring attachments in conjunction with the main frame platform (see Figure 14.47). The equipment consists of a main frame platform which tilts down at the back to form a ramp, up which a vehicle can be driven or winched. Once the car is in position, the platform can be raised hydraulically to a horizontal plane and locked into place.

The platform has 200 tie-down boxes which provide unlimited holding positions for any type of car or light commercial vehicle. Four special anchor clamps are supplied. These lock into tie-down boxes to locate the vehicle. The clamps themselves rotate through 360 degrees and have height adjustment. At the front of the unit are three independent pivoting towers. These between them can be swung in an arc

of 180 degrees, providing a wide range of pulling angles. Each tower is equipped with a 5 tonne hydraulic ram to provide the pulling power, and an adjustable collar to enable precise pulling angles to be achieved. Two portable hydraulic rams are also supplied for lifting, pushing and pulling, thus proividing the means for unlimited pulling capacity at any point on the damaged vehicle. Power for the pulling equipment is supplied by a heavy-duty hydraulic pump operated from a remote control handset.

The universal gauging system gives a fast positive visual alignment analysis before and during repair. Six alignment gauges feature vertical and horizontal calibrated scales, and a variety of hanging attachments. These self-centring gauges form a skeleton image of the damage extended below the vehicle. Each gauge measures width from the centre line as well as interacting with the gauge system. Width is compared with specifications by reading the scale through a convenient window on the top of the centre housing (see Figure 14.48). The attachments are calibrated to show the actual distance from the datum plane to the attachment point. Special colour treatment on each side of the gauges provides easy analysis of the datum plane; yellow stripes on one side and orange stripes on the opposite side can be alternated to provide contrasting colours for siting.

With all the gauges in position, the simultaneous multipull hook-ups are made and the pulling process begins. Throughout the pulling process the gauges show the centre line and datum plane gradually

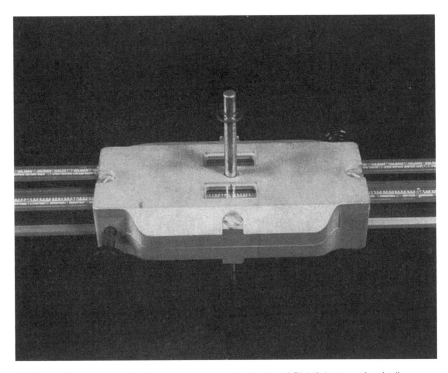

Figure 14.48   Self-centring gauge showing measurement window (*Chief Automotive Ltd*)

becoming aligned until they read central. When the centre line and datum are aligned and the correct length is achieved, the vehicle structure is then in total alignment.

14.2.14 The Genesis electronic measuring system

The Genesis electronic measuring system, developed by Chief Automotive Ltd, integrates the precision of laser scanning with a computerized data base for accuracy in collision repair analysis (Figures 14.49 and 14.50).

The Genesis system begins with the vehicle specification data base. For each vehicle specification there are graphic displays, both overhead and side view, of the vehicle's underbody structure (see Figure 14.51). In addition to showing the extent of collision damage, the Genesis system is designed to monitor progress throughout the repair and verify that the vehicle structure is correctly aligned. Throughout the repair the computer compares vehicle specifications with the manufacturer's current reference point measurements. Computer printouts verify the vehicle's structural

Figure 14.49   The Genesis electronic measuring system (*Chief Automotive Ltd*)

condition by showing overhead and side-view diagrams.

The Genesis system measures using the principle of triangulation. The system's electronic body scanner houses two spinning hubs, each of which emits a laser beam. These two rotating laser lights

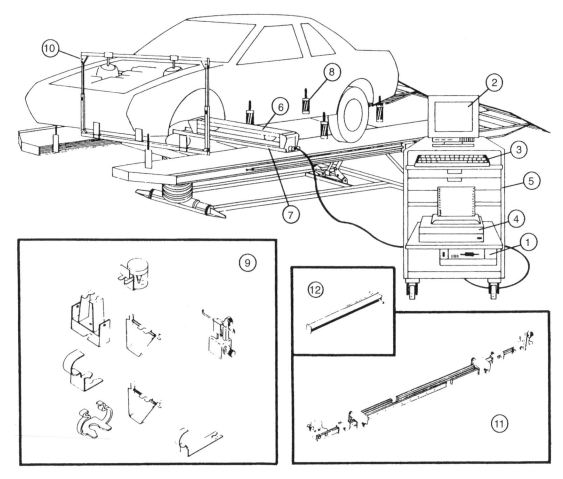

Figure 14.50 Genesis electronic system components (*Chief Automotive Ltd*)

1	Computer	5	Workstation	9	Target attachments
2	Display monitor	6	Body scanner	10	Upper body bar
3	Keyboard	7	Scanner tray	11	Lower body bar
4	Printer	8	Targets	12	Scales

project towards light reflective targets, which are suspended from the vehicle's underbody reference points. Each laser beam strikes a reflective material in the end of the body scanner housing, which activates a counter when the laser light is reflected back to the hub. The beam also strikes a target, which has retro-reflective material on one side, and which is attached at a control point on the underside of the vehicle. The beam is reflected back to its source, the laser hub, which is then reflected up to a detector above the centre of the hub, which again activates or turns off the counter.

The number of counts the counter has made, from when it was turned on by the beam striking the end

of the housing to when it is turned off by the beam striking the target, can be used to calculate the angle of the beam to the housing. The same process occurs with the other hub, which also determines another angle. The body scanner housing and the two laser beams striking the target form a triangle.

Since the spinning hubs are a known distance apart, the counters have determined two angles of the triangle: therefore the length of the sides of the triangle can be mathematically determined by trigonometry (triangulation) (Figure 14.52). This can then be further converted mathematically into length and width measurements for the target locations.

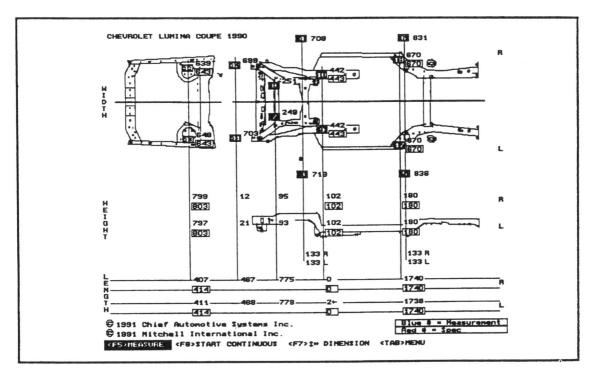

Figure 14.51   Computer measuring screen (*Chief Automotive Ltd*)

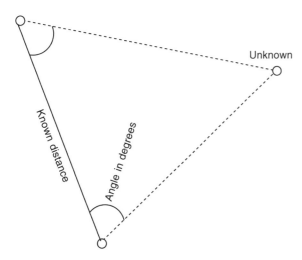

Figure 14.52   Triangulation (laser measurement)
(*Chief Automotive Ltd*)

The body scanner automatically reads all reference points simultaneously, from strategically placed targets positioned on the vehicle underbody. These readings are then compared with published data and are displayed in colour on the monitor screen.

Upper body and strut tower positions are similarly located to provide height, width and length specifications relevant to the lower body. Genesis first displays visual guides, showing where to place each target. It then graphically illustrates how existing positions compare with vehicle specifications and provides direction and distance information for planning the repair. The entire system stays in place while repairs are being made.

Setting up the equipment, hanging the targets in position, entering the vehicle's information, and positioning the body scanner, take a qualified bodyshop technician approximately ten minutes. Calibration of the vehicle is then carried out automatically by Genesis in a few seconds.

To check progress at any moment the computer can provide either a visual display or a printout verification of the vehicle's structural condition. This can be of great benefit for insurance companies for quick approval of additional repairs, and also for customers requiring verification of repairs.

14.2.15 Kroll Autorobot multifunctional system

Kroll Autorobot of Finland have developed a multifunctional system for alignment, repair and straightening. It can be added to as a workshop expands, thus catering for both small and large bodyshops (Figure 14.53).

Autorobot L

For the small bodyshop wishing to invest in a crash repair system, the system begins with the Autorobot L (Figure 14.54). This is easily and quickly loaded using the unique patented sill clamps. The pulling tower can be used 360° around the vehicle to be repaired. Multiple pulling positions around the Autorobot L mean that two or more pulling towers can be used at any one time for both minor and major repairs. This system is mobile and can be moved to any work area in the workshop easily and quickly. The short bed allows access for repair and the engine can be removed while the vehicle is still mounted on the bed.

Autorobot XLS II

By adding to the Autorobot L with a lift, extra pulling towers and a measuring system it becomes the Autorobot XLS II (Figure 14.55). The lifting unit of this system makes many phases of the job comfortable, such as the removal of mechanical parts and the alignment and repair of a damaged chassis. In addition to this the lifting unit assists the mounting of the vehicle. It takes only a few minutes to mount a vehicle on to the Autorobot XLS II. The four operations are as follows:

1. The vehicle is brought up above the bench on the drive-on ramp with or without the aid of a winch.
2. The sill clamps are then moved into place.
3. The straightening bed is hydraulically lifted.
4. Finally the sill clamps are tightened.

The design of the Autorobot allows maximum accessibility under the front and rear of the vehicle; therefore all dismantling, measuring and straightening operations can be carried out with ease.

The measuring system (Figure 14.56) is interchangeable with all Autorobot benches. A damaged vehicle body can be straightened to its original measurements with accuracy and ease using this system. Mechanical measuring rods are used to remeasure the chassis and body. All necessary measuring information for various vehicle models can be found on the data sheets provided. The overhead measuring system can help restore the upper part of the vehicle body to its original shape as on the data sheets. The system traverses the full length

Figure 14.53 Autorobot XLS II showing liftpack servicing four top frames (*Kroll (UK) Ltd*)

of the vehicle, allowing measurements to be taken at any point along its length.

III Super

The **III Super** features five pushing-pulling towers and multifunction robot arm, all operated by remote control electrohydraulics. This system of multitowers allows straightening and counter-pushes combined with holding positions so that the straightening work becomes easier for the operator. The straightening system works on a sliding tower base, allowing the most difficult of repairs to be carried out (Figure 14.57).

Figure 14.54 Autorobot L using lifting crane (*Kroll (UK) Ltd*)

Figure 14.55 Autorobot XLS II system (*Kroll (UK) Ltd*)

Figure 14.56 Autorobot measuring system used on the XLS II (*Kroll (UK) Ltd*)

Figure 14.57 Autorobot III Super complete system (*Kroll (UK) Ltd*)

14.2.16 Choice of systems

The design concept of monocoque construction with its crumple zones which absorb energy through deformation of panel structures, and the increasing use of high-strength steels, have made essential the use of repair and alignment systems even for minor damage repairs. This has been encouraged by the research carried out at the Motor Insurance Repair Research Centre Thatcham, and by the promotion by the VBRA and the AA of improved standards throughout the vehicle body trade.

Owing to the wide range available of repair and alignment systems, it is increasingly difficult to select any one system, as they each have many differing advantages. The following are some of the important criteria to be considered in the selection of suitable equipment.

1 Suitability for the type of repair work normally handled.
2 Suitability for the volume of work handled.
3 Availability of initial training and further training as required.
4 Adequate service and back-up by the manufacturer.
5 Availability of accurate data information which should be constantly reviewed and updated.
6 Expected lifespan of the equipment, and the possibility of it becoming outdated.
7 Mobility, portability and ease of storage.
8 Firm securing of vehicle during repair.
9 Good system of alignment that can be used without too much dismantling of mechanical components (dual systems are sometimes preferable).
10 Multipulling facilities for simultaneous repair over 360 degrees, and enough pulling power to cope with HSS.
11 Adjustable working height positions.
12 Adequate holding facilities to allow the welding in of new panel assemblies.

14.3 Alignment of the modern integral body

14.3.1 Misalignment and vehicle control

Any serious impact on any part of the all-welded body of the mono constructed vehicle will cause distortion throughout the structure, which will not necessarily be adjacent to the damaged area. Since the front and rear suspension units are connected, either directly or through the medium of a front or front and rear subframe, distortion of the body shell means a disturbance of the relationship of the running wheels one with another. When this occurs the result will be crabwise running in which the car seems to travel in a sideways manner, commonly called dogtracking. It is often necessary for the driver of such a vehicle to struggle constantly to maintain a straight course, and this becomes greater as the speed increases.

Misalignment of car underbody can also cause tyre wear and may affect the brakes and steering control, which will not be cured however much checking and adjustment is carrried out. These faults can render a car outside the uniform standards required by law in respect of the fitness of the car for use on the road. A car with a damaged or weakened underbody can be a menace to safety on the road. Aside from the effects it has on steering control and tyre wear, it can also place undue stress on other mechanical parts of the vehicle. The alignment of the engine with the clutch and transmission may be affected. This could cause a manual transmission to jump out of gear and might result in permanent clutch failure. Any radical change in the angle of transmission to the rear axle may cause excessive wear in the universal joints, noise in the rear axle or axle failure. Moreover, whenever the underbody has been damaged by collision there is a possibility of a broken hydraulic brake line which could lead to complete brake failure. Electrical connections also may become broken. There are a number of modern cars where the rear track varies substantially from the front to the back and there are several which have up to 25 mm variation in track; these points are taken into consideration when checking for distortion.

14.3.2 Checking a mono constructed body shell for underframe alignment

In order to make an alignment check of a mono constructed underbody, the first step is to assess the extent of the damage, the probable amount of force involved which caused it, and the direction of that force. This information will act as a guide by indicating where to look for possible distortion. With this knowledge a visual check should be made of doors, bonnets, boot lids, roof and centre pillar positions for any distortion which may be visible. Inconsistent gaps around these panels show that the panel assembly has been moved during the collision.

The underbody should be checked next. This can only be done by jacking the car up and inserting safety stands, or by using lifting equipment in the form of a hoist which will allow a more thorough examination of the underside of the vehicle. The subframes and cross members can now be examined for kinks or buckles with the aid of a portable light, but where sighting is difficult or impossible the members are examined by touch. This is not conclusive evidence of distortion, but will generally be sufficient to convince the repairer whether a more precise check should be made by using the dropline method, the gunsight gauge method, or a jig alignment system.

14.3.3 Drop-line method of checking underbody alignment

For many years in the body repair trade the drop-line method was the basic means of checking the alignment of both composite and mono constructed vehicles. When a car has been damaged the relationship of the vital points, such as rear suspension spring hangers, front hub centres, and front suspension centres, must be checked on a flat floor surface with the vehicle jacked clear of the floor, all four road wheels removed, and safety axle stands in position.

With a plumb bob and length of cord, plot on to the workshop floor a series of points common to both right- and left-hand underbody members. At least eight such points should be selected, four on each side, which could include front and rear spring shackle bolts, bumper bar bracket bolts, or any factory formed holes, rivets, bolts, or intersection of crossmembers common to both side members. On a composite designed body the body bolts provide ideal locating points. Variation in wheelbase can be checked by plotting the centres of each of the four wheels. The plumb-bob line should be held against the centre of each point selected and the position where it strikes the floor should be marked with a pencil cross. The pencil is more easily seen if the floor is previously chalked in the approximate area where the plumb-bob will fall.

After all the points have been plotted and the wheels replaced, the car can then be rolled away, leaving the chalked pattern on the floor. These points are joined diagonally by using a chalked length of cord which is held tight and then flicked to the floor, leaving a chalked straight line between each pair of points. These lines will indicate if any variation has occurred in the diagonal measurements. The next stage is to establish a chalked centre line through the diagonals. This is done by bisecting the lines joining the front and rear pairs of points and using the chalked cord to mark the centre line. If the underbody is in perfect alignment, this centre line should then pass where each diagonal intersects the centre line. When the centre line does not pass within 3 mm to these intersecting points, measurements must be taken to establish which member is distorted. A further check can be made by joining the sets of points transversely. These transverse lines will be at right angles to the centre line if the underbody is in correct alignment (Figure 14.58).

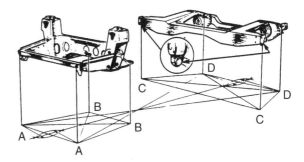

Figure 14.58   Drop-line check on front and rear subframes of a Mini (*Rover Group Ltd*)

A–A	660.40 mm	Front subframe front mounting
	26.0 in	set screws
B–B	412.75 mm	Front subframe rear mounting
	16.25 in	set screws
C–C	1282.70 mm	Rear subframe front mounting
	50.50 in	set screws
D–D	977.90 mm	Rear subframe rear mounting
	38.50 in	block set screws

All dimensions taken at centre line of set screw or set screw hole.

Owing to the inaccuracy of this system, the increase in the complexity of design structure of vehicles, and their critical suspension geometry, the drop-line method has been superseded by the use of alignment jigs, using either a bracket alignment check system or a universal measuring system.

14.3.4 Gunsight gauge method of checking underbody alignment

Another method of checking underbody alignment for distortion is that of using gunsight frame gauges (Figure 14.59). These gauges consist of two sliding bars which can be adjusted by either pulling out or pushing in to make them fit correctly across a chassis frame or underbody. At the middle of the gauge is a sighting pin which always remains central irrespective of the distance to which the gauge is extended, and which is used to establish the datum line (the vehicle's centre line). There are adjustable hanging rods on both ends so that the gauge may be raised or lowered. These hanging rods are also used to attach the gauge to the underbody or frame at symmetrical points on it, which may be factory formed holes, mounting points, or any part of the underframe which has identical positions on both sides of the vehicle (Figure 14.59a). Never hang gauges on movable parts such as control arms, springs, torsion bars, or any mechanical moving parts, as false readings will be obtained. A minimum of three gauges is essential, but the more gauges used the easier it will be to determine exactly where the distortion is located, as sighting along four or five gauges is quicker than relocating three gauges back and forth along the length of the vehicle.

When setting up the gauges to check underbody alignment, the in-line condition will be when all sighting pins are in line and all gauge bars are parallel. If one gauge must be lowered to clear some obstruction on the underbody, all other gauges must be lowered a corresponding distance to maintain a level sighting. The condition where the gauge bars are not parallel (Figure 14.60a) indicates that the underbody has moved up or down at some point along its length. The condition where the sighting pins are not in line, while the gauge bars are still parallel (Figure 14.60b) indicates that the underbody members have moved to one side only (sway). A twisted underbody is recognized by the gauge

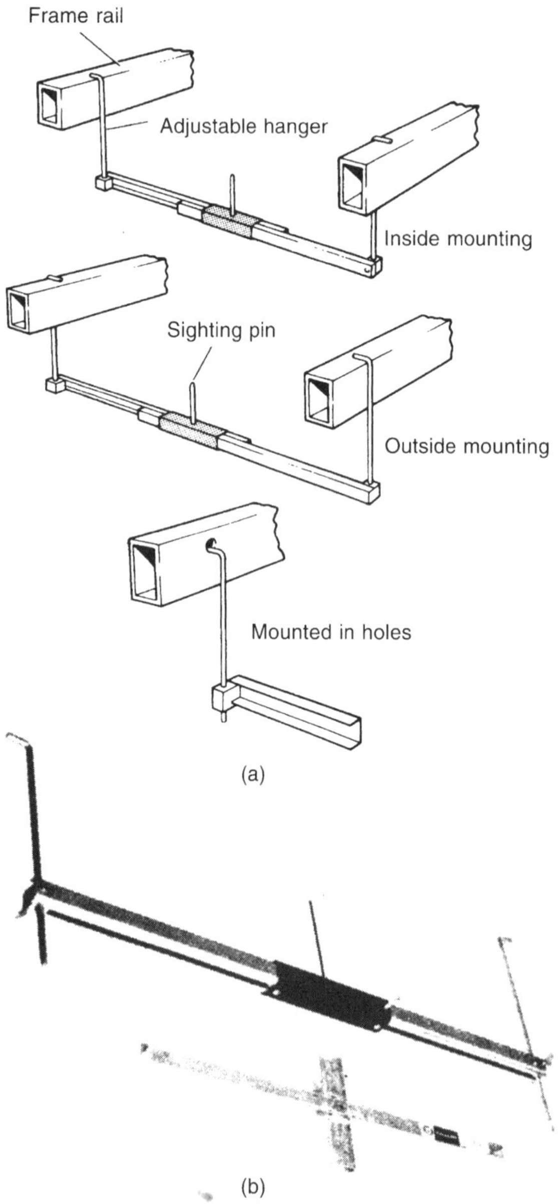

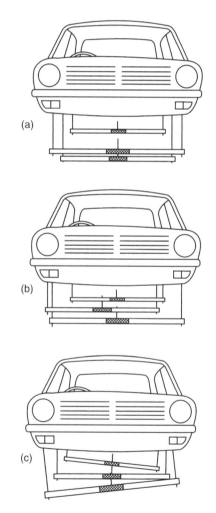

(a)

(b)

(c)

Figure 14.59 (a) Gunsight gauge frame (b) diamond detector gauge (*Blackhawk Automotive Ltd*)

Figure 14.60 Frame gauges showing distortion (*Blackhawk Automotive Ltd*)

condition where sighting pins are out of line in addition to the bars not being parallel (Figure 14.60c).

As well as these basic checks an additional gauge is available, known as a diamond detector gauge (Figure 14.59b), which is specially made to fit on the top or bottom of the self-centring gauges. This provides extra sighting pins to help to detect sideways movement (sway) and diamond damage (when the underbody has been forced into a diamond shape), which may not be evident when using self-centring gauges only. The only indication of a diamond condition would be that the two pins of the diamond detector gauge would be out of line with all other self-centring pins. Having determined the type of damage, the actual location is found by visual inspection. The main advantages gained by using this system of alignment check are the speed at which it can be set up to diagnose distortion, and the convenience it offers in making regular checks during repair procedure.

Chief precision alignment gauge system
The principle of the self-centring siting pin is used by Chief in their alignment gauge system.

These gauges are hung underneath the car from specific data reference points, giving an instant alignment check according to the siting on the pins (see Figure 14.61 and Section 14.2.13).

14.3.5 Vehicle alignment checks using a digital measuring tool

This is a lightweight aluminium measuring tool which can be used to check body alignment on any part of a vehicle body. By extending the tool, its measuring range covers from 2 mm up to 3683 mm. The measurements are displayed on an LCD digital readout, in either metric (millimetres) or imperial (feet and inches), and are accurate to 1 mm. It is supplied with pointers, cones and magnetic bases so that accurate measurements can be taken on various parts of the vehicle without difficulty (Figure 14.62). This measuring tool is used for comparative measurements when checking a vehicle for damage and when estimating a repair. While a vehicle is on a jig, the tool can be used to double check measurements taken by the jig's own measuring system. It also can be used for

Figure 14.61   Chief E-Z Liner II showing hanging gauges, self-centring (*Chief Automotive Ltd*)

Figure 14.62 Digital measuring tool with accessories (*Stanners Ltd*)

measuring a vehicle's wheelbase and comparing it with manufacturer's data (Figure 14.63). In addition it can be used to check the critical suspension points before, during and after repair (Figure 14.64).

Figure 14.63 Checking a vehicle's wheel base with a digital measuring tool (*Stanners Ltd*)

Figure 14.64 Digital measuring tool being used during repair to check vehicle suspension points (*Stanners Ltd*)

14.3.6 Underbody alignment using jig fixtures

Using an alignment and repair system is the only method of guaranteeing a comprehensive and accurate check on the alignment of the underbody of a vehicle. The monocoque body is produced on a production line using a series of quality controlled jigs and fixtures for holding all panels in the correct position for spot welding. To the production engineer this is the only way of consistently reproducing a basic shell fitted with all its mechanical parts in the correct places. When it comes to repairing a body which has been damaged or distorted in an accident it is only logical that accurate alignment jigs should be used to ensure a safe, roadworthy repair and one which has the appearance of a new vehicle. The introduction of systems of body jigs was a major step forward in body repair techniques. These systems were designed and developed to cope with the ever-increasing call on the accident repair trade. Every vehicle presented a different problem but, equally, every body shell had certain points which always could be relied upon to be accurately controlled during factory production (Figures 14.65 and 14.66), and it was around these points that all manufacturers of alignment equipment designed their products. This approach and the enthusiastic cooperation of the vehicle manufacturers at all stages ensured that when a car was mounted on the equipment, the damage of distortion showed up in an unmistakable fashion. The products now available to the body repairer include the following:

1 Alignment systems that incorporate bracket mounting (see Figure 14.67)
2 Alignment systems that incorporate either a mechanical (see Figure 14.68) or a laser light measuring system (see Figure 14.69)
3 Dual equipment, which incorporates both measurement and bracket systems
4 Cassette-type multifunctional alignment and repair systems (see Figure 14.70)
5 Alignment systems that incorporate either a mechanical or an electronic measuring system (Figure 14.71).

The manufacturers are continuing their research to develop appropriate body repair equipment as car design and methods of assembly are influenced by technological progress.

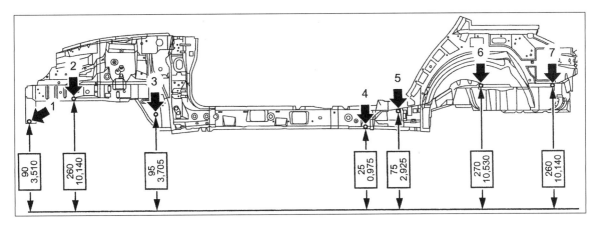

Figure 14.65   Manufacturer's alignment points

No.	Standard measurement point	Hole-Size shape. mm (in)
1	Centre of front bumper position hole – RH	○ – 9 mm (0.354 in)
1a	Centre of front bumper position hole – LH	○ – 9 mm (0.354 in)
2	Centre of front sidemember positioning hole – LH	○ – 25 mm (0.984 in)
2a	Centre of front sidemember positioning hole – RH	○ – 25 mm (0.984 in)
3	Centre of suspension crossmember mounting hole – LH	○ – 16 mm (0.629 in)
3a	Centre of suspension crossmember mounting hole – RH	○ – 16 mm (0.629 in)
4	Centre of front floor sidemember positioning hole – LH	○ – 15 mm (0.59 in)
4a	Centre of front floor sidemember positioning hole – RH	○ – 15 mm (0.59 in)
5	Rear portion of rear seat crossmember positioning hole – LH	⬭ – 25 × 38 mm (0.984 × 1.496 in)
5a	Rear portion of rear seat crossmember positioning hole – RH	⬭ – 25 × 38 mm (0.984 × 1.496 in)
6	Centre of rear floor sidemember drain hole – LH	○ – 10 mm (0.393 in)
6a	Centre of rear floor sidemember drain hole – RH	○ – 10 mm (0.393 in)
7	Rear portion of rear floor sidemember extention positioning hole – LH	⬭ – 18 × 67 mm (0.708 × 2.637 in)

14.3.7 Vehicle body alignment: upper structure

Damage to the inner construction of a car body might be quite severe although not obvious. In some cases the damage to the outer panels of the body can be repaired and the original damage will appear to have been corrected although it has not. Failure to check the correction of the internal damage by alignment methods can result in incorrectly fitting doors, bonnets and boot lids.

The alignment methods are based on comparative measurements in which one distance is compared,

by means of a trammel, tram-track gauge or measuring equipment incorporated in an alignment measuring system (such as in Figure 14.70) to another which should be the same. Comparative measurements are universally used and are regarded as the easiest and quickest method of measuring collision work. These measurements are supplied by most manufacturers on data sheets or in body alignment manuals. Body measurements are made in the same way in both composite and mono constructed vehicles. Checking body measurements is done by a system of diagonal comparisons, commonly called X checking. The

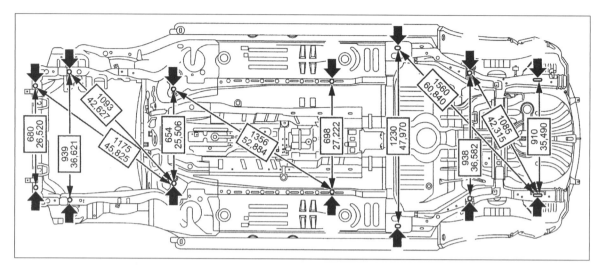

Figure 14.66 Manufacturer's transverse alignment points

Figure 14.67 Alignment and repair using a jig bracket system (*Globaljig, Tri-Sphere Ltd*)

Figure 14.68 Alignment and repair using a measuring bridge system (*Blackhawk Automotive Ltd*)

location and extent of the damage is determined by measuring the body. Further measuring during the repair will indicate when alignment is restored.

In this system of measuring the body is divided into two basic sections: the front section, which is the area from the centre pillar forwards (Figure 14.72); and the centre/rear section, which is the area from the centre pillar backwards to the rear boot compartment (Figure 14.73). In addition to these sections, there are special sets of measurements for the alignment of the bonnet lid, the boot lid, and the front and rear doors. Some alignment systems incorporate

upper body alignment measuring equipment (see Figures 14.74–14.77).

14.3.8 Door alignment

The doors of a car body are said to be misaligned when they do not fit the contour of the body or when they do not fit in the door opening correctly. Doors must provide a good seal against dust, water and air, and in order to do this they must match the contour of the body at all points. When closed, doors must fit correctly into the opening provided for them, because if the door is incorrectly positioned it will

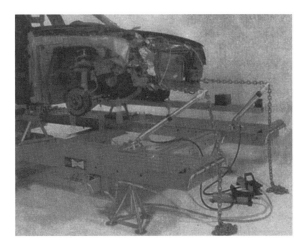

Figure 14.69 Alignment and repair using a laser measuring system (*Dataliner, Geotronics Ltd*)

Figure 14.70 Autorobot L using full measuring system (*Kroll (UK) Ltd*)

Figure 14.71 Alignment system using electronic measuring for both underbody and upper body structures (*Car-O-Liner (UK) Ltd*)

Door alignment is checked by making a visual close inspection of the door and its relation to the surrounding body panels. First open and close the door and observe any movement at the edge of the door at the lock pillar; if there is any up and down movement, the door is out of alignment but is being corrected by the dovetail of the lock as the door closes. Look for signs of rubbing or scraping by the bottom edges of the door on the sill panels or centre pillar. Scrub marks are evidence of a condition known as door sag. In cases where the door is severely damaged and a good deal of straightening is necessary, it is essential to check the door before any attempt is made to reinstall and to align the door with the body. Measurements should be taken on the damaged door, working from points that can be easily established on an undamaged door, and the readings compared to determine whether the damaged door requires further attention. Always be sure to take measurements from the same points on both doors.

not close properly and the gap around the door will not be uniform. In some cases where the door appears to be misaligned the door itself is not the cause of the fault, and so it is advisable to X check the body at the forward and centre positions to determine whether the reason of the failure to match the body contour is due to distortion of the body or distortion of the door itself. In some cases lack of uniform clearance at the front of the door might be due to mispositioning of the front wing.

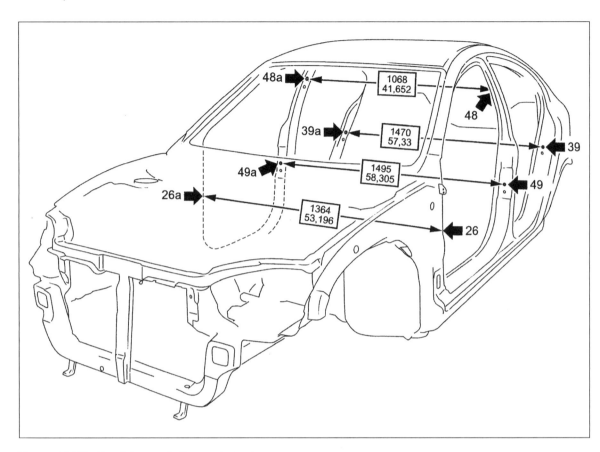

Figure 14.72 Front diagonal alignment check

No.	Standard measurement point	Hole-Size shape. mm (in)
26	Front pillar positioning notch (Inner side) – LH	–
26a	Front pillar positioning notch (Inner side) – RH	–
39	Centre of rear door striker mounting hole (Upper section) – LH	○ – 10 mm (0.393 in)
39a	Centre of rear door striker mounting hole (Upper section) – RH	○ – 10 mm (0.393 in)
48	Centre of hood opener cable routing hole – LH	○ – 5.3 mm (0.208 in)
48a	Centre of hood opener cable routing hole – RH	○ – 5.3 mm (0.208 in)
49	Centre of front door striker mounting hole (Upper section) – LH	○ – 12 mm (0.472 in)
49a	Centre of front door striker mounting hole (Upper section) – RH	○ – 12 mm (0.472 in)

14.3.9 Bonnet alignment

Bonnet alignment can be easily checked by visual inspection. When the bonnet is closed the gap all around it should be uniform; if it is not, misalignment exists. Manufacturers' specifications concerning this gap are available; however, a good general rule is that this clearance should be between 3 mm and 6 mm wide. Correct bonnet alignment with the wings and front grille preserves the streamlined contours of the front end of a car. A uniform gap should exist between the sides of the bonnet and the wings. At the rear end of the bonnet the body is recessed to form the scuttle panel into which the bonnet fits to give a continuous flowing line. If the bonnet is too high or too low a recognizable step occurs at this point. If the bonnet is too far forward

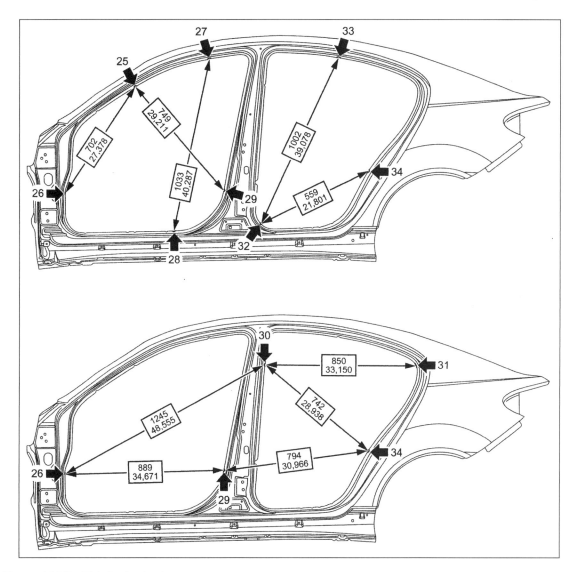

Figure 14.73 Side body check

No.	Standard measurement point	Hole-Size shape. mm (in)
25	Front pillar positioning notch (Upper section)	–
26	Front pillar positioning notch (Lower section)	–
27	Notch of roof side rail	–
28	Notch of front pillar outer and side sill outer	–
29	Centre pillar positioning notch (Lower section)	–
30	Centre pillar positioning notch (Upper section)	–
31	Notch of rear pillar outer and side sill outer	–
32	Notch of centre pillar outer and side sill outer	–
33	Rear pillar positioning notch (Upper section)	–
34	Rear pillar positioning notch (Lower section)	–

Figure 14.74 Complete upper body measuring system (*Kroll (UK) Ltd*)

Figure 14.75 MacPherson strut measuring (*Kroll (UK) Ltd*)

a large gap will exist between the rear of the bonnet and the raised portion of the scuttle panel; however, if the bonnet is too far back this gap will be too small and there is a possibility that when the bonnet is opened it will catch the scuttle panel and chip the paint on either or both of these panels.

When a condition of misalignment exists that affects the bonnet opening, it will be necessary to X check the opening before deciding whether to move either the wings or the grille panel. In some cases this can be done by adjustment; in other cases, where these assemblies are fully welded,

Figure 14.76 Upper body measuring system measuring door aperture (*Kroll (UK) Ltd*)

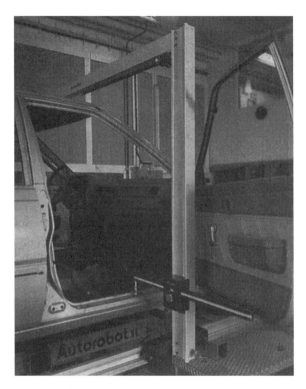

Figure 14.77 Upper body measuring system measuring sill panel (*Kroll (UK) Ltd*)

hydraulic equipment must be used. The bonnet hinges on most cars are also constructed to give adjustment; therefore the bonnet can be moved forwards, backwards and sideways.

14.3.10 Boot lid alignment

Correct boot lid alignment exists when the boot lid matches the contour of the body surrounding panels. Misalignment of the boot lid is checked visually by an inspection of the gap all round the boot lid between it and the body. It is often necessary to check more closely for boot lid alignment, because an incorrectly aligned boot lid could allow the entrance into the luggage compartment of water and dust which might cause damage to anything being carried there. A simple check is to chalk the edges of the body flanges which contact the weatherstrip on the boot lid, then close and reopen the lid, thus transferring the chalk to the weatherstrip at each point where contact is made. If this chalk line is visible around the entire weatherstrip, the boot lid sealing is perfect. Where the chalk line does not appear, the boot lid is not sealing properly and realignment will be necessary. Realignment may be possible by adjustment of the latch or hinges, but in some cases it may be necessary to bend or twist the boot lid back to its correct alignment. When the boot lid opening is at fault it must be carefully X checked, and it can only be realigned by the use of hydraulic equipment.

14.4 Major repair techniques

In major repair work the method of repair is to analyse the crash, establish the order in which the collision damage occurred, and reverse the order when correcting the damage. Distortion of the vehicle's underframe in the case of mono constructed vehicles, or of the chassis in composite constructed vehicles, must be rectified first before any other part of the repair is attempted. Any vehicle body under major repair will never hold its correct alignment unless underframe damage has been completely repaired first. This essential stage in the repair of the vehicle is carried out by first inspecting the damage and checking its alignment by one of the recognized methods to locate the exact position and extent of the damage.

Correction is carried out by using alignment and repair equipment in conjunction with hydraulic pulling and pushing equipment to realign any underframe members which do not align with the specified locating points (see Figure 14.78). Any members which are beyond economical repair can be cut out and replaced with new sections by using a combination of power tools and MIG welding equipment. The use of this hydraulic equipment together with the alignment and repair system will ensure precise accuracy of the complete underbody.

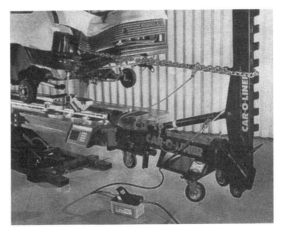

Figure 14.78   Pulling equipment being used to realign underframe members (*Car-O-Liner (UK) Ltd*)

The next stage of the repair is the realigning and reshaping of the body shell structure, and is normally known as roughing out the damage. After alignment checks have been made, hydraulic equipment is used to push and pull the body shell back to its correct shape (Figure 14.79). It is advisable to rough out all damaged sections, even though some of these may eventually need replacing with new panels. The reason for this is that the successful fitting of a replacement section will depend on the correct alignment of the surrounding areas, which can best be achieved by restoring them to as near the true shape as possible before the removal or cutting away of the sections beyond repair. It has been found that the use of multiple hydraulic pulling equipment used simultaneously is the most effective method when dealing with the majority of major collision work, together with standard hydraulic body jack equipment as a

Figure 14.79 Realignment of the body shell using simultaneous push and pull hydraulic equipment (*Kroll (UK) Ltd*)

secondary pushing/pulling facility to aid the repair of the damage. The main advantage is the fact that the pulling is done externally from any angle while the car is anchored securely. In cases where there is a tear in the damaged section, this must be welded up before any pulling or pushing is attempted, otherwise the tear will hinder the rectification and possibly become greater on the application of force.

An important factor is that heat can be used to relieve stress in correcting damaged sections that have been badly creased or buckled as a result of impact. In this case the stress resulting from the buckle helps to retain the metal in its damaged state, and even after it has been restored to its correct shape by the application of pressure from hydraulic equipment, the section will tend to return to its damaged position if pressure is released. Therefore it is of considerable importance that before the pressure is taken off, these buckles or creases are hammered out with the use of heat and hand tools. When using heat for relieving stresses, it is important not to overheat the section beyond dull red, and wherever possible never to heat any point more than once as excess or repeated heating on one spot causes surface oxidation and annealing, both of which weaken the member or panel area. Heat can only be used on low-carbon steel panels and not on high-strength steels, because heat weakens the latter by disturbing the heat treatment which originally strengthened them. This is particularly so in the case of frame members, where part of the strength of the member is derived

from internal stresses set up during the forming operations in the manufacturing processes.

When the roughing out stage is complete and any major body misalignment has been corrected, then the panel assemblies which are beyond repair must be cut out in such a way that the remaining connecting flanges are left intact. These are then straightened with the use of hand tools to facilitate the locating of the new panel section, which after positioning must be checked for alignment with its surrounding panels. When the alignment of the new piece is correct it is welded into position; since the vehicle is still in the repair equipment, correct alignment is maintained (Figure 14.80), check that all doors, the bonnet and the boot lid fit and operate in the appropriate body shell openings. Minor damage such as small dints and scrapes in the original panels should now be repaired using hand tools and the conventional techniques of planishing or filling.

Figure 14.80 Welding in new panel sections while vehicle remains located on alignment and repair bench (*Dataliner, Geotronics Ltd*)

On completion of all repairs to structural and panel damage, the vehicle can be removed from the alignment and repair system. All dismantled mechanical and body trim parts can be reassembled, including any necessary replacements to windscreen, rear window and door windows. It is imperative that after repairing all major accident work the vehicle's steering geometry is checked with optical alignment equipment and a road test carried out to assess its roadworthiness and handling capabilities. When the repairer is satisfied with the vehicle's performance and that all repairs have been completed, the vehicle will need refinishing to return it to its original factory finished condition (Figures 14.81–14.88).

Although every vehicle receiving major collision damage must be individually treated, the following is a basic approach to repair procedure:

1 Inspect and check underframe or chassis for alignment using conventional equipment.
2 Rectify any misalignment of underframe or chassis using hydraulic pulling and pushing equipment, hydraulic body jack, and alignment and repair system.
3 Realign body shell, correcting damage with a combination of pulling equipment and body jack, and the use of heat and hand tools (not HSS). Also cut out and weld in any replaced buckled panels and reinforcing members which are beyond repair.
4 Repair doors, fit glass, and body trim to doors and check for ease of operation and alignment of the door in the opening in the body shell. Also check alignment of bonnet and boot lid and ease of operation.
5 Smooth and finalize minor damage to body panels with hand tools.
6 Replace all mechanical, trim and body parts.

Figure 14.81   Vehicle showing front-end damage being lifted to be positioned on the jig (*Motor Insurance Repair Research Centre*)

Figure 14.83   Vehicle on jig showing damaged sections removed (*Motor Insurance Repair Research Centre*)

Figure 14.82   Vehicle on jig being pulled before cutting out damaged sections (*Motor Insurance Repair Research Centre*)

Figure 14.84   New inner wing and chassis section ready to be welded in place (*Motor Insurance Repair Research Centre*)

Figure 14.85 New inner wing section welded in place (*Motor Insurance Repair Research Centre*)

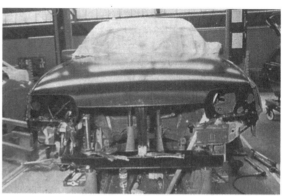

Figure 14.87 Bonnet and wings positioned (*Motor Insurance Repair Research Centre*)

Figure 14.86 Front panel positioned and welded in place (*Motor Insurance Repair Research Centre*)

Figure 14.88 Panel repairs complete (*Motor Insurance Repair Research Centre*)

7 Replace windscreen and rear windows if necessary.
8 Make an optical check on the vehicle's steering geometry.
9 Road test the vehicle.
10 Refinish to original factory condition.
11 Valet car's interior and exterior.
12 Final bodyshop check for quality control of completed work, prior to returning vehicle to customer.

14.4.1 Repairs to monocoque bodies by pulling

The method of correcting damage by pulling from the outside is not new, but the application of this principle has developed into two main methods. These are direct and vector pulling.

Direct pulling

This is achieved by using systems which are designed around a base beam, sometimes mounted on castors for easy movement, with an upright beam pivoted at one end. A hydraulic push ram is mounted across the angle between the two beams. When the ram is extended, the upright beam is pushed in an arc backwards, thus putting tension on to the chain attached to the damage. This produces a direct pull between the repair and the beam (Figure 14.89).

Vector pulling

This is achieved by attaching a chain to the damaged area, then passing it over a hydraulic ram set as near to 45 degrees as possible, and anchoring it to either a rack, a tie-down system or a special clamp. Pressure is applied to the chain by extending the ram; this puts tension in the chain, which in turn

Figure 14.89 Direct pulling (*Kroll (UK) Ltd*)

creates a pulling force on the damage. As pressure is increased the angle will continuously change in response to the pressure applied (Figure 14.90).

Figure 14.90 Vector pulling (*Dataliner, Geotronics Ltd*)

The principle involved in a pulling repair is to apply force to the car body that is opposite to the force that originally damaged it. With a mono-coque body, the engine, suspension and steering rack all contribute to the final strength and shape of the vehicle, so it is important not to remove them unless absolutely necessary for the type of repair being carried out.

Before beginning actual pulling, establish exactly the nature and type of damage sustained. This can only be achieved by the use of alignment and repair equipment; visual inspection is not satisfactory and can lead to inaccuracies. However, the vehicle must still be considered as a unit and the pulling force applied must act on the whole of the body and not just on the damaged parts.

The development of pulling systems has meant that in dealing with major body repair work it is possible to get a direct upward or downward pull to the front, rear or either side of the vehicle using single or multiple pulling equipment.

When repairing mono constructed bodies, basic body repair methods and a knowledge of panel beating is absolutely essential. The combination of methods always necessary in repairs is as follows: heavy external pulling, use of body jack, use of heat (only on low-carbon steels and not on high-strength steels), use of power tools and cutting out buckled sections or reinforcing members. Before any pulling commences, the angle of the pull is already known, but difficulties can arise in setting up the precise angle with the equipment. This problem can be solved by using the principle of parallel lines. First examine the damaged parts carefully and take accurate measurements to see in which direction, and by how much, the damaged parts have moved. This means that both lateral and vertical movement must be determined to make sure the correct reversal of this damage can be achieved. When this is known, the first pulling must always start as near to the major impact as practicable.

Heavy external pulls are a must for all severe mono constructed body damage because the impact has stressed the body metal. It is important when pulling damaged sections back to their original shape not to apply all the force at one place at the same time. Spread the restoring forces by using multipulls, which will allow the pulling forces to be activated from several points on the damaged section simultaneously. This method of pulling is more effective than single pulls on severely damaged vehicles (Figures 14.91 and 14.92). The body jack can also be used to assist by applying additional forces at the same time so they help each other in working the panel back into place.

With all mono constructed bodies, special attention must be given to certain load-bearing or structural members such as frame side members, inner wheel arch panels, cowl areas, cross members and the floor pan that form a part of the cross members or box sections. These must be brought back to the

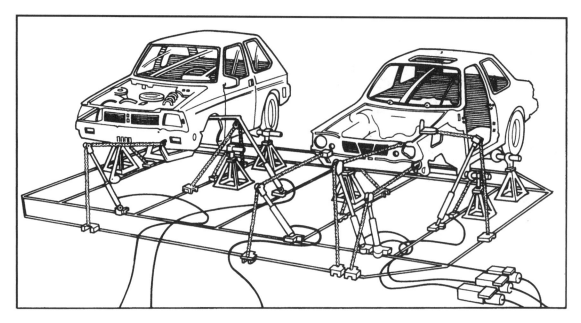

Figure 14.91   Simultaneous multipulls on two vehicles (vector pulling) (*Blackhawk Automotive Ltd*)

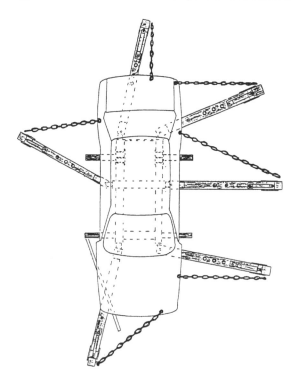

Figure 14.92   Simultaneous multipulling using direct pulling (pulling towers) (*Kroll (UK) Ltd*)

equivalent of their original condition, which means returned to their original shape without wrinkles or excess thinning of the metal.

The first step in making any hook-up is to anchor the pulling equipment into anchor pots, a rack-system, or an alignment and repair system. Next, a pulling attachment such as a sheet metal clamp or pull hook is secured to the damaged part. Work the damage out of the metal with a hammer and dolly while the metal is under tension from the pulling equipment. Use heat sparingly as required to remove severe wrinkles. Where sheet metal damage is widespread, several hook-ups may be required to bring the damage out gradually to prevent tearing and further damage.

With the vector method of pulling, parallel pulling can be carried out in the following way. First connect the chain to the damaged part of the car, then hold the chain tightly in the direction you wish to pull and adjust the pulling unit or ram to the measured angle. It is not necessary to have the chain vertically in line with the direction of pull, but it is important the ram should not be set at an angle lower than 45 degrees; that the chain links are set in line without any twist; and that the chain between the ram and chain anchor point should always lean towards the car. As soon as tension is applied to the chain, safety wires must be fitted in case the clamp or chain tears away from the main structure of the car.

Rectification of various damage types

Although every accident is different and must be treated individually, it is possible to generalize and divide accident damage into several types. Each type can be pulled and pushed straight in a number of ways.

For instance, if a vehicle has been in a head-on collision, and has suffered a straightforward *crumple,* secure the undamaged part of the underbody or chassis leg and pull out the crumple section in the direct line of damage. If the vehicle was struck by another, or if it has hit a low obstacle, it will have *sagged.* To repair this, secure one end of the vehicle, provide a support at the point of the bend, and pull down the raised end. With a *twist,* hold and support the undamaged part and apply a downward force to the front while allowing the other end to pivot up. A *diamond* can be difficult. Hold one side of the vehicle and pull along the line of the diagonal. The tendency will be for the whole of the vehicle to move, so hold the opposite corner to the pull. To repair severe *side damage,* push out and stretch the crushed side, It will also help to pull from the outside; this is because the passenger compartment is very strong, and repairing damage to it requires considerable force.

14.4.2 Repairs to underbody damage of monocoque vehicles

The underbody is the foundation on which the vehicle is built. It is of the utmost importance, therefore, first to determine the extent of the collision damage to the underbody and then to correct it. The repairer must be able to determine what, if any, damage exists before attempting a repair. It is also necessary to determine whether any underbody misalignment exists which requires repair before attempting to correct body alignment and damage. The majority of body repair shops have alignment and repair equipment capable of handling this type of repair. Although at present the majority of British manufacturers are producing motor vehicle bodies designed and built on mono or unitary construction principles, there are still some special sports models which have independent chassis construction, in addition to the available European and American vehicles which are based on composite construction. Consequently body repair establishments must make provisions

to deal with any such vehicles which may be brought in for repair.

The main idea of straightening underbody damage is to exert force in the reverse direction from that of the collision impact. Alignment and repair equipment permits several pulling set-ups to be made at one time. This type of equipment allows the vehicle to be aligned using either brackets or a measuring system. Repairs are carried out using a combination of pulling equipment, body jacks, chains, clamps, wood blocks and welding equipment. The severity of the damage decides the method of repair and techniques to be used. In the case of a vehicle which has received extreme damage, and in which the underbody has twisted badly, the method of repair is that the vehicle must be mounted on to the repair system and the exact amount of distortion established by using either brackets or a measuring system. Whilst the vehicle is still mounted on the system, the underbody must be pulled back into alignment first and then checked against the manufacturer's data, especially the critical suspension and engine mounting positions.

First rough out the damage but do not remove any panels unless it is necessary. Body damage and alignment should, wherever possible, be corrected at the same time as any underbody damage. Work can be made easier by stripping off any body panels or mechanical parts which might interfere with accessibility to the damaged area. As side and cross members are usually formed from low-carbon steel, heat may be used to relieve stresses in badly creased areas. However, when these members are made from high-strength steel, heat must not be used because of its weakening effect; the metal must be repaired in a cold state. For low-carbon steel, heat may be applied with an oxy-acetylene welding torch. Only a very little heat is required to bring the colour of the heated member to a dull red, at which time it is at a proper temperature to be worked. Start to heat the buckled portion or damaged area well out near the edge. Pan the flame over the entire buckled area so that it is heated uniformly. Best results can be obtained by moving the flame in a circular motion until the entire area is heated. Heating well out towards the edges of the damaged area first will remove the chill from the surrounding area, thus preventing the damaged area from cooling too rapidly.

Pulling equipment can be used for heavy external pulling by anchoring it to the under-body members using special bolt-on clamps or chain attachments and pulling the damaged section back into line. Where it is difficult to obtain an anchor point owing to inaccessibility, a substantial metal tab is welded on to the member; this can be used for pulling and later removed. In some cases a combination of pulling and pushing is needed when correcting damage, and this requires a heavy-duty body jack to be used in conjunction with pulling equipment. Wood blocks can be used to spread the load to prevent distortion. As the repair proceeds with a combination of pulling, pushing and heating, continuous checks should be made for alignment, as it is sometimes necessary to overcorrect to allow for springback in the metal structure. A final alignment check should be made after the tension is released.

14.4.3 Wheel alignment (steering geometry)

Over the years there has been increased awareness of the importance of wheel alignment, and particularly four-wheel alignment, as an integral part of the major crash repair procedure. Straightening a vehicle involves two separate alignment operations. First, the body must be aligned, making sure that suspension mounting points are in the correct position. Once the vehicle is jigged, in many cases a mixture of old and new suspension components is fitted. Secondly, a full geometry check must be undertaken to satisfy the repairer that the car is within the manufacturer's tolerances after the repair.

Wheel alignment has an important role to play in the overall safety of the modern vehicle, and proper testing equipment in this area is now vital. Much of the new technology caters for what is now regarded as four-wheel or total alignment. Modern vehicles have become far more sophisticated. They may have active rear suspension, which can also be computer controlled. These set-ups have an element of steering built in to complement the work done by the front wheels: the system is commonly known as rear wheel steering or four-wheel steering. Rear wheel alignment with a fractional variance from the manufacturer's setting can now be critical to a car's handling and steering. It is not normal for cars to veer to left or right: suspension systems are designed and set up by the manufacturer so that the car will travel straight, regardless of road crown.

The main purpose of wheel alignment is to allow the wheels to roll without scuffing, dragging or slipping on the road. The proper alignment of suspension and steering systems centres around the accuracy of six control angles: camber, caster, steering angle inclination (SAI), scrub radius, toe (in and out), and turning radius (Figure 14.93).

Inspection and test drive

Thorough visual inspection and a test drive on a straight flat road should show whether the car needs to go for wheel alignment. The basic checks are:

1 Measure ride height.
2 Does the vehicle appear level or is there any sagging?
3 Are the tyres properly inflated and of correct size and specification?
4 Is there any uneven wear on the tyres?
5 Low-speed wobble may show up a separated tyre tread.
6 High speed may reveal steering wear and odd noises, indicating loose or damaged parts.

Pre-alignment check

This breaks down into two areas: tyre inspection and steering component inspection.

Tyre inspection, front and rear, is as follows:

1 Check for any unusual wear patterns: camber, toe wear, inflation wear, cupping.
2 Check for any physical problems: ply separation, dry rot.
3 Check tyre size and type: same size side-to-side, same brand and tread pattern. Do not mix radial-ply and bias-ply tyres.
4 Tyre pressure and wheels: check pressure and set to specifications; check for bent or egg-shape rims.

Steering component inspection is as follows:

1 Tie rod assemblies.
2 Idler arm and bushings.
3 Centre link and joints.
4 Manual steering gear: check for excessive play and leaks, check U-joint at steering shaft.

5   Power assisted steering (run engine): check for excessive play; inspect all hoses and seals; check fluid level and belt condition; check power assisted balance.
6   Ball-joints and wheel bearings.
7   All bushings, front and rear.
8   Coil springs, torsion bars and shock absorbers.
9   Calibrate alignment equipment.

Wheel alignment procedure

Two-wheel alignment is still used in many workshops as it takes less workshop time, is easier and requires less expensive equipment. However, its efficiency relies on the car having two parallel axles, and front and rear wheels perfectly in line with each other.

Four-wheel alignment owes much to computer technology, which can instantly check and define the alignment of a vehicle. As full wheel alignment starts from a centre-line reference on the rear axle and then checks all steering geometry angles on all four wheels, a simultaneous four-wheel check will speed up the entire operation. Computer-based alignment systems use computer power to carry out the calculations, so that you can be certain of centralizing the steering wheel and alignment of the vehicle. This ensures that the steering wheel spokes are horizontal when the car is travelling straight ahead (Figure 14.94).

14.4.4 Replacement of damaged panels

Damaged panels can be removed from vehicles by using:

1   A hacksaw and metal cutting snips
2   A hammer and thin narrow-bladed chisel
3   Either a conventional drill, a Zipcut tool, or a spot-welding removal tool (air operated) to drill out each spot weld
4   An air chisel with appropriate cutting tools
5   A power saw having either a rotary vibrating blade or a straight reciprocating blade
6   A plasma arc cutting torch
7   A combination of any of the above methods.

A severely damaged inner panel such as a wheel arch can be cut out using the oxy-acetylene torch. However, this method is not recommended because of the very ragged uneven cut edge left by the torch. A better method would be the use of the plasma cutting torch, which leaves a fine cut edge for panel replacement. With both of these methods it is important to be aware of the fire risks.

Lapped and spot-welded joints of panels must be drilled first and carefully separated with a narrow-bladed cold chisel (Figure 14.95). The positions of the spot welds are indicated by small circular discolorations of the metal. These are centre popped, then drilled 4.8 mm or 5.6 mm sufficiently far to break the joint; it is not always necessary to go right through both parts of the panel, as in most cases if the top panel is drilled the joint will break. Alternatively a cobalt drill specially designed for cutting out spot welds can be used, either in a conventional air or electric drilling machine, or fitted to an air attachment designed specifically as a spot-welding removal tool. Another alternative is the Zipcut tool, which again will fit into a conventional drilling machine and which cuts round the spot weld leaving a hole in the top panel surface (see Figure 14.96). In some cases a fine-toothed hacksaw blade, fitted in a padsaw handle, and a pair of snips or shears can be used to remove damaged sections of panels. A tool for removing damaged panels is the power chisel set, which comprises a compressed air gun having interchangeable chisel heads. This sheet metal cutter is capable of speedy and accurate removal of panels or parts of panels that have to be replaced. Power saws can also be used in the removal of damaged sections: either the rotary saw with a vibrating blade, or the straight reciprocating blade (see Figure 14.97). Both of these tools give an excellent fine cut edge suitable for instant rewelding without any dressing.

The bodywork having been straightened and aligned as much as possible, the new panel should be positioned and held in place by two or three clamps and all adjacent panels checked for alignment with the new panel (Figure 14.98). To ensure a perfect fit it is often necessary to trim the adjoining panels of ragged edges, and to straighten out the locating flanges using hand tools (Figure 14.99). When making a close butt weld, exact alignment is ensured by positioning and clamping the new panel outside the old so that a scriber can be run down the edge of the old panel where the joint will come. Excess metal then being carefully trimmed and the edge dressed up with hammer and dolly, the new panel is ready to be

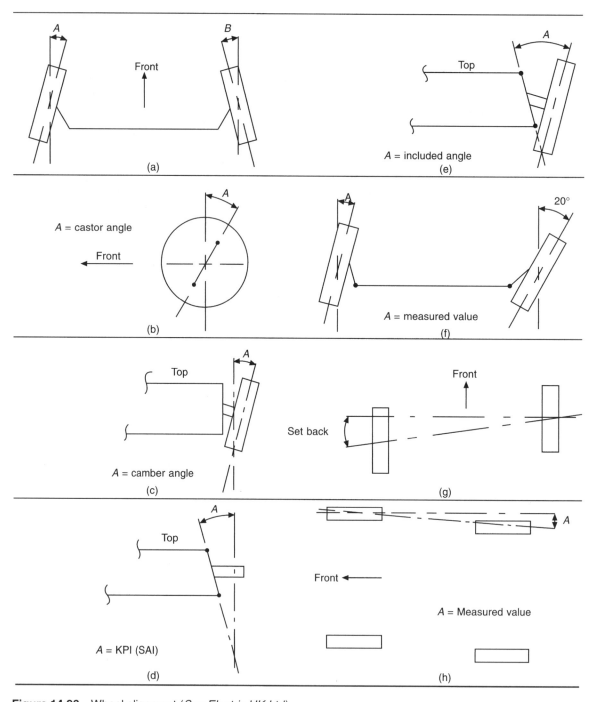

Figure 14.93   Wheel alignment (*Sun Electric UK Ltd*)

(a)   Front wheel toe
   The individual front wheel toe angle *A* is the angle of the front wheel from the straight overhead. The total toe is the sum of angles *A* and *B*. The term 'toe-in' is used when the wheels are closer together at the front, and 'toe-out' is when the wheels are closer at the rear. The purpose of toe is to ensure that the wheels are in the straight ahead position when driving. An incorrect setting would cause excessive tyre wear.

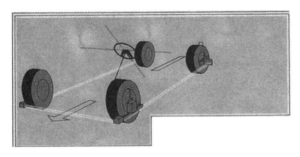

Figure 14.94   Four wheel alignment: steering wheel straight ahead (*Sun Electric UK Ltd*)

welded into place. Always brush on weld-through anti-corrosive primer before welding. The panel should be aligned, clamped and checked. It should next be tack welded at several points; then the clamps should be removed and the panel rechecked for alignment. If this is not satisfactory the tack welds are easily broken. When alignment is correct, further tack welds may be placed between the first set. To avoid distortion only short sections between the tack welds should be welded at a time and the metal allowed to cool (Figure 14.100). In some

---

(Facing page)

Figure 14.93   (*continued*)

(b)   Castor
The castor angle is the forward or backward tilt of the king pin or ball joints, and is measured as an angle from the vertical. A positive castor is when the top ball joint is behind the bottom ball joint. A negative castor is the opposite. The purpose of castor is to cause the front wheels to maintain a straight ahead position and to return to the straight ahead after making a turn. Incorrect castor will not cause tyre wear but would cause the vehicle to wander on the road and not return to the straight ahead after cornering.

(c)   Camber
The camber angle is the inward or outward tilt of the wheel when looking from the front of the vehicle. A positive camber is when the top of the wheel tilts away from the vehicle. A negative camber is when the top of the wheel tilts towards the vehicle. The purpose of camber is to bring the centre point of the steering in the centre of the tyre so the tyre turns on one point. This will decrease the effort required to turn the steering. Incorrect camber will cause the tyres to wear on the inside or outside, heavy steering and poor directional control.

(d)   KPI (SAI)
The KPI or SAI is the angle formed by a line drawn through the king pin or ball joints with a line drawn vertical. The purpose of KPI is the same as for camber. The KPI will reduce the need for large camber angles, and therefore reduce the tyre wear by allowing the wheel to be near vertical.

(e)   Included angle
The included angle is the sum of KPI (SAI) and camber. The point of intersection of KPI and camber gives the point about which the wheel rotates when the steering is turned. This point should be in the centre of the tyre where it contacts the road. Incorrect included angle would cause heavy steering and excessive tyre wear.

(f)   Toe on 20° turns
The toe on turns angle allows the front wheels to follow different arcs when cornering, and is measured when the inner wheel is at 20° from the straight ahead. The purpose of toe on turns is to allow for the greater distance of travel of the outer wheel when cornering. Incorrect toe on turns would cause excessive tyre wear.

(g)   Set back
Front wheel set back is the amount that one front wheel is ahead of the other. Excessive set back would cause the vehicle steering to pull.

(h)   Rear wheel reference toe
The front to rear wheel alignment measures the amount of out of alignment of the front and rear wheels. If the wheels are out of line the vehicle will 'crab', which means that the vehicle will appear to be driving sideways.

Figure 14.95 Drilling out spot welds for the removal of damaged panels (*Motor Insurance Repair Research Centre*)

Figure 14.97 Cutting out damaged panels with a power saw (*Motor Insurance Repair Research Centre*)

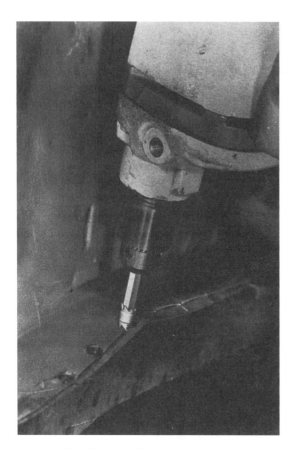

Figure 14.96 Zipcut tool for the removal of spot welds (*Sykes-Pickavant Ltd*)

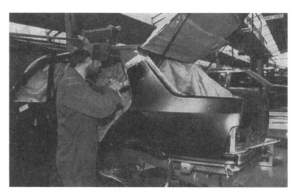

Figure 14.98 Alignment of new panel sections (*Motor Insurance Repair Research Centre*)

Figure 14.99 Cleaning up the locating flanges (*Motor Insurance Repair Research Centre*)

cases it is helpful in confining the heat, and thus reducing the chance of distortion, to place wet cloth packing either side and a little distance away from the weld; alternatively there is a foam type of material which will act as a heat barrier (Section 14.4.7).

Panels that were originally spot welded, but cannot be spot welded back, can be MIG welded in the course of repair; the welds are made in the same place by welding through the holes in the adjacent panel which were left when the original spot welds

Figure 14.100 Welding-in new panel section (*Motor Insurance Repair Research Centre*)

were drilled out. This method is known as plug welding. Another method of securing panels is to MIG braze them in place, but one must remember that although MIG brazing will cause less distortion than MIG welding, it is not as strong. Therefore if the particular panel is a load-bearing panel, it must be welded in fully to obtain the maximum strength for the particular panel assembly. MIG welding has the advantage of reducing distortion to a minimum, owing to the effect of the inert gas shielding the spread of heat around the weld zone. This makes butt joints in panel assemblies more easy to accomplish, as less dressing up and planishing is required than in gas welding. Another advantage arises from the automatic feeding of filler material into the weld pool, which results in better penetration in all weld positions. As this equipment can be used both to weld and to spot weld, it is widely employed as a means of welding in new panel assemblies.

14.4.5 Partial panel replacement

The reason for replacing only part of a panel can vary according to the type of repair. Usually it is quicker and therefore less expensive than full replacement. If sectioning does not save on time or money, then the part is completely replaced: that is, when the repanelling and straightening reaches the cost of the complete new panel. Some manufacturers design repair panels specially for these purposes.

Part panel replacement has long been established as an effective way of repairing body damage on all cars. Depending on the type of damage involved, it used to be common practice to replace an entire sheet metal panel; by means of a part panel repair it is possible to replace only those panel sections which are actually damaged. The possibilities for sectional repairs to load-bearing and non-load-bearing parts have been taken into account during the initial phase of development by manufacturers. All instructions given regarding the locations of cut lines for sectional repairs to load-bearing parts must be strictly adhered to. Sectional repairs to vehicles damaged in an accident should be confined to body parts in which the damage has not produced any loss of strength sufficient to jeopardize safety. Only welding materials reaching the design weld strength should be used. Moreover, a body alignment and welding jig must always be used for repair operations such as sectional replacement of the sidemember/apron-panel assembly. This ensures satisfactory repairs and adherence to the specified dimensions. Manufacturers have also carried out numerous strength tests, and extended trials (crash tests) have proved that part panel replacement is just as effective as replacing complete new panels. Provided repairs are expertly carried out, all repair procedures using part panels will result in the same standards of strength and operational safety being maintained as for a new vehicle.

To carry out this type of repair (see Figures 14.101a–f), the initial procedure is to make certain that the sections to be joined are from identical models. Then determine the most suitable point at which to make the joint: this will be influenced by the length of the weld, the amount of distortion likely to occur in making the weld, and the ease with which it can be dressed up. Remove the bulk of the unwanted section by cutting it away with a power saw, leaving an allowance of about 20 mm from the joint line, which can be trimmed to size when the pieces of the panels are separated. Where it is possible to fit the two sections on the vehicle, it is considered good practice to trim only one edge and to set the panels up with one section overlapping the other, as this enables a check to be made on the alignment of any surrounding or adjacent panels such as bonnet edges and door edges. A good butt joint is achieved by cutting through. When the alignment is correct, the two sections are tack welded along the length of the joint. The joint is then hammered level and fully welded. After this it is planished using normal techniques to achieve a smooth, undetectable joint. If this is impossible owing to inaccessibility, the weld should be hammered in using a shrinking dolly and the weld area filled.

Figure 14.101 Replacement of part panels:
(a) clamping new panel section into place (b) cutting in
(c) tack welding (d) section fully welded in
(e) dressing the welded joints (f) soldering the joint
and filing to a finish (*Motor Insurance Repair
Research Centre*)

Vehicles that have undergone an extended corrosion prevention treatment to prevent body perforation by rust must be repaired using replacement panels coated by the new (cathodic) priming process. These panels are identified by corresponding stickers. Therefore remove as little as possible of the cathodic primer coating on replacement panels during sanding or other abrading operations, e.g. on spot weld flanges or in areas where welds are to be made later. When repairs are being carried out in the underbody area, the zone in question must then be treated with underbody wax. The areas outside the side members and the wheel houses must also be treated with underbody protection material.

14.4.6 Repairing extensively damaged panels

Whenever possible it is best to repair severely damaged panels while they are still attached to the body shell. One advantage of leaving the panel in position is that it remains rigid whilst being processed. As already explained, it is difficult to hold a panel firmly when it is not attached to the body, and to beat a panel that moves with each blow is most unsatisfactory. Second, the shape is maintained while the panel is in position, and this allows the edges of the panel to be lined up with such parts as bonnets, doors and boot lids. In addition the use of the body jack, while the panel is fixed to the body, allows full use of the attachments designed for roughing out such parts.

Having decided whether the panel will be repaired on or off the job the actual repair can then be roughed out, keeping in mind the fact that the shape needs reforming with as little stretching of the metal as possible. The original damaging force could have stretched the panel and, while reshaping, further stretching should be reduced to a minimum by the use of such methods as pushing out the damage with a body jack fitted with suitable rubber flex head; or using a hardwood block between the body jack and panel; or using hardwood blocks in conjunction with the hammer to eliminate the metal-to-metal contact. Where necessary, anneal creased sections with the oxy-acetylene welding torch prior to removing them. As the roughing out proceeds, weld any tears in the metal so as to gain rigid panel as soon as possible. Very stretched sections can be shrunk by the use of heat spots during the roughing

out. A boxwood mallet, in conjunction with a suitable dolly block for the shape of the panel being repaired, should not be used as an initial smoothing or levelling step prior to the final planishing. The actual smoothing with the planishing hammer and dolly should never be started until the correct shape is formed in the panel. At this stage the normal processes of planishing, dressing up, filing and picking up low spots should not be adopted until the job is completed. Either because certain parts of the repair are inaccessible or because of their overstretched condition, it may be necessary to use solder filling or plastic filler to achieve a smooth finish.

14.4.7 Repair methods using heat barrier material

This is a method of repair using a specially formulated compound where heat would cause a problem of distortion or overheating of adjacent materials when welding.

The Cold Front heat sink chemical compound was originally developed for the American space programme. This heat barrier material, based on a chemical formula of magnesium aluminium silicate, prevents the conduction of heat through metal, thus eliminating the risk of distortion to panels, or damage to areas immediately adjacent to any welding operation. When the material is applied to a metal panel on a vehicle body undergoing repair, it prevents the transfer of heat past the Cold Front barrier, giving complete protection to glass, rubber, plastic, paint and any material likely to be damaged by close proximity to the high temperatures which are generated during the processes of welding, brazing or soldering.

The method of application is as follows:

1   Apply generously, as using too little can allow heat damage. Apply to one side of the metal in light gauge work (20 SWG) and both sides of the metal when working with heavy gauge (above 3 mm) for maximum protection.
2   When the mound of material begins to warm up, it becomes saturated with heat. Once saturated, it will allow heat to be conducted through the metal. If there is still work to be carried out, place more Cold Front material on top of the existing mound to delay the conduction of the heat.

3 Always apply the material as close to the area of repair as possible.

4 Damage can be caused by radiant heat as well as conducted heat. The material acts as an effective heat shield against radiant heat when it is applied over the surface to be protected. Be sure not to leave any cavities or depressions when spreading.

5 It is safest to protect the repair area completely around the heat zone. However, if protection is only needed in one direction, the heat barrier should be spread five to six times as wide as the area to which the heat is being applied.

6 Metals have different conducting characteristics: copper, brass and aluminium conduct heat more rapidly than steel. Therefore the better the metal is as a conductor, the more heat barrier should be applied when working with that metal.

Cold Front heat barrier is odourless, non-toxic and harmless to skin. It will not stain surfaces, and any residue can be washed off with cold water once the repair is completed.

The ability of Cold Front to act as a total heat barrier is demonstrated by holding a strip of 20 SWG metal in the bare hand after applying a line of Cold Front halfway along the strip, then heating the portion of the strip above the line to red heat. The part of the strip being held in the hand remains completely cold (Figure 14.102).

Figure 14.102 Application of Cold Front heat barrier material (*Gray–Campling Ltd*)

14.4.8 Door hanging

Door hanging requires particular care, as the fitting of doors is one of the most essential features of crash repair work. The most important aspect when dealing with the hanging of doors is that a door which is known to be the correct shape should never be altered to fit a door opening that is out of alignment. In addition, the resetting of door hinges must not be used as a method of rehanging doors without first checking for and correcting any misalignment.

Hinges used on the modern vehicle can take a good deal of force without altering their settings. In cases where hinges have been subjected to collision damage, the hinge transmits the damaging force to the pillar or door frame, causing twisting and misalignment. Sometimes hinges are so excessively damaged that they need replacing, although this is not the complete answer to correcting misalignment troubles arising from such damage. Where new hinges are necessary a close inspection and comparison should be made with the identical hinge mounting on the opposite side of the car, and any misalignment of this mounting position should be rectified before fitting the new hinge.

When it is considered that the hinge mounting points are correct, the doors should be hung in their respective positions and all screws or hinge fixing bolts placed loosely to allow the door to be realigned according to the car manufacturer's adjustments. The fixing bolts should be tightened if possible while the door is in its closed position. In most cases this is not possible, which means that the doors have to be held open whilst at the same time endeavouring to keep alignment correct while the bolts are tightened. If the door aperture is correct and the door is not twisted, with slight adjustment to the hinge position the door can be made to fit into its appropriate opening.

When the door will not fit into the body opening, the hinges should be tightened in the main adjustment position and the striker plate of the lock mechanism should be removed from the locking pillar to allow the door to swing freely. This gives a truer indication of the swing of the door as the opening is being corrected with the use of hydraulic equipment. The door should then be closed as far as possible, and the points stopping the door from entering the door aperture should be noted. Correction in most cases can be made by diagonal pushing with the body jack within the aperture. Measurements should be taken both before and after applying the pressure so that the effectiveness of the realignment of the door opening can be assessed. This checking and pushing must continue until the door is a good fit inside the door aperture.

14.4.9 Motor vehicle glazing systems

Since the introduction of legislation on the compulsory wearing of seatbelts for front seat passengers in the early 1980s, both the type and style of glass fitted into windscreens have changed. Prior to 1980, toughened glass, which shatters into tiny, relatively harmless pieces when broken, and which can dangerously obscure a driver's vision, was used by many vehicle manufacturers. This glass was indirectly glazed on to the body and secured using a self-tensioning rubber weather strip fitted to the aperture. With the seatbelt legislation in force, however, safety standards were improved, bringing an increased use of laminated glass, which is the type that remains in place when broken and therefore allows the driver to continue the journey.

This resulted in a change in glazing methods, with laminated glass being increasingly secured by the process of direct glazing. This is a method of fitting a glass unit to a vehicle by chemically bonding it to the prepared aperture using specialized compound sealants. This method provides two definite design benefits. It gives an aerodynamic profile, and also increases the strength of the windscreen, thus making it safer and more weather resistant. As well as being easier to fit on the production line, the majority of direct glazing is now done by robots. A direct glazed windscreen can also be replaced to its original specification without affecting the structural integrity of the vehicle.

Indirect glazed windscreen removal and replacement

The two basic types of safety glass used in the car windscreen are laminated and toughened. *Laminated* glass is of a sandwich construction and consists of two pieces of thin glass one on either side of, and firmly united to, a piece of transparent reinforcing known as the interlayer (Figure 14.103). Although such a glass may crack, forming a spider's web pattern (Figure 14.104), it holds together and

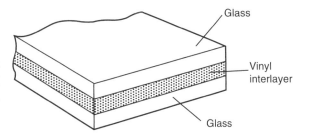

Figure 14.103   Laminated glass

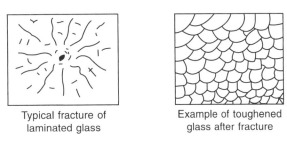

Typical fracture of laminated glass

Example of toughened glass after fracture

Figure 14.104   Characteristic breaking pattern of laminated and toughened glass

the panel remains in one piece except under conditions of the most violent impact. *Toughened* glass is produced by the process of heating a solid piece of glass and then rapidly cooling it so that its liability to fracture is greatly reduced and its strength increased to about six times that of untreated glass. If fracture should take place, the resulting fragments consist of very small comparatively harmless particles (Figure 14.104).

In major accident damage, removal and replacement of the windscreen is an essential feature of the repair. Many windscreens and rear windows are held in place by a rubber weatherstrip which is usually of single-piece construction. Both glass and any outside trim mouldings, if used, are recessed into the weatherstrip. It also has a recess which fits over the body flanges all round the opening for the glass. It is this portion of the weatherstrip that holds the entire assembly in place.

Indirect types of glazing systems can be identified for each specific model by referring to the body section of the relevant manufacturer's repair manual. The rubber weather seals used are either mastic sealed, dry glazed, self-sealing or push fit (Figure 14.105). The self-sealing types are held in

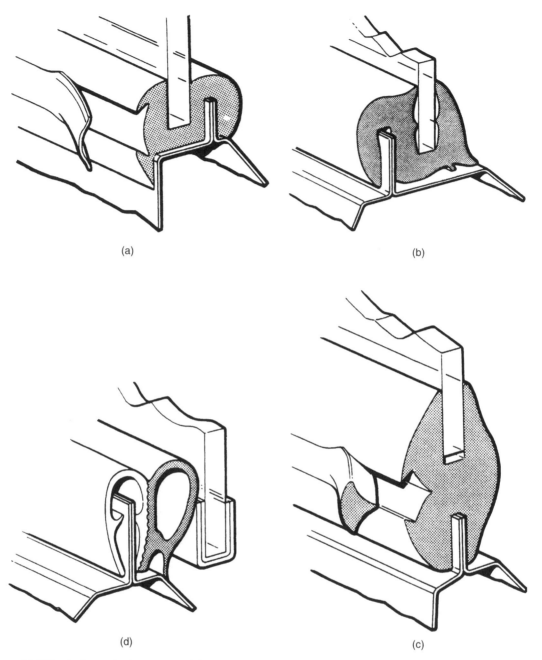

Figure 14.105 Indirect glazing system: (a) mastic-sealed (b) dry glazed (c) self-sealing (d) push fit

place by a removable centre filler strip and can be either mastic sealed or dry glazed (Figure 14.105c). All the systems are used with either a rubber weather strip moulding with a sealing compound (Figure 14.105a) or a rubber weather strip mould-ing without a sealing compound (Figure 14.105b).

The rubber weather strip has two rebates, one for the glass and one to seat over the metal flange of the body aperture. It is the contact area of the aperture that requires careful attention whenever the glazing is disturbed or removed. In all cases it is essential that the aperture is correctly painted prior to refitting any of the weather seals.

In order to remove a windscreen which is fitted with self-tensioning weatherstrip, the following procedure must be carried out. First cover the bonnet to protect the paintwork. Then, working from the inside of the car, remove any instrument panel mouldings or trim which may interfere with the removal of the windscreen. Also, working from the outside of the vehicle, remove any windscreen chrome mouldings which may be fitted into the weatherstrip. Again working from the inside and starting from one of the top corners, pry the rubber weatherstrip off the body flange with a screwdriver. At the same time press firmly against the glass with your hand adjacent to the portion of the rubber lip being removed, or bump the glass with the palm of the hand; this should force the entire assembly over the body flange. The windscreen can then be removed from the body opening.

To replace the windscreen, first clean the glass channel in the rubber weatherstrip to be sure that all traces of sealer and any broken glass are removed. Place a bead of new sealing compound all round the weatherstrip in the glass channel. Work the glass into the glass channel of the weatherstrip and be sure it is properly seated all the way round. Lay the windscreen on a bench, suitably covered to prevent scratching, so that the curved ends are pointing upwards and in this position the pull cords used to seat the rubber lip over the body flange can be inserted into the metal channel on the weatherstrip (Figure 14.106). One or two cords may be used. Where two cords are normally used they can be pulled in opposite directions, which will make the fitting of the windscreen easier. Commence with one cord from the bottom right-hand side of the windscreen, leaving about one foot-spare for pulling. Start the second cord from the bottom left-hand corner and proceed to the right, following round up to the top left-hand corner. Apply sealer all round the metal lip on the body aperture. Place the entire unit in the windscreen opening with the cord ends hanging inside the vehicle. An assistant can help by pressing against the glass from the outside while each cord is pulled out of the groove. As the cord is pulled out, the rubber lip is seated firmly over the body flange. A rubber mallet can be used to tap the unit around the outside so as to seat the entire

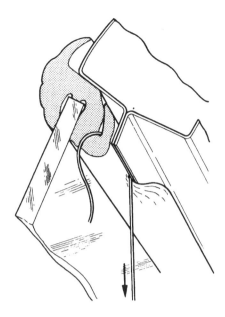

Figure 14.106   Using pull cords to fit windscreens

windscreen assembly firmly in place. Any surplus sealer can be cleaned off with white spirit.

The procedure for removing a windscreen fitted with self-sealing weatherstrip is first to locate the joint in the filler strip which runs right through the centre of the weatherstrip. The strip is then carefully eased up and pulled out slowly around the entire assembly. This relieves the tension on the glass which can then be lifted out of the weatherstrip by easing one corner first. To replace this type of windscreen the metal lip on the body aperture is sealed with a sealer, the appropriate channel in the weatherstrip is then placed over this lip and seated all the way round the body aperture. The windscreen is fitted into the glass channel on the weatherstrip with the aid of a special tool which enables the glass channel lip to be lifted, thus allowing the glass to slip easily into position. A small brush is used to apply a solution of soap and water to the filler channel. This assists the filler strip installation. A specially designed tool is used to insert the filler strip into its channel.

Direct glazed windscreen removal and replacement

The method of direct glazing of windscreens, also known as bonding, is very popular with motor manufacturers. This process involves the bonding of glass into the aperture. Some of the advantages claimed for direct glazing are as follows:

1 Increased rigidity and strength caused by integration of the glass into the body of the vehicle, reducing the need for supporting members and improving visibility with enlarged glass areas.
2 Weight reduction.
3 Improved aerodynamics by the deletion of rubber surrounds, aiding fuel economy.
4 A better seal against the weather, eliminating the need for resealing.
5 Increased protection against car and contents thefts.

Various bonding materials are used in direct glazing. Some require heating to induce a chemical reaction to create adhesion, whilst others will cure at room temperature.

Polyurethane and silicone materials are usually supplied as a pumpable tape. They are cold cured and the material is dispensed on to the glass through a specially formed nozzle out on the end of a cartridge. This can be done with the aid of a hand operated or compressed air cartridge gun. These materials are highly viscous in their uncured state, enabling a high degree of manoeuvrability within the glass aperture to ensure a good seal.

The sealant itself is either one- or two-component polyurethane. The latter contains an accelerator compatible with the adhesive; this is applied evenly to the perimeter of the windscreen (Figure 14.107).

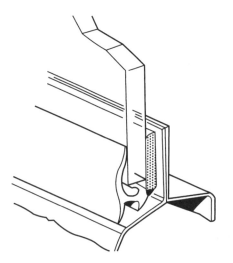

Figure 14.107 Direct glazing with moisture-cured urethane

Once this is achieved, the glass is placed into the aperture using glass suckers and aligned before securing. When a one-component polyurethane is used the car can be driven away in four to six hours: if a two-component polyurethane is used it takes only 30 minutes to cure, allowing the vehicle to be back on the road within an hour of the repair starting.

To remove the glass from the vehicle, the bonding material has to be cut. This can be achieved by using a piano wire, a special cutter or a hot knife, as follows:

Piano wire Before using this method of removal there are certain safety precautions that should be observed. Gloves should be worn to protect the hands, and safety glasses or goggles should be used for eye protection in the event of the wire or glass breaking. When the finishers or mouldings have been removed to expose the bonded area of the glass, the wire has to be fed through the bonding. This is done by piercing a hole through the bonding and feeding the wire through with the aid of pliers. Handles can be fixed to the ends of the wire to allow a pulling action (Figure 14.108).

Pneumatic or electric cutter (oscillating) This is an air or electric powered tool to which special shaped blades to suit specific vehicle models are fitted. The tool removes windscreens that have been bonded with polyurethane adhesives (see Figure 14.109).

Hot knife The hot knife can be used by one person cutting from the outside of the vehicle. It can cut round the average bonded windscreen and clean off the excess remaining adhesive while minimizing the risk of damaging the car bodywork. Before use, all trim around the windscreen, both inside and outside, should be removed (see Figure 14.110). The cutting medium is provided by a heated blade which is placed under the edge of the glass and pulled around the perimeter, melting the bonding to release the windscreen. It is used in conjunction with an air supply, which constantly blows on to the cutting area of the blade. This prevents the heat dissipating along the blade and gives a constant temperature as well as eliminating smoke and fumes emitted from the cutting operation. Overheating can result in toxic fumes being given off and a charcoal filter mask should be worn as a safety precaution. Before glazing can be

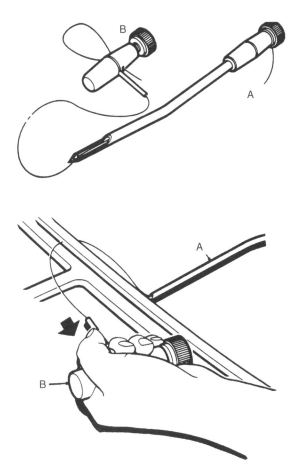

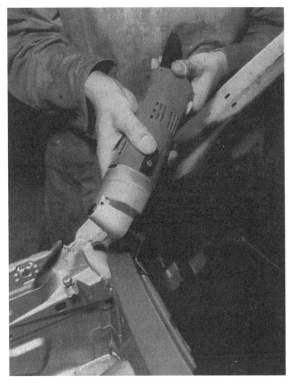

Figure 14.109   Electric windscreen cutter (*Fein/George Marshall (Power Tools) Ltd*)

Figure 14.108   Removing direct glazing using a piano wire

replaced, the bonding surfaces must be prepared correctly. The residue of the original sealer left in the aperture has to be trimmed to ensure there is a smooth layer for the new sealer to adhere to, and to allow the glass to seat in the correct position. Various sealants for windscreen replacement are available.

The surfaces are then prepared with the materials included in the individual manufacturer's glass replacement kit, that is sealer application gun with bonding material, suction cups and leak test equipment, following the instructions provided. There is little variation in the sequence of operations, which is typically as follows:

1   Using a sharp knife, trim and level off sealer remaining in the vehicle windscreen aperture. Remove loose sealer. Areas where bare metal is exposed must be treated with etch primer.

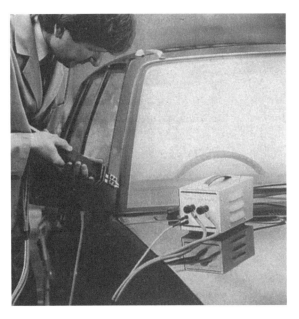

Figure 14.110   Welwyn hot knife (*Welwyn Tool Co. Ltd*)

2 If the original windscreen is to be refitted, level off the remnants of the old sealer. Take care not to damage the black ceramic edging.

3 Apply the cleaning solvent provided in the repair kit to the windscreen aperture and the inboard edges of the screen, then wipe both aperture and screen with a clean cloth.

   Warning: when working with solvents and primers, use PVC gloves or apply barrier cream to hands.

4 Take the spacer blocks from the repair kit, remove the adhesive backing, and fit them in the original position on the windscreen aperture. Ensure that the differently shaped blocks are located correctly.

5 Fit the suction cups to the outer face of the windscreen and place the screen in the vehicle. Centralize the screen in the body aperture. Take strips of masking tape and attach them across the join between the body and the screen. These strips afford a guide when finally fitting the screen after sealant has been applied. Cut the masking tape strips at the glass edges and remove the screen.

6 Shake the tin containing the glass primer for at least 30 seconds. Then, using the felt swab, apply the primer evenly along the screen perimeter. When dry, wipe with a clean cloth. Repeat the above process, this time by cutting the applicator head and using it to apply a coat of primer 17 mm wide (0.7 in) to the windscreen aperture.

7 Immerse a cartridge of sealer in water at a temperature of 60 °C for approximately 30 minutes. Pierce the sealer cartridge in the centre of its threaded end and fit the precut nozzle. Remove the lid from the other end, shake out any crystals present and install the cartridge in the applicator gun.

8 Apply the sealer in a smooth continuous bead around the edge of the windscreen. The sealer bead should be 12 mm (0.5 in) high and 7 mm (0.25 in) wide at its base.

9 Lift the screen using the suction cups and offer it up to the vehicle, top edge leading. Carefully align the masking tape strips, then lower the screen into position. Remove the suction cups. Testing for leaks, using either leak testing equipment or spraying with water, may be carried out immediately.

Caution: if a leak or leaks are detected, apply sealer to the area and retest. Do not remove the screen and attempt to spread the sealer already applied.

10 If leaks are not present, fit the windscreen finishers. Heat the windscreen finishers to 45 °C and, commencing at one end of the screen lower corners, press into position around its entire perimeter. Remove the fascia's protective covering. Trim the lower spacer blocks. Fit retaining clips as necessary.

11 Fit the windscreen wiper finisher panel and air intake mouldings and the wiper arms and blades.

   Caution: do not slam the vehicle's doors with the windows fully closed until the screen is fully cured.

Warning: the integrity of the vehicle's safety features can be impaired if incorrect windscreen replacement bonding materials and fitting instructions are used. The manufacturer's instructions should be adhered to at all times.

14.4.10 Water leak detection

The general principles for searching for the actual location of where water leaks could potentially occur on the vehicle are not specific to any particular model and can apply to all vehicles.

First start by obtaining as much information as possible from the customer as to when, where and how the leak occurs, and also whether the water appears to be clean or dirty. If these facts are not known, considerable time could be spent checking the wrong areas of the vehicle.

Tools and equipment

The following tools and equipment are recommended:

Garden spray (hand pump pressure type)
Wet/dry vacuum cleaner
Torch
Mirror
Seal lipping tool
Trim panel removing tool
Small wooden or plastic wedges
Dry compressed air supply
Hot air blower
Sealant applicator
Ultrasonic leak detector.

Locating the leak

Locating the source of the water leak involves a logical approach together with a combination of skill and experience. For the purpose of locating the leak the vehicle should be considered in three specific areas: the front interior space, the rear passenger space, and the boot space. From the information provided by the customer the body repairer should be able to determine which area is the right one on which to concentrate.

Having identified the area of the leak, the repairer must find the actual point where the water is entering the vehicle. An ordinary garden spray, of the type which can be pressurized and adjusted to deliver water in the form of a very fine spray or a small powerful jet, has been found to be very effective in helping to locate most leaks. Using a mirror and torch will help to see into any dark corners.

Testing

The sequence of testing is particularly important. Always start at the lowest point and work upwards. The reason for this is to avoid masking a leak in one place while testing in another. If for example testing was to commence around the windscreen, water cascading down could leak into the car via the heater plenum or a bulkhead grommet. However, it could be wrongly assumed that the windscreen seal was at fault.

The visible examination of door aperture seals, grommets and weather strips for damage, deterioration or misalignment, as well as the actual shut of the door against the seals, are important parts of identifying an area where water can pass through.

Leak detection using ultrasonic equipment

When the vehicle is in motion, the body shape may produce eddy currents and turbulence which can force air and water through the smallest orifice. When the vehicle is stationary, it can be difficult to reproduce these conditions to a realistic level.

The ultrasonic equipment works on the principle of a transmitter creating ultrasonic waves which penetrate the smallest orifice in the vehicle body and are then picked up by a receiver fitted with a suitable probe, which can in turn pinpoint the exact leak point. The transmitter is placed in a base plate which automatically switches it on via a reed switch in the base of the transmitting body.

Ultrasonic waves will then penetrate out through the unsealed area, including the front windscreen, wiper spindle and washer jet fitment, roof seams, bulkhead seams and grommets, A-post area, door seals, front wheel arch seams and heater air intakes, to the receiver.

An important feature of this equipment is the incorporation of a sensitivity control which enables the operator to check a wide range of fitments. For example, a boot seal is a very light touch seal, whereas a urethane screen fitment is at the other end of the scale setting of the equipment and would need a high setting to determine a leak.

To ensure satisfactory leak location and testing, it is recommended not to use a leak detector on a wet vehicle, which should be blown dry prior to the test. As with any equipment it is important to read the instructions to enable the equipment to be used to its full potential.

Sealing

Having located the point of entry of the water, it is then necessary to carry out satisfactory rectification. Door aperture seals and weather strips should be renewed if damaged or suffering from deterioration. Alternatively the seals can be adjusted by carefully setting the mounting flange after making sure the fit of the door is correct.

Leaks through body seams should be sealed from the outside wherever possible to ensure water is excluded from the seam. First dry out with compressed air or a hot air blower.

When leaks occur between a screen glass and the weather strip, or in the case of direct glazing between the glass and the body, avoid removing the glass if possible. Use recommended sealers to seal between the glass and the weather strip or the body.

If the vehicle is wet then it should be blown dry prior to sealing. It is difficult to seal a wet vehicle because, if a fault is found, it cannot be rectified quickly before moving to another area.

14.4.11 Windscreen repair: Glass-Weld Pro-Vac

This system of repair involves extracting all the air from the damaged glass and replacing the void with a resin which, when cured with ultraviolet light, is optically clear and is stronger than the glass that has been replaced. The system is widely accepted, and repair to Glas-Weld standard would not cause the vehicle to fail the statutory Department of Transport test for windscreens: 80 per cent of stone damaged windscreens can be repaired by this method (Figure 14.111).

When glass breaks, an air gap opens up. Sometimes crushed glass may also be present. There are also flat surfaces present inside the glass on either side of the air gap. These three things result in the visibility of the damaged area being affected. The damage will eventually spread further as the temperature changes and the vehicle flexes in normal use (Figure 14.112).

The system works, as follows:

1 All foreign materials and crushed glass are carefully removed from the centre of the damaged area to open up an airway into the break. Certain types of damage may need to be precision drilled (Figure 14.113).
2 The patent Pro-Vac injector is filled with the appropriate resin, depending on local temperature and humidity, before being mounted on the screen. The injector is then threaded through the stand until the outer seal makes airtight contact with the screen (Figure 14.114).
3 The Pro-Vac injector is capable of creating a total vacuum within the damaged area, which is essential for top quality repairs. Using alternating vacuum and pressure cycles, all of the air in the break is withdrawn and the void is filled with resin. Once the damage is optically clear, curing can commence.
4 Using a special ultraviolet lamp, the resin is cured. Once the curing process is complete the repair area is not only optically clear but also structurally sound; in fact it is stronger than the glass it replaces (Figure 14.115).

14.4.12 Automobile plastics welding

The car manufacturing industry, in its endeavours to reduce weight and improve fuel economy, forecasts a greater use of plastics in future designs (see Section 4.12). Considering the extensive use

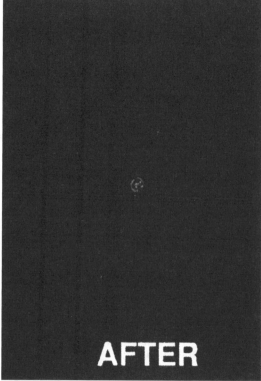

Figure 14.111   Before and after repair (*Glas-Weld Systems (UK) Ltd*)

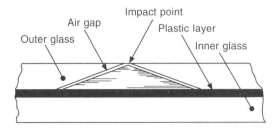

Figure 14.112 Damaged windscreen (laminated safety glass) (*Glas-Weld Systems (UK) Ltd*)

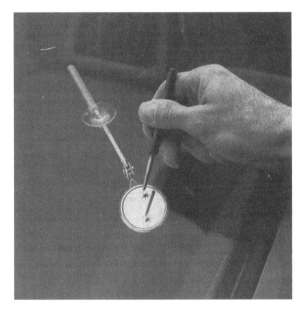

Figure 14.113 Cleaning the damaged area (*Glas-Weld Systems (UK) Ltd*)

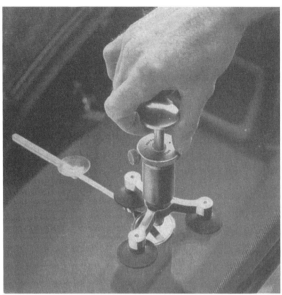

Figure 14.114 Fitting injector to the windscreen (*Glas-Weld Systems (UK) Ltd*)

Figure 14.115 Using ultraviolet lamp to cure resin (*Glas-Weld Systems (UK) Ltd*)

of plastics on current models, coupled with an even greater anticipated usage, it is therefore important that engineers and repairers are knowledgeable in the various types of plastic that are being used and are aware that some plastics can be repaired. Many body components are manufactured in a variety of plastics. Bumpers, grilles, light surrounds and even complete body panels have enabled designers to enhance vehicle aerodynamic styling and cosmetic appeal while retaining impact resistance and eliminating corrosion from these areas.

Plastic offers the same structural strength as steel in a body component because of its greater elasticity. Minor impacts that could deform steel beyond repair can be absorbed by plastic. Where damage is incurred it is capable of repair by welding. Cracks, splits, warping and even the loss of material from plastic components can be remedied with the aid of the Leister Triac hot air welding equipment. Where a steel component with equivalent damage would be replaced at some cost, the repair of the plastic part can save time and expense, particularly when winter accident periods make great demands of the vehicle repairer's parts stock. Welding plastics does not produce fumes when the correct procedure is followed. In a short time a plastic component can be restored to an 'as new' condition without the need for fillers or special treatments. The combination of welding and recommended paint procedures will show no trace of a repair that should last the life of the vehicle.

Identification of the material

There are two basic types of plastic used in cars: thermosetting and thermoplastic (see Section 4.12.3).

Thermosetting plastics are cured by heat during manufacture. They cannot be welded; any more heat applied to them will break down their molecular bonding. The way to repair them is by using adhesive. *Thermoplastic* materials can melt without breaking down, so they can be manipulated like steel. With the right amount of heat they can be bent, softened and welded. The skill is in the selection of the appropriate temperature. The majority of plastics employed in vehicle manufacturing are thermoplastics. There are different types of thermoplastics, each having a specified temperature for welding operations.

Experienced repairers of plastic may recognize broad types of plastic by the degree of hardness or softness of the component surface. A quick test for identifying to which group a plastic belongs can be made by cutting a very thin sliver off the component. The thermoplastic curls right back on itself, but the thermoset stays straight.

In their body repair manuals, some vehicle manufacturers give details of plastic parts, while others mark all plastic parts with code letters. However, when no information or coding is available, identification can be carried out in two ways. The first is the method of combustion testing, where a small strip of material is held in a butane flame and the flame colour is noted. The other method, which works on the majority of the most common plastic materials, is to use an organic solvent test kit. These kits, produced by an automotive paint manufacturer, work by a process of solvent application to the material surface. A normally concealed area should be chosen and degreased, and any paint applied to the surface should be removed from the test area with the aid of an abrasive. Identification is confirmed by reaction between the plastic and test solvent in accordance with the test kit manufacturer's instructions. Gloves and face masks should always be worn when using solvents.

The usual plastic identification codes are as follows:

ABS	acrylonitrile butadiene styrene
ABS/PC	polymer alloy of ABS
PA	polyamide (nylon)
PBT	polybutylene terephthalate (POCAN)
PC	polycarbonate
PE	polyethylene
PP	polypropylene
PP/EPDM	polypropylene/ethylenediene rubber
PUR	polyurethane (not all PUR is weldable)
PVC	polyvinyl chloride
GRP/SMC	glass fibre reinforced plastics (not weldable)

Equipment for plastic welding

The Welwyn Tool Company supply the Leister Triac welding guns (see Figure 14.116). These blow hot air out of their nozzles, and sophisticated electronics make sure that the temperature of this air is exactly right. A potentiometer is set to control the power (Figure 14.117), a photo-electric cell monitors the colour of the heating element and, as a failsafe, a thermocouple double checks the output. The guns can maintain a critical temperature

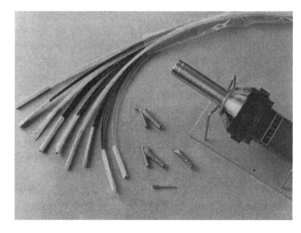

Figure 14.116 Leister Triac electric welding tool and accessories (*Welwyn Tool Co. Ltd*)

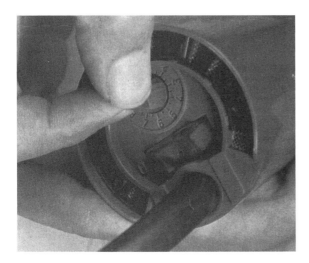

Figure 14.117 Rotary control on rear of hot air tool for temperature setting (*Welwyn Tool Co. Ltd*)

setting to $\pm 5\,°C$, independent of any mains voltage fluctuations. It is important to use a welding rod of the same material as the component: hence the importance of correct identification. The welding rods come in various profiles, and there is a speed-weld nozzle available to suit each rod.

Repair method

Suppose a bumper has suffered a long, straight crack which makes it particularly suitable for repair by welding (see Figure 14.118). When carrying out the repair process, thorough preparation of the

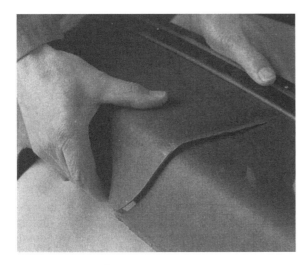

Figure 14.118 Split bumper
(*Welwyn Tool Co. Ltd*)

surfaces is important. When working on a painted bumper it is necessary to sand off any paint in the vicinity of the area to be welded; a mask for facial protection is then necessary. To minimize the risk of damage spreading, the ends of the crack are drilled out with a 2 mm drill. The edges of the crack are then shaped into a V with a burring bit which has a cutting edge on both its circumference and its face (see Figure 14.119). This will allow the welding rod, which in this case is supplied in a V shape, to fit snugly into the crack. Care should be taken not to penetrate more than two-thirds into the depth of the material. The extent of the damage should now be clearly visible.

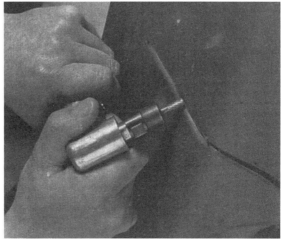

Figure 14.119 Rotary burring tool cutting a V-groove (*Welwyn Tool Co. Ltd*)

In any welding operation it is important that the two surfaces to be joined should remain in alignment, so they should be tacked together using a tacking nozzle (see Figure 14.120). The section of the temperature chart relating to this nozzle must be consulted to determine the correct gun setting; the temperature can then be adjusted on the potentiometer. After allowing the gun to heat up for at least two minutes, the tacking operation can be carried out. The nozzle is held at an angle of about 20 degrees and is run along the crack without applying any pressure.

The speed-weld nozzle is fixed to the gun, and the temperature is again adjusted prior to carrying out the weld. The end of the welding rod is trimmed into a point to ensure a smooth start to the weld (see Figure 14.121). The welding rod is

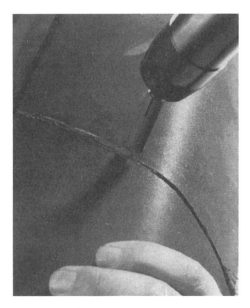

Figure 14.120 Tack welding split in bumper
(*Welwyn Tool Co. Ltd*)

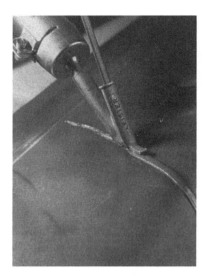

Figure 14.122 Main plastic welding operation
(*Welwyn Tool Co. Ltd*)

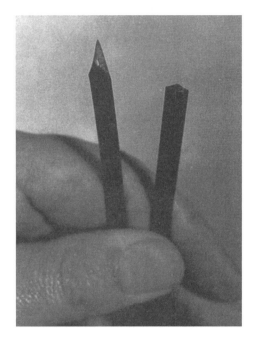

Figure 14.121 Welding rod trimmed to a point
(*Welwyn Tool Co. Ltd*)

inserted into the speed-weld nozzle, and as soon as the edges show signs of melting the weld can be carried out (see Figure 14.122). Contact should be maintained between the nozzle and the workpiece but, as with the tacking operation, no pressure should be applied with the gun. The welding rod, however, should be pushed well into the crack. The rate at which the welding rod is fed into the gun should match the distance travelled. It is also imporant to operate the gun at the correct angle. When the weld is finished, the gun is pulled away leaving the welding rod attached. The excess rod is then trimmed off.

The bumper is allowed to cool, and the welded area can then be sanded down with an 80 grit disc before being passed on to the painter (see Figure 14.123). The bumper has a non-textured painted finish, which will be reinstated. After normal preparation and priming have been carried out a filler is applied. After flatting the filler, two further coats of primer are sprayed on and flatted. The usual cleaning procedures should be carried out between each process. The colour can now be applied. Painting should always be carried out according to the paint manufacturer's instructions.

If the repair method is carried out correctly, the component should achieve up to 90 per cent of its original strength.

When repairing plastic materials there must always be strict observance of health and safety requirements, so the operator should work in a well ventilated environment. Using the gun at an abnormally high temperature does not assist the welding operation, and can cause dangerous fumes.

Figure 14.123 Weld dressing (*Welwyn Tool Co. Ltd*)

Welding defects

A tack weld can be broken and restarted if panel alignment is not achieved on the first attempt. To prove main weld strength, allow it to cool then attempt to pull it from the groove by the attached welding rod. If the weld stays firmly in place then the weld has been successful.

Typical welding defects and their causes are as follows (see Figure 14.124):

Poor weld penetration or bonding Incorrect weld site preparation; welding speed too fast; temperature too low; weld attempted with dissimilar materials; poor technique.
Uneven weld bead width Welding rod stretched; uneven pressure applied to welding rod.
Charred weld Welding speed too slow; temperature too high; repair area overheated.
Warping Parts fixed under tension; poor site preparation.

14.4.13 Plastic repair

Duramix combines the tremendous strength and durability of a plastic weld system with the flexibility of chemical curing products, thus eliminating the need for a filler or stopper to create the complete cosmetic finish. It is compatible with all plastics, fibre-glass and metal surfaces. Repairs can be carried out on flexible plastic bumper bars, semi-rigid plastic panels, ABS plastic grilles, polycarbonate bumper bars, and thermoplastic bumper bars.

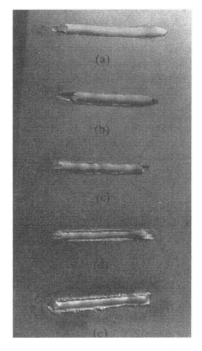

Figure 14.124 Defective plastics welds and their causes (*Welwyn Tool Co. Ltd*)

(a) Weld started correctly but completed too quickly: no wash, indicating haste or welding temperature too low
(b) Hot air tool not allowed to attain the correct welding temperature before starting, and weld finished too soon, leaving a hole
(c) Weld started too late with an unprepared welding rod end: hole and protrusion of plastic result
(d) Too much pressure applied to the welding rod, leaving a low and deformed weld bead: filling may be necessary
(e) Welding temperature too hot for the material, blistering the sides of the weld; repair area may be brittle

Duramix is a urethane composition. It applies easily to vertical surfaces, resists sagging, contours with a spreader to the original surface shape, sets within three minutes, and is easily sanded to a high-quality repair within fifteen minutes. It produces a tenacious bond while maintaining flexibility, even in cold temperatures.

The repair procedure is as follows:

1 Material and equipment required: plastic cleaner, heat gun, reinforcement patches, 3 in grinderette,

DA sander, plastic spreader, masking tape, rubbing-down paper and the Duramix filling material.

2 Identify the plastic to be repaired. Most of the semi-rigid plastics on vertical panels on today's vehicles are urethane, Xenoy, or other repairable materials. A small percentage are thermoplastic polyolefin (TPO), and these must be repaired using a slightly different procedure.

3 Clean the plastic panel surface and back of the repair thoroughly to remove all road dirt, oil and grease.

4 Remove any deformations in the plastic panel. If the panel has undergone a severe impact it may have some residual distortion. To remove any distortion, warm the stretched or distorted area with a heat gun. The warm air will allow the panel to return to its original shape.

5 Apply a structural support to the back of the repair using a Duramix reinforcement patch. This will provide some firmness so that the repair area can be V-grooved and feathered back around the repair.

6 V-groove the area using a coarse 3 in disc, then feather back the paint around the V-groove.

7 Apply Duramix to the plastic panel. First select a length of contouring plastic which will cover the entire repair area. Then mask off the area surrounding the repair to reduce the sanding and clean-up time. Apply Duramix directly to the V-grooved area, slightly overfilling the area; then within 30 seconds overlay a sheet of contouring plastic and use the flat side of a plastic spreader to compress and smooth the Duramix into the V-grooves. Duramix will start to set within 50 seconds after it has been dispensed, so the body repairer should work quickly to apply the contouring plastic.

8 Sand the repair using a DA sander: start with a coarse grit disc and finish with a smooth grit disc, feather edging the whole repair.

9 Apply a sealer, then primer surfacer, then finishing coats to suit the plastic panel or bumper bar being repaired.

14.4.14 Valeting of repaired vehicles

After repair work has been completed, the vehicle should be presented to the customer in a first-class, clean condition. All repaired vehicles should be thoroughly inspected for quality control. The repairs should be guaranteed for a specific mileage or period, as required by the Code of Practice for the Motor Industry and by the VBRA Code. Many insurance companies now make an allowance for valeting within their agreed repair price, and this provides the opportunity for the bodyshop to produce a first-class factory finish to both the exterior and interior of the repair.

Premises, equipment and chemicals used in cleaning and valeting

When cleaning and valeting a vehicle it is essential to organize a standard method of working to ensure that nothing is inadvertently missed. The ideal premises and equipment should be set out in separate purpose designed working areas.

The cleaning process should be undertaken in a wet bay with effective drainage, good lighting and adequate working space all round the vehicle when its doors are open. Compressed air and high-pressure cold and hot water should be available (Figure 14.125).

Figure 14.125 Wet bay (*Autoglym*)

Valeting should be carried out in a well lit, dry bay with adequate working space all round the vehicle. Compressed air, electrical power points, warm water, a workbench and storage cupboards should be available (Figure 14.126).

Equipment, details are as follows:

1 High-pressure hot and cold washer, 70–100 bar (1000–1500 psi), with chemical throughput

Figure 14.126   Dry bay (*Autoglym*)

facility to ensure that engine cleaners and traffic film removers work quickly and effectively

2   Electric/air polisher, 1500–2000 rpm, with polythene foam and lamb's-wool polishing heads

3   Fine grade, 100 per cent cotton, polishing cloths

4   Vacuum cleaner and/or shampoo vacuum cleaner with assorted upholstery brushes and crevice tools

5   Hand-pumped, pressurized sprays for dispensing engine cleaners and traffic film remover

6   Trigger spray dispensers for interior cleaning, together with wheel, engine and carpet brushes

7   Good-quality chamois leather, sponges, polishing cloths (100 per cent cotton, knitted stockinette type), steel wool, spatula, glass scraper, buckets, hot air gun (useful for removing PVC stickers, self-adhesive design trims).

There are many manufacturers and suppliers of valeting and cleaning chemicals. The wide range of valeting and chemical cleaners available is broadly divided up as follows:

Exterior detergents and solvents
Interior detergents and solvents
Interior/exterior rubber, PVC, plastic cleaners and conditioners
Glass cleaners, paints, lacquer and protectorants
Paintwork polishes and conditioners.

Cleaning and valeting process

Prior to commencing work, check that the vehicle is ready. All body repairs, paintwork and mechanical work should be completed. Be aware of areas which have been newly painted as they may be chemically or water sensitive. Individual chemical products carry their own instructions, and health and safety procedures must be adhered to at all times.

Autoglym recommend the following step-by-step procedure for cleaning and valeting.

Cleaning: wet bay

1   Remove spare wheel, rubber mats and wheel trims if fitted.

2   Protect engine air intake and sensitive electrical equipment (distributor cap, fuse box) with plastic sheet.

3   Apply degreaser to engine and compartment. Brush heavy soiling. Alternatively use hot pressure washer with appropriate detergent (Figure 14.127).

Figure 14.127   Degreasing engine (*Autoglym*)

4   Apply degreaser to door apertures and edges. Brush heavy soiling. Alternatively use hot pressure washer with appropriate detergent.

5   Clean wheels, trims, spare wheel with alloy cleaner. Brush brake dust deposits. Treat bright metal and motifs. Rinse all items well (Figure 14.128).

6   Pressure wash engine compartment to remove degreaser. Commence with lower areas. Work methodically upwards (Figure 14.129).

7   Pressure wash door apertures and edges to remove degreaser. Carefully angle water jet away from the vehicle interior.

Figure 14.128 Cleaning wheels (*Autoglym*)

Figure 14.130 Cleaning bodywork (*Autoglym*)

Figure 14.129 Pressure washing engine compartment (*Autoglym*)

12 Apply traffic film remover to rubber mats. Clean with high-pressure washer.
13 Use air line to dry engine. Remove plastic sheeting. Check engine starting. Use water dispersant if necessary. Run engine to aid drying (Figure 14.131).

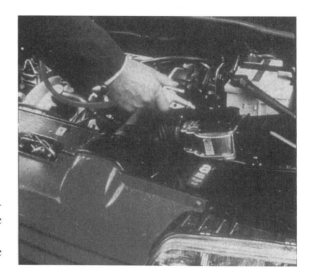

Figure 14.131 Drying engine (*Autoglym*)

8 Apply traffic film remover to engine compartment. If necessary, sponge or brush to remove grime.
9 Pressure wash engine compartment to rinse away traffic film remover.
10 Pressure wash wheel arches to remove mud and debris. Plain water is normally adequate.
11 Pressure wash and rinse bodywork, grilles, tyres, mudflaps. Pay particular attention to difficult to polish areas behind bumpers (Figure 14.130).

14 Use air line and chamois leather to remove excess water from bodywork or trim strips which may trap water (Figure 14.132).

Figure 14.132 Removing water (*Autoglym*)

Valeting: dry bay

1 Finish engine compartment. Clean hoses, wiring, plastic and paintwork with appropriate dressings.

2 Repaint deteriorated black areas. Use fine jet matt black aerosol or small spray gun.

3 Clear lacquer engine if required to enhance and preserve appearance. Close bonnet.

4 Before interior cleaning, remove all loose carpets, tools, ashtrays and personal items to the bench.

5 Use glass scraper to remove all labels from windows. Residual adhesive can be removed with adhesive remover or water.

6 Remove all plastic labels from bodywork using hot air gun. Residual adhesive can be removed with adhesive remover.

7 Vacuum clean all interior surfaces. Slide front seats forward. Use brushes with vacuum nozzle to clean crevices and air vents (Figure 14.133).

8 Clean luggage compartment first. Use interior cleaner by spraying and wiping clean with cloth rinsed frequently. Check body channels and rubbers.

9 Wash interior. Start with headlining and use interior cleaner. Heavily soiled carpets or seats may require shampoo vacuum treatment.

10 Removed carpets should be thoroughly brushed, vacuumed and washed on the bench. Use shampoo vacuum if necessary. Allow to dry (Figure 14.134).

11 Tools and jack should be cleaned, and repainted if necessary. Wash out ashtrays.

Figure 14.133 Brushing and vacuuming (*Autoglym*)

Figure 14.134 Cleaning carpets (*Autoglym*)

12 Plastic coated fibreboard may be painted or stained to cover damage or scrape marks. Check body sides for tar, and top surfaces for industrial fallout. Use appropriate cleaner before polishing.

13 Restore paintwork. Start with roof panel. Use paint renovator with polisher, or appropriate cleaner/polish. Hand polish small areas, corners, edges (Figure 14.135).

14 Apply protective wax coating by hand, ensuring total coverage of all panels. Leave polish applied at this stage.

Figure 14.135 Polishing (*Autoglym*)

15 Clean all external body rubber and plastic mouldings with appropriate dressings.
16 Check all door apertures and rubber seals. Polish door aperture paintwork with clean cloth. Treat rubber seals with appropriate dressings.
17 Check all wheels. If required, clean with steel wool and thinners. Protect tyres with dressing or mask. Respray wheels and clean tyres.
18 Glass cleaning. Lower side windows. Clean top edges completely. Close windows. Polish outside first, then inside. Clean surrounding seals and mirrors.
19 Interior plastic can be dressed to enhance and protect appearance.
20 Replace all carpets, mats, spare wheel, ashtrays, tools, wheel trims.
21 Check interior for remaining imperfections. Check under all seats, glove box, door pockets. Finally vacuum clean. Place protective paper on floor mats.
22 Remove polish. Methodically check all edges, valances, glasses, lights, grilles. Crevices and motifs may be lightly brushed to remove polish.

Checklist

1 Check exterior mirrors, all glass and surrounding seals. Check reverse of interior mirror.
2 Check polish smears, wheel arch edges, front and rear lower pillars, side sills, door handles, grilles. Touch in stone chips.
3 Check windscreen wiper arms and blades, air grilles.
4 Check all light lenses, motifs, number plates and spot lights.
5 Check door edges and apertures, engine and luggage compartment, body channels, and seals.
6 Check control pedals and foot wells, especially under front seats.
7 Check instrument glasses, switches, control levers, interior air vents.
8 Check all ashtrays, glove box, door and seat pockets.

14.5 Estimating and costing

14.5.1 Repair or replacement

The repair of all collision damage can be divided into three categories:

1 Repairing by realigning and reshaping the damaged section which can be economically repaired
2 Repairing by removing sections beyond economical repair and replacing them with new sections
3 Repairing by using a combination of realigning and reshaping together with the replacement of appropriate sections.

Before deciding whether to repair or replace, consideration must be given to the extent and type of damage, whether it is possible to repair a particular panel or not (this fact is influenced by the equipment which is available), and also whether it will be cheaper to repair or replace any one section. A good body repair worker would probably be able to repair almost any part of a car damaged in a collision; however, the labour and time cost involved in attempting to repair extensively damaged panels would be an uneconomical proposition. Only experience in body repair work will give the ability to estimate the time required to do a particular job. Therefore the final decision depends on what is possible to repair and what is economical to repair.

As the construction of the modern vehicle advances, body panel assemblies are becoming very complicated and therefore their cost is constantly increasing. The cost of replacement parts differs between one manufacturer and another, and even between models. In most cases the following parts are usually replaced rather than repaired:

1 All chrome exterior trim. Owing to the intricate design and nature of the materials from which they are made, trims do not lend themselves

to restraightening, and the cost of replating is greater than the cost of a new part.

2 Interior trim which is difficult to clean and repair, especially tears and gashes.

3 Certain mechanical components which it would be ill advised to repair and reuse owing to their critical alignment and strength factors.

On mono constructed vehicles the correct positioning of new panels, which may incorporate reinforcing members, becomes increasingly important and calls for expensive jigging to ensure the accurate alignment of panels which will have to be fitted with major mechanical components and suspension units. The highest proportion of any repair bill is made up of the cost of labour involved in stripping out to gain access to the damaged area, and rebuilding after replacement is effected. In certain cases, therefore, it can be more economical to repair a damaged panel than to replace it because of the time and labour cost involved for the amount of work over and above that required to repair the part. Consequently in any body repair workshop the dividing line between repairing and replacing is determined not only by what it is possible to repair and what it is economical to repair, but also by the availability of the most up-to-date equipment to deal with all types of major collision damage.

14.5.2 Costing

The determining factor when deciding whether to repair or replace lies in the comparison between the cost of repairing the damaged part as against the cost of a new replacement part. When the owner of a car is personally paying the cost of repairs the repairer will need to come to an agreement with him concerning an acceptable standard of repair for the amount of money the owner is able to pay. When it is an older car the extra cost involved in striving for perfection in the repair is not always justified, since these costs could be more than the current value of the vehicle under repair. In cases of this nature a compromise must be arrived at between the repairer and owner as to the quality of repair and cost of repair.

Where an insurance company pays for repairs to the damage of a new or nearly new car, the quality of the work must be such that when the repair is finished there is no indication of the damage having occurred. When an insurance company is paying for repairs to an older car, and the condition of the car (in many cases due to corrosion) affects the quality of the repair, the possibility of achieving perfection of appearance is limited and a mutual agreement must be reached between the owner and the insurance company on a satisfactory quality of repair. Time is always an important factor in the repair of collision damage; however, the majority of car owners are not willing to sacrifice quality of repair for speed of repair.

14.5.3 Estimating

The ability to estimate can only be gained through practical experience in the field of body repair work. An estimate must be competitive to be acceptable to both the private car owner and the insurance company, for the majority of collision jobs are paid for by an insurance company. In estimating, the financial gain for the body shop will depend on the estimator's skill in assessing the damage, his knowledge and experience in repair techniques, and the capabilities of the equipment available.

Small body repair establishments have no separate estimator and therefore it is usual for the man who prepares the estimate to be responsible for carrying out the repair work as well. The larger establishments employ one member of staff who is responsible for all estimating, while the actual repair work is carried out by the tradesmen on the shop floor. Consequently there must be no lack of coordination and understanding between these two and the jobs they perform. The decisions of the estimator are all important as they instigate the organization of work on the shop floor. Moreover, no matter how skilled the tradesmen are, with bad estimating the financial profits can be drastically affected.

In collision work the estimate is considered as a firm commitment to do the work involved for the amount of money shown, and should be detailed so that insurance companies or private owners can determine from the estimate exactly what it is proposed to do to the damaged vehicle. The estimate must include cost of parts and labour costs at the recognized retail rate of the body shop establishment (Figures 14.136 and 14.137).

In the preparation of an estimate, the crash damage should be itemized into the number of damaged panel sections and assemblies, and therefore it is important to have a knowledge of vehicle body

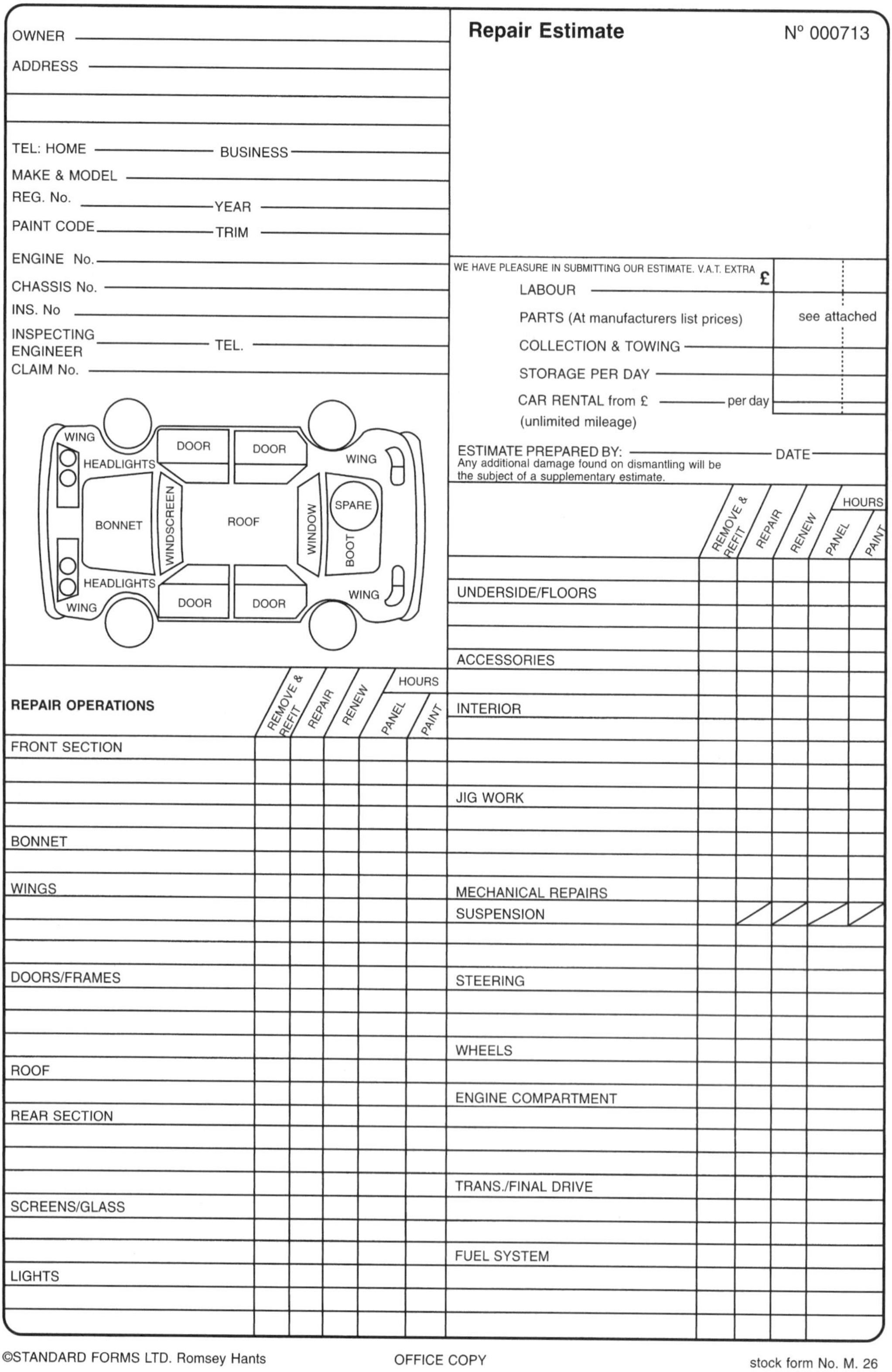

OWNER
ADDRESS

TEL: HOME BUSINESS
MAKE & MODEL
REG. No. YEAR
PAINT CODE TRIM
ENGINE No.
CHASSIS No.
INS. No
INSPECTING ENGINEER TEL.
CLAIM No.

Repair Estimate N° 000713

WE HAVE PLEASURE IN SUBMITTING OUR ESTIMATE. V.A.T. EXTRA £

	£
LABOUR	
PARTS (At manufacturers list prices)	see attached
COLLECTION & TOWING	
STORAGE PER DAY	
CAR RENTAL from £ per day (unlimited mileage)	

ESTIMATE PREPARED BY: DATE
Any additional damage found on dismantling will be the subject of a supplementary estimate.

REPAIR OPERATIONS	REMOVE & REFIT	REPAIR	RENEW	HOURS PANEL	HOURS PAINT
FRONT SECTION					
BONNET					
WINGS					
DOORS/FRAMES					
ROOF					
REAR SECTION					
SCREENS/GLASS					
LIGHTS					

	REMOVE & REFIT	REPAIR	RENEW	HOURS PANEL	HOURS PAINT
UNDERSIDE/FLOORS					
ACCESSORIES					
INTERIOR					
JIG WORK					
MECHANICAL REPAIRS					
SUSPENSION					
STEERING					
WHEELS					
ENGINE COMPARTMENT					
TRANS./FINAL DRIVE					
FUEL SYSTEM					

©STANDARD FORMS LTD. Romsey Hants OFFICE COPY stock form No. M. 26

Figure 14.136 Repair estimate

OWNER

ADDRESS

Cash ☐
Account ☐
Retail ☐
Int. ☐
Wty. ☐

Bodywork Invoice B 01144

TEL. HOME BUSINESS

Is Owner VAT Registered? Yes/No

PART NO.	DESCRIPTION	Qty.	UNIT PRICE	TOTAL

DATE IN	DATE OUT	INVOICE DATE
MAKE MODEL		
SPEEDO READING	FUEL E $^1/_4$ $^1/_2$ $^3/_4$ F	
REG. No.	YEAR	
PAINT CODE	TRIM	
ENGINE NUMBER		
CHASSIS NUMBER		
INSURANCE COMPANY		
INVOICE ADDRESS		
INSPECTING ENGINEER		
TEL.		
ESTIMATE No.	CLAIM No.	

I/We have read and accept your terms of business. I/We agree to pay all sums due on completion of work prior to collection of vehicle. Where applicable I/We agree to pay any Owner's Contribution or Excess Charge specified by the insurance company.

Owner's Signature ———— **Date** ————

LABOUR AS PER ESTIMATE NO.	
TOTAL LABOUR CHARGE	

	INSURANCE COMPANY TO PAY	OWNER TO PAY
LABOUR		
PARTS		
SPECIALIST MATERIALS		
TOTAL EXCL. VAT		
LESS PARTS DISCOUNT %		
SUB-TOTAL		
VAT at %		
SUB-TOTAL		
EXCESS £		
OWNERS CONTRIBUTION £		
DUE FROM INS. Co. →		
DUE FROM OWNER →		
TOTAL INVOICE VALUE £		

COLLECTION NOTE
I Certify that repairs to the above vehicle have been completed and that the vehicle has been made available to me for collection.

Owner's Signature .... Date ....

STANDARD FORMS LTD. Romsey, Hants

Stock Form No. M.16

Figure 14.137 Bodywork invoice

construction. The outer panels which make up the body shell should be referred to by the manufacturer's recognized names, such as:

Near-side front door (NSF door)
Near-side rear door (NSR door)
Off-side front door (OSF door)
Off-side rear door (OSR door)
Roof panel
Boot lid
Bonnet
Near-side centre pillar (NS centre pillar)
Radiator grille
Near-side front wing (NSF wing)
Off-side front wing (OSF wing)
Near-side sill panel (NS sill panel)
Off-side sill panel (OS sill panel)
Front bumper bar assembly with valance
Rear bumper bar assembly with valance.

In addition this can be further broken down to all internal panel structures according to the position of the damage.

It is essential to determine the exact amount of stripping necessary for either the repair or the removal and replacement of a damaged section. To calculate the actual cost of a job it is necessary to establish a set rate per hour for all repairs done in the repair shop, and therefore an accurate total repair time can be calculated on the time taken to repair each section of the body. This then provides the actual cost estimate for repairing each section or part of the job. The retail rate or set rate per hour is the amount of money that the repairer charges the customer for labour, and is made up of; wages paid to the tradesmen; the cost of overheads, which should include such items as supervision, depreciation of equipment, rent, heat, light, electrical power, advertising, telephone accounts, cleaning, office staff, stationery and postage, and workshop materials; and also a reasonable amount of profit.

14.5.4 Computer estimating

Computers are rapidly losing their mystique, and are becoming as indispensable to the bodyshop as the spray booth and chassis jig. Offering benefits in all aspects of bodyshop operation, there are computers and software packages dedicated to quick and easy repair estimates, stock control, invoicing and management accounts, paint and materials, parts list, mixing and colour matching, job cards, work-shop loading, booking in, job efficiency reports, invoice production, reports and letters (Figures 14.138–14.141). Costs have tumbled and the benefits can be reaped by all bodyshops, regardless of size.

The key to success in any bodyshop is the production of an accurate, well presented estimate. An estimate has to be a clear, concise statement of what you consider to be a fair return for the care and work you carry out on a crash damaged vehicle. By utilizing the memory capacity of the computer you will have readily available all necessary details to help you draw up an estimate quickly and accurately.

Your computer system will have the facility to prepare not only a detailed document for the customer, but also a comprehensive breakdown for the insurance company. The documents produced will be tailor made to your own letterhead stationery. The insurance estimate can often be sent direct by fax to the company involved, thereby further reducing the time gap between estimating and authorization by that company. Apart from the major bonus of speed, a computerized estimating system will greatly enhance your company's image as a professional organization. You will have instant access to the progress of any job: most systems include this as standard. The more time and money you spend at the outset, the more control you will have over staff movement and job completion.

14.5.5 Insurance procedure

Under the Road Traffic Act the car owner is obliged to insure his vehicle either under third party insurance cover or fully comprehensive insurance cover. As a direct result of this, most accident damage is covered by insurance and is therefore repaired in body work establishments.

The procedure for dealing with repairs carried out under an insurance claim is in four stages, as described in the following sections.

Claim form
When involved in an accident which has resulted in vehicle damage, the owner should obtain and complete an insurance claim form and immediately return it to his insurance company, or make a written report on the accident and damage received.

BodyMaster House
Wych Cross
Forest Row
East Sussex
RH18 5LJ
Tel: (0342) 826001
Fax: (0342) 826002

ACCIDENT REPAIR CENTRE
FULL JIG FACILITIES
SPRAY BOOTH
ALL INSURANCE WORK

Mrs. M. Jones,
42, Kings Road,
Sharpthorne,
Nr. East Grinstead,
East Sussex,
RH23 4LB.

Estimate No :   10010

Dear Madam,

Re; Ford Escort 1.6L, Registration Number: F376 AFG.

We have pleasure in submitting our estimate for repairs to the above vehicle.

Repairs:
Remove and refit:
Front Bumper, O/S/F Headlamp, O/S/F Indicator, Front Grille, Fog Lamps,
Mouldings, O/S Mirror, Repeater Lamp, "Ford" Decal, O/S/F Door and Trims, Radiator and
O/S/F Tyre.

Load vehicle onto jig.

Repair and reshape:
Front Panel, O/S/F Door and O/S Front chassis Leg end.

Cut out and weld in:
New O/S/F Wing.

Hang and align:
O/S/F Door and New Bonnet.

Check and set wheel alignment.

Paintwork:
Prepare, prime and topcoat:
O/S/F Door, New Bonnet, Front Panel, New O/S/F Wing and
O/S Front chassis leg end.

Labour;   £   571.30

Plus Parts :- (To be supplied at MRP)
New Bonnet   Front Bumper
O/S/F Headlamp   O/S/F Indicator
O/S Mirror   Repeater Lamp
New O/S/F Wing   Clips etc
Radiator (Not discountable)   O/S/F Tyre

Paint, materials and sundry items.

Yours faithfully,

Bodyshop Manager.
Please Note: All Prices Subject To 17.5% VAT.
            Please ensure your radio code is recorded before repair.

Figure 14.138 Customer's estimate (*Bodymaster UK*)

BodyMaster House
Wych Cross
Forest Row
East Sussex
RH18 5LJ

Tel: (0342) 826001
Fax: (0342) 826002

ACCIDENT REPAIR CENTRE
FULL JIG FACILITIES
SPRAY BOOTH
ALL INSURANCE WORK

Engineers Estimate

Mrs. M. Jones,
42, Kings Road,
Sharpthorne,
Nr. East Grinstead,
East Sussex,
RH23 4LB

31st March 1992

Estimate No	: 10010
Vehicle	: Ford Escort 1.6L,
Reg No.	: F376 AFG
Chassis No	: 11-H7 SLBN43/87
Paint Type	: Solid colour

Estimate Details	Hours
Remove and refit:	
Front Bumper	1.00
O/S/F Headlamp	1.00
O/S/F Indicator	0.50
Front Grille	0.20
Fog Lamps	1.00
Mouldings	0.50
O/S Mirror	0.50
Repeater Lamp	0.10
"Ford" Decal	0.10
O/S/F Door and Trims	1.00
Radiator	1.00
O/S/F Tyre	1.00
Repair and make good:	
Front Panel	1.00
O/S/F Door	2.50
O/S Front chassis Leg end	2.00
Cut out and weld in:	
New O/S/F Wing	5.00
Hang and align:	
O/S/F Door	1.00
New Bonnet	1.00
Wheel Alignment	1.50
Load onto Jig	4.00
Prepare and paint:	
O/S/F Door	3.00
New Bonnet	4.00
Front Panel	2.50
New O/S/F Wing	3.00
O/S Front Chassis Leg end	1.00
Total Hours	39.40

Paint & materials   £ 112.61
(Includes dry goods etc.)
Jig bracket hire   £
Recovery charge   £

Comments :
(See separate list for parts)

Total labour cost @ £14.50 per hour £571.30

Figure 14.139 Engineer's estimate (*Bodymaster UK*)

BodyMaster House
Wych Cross
Forest Row
East Sussex
RH18 5LJ

Tel: (0342) 826001
Fax: (0342) 826002

ACCIDENT REPAIR CENTRE
FULL JIG FACILITIES
SPRAY BOOTH
ALL INSURANCE WORK

Vehicle Job Card

Customer   : Mrs. M. Jones,   31st March 1992
Vehicle   : Ford Escort 1.6L,   Reg No : F376 AFG
Chassis No : 11-H7-SLBN43/87
Paint Code : B1
Job/Estimate No: 10010

Repair Format	Est Time	Actual Time
Remove : Front Bumper, O/S/F Headlamp, O/S/F Indicator, Front Grille, Fog Lamps, Mouldings, O/S Mirror, Repeater Lamp, "Ford" Decal, O/S/F Door and Trims, Radiator, O/S/F Tyre.	7.90	
Repair : Front Panel, O/S/F Door, O/S Front Chassis Leg end.	5.50	
Weld in : New O/S/F Wing.	5.00	
Hang and align : O/S/F Door, New Bonnet.	2.00	
Check and set wheel alignment.	1.50	
Load vehicle onto jig.	4.00	
Paint : O/S/F Door, New Bonnet, Front Panel, New O/S/F Wing, O/S Front Chassis Leg end.	13.50	
TOTAL	39.40	

Date	Operation	Initial	On	Off	Total
				TOTAL	

Signed ....

Figure 14.140   Vehicle job card (*Bodymaster UK*)

BodyMaster House
Wych Cross
Forest Row
East Sussex
RH18 5LJ

ACCIDENT REPAIR CENTRE
FULL JIG FACILITIES
SPRAY BOOTH
ALL INSURANCE WORK

Tel: (0342) 826001
Fax: (0342) 826002

Vehicle Checklist

Customer : Mrs. M. Jones,   31st March 1992
Vehicle   : Ford Escort 1.6L,   Reg No: F376 AFG
Chassis No : 11-H7-SLBN43/87
Paint Code : B1
Job/Estimate No : 10010

Workshop :

CHECKLIST	TICK
Lights	
Headlamp beam	–
Door workings	–
Trims and fixings	–
Security of parts	–
Brake fluid level	
Water level/antifreeze	–
Aperture gaps	–
Wax injection	–
Underseal	–
Coachlines	–
Masking tape	–
Blow-ins	–
Overspray	–
Paint finish	–
Cleanliness interior	–
Cleanliness exterior	–
Wheel nuts	–
Roadtest	–

Signed
....

Comments on general condition

Office Use :

Satisfaction Notice Excess Invoice	

Signed
....

Figure 14.141   Vehicle repair checklist (*Bodymaster UK*)

Itemized estimate

According to the extent of the damage, and if the vehicle is still roadworthy, the owner takes the vehicle for the inspection of the damage. The repairer will make a visual inspection of the assessed damage, and from the knowledge gained complete a written itemized estimate which he will submit as a tender to the insurance company. This estimate will show the total cost of repairs, and where the estimated amount is under a certain figure set by the insurance company, the owner has the right to authorize the repairer to do the work. However, in most repair cases this figure is exceeded, and the insurance company's assessor is the only person authorized to allow the repairs to proceed.

One of the important factors when estimating for insurance claim damage is to examine carefully every section of the vehicle, especially those parts which are a known weakness in the construction and therefore liable to be affected directly or indirectly by a collision. A methodical system of estimating is essential to avoid missing any damage, and is usually carried out by noting in order all removal and replacement items, all repair items, all respray items and all items to be supplied new at cost, including mechanical parts and any trim. Supply items are usually difficult to price because of the makers' fluctuating prices, and therefore they should be listed 'at cost'. Spray painting can be quoted either by itemizing each part separately or by a complete price for the total spray operation. While in the owner's presence, the repairer should point out any rusted sections which may affect the repair work, or any previous unrepaired damage which is not covered under this insurance claim.

Authority to repair

It is essential to obtain the authority to repair from the insurance company involved before any work is started on the damaged vehicle. On receiving the repairer's estimate, the insurance company will instruct their own engineer assessor or an independent assessor to examine the vehicle and satisfy himself that the claim is in order and that the estimate submitted by the repairer does not include any labour or material necessary as a result of any other than the accident report. The engineer assessor then agrees the cost of repairs, particularly the labour charges, with the repairer

while inspecting the vehicle, and decides whether or not damaged parts shall be repaired or replaced.

After this inspection the assessor will send written instructions on behalf of the insurance company; these constitute the authority to repair. A condition of these instructions is that the repairer is restricted to repairing the damage as seen and estimated for. Any additional damage disclosed when dismantling the vehicle must have a separate estimate submitted, and work cannot be carried out on this damage until the extra estimate has been agreed and the necessary work authorized by the insurance company.

Clearance certificate

This certificate is provided by the insurance company for the vehicle owner to sign when he has seen that all the agreed repairs have been completed satisfactorily, and the damaged parts reinstated to their original condition. When this certificate has been signed by the vehicle owner, it releases the insurance company from any liability in respect of the claim for damage caused by this particular accident. It is to the repairer's advantage for him to ascertain that the clearance certificate has been duly signed before the vehicle is returned to its owner.

The duty of the repairer is to see that the damaged car is reinstated to its original condition. He is expected to make full use of his skill and knowledge to effect the best possible repair, as the owner relies on his reputed skill.

14.5.6 Motor Insurance Repair Research Centre (Thatcham)

In the 1960s the motor insurance industry became concerned about the escalating cost of vehicle accident damage repair work and its effect on motor insurance premiums. To have some direct influence on these costs it was recognized that motor insurers required a means of researching the cost of repairing accident damaged cars and light commercial vehicles.

In 1969, motor insurance company members of the British Insurance Association and motor syndicates at Lloyd's joined in a scheme to create a research centre. The result was Thatcham, named after the small Berkshire town where it is situated, and the only centre of its type in the UK. Thatcham is unique in that it represents an entire motor insurance market. It has been granted research status by the government, and is funded by levies on the members of the Association of British Insurers and Lloyd's Motor Underwriters Association.

The aims of the centre are as follows:

1 To advise on repair methods and prepare accurate job times for panel replacement
2 To minimize the effects of road accident damage and liaison with manufacturers in the interests of better vehicle design
3 To pioneer quicker, more cost-effective methods of damage repair.

Thatcham exists to promote cost-effective methods for motor vehicle repair. This includes demonstrating the types of equipment and techniques which are available to the repair trade. This effort is backed by a professional approach using skilled workshop personnel and engineers recruited from the repair trade and manufacturers. They are able to provide insurance engineers and repairers with essential repair information, often as soon as new vehicle models become available at the dealers.

Research at Thatcham is divided into two stages. Stage 1 research determines the best method of removing and replacing the outer cosmetic panels of undamaged vehicles, and the time taken to do this. Stage 2 research is more advanced; it relates to vehicles that have been subjected to controlled structural damage, and is concerned with the repair methods and times associated with such damage. Because most research concentrates on prelaunch or recently launched vehicles, accident damage is simulated on the crash test facility. This provides uniformity of impact, irrespective of the model, and provides data based on constant factors, thus making model damage comparison more meaningful. Vehicles can be propelled into a static crash barrier or impacted by a 1000 kilogram mobile crash barrier.

Both stage 1 and stage 2 research is observed by a work study engineer, who determines the most effective method of repair. This is achieved through consultation with the vehicle manufacturer, other specialist engineers and very often the experienced skilled tradesman carrying out the work. Subsequently, repair methods are established before publication. The evolution of manufacturing and repair techniques inevitably means the introduction of new materials, tools and techniques which could have a significant impact on repair methods. Specialized research projects are set up to establish the effect of such materials, tools and equipment on repair methods, times and costs. From time to time, research into specific aspects of vehicle engineering is undertaken for manufacturers.

The results of the Centre's research are published in methods manuals, special reports, newsletters and in the form of parts pricing information. Once the work study engineer has evaluated the information obtained through observation and consultation, the data is published in a methods manual. These contain comprehensive data on repair methods and times, welding tables and diagrams, and technical information supplied by the vehicle manufacturer. The results of the specialized research are published through special reports. Newsletters supplement published data, providing information for immediate action within the vehicle repair industry.

Training is seen as an essential and increasingly important part of the work done. Thatcham provides courses for insurance company staff engineers and independent consultant assessors, concentrating on the latest developments in jig technology, welding and refinishing. Help is also offered to develop potential repair talent by encouraging educational institutes, such as technical training colleges, to visit the Centre.

Overall, the staff are working towards enhanced low-speed impact performances; quicker and more cost-effective repairs; and a better understanding between those who do the repairs and those who pay for them. Thatcham continues to influence and improve repair technology, keeping pace with the ever-increasing sophistication of the modern motor vehicle to the benefit of the motoring public.

14.5.7 Further estimating

The Motor Insurance Repair Research Centre at Thatcham, Berkshire, usually simply referred to as 'Thatcham', have produced repair time schedules for most repairs to most cars. The schedules are divided into replacement time schedules for body panels and methodologies for paint refinishing.

Replacement time schedules

A number of tables are available for each vehicle, these give single panel times and combination panel times. See Figure 14.142. The single panel time is

REPLACEMENT TIME SCHEDULE - NON-METALLIC   **FORD ESCORT 1991 5-DOOR**

FRONT END REPAIRS

SINGLE PANEL TIMES

Fig. Ref.	Panel Description	M.E.T.	Panel	Paint	Comb.	Total Time (inc. Job allowance)
	Front Grille Panel	0.3	M.E.T. Item 2.5	2.5	0.9	3.3
B10:10	Front Wing	1.2RH / 1.4LH	1.0	4.4	1.7	7.1RH / 7.3LH
B10:4	Bonnet	0.5	0.3	3.4	1.8	4.7
B10:3	Bonnet Lock/Headlamp Panel Assembly	3.2	1.8	2.8		8.3
B10:2	Front Crossmember	1.7	1.1	2.0	0.6	5.3
B10:1	Front Panel Assembly	3.3	3.0	3.2	1.7	10.0
B10:5	Front Inner Wing/Chassis Leg Assembly Section (B)	Not applicable				
B10:6	Upper Inner Wing Section (A)	Not applicable				
B10:8	Part Front Chassis Leg	Not applicable				
B10:9	Part Chassis Leg Closing Panel	Not applicable				
B10:7	Inner Wing Reinforcement Section (J)	Not applicable				
	Dashboard Assembly	2.3				2.8
	Front Screen*	2.6				3.1

*See note page B11.

COMBINATION PANEL TIMES

Repair Combinations (each dot denotes one panel)

Panel Description	F1	F2	F3	F4	F5	F6
Front Grille Panel		•	•	•	•	•
Front Wing	•	•	•	•	•	•
Bonnet			•	•	•	•
Bonnet Lock/Headlamp Panel Assembly						
Front Crossmember	•	•	•	•	•	
Front Panel Assembly			•	•	•	•
Front Inner Wing/Chassis Leg Assembly Section (B)			•	•	•	•
Upper Inner Wing Section (A)			•	•	•	•
Part Front Chassis Leg				•	•	•
Part Chassis Leg Closing Panel				•	•	

	F1	F2	F3	F4	F5	F6
M.E.T	4.2RH 4.3LH	4.6	14.4RH 13.8LH	15.9	5.7RH 5.5LH	6.6
Panel	4.0	4.7	8.6	14.0	7.2	11.3
Paint	7.9	9.6	8.9	11.6	8.9	11.6
Job Allowance	1.0	1.3	1.8	2.8	1.8	2.8
Total	17.1RH 17.2LH	20.2	33.7RH 33.1LH	44.3	23.6RH 23.4LH	32.3

A   B   C

Figure 14.142   Replacement time schedules (*Motor Insurance Repair Research Centre*)

given in Section A, the replacement time schedules can be read off horizontally. It should be noted that the total time for each complete replacement includes a time allowance of 0.5 hours for the body repairer to move the vehicle as needed and obtain special tools and parts from the stores, this is called job allowance. The table includes a section for paint refinishing.

If more than one panel is being replaced, the combination panel time is used. For instance, it will take less time to replace both a front wing and a front panel than a separate wing and a front panel. The procedure is to find the nearest combination in Section B, each dot in the column denotes a panel. The time for the combination of panels is read off vertically in Section C. The job allowance is calculated by allowing 0.5 hours for the first panel and 0.25 hours for each subsequent panel or bracketing system. For ease of calculation the allowance is then rounded up to the nearest 0.1 hour.

If the job is a large one which combines several panels and cannot be readily calculated, a paint combination time (paint comb.) is listed in the single panel part of the matrix. The paint combination time is the productive part of actual painting, that is, it does not include paint mixing, test panel spraying, changing overalls or cleaning. So, this time alone can be added to a combination panel time for painting a larger combination of panels.

Paint refinish methods

Thatcham publish refinish guide times for individual vehicles using the common variations in paint systems, these are:

non-metallic
non-metallic, base coat and clear
metallic, base coat and clear
two-coat pearlescent

The times allow for dry-flatting and the use of two-pack materials where appropriate.

Thatcham suggest a set sequence of paint refinishing operations, these are shown in Figure 14.143.

The above is a brief introduction to Thatcham Methods, and it is recommended that the body repairer takes one of the Thatcham short courses.

14.5.8 Glassmatrix II system

Glass's Guide Services, famous for their pocketbook on car prices, offer a computer-based estimating system using Thatcham time schedules.

Glassmatrix II An MS Windows-based, computer-assisted estimating system that is used by insurance companies and vehicle body repairers to produce accurate estimates of repair costs. The estimates are produced using a system of bar coding information from a collision repair estimating guide (CREG). The bar code reader is shown in Figure 14.144. The bar code contains information on the labour time, the cost of replacement parts and the part number for identification. The labour time is mainly supplied by Thatcham, the numbers and the costs of parts is based on manufacturer's recommended retail prices. The Glassmatrix software is supplied on a CD ROM, which is regularly updated – usually once a month. The software is modular, see Figure 14.145.

GlassWord A report and form generator that can be used to design bespoke forms and reports. It works in conjunction with Glassmatrix by importing power fields from an estimate and placing them into a form template. GlassWord can be used to create parts order lists, letters to customers, invoices, and other documents using either plain paper or printed business letterheadings.

GlassImage A system for capturing and storing digital images from video data. A video camera is used to record digital pictures of the damaged vehicle.

Glass's Guide for Windows A PC-based version of the famous valuation booklets; information from this can be imported into the estimate or other forms.

Alternative parts A database of insurance industry approved parts, mainly Veng and Unipart, which may be used instead of the original equipment manufacturer (OEM) parts database.

GlassReport A statistics package for analysing the estimate in terms of common concepts, such as average estimate cost and parts-to-labour ratio.

Glassmatrix communication links Figure 14.146 shows the arrangement of the communication links. It should be noted that e-mail can be sent using either a mobile telephone cell network, or a portable satellite link, both of which are currently more expensive than a land-line link.

Procedure-Sequence of Operations	Method: 1	Method: 2	
Prepare interior and exterior surfaces for primer	▨	▨	P80-180-Scotchbrite
De-oxidene bare metal areas only	▨	▨	
Solvent wash those areas to be masked	▨		
Mask up including fitment of a car cover and dust sheet	▨		
Fit hang-on panel to the spray jig		▨	
Move vehicle/panel into booth. Switch on etc.	▨	▨	
Mix etch primer if necessary	▨	▨	
Mix primer filler/surfacer	▨	▨	
Mix guide coat	▨	▨	
Solvent wash, blow off, tack wipe interior and exterior	▨	▨	
Spray etch primer if necessary to interior and exterior	▨	▨	
Spray primer to interior and exterior	▨	▨	
Spray guide coat to exterior	▨	▨	
Prepare straight colour/straight colour base & clear/metallic base & clear/ready-mix	▨	▨	
Spray a test panel to opacity		▨	
Prepare (minimum) interior surface for straight colour/straight colour base & clear/metallic base & clear	▨		P400-Scotchbrite
Spray straight colour/straight colour base & clear/metallic base & clear to panel interior and test panel to opacity	▨		
Clean spray guns	▨	▨	
Move vehicle/panel and test panel out of booth when dry	▨		
Check test panel for colour match	▨		
Remove hang-on panel from the spray jig		▨	
Remove interior masking only as necessary	▨		
Prepare exterior surface for straight colour/straight colour base & clear/metallic base & clear	▨		P400-Scotchbrite
Prepare interior and exterior surfaces for straight colour/straight colour base & clear/metallic base & clear		▨	P400-Scotchbrite
Seal exterior panel joints as necessary	▨	▨	Polyurethane sealer
Prepare Surfaces for stone chip	▨	▨	Scotchbrite
Mask for stone chip	▨	▨	
Mix stone chip	▨	▨	Solvent or water-based
Spray stone chip	▨	▨	
Clean stone chip spray gun	▨	▨	
Remove masking	▨	▨	
Solvent wash those areas to be masked	▨	▨	
Mask up as necessary	▨	▨	

Figure 14.143   Paint refinishing methods (*Motor Insurance Repair Research Centre*)

PAINT REFINISHING

METHOD DESCRIPTION

Procedure-Sequence of Operations	Method: 1	Method: 2	
Refit hang-on panel to the spray jig		▓	
Move vehicle/panel into booth. Switch on	▓	▓	
Prepare straight colour/straight colour base & clear/metallic base & clear/ready mix	▓	▓	
Solvent wash, blow off, tack wipe exterior	▓		
Solvent wash, blow off, tack wipe interior and exterior		▓	
Spray straight colour/straight colour base & clear/metallic base & clear to panel exterior	▓		
Spray straight colour/straight colour base & clear/metallic base & clear to panel interior and exterior		▓	
Clean spray guns	▓	▓	
Move vehicle/panel out of booth when dry	▓	▓	
Remove masking, car cover and dust sheet	▓		
Prepare for matt black	▓	▓	Scotchbrite
Mask for matt black	▓	▓	
Spray matt black	▓	▓	
Remove masking	▓	▓	
Remove hang-on panel from the spray jig		▓	
ADDITIONAL OPERATIONS FOR STRAIGHT COLOUR BASE & CLEAR			
Prepare lacquer	▓	▓	
Spray lacquer to test panel	▓	▓	
Spray lacquer to panel interior and test panel	▓		
Spray lacquer to panel exterior	▓		
Spray lacquer to panel interior and exterior		▓	
Clean spray guns	▓	▓	
ADDITIONAL OPERATIONS FOR METALLIC BASE & CLEAR/ TWO-COAT PEARLESCENT			
Prepare adjacent panels	▓	▓	Abrasive paste and Scotchbrite
Mask adjacent panels	▓	▓	
Spray metallic base & clear to adjacent panels	▓	▓	
Spray lacquer to adjacent panels	▓	▓	
Remove masking	▓	▓	

The methods shown in the table above and on the previous page are applicable to:

        Method 1–*Welded panels*
        Method 2–*Hang-on panels including doorskins, tailgate skins, etc. where applicable*

Figure 14.143 (*continued*)

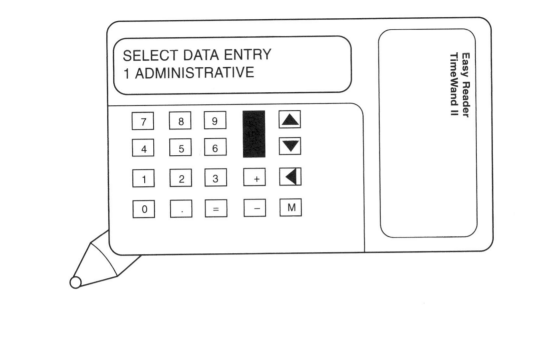

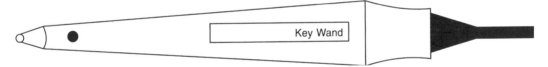

Figure 14.144 Glassmatrix II bar code reader (*Glass's Guide Services*)

MODULAR SOFTWARE

Report/Form
Generator

Veng/Unipart
Parts

Glass
Word

Alternative
Parts

Software
Links Package

Export
Translator

GMX
II

Glass's
Guide for
Windows

Valuation
Data

Digital
Imaging

Glass
Image

Glass
Report

Statistics
Package

Figure 14.145 Glassmatrix II modular software (*Glass's Guide Services*)

Figure 14.135 Polishing (*Autoglym*)

15 Clean all external body rubber and plastic mouldings with appropriate dressings.
16 Check all door apertures and rubber seals. Polish door aperture paintwork with clean cloth. Treat rubber seals with appropriate dressings.
17 Check all wheels. If required, clean with steel wool and thinners. Protect tyres with dressing or mask. Respray wheels and clean tyres.
18 Glass cleaning. Lower side windows. Clean top edges completely. Close windows. Polish outside first, then inside. Clean surrounding seals and mirrors.
19 Interior plastic can be dressed to enhance and protect appearance.
20 Replace all carpets, mats, spare wheel, ashtrays, tools, wheel trims.
21 Check interior for remaining imperfections. Check under all seats, glove box, door pockets. Finally vacuum clean. Place protective paper on floor mats.
22 Remove polish. Methodically check all edges, valances, glasses, lights, grilles. Crevices and motifs may be lightly brushed to remove polish.

Checklist
1 Check exterior mirrors, all glass and surrounding seals. Check reverse of interior mirror.
2 Check polish smears, wheel arch edges, front and rear lower pillars, side sills, door handles, grilles. Touch in stone chips.
3 Check windscreen wiper arms and blades, air grilles.

4 Check all light lenses, motifs, number plates and spot lights.
5 Check door edges and apertures, engine and luggage compartment, body channels, and seals.
6 Check control pedals and foot wells, especially under front seats.
7 Check instrument glasses, switches, control levers, interior air vents.
8 Check all ashtrays, glove box, door and seat pockets.

14.5 Estimating and costing

14.5.1 Repair or replacement

The repair of all collision damage can be divided into three categories:

1 Repairing by realigning and reshaping the damaged section which can be economically repaired
2 Repairing by removing sections beyond economical repair and replacing them with new sections
3 Repairing by using a combination of realigning and reshaping together with the replacement of appropriate sections.

Before deciding whether to repair or replace, consideration must be given to the extent and type of damage, whether it is possible to repair a particular panel or not (this fact is influenced by the equipment which is available), and also whether it will be cheaper to repair or replace any one section. A good body repair worker would probably be able to repair almost any part of a car damaged in a collision; however, the labour and time cost involved in attempting to repair extensively damaged panels would be an uneconomical proposition. Only experience in body repair work will give the ability to estimate the time required to do a particular job. Therefore the final decision depends on what is possible to repair and what is economical to repair.

As the construction of the modern vehicle advances, body panel assemblies are becoming very complicated and therefore their cost is constantly increasing. The cost of replacement parts differs between one manufacturer and another, and even between models. In most cases the following parts are usually replaced rather than repaired:

1 All chrome exterior trim. Owing to the intricate design and nature of the materials from which they are made, trims do not lend themselves

to restraightening, and the cost of replating is greater than the cost of a new part.

2 Interior trim which is difficult to clean and repair, especially tears and gashes.

3 Certain mechanical components which it would be ill advised to repair and reuse owing to their critical alignment and strength factors.

On mono constructed vehicles the correct positioning of new panels, which may incorporate reinforcing members, becomes increasingly important and calls for expensive jigging to ensure the accurate alignment of panels which will have to be fitted with major mechanical components and suspension units. The highest proportion of any repair bill is made up of the cost of labour involved in stripping out to gain access to the damaged area, and rebuilding after replacement is effected. In certain cases, therefore, it can be more economical to repair a damaged panel than to replace it because of the time and labour cost involved for the amount of work over and above that required to repair the part. Consequently in any body repair workshop the dividing line between repairing and replacing is determined not only by what it is possible to repair and what it is economical to repair, but also by the availability of the most up-to-date equipment to deal with all types of major collision damage.

14.5.2 Costing

The determining factor when deciding whether to repair or replace lies in the comparison between the cost of repairing the damaged part as against the cost of a new replacement part. When the owner of a car is personally paying the cost of repairs the repairer will need to come to an agreement with him concerning an acceptable standard of repair for the amount of money the owner is able to pay. When it is an older car the extra cost involved in striving for perfection in the repair is not always justified, since these costs could be more than the current value of the vehicle under repair. In cases of this nature a compromise must be arrived at between the repairer and owner as to the quality of repair and cost of repair.

Where an insurance company pays for repairs to the damage of a new or nearly new car, the quality of the work must be such that when the repair is finished there is no indication of the damage having occurred. When an insurance company is paying for repairs to an older car, and the condition of the car (in many cases due to corrosion) affects the quality of the repair, the possibility of achieving perfection of appearance is limited and a mutual agreement must be reached between the owner and the insurance company on a satisfactory quality of repair. Time is always an important factor in the repair of collision damage; however, the majority of car owners are not willing to sacrifice quality of repair for speed of repair.

14.5.3 Estimating

The ability to estimate can only be gained through practical experience in the field of body repair work. An estimate must be competitive to be acceptable to both the private car owner and the insurance company, for the majority of collision jobs are paid for by an insurance company. In estimating, the financial gain for the body shop will depend on the estimator's skill in assessing the damage, his knowledge and experience in repair techniques, and the capabilities of the equipment available.

Small body repair establishments have no separate estimator and therefore it is usual for the man who prepares the estimate to be responsible for carrying out the repair work as well. The larger establishments employ one member of staff who is responsible for all estimating, while the actual repair work is carried out by the tradesmen on the shop floor. Consequently there must be no lack of coordination and understanding between these two and the jobs they perform. The decisions of the estimator are all important as they instigate the organization of work on the shop floor. Moreover, no matter how skilled the tradesmen are, with bad estimating the financial profits can be drastically affected.

In collision work the estimate is considered as a firm commitment to do the work involved for the amount of money shown, and should be detailed so that insurance companies or private owners can determine from the estimate exactly what it is proposed to do to the damaged vehicle. The estimate must include cost of parts and labour costs at the recognized retail rate of the body shop establishment (Figures 14.136 and 14.137).

In the preparation of an estimate, the crash damage should be itemized into the number of damaged panel sections and assemblies, and therefore it is important to have a knowledge of vehicle body

41   A suspension frame member is to be replaced by welding; explain the correct method of achieving its alignment, before hand.

42   A vehicle is to have the rear boot floor, wheel arch and quarter panel assemblies removed. State the repair procedure to reinstate these panel assemblies.

43   With the aid of a sketch, show the position of a pull-dozer to realign a centre pillar after side impact.

44   Explain the repair procedure necessary for part panel replacement on the quarter panel assembly of a four-door saloon.

45   With the aid of labelled sketches, show the difference between direct glazing and indirect glazing.

46   Outline the procedure for mounting a vehicle on a jig using the MZ bracket system.

47   Show, with the aid of a sketch, how the laser on Dataliner equipment can achieve a 90 degree angle.

48   State the advantages of the use of split measuring bridges on a jig measuring system.

49   Explain what is meant by the term 'universal jig bracket'.

50   State the safety precautions necessary when carrying out pulling in repair.

51   Describe a computer-based estimating system.

52   Describe how to use the Glassmatrix estimating system.

Bodyshop planning

15.1 Initial planning

15.1.1 Preplanning

Setting up a bodyshop capable of carrying out repairs that will meet customers', insurance companies' and motor manufacturers' quality standards requires significant investment in time and money. Therefore to set up or redevelop a bodyshop, professional advice is essential.

There are a number of suppliers, manufacturers and independent consultants offering a range of planning, design and consultancy services, from simple equipment layouts to sophisticated three-dimensional computerized blueprints. The design and planning services provide everything for the re-equipping of new bodyshops in existing premises, extensions to original bodyshops, or the total development of greenfield sites. Preplanning is essential, whether for the building of a new bodyshop or the remodelling of existing premises, because it is necessary to be certain that the finished bodyshop will meet all the operational requirements before construction is started. It is therefore important to conduct as much market research as possible before finalizing your plans. This research should include asking bodyshop staff what improvements they would like to see in the new bodyshop, visiting other bodyshops, and talking to other managers to collect advice and recommendations about good practice and possible problems to be avoided.

15.1.2 Choosing a site

The location of any building or bodyshop is a crucial factor in its profitability: no matter how efficient may be the management and staff, customers will not be encouraged to bring their vehicles for repair if access to the site is difficult.

A new bodyshop located on an industrial estate usually presents few problems. It also usually means a lack of sensitive residential neighbours. Access is generally good, and the essential services of gas, electricity, water, drains and telephone are already available. On sites lacking these services it could prove costly to have them installed.

A site that is relatively cheap may have distinct disadvantages such as a narrow road, steep hills, a particularly bad surface, sharp corners, one way in and one way out only; all these affect access to the premises. Ideally the customer should be able to see the site before he or she is on top of it. Your outside sign should be simple, easy to read and brightly coloured, otherwise you run the risk of customers missing it, driving past your bodyshop and ending up frustrated even before walking through the door.

The premises should be easy to drive in and out of, and there should be adequate space for vehicles to turn safely. These requirements apply equally to suppliers delivering bodyshop materials, especially with large commercial vehicles. Parking facilities should be adequate and made attractive by either landscaping or the addition of planters. A prospective customer will not be impressed if he has to park halfway down the street to walk to your bodyshop.

The choice of site will of course depend to a large extent on the amount of capital available. The operator who buys a leasehold site pays proportionally less but may not be able to make the necessary profit within a set number of years. Freehold land is more expensive but a much better investment. Another option is renting an existing bodyshop. Alternatively, converting a purpose-built building might be considered, although this could be costly and may not ultimately fulfil the necessary criteria for an efficient bodyshop, although it is probably the most popular option.

construction. The outer panels which make up the body shell should be referred to by the manufacturer's recognized names, such as:

Near-side front door (NSF door)
Near-side rear door (NSR door)
Off-side front door (OSF door)
Off-side rear door (OSR door)
Roof panel
Boot lid
Bonnet
Near-side centre pillar (NS centre pillar)
Radiator grille
Near-side front wing (NSF wing)
Off-side front wing (OSF wing)
Near-side sill panel (NS sill panel)
Off-side sill panel (OS sill panel)
Front bumper bar assembly with valance
Rear bumper bar assembly with valance.

In addition this can be further broken down to all internal panel structures according to the position of the damage.

It is essential to determine the exact amount of stripping necessary for either the repair or the removal and replacement of a damaged section. To calculate the actual cost of a job it is necessary to establish a set rate per hour for all repairs done in the repair shop, and therefore an accurate total repair time can be calculated on the time taken to repair each section of the body. This then provides the actual cost estimate for repairing each section or part of the job. The retail rate or set rate per hour is the amount of money that the repairer charges the customer for labour, and is made up of; wages paid to the tradesmen; the cost of overheads, which should include such items as supervision, depreciation of equipment, rent, heat, light, electrical power, advertising, telephone accounts, cleaning, office staff, stationery and postage, and workshop materials; and also a reasonable amount of profit.

14.5.4 Computer estimating

Computers are rapidly losing their mystique, and are becoming as indispensable to the bodyshop as the spray booth and chassis jig. Offering benefits in all aspects of bodyshop operation, there are computers and software packages dedicated to quick and easy repair estimates, stock control, invoicing and management accounts, paint and materials, parts list, mixing and colour matching, job cards, work-shop loading, booking in, job efficiency reports, invoice production, reports and letters (Figures 14.138–14.141). Costs have tumbled and the benefits can be reaped by all bodyshops, regardless of size.

The key to success in any bodyshop is the production of an accurate, well presented estimate. An estimate has to be a clear, concise statement of what you consider to be a fair return for the care and work you carry out on a crash damaged vehicle. By utilizing the memory capacity of the computer you will have readily available all necessary details to help you draw up an estimate quickly and accurately.

Your computer system will have the facility to prepare not only a detailed document for the customer, but also a comprehensive breakdown for the insurance company. The documents produced will be tailor made to your own letterhead stationery. The insurance estimate can often be sent direct by fax to the company involved, thereby further reducing the time gap between estimating and authorization by that company. Apart from the major bonus of speed, a computerized estimating system will greatly enhance your company's image as a professional organization. You will have instant access to the progress of any job: most systems include this as standard. The more time and money you spend at the outset, the more control you will have over staff movement and job completion.

14.5.5 Insurance procedure

Under the Road Traffic Act the car owner is obliged to insure his vehicle either under third party insurance cover or fully comprehensive insurance cover. As a direct result of this, most accident damage is covered by insurance and is therefore repaired in body work establishments.

The procedure for dealing with repairs carried out under an insurance claim is in four stages, as described in the following sections.

Claim form

When involved in an accident which has resulted in vehicle damage, the owner should obtain and complete an insurance claim form and immediately return it to his insurance company, or make a written report on the accident and damage received.

```
BodyMaster House                        ACCIDENT REPAIR CENTRE
Wych Cross                                 FULL JIG FACILITIES
Forest Row                                    SPRAY BOOTH
East Sussex                               ALL INSURANCE WORK
RH18 5LJ
Tel: (0342) 826001
Fax: (0342) 826002


Mrs. M. Jones,
42, Kings Road,
Sharpthorne,
Nr. East Grinstead,
East Sussex,                   Estimate No :    10010
RH23  4LB.

Dear Madam,

Re; Ford Escort 1.6L, Registration Number: F376 AFG.

We have pleasure in submitting our estimate for repairs to the above vehicle.

Repairs:
Remove and refit:
Front Bumper, O/S/F Headlamp, O/S/F Indicator, Front Grille, Fog Lamps,
Mouldings, O/S Mirror, Repeater Lamp, "Ford" Decal, O/S/F Door and Trims, Radiator and
O/S/F Tyre.

Load vehicle onto jig.

Repair and reshape:
Front Panel, O/S/F Door and O/S Front chassis Leg end.

Cut out and weld in:
New O/S/F Wing.

Hang and align:
O/S/F Door and New Bonnet.

Check and set wheel alignment.

Paintwork:
Prepare, prime and topcoat:
O/S/F Door, New Bonnet, Front Panel, New O/S/F Wing and
O/S Front chassis leg end.

Labour;                                    £    571.30

Plus Parts :- (To be supplied at MRP)
New Bonnet                      Front Bumper
O/S/F Headlamp                  O/S/F Indicator
O/S Mirror                      Repeater Lamp
New O/S/F Wing                  Clips etc
Radiator (Not discountable)     O/S/F Tyre

Paint, materials and sundry items.

Yours faithfully,

Bodyshop Manager.
Please Note: All Prices Subject To 17.5% VAT.
          Please ensure your radio code is recorded before repair.
```

Figure 14.138   Customer's estimate (*Bodymaster UK*)

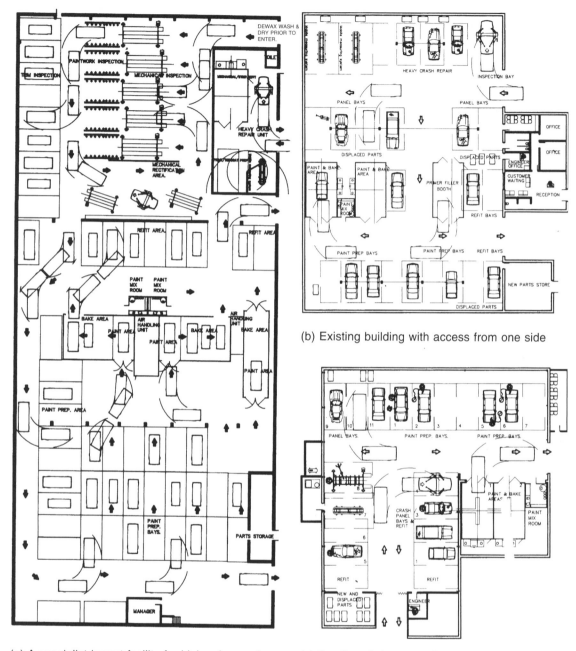

(a) A specialist import facility for high volume minor repairs

(b) Existing building with access from one side

(c) Small workshop complex

Figure 15.1 Bodyshop layout (*ICI Autocolor, Ernest W. Godfrey, Pickles Godfrey Design Partnership*)

Parts department for trade and retail This could include parts manager's office, parts store area, and storemen.

Staff facilities These could include toilets, washroom and showers, rest room, canteen, first aid room.

Fire fighting equipment This must be approved by the local fire officer.

Working areas should be tailor-made to suit the needs of the particular bodyshop. Typical requirements are as follows:

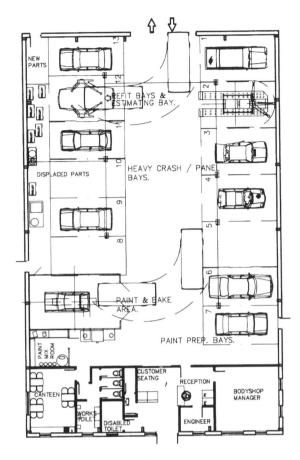

(d) Greenfield site development

Figure 15.1 (*continued*)

Vehicle damage assessment area This is where the vehicle is first assessed for the degree of damage in order that an estimate may be written for customers and insurance companies.

Stripping area This is where damaged vehicles are, prior to repair, stripped of body panels, body trim, electrical and mechanical assemblies.

General body repair area This is where minor repairs, general repair work, and work not requiring a jig can be carried out.

Specialist repair area or jig area This is designed for major accident damage where the vehicle must be repaired on a jig system.

Preparation area This is for the pretreatment of the repair in the form of sanding, priming and masking before painting.

Refinishing area This is where the vehicles are repainted and baked to original factory condition.

Refit area This is where interior and exterior trim, as well as mechanical and electrical assemblies, are refitted after the vehicle has been painted.

Valeting and wax injection area This is designed for the cleaning of the vehicle's interior and exterior back to showroom condition as well as for replacing the corrosion prevention treatments of undersealing and injection wax treatment.

Final inspection area This is where the vehicle can be checked for any minor defects after the repair process is completed.

15.2.1 Reception area

The reception area is very important. Here customers get their first impressions by meeting the receptionist, who represents the company. Reception staff play a very important part by instilling confidence into customers through their courteous approach and knowledge of vehicle repair. The area should be clearly signed and the customer should walk into a pleasant, clean and tidy reception office. Natural colours are probably the best for decorating the walls, and wall space should be used to promote the bodyshop by displaying framed certificates of the staff's qualifications, association memberships, and insurance company approvals. The area should be large enough to accommodate customers and the reception engineer, and to hold customers' records, bodyshop loading display boards, telephone and workshop intercom systems. Lighting should be efficient, and floor covering attractive but practical and easy to clean. Furniture should be serviceable and comfortable without being elaborate, and sufficient to seat at least six people. The area must be sited near an entrance door, and ideally provide customers' toilets and a drinks vending machine. A window installed between the work area and the reception area allows customers to view tasks being carried out. Professional technicians at work on vehicles are a good selling point, as customers are always interested in seeing work being carried out. However, if customers can see into the bodyshop, it must be kept tidy at all times.

15.2.2 Administration offices

The bodyshop manager's office should also be adjacent to the reception office for easy access to liaise with the customers. He or she needs to be easily accessible to the secretarial staff employed by the

BodyMaster House
Wych Cross
Forest Row
East Sussex
RH18 5LJ

Tel: (0342) 826001
Fax: (0342) 826002

ACCIDENT REPAIR CENTRE
FULL JIG FACILITIES
SPRAY BOOTH
ALL INSURANCE WORK

Vehicle Checklist

Customer   : Mrs. M. Jones,
Vehicle   : Ford Escort 1.6L,
Chassis No : 11-H7-SLBN43/87
Paint Code : B1
Job/Estimate No : 10010

31st March 1992
Reg No: F376 AFG

Workshop :

CHECKLIST	TICK
Lights	
Headlamp beam	–
Door workings	–
Trims and fixings	–
Security of parts	–
Brake fluid level	
Water level/antifreeze	–
Aperture gaps	–
Wax injection	–
Underseal	–
Coachlines	–
Masking tape	–
Blow-ins	–
Overspray	–
Paint finish	–
Cleanliness interior	–
Cleanliness exterior	–
Wheel nuts	–
Roadtest	–

Signed
....

Comments on general condition

Office Use :

Satisfaction Notice Excess Invoice	

Signed
....

Figure 14.141 Vehicle repair checklist (*Bodymaster UK*)

Itemized estimate

According to the extent of the damage, and if the vehicle is still roadworthy, the owner takes the vehicle for the inspection of the damage. The repairer will make a visual inspection of the assessed damage, and from the knowledge gained complete a written itemized estimate which he will submit as a tender to the insurance company. This estimate will show the total cost of repairs, and where the estimated amount is under a certain figure set by the insurance company, the owner has the right to authorize the repairer to do the work. However, in most repair cases this figure is exceeded, and the insurance company's assessor is the only person authorized to allow the repairs to proceed.

One of the important factors when estimating for insurance claim damage is to examine carefully every section of the vehicle, especially those parts which are a known weakness in the construction and therefore liable to be affected directly or indirectly by a collision. A methodical system of estimating is essential to avoid missing any damage, and is usually carried out by noting in order all removal and replacement items, all repair items, all respray items and all items to be supplied new at cost, including mechanical parts and any trim. Supply items are usually difficult to price because of the makers' fluctuating prices, and therefore they should be listed 'at cost'. Spray painting can be quoted either by itemizing each part separately or by a complete price for the total spray operation. While in the owner's presence, the repairer should point out any rusted sections which may affect the repair work, or any previous unrepaired damage which is not covered under this insurance claim.

Authority to repair

It is essential to obtain the authority to repair from the insurance company involved before any work is started on the damaged vehicle. On receiving the repairer's estimate, the insurance company will instruct their own engineer assessor or an independent assessor to examine the vehicle and satisfy himself that the claim is in order and that the estimate submitted by the repairer does not include any labour or material necessary as a result of any other than the accident report. The engineer assessor then agrees the cost of repairs, particularly the labour charges, with the repairer while inspecting the vehicle, and decides whether or not damaged parts shall be repaired or replaced.

After this inspection the assessor will send written instructions on behalf of the insurance company; these constitute the authority to repair. A condition of these instructions is that the repairer is restricted to repairing the damage as seen and estimated for. Any additional damage disclosed when dismantling the vehicle must have a separate estimate submitted, and work cannot be carried out on this damage until the extra estimate has been agreed and the necessary work authorized by the insurance company.

Clearance certificate

This certificate is provided by the insurance company for the vehicle owner to sign when he has seen that all the agreed repairs have been completed satisfactorily, and the damaged parts reinstated to their original condition. When this certificate has been signed by the vehicle owner, it releases the insurance company from any liability in respect of the claim for damage caused by this particular accident. It is to the repairer's advantage for him to ascertain that the clearance certificate has been duly signed before the vehicle is returned to its owner.

The duty of the repairer is to see that the damaged car is reinstated to its original condition. He is expected to make full use of his skill and knowledge to effect the best possible repair, as the owner relies on his reputed skill.

14.5.6 Motor Insurance Repair Research Centre (Thatcham)

In the 1960s the motor insurance industry became concerned about the escalating cost of vehicle accident damage repair work and its effect on motor insurance premiums. To have some direct influence on these costs it was recognized that motor insurers required a means of researching the cost of repairing accident damaged cars and light commercial vehicles.

In 1969, motor insurance company members of the British Insurance Association and motor syndicates at Lloyd's joined in a scheme to create a research centre. The result was Thatcham, named after the small Berkshire town where it is situated, and the only centre of its type in the UK. Thatcham is unique in that it represents an entire motor insurance market. It has been granted research status by

15.2.6 Preparation and refinishing area

When planning your bodyshop, productivity, health, safety, and environment regulations must all be taken into consideration.

Start with the preparation area. There must be space to pressure wash vehicles to get cleaner paintwork before they enter the paint shop.

The dust created by sanding in the preparation stage must be prevented from being drawn into the paint area. The best solution is an efficient extraction system or extraction tools. Avoid walls or curtains as they lead to rigid space allocation and are inflexible because they prevent bodyshops reacting to variable workloads. If possible, design the paint shop so that the vehicle moves forward in a relatively straight line or an easy loop, and try to keep it inside the workshop to avoid the problem of coming into contact with dirty, dusty air.

Ideally the refinishing area should form a self-contained area separate from the mechanical and repair side of the workshop and the preparation area. The main objective is to minimize the amount of handling of the vehicles by arranging for a flow system from the vehicle's entry to its exit from the spray booth, thus preventing any dust entering the spray booth. Consequently spray booths and ovens can be used to their full capacity throughout the working day.

In the paint shop, health and safety regulations will dictate many decisions and must be taken into account in your plans. The ramifications of the EPA and COSHH legislation are far reaching for all bodyshops. In effect, any bodyshop currently operating without a properly constructed fully operational and well maintained spray booth is operating outside the law and risks summary closure.

A spray booth is in most cases the single biggest purchase decision made by the bodyshop owner. Most bodyshops are now fitting combi-booths. A combi-booth is a combined spray booth and low-bake oven with an electrically controlled operational cycle which provides the correct temperature, filtration and lighting requirements for the spraying and baking of paint. The purpose of the combi-booth is to provide an efficient, clean area, free of humidity and within the temperature band 20–25 °C. The booth must be sealed off to prevent overspray escaping into the workshop, and must be force ventilated to remove solvent fumes and microscopic paint particles whilst drawing in clean and dust-free air. Once spraying is completed, the combi-booth can be put into the bake cycle in which the internal booth temperature is raised to promote rapid curing of the paint. One important point to consider is the choice of spray booth. This must meet all the requirements of the EPA in relation to the throughput of air and the placement of the exhaust stack. This point is particularly important in residential areas. The booth must operate under negative pressure and to comply must be fitted with a pressure gauge. As a further precaution, emissions from spray booths should be tested at least once a year to monitor particulate matter emitted. The EPA stipulates that concentration of total particulate matter in final discharge to air from the booth should not exceed 10 mg/m^3.

When constructing the paint store room, remember to provide space for the mixing system and for the safe storage of paint. Again, follow all the necessary regulations.

15.2.7 Valeting

The vehicle should now be back at the valeting bay for both interior and exterior valeting, to emerge as good as new with the minimum wasted time. Careful valeting work can transform the final appearance of the vehicle, thus creating customer satisfaction by returning the car looking like it did when new. If the customer's first impressions on collecting the vehicle exceed expectations, he will be more than willing to recommend the comapany's quality of service.

15.2.8 First aid

First aid equipment should be supplied at set points throughout the workshop, and staff should be encouraged to attend first aid classes. The Health and Safety at Work Act requires that where more than 50 personnel are employed, one qualified first aid person must be appointed.

15.2.9 Fire fighting equipment

Consult the local fire officer regarding fire points and types of fire appliances, exits, and fire doors. Organize staff fire drill and training procedures, and make sure that all fire extinguishers are maintained.

15.3 Bodyshop heating

The body repair shop can be a difficult building to heat because of the very nature of the work that is carried out. First, it needs to be a reasonable height so that vehicle bodies can be elevated and moved about easily; and secondly, doors are constantly being opened and shut to allow for the movement of vehicles in and out of the workshop.

Careful consideration of bodyshop heating requirements will not only result in energy consumption savings but also make the bodyshop more comfortable. Whatever type of heating systems are used, care should be taken to ensure that heaters are placed where they will not interfere with work in progress, will not take up unnecessary floor space, and will provide employees with the maximum benefit.

Most of the conventional methods of space heating can be found in bodyshops, ranging from portable gas fired units through wet systems to radiant systems.

15.3.1 Radiant heating

Infrared (IR) heating is the transfer of energy by IR electromagnetic radiation; this is the portion of the electromagnetic spectrum between visible light and the top section of the radio-radar wavebands.

Any object will emit IR energy; the higher the temperature, the more IR radiation is emitted. Therefore when two objects are adjacent to one another and one is hotter than the other, there will always be an exchange of heat from the hotter object to the cooler one. Radiant heaters operate by the transfer of heat to solid objects rather than by heating the air itself; therefore high-ceiling buildings can be heated with a high degree of efficiency. By installing radiant heaters above a cold concrete floor, there will be a transfer of IR radiant energy downwards towards the cold concrete, which will absorb energy until it becomes warm. Other objects at floor level, such as vehicles, benches and machinery, as well as the walls, will become warm either by receiving direct IR energy or by the transfer of energy from the floor. All of these items become secondary heat sources, transmitting energy into the surroundings of

the workshop. Ceiling-mounted radiant heating units are ideal for bodyshop use in that they do not take up usable space and encroach into the working area; however, they are not suitable where the ceiling height is less than 3.5 metres. Also radiant heat absorbed by its surroundings does not easily escape through service doors: therefore heat loss is reduced and a saving is made on energy costs. Furthermore, heating the floor keeps it dry, and the fact that vehicle bodies become heated reduces problems associated with condensation as the cold vehicle bodies are brought into a heated environment.

15.3.2 Electric heating

Electric spot heating allows comfortable heat to be concentrated where and when it is wanted. Linear quartz heaters can be used for the spot heating of specific work areas, for zone heating of a particular area of the bodyshop, or for full coverage heating.

For spot heating, the heaters should be positioned about 3–5 metres apart, at a height of approximately 2.5–3.5 metres from the floor, depending upon the heater rating. Heaters should be rated at 1.5 kW for sheltered locations, rising to 3 kW or even 4.5 kW for more adverse conditions. For zonal heating, the heating requirement should be calculated on the basis of 150–300 W/m^2. The actual value selected will depend upon the state of the building, internal draughts, etc. If the zone is more than about 15 metres wide, then additional rows of heaters may be required. For full coverage, the heaters should be positioned around the perimeter of the building at spaces of about 4–6 metres, using load densities of 100–200 W/m^2 based upon heat loss calculations. If the building is more than about 17 metres wide, additional heaters can be installed.

When installing these heaters, the general principle is for the heater to be angled down at about 45 degrees when wall mounted and up to 90 degrees when mounted overhead.

As with all forms of heating, care should be taken with these heaters if there is a risk of combustible dusts or flammable gas hazards. The lamp/filter assembly is the hottest part of the heater, with surface temperatures of 750–800 °C, broadly similar to the surface temperature of the lamp in a tungsten halogen floodlamp. In case of concern over this point, consultations should be held with the factory inspector.

given in Section A, the replacement time schedules can be read off horizontally. It should be noted that the total time for each complete replacement includes a time allowance of 0.5 hours for the body repairer to move the vehicle as needed and obtain special tools and parts from the stores, this is called job allowance. The table includes a section for paint refinishing.

If more than one panel is being replaced, the combination panel time is used. For instance, it will take less time to replace both a front wing and a front panel than a separate wing and a front panel. The procedure is to find the nearest combination in Section B, each dot in the column denotes a panel. The time for the combination of panels is read off vertically in Section C. The job allowance is calculated by allowing 0.5 hours for the first panel and 0.25 hours for each subsequent panel or bracketing system. For ease of calculation the allowance is then rounded up to the nearest 0.1 hour.

If the job is a large one which combines several panels and cannot be readily calculated, a paint combination time (paint comb.) is listed in the single panel part of the matrix. The paint combination time is the productive part of actual painting, that is, it does not include paint mixing, test panel spraying, changing overalls or cleaning. So, this time alone can be added to a combination panel time for painting a larger combination of panels.

Paint refinish methods

Thatcham publish refinish guide times for individual vehicles using the common variations in paint systems, these are:

non-metallic
non-metallic, base coat and clear
metallic, base coat and clear
two-coat pearlescent

The times allow for dry-flatting and the use of two-pack materials where appropriate.

Thatcham suggest a set sequence of paint refinishing operations, these are shown in Figure 14.143.

The above is a brief introduction to Thatcham Methods, and it is recommended that the body repairer takes one of the Thatcham short courses.

14.5.8 Glassmatrix II system

Glass's Guide Services, famous for their pocketbook on car prices, offer a computer-based estimating system using Thatcham time schedules.

Glassmatrix II An MS Windows-based, computer-assisted estimating system that is used by insurance companies and vehicle body repairers to produce accurate estimates of repair costs. The estimates are produced using a system of bar coding information from a collision repair estimating guide (CREG). The bar code reader is shown in Figure 14.144. The bar code contains information on the labour time, the cost of replacement parts and the part number for identification. The labour time is mainly supplied by Thatcham, the numbers and the costs of parts is based on manufacturer's recommended retail prices. The Glassmatrix software is supplied on a CD ROM, which is regularly updated – usually once a month. The software is modular, see Figure 14.145.

GlassWord A report and form generator that can be used to design bespoke forms and reports. It works in conjunction with Glassmatrix by importing power fields from an estimate and placing them into a form template. GlassWord can be used to create parts order lists, letters to customers, invoices, and other documents using either plain paper or printed business letterheadings.

GlassImage A system for capturing and storing digital images from video data. A video camera is used to record digital pictures of the damaged vehicle.

Glass's Guide for Windows A PC-based version of the famous valuation booklets; information from this can be imported into the estimate or other forms.

Alternative parts A database of insurance industry approved parts, mainly Veng and Unipart, which may be used instead of the original equipment manufacturer (OEM) parts database.

GlassReport A statistics package for analysing the estimate in terms of common concepts, such as average estimate cost and parts-to-labour ratio.

Glassmatrix communication links Figure 14.146 shows the arrangement of the communication links. It should be noted that e-mail can be sent using either a mobile telephone cell network, or a portable satellite link, both of which are currently more expensive than a land-line link.

Procedure-Sequence of Operations	Method: 1	2	
Prepare interior and exterior surfaces for primer	●	●	P80-180-Scotchbrite
De-oxidene bare metal areas only	●	●	
Solvent wash those areas to be masked	●		
Mask up including fitment of a car cover and dust sheet	●		
Fit hang-on panel to the spray jig		●	
Move vehicle/panel into booth. Switch on etc.	●	●	
Mix etch primer if necessary	●	●	
Mix primer filler/surfacer	●	●	
Mix guide coat	●	●	
Solvent wash, blow off, tack wipe interior and exterior	●	●	
Spray etch primer if necessary to interior and exterior	●	●	
Spray primer to interior and exterior	●	●	
Spray guide coat to exterior	●	●	
Prepare straight colour/straight colour base & clear/metallic base & clear/ready-mix	●	●	
Spray a test panel to opacity		●	
Prepare (minimum) interior surface for straight colour/straight colour base & clear/metallic base & clear	●		P400-Scotchbrite
Spray straight colour/straight colour base & clear/metallic base & clear to panel interior and test panel to opacity	●		
Clean spray guns	●	●	
Move vehicle/panel and test panel out of booth when dry	●	●	
Check test panel for colour match	●	●	
Remove hang-on panel from the spray jig		●	
Remove interior masking only as necessary	●		
Prepare exterior surface for straight colour/straight colour base & clear/metallic base & clear	●		P400-Scotchbrite
Prepare interior and exterior surfaces for straight colour/straight colour base & clear/metallic base & clear		●	P400-Scotchbrite
Seal exterior panel joints as necessary	●	●	Polyurethane sealer
Prepare Surfaces for stone chip	●	●	Scotchbrite
Mask for stone chip	●	●	
Mix stone chip	●	●	Solvent or water-based
Spray stone chip	●	●	
Clean stone chip spray gun	●	●	
Remove masking	●	●	
Solvent wash those areas to be masked	●	●	
Mask up as necessary	●	●	

Figure 14.143 Paint refinishing methods (*Motor Insurance Repair Research Centre*)

Figure 15.5 Lighting for the spray booth (*Fifth Generation Technology Ltd*)

existing booth. Lighting is certainly an area worthy of meticulous attention by the bodyshop operator (Figure 15.5).

15.4.1 Light sources in bodyshops

There are several general types of lamp found in bodyshops:

Tungsten bulbs (general lighting service (GLS) lamps) These are still used for hand lamps and toilets, but are steadily being replaced by low-energy compact fluorescent lamps.

High pressure sodium (SON) These give a warm golden light (SON DL) or a pleasant, less yellow light (white SON). Both are good for mechanical areas and forecourts, but their colour rendering makes them totally unsuitable for bodyshops.

Mercury lamps (MBF) These give white light similar to fluorescent, although not good enough for colour matching. They are suitable for general areas.

Metal halide (MBIF) These lamps are now both long life and suitable for colour matching.

Fluorescent tubes (MCF) These are the industry standard, and provide an excellent form of low-cost lighting. There are many different types of fluorescent tube on the market, and it is important that the correct type of tube is fitted in each area of the bodyshop. Spray booths should be fitted with colour matching tubes and it is advisable that this type of tube is used in the preparation area too. It is vital that all tubes fitted in a spray booth should be the same colour. If one tube in the spray booth should fail and a direct matching replacement is not available immediately, it is actually better to run

with a tube missing than to temporarily substitute a non-matching tube. In areas where critical colour matching is regularly performed, it is definitely worth while to change lamps on a regular basis about once a year and so to avoid the problems of lamp failures and the colour variations which occur with age.

15.5 Essential equipment for the bodyshop

In the utilization of equipment there are two important factors:

Return on investment

This is simply the profit made from the amount invested in the workshop and equipment.

Return on assets

This is the assessment of how the workshop and equipment are being used. This relates to net profit achieved against an asset, which would be the work being produced in the workshop. This base is divided between fixed assets and current assets. Fixed assets are money in buildings and equipment. Current assets are money tied up in the work in progress and in stocks held within the stores department. To achieve the necessary return it is therefore absolutely essential to use the site, and the expensive equipment required for vehicle repair, to the fullest extent.

15.5.1 Essential equipment guide

Before purchasing and installing expensive equipment an analysis of any past work and a forecast of the future are essential to forecast the probable availability of the work.

The equipment requirements can be categorized as follows:

Specialist equipment: essential and desirable
General workshop equipment: essential
Hand tools and expendable items: essential.

Specialist equipment needed for stripping, repairing and painting

If a full and efficient service is to be given to the customer, certain equipment is required in the workshop. All equipment used in the workshop must meet all current legislation.

Alignment and repair jigs for measuring and straightening

Pulling equipment for reshaping damaged body sections

Hydraulic body jack equipment for panel repair

Welding equipment: oxy-acetylene, MIG/MAG welding, TIG welding, spot welding (double and single sided)

Plasma arc cutting equipment

Dust and fume extraction systems (portable and static)

Wheel alignment equipment (four-wheel alignment)

Combi-booth

Panel booth

Mixing room with mixing system

Infrared driers

Wall-mounted breathing air filters

Air regulators

Respirators and air-fed visors

Spray guns (standard and/or HVLP guns)

Gun cleaning tank, extracted

Combined paper baler and can crusher

Pedal waste bin for solvent contamination waste

Plunger cans to dispel solvent

Screw compressor (Suitable for the total capacity of air requirement of the workshop).

See Chapters 13, 14 and 17 for further information about the equipment.

General workshop equipment

Hoist (wheel-free type)

Headlight focusing equipment

Axle stands and trolley jacks

Folding crane (1 tonne)

Battery charger (portable)

Parts trolleys

Masking machine

Filler dispensers

Solvent dispensers

Panel stands

Impact wrenches

Torque loading spanners for wheel nuts

Waste oil dispensers

Wet and dry vacuum

Valeting machine

Polisher

Wax injection equipment

Vehicle moving skates

Benches (mobile).

See Chapters 13, 14 and 17 for further information about the equipment.

Hand tools and power tools

Panel tools and wallboards

Air tools, electric tools: grinders, chisels, drills, saws

Windscreen cutting tool

Spottle spot welding cutter

Spot welding dresser

Random orbit sanders and dust extraction

Block sanders and dust extraction

See Chapter 3 for further information about the equipment.

Miscellaneous and expendable items

Welding goggles approved to BS

Welding headscreens approved to BS

Fire blankets approved to BS

Welding curtain/screen approved to BS

Gloves: disposable, plastic, leather, rubber, canvas

First aid equipment

Fire extinguishers

Storage cabinet/lockers.

See Chapter 2 for further information.

15.5.2 Dust and fume extraction (extraction and arrestment systems)

Polluted air is often invisible to the naked eye. However, the effect it can have on the health of a workforce and the overall efficiency of an organization can be dramatic. The most effective way to purify air is to capture airborne pollutants at source and, depending on individual applications, to either recirculate fresh and preheated air or vent the pollutants away from the working environment to a safe collection point. The systems should actually be extraction and arrestment systems and should extract the pollutant materials and collect them in a safe and manageable form (Figure 15.6).

In the context of bodyshops, the main problems are fillers and paint dust from the rubbing down and flatting processes, paint and solvent fumes from the wiping down and painting processes, and welding fumes.

Filler and paint dust generated in the preparation area is best collected as soon as it is produced by

Figure 15.6 Portable dust extraction system used in the workshop (*Nederman Ltd*)

Figure 15.7 Tool extraction system (*Minden Industrial Ltd*)

using off-the-tool extraction (Figure 15.7). This can either be by portable units serving one or two operators, or by fixed systems with a central extraction unit serving a number of fixed extraction points located in the workshop (Figure 15.8).

Specially designed extraction equipment can be tailor-made to an individual bodyshop for the removal of dust and fumes. Gases, powders and chemical vapours are all types of hazardous elements to which fume extraction can be applied. The range of self-supporting arms, combined with the versatility to mount the system on ceilings, floors, benches and walls, make the access to applications unlimited. Furthermore, the easily manoeuvred suction hoods create extraction right at the source of the problem. The fans are designed to draw the polluted air through the extraction arms, dispersing the fumes via the assembly ducting. Alternatively, to recirculate the purified air, an electrostatic unit can be employed to eliminate harmful particles and utilize existing preheated air.

A range of vehicle exhaust extraction systems is available: a choice can be made from a simple drop system through to the rail system (Figure 15.9), which allows vehicles to be driven whilst maintaining at-source extraction with both advanced infrared remote controlled and electrically motor driven reels.

At-source extraction of welding fumes is far more energy efficient than using central ventilation systems. Harmful fumes can be captured and disposed of in a safe and simple way regardless of the welding environment. Irritation, fever, poisoning and fibrosis are a few of the effects that can be minimized with the extraction of fumes from the welding operations. Many welding processes create noxious and harmful fumes which can be eliminated with portable welding smoke eliminators (Figure 15.10). These provide complete extraction where confined areas pose a problem, especially in body repair workshops. There is a wide range of light-weight smoke eliminators which take up very

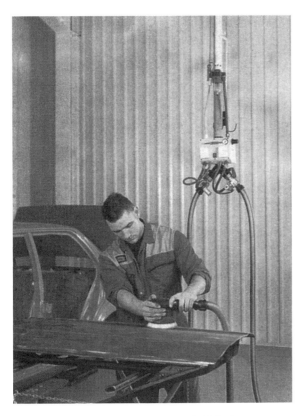

Figure 15.8   Central extraction unit (*Minden Industrial Ltd*)

Figure 15.10   Portable fume and smoke extraction system (*Nederman Ltd*)

Figure 15.9   Vehicle exhaust extraction using a rail system (*Nederman Ltd*)

little floor space and can be carried from job to job. The filters in these portable extraction systems can be changed very quickly, and some have an alarm light fitted to warn the user that the filter needs replacing.

It is a specific requirement of the COSHH Regulations that equipment is maintained in efficient working order. The physical system must be regarded as part of a broader health and safety housekeeping policy aimed at keeping the whole area dust and fume free.

15.6 Bodyshops and legal requirements

Health and safety legislation has made the vehicle body repair industry increasingly aware of the need to provide adequate facilities for employees, both as a legal duty and to improve the working environment. Within this framework of a safe working environment the employer must also promote efficient work methods, which together should result in improved productivity.

15.6.1 Statutory legislation

Petroleum (Consolidation) Act 1928
Petroleum (Mixtures) Orders 1929 and 1947
Factory Act 1961
Weights and Measures Act 1963
Fire Precautions Act 1971
Highly Flammable Liquids and Liquefied Petroleum
  Gases Regulations 1972
Road Traffic Act 1972 (MOT)
Abrasive Wheel Regulations 1974
Control of Pollution Act 1974
Protection of Eyes Regulations 1974
Health and Safety at Work Act 1974
Fire Precautions (Factories, Offices, Shops and
  Railway Premises) Order 1976
Motor Vehicle Construction and Use
  Regulations 1978
Control of Pollution (Special Wastes)
  Regulations 1980
Classification, Packaging and Labelling of
  Dangerous Substances Regulations 1984
Control of Substances Hazardous to Health
  (COSHH) Regulations 1988
Environment Protection Act 1990
Management of Health and Safety at Work
  Regulations 1992
Provision and Use of Work Equipment
  Regulations 1992
Personal Protection Equipment at Work (PPE)
  Regulations 1992
Manual Handling Operation Regulations 1992
Workplace (Health, Safety and Welfare)
  Regulations 1992

The bodyshop is most affected by the legislation described in the following sections.

15.6.2 Regulations covering paint storage, fire precautions and fire fighting

The storage of paints with a flashpoint below 22 °C is governed by the Petroleum (Consolidation) Act and the statutory rules and orders 1929. Those with a flash point of 22–32 °C are governed by the Highly Flammable Liquids and Liquefied Petroleum Gases Regulations 1972.

Requirements of the Petroleum (Consolidation) Act 1928

Paints, thinners, and other products governed by this Act will either be labelled 'Petroleum mixture giving off an inflammable heavy vapour' or have a flame symbol and the words 'Highly flammable', indicating a flashpoint below 22 °C.

The Act requires that the storeroom should be licensed and constructed to approved standards.

Requirements of the Highly Flammable Liquids Regulations 1972

Paints governed by these Regulations will be labelled 'Flammable'. More precise flashpoint details can be obtained from the relevant product data sheet, which indicates products with a flashpoint range between 22 and 32 °C.

Products with a flashpoint greater than 32 °C do not fall within the Regulations, but other Regulations require that products with a flashpoint up to 55 °C have a 'Flammable' warning.

15.6.3 Health and safety legislation

The Health and Safety at Work Act 1974 places a duty upon employers to ensure, so far as is reasonably practicable, safe working conditions and the absence of risks to health in connection with the use, handling, storage and transport of articles and substances.

The Act places a statutory duty on employers to have a declared safety policy for a business in which more than five persons are employed. The Act does not specify what you should do; it merely provides the framework in which you should operate, together with establishing the Health and Safety Executive and the Health and Safety Commission. The Act regulates all working methods. The importance of this Act cannot be over-emphasized; no working methods may be employed that can be seen to be a health or safety hazard to employees.

First a company safety policy must be established and a safety committee formed. The committee should consist of members with specialized knowledge of the risks of a particular area, i.e. bodyshop, paint shop, parts department, offices. The chairman of the committee should be a senior member of the company.

Safety committee's main objectives

1   To ensure that the company's premises outside and inside are safe and healthy.
2   To set up an administrative system to maintain a safe working environment.

3 To set up some kind of information system on all matters relating to health and safety.
4 To ensure that employees receive adequate training in health and safety matters.

Regular inspections of all equipment as outlined in the safety practice code are now a legal requirement. Failure to meet this requirement can result in imprisonment or unlimited fines or both.

Employer's duties

To provide and maintain:

1 Safe equipment.
2 Safe place of work with safe access and exit.
3 Safe systems of work.
4 Healthy working environment.
5 Welfare arrangements (washing facilities, eating facilities, first aid).
6 Proper training with time off for this purpose.

Employee's duties

1 To take reasonable care for his/her own safety and the safety of others who may be affected by his/her acts.
2 To cooperate with his/her employer or any person on whom a duty or requirement is imposed.

The company should establish, in conjunction with the safety committee, its own code of practice on safety procedures within its establishment, and all personnel should receive a copy.

Objectives of the Act

1 To repeal, replace or modify existing regulations
2 To maintain or improve standards of health and welfare of people at work
3 To protect people or other workers against any risk to health or safety arising out of the activities of people at work
4 To control the keeping and use of dangerous substances including explosive and highly flammable materials (paints, solvents, resins)
5 To control the emissions of toxic substances into the atmosphere (dust, fumes, smoke, gases).

15.6.4 Control of Substances Hazardous to Health (COSHH) Regulations 1988

Since 1974 a number of Regulations have been introduced which describe in detail the require-ments for specific safe working practices. One of the most recent is the Control of Substances Hazardous to Health (COSHH) Regulations 1988.

COSHH is a major piece of health and safety legislation. It tightens the general obligations of an employer under Section 2 of the Health and Safety at Work Act by specifying comprehensive rules on how substances should be controlled. COSHH applies to all substances in all forms, including gases, vapours, solids, dusts, liquids, and even micro-organisms. It also covers mixtures and preparations.

A substance may be a 'hazard to health' if the substance itself is harmful or if:

1 There are hazards from impurities.
2 Dust or fumes are generated during use.
3 It is dangerous when used in combination with other materials.

COSHH attempts to follow the principle of good occupational hygiene practice:

1 Assess the hazard (Regulation 6)
2 Control it (Regulations 7, 8, 12)
3 Maintain the control (Regulations 9, 10, 11).

Assessment: Regulation 6

An employer shall not carry on any work which is liable to expose any employees to any substance hazardous to health unless he has made a suitable and sufficient assessment of the risk created, and the steps needed to achieve and maintain control.

Control of exposure: Regulation 7

Exposure must be prevented or adequately controlled. 'Adequately controlled' means that repeated exposure must not cause damage to health. Maximum exposure limits must not be exceeded or employers may be prosecuted.

Use of controls: Regulation 8

Employers should take all reasonable steps to ensure that control measures are properly used, and employees shall make full and proper use of the controls provided.

Maintenance of controls: Regulation 9

Controls must be maintained in efficient working order and good repair. Thorough examinations and tests must be made at suitable intervals. Records of tests and repairs must be kept for five years.

Exposed monitoring: Regulation 10

Air sampling is required to monitor exposure every twelve months if:

1 It is needed to maintain control.
2 It is needed to protect health.
3 Listed carcinogens are in use.

Health surveillance: Regulation 11

Health surveillance is required at least every twelve months if:

1 Valid health surveillance techniques exist.
2 Specified substances are in use.

Surveillance may include biological monitoring, clinical examinations and review of health records. Records must be kept for 30 years (Figure 15.11).

HEALTH SURVEILLANCE RECORD

EMPLOYEE'S NAME	TYPE OF SURVEILLANCE	RESULT	DATE CARRIED OUT
MR JOHN SMITH	RESPIRATORY FUNCTION SCREENING.	WITHIN NORMAL LIMITS.	30-1-93
MR DEREK BROWN	CHEST X-RAY	WITHIN NORMAL LIMITS.	6-3-93
MR WILLIAM BLACK	AUDIOMETRIC HEARING TEST.	CATEGORY 5 (NORMAL)	20-4-93

Figure 15.11 Health surveillance record (*Akzo Coatings PLC*)

Information, instruction and training: Regulation 12

The employees must be told:

1 The risk to health
2 The results of the tests
3 The control measures to be used.

They must be given appropriate instruction and training. Any person carrying out work under the Regulations must be competent to do so (Figure 15.12).

Carrying out a COSHH assessment Gather information

1 What is used, handled, stored
2 Intermediates, by-products, wastes.

TRAINING RECORD

EMPLOYEE'S NAME	TYPE OF TRAINING	DATE CARRIED OUT
MR JOHN SMITH	SAFETY TRAINING ON THE USE OF AIR-FED MASKS	1-3-93
MR DEREK BROWN	TRAINING IN THE USE OF PORTABLE FIRE-FIGHTING EQUIPMENT FOR THE WORKSHOP.	10-4-93
MR WILLIAM BLACK	TRAINING IN THE SAFE USAGE OF MIG WELDING EQUIPMENT.	5-5-93

Figure 15.12 Training record (*Akzo Coatings PLC*)

Identify hazards (a hazard is the potential to cause harm)

1 Nature of hazard
2 Route of exposure
3 Possible interactions.

Evaluate the risk (risk is the likelihood that a substance will cause harm in the actual circumstance of use)

1 How is the substance used?
2 How is it controlled?
3 What is the level of exposure?

Decide on controls

1 Type of controls (substitution, local exhaust ventilation (LEV))
2 Maintenance and testing procedures
3 Air monitoring
4 Health surveillance
5 Record keeping.

Record assessment A record must be kept of all assessments other than in the most simple cases that can be easily repeated (Table 15.1, Figure 15.13).

Review Assessments should be reviewed for example when:

1 There are any significant changes to the process, e.g. plant changes, volume of production.
2 New substances are introduced.
3 Ill-health occurs.
4 New information becomes available on hazards or risks.

Table 15.1 Guide to safe working practices: to be filled in by the bodyshop. The example gives a typical process carried out in most paint shops every day. The initial assessment has identified there is a serious risk to peoples' health from the way the hazardous substance is being used. Immediate steps must be introduced to control and reduce the risks. Failure to comply to the approved code of practice by an employer may lead to prosecution

| Activity | Hazard | Risk | Precautions | | Products used | Products contain | Remarks |
			Personal	General			
Wet sanding of fillers and stoppers	Wet sludges of unknown composition	Skin contact with unknown materials Eye contact with splashes of unknown materials Remarks:	Gloves Goggles or visor Conclusions:	Carry out in a properly ventilated area Conclusions:			
Priming Mixing (hardener addition and or thinning)	Solvents and, depending on the particular product, isocyanates, acids or polyamides may be present	Inhalation of solvent vapours Skin contact with solvents Eye contact with vapours, solvents or paint splashes Remarks: High risk Carried out three times daily	Respirator suitable for solvent vapours where extraction inadequate Gloves Goggles or visor Conclusions: Strong smell of solvent from acid hardener No respirator or eye protection used whilst mixing	Carry out in a well ventilated area or with local exhaust ventilation to minimize vapour build-up. NB air-fed respirators are not required when mixing isocyanate containing products at ambient temperatures under conditions of good ventilation Conclusions: No LEV in mixing area	Washfiller 580 Washfiller phosphoric hardener	Zinc chromate <5% MEL Butyl acetate 10–25% OES Plus blends of additional solvents Ethanol 15–30% OES Methyl isobutyl ketone 30–45% OES+ Plus blend of additional solvents including phosphoric acid	Product substitution not possible. Issue staff with respirator and eye protection Arrange for extraction to be installed in mixing room Reduce mixing from three times a day to once daily

(Continued)

Table 15.1 (Continued)

Activity	Hazard	Risk	Precautions		Products used	Products contain	Remarks
			Personal	General			
Application	Spray mist containing solvent and, depending on the particular product, isocyanates, zinc or strontium chromates, polyamides or acids	Inhalation of solvent vapours or spray mists	Respirator suitable for solvent vapours and particulates	All spraying *must* be carried out in a properly ventilated spray enclosure	As above	As above	Issue replacement charcoal respirator, set up inspection and maintenance records. No spraying to be done outside spray booth
			Air-fed respirators *must* be worn (i) whenever isocyanates are present or (ii) when zinc or strontium chromate containing primers are used and the ventilation in the spray enclosure is inadequate for the particular job				
		Skin contact with solvent vapour or spray mists	Gloves				
		Eye contact with solvent vapour or spray mists	Goggles or visor				
		Remarks: High risk Process carried out daily	Conclusions: Cartridge mask used but in poor condition	Conclusions: Spraying in open workshop Poor extraction			

COSHH ASSESSMENT RECORD		RECORD NO.

PROCESS IDENTIFICATION ....

....

PROCESS LOCATIONS ....

FREQUENCY OF USE ....

NUMBER OF OPERATORS ....

SUBSTANCES/PRODUCTS USED	SUPPLIER	HAZARDS	OES	MEL

CONTROLS

PRECAUTIONS CURRENTLY USED ....

HOW EFFECTIVE ARE THEY ....

LIKELY HAZARD ....

– INHALATION ....

– SKIN CONTAMINATION ....

– EYE CONTACT ....

ACTION REQUIRED ....

....

NEW CONTROLS INTRODUCED ....

....

....

IS A MAINTENANCE SCHEDULE REQUIRED	YES	NO

....

HEALTH SURVEILLANCE REQUIRED	YES	NO

....

CARRIED OUT BY ....

SIGNATURE .... DATE ....

Figure 15.13 COSHH assessment record (*Akzo coatings PLC*)

15.6.5 Environment Protection Act 1990

The Environment Protection Act 1990, like the Health and Safety at Work Act, allows the Secretary of State to introduce Regulations to control, amongst other things, the release of harmful substances to air, water and land. The legislation is wide in scope, with the long-term aim of minimizing environment pollution. The paint application industry is among the third largest group of environment polluters, and it is because of this that the EPA introduces many refinishing industry specific controls. The object of the Act is to reduce and eventually eliminate the use of volatile organic compounds (VOCs) in surface coatings, using the best available techniques not entailing excessive costs (BATNEECs). VOCs are substances which react with nitrous oxides in sunlight to create low-level ozone. This photochemical air pollution causes damage to vegetation and can also cause serious breathing difficulties in humans and animals. All users will have to use BATNEECs to prevent, minimize and render harmless any release of VOCs. If there is alleged contravention, the user will have to prove that no better techniques were available.

Bodyshops purchasing more than 2 tonnes of solvent per year will be required to register under the classification of Part B 'The Respraying of Motor Vehicles' of the Act. Any bodyshop currently operating below these levels is not required to register.

Existing bodyshops who registered by the deadline set for their area of the country (in England and Wales by 30 September 1992, and in Scotland by 31 March 1993) have until October 1998 to fulfil all their legal obligations under the Act.

Two separate pollution control regimes are established in Part 1 of the Act to control industrial processes falling into the categories A and B. Category A is integrated pollution control (IPC) operated by HMIP in England and Wales and HMIPI in Scotland. Category B is local authority air pollution control.

Environmental Protection (Prescribed Processes and Substances) Regulations 1991

These identify the respraying of road vehicles as a category B process (local authority air pollution control) if the process may result in the release into the air of any particular matter or of any volatile organic compound (solvents) where the process is likely to involve the use of 2 tonnes or more of organic solvents in any 12 month period. Cleaning agents, fillers, stoppers, primers, gunwash and many other products in the refinishing of motor vehicles all contribute to this figure. The figure should represent the amount consumed in the process, i.e. lost through evaporation, spillage and transfer.

If the process involves the use of 2 tonnes of organic solvents in any 12 month period then the bodyshop operator must seek authorization from his or her local authority to carry out the process. If the operator of the bodyshop calculates the solvent usage to be less than 2 tonnes, then the best advice would be to monitor solvent usage in order to demonstrate the fact to the local authority.

First check with the local authority, who will require information about your operation such as:

1   Who and where you are
2   What you do and how you do it
3   What is released to air, how much, and where-from
4   What you do to prevent, control and monitor such releases
5   Evidence or proposals that the objectives of BATNEEC will be met.

Upgrading of existing bodyshops to comply with the Act

The engineering controls that the EPA imposes upon equipment are very significant, particularly regarding spray booths if emission exceeds the 2 tonnes VOCs limit. For spray booths already working there is a requirement to submit an upgrading programme to the local authority within 12 months of the initial registration. That programme must be implemented by 1 April 1998.

The requirements for upgrading are as follows:

1   Filtration systems have to ensure that the concentration of paint particulate matter in the final discharge to atmosphere from the spray booth does not exceed 10 mg/m^3.
2   Manufacturers or companies upgrading booths will be required to provide a guarantee confirming that the equipment conforms to the emissions limit.
3   The vent velocity of the extract duct must achieve a minimum of 15 m/s for dry filter systems.
4   The vent velocity of the extract duct must achieve a maximum of 9 m/s where a wet method is used.
5   Pressure control systems must be provided which will shut down the spray booth if it is over-pressurized and activate an audio alarm.

6   Exhaust duct openings must not be fitted with plates, caps or cowls that could act as restrictors or deflectors.

7   Exhaust ducting (the chimney) must be a minimum of 8 metres above ground level, but it also has to be a minimum of 3 metres higher than the roof height of any nearby building which is within a distance of five times the uncorrected chimney height.

8   Preventive measures must be taken for fugitive emissions of odour, fumes and particulate matter from mechanical operations like welding, grinding or sanding. All these activities should take place in a building.

9   Shot blasting emissions must not exceed 50 mg/m^3.

10 Paint mixing and equipment cleaning are to take place in an adequately ventilated area.

11 An automatic totally enclosed machine for cleaning spray guns and equipment must be provided.

12 Spray gun testing must be into an extracted area.

Waste management: EPA Part 11

Section 34 of the EPA makes it a criminal offence to 'treat, keep or dispose of controlled waste in a manner likely to cause pollution of the environment or to harm human health'. In order to meet this stated aim, the Part 11 provisions impose a 'duty of care' on everybody involved in the chain of waste management. This means that if you are involved in the creation of waste, then you are responsible for its safe and proper disposal. Typical relevant bodyshop waste includes dirty solvent, paint residues, empty cans and dirty rags.

   Those subject to the duty of care must try to achieve the following four things:

1   To prevent any other person committing the offence of disposing of 'controlled waste', or treating it, or storing it without a waste management licence; or breaking the conditions of a licence; or in a manner likely to cause pollution or harm to health.

2   To prevent the escape of waste which is, or at any time has been, under their control (this has implications for waste storage facilities and waste containment).

3   To ensure that if waste is transferred it goes only to an 'authorized person' or to a person for 'authorized transport purposes'.

4   When waste is transferred, to make sure that there is also transferred a written description of the waste, which is good enough to enable each person receiving it to avoid committing any offence under 1, and to comply with the duty at 2 to prevent the escape of waste.

Under the code of practice, the waste must be:

1   Identified (paint, solvent, paper and tape, dust, loaded extract filter media, scrap metal, tyres and batteries)

2   Categorized

3   Kept in appropriate containers (external skip, internal container, paper baler, can crusher, solvent/paint 'closed loop' system, parts for recycling)

4   Collected and disposed of by a registered operator (transfer note signed by the disposer and collector, and a written description of the waste)

5   Documented at all stages.

15.6.6 Management of Health and Safety at Work Regulations 1992

These Regulations set out broad general duties which apply to almost all work activities in Great Britain. They are aimed mainly at improving health and safety management, and can be seen as a way of making more explicit what is required of employers under the HSW Act. The main provisions are designed to encourage a more systematic and better organized approach to deal with health and safety.

   The Regulations require that employers should:

1   Assess the risk to health and safety of employees and of anyone else who may be affected by the company's work activity. This is so that the necessary preventive and protective measures can be identified. Employers with five or more employees will have to record the significant findings of the assessment.

2   Make arrangements for putting into practice the health and safety measures that follow from the assessment.

3   Provide appropriate health surveillance for employees where the risk assessment shows it to be necessary.

4   Appoint competent people either from inside the organization or from outside to help devise and apply measures needed to comply with duties under health and safety law.

5 Set up emergency procedures.
6 Provide employees with information they can understand about health and safety matters.
7 Cooperate with other employers sharing the work site.
8 Make sure that employees have adequate health and safety training and are capable enough at their jobs to avoid risk.
9 Provide temporary workers with some particular health and safety information to meet special needs.

The regulations will also:

1 Place duties on the employees to follow health and safety instructions and report danger.
2 Extend the current law which requires the employer to consult employees' safety representatives and provide facilities for them.

15.6.7 Provision and use of Work Equipment Regulations 1992

These regulations will place general duties on employers, and list minimum requirements for work equipment to deal with selected hazards whatever the industry.

In general the Regulations will make explicit what is already somewhere in the law or is good practice. If equipment has been well chosen and well maintained there should be little need to do more than follow the guidance on the Regulations. Some older equipment may need to be upgraded to meet the minimum requirements, but the company will have until 1997 to do any necessary work.

Work equipment will be broadly defined to include everything from a hand tool, through machines of all kinds, to a complete plant. Use of the equipment will include starting, stopping, repairing, modifying, installing, dismantling, programming, setting, transporting, maintaining, servicing and cleaning.

The general duties will require the employer to:

1 Make sure that equipment is suitable for the use that will be made of it.
2 Take into account the working conditions and hazards in the workplace when selecting equipment.
3 Ensure that equipment is used only for operations for which and under conditions for which it is suitable.

4 Ensure the equipment is maintained in an efficient state, in efficient working order and in good repair.
5 Give adequate information, instruction and training.
6 Provide equipment that conforms with EC product safety Directives (see below).

Specific requirements will cover:

1 Guarding of dangerous parts of machinery (replacing the current law on this)
2 Protection against specific hazards, i.e. falling/ejected articles and substances, rupture/disintegration of work equipment parts, equipment catching fire or overheating, unintended or premature discharge of articles and substance, explosion
3 Working equipment parts and substances at high or very low temperatures
4 Control systems and control devices
5 Isolation of equipment from the source of energy
6 Stability of equipment
7 Lighting
8 Maintaining operations
9 Warnings and markings.

The Regulations implement an EC Directive aimed at the protection of workers. There are other Directives setting out conditions which much new equipment (especially machinery) will have to satisfy before it can be sold in the EC member states. They will be implemented in the UK by Regulations made by the Department of Trade and Industry.

15.6.8 Personal Protective Equipment (PPE) at Work Regulations 1992

These Regulations set out in legislation sound principles for selecting, providing, maintaining and using PPE. They do not replace recently introduced law dealing with PPE, for example COSHH or noise at work regulations.

PPE is defined as all equipment designed to be worn or held to protect against a risk to health or safety. This includes most types of protective clothing and equipment such as eye, foot and head protection, safety harnesses, life jackets and high-visibility clothing. PPE should be relied upon only as a last resort, but where risks are not adequately controlled by other means the employer will have a duty to ensure that suitable PPE is provided free of charge for employees exposed to these risks. The Regulations say what is meant by suitable PPE, a key point in making sure that it effectively protects the wearer. PPE will only be suitable if it is appropriate for the risks and the working conditions, takes account of workers' needs, and fits properly, gives adequate protection, and is compatible with any other item of PPE worn.

Employers also have duties to:

1   Assess the risks and PPE intended for issue, to ensure that it is suitable.
2   Maintain, clean and replace PPE.
3   Provide storage for PPE when it is not being used.
4   Ensure that PPE is properly used.
5   Give training, information and instruction to employees on its use and how to look after it.

PPE is also subject to a separate EC Directive on design, certification and testing, and will be marked by the manufacturer with a CE mark.

15.6.9 Manual Handling Operations Regulations 1992

These will apply to any manual handling operations which may cause injury at work. These operations will be identified by the risk assessment carried out under the Management of Health and Safety at Work Regulations 1992. They will include not only lifting of loads, but also lowering, pushing, pulling, carrying, or moving them, whether by hand or other bodily force.

Employers will have to take three key steps:

1   Avoid hazardous manual handling operations wherever reasonably practicable. Consider whether the load must be moved at all and, if it must, whether it can be moved mechanically.
2   Assess adequately any hazardous operations that cannot be avoided. An ergonomic assessment should look at more than just the weight of the load. Employers should consider the shape and size of the load, the way the task is carried out, the handler's posture, the working environment (cramped or hot), the individuals's capacity, and the strength required. Unless the assessment is very simple, a written record will be needed.
3   Reduce the risk of injury as far as reasonably practicable. A good assessment will not only show whether there is a problem but will also point to where the problem lies.

15.6.10 Workplace (Health, Safety and Welfare) Regulations 1992

The Regulations will cover many aspects of health, safety and welfare in the workplace. Some of them are not explicitly mentioned in the current law, though they are implied in the general duties of the Health and Safety at Work Act.

The regulations will set out general requirements in four broad areas.

Working environment

1   Temperature in indoor workplaces.
2   Ventilation.
3   Lighting including emergency lighting.
4   Room dimensions and space.
5   Suitability of workstations and seating.

Safety

1   Safe passage of pedestrians and vehicles (traffic routes must be wide enough and marked where necessary, and there must be enough of them).
2   Windows and skylights (safe opening, closing and cleaning).
3   Transparent and translucent doors and partitions (use of safety material and marking).
4   Doors, gates, escalators (safety devices).
5   Floors (construction and maintenance, obstructions and slipping and tripping hazards).
6   Falling a distance into a dangerous substance.
7   Falling objects.

Facilities

1   Toilets.
2   Washing, eating and changing facilities.
3   Clothing storage.
4   Drinking water.
5   Rest area (and arrangements to protect people from the discomfort of tobacco smoke).
6   Rest facilities for pregnant women and nursing mothers.

Housekeeping

1   Maintenance of workplace, equipment and facilities.
2   Cleanliness.
3   Removal of waste materials.

Employers will have to make sure that any workplace within their control complies with the Regulations. Existing workplaces will have until 1996 to comply.

15.6.11 Risk assessment

For every task which is carried out in the workshop a risk assessment should be carried out. Risk assessment is about identifying hazards and the severity of the hazard, deciding on the level of likelihood of the occurrence and setting control to deal with them. For instance a can of solvent with a secure top in a cupboard is relatively safe. The same can of solvent left without a top and placed on the floor is likely to get kicked over, or the vapour could cause a fire. That is, it now has a potential to cause a major injury and is very likely to happen.

UNISON's 25 steps to risk assessment

UNISON have produced a step-by-step approach to risk assessment, the 25 steps are:

1   Set up a programme of risk assessments.
2   Consult safety representatives about the appointment of competent persons.
3   Appoint competent people.
4   Decide on methods and approach.
5   Identify any other specific health and safety legislation which applies.
6   Collect information.
7   Consult safety representatives and employees about work and perceived hazards.
8   Observe what happens in practice.
9   Identify hazards.
10 Identify the harm that could arise from hazards.
11 Identify those at risk.
12 Identify how they may be harmed.
13 Evaluate the likelihood of harm occurring.
14 Evaluate the likely severity of the harm.
15 Evaluate the likely numbers who could be harmed.
16 Identify the control measures already in place, including information, instruction and training.
17 Evaluate the effectiveness of the control measures.
18 Decide what more needs to be done to eliminate or control risks, in accordance with the accepted priorities of risk prevention and control measures.
19 Record the assessments.
20 Provide safety representatives with copies of the assessments and supporting information.
21 Draw up an action plan and prioritize risks to be tackled.
22 Draw up a timetable for completion of action (action plan or development plan).
23 Allocate financial and staff resources for carrying out the action plan.
24 Implement measures.
25 Monitor the effectiveness of control measures and review the risk assessments at agreed regular intervals and whenever changes require it.

Risk grid

One way to assess risks is to give both the severity of the hazard and the likelihood of occurrence, a factor between one and five. Multiply the factors to get a severity of risk. This will give you a matrix from 1 to 25 (Table 15.2). That is one is low risk; 25 means somebody is likely to be killed.

Possible hazards

Hazards come in different forms. The following lists are not exhaustive, but give a good framework to start from.

Table 15.2   Risk grid

		Hazard security index				
		Death	Major injury	Significant injury	Minor injury	Trivial injury
Likelihood of occurrence	Very likely	25	20	15	10	5
	Likely	20	16	12	8	4
	Quite possible	15	12	9	6	3
	Possible	10	8	6	4	2
	Not likely	5	4	3	2	1

Physical hazards

- Asbestos
- Awkward posture
- Chemicals
- Display screen equipment
- Drugs
- Electricity
- Fire
- Lead
- Machinery
- Manual handling
- Noise
- Paint
- Petrol/diesel
- Radiation
- Slips, trips and falls
- Solvents
- Transport
- Vibration

Biological hazards

- Animal allergens
- Hepatitis

- HIV
- Legionnaires disease
- MSRA
- Plant allergens
- Tuberculosis
- Weil's disease

Psychosocial hazards

- Boredom
- Bullying
- Isolation
- Lack of control
- Lack of support
- Long hours
- Monotonous work
- Shift work
- Stress
- Violence
- Work overload

Using this information the Risk Assessment Form (Figure 15.14) can be completed. Such a form should be completed for each discrete task in the

Brooklandsgreen Motor Company

BODYSHOP RISK ASSESSMENT

Support Team:		SAFETY MANAGER:		DATE:		
ACTIVITY	HAZARDS	RISK FACTOR High/Med/ Low	EXISTING OR PLANNED CONTROLS	THOSE AT RISK	IF NOT CONTROLLED, NOTE ACTION REQUIRED	

Figure 15.14   Risk assessment form

bodyshop. Remember, if in doubt, don't do it no matter what the reward.

15.6.12 Further reading

One thing that any person working in a bodyshop must always bear in mind is that ignorance is no excuse in the eyes of the law. That is, it is, whether you are the newest apprentice or the oldest manager, you have a legal requirement to comply with the relevant regulations. Most prosecutions by the Heath and Safety Executive (HSE) are under the 1974 Health and Safety at Work Act; this is because it is a catch-all. However, it is important to keep up to date with the latest regulations. The HSE publishes books on all the regulations, obviously not all of them are relevant to the bodyshop, current ones, other than those already mentioned, which have relevance to the bodyshop of a general nature are:

- Health and Safety in Engineering Workshops (2002) ISBN 0 7176 1717 3
- Safety Representatives and Safety Committees (1996) ISBN 0 7176 1220 1
- Risk Management (1995) ISBN 0 7176 0905 7
- Safe Use of Work Equipment (1999) ISBN 0 7176 1626 6
- The Law on VDUs (2003) 0 7176 2582 6
- Work with Display Screens (2002) 0 7176 2582 6
- Management of Health and Safety at Work (2002) 0 7176 2488
- Personal Protective Equipment at Work (2000) 0 7176 0415 2

The HSE website can be found at www.hse.gov.uk. On the HSE website can be found lists of all the latest publications and news items on current developments as well as other relevant information.

15.7 Quality Management for the bodyshop: BS EN ISO 9001–2000

The motor industry is under increasing pressure from its customers to show that it is able to guarantee a high standard for its repairs. Increasingly, customers are expecting their suppliers to have formal and verifiable quality management systems as a guarantee that work will be completed with an acceptable level of assurance of quality. This expectation applies as much to smaller firms as it does to large.

Some of Britain's largest organizations now include BS EN ISO 9001–2002, or other acceptable quality management systems, as a contractual requirement for their suppliers. Many other organizations give preferential treatment to suppliers who show a visible commitment to quality by pursuing this British Standard, BS EN ISO 9001–2000.

An increasing number of companies in the body repair industry are now seeking registration to BS EN ISO 9001–2000, the national standard for quality management systems. Registration is being sought by vehicle manufacturers and suppliers of parts and equipment, by car dealers, by garages offering service and repair and by body repair shops.

The assurance of quality is needed to eliminate failures and complaints and their associated costs as far as possible. By operating a management system according to the defined criteria of a standard such as BS EN ISO 9001–2000, you have confidence that you have taken all reasonable steps to guarantee quality to your customer, and you provide evidence of your commitment to do so. Quality assurance systems provide the mechanism for ensuring that a business satisfies the needs of its customers. BS refers to British Standards Institute, other countries may just use the term ISO 9001.

15.7.1 Requirements of BS EN 9001–2000

BS EN 9001–2000 identifies basic disciplines, procedures and criteria to ensure that products and services meet the customer's requirements. It is the British Standards Institution's (BSI) assurance of quality and can be applied to all industries. All accredited companies are permitted to display the easily recognizable BSI symbol.

The main requirement of BS EN ISO 9001–2000 is the procedures manual. All companies have to prepare a manual which meets the 18 working procedures required by BSI in order to conform:

1 Management responsibilities
2 Quality systems
3 Contract review
4 Document control
5 Purchasing

6 Purchaser supplied products
7 Product identification and traceability
8 Process control (workshop procedures for body repair, welding and painting)
9 Inspection and testing
10 Inspection, measuring and test equipment (to include spray booths, welding equipment, body jigs, measuring equipment)
11 Inspections and test status
12 Control of non-conforming products
13 Corrective action
14 Handling, storage, packing and delivery
15 Quality records
16 Internal quality audits
17 Training
18 Statistical techniques.

Service departments and bodyshops **must** establish:

(a) What repairs each customer wants on the vehicle
(b) a procedure for the ordering of necessary parts
(c) A check that all tools are calibrated for use by trained and authorized staff
(d) The production of a workshop manual covering repairs and welding techniques used in the bodyshop
(e) The production of a quality systems manual which will include all 18 procedures.

In order to enable this system to work, documented records must be kept at all stages of the repair from the car being booked in for repair until it is returned to the customer.

15.7.2 Benefits of BS EN ISO 9001–2000 for bodyshops

1 BSI certification.
2 The use of the BSI logo on publicity material.
3 Customers accept BSI certification as proof of quality commitment.
4 Customers become less likely to look for a special assessment of work done.
5 Quality performance will improve customer satisfaction.
6 The company will have the confidence of knowing that its quality system has been externally assessed and approved.
7 Staff will work better in an organized working situation.

15.7.3 The bodyshop's action plan

1 Accept the opportunity to move towards BSI approval.
2 Provide the environment in which the participants can work together towards the required standard.
3 Provide the expertise and equipment needed to work through all the necessary stages to produce a documented quality system which will include a quality manual, supporting procedures, working instructions and other associated documents.
4 Provide quality training as required by the participants.
5 Provide support throughout the whole procedure.

To achieve the action plan a company can:

1 Undertake the development work itself and conduct an internal quality improvement project.
2 Collaborate with other local businesses and work together to achieve the standard with external support.
3 Pursue BS EN ISO 9001–2000 using external consultancy support.

If it is decided to develop and implement a formal quality system in the company, a controlled, logical action plan should be followed. There is always a tendency to make a system too elaborate; the aim should be to develop a system which consistently gives the degree of assurance your business requires.

A suggested plan of action for the development, implementation and certification of any quality management system is as follows:

1 Establish the company's commitment to quality and the achievement of BS EN ISO 9001–2000 standards.
2 Review the existing arrangements for quality management, and identify a company quality development plan with a controlled implementation schedule with target dates for the completion of essential stages.
3 Define and document the system.
4 Apply for assessment and certification to the British Standards Institution.
5 Implement and monitor the system over a period of at least three months.
6 Maintain certification by reviewing the system at predetermined intervals, conducting system audits, and maintaining a disciplined approach to quality assurance.

15.7.4 Implementation and assessment of BS EN ISO 9001–2000

Quality assurance personnel, line managers and supervisors will be required to document their quality systems and practices into manuals and procedures to comply with BS EN ISO 9001–2000. Within the standard, the following areas are addressed through an organized approach to the task:

Background on documenting a system
Consideration of all the elements
Understanding the written task
Support base necessary
Controls on procedures
Getting everyone involved
Contents of procedures.

The bodyshop's company programme should include:

Quality policy manual
Quality procedures manual
Workshop procedures manual
Welding test procedures manual.

BS EN ISO 9001–2000 also supplies the basic management controls upon which the business can build a documented management system, allowing recognition and acceptance by a wider number of people within the vehicle industry and by those wishing to either supply or purchase from the company.

To meet the requirements of BS EN ISO 9001–2000, the business must ensure that complete controls exist for such areas as workshop loading and receiving customer orders. Systems must also be in place to show that if faults do occur, positive actions are taken to ensure that they do not occur again.

Those who have already met the requirements of BS EN ISO 9001–2000 and passed assessment by BSI Quality Assurance agree that a considerable effort is required. However, the real benefits in efficiency and staff morale have made it worthwhile.

BS EN ISO 9001–2000 is about the maintenance of systems which work to ensure the provision of a documented quality service. The procedures manual is the starting point for this:

1 Are you doing what your manual says you should do?
2 Can you prove that your procedures are correct?
3 Are your staff trained professionals who take quality seriously?

The BSI needs evidence and history, and so the assessors look at all the paperwork in the relevant departments to search for major or minor non-conformances. A minor non-conformance would be a small error which would be acceptable in isolation. A major non-conformance would be an error with possible adverse effect on quality. The assessors are looking for a system that will identify problems at an early stage and take action to solve them.

15.7.5 What help do consultants offer?

Consultants offer a support and advisory service to companies pursuing BS 5750 or looking to implement a quality improvement programme. This includes:

1 Consultancy support to develop company action plans and the production of quality manuals including standard operating procedures, policy statements and all other mandatory documentation required by the BSI.
2 Help with implementation, including training for management and staff, production of company literature, contact with suppliers and customers.
3 Support for the company and liaison with the BSI and/or independent assessors and certification bodies.
4 Advice on the maintenance of the system once it is installed.
5 Estimate of cost including running costs.

Questions

1 Explain the role of a bodyshop planning consultant in the planning of a new workshop.

2 Explain the importance of the choice of site in locating a new workshop.

3 State the legal requirements necessary when building a new workshop.

4 Define the following abbreviations: COSHH, HASAWA, EPA.

5 Explain the importance of entrances, roof supports and floor space when planning a workshop.

6 List the necessary areas to be considered for inclusion in a new workshop.

7 Give reasons for the difference in floor areas between a panel bay and a bay with a static jig.

8 Discuss the importance of the reception area as part of the workshop.

9 Why should a new workshop have a separate vehicle assessment area?

10 What are the four types of heating system that could be used in workshops?

11 Explain why bodyshop lighting plays a crucial role in the repair and final finish of a vehicle.

12 Make a list of the essential equipment needed for the stripping, repairing and painting areas.

13 Make a list of hand tools, power tools and safety equipment needed in a workshop.

14 Explain the importance of dust and fume extraction systems.

15 Name five of the main health and safety legislation Acts which affect the workshop.

16 Discuss volatile organic compounds (VOCs) and their effect on the environment.

17 Which Act applies to waste management?

18 Explain the importance of the Personal Protective Equipment (PPE) at Work Regulations 1992.

19 Explain what is meant by the Manual Handling Operation Regulations 1992.

20 Explain the term BS EN ISO 9001–2000 and its relevance to bodyshops.

Reinforced composite materials

16.1 Development of reinforced composite materials

Glass fibres have been known in one form or another since 1500 BC, and some glass fabrics have been produced as far back as the beginning of the eighteenth century, but all of these were far too coarse to be of any industrial value. During the late 1930s glass fibre became available on a commercial scale and of a quality and smallness of diameter which enabled it to be manufactured into textile products. This was due to the development of the continuous filament process. It was soon apparent that the fibres made by this new process had many desirable properties such as strength, smallness of diameter, high elasticity, and the ability to withstand high temperatures.

Early attempts to use glass fibre as a reinforcement were disappointing. The resin then being used required high moulding pressures and this led to a crushing of the fibres with a resulting loss in strength. During the early 1940s an entirely new group of resins was introduced; these were known as contact resins as they could be used without pressure. Of the contact resins the polyesters are undoubtedly the best known, and they are the most widely used resins in the preparation of glass reinforced laminates because of their combination of low cost and ability to be moulded without pressure, and because their conditions of use fitted into normal workshop practice. With glass fibre as a reinforcement they could be used by relatively unskilled labour to produce strong, lightweight structures in complicated shapes which would be either too expensive to produce by panel beating or too complicated for pressing. The glass fibre reinforced plastic,

having characteristics of high strength, low weight, resistance to corrosion, good electrical properties and design flexibility, together with its suitability for a wider range of moulding methods, has resulted in a rapid growth of its use. A big advantage was the low tooling cost needed for limited production.

In the automobile field the availability of resins of various types and an extensive range of reinforcing materials has widened the scope of the designer. Increased production of polyester, although small compared with many other resins, has permitted raw material costs to be reduced sufficiently to make the polyester/glass combination competitive with traditional materials for the manufacture of many components.

Although glass reinforced laminates are characterized by high strength/weight ratio, it is well known that they do not offer the same flexural strength as steel and aluminium. Thus for bodywork the successful application of the materials is largely dependent upon careful design and a full understanding of their properties. To the body designer one of the most outstanding characteristics of these materials is the ease with which complex curves can be produced as compared with methods of panel beating and wheeling, while from a production point of view the fact that complex structures can be produced as one-piece mouldings can greatly reduce assembly time and the need for complicated assembly fixtures.

Although this system would appear advantageous by reducing the large number of assembly operations required for building a steel body, it must be realized that in the layout of a steel body plant little or no allowance is made for processing time and that the speed of production is determined only by the time required for handling and transporting panels from one stage to the next. When time has to be allowed for curing, and with resins this may be considerable, production speed is greatly reduced and the floor area requirements may make the use of these materials prohibitive for quantity production.

The last few years have seen a rapidly increasing acceptance of the material as an ideal alternative to steel in the motor industry. Use of glass fibre as a replacement for steel in body panels is economically possible at comparatively low volume levels, say 1000 vehicles a year. The reason is that, unlike steel bodies, glass fibre panels do not require a large investment in presses and dyes, and the moulds to produce the panels, as well as the jigs and fixtures to manufacture the vehicle, can also largely be produced from glass fibre.

From the manufacturer's viewpoint, glass fibre production means low tooling costs, comparatively small initial investment, flexibility of design, and strength with lightness (Table 16.1). The user benefits by the complete resistance of glass fibre vehicles to corrosion, their sturdiness in a collision, and ease of repair.

16.2 Basic principles of reinforced composite materials

The basic principle involved in reinforced plastic production is the combination of polyester resin and reinforcing fibres to form a solid structure (Figure 16.1). Glass reinforced plastics are essentially a family of structural materials which utilize a very wide range of thermoplastic and thermosetting resins. The incorporation of glass fibres in the resins changes them from relatively low-strength, brittle materials into strong and resilient structural materials. In many ways glass fibre reinforced plastic can be compared to concrete, with the glass fibres performing the same function as the steel reinforcement and the resin matrix acting as the concrete. Glass fibres have high strength and high modulus, and the resin has low strength and low modulus. Despite this the resin has the important task of transferring the stress from fibre to fibre, so enabling the glass fibre to develop its full strength.

Polyester resins are supplied as viscous liquids which solidify when the actuating agents, in the form of a catalyst and accelerator, are added. The proportions of this mixture, together with the existing workshop conditions, dictate whether it is cured at room temperature or at higher temperatures and also the length of time needed for curing. In common practice pre-accelerated resins are used, requiring only the addition of a catalyst to affect the cure at room temperature. Glass reinforcements are supplied in a number of forms, including chopped strand mats, needled mats, bidirectional materials such as woven rovings and glass fabrics, and rovings which are used for chopping into random lengths or as high-strength directional reinforcement. Other materials needed

Table 16.1 Advantages of reinforced composite plastics (*Owens-Corning Fiberglas*)

Compared with metals	Compared with laminated thermoplastics	Compared with injection moulded thermoplastics	Compared with injection moulded thermosets	Compared with wood
1 Higher strength weight ratio	1 Greater scope in mouldable shapes	1 Far higher strength	1 Far higher strength	1 Much higher strength
2 Easier and cheaper manufacture of complex shapes	2 Higher strength	2 Improved dimensional stability	2 Ability to be formed into thin flat sections or panels	2 Greatly increased strength/weight ratio
3 Good corrosion resistance	3 Comparable or better electrical properties	3 Higher temperature resistance		3 Improved dimensional stability
4 Ability to incorporate self colours				4 Better weathering properties
				5 Higher water resistance
				6 Ease of fabricating complete structures

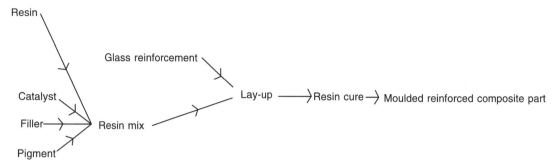

Figure 16.1   Flow chart showing the principles of reinforced composite materials

are the releasing agent, filler and pigment concentrates for the colouring of glass fibre reinforced plastic.

Among the methods of production, the most used method is that of contact moulding, or the wet laying-up technique as it is sometimes called. The mould itself can be made of any material which will remain rigid during the application of the resin and glass fibre, which will not be attacked by the chemicals involved, and which will also allow easy removal after the resin has set hard. Those in common use are wood, plaster, sheet metal and glass fibre itself, or a combination of these materials. The quality of the surface of the completed moulding will depend entirely upon the surface finish of the mould from which it is made. When the mould is ready the releasing agent is applied, followed by a thin coat of resin to form a gel coat. To this a fine surfacing tissue of fibre glass is often applied. Further resin is applied, usually by brush, and carefully cut-out pieces of mat or woven cloth are laid in position. The use of split washer rollers removes the air and compresses the glass fibres into the resin. Layers of resin and glass fibres are added until the required thickness is achieved. Curing takes place at room temperature but heat can be applied to speed up the curing time. Once the catalyst has caused the resin to set hard, the moulding can be taken from the mould.

16.3 Manufacture of reinforced composite materials

When glass is drawn into fine filaments its strength greatly increases over that of bulk glass. Glass fibre is one of the strongest of all materials. The ultimate tensile strength of a single glass filament (diameter 9–15 micrometres) is about 3447000 kN/m^2. It is made from readily available raw materials, and is non-combustible and chemically resistant. Glass fibre is therefore the ideal reinforcing material for plastics. In Great Britain the type of glass which is principally used for glass fibre manufacture is E glass, which contains less than 1 per cent alkali borosilicate glass. E glass is essential for electrical applications and it is desirable to use this material where good weathering and water resistance properties are required. Therefore it is greatly used in the manufacture of composite vehicle body shells, both for private and for commercial vehicles.

Basically the glass is manufactured from sand or silica and the process by which it is made proceeds through the following stages:

1   Initially the raw materials, including sand, china clay and limestone, are mixed together as powders in the desired proportions.
2   The 'glass powder', or frit as it is termed, is then fed into a continuous melt furnace or tank.
3   The molten glass flows out of the furnace through a forehearth to a series of fiberizing units usually referred to as bushings, each containing several hundreds of fine holes. As the glass flows out of the bushings under gravity it is attenuated at high speed.

After fiberizing the filaments are coated with a chemical treatment usually referred to as a forming size. The filaments are then drawn together to form a strand which is wound on a removable sleeve on a high-speed winding head (Figure 16.2). The basic packages are usually referred to as cakes and

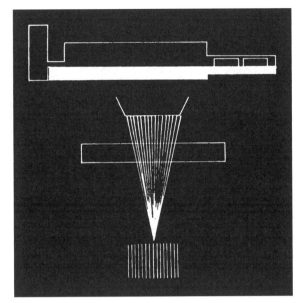

Figure 16.2 Manufacture of glass fibre. The story starts with molten glass, heated in a modern furnace. As it is pulled through tiny holes, that liquid mass is transformed into fibres smaller in diameter than a human hair. These fine fibres are then assembled into textile yarn, or into reinforcement products, including glass fibre roving, mat, chopped strands, glass web and milled fibres material (*Owens-Corning Fiberglas*)

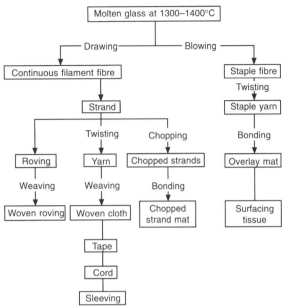

Figure 16.3 Derivation of glass fibre reinforcement

form the basic glass fibre which, after drying, is processed into the various reinforcement products (Figure 16.3). Most reinforcement materials are manufactured from continuous filaments ranging in fibre diameter from 5 to 13 micrometres. The fibres are made into strands by the use of size. In the case of strands which are subsequently twisted into weaving yarns, the size lubricates the filaments as well as acting as an adhesive. These textile sizes are generally removed by heat or solvents and replaced by a chemical finish before being used with polyester resins. For strands which are not processed into yarns it is usual to apply sizes which are compatible with moulding resins.

Glass reinforcements are supplied in a number of forms, including chopped strand mats, needled mats, bidirectional materials such as woven rovings and glass fabrics, and rovings which are used for chopping into random lengths or as high-strength directional reinforcements (Figures 16.4–16.7).

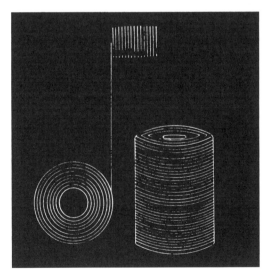

Figure 16.4 Manufacture of rovings. Rovings can be supplied to suit a variety of processes, including projection moulding, continuous laminating, filament winding, pultrusion and as reinforcement for sheet moulding compound. These rovings consist of continuous glass strands, gathered together without any mechanical twist and wound to form a tubular, cylindrical package (*Owens-Corning Fiberglas*)

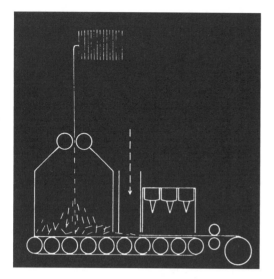

Figure 16.5 Manufacture of mats. Glass fibre mats are used as resin reinforcement in contact and compression moulded applications. Chopped strand mats, made from fine chopped glass strands bonded with a powder of emulsion binder, are used in both areas (*Owens-Corning Fiberglas*)

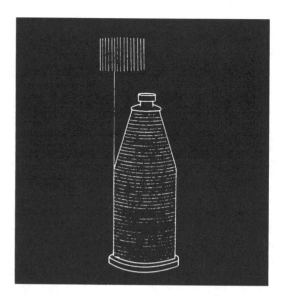

Figure 16.7 Manufacture of yarn (*Owens-Corning Fiberglas*)

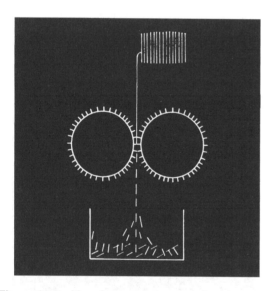

Figure 16.6 Manufacture of chopped strands. Chopped strands are widely used to reinforce thermo-plastic compound (GRTP), polyester bulk moulding compounds (BMC) and in the manufacture of wet-laid glass webs. Chopped strands consist of continuous glass strands chopped to a desired length and are available with a wide veriety of surface treatments to ensure compatibility with most resin systems. They are generally solvent and heat resistant and offer excellent flow properties (*Owens-Corning Fiberglas*)

16.4 Types of reinforcing material

16.4.1 Woven fabrics

Glass fibre fabrics are available in a wide range of weaves and weights. Lightweight fabrics produce laminates with higher tensile strength and modulus than heavy fabrics of a similar weave. The type of weave will also influence the strength (due, in part, to the amount of crimp in the fabric), and usually satin weave fabrics, which have little crimp, give stronger laminates than plain weaves which have a higher crimp. Satin weaves also drape more easily and are quicker to impregnate. Besides fabrics made from twisted yarns, it is now the practice to use woven fabrics manufactured from rovings. These fabrics are cheaper to produce and can be much heavier in weight (Figure 16.8).

16.4.2 Chopped strand mat

Chopped strand glass mat (CSM) is the most widely used form of reinforcement. It is suitable for moulding the most complex forms. The strength of laminates made from chopped strand mat is less than that with woven fabrics, since the

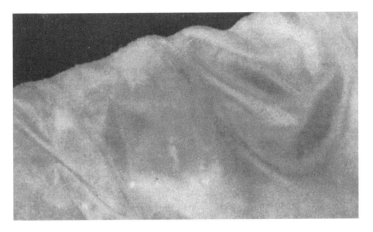

Figure 16.8 Woven fabric (*Scott Bader Co. Ltd*)

glass content which can be achieved is considerably lower. The laminates have similar strengths in all directions because the fibres are random in orientation. Chopped strand mat consists of randomly distributed strands of glass about 50 mm long which are bonded together with a variety of adhesives. The type of binder or adhesive will produce differing moulding characteristics and will tend to make one mat more suitable than another for specific applications.

16.4.3 Needle mat

This is machanically bound together and the need for an adhesive binder is eliminated. This mat has a high resin pick-up owing to its bulk, and cannot be used satisfactorily in moulding methods where no pressure is applied. It is used for press moulding and various low-pressure techniques such as pressure injection, vacuum and pressure bag.

16.4.4 Rovings

These are formed by grouping untwisted strands together and winding them on a 'cheese'. They are used for chopping applications to replace mats either in contact moulding (spray-up), or translucent sheet manufacture of press moulding (preform). Special grades of roving are available for each of these different chopping applications. Rovings are also used for weaving, for filament winding and for pultrusion processes. Special forms are available to suit these processes (Figure 16.9).

Figure 16.9 Rovings (*Owens-Corning Fiberglas*)

16.4.5 Chopped strands

These consist of rovings prechopped into strands of 6 mm, 13 mm, 25 mm or 50 mm lengths. This material is used for dough moulding compounds, and in casting resins to prevent cracking (Figure 16.10).

16.4.6 Staple fibres

These are occasionally used to improve the finish of mouldings. Two types are normally available, a compact form for contact moulding and a soft bulky form for press moulding. These materials are frequently used to reinforce gel coats. The weathering properties of translucent sheeting are considerably improved by the use of surfacing tissue (Figure 16.11).

Figure 16.10 Chopped strands (*Owens-Corning Fiberglas*)

Figure 16.11 Staple fibres (surface mat) (*Owens-Corning Fiberglas*)

16.4.7 Application of these materials

Probably chopped strand mat is most commonly used for the average moulding. It is available in several different thicknesses and specified by weight: 300, 450 and 600 g/m². The 450 g/m² is the most frequently used, and is often supplemented with the 300 g/m². The 600 g/m² density is rather too bulky for many purposes, and may not drape as easily, although all forms become very pliable when wetted with the resin.

The woven glass fibre cloths are generally of two kinds, made from continuous filaments or from staple fibres. Obviously, most fabricators use the woven variety of glass fibre for those structures that are going to be the most highly stressed. For example, a moulded glass fibre seat pan and squab unit would be made with woven material as reinforcement, but a detachable hard top for a sports car body would more probably be made with chopped strand mat as a basis. However, it is quite customary to combine cloth and mat, not only to obtain adequate thickness, but because if the sandwich of resin, mat and cloth is arranged so that the mat is nearest to the surface of the final product, the appearance will be better.

The top layer of resin is comparatively thin, and the weave of cloth can show up underneath it, especially if some areas have to be buffed subsequently. Chopped fibres do not show up so prominently, but some fabricators compromise by using the thinnest possible cloth (surfacing tissue as it is known) nearest the surface, on top of the chopped strand mat. When moulding, these orders are of course reversed, the tissue going on to the gel coat on the inside of the mould, followed by the mat and resin lay-up.

It is important to note that if glass cloths or woven mat are used, it is possible to lay up the materials so that the reinforcement is in the direction of the greatest stresses, thus giving extra strength to the entire fabrication. In plain weave cloths, each warp and weft thread passes over one yarn and under the next. In twill weaves, the weft yarns pass over one warp and under more than one warp yarn; in 2×1 twill, the weft yarns pass over one warp yarn and under two warp yarns. Satin weaves may be of multishaft types, when each warp and weft yarn goes under one and over several yarns. Unidirectional cloth is one in which the strength is higher in one direction than the other, and balanced cloth is a type with the warp and weft strength about equal. Although relatively expensive the woven forms have many excellent qualities, including high dimensional stability, high tensile and impact strength, good heat, weather and chemical resistance, how moisture absorption, resistance to fire and good thermo-electrical properties. A number of different weaves and weights is available, and

Table 16.2 Advantages and disadvantages of glass fibre reinforcement

	Advantages	*Uses*	*Disadvantages*
Rovings	Unidirectional strength	Spray-up Local longitudinal strength Mechanical bond for bulkheads Making tubes	Limited use in hand lay-up
Woven rovings	Easy to handle Bidirectional strength High glass content High impact resistance	To increase longitudinal and transverse strength	Poor interlaminar adhesion Traps air, causes voids High cost

Chopped strand mat	Multidirectional strength
	Low cost
	Good interlaminar bond
	Can be moulded into
	  complex shape
		Various
		General-purpose
		  reinforcement	

Woven cloth	High strength Smooth finish	Sheathing As a fire barrier	Very high cost difficult to 'wet out'
Surface tissue	Fine texture	Reinforcing gelcoat	Low strength

thickness may range from 0.05 mm to 9.14 mm, with weights from 30 g/m^2 to 1 kg/m^2, although the grades mostly used in the automotive field probably have weights of about 60 g/m^2 and will be of plain, twill or satin weave.

The advantages and disadvantages of glass fibre reinforcement materials are indicated in Table 16.2.

16.4.8 Carbon fibre

This is another reinforcing material. Carbon fibres possess a very high modulus of elasticity, and have been used successfully in conjunction with epoxy resin to produce low-density composites possessing high strength.

16.5 Resins used in reinforced composite materials

The first man-made plastics were produced in this country in 1862 by Alexander Parkes and were the forerunner of celluloid. Since then a large variety of plastics has been developed commercially, particularly in the last twenty-five years. They extend over a wide range of properties. Phenol formaldehyde is a hard thermoset material; polystyrene is a hard, brittle thermo-plastic; polythene and plasticized polyvinyl chloride (PVC) are soft, tough thermoplastic materials; and so on. Plastics also exist in various physical forms. They can be bulk solid materials, rigid or flexible foams, or in the form of sheet or film. All plastics have one important common property. They are composed of macro-molecules, which are large chain-like molecules consisting of many simple repeating units. The chemist calls these molecular chains polymers. Not all polymers are used for making plastic mouldings. Man-made polymers are called synthetic resins until they have been moulded in some way, when they are called plastics (Figure 16.12).

Most synthetic resins are made from oil. The resin is an essential component of glass fibre reinforced plastic. The most widely used is unsaturated polyester resin, which can be cured to a solid state either by catalyst and heat or by catalyst and accelerators at room temperature. The ability of polyester resin to cure at room temperature into a hard material is one of the main reasons for the growth of the reinforced plastics industry. It was this which led to development of the room temperature contact moulding methods which permit production of extremely large integral units.

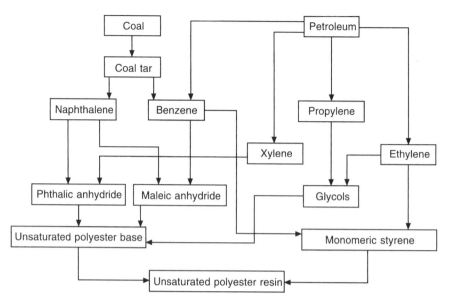

Figure 16.12 Production of unsaturated polyester resin (*Scott Bader Co. Ltd*)

Polyester resins are formulated by the reaction of organic acids and alcohols which produces a class of material called esters. When the acids are polybasic and the alcohols are polyhydric they can react to form a very complex ester which is generally known as polyester. These are usually called alkyds, and have long been important in surface coating formulations because of their toughness, chemical resistance, and endurance. If the acid or alcohol used contains an unsaturated carbon bond, the polyester formed can react further with other unsaturated materials such as styrene or diallyl phthalate. The result of this reaction is to interconnect the different polyester units to form the three-dimensional cross-linked structure that is characteristic of thermosetting resins. The available polyesters are solutions of these alkyds in the cross-linking monomers. The curing of the resin is the reaction of the monomer and the alkyd to form the cross-linked structure. An unsaturated polyester resin is one which is capable of being cured from a liquid to a solid state when subjected to the right conditions. It is usually referred to as polyester.

16.5.1 Catalysts and accelerators

In order to mould or laminate a polyester resin, the resin must be cured. This is the name given to the overall process of gelation and hardening, which is achieved either by the use of a catalyst and heating, or at normal room temperature by using a catalyst and an accelerator. Catalysts for polyester resins are usually organic peroxides. Pure catalysts are chemically unstable and liable to decompose with explosive violence. They are supplied, therefore, as a paste or liquid dispersion in a plasticizer, or as a powder in an inert filler. Many chemical compounds act as accelerators, making it possible for the resin-containing catalyst to be cured without the use of heat. Some accelerators have only limited or specific uses, such as quaternary ammonium compounds, vanadium, tin or zirconium salts. By far the most important of all accelerators are those based on a cobalt soap or those based on a tertiary amine. It is essential to choose the correct type of catalyst and accelerator, as well as to use the correct amount, if the optimum properties of the final cured resin or laminate are to be obtained.

16.5.2 Pre-accelerated resins

Many resins are supplied with an in-built accelerator system controlled to give the most suitable gelling and hardening characteristics for the fabricator. Pre-accelerated resins need only the addition of a catalyst to start the curing reaction at room temperature. Resins of this type are ideal for production runs under controlled workshop conditions.

The cure of a polyester resin will begin as soon as a suitable catalyst is added. The speed of the reactions will depend on the resin and the activity of the catalyst. Without the addition of an accelerator, heat or ultraviolet radiation, the resin will take a considerable time to cure. In order to speed up this reaction at room temperature it is usual to add an accelerator. The quantity of accelerator added will control the time of gelation and the rate of hardening.

There are three distinct phases in the curing reaction:

Gel time   This is the period from the addition of the accelerator to the setting of the resin to a soft gel.

Hardening time   This is the time from the setting of the resin to the point when the resin is hard enough to allow the moulding or laminate to be withdrawn from the mould.

Maturing time   This may be hours, several days or even weeks depending on the resin and curing system, and is the time taken for the moulding or laminate to acquire its full hardness and chemical resistance. The maturing process can be accelerated by post-curing.

16.5.3 Fillers and pigments

Fillers are used in polyester resins to impart particular properties. They will give opacity to castings and laminates, produce dense gel coats, and impart specific mechanical, electrical and fire resisting properties. A particular property may often be improved by the selection of a suitable filler. Powdered mineral fillers usually increase compressive strength; fibrous fillers improve tensile and impact strength. Moulding properties can also be modified by the use of fillers; for example, shrinkage of the moulding during cure can be considerably reduced. There is no doubt, also, that the wet lay-up process on vertical surfaces would be virtually impossible if thixotropic fillers were not available.

Polyester resins can be coloured to any shade by the addition of selected pigments and pigment pastes, the main requirement being to ensure thorough dispersion of colouring matter throughout the resin to avoid patchy mouldings.

Both pigments and fillers can increase the cure time of the resin by dilution effect, and the adjusted catalyst and promoter are added to compensate.

16.5.4 Releasing agents

Releasing agents used in the normal moulding processes may be either water-soluble film-forming compounds, or some type of wax compound. The choice of releasing agent depends on the size and complexity of the moulding and on the surface finish of the mould. Small mouldings of simple shape, taken from a suitable GRP mould, should require only a film of polyvinyl alcohol (PVAL) to be applied as a solution by cloth, sponge or spray. Some mouldings are likely to stick if only PVAL is used. PVAL is available as a soultion in water or solvent, or as a concentrate which has to be diluted, and it may be in either coloured or colourless form.

Suitable wax emulsions are also available as a releasing agent. They are supplied as surface finishing pastes, liquid wax or wax polishes. The recommended method of application can vary depending upon the material to be finished. Hand apply with a pad of damp, good quality mutton cloth or equivalent, in straight even strokes. Buff lightly to a shine with a clean, dry, good quality mutton cloth. Machine at 1800 rpm using a G-mop foam finishing head. Soak this head in clean water before use and keep damp during compounding. Apply the wax to the surface. After compounding, remove residue and buff lightly to a shine with a clean, dry, good quality mutton cloth.

Wax polishes should be applied in small quantities since they contain a high percentage of wax solids. Application with a pad of clean, soft cloth should be limited to an area of approximately 1 square metre. Polishing should be carried out immediately, before the wax is allowed to dry. This can be done either by hand or by machine with the aid of a wool mop polishing bonnet.

Frekote is a semi-permanent, multi-release, gloss finish, non-wax polymeric mould release system specially designed for high-gloss polyester mouldings. It will give a semi-permanent release interface when correctly applied to moulds from ambient up to 135 °C. This multi-release interface chemically bonds to the mould's surface and forms on it a microthin layer of a chemically resistant coating. It does not build up on the mould and will give a high-gloss finish to all polyester resins, cultured marble and onyx. It can be used on moulds made from polyester, epoxy, metal or composite moulds. Care should be taken to avoid contact with the skin, and the wearing of suitable clothing, especially gloves, is highly recommended. These products must be used in a well ventilated area.

16.5.5 Adhesives used with GRP

Since polyester resin is highly adhesive, it is the logical choice for bonding most materials to GRP surfaces.

Suitable alternatives include the Sika Technique, which is a heavy-duty, polyurethane-based joining compound. It cures to a flexible rubber which bonds firmly to wood, metal, glass and GRP. It is ideal for such jobs as bonding glass to GRP or bonding GRP and metal, as is often required on vehicles with GRP bodywork. It is not affected by vibration and is totally waterproof. The Araldite range includes a number of industrial adhesives which are highly recommended for use with GRP. Most high-strength impact adhesives (superglues) can be used on GRP laminates.

Most other adhesives will be incapable of bonding strongly to GRP and should not be used when maximum adhesion is essential.

16.5.6 Core materials

Core materials, usually polyurethane, are used in sandwich construction, that is basically a laminate consisting of a foam sheet between two or more glass fibre layers (Figure 16.13). This gives the laminate considerable added rigidity without greatly increasing weight. Foam materials are available which can be bent and folded to follow curved surfaces such as vehicle bodies. Foam sheet can be glued or stapled together, then laminated over to produce a strong box structure, without requiring a mould. Typical formers and core materials are paper rope, polyurethane rigid foam sheet, scoreboard contoured foam sheet, Termanto PVC rigid foam sheet, Term PVC

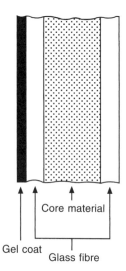

Figure 16.13 Typical sandwich construction (*Scott Bader Co. Ltd*)

contoured foam sheet, and Termino PVC contoured foam sheet.

16.5.7 Formers

A former is anything which provides shape or form to a GRP laminate. They are often used as a basis for stiffening ribs or box sections. A popular material for formers is a paper rope, made of paper wound on flexible wire cord. This is laid on the GRP surface and is laminated over to produce reinforcing ribs, which give added stiffness with little extra weight. The former itself provides none of the extra stiffness; this results entirely from the box section of the laminate rib. Wood, metal, or plastic tubing and folded cardboard can all be used successfully as formers. Another popular material is polyurethane foam sheet, which can be cut and shaped to any required form (Figure 16.14).

16.6 Moulding techniques for reinforced composite laminates

16.6.1 Contact moulding

This is the oldest, simplest and most popular fabrication technique for the automotive, reinforced plastic body industry. It is normally used for relatively short runs, but it has also been adapted successfully for series production. It is the only production method which takes full advantage of the two

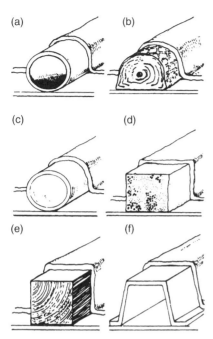

Figure 16.14 Typical formers: (a) metal tube
(b) paper rope (c) cardboard tube (d) foam strip
(e) wood (f) folded cardboard (*Scott Bader Co. Ltd*)

most important characteristics of polyester resin, namely that it can be set without heat and without pressure. A considerable industry has been built around contact moulding, which has facilitated the cost effective production of large one-piece mouldings, particularly for low production runs. Contact moulding advantages are that a minimum of equipment is required, tooling is inexpensive, there are practically no size restrictions, and design changes are easily made. Disadvantages are that the labour context is high and the quality of the moulding depends on the skill of the operators. The lay-up and curing times are comparatively slow, and only one good surface finish is achieved.

The contact moulding process is carried out in the following manner. A master pattern or model is made, representing in all its dimensions the finished product. This could be, for example, a full-size wood model of the type of body shell or cab shell required, or it might equally well be a steel or aluminium panel-beaten structure of composite type, or even a plaster model reinforced with wire mesh. From this is made a master mould, which must be female or concave for the most part, and this would in all probability be made in reinforced plastics

similar to those used for the final product. It is important to differentiate, as a matter terminology, between 'mould' and 'moulding', one being the production tool and the other the product itself.

An important aspect of the process is that the surface of the mould will inevitably be faithfully reproduced in the moulding, and accordingly if the mould is bumpy or rough so will be the final article. It does not usually matter if the unseen or partly hidden side of the moulding is rough (indeed, it usually is), but for the displayed surface to be unsightly is not normally tolerable. This is why a female mould is usually used. If the original pattern was very smooth, so will be the inside of the mould and, therefore, the outside of the moulding. This is why so much trouble is taken over the pattern. If in wood it is tooled and sanded to perfection, and if in metal it is panel beaten with the greatest possible skill, then ground and polished if necessary. It is of the greatest importance that separation of the moulding from the mould be easy. A special compound or a polish such as carnauba wax and/or silicone lubricant can be used; plastics film is occasionally used as a separating membrane, as it will not adhere strongly to either the mould or the moulding. For contact moulding, the equipment is relatively simple and inexpensive. Contact moulding can be further subdivided into hand lay-up and spray lay-up.

The application of the release agent to the mould is followed by brush or spray application of a gel coat of resin. There has been a constant improvement and development in polyester resins, and among other things this has led to the introduction of successful semi-flexible gel coats, which are of particular interest to the motor industry. The gel coat is a continuous skin on the working face of a moulding. It is almost pure resin and its object is to give a good finish as well as to protect the working surface from corrosion and other damage. It also hides the fibre pattern of the reinforcement. The gel coat can be colour impregnated or otherwise specially formulated, e.g. for extra abrasion or impact resistance. It should be as even in thickness as possible, as thicker areas are very susceptible to accidental damage, while thin patches can lower the resistance of the structure to moisture and to chemical attack.

Even in hand lay-up the spray method may be used for this stage and for the application of the separating agent, so that there is a small element of mechanization. A fine surfacing tissue may be applied to the gel coat while wet, or it may simply be allowed to gel as it is. Further resin is sprayed on or brushed on, and mat or woven cloth, which has been carefully cut to patterns, is laid in position. Consolidation and air removal are then effected by manual means. It is customary to use rollers made up of split washers for this operation, which is an extremely important one if consistency and strength of the moulding are to be obtained. More mat or cloth is added in order to build up the requisite thickness of reinforced plastics, and the moulding is then allowed to set or cure. Curing normally takes place at room temperature, but sometimes under a certain degree of heat if the process is to be accelerated. It should be remembered that curing is itself a heat-producing process. Contact moulding can also be carried out by simple mechanical means, but the general principle is always that of bringing the materials into contact with the mould, without the use of any dies.

The following is a summary of the contact moulding process:

1   The master mould must be spotlessly clean.
2   A release agent is applied to the entire surface of the mould face (Figure 16.15).

Figure 16.15   Application of release agent to mould surface (*Scott Bader Co. Ltd*)

3   Gel coat covering is applied by brush or spray (Figure 16.16).
4   Catalyst is added to the resin and the catalysed resin is smoothed over the gel-coated mould.
5   Glass fibre mat, precut to exact size, is laid on the mould and a further small quantity of resin is poured on to the mat. With brushes and hand roller, the resin is drawn through the mat (Figure 16.17). A second layer of mat is applied

and the drawing-through process is repeated until the required thickness is achieved. It is critical that all air bubbles be removed by the brushing and rolling (Figure 16.18).

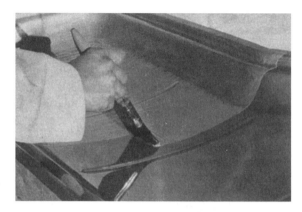

Figure 16.16   Application of gel coat covering (*Scott Bader Co. Ltd*)

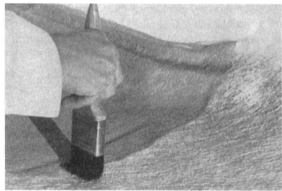

Figure 16.17   Impregnation of glass mat with resin (*Scott Bader Co. Ltd*)

Figure 16.18   Rolling impregnated mat (*Scott Bader Co. Ltd*)

6 The mould is allowed to cure naturally or heat is used to speed up the curing process.

7 After curing, the moulds are broken and the completed sections are removed (Figure 16.19).

8 They are then trimmed and ready for use.

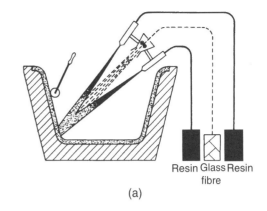

Resin Glass Resin
fibre
(a)

Figure 16.19 Final mould and moulding (*Scott Bader Co. Ltd*)

16.6.2 Spray-up technique

A development from the basic manual contact process which is employed with increasing frequency in the automotive body industry is known as spray-up (Figure 16.20). In this method, rovings are automatically fed through a chopping unit and the resultant chopped strands are blown or carried by the sprayed resin stream on to the mould. The glass and resin mix applied in this way is consolidated, and the air pockets or bubbles are removed by manual rolling, as in simple hand lay-up.

There are several commercial spraying systems available, where the glass fibre and resin are deposited simultaneously on the mould face. They consist of two principal types. In the *twin-pot system* a twin-nozzle spray gun is used, and in order to prevent gelation in the gun the resin is divided into two parts, one of which is catalysed and the other accelerated. The two streams of resin spray converge near the surface of the mould simultaneously with a stream of glass fibre ejected by a glass rovings chopper. In the other type, the *catalyst injector system*, accelerated resin is sprayed from a

(b)

Figure 16.20 The spray-up technique (*BP Chemicals (UK) Ltd*)

single-nozzle gun, but liquid catalyst is metered into the resin in the gun itself. A glass rovings chopper delivers the reinforcement to the mould surface as in the former system.

Although much of the manual labour of the hand lay-up is eliminated by using the spray process, thorough rolling is still necessary, not only to consolidate the deposited glass resin mixture, but also to ensure that the accelerated and catalysed portions of the resin are adequately mixed. Considerable skill is needed to control the thickness of the laminate when using the resin glass spray gun. Spraying reduces labour costs, especially when the volume of production is large enough to keep the equipment in constant use.

16.6.3 Hot press moulding

This process involves the use of chopped strand glass mat, pre-impregnated with polyester resin, which is then in general principle formed in presses in a similar manner to that used for forming steel sheet (Figure 16.21). In this case, however, the dies, which are preheated, have to remain closed for the curing cycle of the pre-impregnated mat, which may involve a period from 15 to 30 seconds. Hot press moulding using matched dies have good finish on both surfaces. Further, this method enables high glass content and uniform dimensional properties and appearances to be achieved at lower cost than by other methods for runs above 1000 units. Such a process reduces the labour content of producing panels, but much increases the initial tooling charge.

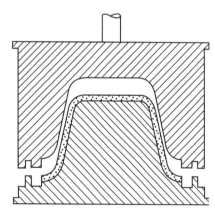

Figure 16.21 Hot press moulding

16.6.4 Production composite moulding processes

There are many fibreglass reinforced plastic moulding processes available to the designer. Each process has its own characteristics, as well as its own limitations as to part size, shape, production rate, compatible reinforcements, and suitable resin

systems. The most common moulding processes are as follows (Figure 16.22): hand lay-up, spray-up, resin transfer moulding, compression moulding, injection moulding thermoplastics, injection moulding thermosets, pultrusion, and reinforced reaction injection moulding.

The three techniques used in the production of body panels are as follows:

Cold press mouldings This is used in the manufacture of the boot lid. The boot lid is formed by cold pressing a mineral reinforced resin-coated glass fibre mat in a gel-coated mould, forming a component which is very stiff in relation to its weight (see Figure 16.23).

Reinforced reaction injection moulding The RRIM technique is used for all the vertical body panels such as the front and rear wings, front grille and bumper assembly, and rear panel and bumper assembly. RRIM polyurethane has the properties of good recovery from deformation, outstanding resistance to wear, impact and abrasion, and a fast cycle time in manufacture. The use of this material for all exposed corners of the car helps to reduce minor body damage repair (see Figure 16.22h).

Pressure assisted resin injection moulding The bonnet is a pressure assisted moulding of sandwich construction with polyester resin exterior on either side of a rigid urethane core. The underside of the moulding is an intumescent fire barrier which is a major safety factor for an engine compartment cover (see Figure 16.22e).

Lotus Cars Ltd
Lotus have been producing reinforced composite motor cars since 1957. In 1973 the company introduced the vacuum assisted resin injection (VARI) system for vehicle body manufacture. The first VARI moulded production car was the Lotus Elite introduced in 1974; since then developments have continued in the processes, tooling and techniques of producing composite vehicle bodies.

The Lotus VARI process provides a method of moulding fibre reinforced composite panels from matched tooling. The process can be used to manufacture large body panels with integrated foam structures and captive metal fixings, using relatively low-cost tooling. As there is no dependency upon platen size and press tonnage – an obvious limitation of other processes – there are no panel size limitations. This means that the vehicle body structures can be moulded in much fewer sections, minimizing panel assembly times and costs. Lotus mould their complete bodyshell in a few pieces, each of which has an integral moulded structure. This structure uses isophthalic polyester resin in conjunction with continuous filament mat. In addition, woven and unidirectional glass reinforcements are used in areas where more specific loadings are required. Kevlar aramid materials are also included where particularly high strengths are needed.

The VARI tooling, which can be either metal faced or constructed entirely from composite materials, is designed so that each tool becomes its own press. The pressure is created by vacuum which draws the tools together and holds the male and female throughout the moulding cycle. Therefore no mechanical clamping mechanisms are involved. After closing the tools the resin is injected using a machine which dispenses precise quantities of catalysed materials. After the curing cycle the vacuum is reversed and the tools open, releasing the moulded panel or body section.

16.7 Designing reinforced composite materials for strength

The ability to design and fabricate large structures as a whole, rather than as an assembly of components, is one of the chief advantages that glass resin laminates bring to the designer. This is supported by the ease of modifying the material thickness at specific locations and, taking advantage of the properties of the various types of reinforcement, by building in extra strength at any point and in any direction (Figure 16.24). The skill of the operator is an important factor.

Strength of glass fibre reinforced plastic laminates in any direction is dependent on the orientation of fibre reinforcement to that direction (Figure 16.25). When chopped strand mat is used, its random fibre arrangement can be expected to give roughly equal mechanical properties in all planes; however, maximum strength will in practice be parallel to the plane of the laminate. Plain woven roving gives optimum mechanical properties at right angles, while unidirectional roving mat shows highest strength along the roving; as the roving is continuous and uncrimped, this last type will be stronger than other types of reinforcement.

There is still an unfortunate tendency on the part of designers to use a traditional design, known to be satisfactory for wood or metal, for reinforced plastics which have, of course, completely different properties and processing characteristics (Tables 16.3 and 16.4). This may give glass fibre reinforced plastic mouldings of incorrect shape, since although conventional materials are well suited to straight lines and flat surfaces, the properties of glass fibre reinforced plastic components are improved by the introduction of curvature, and if possible double curvature, to the design. The specific strength (tensile strength weight ratio) of glass-polyester laminates is high, but rigidity tends to be on the low side. It may therefore be necessary to design for additional rigidity rather than for optimum tensile strength. This can be effected by various means, of which increased overall moulding thickness is perhaps the least desirable as it is wasteful and may well add unnecessary weight. The use of simple or compound curves in the design may be the answer, or perhaps local corrugations can be introduced as in metal designs; these can often be incorporated into the overall styling, especially in vehicle bodies. Localized thickening, particularly towards the edges of a panel, will contribute usefully to the stiffness of the moulding. A common practice to achieve extra rigidity is the integral moulding of ribs into the reverse face of the laminate. These ribs, which are often used in large boat hulls, can be solid or hollow; for solid ribs a permanent core of glass fibre, wood or plastic foam can be laminated in, while hollow ribs are achieved by use of removable tube or simple former of cardboard or similar material (see Figure 16.14).

Large panels can be made considerably more rigid by the employment of a sandwich form of construction, whereby two layers of glass fibre reinforced plastic are separated by a thick but relatively weak lightweight material. The benefit here derives from the fact that stiffness is a direct function of thickness. Other advantages of this method are increased heat and sound insulation. Stress analysis is usually based upon the tensile strength at which crazing of the resin matrix occurs. This corresponds to the yield point of conventional materials. It is usual to allow a safety factor of between 1:3 and 3:5, depending upon the conditions of service. Frequently the stressing of a moulding is so

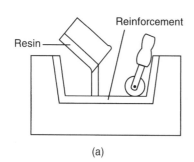

(a)

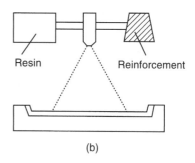

(b)

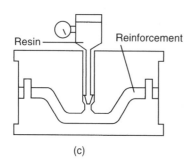

(c)

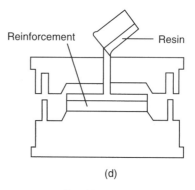

(d)

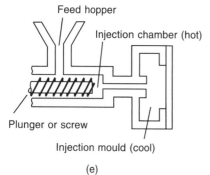

(e)

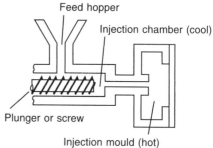

(f)

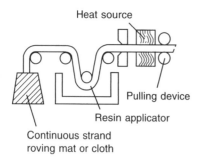

(g)

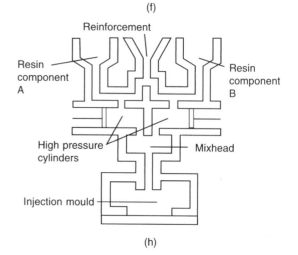

(h)

complex that it defeats analysis. In such cases it may be necessary to base the design upon the known performance of a similar structure and to produce a prototype moulding for testing under service conditions. Where metal inserts are to be incorporated in the moulding, allowance should be made in the design of the joint for the greatly differing yields of the two materials under similar loading. A wider insert will result in lower stress per inch at the load transfer point. The stress in the resin laminate can be taken up to a point where strain in both materials is similar by thickening the laminate where it approaches and surrounds the metal insert. Ideally the base of the insert should be about four times as wide as it is long.

It is better to achieve load transfer by adhesion rather than by mechanical interlocking; mechanical methods are satisfactory, however, if only small loads are involved. Joining of two glass fibre reinforced plastic componenets to each other can be effected by adhesive or mechanical means, or by a combination of both. In adhesive joint design the bonding area should be as large as possible. For ordinary butt or scarf joints, extra reinforcement should be provided by lamination of extra layers of resin and glass over the joint on both sides; in general it is better to use overlapping joints (Figure 16.26). Exposure of glass fibre by roughening in each case will enhance bond strength. Cured laminates are commonly bonded by means of epoxy-type adhesives. A polyester resin adhesive may be used, but here it is necessary to ensure that the adhesive films are thick enough to avoid problems of undercure. This can be done by including a single sheet of glass cloth or mat with open texture in the joint. Most kinds of mechanical fastener (nuts and bolts,

---

(Facing page)

Figure 16.22 Composite moulding processes (*Owens-Corning Fiberglas*)

(a) Hand lay-up is a low-to-medium volume moulding method suitable for making boats, tanks, housings and building panels and other large parts requiring high strength. The process provides only one finished surface.

(b) Spray-up is a low-to-medium volume moulding method similar to hand lay-up in its suitability for making boats, tanks, tub/showers and other medium to large size shapes. Greater shape complexity is possible with spray-up than with hand lay-up.

(c) Resin transfer moulding (RTM) is suitable for medium-volume production and may be regarded as an intermediate process between spray-up and faster compression moulding methods using SMC and BMC. RTM provides two finished surfaces. The reinforcement is placed in the bottom half of the mould. The mould is then closed and clamped, and catalysed resin is pumped in under pressure until the mould is filled. Moulds are usually made of reinforced plastics.

(d) Compression moulding is a high-volume, high-pressure process suitable for moulding complex, high-strength fibreglass reinforced plastic parts using sheet moulding compound, bulk moulding compound or preforms. Fairly large parts can be moulded in medium to high volumes with excellent surface finish.

(e) Injection moulding is the highest volume method of any of the fibreglass reinforced plastic processes using single or multi-cavity moulds to produce large volumes of complex parts at high production rates. Advanced fibreglass technology makes it possible to injection mould glass reinforced thermoplastics that provide a wider variety of mechanical, chemical, electrical and thermal properties than previously available.

(f) Thermoset moulding compounds are injection moulded with a low-temperature injection screw/plunger and chamber, and a high temperature mould. This cures the thermoset material under heat and pressure. Injection moulding of thermoset resins offers the capability to produce high volumes of very complex parts with good mechanical properties and impact strengths.

(g) Pultrusion is a continuous process for the manufacture of products having a constant cross-section, such as rod stock, structural shapes, beams, channels, pipe, tubing and fishing rods.

(h) RRIM is emerging as a leading process. RRIM provides low-cost tooling, low-pressure moulding and design flexibility to accommodate inserts or encapsulations of structural supports.

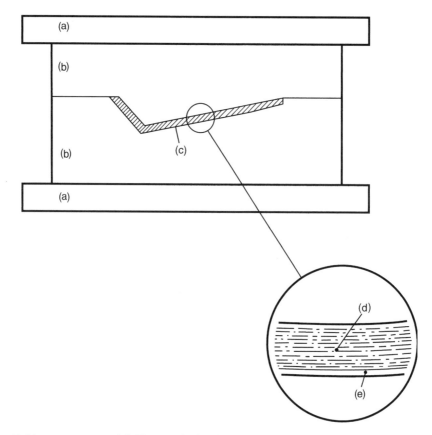

Figure 16.23   Cold press process: (a) 50 tonne hydraulic press (b) Kirksite metal mould (c) moulding (d) continuous filament glass (e) polyester surface finish (*Reliant Motor PLC*)

self-tapping screws, rivets and others) can usually be employed without any trouble, provided the load is spread by means of large heads and/or large washers. To prevent laminate crushing it is a good plan to use spacers for bolted connections. Provided the extent of the bonded area is taken into account, and thorough cleaning and roughening are carried out, there is no reason why drilled and tapped metal plates, or special tapped inserts, should not be laminated into the moulding or on the reverse side.

16.7.1 Composite theory

In its most basic form a composite material is one which is composed of two elements working together to produce material properties that are different to the properties of those elements on their own. In practice, most composites consist of a bulk material called the matrix, and a reinforcement

material of some kind which increases the strength and stiffness of the matrix.

Polymer matrix composites (PMC) is the type of composites used in modern vehicle bodywork. This type of composite is also known as Fibre reinforced polymers (or plastics) (FRP). The matrix is a polymer-based resin and the reinforcement material is a fibrous material such as glass, carbon or aramid. Frequently, a combination of reinforcement materials is used.

The reinforcement materials have high tensile strength, but are easily chaffed and will break if folded. The polymer matrix holds the fibres in place so that they are in their strongest position and protects them from damage. The properties of the composite are thus determined by:

* The properties of the fibre.
* The properties of the resin.

- The ratio of fibre to resin in the composite – fibre volume fraction (FVF).
- The geometry and orientation of the fibres in the composite.

Resin

The choice of resins depends on a number of characteristics, namely:

- Adhesive properties – in relation to the type of fibres being used, and if metal inserts are to be used such as for body fitting.
- Mechanical properties – particularly tensile strength and stiffness.
- Micro-cracking resistance – stress and age hardening causes the material to crack, the micro-cracks reduce the material strength and eventually lead to failure.
- Fatigue resistance – composites tend to give better fatigue resistance than most metals.
- Degradation from water ingress – all laminates permit very low quantities of water to pass through in a vapour form. If the laminate is wet for a long period, the water solution inside the laminate will draw in more water through the osmosis process.
- Curing properties – the curing process alters the properties of the material. Generally oven curing at between 80 °C and 180 °C will increase the tensile strength by up to 30%.
- Cost – the different materials cost different prices.

The main types of resins are polyesters, vinylesters, epoxies, phenolics, cyanate esters, silicones, polyurethanes, bismaleides (BMI) and polyamides. The first three are the ones mainly used for vehicle body work as they are reasonably priced (Table 16.5). Cyanates, BMI and polyamides cost about 10 times the price of the others.

Reinforcing fibres

The mechanical properties of the composite material are usually dominated by the contribution of the reinforcing fibres. The four main factors which govern this contribution are:

1 The basic mechanical properties of the fibre.
2 The surface interaction of the fibre and the resin – called the interface.

3 The amount of fibre in the composite – FVF.
4 The orientation of the fibres.

The three main reinforcing fibres used in vehicles are: glass, carbon and aramid. In addition, the following are used for non-body purposes: polyester, polyethylene, quartz, boron, ceramic and natural fibres such as jute and sisal.

Glass is discussed separately in Section 16.4

Aramid   Aramid fibre is a man-made organic polymer, an aromatic polyamide, produced by spinning fibre from a liquid chemical blend. The bright golden yellow fibres have high strength and low density giving a high specific strength. Aramid has good impact resistance. Aramid is better known by its Dupont trade name Kevlar.

Carbon   Carbon fibre is produced by the controlled oxidation, carbonization and graphitization of carbon-rich organic materials – referred to as precursors – which are in fibre form. The most common precursor is polyacrylonitrile (PAN); pitch and cellulose are also used (Table 16.6).

16.7.2 Fabric types and constructions

To weave any material warp threads and weft threads are needed. The warp is the longest part, the part which is wound around the roller or folded. The weft is the shorter part, running at 90° across the warp. The spun warp threads are wound onto a giant roller and the weft threads are woven into the warp threads in different patterns for different purposes.

Plain

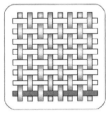

(Courtesy of SPSystems)

Each weft thread passes alternately under and over each warp fibre. The resultant weave is symmetrical with good stability and reasonable porosity. However, because of the high level of crimp of the fibres the mechanical properties are low compared

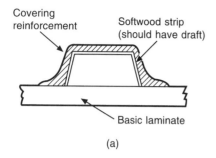

Covering reinforcement

Softwood strip (should have draft)

Basic laminate

(a)

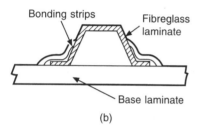

Bonding strips

Fibreglass laminate

Base laminate

(b)

Cardboard form

Base laminate

(c)

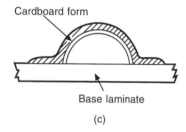

Countersunk screw

Fibreglass flange

Base laminate

(d)

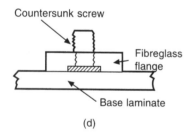

Coupling

Tack weld

Fibreglass bonded layers

Metal screen

Base laminate

(e)

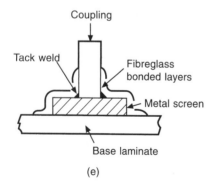

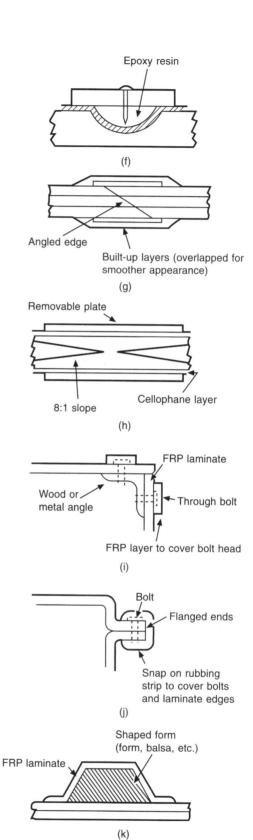

Epoxy resin

(f)

Angled edge

Built-up layers (overlapped for smoother appearance)

(g)

Removable plate

8:1 slope

Cellophane layer

(h)

FRP laminate

Wood or metal angle

Through bolt

FRP layer to cover bolt head

(i)

Bolt

Flanged ends

Snap on rubbing strip to cover bolts and laminate edges

(j)

Shaped form (form, balsa, etc.)

FRP laminate

(k)

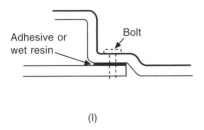

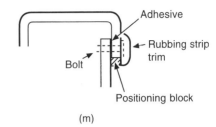

(l)

(m)

(Facing page)

Figure 16.24 Design consideration for reinforced composite mouldings (*Owens-Corning Fiberglas*). Designs may incorporate large areas of sandwich construction with cores for vibration dampening, stiffening, thermal inculation or flotation. The type of core material will depend upon the specific property required. Care must be exercised to insure adequate adhesion between the laminate and the core material used. Unusual and acute contours, expensive to cut in metal dies, are both practical and economical in fibreglass spray-up.

(a) Encased. Wood strips, plywood or metal stiffeners can be pressed into the part while wet and covered with resin and reinforcement to anchor and protect from deterioration. Softwoods are normally used since hardwoods are difficult to bond. Note: if stiffeners are added after cure, bonded surfaces should be sanded to assure good adhesion.

(b) Separately fabricated fibreglass stiffeners can be pressed into layers to be reinforced while it is still wet. While fibreglass adheres well to itself, additional spray-up is needed to tie into part.

(c) Integral. Light, low-cost forms (cardboard mailing tubes, folded cardboard, balsa wood) pressed into the mould during spray-up give stiffening shape to fibreglass. Thickness of spray-up over the form should be at least 3 mm. Exact thickness will depend upon strength and stiffness requirements of the part.

(d) Bolting flange. This can be made by cutting a section of a 13 mm laminate. This laminate can then be tapped for studs or countersunk for cap screws. The assembly is then placed into a wet layer of spray-up material. The flange can be strengthened by adding layers.

(e) Threaded coupling. A collar of expanded metal mesh (for coarse metal screen) is tack-welded to a threaded pipe coupling. Then, this assembly is pressed into a wet spray-up layer and cover layers are added for increased torque resistance. Large beads on the tack welds will act to 'key' the laminate to prevent twisting.

(f) Nailing or screw insert. Into a depression in the spray-up part, a shrink-resistant epoxy resin is cast to hold nails or other mechanical fasteners. Mould surface in depression should be purposely left rough to give maximum gripping to insert. Additional layers can then be sprayed over insert.

(g) Simple butt. When premoulded sheets are joined, edges are angled to give more bonding area. Then laminate layers are built up in thickness on both sides to desired strength.

(h) Optimum appearance. For smooth surface a double-tapered joint cured between removable plates is required. Slope of angle should be 8:1. Cellophane placed between plate and laminate will facilitate plate removal.

(i) The simple corner uses a premoulded sheet of *Fibreglas* reinforcement and resin. For strength, a wood or metal angle may be incorporated or the butt itself carried around corner.

(j) Flanged edges bolted together require trim to hide joint. Though bolts are shown. Self-tapping screws or rivets may be used to hold parts together.

(k) Built-up stiffeners incorporated in laminate may be cardboard, balsa, various foams, plastic, or sheet metal. They should be designed to spread loads over wide area.

(l) Integral stiffeners are most easily incorporated as flanges at a joint to provide dual function. Pads or gussets are desirable where stiffeners intersect a flat, flexible area.

(m) Lapped sections are held in place mechanically with adhesive or wet resin between sections as seal. Trim hides edges. Fastening device may be bolts, screws or rivets.

The best procedure is a corner integral with the structure, with no joints. This provides optimum performance opportunity.

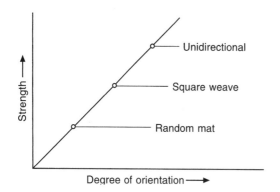

Figure 16.25   Orientation of fibre reinforcement

to other weave patterns; this also means that it does not drape very well.

Twill

(Courtesy of SPSystems)

One or more weft fibres alternate by passing over or under two or more warp fibres. This gives the visual appearance of a diagonal rib. The threads therefore have less crimp giving higher mechanical properties than the plain weave. The fabric looks smoother and drapes better.

Satin

(Courtesy of SPSystems)

Satin weaves are fundamentally twill weaves modified to produce fewer intersections of warp and weft. The harness number used in the designation, typically 4, 5 and 8 is the total number of fibres crossed and passed under before the fibre repeats the pattern. A crowsfoot weave is a form of satin weave with a different stagger in the repeat pattern. Satin weaves are very flat and have low crimp which means that they have excellent mechanical properties and drape well. Because of the pattern shape one face of the material will have sections of fibre running open in one direction, giving an asymmetric imbalance. Care must be taken when laying up several layers, to ensure the pattern is set to minimize stress levels.

Basket

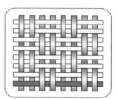

(Courtesy of SPSystems)

Basket weave is fundamentally the same as plain weave except that two or more warp fibres alternately interlace with two or more warp fibres. An arrangement of two wefts crossing two warps is designated 2X2 basket. The arrangement of fibres need not be symmetrical. Basket weave is flatter and through less crimp is stronger than plain weave. Basket weave is less stable and is best used with thicker fibres, called high tex, to avoid excessive crimping.

Leno

(Courtesy of SPSystems)

Table 16.3 Properties of fibreglass composites and alternative materials

Material	ASTM test method	Glass fibre by weight %	Specific gravity D792	Density lb/in³	Tensile strength 10³psi D638	Tensile modulus 10⁶psi D638	Elongation % D638	Flexural strength 10³psi D790	Flexural modulus 10⁶psi D790	Compressive strength 10³psi D695	Impact strength Izod ft lb/in notched at 73°F D256	Hardness Rockwell (except where noted) D785	Flammability UL-94
Glass fibre reinforced thermosets	Polyester SMC (compression)	30.0	1.85	0.066	12.00	1.70	1.0	26.00	1.60	24.00	16.00	Barcol 68	5V
	Polyester SMC (compression)	20.0	1.78	0.064	5.30	1.70	.4	16.00	1.40	23.00	8.20	Barcol 68	5V
	Polyester SMC (compression)	50.0	2.00	0.072	23.00	2.27	1.7	45.00	2.00	32.00	19.40	Barcol 68	5V
	Polyester BMC (compression)	22.0	1.82	0.065	6.00	1.75	.5	12.80	1.58	20.00	4.26	Barcol 68	5V
	Polyester BMC (injection)	22.0	1.82	0.065	4.86	1.53	.5	12.65	1.44	—	2.89	Barcol 68	VO
	Epoxy (filament wound)	80.0	2.08	0.061	80.00	4.00	1.6	100.00	5.00	45.00	45.00	M98	VO
	Polyester (pultruded)	55.0	1.69	0.060	30.00	2.50	—	30.00	1.60	30.00	25.00	Barcol 50	VO
	Polyurethane, milled fibres (RRIM)	13.0	1.07	0.038	2.80	—	140.0	—	0.037–0.053	—	—	SD 65–75	VO
	Polyurethane flaked glass (RRIM)	23.0	1.17	0.042	4.41	—	38.9	—	0.15	—	2.10	—	VO
	Polyester (spray-up/lay-up)	30.0	1.37	0.049	12.50	1.00	1.3	28.00	0.75	22.00	13.00–15.00	Barcol 50	VO
	Polyester, woven roving (lay-up)	50.0	1.64	0.059	37.00	2.25	1.6	46.00	2.25	27.00	33.00	Barcol 50	VO
Glass fibre reinforced thermoplastics	Acetal	25.0	1.61	0.058	18.50	1.25	3.0	28.00	1.10	17.00	1.80	M79	HB
	Nylon 6	30.0	1.37	0.049	24.00	1.05	3.0	29.00	1.11	24.00	2.20	R121	HB
	Nylon 6/6	30.0	1.48	0.053	23.00	1.20	1.9	35.00	0.80	26.50	2.20	M95	HB
	Polycarbonate	10.0	1.26	0.045	12.00	.75	9.0	16.00	0.60	14.00	2.00	M80	V-1
	Polypropylene	20.0	1.04	0.037	6.50	.54	3.0	8.30	0.52	25.00	1.10	R103	HB
	Polyphenylene sulphide	40.0	1.64	0.059	22.00	2.05	3.0	37.00	1.90	21.00	1.50	R123	V-O/5V
	Acrylonitrile butadiene styrene (ABS)	20.0	1.22	0.044	11.00	.90	2.0	15.50	0.87	14.00	1.20	R107	HB
	Polyphenylene oxide (PPO)	20.0	1.21	0.043	14.50	.92	5.0	18.50	0.75	17.60	1.80	R107	HB
	Polystyrene acrylonitrile (SAN)	20.0	1.22	0.044	14.50	1.25	1.8	19.00	1.10	17.50	1.10	R122	HB
	Polyester (PBT)	30.0	1.52	0.054	19.00	1.20	4.0	28.00	1.17	18.00	1.80	R118	HB
	Polyester (PET)	30.0	1.56	0.056	21.00	1.30	6.6	32.00	1.25	25.00	1.80	R120	HB

(Continued)

Table 16.3 (*Continued*)

Material	ASTM test method	Glass fibre by weight %	Specific gravity D792	Density lb/in³	Tensile strength 10³psi D638	Tensile modulus 10⁶psi D638	Elongation % D638	Flexural strength 10³psi D790	Flexural modulus 10⁶psi D790	Compressive strength 10³psi D695	Impact strength Izod ft lb/in notched at 73° F D256	Hardness Rockwell (except where noted) D785	Flammability UL-94
Unreinforced thermoplastics	Acetal	—	1.41	0.051	8.80	.41	40.0	13.00	0.38	16.00	1.00	M78–M80	HB
	Nylon 6	—	1.12	0.040	11.80	.38	30.0	15.70	0.39	13.00	0.60	R119	HB
	Nylon 6/6	—	1.13	0.041	11.50	.40	60.0	17.00	0.42	15.00	0.80	R120, M83	V-2
	Polycarbonate	—	1.20	0.043	9.50	.34	110.0	13.50	.34	12.50	16.00	M70	V-2
	Polypropylene	—	.89	0.032	5.00	.10	200.0	5.00	0.13–0.20	3.50	1–20	R50–96	HB
	Polyphenylene sulphide	—	1.30	0.045	9.50	.48	1.0	14.00	0.55	16.00	0.50	R123	V-O
	Acrylonitrile butadiene styrene (ABS)	—	1.03	0.037	6.00	.30	5.0	11.00	0.35–0.40	10.00	3–6	R107–115	HB
	Polyphenylene oxide (PPO)	—	1.10	0.039	7.80	.38	50.0	12.80	0.33–0.40	12.00	5.00	R115	V-1
	Polystyrene acrylonitrile (SAN)	—	1.05	0.038	9.50	.40	.5	14.00	0.55	14.00	0.30–0.45	M80–85	HB
	Polyester (PBT)	—	1.31	0.047	8.20	.28	50.0	12.00	0.33–0.40	8.60	.80	M68–78	HB
	Polyester (PET)	—	1.34	0.048	8.50	.40	50.0	14.00	0.35–0.45	11.00	0.25–0.65	M94–101	HB
Metals	ASTM A-606 HSLA steel (cold rolled)	—	7.75	0.280	65.00	30.00	22.0	—	—	65.00	—	B80	—
	SAE 1008 low-carbon steel (cold rolled)	—	7.86	0.280	48.00	30.00	37.0	—	—	48.00	—	B34–52	—
	AISI 304 stainless steel	—	8.03	0.290	80.00	28.00	40.0	—	—	80.00	—	B88	—
	TA 2036 aluminium (wrought)	—	2.74	0.099	49.00	10.20	23.0	—	—	49.00	—	R80	—
	ASTM B85 aluminium (die cast)	—	2.82	0.102	48.00	10.30	2.5	—	—	48.00	—	Brinell 85	—
	ASTM AZ91B magnesium (die cast)	—	1.83	0.066	33.00	65.00	3.0	—	—	33.00	—	Brinell 85	—
	ASTM AG40A zinc (die cast)	—	6.59	0.238	41.00	10.90	10.0	—	—	41.00	—	Brinell 82	—

*Shore D

Table 16.4   Compatibility of materials and processes for fibreglass composites

	Thermosets					Thermoplastics									
	Polyester	*Polyester SMC*	*Polyester BMC*	*Epoxy*	*Polyurethane*	*Acetal*	*Nylon 6*	*Nylon 6/6*	*Polycarbonate*	*Polypropylene*	*Polyphenylene sulphide*	*ABS*	*Polyphenylene oxide*	*Polystyrene*	*Polyester PBT*
Injection moulding	•		•	•	•	•	•	•	•	•	•	•	•	•	•
Hand lay-up	•				•										
Spray-up	•				•										
Compression moulding	•	•	•		•						•				
Preform moulding	•				•										
Filament winding	•				•										
Pultrusion	•				•										
Resin transfer moulding	•													•	•
Reinforced reaction injection moulding	•			•	•		•								

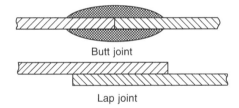

Butt joint

Lap joint

Figure 16.26   Bonded joints

Figure 16.28   Carbon fibre composite front nose of Mercedes Mchaven SLR

Figure 16.27   Mercedes Mchaven SLR body-in-white

Leno weave is used to improve the stability in open fabrics which have a low fibre count. It is a form of plain weave in which the adjacent weft fibres are twisted around consecutive warp fibres. Fabrics in leno weave are usually used in conjunction with other weave style fabrics as leno weave alone is unlikely to produce sufficient strength because of its openness.

Table 16.5 Resin comparison (Courtesy of SPSystems)

Material	Advantages	Disadvantages
Polyester	Easy to use; lowest cost resin	Only moderate mechanical properties; high styrene emissions on open moulds; high cure shrinkage; limited range of working times
Vinylester	Very high chemical/environmental resistance; higher mechanical properties than polyesters	Postcure generally required for high mechanical properties; high styrene content; double cost of polyester; high cure shrinkage rate
Epoxy	High mechanical and thermal properties; high water resistance; long working times available; temperature resistance up to 220 °C; low cure shrinkage	Critical in mixing; corrosive; dangerous to handle; triple cost of polyester

Table 16.6 Properties of reinforcement fibres and other materials (Courtesy of SPSystems)

Material type	Tensile strength (MPa)	Tensile modulus (GPa)	Density (g/cc)	Specific modulus
Carbon HS	3500	160–270	1.8	90–150
Carbon IM	5300	270–325	1.8	150–180
Carbon HM	3500	325–440	1.8	180–240
Carbon UHM	2000	440+	2.0	200+
Aramid LM	3600	60	1.45	40
Aramid HM	3100	120	1.45	80
Aramid UHM	3400	180	1.47	120
Glass – E glass	2400	69	2.5	27
Glass – S2 glass	3450	86	2.5	34
Glass – quartz	3700	69	2.5	31
Aluminium alloy (7020)	400	1069	2.7	26
Titanium	950	110	4.5	24
Mild steel (55 Grade)	450	205	7.8	26
Stainless steel (A5–80)	800	196	7.8	25
HSS steel (17/4 H900)	1241	197	7.8	25

Mock leno

(Courtesy of SPSystems)

Mock Leno is a version of plain weave in which the fibres under–over interlace every two or more fibres apart. This give a thicker fabric which has a rougher surface texture and very good porosity.

16.7.3 Pre-impregnated material (Pre-preg)

Woven material is available pre-impregnated with resin. It is referred to as Pre-preg. This means that the material has exactly the right amount of resin applied to it. The resin is fully coating the material – so that there are no dry spots which could lead to component failure. Pre-preg is therefore quicker to use and the resin density is accurate.

Pre-preg has a limited shelf life which is compounded by the fact that it must be stored at −18 °C. A deep freeze cabinet is therefore needed for storage. The pre-preg can not be unrolled nor cut when it is in the frozen state, so it must be removed from the freezer and brought up to normal room temperature. It is only possible to freeze and de-frost the pre-preg a limited number of times so the material must be managed carefully. The usual way to do this is by means of a control card. The dates and times of defrosting are recorded as is the amount of material taken off the roll. That way the life of the roll and the amount of material left can be seen without removing the roll from the freezer.

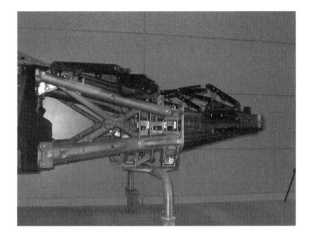

Figure 16.29 Front nose attached to main tub of Mercedes Mchaven SLR with aluminium frame

Curing

The resin, whether it is by wet lay-up or pre-preg needs time and heat to dry it out and make it hard. When the hardener is added to the resin it will generate heat chemically. Be careful, this heat can cause fire and other damage. However, at normal temperature, 20 °C, it will take about 5 days for the resin to become fully hard. During this time period the component should not be moved nor should any stress be applied. To speed up the hardening process and to add extra strength to the component it is normal to use an oven. The oven may be a simple box with an heating element, or an autoclave which is a cylindrical shaped oven that can be pressurized or evacuated inside. The normal procedure is to place the newly made component in the oven, or autoclave, then rack up the temperature gently, over a period of about 30 minutes. Maintain the temperature typically at 150 °C for about 5 hours, then gradually lower the temperature, again over about a 30-minute period. The best way to do this is with a computer control system.

Core materials

Engineering theory tells us in most cases that the stiffness of a panel is proportional to the cube of its thickness. That is, the further apart that we can keep the outer fibres the stiffer the panel will be. Putting a low density core between two layers of composite material will add stiffness with minimum weight and at reasonable cost.

Foam

A variety of materials are used, one of the most common is foam. Foam can be made from a variety of synthetic polymers (Table 16.7) Densities of foam can vary between 30 and 300 kg/m^3 and thicknesses available are from 5 to 50 mm.

Honeycomb

Honeycombs are made from a variety of materials, including extruded thermoplastic – ABS, polycarbonate, polypropylene and polyethylene – bonded paper, aluminium alloy and for fire resistant parts, Nomex. Nomex is a paper-like material based on Kevlar fibres.

Table 16.7 Synthetic polymers and their characteristics for use in foam making

Foam material	Abbreviation	Characteristics
Polyvinyl chloride	PVC	Good resistance to water; available in plain sheet or grid scored
Polystyrene	PS	Very light; low mechanical strength
Polyurethane	PU	Good for thermal and acoustic insulation; can be used at up to 150 °C
Polymethyl methacrylamide	Acrylic	Very strong, but expensive
Polyetherimide	PEI	Good fire resistance, used for interior trim on public carrying vehicles
Styreneacrylonitrile	SAN	High impact strength

Heat

A point to be noted is that most carbon fibre materials are affected by heat. Thermal expansion can lead to micro-cracking. A carbon fibre panel which is painted black will absorb a lot of heat if left in the sun for a long period. This can cause the panel to expand which could lead to micro-cracks in the panel and cracks in the paint work. This will then allow in moisture which will cause further deterioration of the panel.

16.8 Body production in reinforced composite plastic (Lotus)

At present composite reinforced plastic finds its use in road transport applications, where in some cases complete cabs and bodies are manufactured using the material. It is also on the increase in the manufacture of public service vehicles, luxury coaches and caravans. The manufacture of car bodies in this material is still somewhat limited, although some of the British car manufacturers, particularly Lotus Cars Limited, are developing the use of this material in their fibre reinforced composite constructed bodies.

Shapes and forms which are acceptable in steel vehicle bodies can also be produced in composite materials. These materials, such as Kevlar, carbon fibre, glass fibre, non-woven, unidirectional, diagonal and bidirectional forms, can produce moulded structures with a variety of properties. The performance of body panels can be changed whilst retaining the panel thickness, simply by altering the type of reinforcement used within any given panel thickness. The ability to create these effects in composite vehicle design depends on the skill of the designer. Advanced structures are made by incorporating premoulded rigid foam and metallic inserts in the fibre reinforced resin during the injection moulding process, ensuring flexibility that in turn allows reinforcement properties to be accurately tailored to a specific design requirement. This can be done without compromising the original design concept.

Lotus chassis design and construction

Lotus has modified its practice of using a pure backbone chassis for the Elan and opted for a unique, composite platform and backbone type of construction. A major factor behind this decision

was the engineering requirements to manufacture a very taut, rigid open sports car.

The Elan body platform is a one-piece 3 mm nominal thickness vacuum assisted resin injection (VARI) moulding which is riveted and bonded to the welded steel reinforcing outriggers comprising: inner sill, toe board, heel board, A-post and B-post. When bolted to the backbone chassis this results in high torsional stiffness which gives the car exceptional handling characteristics (Figures 16.30 and 16.31). The floor pan is manufactured from isophthalic polyester

Figure 16.30 First stages in building up chassis details on the Lotus Elan (*Lotus Engineering*)

Figure 16.31 Final stages of building up chassis details on the Lotus Elan (*Lotus Engineering*)

resin continuous filament glass fibre with additional local reinforcements in high-load areas such as the body to chassis attachment points and the fuel tank mounting area. The outrigger and A- and B-posts are manufactured from 18 gauge steel, E coated and wax injected for maximum corrosion resistance prior to assembly. Elastomeric polyurethane adhesives are used throughout the construction. These steel components not only contribute to the bending and torsional stiffness of the vehicle but also provide rigid attachment points for seat runners, lower seatbelt mountings and door hinges. Additional structural rigidity and side impact protection is provided by steel cross-braces between the A-posts at the front and the B-posts at the rear (Figure 16.32). The backbone chassis extends rearwards from the front bulk-

Figure 16.32 Floor pan and outrigger attachments, undertray and bulkheads (*Lotus Engineering*)

head and incorporates the rear suspension pick-up points, while the front longeron/underframe assembly bolts on to the front of the backbone frame. This incorporates the front suspension pick-up points, engine mountings and front energy absorbing structures. The complete subframe assembly, including the power train, is detachable to ease both manufacturing and service. High-strength cast aluminium is used for the windscreen pillars, which bolt directly on to the top of the A-posts and are joined by an extruded and formed aluminium header rail.

Lotus body design and construction

The body structure comprises a moulded composite floor pan reinforced with steel in key areas to form stiff box sections. The floor pan is bolted at 16 points to the box section steel back-bone chassis, with further rigidity and occupational protection provided by a high-strength aluminium alloy windscreen frame, a tubular steel scuttle beam, and steel beams in the

doors and B-posts. Most composite exterior panels are bonded to this structure using a flexible polyurethane adhesive, but the frontal panels are secured by threaded fasteners for ease of service access and collision repair. The front bumper/spoiler and rear bumper valance are flexible reinforced polyurethane mouldings resistant to damage from minor knocks. Composite structures have the ability to absorb high impact loads by progressive collapse, with impact damage being localized. In accidents this feature protects the occupants from injurious shock loads and greatly reduces the danger of entrapment by deformation of body panels. This behaviour also facilitates repair by replacing the damaged bolt-on or bonded panels using recognized approved methods.

All the outer body panels are a nominal 2 mm as they are cosmetic and not load bearing. However, there are some exceptions: the undertray, bulkheads, bumper armatures and door inners are thicker to contribute to the structural performance (Figure 16.33).

Figure 16.33 Vehicle undergoing interior trim (*Lotus Engineering*)

All Elans have RRIM bumpers front and rear to comply with US federal regulations, and energy absorbing front bumper construction is used. The door's outer panel shape does not allow conventional hinges: a unique design allows the door to swing in an arc outside, instead of the more traditional inside, of the front A panel. At the latch end of the door a tapered interlock bar has been designed so that during side impact the load path of the low mounted side intrusion beam is through the latch and hence into the main vehicle structure.

The body panels are produced from composite materials, which include a low-profile non-shrink polyester polymer system which has been developed to suit the Lotus VARI process requirements. A patent fibreform process has been developed by Lotus to provide preformed fibre reinforcement which is self-locating inside the VARI tools during the moulding process. An added sophistication is that the production moulds have an electroplated nickel shell surface, which not only extends tool life but also gives a high standard of finish to the body panels, allowing minimal preparation for the painting process (Figure 16.34).

Figure 16.34 Final inspection of the completed Lotus Elan (*Lotus Engineering*)

16.9 Repair of reinforced composite bodies

The repair of reinforced composite bodies and component panels is not difficult, and unless major damage has been sustained it can quite readily be carried out by a competent body repair shop. If the correct materials and tools are used and operations are carried out in the correct sequence, repair is usually considerably easier than to a metal body damaged in an accident of similar severity. The equipment required is simple, the only two items not generally found in the average workshop being a split washer roller and a graduated measuring cylinder. Power tools can speed preparation, but hand tools will be just as effective. Much of the success of the repair depends on the correct preparation of the resin mix. If polyester resin is supplied from a manufacturer, the correct proportions of resin, catalyst and accelerator will be stipulated. In calculating the amount of resin required for the repair, first weigh the cut patches of chopped

strand mat and activate three times this weight of resin. If a large repair is being attempted, do not prepare more than half a pound (0.2 kg) of resin at a time to avoid the mix curing before it can be used.

The first essential when considering the repair of a reinforced composite moulding is to ensure that the area to be repaired is clean and dry, including the rough edge of any torn portion. In many cases a vehicle with minor damage will be driven back from the site of the accident and road moisture and dirt is deposited on the damaged area. The dirt must be washed off and the area dried using some convenient form of heating, but care must be taken that the moulding is not further damaged by too high a temperature. The area of the damage must then be checked and marked for cutting. It is usually found that any break in a moulding is surrounded by an area of bruising where the resin is crushed, and this must be removed. It is unusual for this to extend more than about 8 cm from a break, and the easiest way of checking the exact limits is to shine a powerful lamp through the laminate. The bruising will then show up as a dark or light patch depending on the colour and content of the laminate, and can be marked out for removal.

After all damaged areas have been removed, the cut edges should be feathered on the non-weathering side and then the surface is roughened for about two inches back from the cut. A single layer of chopped strand mat and polyester resin is then laminated into the roughened area. While the area is still wet, a sheet of cellophane large enough to cover this material and the hole is applied and pressed into the uncured resin. The cellophane is then supported by sheet metal, hardboard or card which can be fastened to the moulding with adhesive tape and struts fixed to hold it to contour. The hollow is then almost filled by applying polyester resin and chopped strand mat, continuing to build it up until it is nearly flush with the original surface. The laminate is allowed to cure and then finished off flush and smooth with a standard two-compound gap filler.

16.9.1 Accident damage assessment

All damage to the body is covered by the following categories:

Highly stressed
Moderately stressed
Lightly stressed.

On that definition depends the original construction and therefore the repair method to be employed.

As a general rule there should be a bonded joint wherever two panels touch, or wherever they close on important points. It is usually possible to check these bonds both visually and physically for fractures and breaks. Ascertain the cause of damage and the direction of impact and examine all panels or bonds which may have been affected.

A front-end impact, for example, may easily cause the bonds at the bulkhead to split without the defect being easily visible. To facilitate a closer examination, it may be necessary to remove parts and mechanical components to determine the extent of the damage. When determining the replacement underbody sections and panels to be ordered, make sure the new underbody section and panels will be attached to firm composite materials: avoid badly crazed areas and badly burnt areas.

Fire damage is the most difficult to assess, but generally only the obviously burnt or charred sections will need to be replaced or reinforced.

After a moderate or severe accident, checks should be carried out on the A- and B-posts, chassis datum points, and suspension and wheel alignment.

16.9.2 Accident repair

For repair purposes, accidents may be defined as

Front section
Centre section
Rear section.

This definition determines the original construction and therefore the repair method employed.

Resin and gel mix

The panels on GRP vehicles are manufactured using different types of materials. Therefore, when carrying out repair, the type of resin and gel mixes used will depend on the original materials used in manufacture.

Superficial defect repairs

Pin holes or air voids Either drill or rout out so as to leave a larger hole with near vertical walls, or enlarge by gouging or picking out. Then fill the holes with polyester stopper or filler.

Surface crazing There are various causes of surface crazing, but most results from sharp impacts or accidental damage. During an accident some panels may

flex sufficiently to cause the surface to craze without immediate apparent damage to the painted surface. The crazing may not work its way through the paint surface for some weeks, so it may be necessary when assessing accident damage to carefully examine all panels, particularly near cracks or split bonds. In case of doubt it may be possible to promote the appearance of the crazing by applying gentle heat.

Surface crazing itself generally stops at the first layer of glass fibre and is consequently not structurally serious. However, the extensive crazing near damaged areas should be taken as an indication of over-stressing and the panel should be reinforced or replaced. It is not possible to remedy crazing by simply resurfacing with a further layer of resin.

Replacement sections and panels

Where the repair of a damaged vehicle calls for replacement sections and panels, they may be obtained in most cases direct from the manufacturer of the vehicle under repair.

Where severe damage has been sustained, the damaged section or panel may be cut away, and a replacement section or panel grafted in. Before cutting away any damaged section, or before ordering replacements, the repairer should ascertain the proposed method of repair and the positioning of joint lines. Determine a method of correct positioning of replacement panels and sections such as B-posts, A-posts, or any prominent feature from which measurements can be made, and scribe these clearly on to the section which is to be used. Use a marker to define the lines on which it is proposed to cut the section or panel. Study these lines to see that any damaged section or panel removed will allow adequate pick-up points on the replacement section to ensure it can be accurately positioned, and that the proposed outline transverses longitudinal, lateral and horizontal definition points to assist easy lining up of the new section or panel in all three planes.

16.9.3 Repair of a damaged wing

1   Investigate the damaged area to find its extent. A powerful lamp held on the reverse side of the panel helps. Damage then shows as a dark or light patch depending on the colour of the car. All road dirt, grit and moisture must be cleaned from the area of repair to allow effective adhesion of the new material (Figure 16.35).

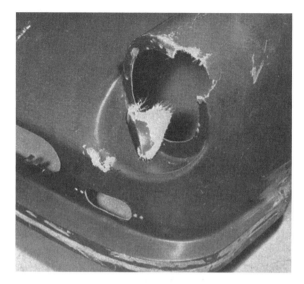

Figure 16.35 Damage area on composite body

2 Decide on the most effective way of repairing the damage, bearing in mind ease of access and the degree of surface finish required.
3 Cut away the damaged area surrounding the point of impact, ensuring that the cut is taken into sound, undamaged laminate (Figure 16.36).
4 Finish preparation of area to be repaired by sanding or filing a scarf edge to the laminate. Cut the scarf with the slope on the opposite face to that from which work will be carried out. This assists in locking the repair in place (Figure 16.37).
5 Temporary moulds are prepared, usually in metal, but before fixing these in place a layer of cellophane is applied to the working area to prevent adhesion of the repair to the mould. Wax polish and polythene sheet are other suitable release media (Figure 16.38).

Figure 16.37 Scarfing the laminate

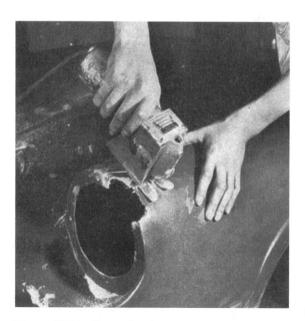

Figure 16.36 Removing damaged laminate

Figure 16.38 Applying release medium and temporary moulds

6 The temporary moulds can be held in place with self-adhesive tape, by wedging, or by packing to a curve with struts and wadding.

7 Prepare the resin and, if working from the inside of the laminate, first paint the temporary moulds with a gel coat of activated resin and allow this to cure.

8 Cut patches of chopped strand mat to fit the area of repair. Two or more layers will be required, and as work proceeds they should be made larger to bond well to the original laminate surrounding the repair (Figure 16.39).

Figure 16.40 Rolling

Figure 16.39 Applying chopped strand mat

9 Impregnate each layer with resin, and roll with a split washer roller to pack down the fibre glass and resin (Figure 16.40).

10 Allow the repair to cure and then remove the temporary moulds (Figure 16.41).

11 Clean up the area of repair, removing all surplus resin and glass and trimming and smoothing any raw edges.

12 File and sand the repair smooth and, if necessary, finish with a proprietary polyester filler. Normal car painting procedure is then carried out (Figure 16.42).

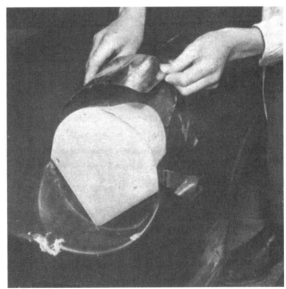

Figure 16.41 Removing temporary moulds

16.9.4 Repairing a body scratch or gash

1 Where damage takes the form of a scratch which may or may not penetrate the skin of the car, the two main requirements are to replace the lost panel strength and fill the damage flush with the original surface.

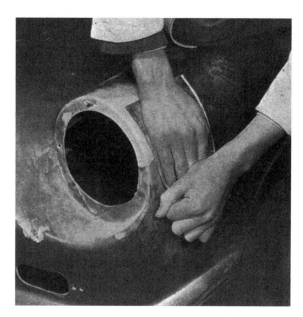

Figure 16.42   Finishing the laminate repair

2   The damage will generally extend for some distance around the scratch, and this must be removed.
3   Again the edge of the cut-out should be scarfed and, as the repair will be carried out from the inside of the body, scarfing should be on the outside.
4   When repairing relatively flat surfaces which are near vertical, it is easier to impregnate the chopped strand mat with polyester resin on the bench. Lay down a piece of cellophane slightly larger than the patch and paint this with resin. Place the glass fibre mat into the resin and complete impregnation by stippling on more resin.
5   Extract air from the laminate while still on the bench, using the split washer roller. When removing them from the bench, lift the cellophane and the laminate together.
6   Use self-adhesive tape to cover the cut-out area from the outside of the body. A smoother surface will be obtained if this is backed by a solid flexible former. Paint resin on to the mould from inside the body, ensuring that it fills the scarfed edge completely.
7   Apply the laminate complete with cellophane to the inside of the repair and press into contact through the cellophane (the cellophane should be left in position until the repair is cured).

8   After curing, strip off the cellophane and the former and adhesive tape from the outside of the body. The repair is completed by filling imperfections with polyester resin, after this has cured, by normal smoothing, flatting and painting.

16.9.5 Repairs to a blind panel

On occasion it is impossible to reach both sides of a damaged panel. In such a case repair is more difficult but can be still carried out. When removing the damaged section of laminate, cut the panel back to form a rectangle. Prepare a separate moulding larger all round than the rectangle but still sufficiently small to pass through the panel.

When this patch is cured, fix a loop of wire through two holes in the centre. Apply resin and chopped strand mat to the rough surface of the cured laminate, leaving the centre free. Insert the prepared patch through the panel, keeping hold of the wire loop. Using a convenient lever, apply pressure to pull the patch into contact with the reverse side of the body panel and allow to cure. Remove the wire loop and fill the repair flush with the original body panel using chopped strand mat and polyester resin. Finish off resin rich, and cover with cellophane and squeegee to give a smoother finish. After cure, remove the cellophane and proceed with normal filling, flatting and painting.

16.9.6 Replacing part components

The techniques for freeing composite body components and for replacing part of a component are shown in Figures 16.43 and 16.44. Special clamps are available for holding composite body panels in alignment while they are under repair and being bonded in place (Figure 16.45).

16.9.7 Important points in reinforced composite material repair

1   The strength and ultimate surface finish of a repair will depend to a great extent on thorough preparation of the damaged area.
2   If either the damaged area or the repair materials are at all damp, the strength and life of the repair will be seriously affected.

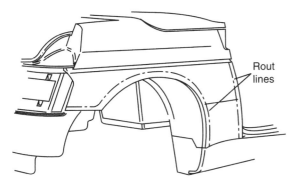

Figure 16.43 Removing lower quarter panel on Lotus Elan (*Lotus Engineering*). Before removing the lower quarter panel it will be necessary to remove the boot lid, rear top shell and rear transom.

1 Using a router, cut through the rear of the lower quarter panel just above the undertray.
2 Carefully rout along a line 40 mm above the undertray and following the rear wheel arch to the B-post cover.
3 Using the appropriate Lotus panel tool, separate the lower quarter panel from the B-post cover and discard.
4 Using the Lotus panel tool, separate the remaining section of the lower quarter panel from the undertray.
5 Using a 36 grit paper and mechanical sander, remove all old adhesive from the undertray and B-post cover.

3 Many repairs are ruined by attempting to smooth off or polish before the resin is thoroughly cured. Any repair should be allowed to stand at least overnight, or longer if the weather is particularly cold.

4 The glass fibre reinforcement provides the strength of the repair, and either chopped strand mat or cloth must be used to achieve the required strength. It is not sufficient to embed layers of glass tissue in polyester resin.

16.10 Common faults in moulded laminates

Many complaints concerning the appearance and performance of these mouldings stem from the basic cause that the resin is undercured. There are, however, several problems in the form of visible flaws or other defects, the remedies of which will become apparent from the analysis of the causes.

16.10.1 Wrinkling

This is caused by solvent attack on the gel coat by the monomer in the laminating resin due to the fact that the gel coat is undercured. Wrinkling can be avoided by ensuring that the resin formulation is correct and the gel coat is not too thin, and by controlling temperature and humidity and keeping the work away from moving air, especially warm air. If the workshop is equipped with hot air blowers, these should be directed away from the moulds (Figure 16.46).

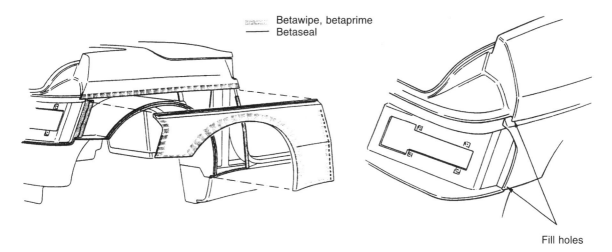

······ Betawipe, betaprime
——— Betaseal

Fill holes

Figure 16.44 Refitting lower quarter panel by bonding (*Lotus Engineering*)

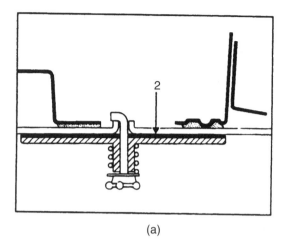

(a)

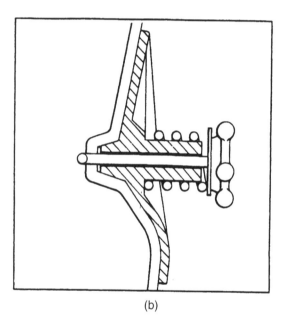

(b)

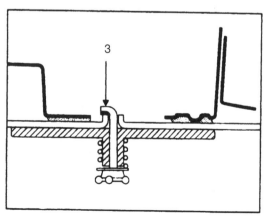

(c)

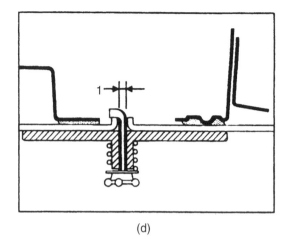

(d)

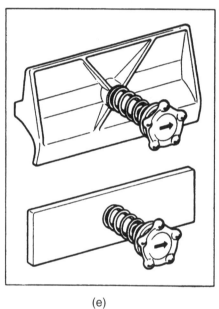

(e)

Figure 16.45 Centring and aligning composite components using special clamps (*Lotus Engineering*)

(a) Side components (front wing, front and rear door external panels, rear wing) are centred and aligned with respect to each other, using tools available in the form of a kit.

(b) Kit comprises: six flat plates, two profiled plates.

(c) Function: vertical adjustment by means of the shape of the profiled plate fitting in the recess between the parts.

(d) Clearance 5.5 mm, obtained by the diameter of locking pin (1) with the components in contact with one another.

(e) Alignment obtained by the flat of plates (2).

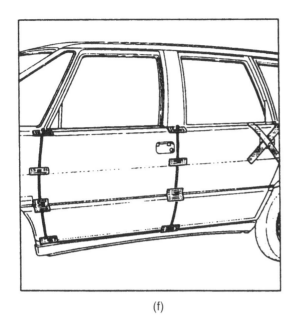

(f)

Figure 16.45 (*continued*)

(f) Fitting and locking: each component is locked in place on the side of the component which is not replaced (3) (except when replacing a rear door panel with a rear wing). Leave the adhesive to harden for 30 minutes before removing the tools.

Do not open the door or doors when the tools are fitted on them: there is a risk of damaging the panel or the tools.

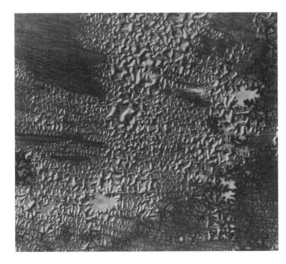

Figure 16.46 Wrinkling (*Scott Bader Co. Ltd*)

16.10.2 Pin-holing

Surface pin-holing is caused by small air bubbles which are trapped in the gel coat before gelation. It occurs when the resin is too viscous, or has a high filler content, or when the gel coat resin wets the release agent imperfectly (Figure 16.47).

Figure 16.47 Pinholing (*Scott Bader Co. Ltd*)

16.10.3 Poor adhesion of the gel coat resin

Unless the adhesion of the gel coat to the backing laminate is very poor, this defect will be noticed only when structure is being handled and pieces of gel coat flake off. Areas of poor adhesion can be detected sometimes by the presence of a blister, or by local undulations in the surface when it is viewed obliquely. Poor gel coat adhesion can be caused by inadequate consolidation of the laminate; by contamination of the gel coat before the glass fibre is laid up; or, more generally, by the gel coat being left to cure for too long (Figure 16.48).

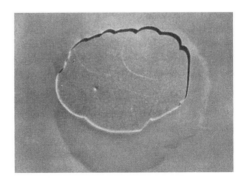

Figure 16.48 Poor adhesion of the gel coat (*Scott Bader Co. Ltd*)

16.10.4 Fibre pattern

The pattern of the composite reinforcement is sometimes visible through the gel coat or prominently noticeable on its surface. This usually occurs when the gel coat is too thin or when the reinforcement has been laid up and rolled before the gel coat has hardened sufficiently, or when the moulding is removed too early from the mould (Figure 16.49).

Figure 16.49   Fibre pattern (*Scott Bader Co. Ltd*)

16.10.5 Fish eyes

On a very highly polished mould, particularly when silicone modified waxes are used, the gel coat is almost non-existent. This shows up as patches of pale colour usually up to 6 mm in diameter. It can also occur in long straight lines following the strokes of the brush during application. This fault is rarely experienced when a PVAL film is correctly applied.

16.10.6 Blisters

The presence of blisters indicates that there is delamination within the moulding and that the air or solvent has been entrapped. Blisters which extend over a considerable area may also indicate that the resin is undercured, and this type of blister may not form until some months after moulding. Blisters can also occur if the moulding is subjected to an excessive amount of radiant heat during cure. A possible cause of this defect is the use of MEKP rather than cyclohexanone peroxide paste. If, on the other hand, the blister is below the surface, the cause is likely to be imperfect wetting of the glass fibre by the resin during impregnation. This would be due to the fact that

insufficient time had been allowed for the mat to absorb the resin before rolling. Blisters of this kind can usually be detected by inspection as soon as the moulding has been removed from the mould (Figure 16.50).

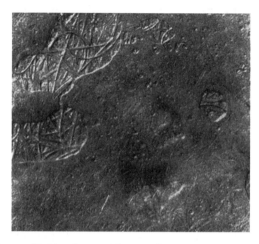

Figure 16.50   Severe blisters (*Scott Bader Co. Ltd*)

16.10.7 Crazing

Crazing can occur immediately after manufacture or it may take some months to develop. It appears as fine hair cracks in the surface of the resin. Often the only initial evidence of crazing is that the resin has lost its surface gloss. Crazing is generally associated with resin-rich areas and is caused by the use of an unsuitable resin or resin formulation in the gel coat. The addition of extra styrene to the gel coat resin is a common cause. Alternatively the gel coat resin may be too hard with respect to its thickness. In other words, the thicker the gel coat the more resilient the resin needs to be. Crazing which appears after some months of exposure to the weather or chemical attack is caused by either undercure, the use of too much filler, or the use of a resin which has been made too flexible (Figure 16.51).

16.10.8 Star cracking

This is the result of having an over-thick gel coat, and occurs when the laminate has received a reverse impact. Gel coats should never be more than 0.5 mm thick (Figure 16.52).

Figure 16.51   Crazing (*Scott Bader Co. Ltd*)

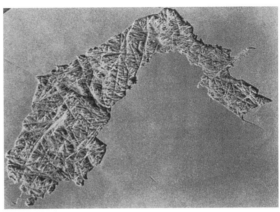

Figure 16.53   Internal dry patch (*Scott Bader Co. Ltd*)

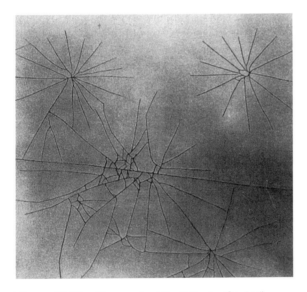

Figure 16.52   Star cracks (*Scott Bader Co. Ltd*)

16.10.9 Internal dry patches

These can be caused by attempting to impregnate more than one layer of mat at a time. The presence of internal dry patches can be readily confirmed by tapping the surface with a coin (Figure 16.53).

16.10.10 Leaching

This is a serious fault. Leaching occurs after exposure of the laminate to the weather, and is characterized by a loss of resin from the laminate, leaving the glass fibres exposed to attack by moisture. Leaching indicates either that the resin used

has not been adequately cured, or that it is an unsatisfactory resin for that particular application.

16.10.11 Cracking, shrinking and discolouring

The identification of this fault is visual. Deep cracks appear in the resin and the colour changes from a green to a mauve purple and hot to the touch. The cause of this fault can be large build-up of resin due to drainage or to excess application; both lead to extreme exotherm. Alternatively, incorrect amounts of catalyst or accelerator may be used, usually an excess. The effect of the build-up of exotherm is to cause excessive shrinking of the moulding and internal stress. The extreme exotherm can also damage and distort the mould itself. This can be prevented by: avoiding large amounts of resin (but where this is necessary the build-up should be gradual); using the correct amount of ingredients; being aware of any increase in workshop temperatures; and noting any variations in the correct percentages of catalyst or accelerator used.

16.10.12 Low-rigidity laminate

The first identification is by touch; the laminate will feel spongy and more flexible than usual. Alternatively, check by applying a Barcol hardness tester. The common causes of this fault are low resin content or undercure of the resin. To prevent this fault, ensure the correct ratio of resin to glass and eliminate any draughts. Large areas should be made to gel quicker to cut down styrene loss, and correct proportions of catalyst and accelerator must be used.

16.11 Safety precautions

The handling of polyester resin, glass fibre, and ancillary materials such as catalysts, presents several hazards which can be reduced to a minimum if the correct precautions are taken. Most glass fibre materials and resins are perfectly safe to use provided the potential hazards are recognized and reasonable precautions are adopted. Normally you will have no problems if you follow these rules:

1 Do not let any materials come into contact with the skin, eyes or mouth.
2 Do not inhale mist or vapours, and always work in a well ventilated workshop.
3 Do not smoke or use naked flames in the workshop.

16.11.1 Storage precautions

Liquid polyester resins are flammable but not highly flammable, most of them having a flashpoint of 31 °C. Resins and accelerators should preferably be kept in a brick-built store conforming to the normal fire regulations for a paint store. The storage life of polyester resin is about six to twelve months provided the resin is kept below 20 °C in the dark (in metal drums). Storage at higher temperatures, even for only a few days, will considerably reduce the shelf life.

Catalysts are organic peroxides and present a special fire hazard. They should be stored in a separate area, preferably in a well ventilated fire-resisting compartment. If kept reasonably cool they will not burn or explode. In case of fire in the vicinity, they can be kept safe by drenching the containers with water.

16.11.2 Operating precautions

Most polyester resins contain monomeric styrene, which is a good grease solvent and may cause irritation to the skin. The most effective method of protecting hands is the use of a barrier cream or rubber gloves, and this is strongly recommended. Resin can be removed from the hands with proprietary resin removing creams, or with acetone followed immediately by a wash in warm soapy water. In sufficient concentration styrene vapour is irritating to the eyes and respiratory passages, and therefore workshops should be well ventilated. When resin is sprayed a gauze mask should be worn to protect the mouth and nose. This also applies to

trimming operations when resin and glass dust can cause irritation.

Catalysts are extremely irritating to the skin and can cause burns if not washed off immediately with plenty of warm water. Particular care must be taken with liquid catalysts to avoid splashing, spilling or contact with the eyes. Protective goggles should be worn as a necessary precaution.

16.11.3 Workshop conditions

The building should not be damp and it should be adequately heated and ventilated. Good head room is desirable and sufficient space should be allowed for all operations. The floor area should be divided into sections as follows: preparation of reinforcement, mixing of resins, moulding, trimming and finishing. Resin and curing agents should be stored away from the working area in a cool place, observing the necessary precautions for flammable liquids and keeping in mind the special hazards associated with organic peroxides. Glass fibre should be stored and tailored under dry conditions and separately from the moulding area. The temperature of the building should be controlled between 15 °C and 25 °C. Ventilation should be good by normal standards, but draughts and fluctuations in temperature must be avoided, so doors and windows should not be used for ventilation control. Dust extraction in the trimming section should be of the down-draught type. Cleanliness is important both for health of the operators and for preventing contamination of resin and reinforcement.

As far as possible, health and comfort depend in the first place on planned extraction, and in the second place on workshop education in the nature of the materials used. Almost all the offence comes from the styrene vapour and glass filaments, both of which advertise their presence before the concentration reaches a danger level. As far as is known the only real source of physical harm is the dust produced in grinding, but all the materials and byproducts contribute to discomfort, and sensible evasive action is essential.

16.11.4 Spillage and disposal

Most of the following products are covered by the terms of the Deposit of Poisonous Wastes Act:

Polyester resin Absorb spillages in dry sand and dispose by landfill or controlled incineration.

Furane resin Extinguish all naked lights, open doors and windows. Absorb spillage into sand or chalk. Pack into drums, seal and store prior to collection by specialized chemical disposal company.

Catalyst Absorb into vermiculite, remove to landfill or controlled incineration. Wash down remaining traces with copious water.

Accelerator Absorb into dry sand and dispose by landfill or controlled incineration.

Release agents Wash down with water.

Mould cleaner Absorb into sand or earth, remove to landfill or controlled incineration. Flush contaminated area with water.

16.11.5 Fire hazards

Many resins and associated products are either flammable or contain flammable additives. Styrene, catalyst and acetone are particularly dangerous. Do *not* smoke or use naked lights, oil burners or similar heating devices in the working area. If a fire does start, do not attempt to put it out with water unless it is a catalyst. Dry powder extinguishers can be used on accelerator, mould cleaner, acetone, resins and release agents.

Fires can be started if catalysed but uncured resins are thrown away. The waste resin will continue to cure and the heat generated by the curing process can ignite other waste materials. Therefore unwanted resin should be left in a safe place until it is fully cured; it can then be discarded without risk of fire.

Questions

1 Explain the following abbreviations: RRIM, VARI, PVAL.

2 State three advantages of reinforced composite materials when used in vehicle body construction.

3 List the main physical properties of glass fibre composite materials.

4 Explain the advantages and disadvantages of reinforced composite materials when used as an alternative to low-carbon steel.

5 Explain briefly the function of (a) the releasing agent (b) the gel coat (c) the catalyst.

6 Describe the process of contact moulding.

7 Explain the purpose of pre-accelerated resins.

8 With the aid of a sketch, show the lay-up of a laminate in the mould.

9 Describe the type of tools that would be used to trim the edges of reinforced laminates.

10 Describe the sequence of repair to a damaged glass fibre composite body panel.

11 Describe how a patch mould can be used during the repair to damage of a GRP laminate.

12 Describe the likely damage that would occur to a panel made from GRP and which had been subjected to a heavy blow.

13 Name the types of materials that could be used to reinforce polyester resin.

14 Name three reinforcing materials that could be added to a GRP moulding to give strength to the laminate.

15 Explain why a GRP panel may offer better resistance to minor damage when compared with a low-carbon steel panel.

16 In the automobile industry, why is GRP limited in use to a small specialist sector?

17 Describe, with the aid of a sketch, a test to show that GRP is more elastic and less ductile than aluminium sheet.

18 State the reasons why GRP has not replaced low-carbon steel as the material used to manufacture vehicle bodies.

19 State the reasons why certain manufacturers of sports cars prefer to make their vehicle bodies from GRP.

20 Explain how to repair a deep scratch in a GRP body panel.

Automotive finishing and refinishing

17.1 History of automotive finishing

No repair to a vehicle is complete until it has been painted to match the rest of the vehicle and is rendered undetectable. This part of the operation is carried out by the spray painter, who must have a knowledge of the type of materials used in the repair shop in order to help him select the best process for refinishing the vehicle concerned. Nowadays the spray painter has the help of the paint manufacturers, who can supply him with literature to cover every painting process, but his predecessor, the coach painter, had to have a very solid understanding of the materials at his disposal.

As the term coach painting implies, this is a craft which dates back long before the days of the motor car. Reference is made to coach painting in the diary of Samuel Pepys, in which he makes mention of the buying and repairing of a second-hand horse-drawn carriage. The amount charged to him for the repainting was sixty pounds which, four hundred years ago, would be a fairly large sum of money. The time involved for the repainting was one month which, bearing in mind the type of paint used, was not unreasonable.

In those days the painter not only mixed his own colours but actually manufactured his own paints, using a pestle and mortar to grind the pigment and oil together. The choice of materials available was rather limited, but it is to the credit of the craftsmen of those days that the finished appearance was of a very high standard and extremely durable. Perhaps the best protective pigment available to the craftsman was white lead, and he made full use of it. The

lead paste was mixed with linseed oil and applied by brush, one coat every second day, being too slow in drying to allow for more frequent coatings. Several coats were applied, each being rubbed down smooth prior to the application of the next coat. When the work was judged to be ready for the colour coats, these were also applied in several layers, being too transparent to cover in one solid coat.

When the painting was completed, the sign writer took over and embellished the coach with line work and heraldic emblems. One of the most widely used materials was silver leaf; gold leaf was not then available. Following this part of the work, as many as seven coats of varnish were applied. The varnish, being rather yellow in colour, tended to enhance the silver leaf by tinting it amber and giving it the appearance of gold. This slow, laborious and costly painting process continued almost without alteration right up to the end of the nineteenth century and the birth of the motor car. Paint manufacturers had, however, come into being, with a consequent improvement in the range and quality of materials at the disposal of the painter.

The early motor car, like its predecessor the coach or carriage, was of coach-built construction, and the existing methods of painting were suitable for this type of vehicle. However, as the motor car increased in popularity and demand for it grew, a new and faster method of production had to be found. This was achieved by the advent of pressed steel construction, but the paint process caused a bottleneck to production and so research was carried out to solve this problem.

The answer came with the development of cellulose lacquer which, though not a complete answer, was nevertheless an extremely fast drying material which allowed for several coats to be applied in one day (speed, of course, being the main criterion). Being so rapid in drying, cellulose was not suitable for brushing purposes and so the application of paint using a spray gun came into its own. By 1930 all new motor cars were being finished by this method. The material, however, was lacking in solid content, and consequently several coats had to be applied to achieve a coating of worthwhile thickness. Another time consuming factor was that the finished vehicle had to be burnished to obtain a high gloss.

Around 1935 cellulose-lacquer based paints were combined with other synthetic materials to produce a paint which dried in thirty minutes, had better 'build' qualities and thus required fewer coats. It also reduced the burnishing time and so eased the bottleneck which still existed in the paint section of the production line.

During the Second World War a great deal of research was carried out, and success achieved, with thermosetting paints which could be force dried at elevated temperatures. These paints provided a hard glossy finish, required fewer coats than the cellulose materials and were more chemically resistant. A further advantage was that the finish required no burnishing or polishing. In all, this was a paint ideally suited to the expanding motor vehicle industry, and by the early 1950s all new motor cars were being finished in these stoving synthetics.

As well as improvements in finishing materials, changes in painting techniques were being evolved. Perhaps the most revolutionary changes were introduced in the application of the priming paints, mainly in the field of dip application. In this method the entire body shell is completely immersed in a tank of priming paint (which is specially formulated for this purpose), is withdrawn, allowed to drain, and is then passed on to a stoving oven for baking.

Stoved synthetic finishes became the accepted finish on new motor vehicles, but difficulties were experienced in refinishing damaged areas as a result of colour fading. Though the colours did not fade drastically, they did, however, fade sufficiently to give the refinisher a difficult job to obtain a perfect match.

In 1963 Vauxhall introduced a finish on their new Viva model which the paint manufacturers claimed had better colour stability. This was the acrylic resin stoving finish which was produced with the cooperation of the plastics industry (being of the thermoplastic type). By 1965 Ford had changed all of their colours to a high-bake acrylic finish, which was a product of the paint industry only, being thermosetting. BLMC, Rootes, Standard, Triumph and Rover followed suit by changing most of their colours to the acrylic range. Acrylic paints, as well as possessing good colour stability, are durable, have good gloss and are easily polished.

The method by which the priming coat on modern vehicle body shells is applied is known as electrodeposition (Figure 17.1). A large dip tank containing 2500 litres of a water-borne paint is included in the production line. An overhead conveyor carries the body shells from the pre-cleaning area to the dip tank. The paint is charged with electricity and the shell is earthed through the conveyor. The thinner of the paint, being water, acts as an electolyte; the paint solids, i.e. pigment and

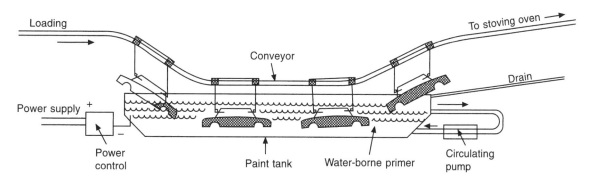

Figure 17.1 Electrodeposition of priming coat

binder, are ionized and are attracted to the earthed car body. An even coating of paint is thus applied, even on thin metal edges. The thickness of the coating can be varied according to the electrical potential introduced. When the car body moves out of the tank, surplus paint drains out of it and the shell is then rinsed off under sprinklers, which does not affect the electrodeposited coating. The car body is then dried off and baked.

As to the future of motor vehicle finishing, it seems reasonable to expect water-borne paints to be developed to such an extent that they will become the accepted finish on new motor vehicles. Looking even further ahead, it could be that the car body will be formed entirely of a moulded plastic which could be self-coloured. Should this come about, damaged areas could be removed and replaced with a new section which is already coloured to match the rest of the car. However, there will still be a place for the refinisher, as car owners will, in all probability, desire the occasional colour change on their car. In all, there have been many developments since the days of the coach painter and his homemade paints.

17.2 Glossary of terms used in spray painting

In order to be able to appreciate more fully the descriptions of processes and practices in the paint shop, the reader should make himself acquainted with the following trade terms and items of equipment.

Air delivery   The actual volume of compressed air delivered by the compressor after making allowances for losses due to friction. It is measured in litres per second.

Air duster   A tool which, when fitted to an air line, is useful for blowing water from recesses and for drying a surface quickly prior to painting.

Air pressure   The pressure of air which has been mechanically compressed. It is measured in bars or pounds per square inch (psi).

Air receiver   A reservoir or storage tank to contain compressed air.

Atomization   The breaking up of paint or other materials into very fine particles. Good atomization is essential in spray painting.

Double-header coating   This results from the practice of spraying one coat immediately after another without allowing a flash-off period.

Dry coating   Several thin coats of paint can be applied fairly rapidly if they are sprayed 'dry'. This is done by increasing the ratio of atomizing air to paint at the spray head of the gun. Dry coating is particularly useful when carrying out local repairs to paintwork.

Feather edging   The rubbing down of a damaged area of paintwork until there is no perceptible edge between the paint and the substrate.

Feathering the gun   To ease the pressure on the gun trigger whilst spraying, thereby reducing the volume of paint passing through the fluid tip. This is done mainly when spraying local areas to achieve a feather edge.

Flash off   To allow the greater part of the more volatile solvents in a sprayed coat of lacquer or enamel to evaporate before proceeding with the application of another coat or with stoving.

Fluid cup   A container for the paint attached to the spray gun in conventional spray painting, or a separate item in pressure feed systems connected to the spray gun by means of a fluid hose.

Fluid nozzle   The orifice in the fluid tip.

Ground coats   The paint coats between the primer and finishing coats. Ground coats are usually of a similar colour to the enamel.

Guide coat   A thin coating applied as evenly as possible over a surface to be rubbed down. Following rubbing down, no trace of the guide coat should remain, so that complete flattening of the surface is achieved. The guide coat should obviously be of a contrasting colour to that over which it is applied.

Hold-out   The degree of imperviousness of a dried paint film. Some filler coatings in particular are porous and so they tend to absorb the binder or medium of finishing coats, thus reducing their effectiveness as a glossy finish.

Matt finish   A surface finish which has no glossy effect.

Shrinkage   This refers to the manner in which some paints decrease in size not only vertically but also horizontally. As with sinkage, this is caused by solvent evaporation. Nitrocellulose materials are particularly prone to this phenomenon, which can affect the adhesion to the substrate.

Sinkage   This trade term can have two interpretations. When paint is applied over a particularly porous surface it will sink into it, and if this paint is a finishing material the gloss will be impaired. The other explanation of the term brings in solvent evaporation. The solvents or thinner added to paints to reduce the viscosity evaporate during the drying process and consequently some of the liquid content of the paint vanishes. When this happens the paint film becomes thinner and projections on the substrate come through it.

Tacking off   To wipe over a surface with a specially treated cloth, which is slightly sticky, to remove dust. Tacking off is essential before applying finishing coats.

Viscosity   The degree of resistance to flow of a liquid. More simply, it refers to the thickness or otherwise of a fluid such as paint.

17.3 Basic composition of paint

Pigments   Fine solid particles which do not dissolve in the binder. They give colour and/or body to the paint. Some pigments possess good anti-corrosive properties and are used in paints designed to give protection to the substrate. Extenders are cheaper than pigment, but when used in the correct proportions they carry out many useful functions such as improvement of adhesion and ease of sanding.

Binder   Reacts to form a film, and binds the pigments together and to the surface. The binder is often referred to as the medium of the paint.

Thinner   Some of the liquid of the paint is often withheld from the paint container and supplied separately as a *thinner*. The user adds thinners to adjust the viscosity to suit his requirements.

Additives   Small quantities of substances which are added to carry out special jobs. Wax in varnish creates a matt finish, and silicones in metallic paint give a hammer finish.

Figure 17.2 illustrates the composition of paint.

17.4 Types of paint

Cellulose synthetic

This dries by the evaporation of the solvent. The main advantage of this material is rapid air drying. However, there are a number of disadvantages. The coating dries rapidly only when thin films are applied, otherwise drying is delayed by solvent retention. The high proportion of solvent used (60 per cent in most cases) results in shrinkage which causes the film to adhere poorly to the substrate. The absence of chemical change means that the dried film does not increase in chemical resistance, and is readily softened by the original solvent.

Oil paints

The drying of an oil paint depends on the ability of certain drying oils to dry by a reaction that involves atmospheric oxygen, a process which is confined to relatively thin films.

Synthetic paints

These are mixtures of drying oils and synthetic resins. The most obvious limitation of a paint based solely on a drying oil is slow drying. To improve this property and to give tougher films and improve the gloss, a resin is added to the oil and they are cooked together for a period so that they chemically combine. The varnishes produced can be divided into two main classes based on their oil to resin content: long oil and short oil.

Stoving paints

These are also mixtures of oils and resins that require exposure to an elevated temperature to

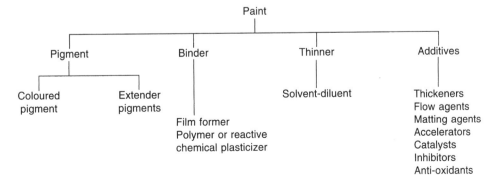

Figure 17.2   Composition of paint

produce a cure (dried film). The time of exposure is mainly dependent on the temperature: 60 minutes at a temperature of not less than 138 °C; and 10 minutes at a temperature of not less than 205 °C.

Blacking paints

Chassis black is a cheap black paint generally based on bitumen. It has good adhesive qualities on bare metal and is a good rust inhibitor.

Tyre black is also a cheap black paint, being of low viscosity. Several proprietary brands are available.

Two-pack paints

These are probably the most widely used paints in the vehicle refinishing trade, with more than 80 per cent of refinishers preferring them. They present special health hazards, and the user should be equipped with an air-fed mask and face visor to prevent inhalation of the vapours when spraying (Figure 17.3). A canister mask of the CC type can be used as an alternative, but these can prove to be expensive as the canister is only useful for 15 minutes continuous use and should then be discarded. Precautions should also be taken to prevent the mixed material or spray vapours making contact with the skin, as this can cause dermatitis.

These paints consist of a base material and a catalyst or activator. When they are mixed together, a chemical reaction takes place which results in complete polymerization. Two-pack (or 2K) paints have a limited pot life after mixing, but when curing is complete they can equal stoving paints in hardness and durability. They are characterized by high solids content and low solvent content, which results in high build and good scratch filling with the minimum number of coats, thus resulting in savings on labour time and overspray wastage.

The gloss from the gun is good and no burnishing or polishing is needed unless dirt is present in the finish. Should this be the case, the finish can be wet flatted with P1200 paper, using soap as a lubricant, and burnishing can be carried out using fine rubbing compound and a polishing machine of 6000 rev/min. A clean, dry lambswool mop is recommended for best results.

Materials included in this group are: acrylic and polyurethane primer undercoats and finishes, including base-coat-and-clear finishes; epoxy resin primers and finishes; and polyester spraying fillers.

Figure 17.3 DeVilbiss Pulsafe breathing air kit showing half-mask and visor outfits (*DeVilbiss Automotive Refinishing Products*)

It is common practice in most refinishing paint shops to force dry these materials using low-bake ovens on large areas for 30 minutes of 60 °C, or using infrared lamps on small areas.

Low-bake finishes

These are modified stoving paints which can be completely cured at a temperature between 66 and 93 °C. The material was formulated for the refinisher to enable him to match more closely the original finish of the car manufacturers. These are now losing favour to two-pack materials.

17.5 Materials used in refinishing

17.5.1 Primers and dual-purpose primers

Primer

The priming coat is the first coat of paint on any surface. Its functions are to gain maximum adhesion to the substrate, to provide a sound base for subsequent coatings, and on metals to act as a corrosion inhibitor. The priming paint should be selected to suit the type of surface on to which it is to be applied.

Etch primer

The best types of etch primer are two-pack materials, the base and activator being supplied separately. They are mixed ten minutes before use, but have a limited pot life (about six hours), although longlife etch primers are now available. Brushing activators are available for those shops where spray painting is not practical. Etch primers have a fairly high water absorption characteristic and should be coated with surfacer or filler after the appropriate drying time, to avoid moisture absorption from the surrounding atmosphere. Special thinners are provided for etch primers and should be kept for this purpose only.

The pigment for this type of paint is zinc chromate which makes it an ideal primer for aluminium, although it can be used also with good effect on most other metals. The activator contains phosphoric acid which etches the surface, thereby ensuring good adhesion. An extremely thin coating gives best results. Once the base and activator have been mixed together they should on no account be returned to the tin of base.

Primer surfaces

A primer surfacer does the work of both a primer and a surfacer.

Self-etching primer surfacers are available which eliminate the need for using an etching primer and then coating over with surfacer or filler. Self-etching primer surfacers have gained in popularity, the main advantage being improved adhesion to metal substrates when compared with standard primer surfacers.

Primer filler

A primer filler is similar in function to a primer and a filler.

Polyester primer filler is a two-pack paint, first used extensively on the European continent. When introduced into Britain it was viewed with a certain amount of distrust, as the claims made for it (primer, stopper, surfacer rolled into one) appeared too good to be true. There was good cause for this distrust, as one of the difficulties encountered was that blistering occurred when it was used with nitro cellulose materials. Another factor against it was that it required four hours to cure. However, modifications have been made to it which have resolved the problem of blistering and reduced the curing time to two hours at an ambient temperature of 25 °C, which can be still further reduced to fifteen minutes when force dried.

This paint is particularly suitable for use on rough surfaces where heavy coatings are required to level up the surface with the minimum of effort. Several coatings can be applied wet-on-wet as it dries by catalyst action. The amount of sinkage after drying is virtually nil. Polyester primer/filler possesses exceptionally good build qualities, and when dry can be very easily smoothed down with abrasive paper used either wet or dry. Face masks should be worn when dry rubbing is carried out.

Best results are obtained if the paint is applied with a gravity-fed spray gun. When spraying local repairs the paint can be applied at the supplied viscosity, but for a large area a small quantity of the appropriate thinner may be added to obtain better atomization and to provide a smoother coating. Spray guns must be thoroughly cleaned immediately after being used with polyester primer filler. As the paint solvent is extremely volatile, the cap must be screwed down tightly on the tin when not in use. Polyester primer filler can be used with most car refinishing processes and is very useful when used with low-bake enamels.

17.5.2 Fillers and levellers

Surfacer

A surfacer is applied over the primer. Its function is to build up the coating thickness, whilst filling up minor defects such as scratches.

Filler

Filler is a heavy-bodied material used for levelling defects which are too deep to be filled economically with surfacer. Fillers are manufactured to suit the method of application: there are spraying, brushing and knifing types. Fillers are available as both single-pack or two-pack, cellulose and synthetic types.

Stopper

Stopper is a putty-like substance used for filling up defects too deep for satisfactory levelling with either filler or surfacer.

Though deep indentations are normally filled up by the body repair worker, the painter is sometimes required to carry out levelling work. A two-pack stopper, usually based on polyester resin and a catalyst, is used for this purpose. It dries rapidly in heavy layers, unlike cellulose or oil-based stoppers which must be applied in thin layers with a drying period between applications. Polyester stopper is intended for use on bare metal or over high-baked primers. It cannot be used as an intercoat stopper over etching primers, or between enamel coats. It can, however, be coated over with most standard paint systems used in refinishing, including cellulose synthetic, coach finish and low-bake synthetics. The normal curing time of a polyester stopper is about thirty minutes at 20 °C.

17.5.3 Sealers

There are three types of sealers used by the refinisher: standard, isolators and bleed inhibitors.

Standard sealers have a low pigment and high binder content. They are supplied ready for use and are applied over the final coat of surfacer or spray filler to provide hold-out of the finishing material and promote higher gloss. They also reduce the risk of crazing when applying acrylic lacquer-type finish over cellulose-based undercoats.

Isolating sealers are more heavily pigmented and are used to avoid a reaction between different types of paint systems, e.g. when applying a lacquer-type paint over a synthetic enamel.

Bleed-inhibiting sealers contain carbon black pigment which is able to absorb floating colour. These sealers are recommended when carrying out a colour change over a colour which is suspected of being a bleeder, i.e. some of the pigment or dyestuff will float into the new coating and discolour it. Several reds, particularly those containing organic pigments, are prone to this behaviour.

Whichever sealer is used, it can only do its job if it is a continuous film. For this reason it must not be flatted down, through a light denib, carefully done, is permissible.

17.5.4 Finish

Finish is the term used to describe the finishing colour coats. They have a comparatively low pigment content as opposed to surfacers and fillers. The high percentage of the vehicle or binder provides the glossy effect.

17.5.5 Abrasive papers

The type of abrasive paper mainly used by the spray painter is known as wet-or-dry paper. The abrasives used are silicon carbide and aluminium oxide, and these are attached to a treated paper backing by means of waterproof resin glue.

Wet-or-dry paper is available in various grades ranging from 80D (coarse) to P1200 (very fine): low numbers denote coarse grades and high numbers identify the finer grades. This type of abrasive paper is normally used in conjunction with water to avoid a build-up of paint particles which would affect the abrading effectiveness of the paper.

In addition to wet-or-dry paper, the spray painter should have in his stock emery paper and production papers. These are much coarser than wet-or-dry papers and are used for rougher work such as removing rust or mill scale. These two papers are most effective when used with power tools in the form of circular discs, being attached to the pad with specially formulated disc adhesives. As they are normally used dry, the particles are attached to the paper or cloth backing with a good quality hide glue.

Abrasive papers fall into two categories, based on the amount of space between the particles. If these are widely spaced the paper is referred to as open coated and is used to remove paint or rust which tend to fill up the spaces on finer grades.

When the particles are tightly packed the paper is known as close coated and is used to rub down smoother surfaces.

17.5.6 Masking

Masking tape

This is a paper tape, one side of which is coated with an adhesive of a non-drying composition. It is supplied in rolls in a variety of widths, but the most widely used are those measuring 20 mm and 25 mm. Wider tapes are considered to be uneconomical, though narrower tapes such as 13 mm do have limited uses. Where two-tone work is carried out, a finer edge can be achieved by the use of gummed paper than is possible with masking tape, but the time spent on removing it makes it less popular.

Masking machines are also available in which a roll of masking paper and various widths of masking tape are mounted. As the paper is pulled out of the machine, a strip of the tape is automatically attached to one of the edges (Figure 17.4).

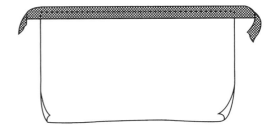

Figure 17.5 Method of fixing masking tape to paper

mind that newspaper is extremely porous and subject to solvent penetration, and it is therefore recommended that at least two layers be used over existing paintwork. It is common practice in some shops to oil or grease the newspaper, but this cannot be recommended because of the risk of grease contamination to those areas to be sprayed.

17.5.7 Burnishing

Burnishing compound

There are both burnishing pastes and polishing liquids. These are generally emulsions of mineral oils and water with the addition of an emulsifying agent. They also contain mild abrasives to 'cut down' the final coat of enamel and promote a good lustre. Burnishing compounds are also used during the carrying out of local paint repairs to remove overspray.

Standard compounds and creams or liquids contain ammonia to keep them fresh in the tin, but ammonia can cause staining of clear coatings. Special ammonia-free compounds and liquids are available for the burnishing of clear-over-base finishes.

Mutton cloth

In days gone by, butchers and slaughterers wrapped joints of meat in this material – hence its name. Mutton cloth is available in different grades: coarse, medium and fine. The coarser grades are used with rubbing compound to remove overspray when carrying out localized paint repairs. The fine grades are used with fine compounds and creams for final burnishing to promote a deep gloss.

Paper roll

Masking tape

Figure 17.4 Masking machine

Masking paper

Brown paper (kraft paper) is an ideal masking material, though newspaper is very widely used for this purpose (Figure 17.5). It should be borne in

17.5.8 Solvents

The paint shop should be well stocked with the appropriate solvents for the types of paint to be used. The importance of using the correct thinner for a particular type or make of paint cannot be over-stressed, as many painting defects can be traced to the incorrect use of solvents. Solvents can prove to be an expensive item, and in consequence a cheaper form of cleaning solvent should be stocked for the purposes of cleaning spray guns and equipment.

Solvents fall into two categories, high boilers and low boilers. Those which require a high temperature to bring them up to their boiling point tend to be slow in evaporating and consequently slow in drying. Low boilers, on the other hand, evaporate and dry quickly. Low boilers are best used with primers, surfacers and fillers where it is required to build up the coating thickness fairly rapidly. High boilers can be used in finishing coats to promote better flow, helping to eliminate an orange-peel defect. A good quality solvent, however, will contain both high and low boilers in well balanced proportions.

17.6 Spray painting equipment

The items of equipment in a spray painting shop are basically as follows: air compressing unit, air line, air filter, pressure regulator; air hose and finally spray gun.

17.6.1 Air compressor

When selecting an air compressor for a particular workshop, one should calculate the volume of compressed air that will be required to operate the various tools throughout the workshop, such as rubbing-down tools and spray guns. The size of the compressor chosen should be capable of giving a higher free air delivery than is required.

In a workshop where more than one activity is carried out, such as panel beating and spray painting, it is advisable to install a stationary two-stage compressor. This should be bolted to the floor at least 300 mm from any wall, and sited where it can receive an ample supply of clean dry air. A two-stage compressor consists of an electric or fuel oil motor, the compression unit itself, and a storage tank. In addition to these basic components there can also be various refinements in the way of safety devices and operating switches.

Taking a piston-type compressor as an example, the operational sequence is as follows. The motor drives the compressing pistons, which are situated within two cylinders set above the storage tank. Air

at normal atmospheric pressure is drawn into the first cylinder via the air intake, to which is attached a filter. The air is then compressed to an intermediate pressure by the action of the piston moving upwards and reducing the volume of the cylinder (Figure 17.6). When a sufficient pressure has been attained, the air is pumped through a chamber (or intercooler) into the second smaller cylinder, where it is compressed even further by piston action and finally discharged into the storage tank for subsequent use (Figure 17.7).

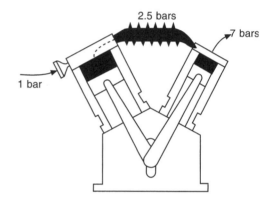

Figure 17.6 Single-acting two-stage compressor

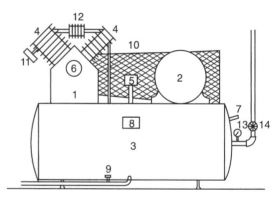

1. Compressor	7. Safety valve
2. Motor	8. Starter
3. Air receiver or storage tank	9. Drain cock
4. First and second stage valves	10. Safety guard
5. Pressure switch automatically controls volume of air in tank	11. Air intake
	12. Intercooler
6. Centrifugal pressure release on large models relieves motor starting against load	13. Pressure gauge
	14. Release cock

Figure 17.7 Principal parts of a stationary two-stage compressor

The compressed air is stored in the receiver (or storage tank) for a cooling-off period, during which moisture and vapours within the air will condense and collect on the floor of the tank. The very act of compressing air will generate heat, and so the cooling period is essential. In a further effort to keep the air cool, the compression cylinders have external fins to provide a greater surface area for the heat generated within them to dissipate into the atmosphere. In addition to storing the compressed air and providing a cooling-off period, the air receiver acts as a buffer between the compressor and the spray gun, blanketing out pulsations so the compressed air can be drawn from it at a steady even pressure.

It is important that the size of the air receiver be in direct proportion to the size of the compressor. Contrary to popular belief, air volume is more important to the spray painter than is air pressure. A typical suction-fed spray gun will require 3–4.5 litres of free air delivered (FAD) per second to allow it to operate satisfactorily. The gun must have this volume of air from the air receiver regardless of the pressure. In a typical two-stage piston-type compressor, the free air delivered will only be 70 to 75 per cent volumetric efficiency. Free air delivery is in consequence the prime factor to consider when purchasing a compressor.

17.6.2 Air lines

The compressed air is drawn from the air receiver and led to the air transformer through a galvanized tube known as an air line. The three main requirements of an air line installation are:

1 Low pressure drop between the compressor plant and the points of air consumption
2 Minimum of air leakage
3 High degree of contamination filtering throughout the system.

Pressure drop in an air line is caused by the frictional action of the air molecules on the inside surface of the pipe, as well as pressure build-up at angles. The greater the distance between the compressor and consumption points, the greater will be the pressure drop. For this reason the air line must be of sufficient internal diameter to carry a satisfactory volume of compressed air into the spray room. For the average refinishing shop an internal diameter of 25 mm will suffice. Pressure drops can

be increased where angles are introduced into the system, and in consequence the number of angles should be kept to a minimum. Also the more couplings there are in the installation, the greater is the risk of air leakage with a consequential drop in pressure.

It is inevitable that some moisture vapour will be drawn into the air line from the air receiver, and for this reason the air line should slope down towards the air receiver to enable condensed moisture to run into it rather than to the air transformer. The service line leading to the transformer should be tapped from the top of the main air line, and should preferably be in the form of a U bend rather than elbow joints with right-angled bends on them (Figure 17.8).

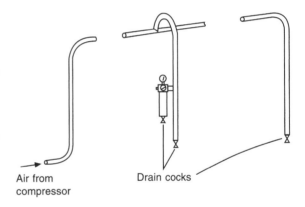

Air from compressor   Drain cocks

Figure 17.8   Air line installation

17.6.3 Air transformer

This consists of two units: a condenser or filter, and a pressure regulator. The *condenser* allows the compressed air to expand into a chamber, thus assisting cooling, and then removes moisture from it by means of a removable filter. A drain cock is situated at the base of the chamber to drain off accumulated impurities periodically. The *regulator* is a reducing valve with which to reduce the air pressure from the compressor to that required for spraying. The air transformer should be fitted with a pressure gauge giving an accurate reading of the pressure of air passing through the regulating valve (Figure 17.9). A small lightweight filter can be fitted to the handle of the spray gun as an additional safeguard in conditions where exceptional humidity exists.

Figure 17.9 DeVilbiss DVFR-1 filter regulator assembly and DVF2-2 filter regulator coalescer (*DeVilbiss Automotive Refinishing Products*)

17.6.4 Air hose

The compressed air is led from the transformer to the spray gun by means of an air hose. This consists of a rubber tube covered with cotton braid enclosed within a rubber covering; the three layers are vulcanized into one (Figure 17.10). Multibraid hoses are available for high-pressure work. At each end of the hose are couplings for attachment to the transformer and the spray gun.

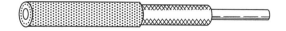

Figure 17.10 Single-braid air hose

Though the interior wall of the air hose is smooth, it will still create a certain amount of resistance to the flow of air, particularly when long lengths are used. Hoses with an interior diameter of 6 mm should never be used in lengths exceeding 4 m because of the high pressure drop encountered. For example, if the air transformer is delivering air at a pressure of 3.5 bar (50 psi) the pressure drop over 4 m would be 0.7 bar (10 psi). However,

using an air hose of diameter 8 mm over the same length and at the same supplied pressure, the drop would only be 0.2 bar (3.5 psi). From this it can be seen that a hose of 8 mm diameter will give best results on lengths exceeding 4 m. When working in a spray booth, the spray painter rarely, if ever, requires an air hose greater than 6 m in length.

17.7 Types of spray gun

Of all the tools and techniques used in paint shops, the spray gun and spray painting have provided the most satisfactory method of applying paint. Unless there is a complete change in the design and construction of the motor vehicle, the spray gun will be used for many years to come. The spray gun, like all tools, is only effective in the hands of a skilled operator, and therefore a painter should know as much as possible about what has become the main tool of the trade.

A spray gun is a precision instrument which uses compressed air to atomize the fluid paint and break it up into small particles. The air and paint enter the gun through separate passages, mix, and are then ejected at the front of the gun.

Spray guns can be divided into groups:

1 By methods of paint supply, such as suction feed, gravity feed or pressure feed
2 Those with detached or attached paint containers
3 Internal mix or external mix types
4 Bleeder or non-bleeder types.

The most widely used spray gun is the suction-feed, external-mix, non-bleeder model.

17.7.1 Paint supply methods

Suction-feed gun
With this type, a stream of compressed air creates a vacuum at the fluid tip which allows atmospheric pressure within the fluid cup to force the paint up the fluid tube to the fluid tip and air cap. The paint container (fluid cup) is limited to one litre (1000 cm^3) capacity to enable the gun to be handled without fatigue. The suction-feed spray gun is easily identified, as the fluid tip protrudes slightly beyond the air cap (Figure 17.11).

Gravity-feed gun
The fluid cup is mounted above the spray head and paint is fed to the gun by the force of gravity. The fluid cup is usually limited to 0.5 litre capacity,

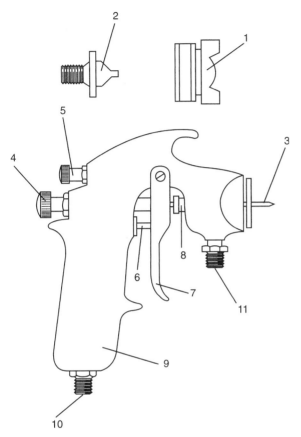

Figure 17.11 Typical suction-feed spray gun

1	Air cap	7	Trigger
2	Fluid tip	8	Fluid packing nut
3	Fluid needle	9	Gun body
4	Fluid control screw	10	Air inlet
5	Spreader control	11	Fluid inlet
6	Air valve		

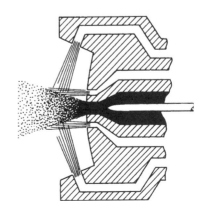

Figure 17.12 Gravity-feed spray gun (*DeVilbiss Automotive Refinishing Products*)

which makes the gravity-feed gun unsuitable for the spraying of large areas. However, it is very useful for painting local repairs where heavy-bodied fillers are applied, and where rapid colour changes are necessary (Figure 17.12).

Pressure-feed gun

This type of gun sprays paint that has been forced from the paint container by compressed air. The air cap of this gun is not designed to create a vacuum, and the fluid tip is flush with the front of the air cap (Figure 17.13). Pressure-feed guns are used where a large quantity of a particular paint is to be sprayed, or where the material is too viscose or

Figure 17.13 Pressure-feed spray head

heavy to be siphoned from the fluid container as in suction-feed guns.

The fluid container or pressure vessel is connected to the gun by means of a reinforced fluid hose, and normally ranges in size from 2 to 25 litres. The smaller pressure vessels can be carried in the operator's free hand (Figure 17.14), but ones from 10 litres capacity upward can be mounted on wheels for easy portability (Figure 17.15).

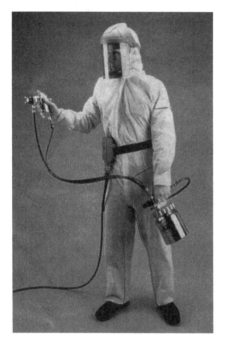

Figure 17.14 Remote cup with gun (2.3 litre capacity) (*DeVilbiss Automotive Refinishing Products*)

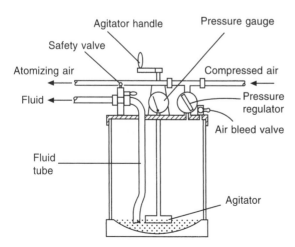

Figure 17.15 Typical pressure-feed tank

17.7.2 Mix methods

Internal-mix gun

This gun mixes air and paint inside the air cap, and is used with low air pressure to apply show-drying materials (Figure 17.16).

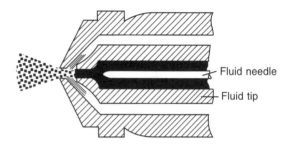

Figure 17.16 Internal-mix spray head

External-mix gun

This is the most widely used type of gun, and can be used to spray most types of paint. It is the best type of gun for spraying quick-drying materials. The air and paint mix beyond the air cap, and perfect atomization can be achieved (Figure 17.17).

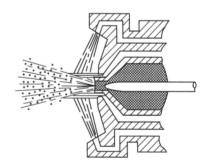

Figure 17.17 External-mix suction-feed spray head

17.7.3 Bleed methods

Bleeder gun

This is designed without an air valve. Air continually passes through the gun, thus preventing a build-up of pressure in the air hose. The gun is usually used with small compressors having a limited output and having no unload or pressure switch. The trigger on the gun controls the flow of paint only.

Non-bleeder gun

These are equipped with an air valve. The trigger controls both the flow of compressed air passing through the gun and the flow of paint.

17.7.4 High-volume low-pressure (HVLP) spray guns

Legislation which is part of the Environmental Protection Act requires that spray painters must reduce the amount of paint vapours being released into the atmosphere as a result of their working activities (see Chapter 15).

In conventional air-atomized spray painting, about 35 per cent of the mixed paint actually reaches the job surface. The remaining 65 per cent is either extracted to the atmosphere or collects on the workshop floor and walls. Of this waste material, 30 per cent is solid whilst 70 per cent is classified as volatile organic compounds (VOCs). It is with these VOCs that the legislation is concerned.

In an effort to conform to the EPA, spray gun manufacturers have developed spray guns which atomize the paint at low air pressure, that is 0.6 bar (10 psi) as opposed to the usual 4–5 bar (60–75 psi). This reduced air pressure results in greatly reduced overspray, and as a further bonus 65 per cent of the mixed paint reaches the surface. Savings on paint wastage are obvious.

Though atomization is achieved with much lower pressure, these spray guns still require large volumes of compressed air to operate them. A typical air pressure at the air transformer may be 4 bar (60 psi), but this is reduced in the gun body by means of an air restrictor which reduces the air velocity at the gun outlet in the ratio of about 6:1. Other manufacturers like DeVilbiss do not have restrictors in the guns and only need 25 psi at the inlet to give 10 psi at the cap (Figures 17.18 and 17.19).

A special air cap is provided with the gun. This has a pressure gauge attached to it which enables the spray painter to adjust the outlet pressure. When the pressure has been adjusted to 10 psi, the air cap is removed and a more conventional type fitted (Figure 17.20). The fluid cup is pressurized and can be either attached to the gun (Figure 17.18) or remote from it (Figure 17.21). Gravity types are also available (see Figure 17.22). The basic parts of these types of spray guns are seen in Figure 17.23.

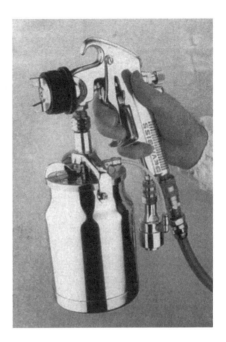

Figure 17.18 High-volume low-pressure (HVLP) cup gun (*DeVilbiss Automotive Refinishing Products*)

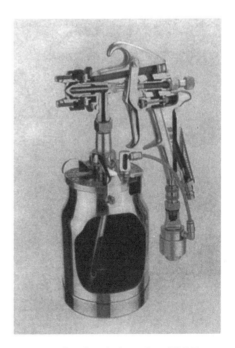

Figure 17.19 Sectional view of an HVLP cup gun (*DeVilbiss Automotive Refinishing Products*)

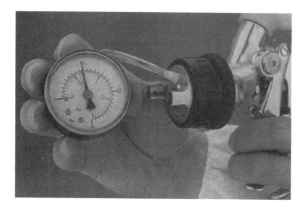

Figure 17.20 Air cap with pressure gauge attached (*DeVilbiss Automotive Refinishing Products*)

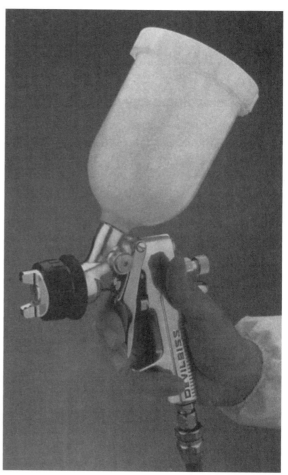

Figure 17.22 HVLP gravity-fed spray gun (*DeVilbiss Automotive Refinishing Products*)

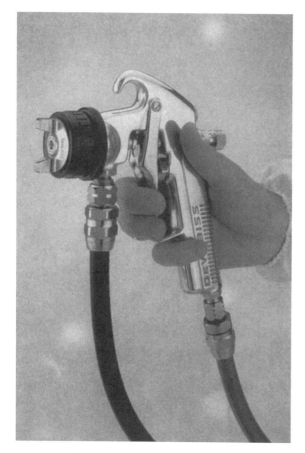

Figure 17.21 HVLP spray gun using a remote fluid cup (*DeVilbiss Automotive Refinishing Products*)

17.8 Basic parts of a standard spray gun

Air cap This is the nozzle at the front of the gun that directs compressed air into the stream of paint, thus atomizing it and forming the spray pattern. Air caps are designed to give a wide variety of spray patterns and sizes, and in addition ensure perfect atomization over a wide range of paint viscosities. The choice of an air cap (Figure 17.24) depends on the following:

1 Volume of compressed air available
2 Type of paint feed system being used
3 Type and volume of paint to be sprayed
4 Size of the fluid tip (air caps are usually designed to operate with a particular fluid tip)
5 Nature and size of the surface to be painted.

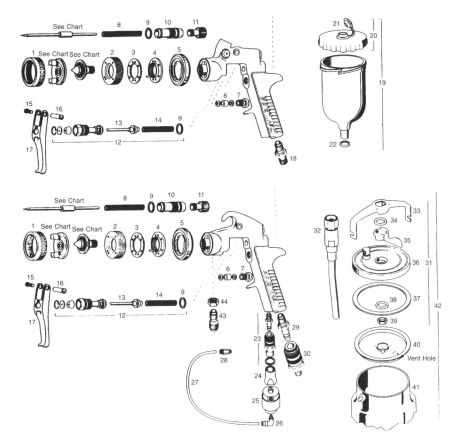

Figure 17.23   Basic parts of the HVLP spray guns (*DeVilbiss Automotive Refinishing Products*)

1 Nylon air cap retaining ring	23 Mini QD female valve and male stem
2 Front air cap baffle (nylon)	24 Adaptor to fit regulator to QD valve
3 Kit of 2 baffle seals	25 Cup pressure regulator
4 Rear baffle assembly	26 Kit of 2 regulator elbow connectors
5 Spreader/fan adjustment ring	27 Kit of 5 regulator to cup tubes
6 Fluid needle packing set	28 Cup pressure tube connector
7 Fluid needle packing gland	29 QD stem
8 Kit of 5 fluid needle springs	30 QD valve
9 Kit of 5 gun body bushing gaskets	31 Complete cup lid assembly $\frac{3}{8}$ in BSP
10 Fluid needle body bushing	32 Fluid tube $\frac{3}{8}$ in BSP female fitting
11 Fluid needle adjusting screw	33 1 quart KR cup yoke
12 Complete air valve assembly	34 Kit of 5 washers
13 Air valve stem	35 Cam lever
14 Kit of 3 air valve (trigger) springs	36 KR 1 quart pressure cup lid
15 Kit of 5 trigger pivot screws	37 Kit of 3 KR cup lid gaskets
16 Kit of 5 female trigger pivot studs	38 Washer
17 Chrome plated trigger	39 Fluid tube retaining nut
18 Male/male air connector $\frac{1}{4}$ in BSP	40 Kit of 5 drip free diaphragms
19 1 pint nylon gravity cup assembly	41 1 quart (1.14 litre) PTFE lined pressure cup
20 Gravity cup lid assembly	42 KR cup and lid assembly (PTFE lined)
21 Kit of 5 drip check lids	43 Fluid inlet connector $\frac{3}{8}$ in BSP
22 Kit of 5 O-rings for gravity cup	44 Fluid inlet locknut

Figure 17.24 Air cap, fluid tip and fluid needle (*DeVilbiss Automotive Refinishing Products*)

Fluid tip This is situated behind the air cap and meters out the paint. The volume of paint passing through the fluid tip and into the stream of compressed air is governed by the diameter of the orifice in the fluid tip. The choice of fluid tip depends, in the main, on the type of material to be sprayed. Heavy, coarse or fibrous materials require large nozzle sizes to prevent clogging, whilst thin materials, which are applied at low pressures, require small nozzle sizes to prevent an excessive flow of paint. The fluid tip provides a seating for the fluid needle (Figure 17.24).

Fluid needle This seats in the fluid tip, its function being to start and stop the flow of paint. For the gun to operate efficiently, the fluid tip and fluid needle should be selected as a pair, and should be of the same size (Figure 17.24).

Fluid control screw This is an adjustment control which limits the length of travel of the fluid needle, governing the flow of paint from the fluid tip.

Trigger The function of the trigger is to operate the air valve and also the fluid needle.

Air valve This is situated in the handle of the spray gun (or gun body) directly behind the trigger by which it is operated. Its function is to control the passage of air through the gun.

Gun body This can be regarded as a basic frame on to which the spray painter will mount a suitable set-up to suit his requirements.

Spreader control This is of great importance in controlling the volume of air passing to the horn holes of the air cap. If air is cut off from the horn holes a narrow jet of paint giving a spot pattern is ejected, but when air is allowed to pass through the horn holes a fan spread is obtained, the width of fan varying according to the volume of air (Figure 17.25).

17.9 Spray gun maintenance and cleaning

17.9.1 Cleaning a suction-feed gun

First, loosen the gun from the paint container, allowing the fluid tube to remain in the container, and unscrew the air cap a few turns. Holding a piece of cloth over the air cap, pull the trigger to divert air down the fluid tube and drive the paint back into the container. Next, empty paint from the container and rinse it out with the appropriate solvent and an old paint brush. Pour a small quantity of the solvent into the container and spray it through the gun to flush out the fluid passages. Remove the air cap and immerse it in clean solvent, then dry it out by blowing with compressed air.

If the holes in the air cap become blocked with dried paint, a stiff brush moistened with solvent will usually remove the obstruction. If not, a toothpick or sharpened matchstick can be used to clean out the hole. On no account must wire or a nail be used, for this can distort the holes and permanently damage the air cap, resulting in a distorted spray pattern.

17.9.2 Cleaning a gravity-feed gun

Remove the cup lid, empty out surplus paint and replace it with a small quantity of solvent. Replace the lid, unscrew the air cap a few turns and, holding a piece of cloth over the air cap, pull the trigger. Air will be diverted into the fluid passage, causing a boiling action and flushing it clean. Spray solvent through the gun and clean the air cap. Finally, with a solvent-soaked rag wipe the outside of the paint container clean.

17.9.3 Cleaning a pressure-feed set-up

Shut off the air supply to the pressure tank, release pressure in it and loosen the lid. Unscrew the air cap a few turns, hold a piece of cloth over it and pull the trigger. The pressure will force the paint

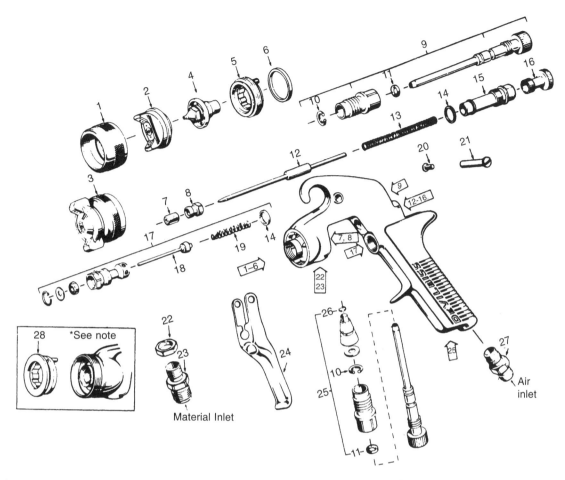

Figure 17.25 Basic parts of a standard spray gun (*DeVilbiss Automotive Refinishing Products*)

1	Retaining ring for air cap	15	Gun body bushing
2	Air cap	16	Fluid needle adjusting screw
3	Air cap and retaining ring	17	Air valve assembly
4	Corrosion resistant fluid tip and gasket	18	Air valve
5	Baffle	19	Kit of three springs
6	Kit of five seals	20	Kit of five screws
7	Kit of five JGA-7 fluid needle packings	21	Kit of five trigger bearing studs
8	Fluid needle packing nut	22	Locknut
9	Valve assembly	23	Connector $\frac{3}{8}$ in BSP
10	Kit of five circlips	24	Trigger
11	Kit of five O-rings	25	Air flow valve
12	Fluid needle	26	Retaining ring
13	Spring	27	Connector $\frac{1}{4}$ in BSP
14	Kit of five gaskets	28	Baffle

from the gun and fluid hose back into the tank. Empty surplus paint from the tank and pour in a small quantity of solvent, replace the lid firmly and pressurize the tank at about 0.3 bar (5 psi). The pressure will force the solvent to the gun. Hold a piece of cloth over the air cap and pull the trigger, and the higher atomizing pressure will force the solvent back into the pressure tank. If this is repeated about a dozen times, the purging action will clean out the fluid passages. Disconnect the fluid hose and blow it out with compressed air. Steep and clean the air cap and finally dry out the tank with a piece of cloth.

Whichever type of gun is used it must not be immersed in solvent, as this causes the lubrication oil to be washed away and will cause paint leakage from the gun. In addition to this, the air passage could become blocked with pigment sludge.

17.9.4 Lubrication

After the gun has been cleaned, a drop of oil should be applied to the fluid needle packing, air valve packing, and trigger fulcrum screw.

A spray gun will function efficiently provided that it is clean and well maintained, but neglect will eventually cause the gun to malfunction.

17.10 Spray gun motion study

The spray gun is not a difficult tool to master, but a study of the following text and the accompanying diagrams will be invaluable to the inexperienced sprayer. Any person who is using a spray gun for the first time should obviously spray a few practice panels such as disused car doors, wings, bonnets and so on to get the feel of the gun before undertaking actual work.

17.10.1 Spraying flat surfaces

The gun should be held at right angles to the work and at a distance of 150–200 mm. Should the gun be held too close to the surface this will result in too much paint being applied, causing runs and sags. Holding the gun too far from the surface creates excessive overspray and a sandy finish (Figure 17.26). The relationship of gun distance and stroke

speed is easily understood, and with a little practice the sprayer is able to adjust his speed of movement to suit the distance between the gun and the surface. The gun distance should be kept as constant as possible, and arcing of the gun must be avoided to obtain an even coating thickness. The correct gun action is acquired by keeping the wrist flexible (Figure 17.27). Do not tilt the gun; hold it perpendicular to the surface. Tilting will give an uneven spray pattern resulting in lines across the work (Figure 17.28).

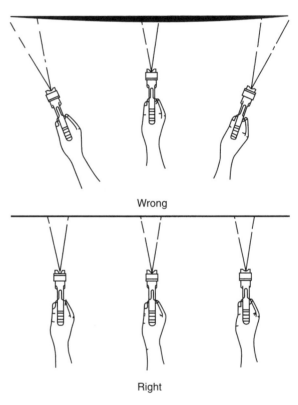

Wrong

Right

Figure 17.27　Maintaining spray distance

When spraying a panel, the technique of 'triggering' the gun must be mastered. The stroke is started off the panel and the trigger is pulled when the gun reaches the edge of the panel. The trigger is released at the other edge of the panel but the stroke is carried on for a short distance before reversing for the second stroke. This triggering action must be practised and perfected to avoid a build-up of paint at the panel edges and to reduce paint wastage due to overspray.

The method of spraying a panel is shown in Figure 17.29. Note that the gun is aimed at the top

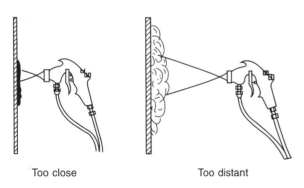

Too close　　　　　　　Too distant

Figure 17.26　Spray distance

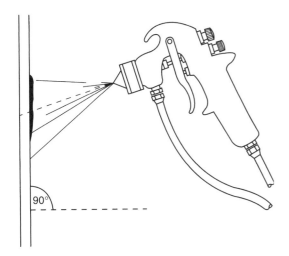

Figure 17.28 Tilting spray gun

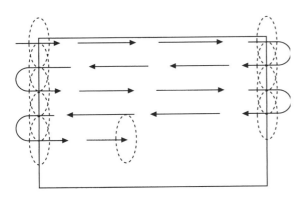

Figure 17.29 Panel spraying method

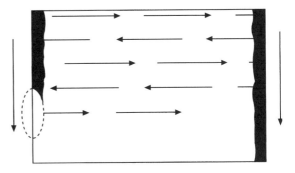

Figure 17.30 Alternative panel spraying method

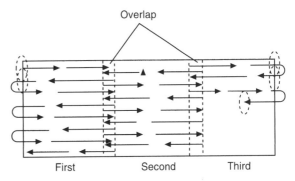

Figure 17.31 Spraying large panels

edge of the panel, and from then on the aiming point is the bottom of each previous stroke. This gives the 50 per cent overlap necessary to obtain a wet coating. An alternative method is shown in Figure 17.30. In this method the ends of the panel are first sprayed with single vertical strokes, the panel then being completed with horizontal strokes. This technique reduces over-spray and ensures complete coverage of the surface.

Long panels such as those encountered on furniture vans require a different approach. A certain amount of arcing is permitted to avoid a build-up of paint where the strokes overlap, and the triggering of the gun is very important. The length of each horizontal stroke is 450–900 mm approximately, or whatever the sprayer can manage comfortably. Figure 17.31 shows the method of overlapping with

the panel being sprayed in separate sections, each section overlapping the previous one by about 100 mm.

When spraying level surfaces such as car roofs and bonnets, always start on the near side and work to the far side to redissolve any overspray. A certain amount of gun tilting is usually unavoidable when reaching across a car roof and overspray is thus created.

17.10.2 Spraying curved surfaces

As previously stated, the gun should be kept at right angles to the surface and as near a constant distance from it as possible (Figure 17.32).

17.10.3 Spraying external corners

Figure 17.33 shows the method of spraying the edges and corners of a panel, the centre being sprayed like a plain panel. Figure 17.34 shows the technique used to paint vertical corners, the point to watch being to half trigger the gun to avoid applying too much paint.

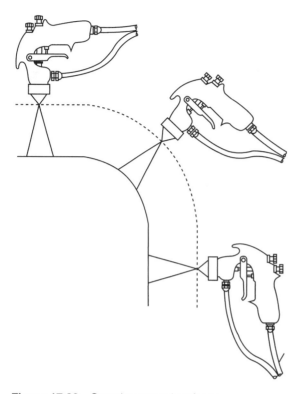

Figure 17.32 Spraying curved surfaces

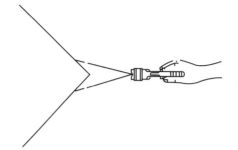

Figure 17.34 Spraying vertical corners

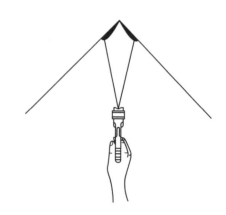

Figure 17.35 Spraying internal corners

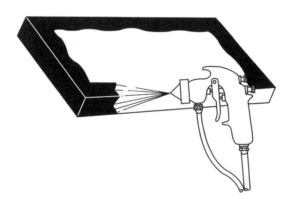

Figure 17.33 Spraying external corners

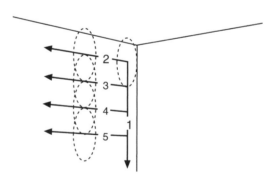

Figure 17.36 Spraying internal corners: method for better finish

17.10.4 Spraying internal corners

Spraying directly into a corner (Figure 17.35) gives an uneven coating but is satisfactory for most work. When an even coating is necessary, such as with metallic finishes, it is better to spray each face separately, starting with a vertical stroke at the edge of the panel (Figure 17.36). The vertical stroke should be followed with short horizontal strokes in order to avoid overspraying or double coating the adjoining surface.

17.10.5 Spraying sequence

An automobile should be sprayed in sections, spraying one section at a time before moving on to the next one. Figure 17.37 shows a typical method, but this may vary depending on the size and shape

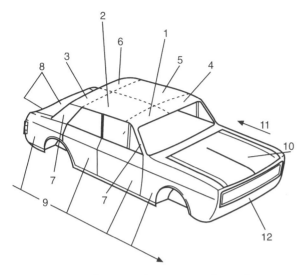

Figure 17.37 Suggested sequence of spraying a car

of the vehicle concerned (Figures 17.38 and 17.39). The painter must decide on his approach before commencing to spray and then work methodically round the car, finishing at a point where an overlap is least noticeable. As the bonnet of the car is the

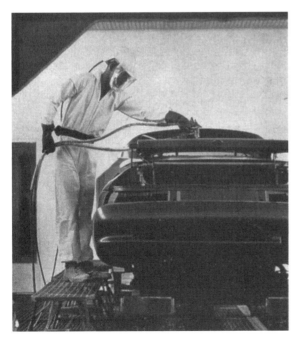

Figure 17.38 Spraying a vehicle roof (*Racal Safety Ltd*)

Figure 17.39 Spraying a vehicle wing (*DeVilbiss Automotive Refinishing Products*)

panel which attracts most attention by the customer, most spray painters prefer to spray this last to avoid the risk of overspray falling on it.

17.11 Spraying defects

No matter how excellent a spraying equipment is, sooner or later some small trouble shows itself which, if it were allowed to develop, would mar the work done. However, this trouble can usually be very quickly rectified if the operator knows where to look for its source. The following sections contain the causes and remedies of all the troubles most commonly encountered in spraying.

17.11.1 Fluttering spray

Sometimes the gun will give a fluttering or jerky spray (Figure 17.40), caused by an air leakage into the paint supply line. This may be due to the following (numbers correspond to those on the figure):

1 Insufficient paint in the cup or pressure feed tank so that the end of the fluid tube is uncovered
2 Tilting the cup of a suction-feed gun at an excessive angle so that the fluid tube does not dip below the surface of the paint
3 Some obstruction in the fluid passageway which must be removed

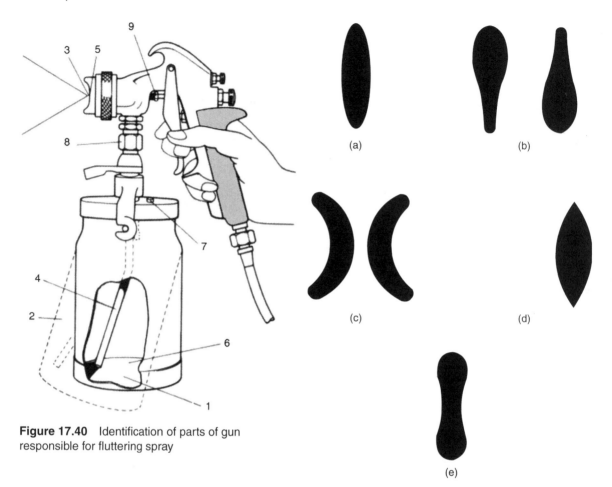

Figure 17.40 Identification of parts of gun responsible for fluttering spray

Figure 17.41 Faulty spray patterns: (a) correct pattern (b) top or bottom heavy (c) right or left sided (d) heavy centred (e) split

4 Fluid tube loose or cracked or resting on the bottom of the paint container
5 A loose fluid tip on the spray gun
6 Too heavy a material for suction feed
7 A clogged air vent in the cup lid
8 Loose nut coupling the suction feed cup or fluid hose to the spray gun
9 Loose fluid needle packing nut or dry packing.

17.11.2 Faulty spray patterns

The normal spray pattern produced by a correctly adjusted spray gun is shown in Figure 17.41a, and defective spray patterns can develop from the following causes:

1 Top or bottom heavy pattern (Figure 17.41b) caused by:
   (a) Horn holes in air cap partially blocked.
   (b) Obstruction on top or bottom of fluid tip.
   (c) Dirt on air cap seat or fluid tip seat.

2 Heavy right or left side pattern (Figure 17.41c) caused by:
   (a) Right or left side horn hole in air cap partially clogged
   (b) Dirt on right or left side of fluid tip.

3 Heavy centre pattern (see Figure 17.41d) caused by:
   (a) Too low a setting of the spreader adjustment valve on the gun
   (b) Atomizing air pressure which is too low or paint which is too thick
   (c) With pressure feed, too high a fluid pressure or a flow of paint which exceeds the normal capacity of the air cap
   (d) The wrong size fluid tip for the paint being sprayed.

4   Split spray pattern (Figure 17.41e) caused by the atomizing air and fluid flow not being properly balanced.

To correct defects 1 and 2 (top or bottom heavy pattern, or heavy right or left side pattern) determine whether the obstruction is in the air cap by spraying a test pattern; then rotate the air cap half a turn and spray another test. If the defect is inverted the obstruction is obviously in the air cap, which should be cleaned as previously instructed. If the defect has not changed its position, the obstruction is on the fluid tip. When cleaning the fluid tip, check for fine burr on the tip, which can be removed with P1200 wet-or-dry sandpaper. To rectify defects 3 and 4 (heavy centre pattern, or split spray pattern), if the adjustments are unbalanced readjust the atomizing air pressure, fluid pressure, and spray width control setting until the correct pattern is obtained.

17.11.3 Spray fog

If there is an excessive mist or spray fog, it is caused by:

1   Too thin a paint.
2   Over-atomization, due to using too high an atomizing air pressure for the volume of paint flowing.
3   Improper use of the gun, such as making incorrect strokes or holding the gun too far from the surface.

17.11.4 Paint leakage from gun

Paint leakage from the front of the spray gun is caused by the fluid needle not seating properly (Figure 17.42). This is due to the following (numbers correspond to those on the figure):

1   Worn or damaged fluid tip or needle
2   Lumps of dried paint or dirt lodged in the fluid tip

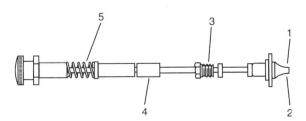

Figure 17.42   Fluid needle assembly

3   Fluid needle packing nut screwed up too tightly
4   Broken fluid needle spring
5   Wrong size needle for the fluid tip.

17.11.5 Faulty packing

Paint leakage from the fluid needle packing nut is caused by a loose packing nut or dry fluid needle packing. The packing can be lubricated with a drop or two of light oil, but fitting new packing is strongly advised. Tighten the packing nut with the fingers only to prevent leakage but not so tight as to bind the needle.

17.11.6 Air leakage from gun

Compressed air leakage from the front of the gun (Figure 17.43) is caused by the following (numbers correspond to those on figure):

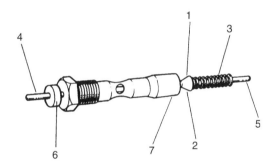

Figure 17.43   Air valve assembly

1   Dirt on the air valve or air valve seating
2   Worn or damaged air valve or air valve seating
3   Broken air valve spring
4   Sticking valve stem due to lack of lubrication
5   Bent valve stem
6   Air valve packing nut screwed too tightly
7   Air valve gasket damaged.

17.11.7 Oil in air line

If the air compressor pumps oil into the air line, it can have the following causes:

1   The strainer on the air intake is clogged with dirt.
2   The intake valve is clogged.
3   There is too much oil in the crank case.
4   The piston rings are worn.

17.11.8 Compressor overheating

An overheated air compressor is caused by:

1 No oil in the crankcase
2 Oil which is too heavy
3 Valves which are sticking, or dirty and covered with carbon
4 Insufficient air circulating round an air-cooled compressor due to it being placed too close to a wall or in a confined space
5 Cylinder block and head being coated with a thick deposit of paint or dirt
6 Air inlet strainer clogged.

17.12 Sanding and polishing machines

Sanding and polishing by hand can prove to be both laborious and expensive. Unfortunately there are many parts of vehicle surfaces where there is no alternative but to carry out these processes without the aid of power tools. However, surfaces such as the roof, bonnet, boot lid and parts of the doors and wings of cars can be rubbed down or polished more economically and efficiently with power tools. Damaged areas of paintwork can be rubbed down very quickly with these machines, but final feather edging is best done by hand. Two types of machine are favoured by the refinishing painter, the rotary sander and the orbital sander. Both are obtainable as either compressed air operated or electrically driven machines (Figures 17.44 and 17.45). In addition,

Figure 17.45 Electrically driven palm grip orbital sander with dust extraction (*Black and Decker Ltd*)

orbital sanders are available which operate with compressed air and can be connected to a water supply so that the worked surface is continually washed with clean water whilst rubbing down takes place.

17.12.1 Sanding processes

In order to produce smooth, glossy finishes the substrates and undercoats must be levelled down without leaving deep scratches. Non-sand primer surfaces have virtually eliminated many of the problems associated with scratch swelling, but they must be applied over staisfactorily prepared surfaces. Where repair work has been carried out which includes the use of polyester fillers and/or surfacers, the repaired area must be carefully sanded prior to applying the non-sand coatings.

The wet sanding of high-build undercoats and fillers has long been the accepted method of levelling these materials in order to avoid the creation of excessive dust. In addition, this method of sanding helps to clean the surface but also presents certain problems. Water penetration behind window and windscreen rubbers, door checks, etc. can lead to lengthy drying-out times. All undercoats are slightly porous and, if the moisture from the sanding process is not completely dried out, problems of micro-blistering and faulty adhesion may result.

The two main objections to dry sanding have always been that the abrasive paper clogs up and the process creates excessive dust in the workshop.

Figure 17.44 Air driven orbital sander with dust extraction (*Desoutter Automotive Ltd*)

However, clogging up of the abrasive paper is no longer a big problem since the introduction of coated abrasives. These abrasive papers have a coating over the grit particles, of a material which is based on zinc stearate. This coating allows the sanded-off residue to be shaken from the paper to produce a fairly clean sanding surface, thus creating more mileage from each sheet or disc of the abrasive paper. Coated abrasive papers are available in production paper grades (e.g. P40 to P120) and also in finer grades of lubricoat papers (e.g. P150 to P600).

It should be noted that, following sanding down with these coated abrasives, the surface should be cleaned with a proprietary spirit cleaner to remove particles of the stearate coating from the surface before applying paint. Failure to do so may result in the appearance of 'fish eyes'.

Dry sanding of painted surfaces, and surfaces to be painted, has increased in popularity as a result of the development of sanding tools which incorporate dust extraction methods. These dust extractors can be in the form of either a vacuum bag attached to the sanding tool, or a large vacuum dust collecting unit remote from the tool and to which two or more sanders can be connected. An added bonus with the remote machines is that they can also be used as a vacuum cleaner for the workshop. The sanding tools have a pad with eight or more holes in them to which are attached sanding discs with similar holes. In operation, the dust created by sanding is drawn through the holes and deposited in the collecting unit or bag.

Sanding machines available include types which are dual acting: that is, they can be set to either rotary or orbital (eccentric) actions (see Figures 17.46 and 17.47). The rotary action is in the region of 420 rev/min, whilst the eccentric action moves at 12 000 strokes/min. The machines can be either electric or driven by compressed air. The rotary action is suitable for removing old paint films and surface rust and, when fitted with a polishing head, the machine can be used for final polishing of paintwork. The eccentric action is suitable for feather edging, flatting surfaces prior to painting, and levelling surfacers and fillers. The sanding discs can be either self-adhesive (synthetic resin adhesive) or of a type which has a looped velvet reverse side which is simply pressed on to a special pad. These pads themselves are available as hard,

Figure 17.46 Festo orbital sander (*Minden Industrial Ltd*)

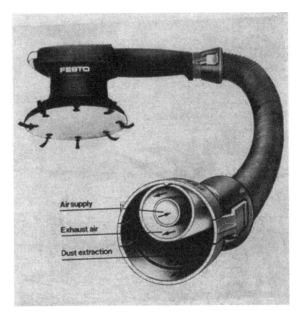

Figure 17.47 Festo random orbital sander (*Minden Industrial Ltd*)

soft and supersoft types depending on the type of work for which they are required.

The Festo sanding systems for body repair and paint shops can be of the mobile type which has the extraction unit mounted on to castors (see Figure 17.48), or of the fixed position type. In the latter type a boom, which can be moved over the car body, has all the facilities that the worker requires, e.g. compressed air supply hoses, extraction hoses, and electrical supply for both 240 V and 110 V tools.

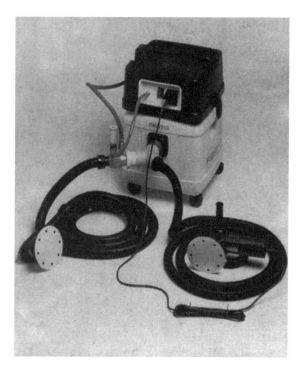

Figure 17.48 Festo dust extractor (*Minden Industrial Ltd*)

Figure 17.49 Air driven polisher with foam pad (*Desoutter Automotive Ltd*)

17.12.2 Polishing machines

There are many types of machine suitable for polishing, and they all operate with a common circular movement. To be suitable for polishing, however, the machine must not rotate with too many revolutions per minute as scorching of the film will easily result. The pad is covered with a lambswool disc or foam pad which must be kept clean at all times (Figure 17.49). Remember that polishing is the final process; should grit be picked up on the disc, scratches will result which can ruin the finish on a vehicle. Polishing machines are best used with liquid polishes and should be used with a sweeping movement. Even with slower revolving machines, polishing in one spot for too long can cause scorching and blistering of the paint film.

17.13 Preparation of a motor vehicle for repainting

Two methods of preparing a vehicle for repainting are open to the painter: (a) completely strip the existing finish down to the metal substrate, or (b) prepare the surface by rubbing down with abra-

sives. The main point to consider when deciding which process to use is the condition of the existing finish allied to the extent of damage in the case of a local repair. Which process to choose can only be decided upon by a close inspection of the vehicle.

On a motor car that has been involved in a collision and received damage to its front wing, but not serious enough to warrant replacement, the panel beater will carry out his work first, bringing the wing as near as possible to its original shapes. The vehicle then becomes the responsibility of the painter, who will carry out a close inspection of the condition of the paintwork. Should the paint show signs of poor adhesion to the substrate, it is advisable to strip it off completely using paint remover. However, if the damaged area is not extensive and the paintwork on the rest of the wing is in sound condition, the work can be carried out, without the use of paint remover, by rubbing down the area.

17.13.1 Preparation by using paint remover

If the existing paintwork is in poor condition, or if the damaged area is so extensive that the remaining paintwork on the damaged panel is not worth rubbing down, the existing finish can be removed with paint remover. The first task is to remove any flashes or chrome strips that can easily be removed. If the work is being carried out in a shop where several vehicles are being refinished, it is advisable to store these fittings in boxes which should be labelled with details of the car such as customer's name, registration number and type of car. On a complete refinishing job it is also advisable to label each part to indicate from which side of the vehicle it was removed. Should it prove to be too difficult or impractical to remove any parts, these should be masked completely. Though it is not always necessary to mask the surrounding areas when stripping, it is advisable to do so because if any paint remover accidentally came into contact with an adjoining door or bonnet, damage to these areas would occur. However, if masking paper is applied it should be of a fairly stout nature and laid in two or three thicknesses to ensure that any paint remover accidentally deposited on it does not immediately penetrate to the paintwork beneath.

Paint removers were formerly of a highly flammable nature and constituted a fire risk, but nowadays non-flammable types are available and are widely used. As well as removing fire risks they also evaporate at a much slower speed; hence they remain wet for a longer period and so have better penetration properties. Though they are quite easy to use, some precautions should be taken when working with these paint removers. They can be a cause of dermatitis by penetrating the pores of the skin, dissolving the natural oils beneath and so leaving the pores open to attack by bacteria. A simple precaution is to apply a barrier cream to the hands before commencing paint stripping, and to rinse off with water any remover which comes into contact with the skin. Strong rubber gloves are very useful when using liquid paint removers. The operator should not smoke when using paint remover, even though it is not flammable, as it gives off a strong vapour which, on coming into contact with a naked light, becomes toxic in nature. This toxic vapour if inhaled into the lungs will cause dizziness, nausea, and even vomiting. Adequate ventilation should be provided in any shop where paint remover is used, for the same reasons.

A fairly liberal coat of stripper should be applied to the surface and allowed sufficient time to penetrate through the various layers of paint. The softened paint is then removed with a stripping knife and the surface washed down with a generous quantity of water. Both the laying on and the washing off of the stripper can be done with old or cheap paint brushes. To assist the removal of paint at awkward places, a wire brush is an invaluable tool. Very often a layer of scum or a fine stain of primer is left on the surface but this can be quite easily removed with steel wool and water. It is imperative that all traces of paint remover be thoroughly rinsed off, otherwise it will cause damage to any subsequent paint system applied to the job.

17.13.2 Preparation by using abrasive paper

Before the rubbing down of a local repair commences, the whole panel should be thoroughly washed down with a detergent to remove all traces of wax polish. Nowadays nearly all wax polishes contain silicones, which can create 'pinholing' in any paint applied over them. Water miscible cleaning solutions, which are normally used for degreasing, are suitable for wax removal. Thorough rinsing with water and drying should follow. The grade of wet-or-dry paper chosen to rub down the damaged area depends very greatly on the degree of damage, but normally 180 grade is suitable. The paper should be folded and torn so that one-quarter of the length can be removed (Figure 17.50). This is then attached to a rubber rubbing block and the damaged area abraded using a liberal quantity of water to keep the work clean so that continual inspection can take place. The rubbing down is normally carried out with a forward and backward motion until the paint is removed and a feather edge achieved. Should paint remain in recessed areas, it can be removed with a wire brush, coarse steel wool, or abrasive paper without the rubbing block.

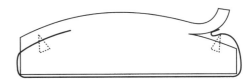

Figure 17.50 Method of fixing abrasive paper into rubbing block

As an alternative to rubbing down by hand, power tools can be used to speed up the process. These can be either electric or driven by compressed air. Those which are electrically driven are used dry, but compressed air tools can be used either wet or dry. The latter types are generally favoured for lighter work where damage to the paintwork is not extensive, but the electric tools are most suitable where severe rusting of the metal has occurred.

17.13.3 Treatment with phosphating liquid

Following the removal of the paint by either of the methods described, the bare metal should be treated with a proprietary phosphating liquid. These liquids (there are several brands available) have an acid content, mainly phosphoric, which etches the metal and completes the derusting of it. In addition to etching the metal, phosphating solutions deposit a layer of iron or zinc phosphates on to the metal which inhibits corrosion. These liquids should be used according to the manufacturer's instructions, but are generally diluted in the ratio 1:1 with water, applied with an old paintbrush and washed off with water after about fifteen minutes. The liquid, being of acid content, should not be allowed to come into contact with the skin, eyes or clothing of the operator, and rubber gloves should be worn to protect the hands. In the event of its accidentally doing so, the affected part should be rinsed thoroughly in running water. Following drying of the bare metal with an air duster and wash leather, the area is now ready for repainting.

17.14 Finishing and refinishing processes

There are four main car refinishing paints at the refinisher's disposal: cellulose synthetic (half-hour enamel) paint; acrylic resin paint; low-bake synthetic paint; and two-pack paints. Cellulose synthetic and low-bake synthetics are used as repair materials over the high-bake synthetics applied by the car manufacturers; in addition, of course, they can be used for complete resprays. Acrylics are best used on repair work over an original high-bake acrylic finish, but can also be used as a refinishing material on complete resprays. To determine whether a paint is suitable for repairing a particular job, a flat area should be chosen and a wet coat of the paint sprayed on to it. If wrinkling or lifting occurs this is proof that the solvent is too strong and will probably lift the existing finish.

The following paint systems are typical of those carried out in refinishing shops. They do not follow any one particular paint manufacturer's specifications but are intended as a guide to the use of the various paint types. When using a particular brand of paint, the operator should always follow the maker's instructions as to viscosities, drying times and temperatures. The processes outlined are based on the assumption that the vehicle has been prepared for refinishing as described in Section 17.13, followed by the necessary masking up.

17.14.1 Coach finish

This is the traditional material, used in coach painting workshops for many years, but it has lost favour to more modern materials mainly because of its lengthy drying time. However, in paint shops where spraying equipment is not available, this material still has a use. It is essential that the workshop should be kept clean, otherwise dirt in the finish is a certainty owing to the prolonged period of paint film wetness.

The finishing material can be sprayed either hot or cold and can also be brushed. It is suitable for application on all vehicle construction materials including wood. Undercoats may also be sprayed hot when thinned 6:1 with white spirit. The paint system is of the simplest type. For example, for metal surfaces the procedure is as follows:

1 Prepare the surface.
2 Carry out any necessary masking.
3 Wipe the surface down with a proprietary spirit wipe or a mixture of methylated spirit and water.
4 Apply one coat of self-etching primer, either spraying or brushing type. Allow to dry and then apply ground coats as soon as possible.
5 Apply two coats of ground coat:
    (a) If cold sprayed, allow 2 hours between coats.
    (b) If hot sprayed, heat to 60 °C and allow 1 hour between coats.
    (c) When brushed, allow to dry for 6 hours between coats.
6 Should stopping up be necessary, this should be done after the first ground coat which should, of course, be allowed to through-dry completely. A synthetic resin stopper should be used and, when hard-dry, wet flatted with P360 wet-or-dry paper.
7 Should sanding down of the ground coats be necessary, allow to dry for at least 6 hours, and dry-sand using P400 lubricoat paper. Dust off and tack off. If sanding down is not necessary, the finishing material can be applied following the flash-off times listed in point 5.

8   Apply finishing enamel:
(a) When brushed, no thinning is required.
(b) When not sprayed, heat to 60 °C without thinner.
(c) When cold sprayed, thin 6:1 with white spirit or preferably with a thinner supplied by the paint manufacturer.
Apply two coats as follows:
(a) Brushed: allow at least 6 hours between coats, the time depending on workshop temperature.
(b) Hot sprayed: one light coat; allow 30 minutes and apply a double-header coating.
(c) Cold sprayed: one light coat; allow 60 to 90 minutes before applying a double-header coat, but take care not to apply too wet a coating otherwise runs will result.
Spraying pressures should be in the order of 3–4 bars (45–60 psi). Spray gun fluid tips should be of the smaller types (say 1.25 mm diameter) to avoid flooding the job. Spray gun distance from job is best about 200 mm.

9   Remove masking whilst the paint is at the tacky stage. This will allow the edges to settle down into place.

10   After overnight hardening, transfers, lining and lettering may be applied.

17.14.2 Cellulose synthetic (half-hour enamel): complete respray from bare metal

The process summary is as follows:

1   Spray a thin coat of etching primer, mixed 1:1 with activator, thinned to 20–25 seconds viscosity.
2   Allow to dry for 15 minutes.
3   Spray cellulose primer surfacer, thinned 1:1 to 19–22 seconds viscosity. Apply three full coats, allowing 10–15 minutes between coats.
4   Allow to dry for 1 hour.
5   Wet flat with P400 or P600 wet-or-dry paper.
6   Apply cellulose stopper where necessary, allowing 15–20 minutes between layers.
7   Allow to dry for $1-1\frac{1}{2}$ hours.
8   Wet flat stopper with 320 wet-or-dry paper.
9   Spray cellulose primer surfacer to stopped up areas, and flat with P600 grade paper.
10   Blow off vehicle with air gun and tack off.

11   Spray finishing material thinned 1:1 to a viscosity of 21–23 seconds. Apply one coat and allow to dry for 15–30 minutes. Apply second coat.
12   Allow overnight drying.
13   Wet flat with P800 grade paper, dry with air gun, tack off.
14   Spray double-header coat, thinned as before.

Alterations to this standard method can be made where necessary. For example, an alternative to using cellulose primer surfacer at stage 3 would be to use a synthetic resin primer surfacer which has better build qualities and may be preferable on rough surfaces. It is thinned 4:1 giving a viscosity of 26–30 seconds. Spray two or three coats, allowing a flash-off period of 30 minutes between coats, and the surface will appear completely matt. Allow the final coat to dry for 4 to 6 hours before dry scuffing, or to dry overnight before wet flatting with P600 or P400 grade paper. Cellulose stopper cannot be applied between coats of this material, and when used after the third coat must be oversprayed with cellulose primer filler. Cellulose finishing paints can be applied directly over synthetic resin primer surfacers, but if drying conditions are less than ideal a certain amount of lifting of the surfacer could be experienced. This is caused by the strong solvent used in cellulose finishes penetrating the still soft synthetic primer surfacer and acting as a paint remover. Therefore when adverse drying conditions prevail, such as low temperature, dampness or high humidity, it is advisable to spray a coat of sealer or cellulose surfacer over the job before applying the cellulose-type finish.

At stage 11 in the process, and alternative method is to apply one coat of the finishing paint, allow to dry for 15 to 30 minutes, then spray a double-header coating. This is a quicker method than that mentioned in the process summary, but does not produce the same standard of finish.

It should be noted that with the exception of tacking off and the actual application of the paint, all the process stages should be done outside the spray booth. Cleanliness within the booth or spray room is essential if high-class finishes are to be obtained. Rubbing down, even though done with water, will leave a scum in the workshop floor, which when dried out will leave behind a powdery residue. When disturbed this will cause air contamination. Burnishing and polishing cloths also cause contamination of the atmosphere, and these processes should be carried out in a part of the workshop away from the spray booth. These polishing cloths must not be put down on the floor or dirty work benches, for if grit should be picked up on them, damage to the new paint finish will occur.

As to the correct air pressure to use when spraying, this will vary according to the type of spray gun being used, the paint viscosity, and the type of paint. For cellulose synthetic paints and air pressure of about 4 bars (60 psi) is normal, though this can be adjusted by the spray painter to suit his own requirements. The golden rule is to use the lowest air pressure that will give satisfactory atomization.

Stage 6 in the process summary refers to the application of cellulose stopper. It should be borne in mind that this material dries by the evaporation of the solvent content, and if it is applied in heavy layers the inside of the coating will remain wet for a considerable period. In addition to this, when drying does eventually take place the stopper will contract and sink. A much more satisfactory repair will be obtained if the material is applied in thin layers with a suitable drying period between applications. Plastic spreaders are to be preferred to metal knives when laying on the stopper, as these cause less damage to surrounding areas of paintwork. The lid must always be firmly replaced on the tin or tube of the stopper when not in use, otherwise the stopper will harden in the container and cannot be satisfactorily resoftened.

A coat of sealer could be added to the process, where costs permit, between stages 10 and 11. This would reduce absorption of the finishing coats by the surfacer and promote a high gloss.

Following the finishing coats, the vehicle may, if required, be burnished and polished; the method for this is described in Section 17.15.

Finally, it should be borne in mind that the use of a guide coat at each of the stages of sanding down is invaluable for locating surface defects.

17.14.3 Cellulose synthetic (half-hour enamel): refinishing over an existing finish

The majority of cars or vans that are refinished today are first patch repaired, followed by refinishing of the entire body. Only when the paintwork is in very bad condition is it considered necessary to strip it all off and it is usually only essential to cut back to bare metal those parts where the coating is damaged or corrosion has set in. However, it is not always necessary to repaint the whole car or van. A typical example would be that of a comparatively new car with paintwork in excellent condition but with local damage such as would be sustained in a collision. In this case it would only be necessary to repaint the repaired areas.

Following careful inspection of the vehicle, it can be brought up to the painting stage by using the method described in Section 17.13.2. A process summary would be:

1 Apply etch primer to bare metal parts only.
2 Spray one or two coats of cellulose primer surfacer, allowing 10–15 minutes between coats, on to the etch primer.
3 Apply cellulose stopper.
4 Rub down stopper with 360 paper.
5 Spray sufficient coats of primer surfacer or filler to bring the repaired area up to the level of the surrounding surface; use air pressure of 3 bars (45 psi).
6 Wet flat these areas with 360 paper.
7 Apply three full coats of primer surfacer to the entire body. Alternatively, synthetic resin primer could be used, providing more build to the system and allowing for easier flatting at stage 8.
8 Wet flat with P600 grade paper.
9 Apply cellulose stopper where necessary.
10 Rub down stopper with P400 grade paper, and spray locally with cellulose surfacer. Wet flat these areas with P600 paper.
11 Dry off and tack off.
12 Spray finishing coats as described in Section 17.14.2.

17.14.4 Cellulose synthetic (half-hour enamel): local repair

1 Rub down damaged area with P180 grade paper (wet) to obtain a feather edge. Wet flat surrounding paintwork with P400–P600 grade paper. Alternatively, dry sand with a dual-acting sanding tool using P80–P150 grade paper.
2 Treat exposed metal with phosphating solution.

3   Apply thin coat of etch primer to bare metal only; alternatively, spray one coat of primer surfacer directly on to bare metal, thinned 1:1.

4   Fill up defects with cellulose stopper in thin layers.

5   Wet flat stopper with 320–360 grade paper.

6   Spray sufficient coats of primer surfacer or filler (thinned 1:1) at 3 bars (45 psi) to bring repair up to level of surrounding surface.

7   Wet flat with P600 grade paper.

8   Burnish surrounding area with rubbing compound and a damp cloth to ensure a good colour match and better blending in.

9   Tack off.

10   Spray colour coats, thinned 1:1 to a viscosity of 19–23 seconds, at a pressure of 3 bars (45 psi). Spray several coats lightly until a good colour match is achieved. Allow to dry hard.

11   Wet flat with P800 grade paper, dry off and tack off.

12   Overspray the repair with a mixture of 75 per cent thinner, 25 per cent colour, carrying the spraying beyond the edge of the repair to obtain a soft blend.

13   Allow to dry hard (preferably overnight) and wet flat with P200 grade paper.

14   Burnish and polish.

Any masking off that may be required should be done between stages 2 and 3. A coat of sealer may be applied just prior to the colour coats to provide better hold-out, thus obtaining a smoother finish. Synthetic resin primers have not been included in this process as they have a tendency to peel back from the edges when rubbing down takes place. Overspray from the spray gun can create unnecessary work when carrying out local paint repairs, but this can be restricted by using a narrower fan pattern than that used for spraying a whole panel. When this is done, a higher volume of paint will be applied, increasing the risk of runs, and so the fluid needle adjusting screw should be turned to the right until a satisfactory volume of paint issues from the fluidnozzle.

   Occasionally the spray painter may be called upon to repaint a motor car on which the paintwork is in excellent condition, the customer simply desiring a change of colour. In this case he may only require to wet flat the surface with P600 grade paper using a weak solution of water miscible

cleaning solution to remove any wax polish. Then following thorough drying off and masking up, three coats of half-hour enamel will produce quite a good finish which can be further improved by flatting, burnishing and polishing.

17.14.5 Acrylic lacquer

Though cellulose synthetic finishes are best left to dry in their own time to obtain best results, acrylic lacquer can be force dried without damage to the paint film, with a consequent speeding up of the process. This can be done with infrared lamps or in a heated booth (though not in excess of 50 °C). The filler materials possess better build and flowout than the cellulose-based materials, thus providing better surfaces for the finishing coats. The spraying viscosity of the finishing enamel is more critical than the half-hour enamel, and only the thinner recommended by the paint manufacturer must be used. Solvent evaporation from the wet paint film is governed to some extent by the workroom temperature, and should this be below 15 °C a special quick repair thinner should be used. This thinner evaporates very quickly but tends to produce a low gloss which will require burnishing and polishing. However, should the ambient temperature be 15 °C or above, a good hard glossy finish can be obtained straight from the gun which does not require polishing. As acrylic primer fillers have good adhesion properties the use of etch primer, though recommended, is not essential provided that the metal substrate has been properly prepared and treated with a phosphating solution.

17.14.6 Acrylic lacquer: complete respray from pretreated bare metal

1   Spray one coat of acrylic primer filler thinned to a viscosity of 21 to 23 seconds at 25 °C. Allow 5 to 10 minutes to flash off.

2   Spray two coats of primer filler thinned 1:1, 26 to 29 seconds at 25 °C. Allow 5 to 10 minutes between coats and 1 to 2 hours after second coat.

3   Apply cellulose stopper where necessary in thin layers, allowing 15 to 20 minutes between layers.

4   Wet flat stopper with 320–P400 grade paper, dry off and tack off.

5   Spray stopper locally with primer surfacer, allow 5 to 10 minutes, then apply a full coat over the entire surface. Leave to dry for 1 to 2 hours.

6 Wet flat with P600 grade paper, rinse, dry off and tack off.

7 Apply one wet coating of acrylic sealer, allow to dry for 30 minutes, denib and dry. Do not wet flat. The lacquer coats should be applied within 2 hours. Tack off before spraying the enamel.

8 Spray lacquer coats as necessary using one light coat followed by a double header. Thin the enamel 2:3 to a viscosity of 16 to 19 seconds at 25 °C. Use the appropriate thinner only. Should more than one coating of the lacquer be considered necessary, allow a flash-off time of 5 to 10 minutes between coats. The lacquer is touch dry after 15 minutes and can be safely handled after 1 to 2 hours, depending on room temperature. If necessary it can be burnished and polished after overnight drying, though this can be done after 4 to 6 hours if the enamel is force dried. The air pressure used when applying acrylic lacquer is between 3 bars (45 psi) and 4 bars (60 psi) depending on the make of spray gun employed.

17.14.7 Acrylic lacquer: complete respray over an existing finish

The method chosen here must obviously depend on the condition of the paint film. If the surface requires filling and stopping up, it should be wet flatted with 280 grade paper using a water miscible solution or liquid detergent. After rinsing and drying off, the system described for bare metal can then be used. If filling and stopping is not considered necessary and the paint is sound, it can be wet flatted with P600 grade paper and a solution of liquid detergent. Following rinsing and drying off, tack off and apply the sealer and lacquer coats as described.

17.14.8 Acrylic lacquer: local repair

Acrylic lacquer can be used to repair high- and low-bake enamels, but is not recommended for repairs to half-hour enamels or nitrocellulose-based air drying finishes. Nor is it suitable for use on wood or the repair of synthetic coach finishes.

A typical system for a local or spot repair is as follows:

1 Degrease with a solution of liquid detergent.
2 Wet flat damage area with 180 grade paper and feather edge.

3 Treat bare metal with metal conditioner or phosphating liquid, rinse and dry off.
4 Spot prime with acrylic primer filler.
5 Stop up with cellulose stopper where required.
6 Wet flat with 280 grade paper and rubbing-down block. Finish off with 320 grade paper, rinse and dry off.
7 Spray in with acrylic primer filler sufficient coats to level up the surface.
8 Wet flat with P600 grade paper, rinse, dry off and tack off.
9 Spray over the repair with acrylic sealer.
10 Denib and dry. Wet flat around the edge of repair with P800–P1000 grade paper. Burnish surrounding panel.
11 Apply acrylic finish, thinned with quick-drying repair thinner, in light coats. Finish off with a double-header coat to obtain a smooth finish.
12 When dry, wet flat with P1200 grade paper, burnish and polish.

Should the damage to the panel be too severe for satisfactory or economical levelling up with cellulose stopper, the two-pack polyester resin stopper described in Section 17.5.2 could be used. This is applied to the bare metal prior to the paint system. It is best rubbed down and, after dusting off and tacking off, coated with primer filler.

The use of a sealer coat at stage 9 may be eliminated if the surface is carefully prepared. Sealers are supplied ready for use, and it is difficult to spray them without leaving an edge which is difficult to remove. Providing that the original finish is flatted 50–75 mm beyond the repair area, a satisfactory job can be produced by spraying the finishing coats on to the filler to slightly beyond its edge. Care should be taken not to overlap the colour on to unflatted enamel.

17.14.9 Metallic finishes

Practically without exception, metallic colours being applied by car manufacturers in Britain are based on acrylic resins. However, they present problems not experienced when applying straight colours, which are caused by the metallic particle content. Without delving too deeply into the realms of paint technology, a metallic finishing paint could be described as a tinted, semi-transparent varnish containing finely ground metallic particles such as aluminium, bronze and copper. Polished aluminium flakes, because of their silvery metal appearance, are the pigments most widely used. Because they are lacking in opacity, these paints are generally applied in several layers to achieve the desired effect of colour depth and an even distribution of the metal flakes. The best method is to apply a single coat followed by wet on wet (double-header) coats. The coverage may vary from colour to colour. Apply as many coats as may be necessary.

Though car manufacturers favour acrylic-based metallic paints, they are, however, available in cellulose synthetic, slower drying types of spraying synthetics, and, most widely used nowadays, two-pack synthetics. They are not suitable for brush application. It is not necessary to outline a complete paint system for metallic finishes; this section deals with the application of the finishing coats. The actual spraying of these has a great influence on the finished appearance as regards colour. Spray gun technique is very important. The gun should be held at right angles to the surface and at a distance of 150–200 mm approximately. If the distance between the gun and the painted surface varies, dark and light patches will result. A 50 per cent overlap of gun stroke is essential to obtain an overall even colour and texture.

When too wet a coating is applied, the metallic flakes move freely within the wet film, and when solvent evaporation takes place they are generally in a fairly upright position. This tends to darken the final effect, as light does not reflect too well from the flakes in this position (Figure 17.51). Opacity is also reduced and when sinkage takes place as a result of solvent evaporation, the particles tend to stick through the top skin of the paint film, causing the finish to have a seedy appearance.

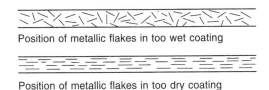

Position of metallic flakes in too wet coating

Position of metallic flakes in too dry coating

Figure 17.51   Coatings too wet and too dry

Too dry an application will create a dusty effect, too pale a colour when viewed head on, and too deep a colour when viewed from the side. This is because the metallic flakes have a tendency to lie parallel to the surface when a dry coating is applied.

When repairing a metallic finish, two options are available to the refinisher. One is to extend the repair to a natural break in the car body lines and refinish the entire panel. Using this method there is always the danger of ending up with a mismatch when the masking is removed. A better method is to apply the colour coats to the whole panel, remove the masking and then recoat but extend the colour further still, using a fade-out technique. Overspray thus created should be overcoated with a clear blend-in material to protect the overspray when burnishing is done. When using this second method with the base-coat-and-clear materials, it is best to coat the panel with the clear coating first and, whilst this is still wet, to apply the colour; this should then be overcoated with further coats of clear.

There are too many variables involved to make it possible to lay down a hard and fast system in obtaining a perfect match on repair work. No matter how much technical data is available, in the final analysis it becomes a system of trial and error, with the spray painter trying out various spray gun set-ups, varying the air pressure, speed of gun stroke, and distance from the surface. The spraying viscosities vary according to the type of material being used but a rough guide is as follows: acrylic 16–19 seconds, cellulose synthetic 19–23 seconds, synthetic 23–28 seconds.

The damaged area should be levelled up as previously described, and the whole of the surrounding panel wet flatted with P400 grade paper. Following drying and tacking off, the panel should be sprayed with primer surfacer or filler and flatted with P600 grade paper. After drying and tacking, the colour coats are applied. The function of this final coat of surfacer is to equalize solvent penetration from the finishing material. Surfacers and fillers are more porous than enamels, and when the surrounding panel is sprayed with the finishing colour only the repaired area can be detected by a slight variation in colour and texture. Where costs permit, a coat of sealer could be applied prior to the finishing colour.

Light burnishing and polishing of metallic finishes can be done after overnight drying in the case of acrylic- and cellulose-based materials. With the slower drying synthetics this process is best left for about four days to allow for complete solvent evaporation and to give the paint film time to harden off. When using two-pack materials, burnishing can be carried out after 16 hours at a workshop temperature of 15–20 °C.

17.14.10 Quick air drying synthetics

This type of material offers special advantages for low-cost resprays and rapid finishing of commercial vehicles. The main advantage is that it is suitable for application over all types of existing finishes, and is suitable for hot or cold spray application.

The rapid respray system is as follows:

1　Clean off all traces of traffic dirt, grease, wax and silicone polishes by using water miscible cleaning solution.
2　Feather edge all damaged areas including parts damaged by stone chips (generally on sills). Treat bare metal with derusting liquid.
3　Spot prime any bare metal areas with cellulose-based primer surfacer, and allow to dry for half to one hour. Stopping, if necessary, is best done before priming with two-pack polyester stopper.
4　Wet flat the whole vehicle with P600 grade paper, using liquid detergent or water miscible solution as a lubricant and to remove wax. Rinse, dry and tack off.
5　Thin the enamel to a viscosity of 23–26 seconds (4:1). Spray one coat and allow to flash off for 10 minutes, then apply a double-header coating. The finish may also be applied hot, in which case it requires no thinning. The paint is heated to a temperature of 70–80 °C, and one coat will be sufficient to give the required film thickness. The flash-off times between coats recommended by the paint manufacturer must be strictly observed. If too great a time lapse is allowed, then problems of lifting or wrinkling may be encountered.

17.14.11 Low-bake enamels

These are modified high-bake enamels rather similar in composition to those used by motor vehicle manufacturers. Though developed primarily for the refinishing trade, they are in fact used by vehicle manufacturers in the rectification of damaged or faulty finished vehicles. One of the problems encountered by the refinisher has always been to obtain a close match to the original finish. When exposed to sunlight some colours have a tendency to fade, whilst a certain amount of discoloration takes place in others; some whites, for instance, tend to yellow with time. In addition to this, some pigments combine very well with a synthetic medium but not so well with cellulose-based vehicles. As the latter have always been the most widely used materials in refinishing, colour matching thus presented problems. These would have been less if refinishing shops were equipped with equipment similar to that of the vehicle makers, allowing them to use the same finishing materials. This, of course, would require a tremendous capital outlay which would be well beyond the reach of the refinisher, and would in all probability never be recovered.

In an effort to improve refinishing techniques, to reduce drying times and to obtain faster production, low-bake enamels (i.e. paints that cure at lower temperatures than the original high-bake finish) were developed. Being similar in structure to the high-bake material, these paints offer a closer match in colour and texture than do cellulose synthetic materials. Burnishing and polishing is seldom required as these paints provide a good gloss from the gun.

Equipment

An obvious requirement for the drying of low-bake enamels is a stoving room, and various types are available. A stoving room can be acquired which can be connected to an existing spray booth, or where workshop space is limited a combined spray booth and low-bake oven can be installed (Figure 17.52). The obvious limitation of a combined spray booth and low-bake oven is that no further paint spraying can be done during the stoving schedule, in addition to which valuable working time can be lost whilst waiting for the booth to return to a comfortable working temperature.

Where workshop space permits, a better proposition would be to install a low-bake oven which can be joined up to an existing spray booth with sliding shutter doors to seal the two areas (Figure 17.53). This can be further improved upon by having entrance/exit doors at each end of the unit which would provide a flow-line system of painting and stoving (Figure 17.54), thus reducing time wastage on vehicle movement. The type of installation must obviously be governed by the size and shape of the workshop (Figures 17.55–17.59). A further variation

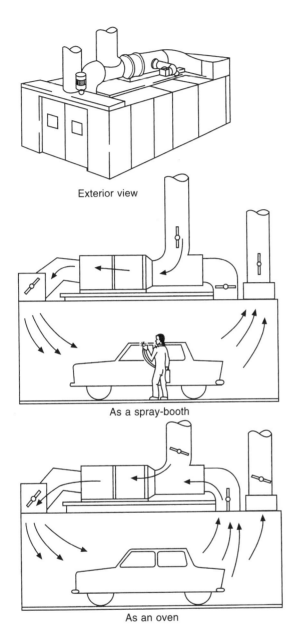

Exterior view

As a spray-booth

As an oven

Figure 17.52 Spray booth combined with low-bake oven

is shown in Figure 17.60, in which the vehicle being painted is moved sideways on rails and bogies.

Apart from the stoving room, the usual spray painting equipment used in refinishing shops is required.

Preparatory work

Deep indentations in the vehicle body can be filled with polyester stoppers, which cure very quickly

and have good adhesion properties when baked. Following rubbing down and cleaning off, the surface can be sprayed with synthetic resin primer surfacer if it is of a rough nature. As previously stated, these primers have exceptional filling properties, but when used with low-bake finishes the stoving schedule ($1-1\frac{1}{2}$ hours at 82–93 °C) can prove to be too long to be economical. Where the surface is not too rough, a better method would be to use cellulose primer surfacer thinned 1:1. Slight imperfections should be levelled with cellulose stopper, which must be over-coated with surfacer. The surfacer can then be stoved for thirty minutes at 70–80 °C. A third material finding favour is the polyester primer filler described in Section 17.5.1, which obviates the need for stopper on small imperfections and provides a single coat build-up to the finishing stage. As previously stated, several coats can be applied wet on wet without flash-off periods, and force dried for 15 minutes at about 75 °C. A further advantage is that polyester stopper can be used, if necessary, between coats of polyester primer filler. This material should be over-coated with cellulose primer surfacer before applying the finishing material. Whichever of the above materials is chosen, the final coating must be rubbed down wet, and following drying and tacking off, the colour coats can then be applied.

Spraying

The finishing material is thinned 7:2 and sprayed in one single coat followed by a double-header coat, with a 15 minute flash-off between coats. Another flash-off period of 15 minutes should be allowed before stoving.

Preparation for stoving

It is current practice in the motor industry to repair areas damaged during assembly with low-bake enamels and then to stove the whole body fully trimmed. The temperature inside the vehicle does not rise sufficiently to damage the trim provided that the doors and windows are properly closed. Before stoving a vehicle refinished with low-bake enamel, check that doors and windows are closed, remove all exterior plastic fittings, increase tyre pressure by 0.5 bar (5 psi), check that the petrol tank is not too full and, if duotone work is being done, remove any masking. It is not necessary to remove the battery.

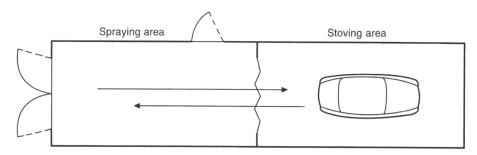

Figure 17.53 Low-bake oven joined to spray booth

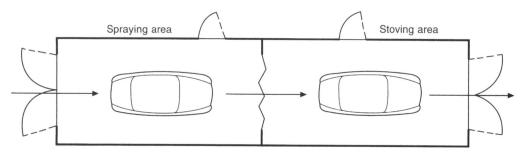

Figure 17.54 Oven joined to spray booth, with through doors for flow line

Stoving

The usual stoving schedule is 30–40 minutes at 80–100 °C, following which the body should be allowed to cool for about 45 minutes (longer if necessary) before further masking up for duotone work is carried out. Certain parts of the car will be sheltered from the heat, and consequently paint applied to these parts will not be fully cured. A converter liquid is available from most manufacturers of low-bake paints which can be added to the enamel for brushing in door edges and insides of boot lids and bonnets, or these areas can be

Figure 17.55 Spraybake exterior units, height 3 m (*Spraybake Ltd*)

Figure 17.56 Multiple installation showing drive through and paint mix room (integral) between units (*Spraybake Ltd*)

Figure 17.57 Multiple installation showing side loading doors and track, and housing IRT arch (*Spraybake Ltd*)

Figure 17.58 Corner installation showing two low-bake units, height 2.5 m (*Spraybake Ltd*)

Figure 17.59 Interior shot of Green Booth designed to meet EPA regulations (*Spraybake Ltd*)

sprayed if the viscosity is reduced to 22 seconds with the appropriate thinner. Any areas touched in thus will dry quite hard at the normal paint shop temperature.

Finishing

Polishing of low-bake enamels is not normally necessary as a good gloss from the gun is obtainable, but it should be borne in mind that these paints remain more open, i.e. wetter, than more coventional materials during the flash-off period prior to stoving. For this reason, a clean workshop is a necessity; otherwise dust may settle on the

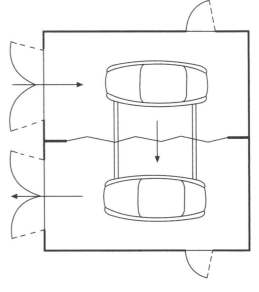

Figure 17.60 Joined oven and spray booth, with sideways flow

still wet film which will necessitate flatting and burnishing to obtain a smooth finish. The nibs should be removed with P800 grade wet or dry paper and the flatted area brought up to a high gloss with burnishing compound, followed with a good-quality car polish. Low-bake finishing can be carried out over both stoving and nitrocellulose finishes providing they are in a sound condition. Red or maroon cellulose finishes which might bleed are best coated with a sealer before commencing the spraying of the surfacer coatings.

Developments

Apart from the cost of the stoving equipment, one of the reasons why low-bake finishes were slow to find popularity with refinishers was that it meant having to increase the already extensive stock of paints normally carried in refinishing shops. With this in mind, paint manufacturers have developed a type of thinner which, when added to their half-hour enamel (nitrocellulose based), will convert it to a low-bake cellulose enamel. Over 2000 car colours are available in this material, and so colour matching does not present too many problems. The cellulose stoving enamel should be thinned with the converter thinner to a viscosity of 23 to 26 seconds with a BS B4 flow cup. Spray either three single coats or one single and one double-header coat at 4–5 bars (60–75 psi), allowing 15 minutes flash-off between coats and another 15 minutes before stoving.

The stoving schedule recommended is 40 minutes at 91–93 °C, during which the car body attains a metal temperature of 80 °C for 30 minutes. This temperature is not high enough to cause damage to the car interior but is high enough to effect a complete cure of the paint film.

Advantages

Low-bake refinishing, though costly to set up initially, undoubtedly increases the potential productivity of refinishing shops. Faster drying times are obtained at almost every stage of the job, with the result that vehicles are refinished much more quickly than when using air-drying materials. This leads to faster delivery of customers' vehicles, which is a very good selling point, and as the finished article can be moved out of the working space after cooling of the body, freer movement within

the workshop is obtained. Used properly, low-bake finishes lead to a higher turnover of work, which in turn increases the need for good organization within the workshop (see Section 17.18).

17.14.12 Short-wave infrared paint drying

Infrared drying lamps have been used in finishing shops for many years to force dry localized areas of paintwork and to accelerate the curing of two-pack polyester stoppers and fillers. A more recent development is the short-wave infrared heating module which, instead of easily breakable lamps, has heating elements mounted into an aluminium casette.

One firm specializing in this type of equipment is Infrarödteknik AB, whose IRT 100 unit has a reflector coated with a thin layer of gold to give maximum reflectivity and long life. In addition it is equipped with a fan for solvent vapour removal, and a control panel which guarantees efficient control of the drying process according to the type of paint being used. The unit is mounted on a stand with a flexible support lever which makes it possible to locate the heating module in any position without locking devices (see Figure 17.61). A variation on this type of heater is a vertical heater mounted on castors for force drying of doors, quarter panels and so on.

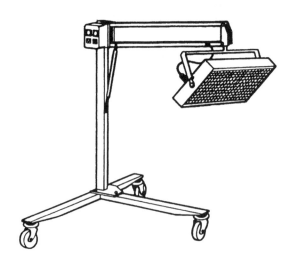

Figure 17.61 Short-wave infrared heater IRT 100 (*Infrarödteknik AB – Stanners*)

A further development is the infrared drying arch, designed to be mounted inside an existing spray booth. The arch consists of a number of heating modules mounted on to a steel frame in the form of an arch. This arch is stored at the rear end of the booth and is sheltered from overspray. Following the spraying process, it is moved quickly to the vehicle and then moves along the length of the vehicle at a slower speed. After the stoving of the vehicle, it then switches off automatically and moves back to the rear of the booth (see Figure 17.62). Being computer controlled, the arch is capable of providing the refinisher with the option of drying a complete vehicle or one or more panels at various positions on the vehicle body. It also caters for the various types of paint and colours.

Short-wave infrared units require no preheating, which makes for fairly high savings where energy costs are concerned. They also cool in seconds, which means the vehicle can be moved out of the booth almost immediately. When used to force dry two-pack materials, the drying times claimed are quite extraordinary: for example, a complete respray can be dried in 5 to 10 minutes, and minor touch-ups in 1 to 2 minutes. Single panels finished in a material such as Bergers Standox 2K are baked in 4 to 5 minutes, whilst primers and fillers bake even faster. With drying times like this, the savings on energy costs and the gain in turnover of work are obvious.

17.14.13 Two-pack paint system

There are many variations available to the refinisher with these materials, and the paint manufacturers' literature must be referred to. It is inadvisable to mix one manufacturer's materials with another.

Although it is common practice in some workshops to use cellulosc-based primer surfacers under two-back finishes, better build and intercoat adhesion is obtained when two-pack undercoats are used. These undercoats are multipurpose materials which can be used as primers or non-sand surfacers. Selfetch 2K primer surfacers are available for use on bare, prepared substrates such as steel, aluminium and glass fiber, and can even be applied over vehicle manufacturers' finishes. Transparent adhesion primers are available which can be applied on to manufacturers' finishes which have merely been cleaned with a cleaning spirit and a Scotchbrite pad. These primers can then be overcoated with the finishing material after 15 minutes as a wet-on-wet process. As previously stated, there are many variations of materials with these products, and so there is no basic system or process which can be described. A variety of thinners are available for use at various workshop temperatures. Various hardeners may be used depending on workshop temperature and humidity.

Two-pack paints are available as straight colours, metallics and base-coat-and-clear systems. To give some idea of the range of choice in two-pack base-coat-and-clear materials, the following list will be helpful. Each of the clear coatings contains an isocyanate hardener:

1 Base coat: acrylic/polyurethane synthetic
   Clear coat: two pack (2K)
2 Base coat: acrylic synthetic
   Clear coat: 2K modified acrylic synthetic

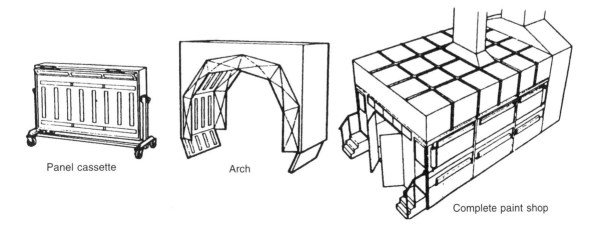

Panel cassette   Arch   Complete paint shop

Figure 17.62 Infrared cassette, arch and paint shop (*Infrarödteknik AB – Stanners*)

3 Base coat: acrylic lacquer
  Clear coat: 2K modified acrylic synthetic
4 Base coat: cellulose
  Clear coat: 2K acrylic synthetic or polyester/acrylic.

It must be stressed that the time lapse between applying the base coat and the clear coatings recommended by the paint manufacturer must be strictly observed if intercoat adhesion problems are to be avoided.

17.14.14 Water-based paint

Based paints use, as their name suggests, water instead of solvent. The use of water means that the emission of volatile organic compounds (VOCs) is eliminated. VOCs generate photochemical oxidants and suspended particulate matter (SPM); these are sources of air pollution which most countries are seeking ways to reduce.

Water-based paints are applied in the normal way, but care must be taken with the volume of paint delivered. The size of the fluid tip is therefore very important (Figures 17.63a, b).

The use of water-based paint also allows the reduction of carbon dioxide (CO_2) by giving a faster drying time. Toyota have developed a system which does not need drying after the second coat, the third coat is applied straight away. This also saves painting time. This system reduces VOCs by 70% and CO_2 by 15% compared to conventional techniques. Water-based paint also offers easy clean-up and is therefore better in terms of waste management.

However, currently solvent clear lacquer coats are still needed as the final finish needs to withstand the elements of the weather.

17.15 Burnishing, polishing and final detail work

Though brief reference to burnishing and polishing has been made earlier, these subjects are important enough to warrant fuller description. Even though most car refinishing paints are nowadays formulated to provide a good gloss from the gun, the final appearance of the vehicle can be further enhanced by careful burnishing and polishing. Burnishing helps to smooth out the surface whilst imparting a fuller gloss and revealing depth to the colour (Figure 17.64). Polishing, if carefully done,

(a)

(b)

Figure 17.63 Application of water-based paints is strictly controlled by the size of the fluid tip on the spray gun

Figure 17.64 Burnishing using rubbing compound and a sponge polishing pad (*Faréncla Products Ltd*)

will improve the lustre still further and provide a protective coating over the paint film.

17.15.1 Burnishing

Where the final coat of enamel contains small particles of foreign matter, it is best flatted with P1200 grade wet-or-dry paper with soap as a lubricant. The soap must be rubbed into the paper and not into the painted surface. Care must be taken not to rub through the colour coats, especially where projections exist in the vehicle construction.

Following washing down, the car or van is now ready for burnishing. The mutton cloth (or any suitable soft cloth) is first wetted in clean water and wrung out. A small quantity of burnishing compound is applied to the cloth and rubbing can commence, working in straight lines over a small area. A fairly firm pressure should be applied at first, but as the area shows sign of glossiness the pressure should be reduced. The friction caused by rubbing will generate heat, and when this takes place the cloth must be turned over and the bur-

nishing continued until a smooth glossy surface is achieved. Some of the colour will come away on the cloth but, provided that a sufficient coating thickness has been applied, this is unimportant. The cloths used for burnishing should be washed out periodically in clean water to remove the build-up of pigment which will hinder the action of the cloth.

Machines can be used for burnishing large areas of panel using liquid burnishing materials. These liquids, which contain milder abrasives, reduce the risk of swirls appearing on the surface as would be the case should burnishing compounds be used with machines. Hand burnishing, however, will produce the better finish should costs permit. When using machines for burnishing, the precautions set out in Section 17.12.2 should be observed. Following the burnishing process comes the final polishing.

17.15.2 Polishing

This is where the rotary-type polisher fitted with a lambswool pad (Figure 17.65) comes into its own.

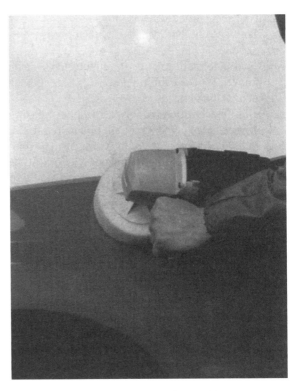

Figure 17.65 Rotary polisher fitted with a sponge polishing pad (*Faréncla Products Ltd*)

These machines are best used with liquid car polishes, but are also very effective when used with the more viscous conventional polishes. The polish is spread lightly over the surface with soft clean cloth, following which the finish is brought up to a high lustre with the machine. Heavy pressure is not required when polishing with a machine. Some parts of the vehicle body are obviously not suitable for machine polishing and must be done by hand. This is best done with soft yellow dusters rather than mutton cloth, which can leave striations on the surface and spoil the finished effect. The same care and precautions should be practised when polishing as when burnishing (see Section 17.12.2). The importance of working cleanly cannot be overemphasized.

17.15.3 Final detail work

Attention to detail when completing a respray can make the difference between a professionally done job and the amateur one. The sequence of operations is usually as follows:

1 Remove overspray from windows with a razor-blade scraper and wash leather.
2 Clean chrome bumper bars with rubbing compound.
3 Clean and refit all parts which have been removed for the painting process.
4 Vacuum clean inside of car.
5 Paint tyres with tyre black. Rubber glazing strips can also be painted with this material.
6 Finally carry out any lining work, using either lining tape, fine lining brushes or roller-type lining tools. The type of paint used is usually of a synthetic resin nature.

17.16 Rust-proofing

A vehicle body contains many recessed areas and box sections which are difficult, and in some cases impossible, to paint properly with a spray gun and conventional paints. The only time that they receive a coat of paint is in the electropriming process carried out by the vehicle manufacturer. All primers are porous, and this material is no exception. Should moisture be allowed to penetrate the primer, it will eventually break down and corrosion of the substrate will result. It is estimated that 90 per cent of corrosion on vehicle bodies is of the 'inside-out' type.

In an effort to prolong the life of a car body and provide an additional selling point, most manufacturers now carry out a rust-proofing process as part of the finished product or as a chargeable extra. A warranty is provided to the customer, but usually on condition that he or she has the vehicle checked for signs of corrosion at regular intervals and is prepared to pay for any rectification work that may be required.

The material used to treat these cavity areas is based on wax dissolved in a solvent with rust inhibitors added. A compound of this type has good capillary attraction and can filtrate into joints, seams and otherwise inaccessible areas. Different types of wax coatings are available, some classed as penetrants and others as heavy-duty coatings. There are several firms worldwide who specialize in these types of materials and the equipment for applying them. One such company is Tuff Kote Dinol (TKD).

Measures employed by car manufacturers in an effort to combat corrosion include the use of pre-coated steel; washing and spraying the assembled body; and immersion in anti-rust primer, which is then baked. Spot-welded seams, which would be prime sites for corrosion, are sealed; and wheel-arches, along with the underbody, are protected with anti-chip coatings. After a further baking period the car body is painted and some of the critical cavities can be flooded with hot wax. However, a car is subjected to a great deal of stress throughout its life; it is scratched, scraped and banged, and panels may flex, seams move and joints vibrate. If untreated metal becomes exposed, rust will get in; therefore cars need to be treated regularly, and so rust-proofing becomes a continuous process.

The basic anti-corrosion tool is the spray gun with its four attachments. The rigid and flexible lances are best for all enclosed areas; the hook nozzle is designed for more accurate, directional work; while the fan spray tool is for coating open underbody surfaces.

17.16.1 Spray nozzles

Rigid lance (1100 mm)
This produces a 360 degrees spherical spray pattern at right angles to the lance, combined with a forward and backward directed spray that allows all surfaces – front, back, sides, top and bottom – to be coated in one single sweep or stroke. It is highly effective in places where straight structures

such as doors, tailgate panels and long channels exist, or where the operator needs to control the position of the lance, such as enclosed front or rear wing box sections.

It is advisable to place the tip of the lance into the section being processed and to fog spray on both inward and outward strokes. Do not force into access holes. Test penetration before spraying to ensure adequate clearance for lance.

Flexible lance (1100 mm)

This is basically a flexible version of the rigid lance. It produces a 360 degrees spherical spray at right angles to the lance combined with a straight-ahead jet. It is highly effective in long narrow sections which would otherwise be inaccessible or awkwardly positioned and where its flexibility enables it to operate even where there are bends and restrictions which would prohibit the use of the rigid lance.

The main use of the flexible lance is for treating sills, underbody box sections, strengtheners and pillars. Its limitations are in narrow sections where it cannot gain entry or in very large box sections where it is desirable to control the position of the tip of the lances, such as in doors.

It is advisable to work the lance into the desired position and to fog on the outward stroke. In larger sections, retract the lance slowly so that one sweep will sufficiently cover all surfaces.

To extend the life of the nylon tube, the sides of the lance should be held away from the edges of the hole as it is entered or withdrawn. To eliminate snagging on the edge, always hold the exposed part of the lance at right angles to the hole. Feed the lance into the section gradually, keeping thumb and forefinger around the tube, close to the hole. Use the same technique when withdrawing the lance. The tube will go slack just before the tip emerges from the opening. This reduces the risk of accidentally spraying the interior trim and fittings by withdrawing the lance too rapidly.

Hook nozzle

This produces a highly atomized forward-directed full-cone jet which gives a powerful long range and, at the same time, good dispersion. Its value lies in its long range and directional capability, so that the product can be directed precisely where pointed.

This makes it suitable for treating narrow box sections such as door pillars, boot and bonnet lid reinforcements, and areas with difficult access such as door hinge areas, head and tail lamp housings, wing supports, door sills and underbody box sections, reinforcements, and suspension mountings. It may also be used for spray coating the wheel arches and underside of the vehicle.

The thumb should always be positioned on the machined flat portion of the neck to direct the nozzle where required. When spraying into narrow sections, the nozzle should enter the section with the spray directed towards the surface immediately opposite the opening into the section. As soon as the trigger is applied, the nozzle should be moved in an arc until the spray is being directed towards the end of the section. This should be repeated in the opposite direction. Where seam penetration is required, the nozzle should be specifically directed toward the seam.

Straight nozzle

This produces a forward spray with a restricted spray width. Ideally this nozzle should only be used with underbody sealants on to easily accessible panels. More obscure areas such as ledges within wheel arches should be processed with the hook nozzle.

Rules for nozzle selection

To minimize operator error when processing enclosed sections, always use the flexible or rigid lances first. Only use the hook nozzle where openings are insufficient for access with the 360 degree tooling. An exception would be where a very narrow section is to be processed and it would be preferable to spray with a powerful directional jet, such as a bonnet or boot lid strengthener.

17.16.2 Materials

Various anti-rust corrosion compounds include heavy-duty waxes and sealants for wings, wheel arches and underneath the vehicles; high-performance cavity waxes and penetrants for all enclosed areas, box sections, doors and pillars; special engine compartment waxes; and waxes for special purposes. These compounds are waxes and inhibitors dissolved in a solvent. They are applied as a liquid, but solidify as the solvent evaporates. For ease of application make sure the materials and the vehicle to be treated are at room temperature (about 15 °C). This is particularly important in winter.

17.16.3 Process

To assemble the gun, put a can of wax into the pressure pot, connect the head to the air supply, and regulate the pressure. In general, work with pressures of 5–8 bars (75–120 psi). Attach the appropriate tool and then spray. To change materials, release the pressure and replace the can. Try not to let the gun get too dirty because, although it is robust, dust and dirt will clog the nozzle.

The principle of rust-proofing is to prevent the atmosphere and corrosive substances from attacking the metal of the car body by applying an impermeable layer to the metal. In this case it is a layer of a penetrant or a sealant. Applying these materials is straightforward enough, but it does need care and attention.

Engine compartment

When protecting the engine compartment, the only preparation necessary is thoroughly to clean and dry the engine unit and compartment. Assemble the gun and load the engine compartment wax, which is designed to withstand high temperatures. The hook nozzle is best here, as the spray from the flexible lance would go everywhere and cover everything with wax. A face mask is necessary when working in confined areas or on open surfaces.

Putting your thumb on the flat section at the base of the hook will help to direct the spray. Beware of one piece of metal getting in the way of another, and by obscuring it prevent it from getting fully coated. The gun is two stage: the first pressure delivers just air, and the second delivers both air and material. The quantity of material is proportional to the distance you pull back the trigger. Match the speed of your sweep to the amount of spray to avoid build-up of wax. Work in short bursts rather than spraying continuously, and always remember to hold the gun upright.

Using the flexible lance, spray inside the front cross member and any other reinforcing sections. Unprotected cavities are prone to rust. Watch out for the join between sheets of metal, as dirt and moisture can gather here and start rust. Use the hook spray, since it is directional. Always wipe away any surplus anti-corrosion fluid as you go. Make sure you do inside all the reinforcements. Spray with reduced pressure, otherwise anti-corrosion fluid will go everywhere. Work from the bottom up, so that anti-corrosion fluid will not drip on you. When you rust-proof the chassis legs you should use a penetrating fluid and the flexible lance for all enclosed areas.

Pillars

Whenever possible remove the courtesy input switch for access, otherwise extra holes may have to be made. Spray as far up the pillar as possible, then down to the sills. Finally extend and secure the webbing on inertia reel seatbelts and spray the surrounding areas with great care. Spray into the D-post. Pay attention to fully coating the wheel arch seam area.

Door panels and rear quarter panels

Use the existing openings to spray into door panels, or drill extra access holes as required. Take care to thoroughly protect all door hinges. Remove the trim where necessary, to gain access to the rear quarter panel. Reach over the rear wheel arch and spray coat the wheel arch seam, the lower rear wing, around the tail lights and into all the inaccessible areas. Remove the existing plugs for access to the tailgate panel. Spray up beside the window of the tailgate; if unprotected this is particularly vulnerable to rust.

Sill panels and underbody sections

It is most important to achieve thorough coverage of the sills. Use the manufacturer's existing access hole, or open extra access holes. Look out for double sill sections, which must both be treated. Treat the underbody box sections and cavities in a systematic manner. Underbody seams, joints, brackets and attachments must all be coated with the fluid. Thoroughly apply a coating of rust-proofing fluid to the underside of the car, paying special attention to exposed seams and joints well away from the spray.

Wheel arches and underbody

Before applying a coat of sealant or heavy-duty wax, first remove any loose or flaking materials. When spraying, avoid blocking the manufacturer's drain holes; they are a vital part of anit-rust protection. If they do get blocked, clear out the surplus.

The wheel arches must be done carefully (Figure 17.66). Apply an extra coating to all forward-facing surfaces. To give the best possible abrasion resistance, pay particular attention to all joints and recesses. These are the places where mud can easily accumulate. When spraying the chassis legs try not to get any overspray in the engine compartment. Using the hook nozzle, deposit a generous

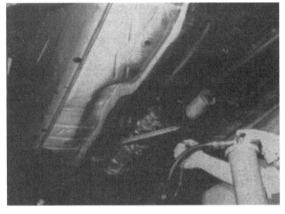

Figure 17.66 Sealing wheel arches and floor pan (*Tuff Kote Dinol*)

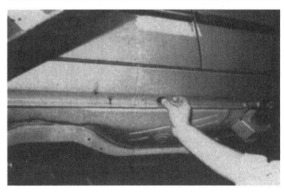

Figure 17.67 Sealing underbody (*Tuff Kote Dinol*)

coating of sealant all the way round. Make sure that the ledges are fully coated; not all of these are visible from underneath, so check them when the vehicle is back on the ground.

Spray the underbody from every direction to prevent shadow areas (Figure 17.67). Check your work as you go. Use the fan spray nozzle for rapid coverage of large open areas. Use the hook nozzle to get a good finish around the edges and to reach into awkward corners, then wipe away any excess sealant before it sets.

Clean any excess rust-proofing fluids off the car body with the recommended solvent, and remember to always clean the gun thoroughly after using the sealant or other heavy-bodied products.

In some situations you *must* rust-proof: for instance, when fitting a new body panel which is supplied untreated or when you do service bodywork repairs. But rust can also develop as a result of any accidents. Check the vehicle when it comes in for signs of external corrosion or paint damage;

if neglected it risks forfeiting the protection of any body warranty given.

17.17 Comparison of hot and cold spraying

The true value of a paint film is the amount and character of the solids deposited that remain after drying or curing. Generally it is not possible to apply paint by spray as supplied by the paint manufacturers. To obtain the necessary consistency for spray application and good flow-out, thinners are normally used. The function of a solvent is merely to reduce the viscosity of the paint to assist its application to the work.

A paint at a viscosity of 35 seconds BS B4 flow cup for conventional spray application requires a high air pressure to produce good atomization (Figure 17.68). The high pressure and the expanding compressed air evaporate a lot of the solvent and carry off some of the finely atomized paint as overspray. This is a recognized loss, and therefore less solids are deposited per coat. The basic principle of hot spray is to reduce the initial viscosity of the paint by heat, in contrast to cold spray where the viscosity is reduced by the addition of a thinner which subsequently evaporates. The underlying principle of a spray system involves splitting up the fluid paint into tiny droplets. This atomization is only possible if the viscosity of the liquid is sufficiently low. In conventional cold spray application this low viscosity is achieved by the addition of a solvent, whereas in hot spraying the necessary reduction in viscosity is achieved by heat. Heating of enamels etc. to 60–80 °C generally results in the viscosity becoming one-third to one-quarter that at ambient temperature. Further heating does not usually cause any further fall (Figure 17.69).

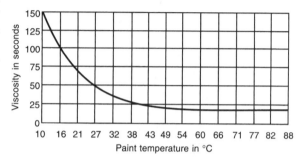

Figure 17.69 Typical viscosity curve

The viscosity of paints can vary in the paint shop owing to changes in the atmospheric and shop temperatures throughout the working day. This can be a problem to the painter, as work can be subject to uneven build and runs and sags can occur in spite of gun adjustments. Even if wide temperature changes do occur in hot spray, the atomization and spray pattern is unaffected because the paint heater delivers the fluid at a controlled temperature and therefore the viscosity remains constant. At higher temperatures temperature viscosity is negligible, but there is a sharp variation within normal atmospheric temperature range (Figure 17.70).

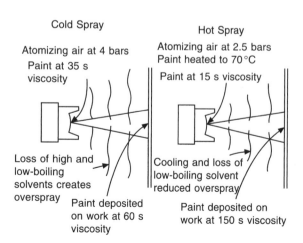

Figure 17.68 Comparison of cold spraying and hot spraying

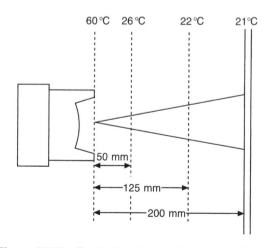

Figure 17.70 Rapid drop in paint temperature from fluid tip to job surface

In hot spray application, low boiling solvents are released more quickly during atomization because of the heat. These are therefore not present in the applied liquid film to cause chilling and resulting condensation, called 'blushing', which occurs with certain paints on applications under humid conditions. The effect is further checked because the material does not reach the low temperature when chilled by the evaporating solvent at the spray tip of the gun.

17.18 Movement of vehicle in the paint shop

In any refinishing shop, the movement of vehicles is of primary importance. Much time and effort may be wasted in moving vehicles from one point to another as dictated by the finishing process. This is particularly true in those shops where expensive air conditioned spray booth and/or spray booth ovens have been installed. To recoup the high capital outlay in such units it is essential that they should be used at full capacity for the whole of the working day. The simplest spraying/stoving unit, i.e. the combined spray booth and oven, is capable of handling six to seven vehicles per working day of eight hours, providing it is kept in use over the lunch hour. Furthermore, with soaring building costs and high rateable values, coupled in some cases with restricted space for expansion, it is important that floor space be kept to a minimum. Whilst it is not possible to achieve a factory-type flow line for refinishing jobs that may vary tremendously in their requirements, e.g. size, extent of repair and amount of work to be carried out, the desired continuous supply of vehicles to the spray booth may be achieved. Careful consideration of the problems of the vehicle movement will enable this supply to be obtained with the minimum of time, labour and space.

There are several methods of moving vehicles within the refinish shop:

Under own power This method is the most economical of labour, but requires the maximum of space. Also it gives rise to additional dirt, air pollution and fire hazards.

Manually Space requirements are as above but demand on labour is heavier. Three men may be required to move an average car.

Mobile hydraulic jacks These permit some saving of space in that they allow a tight turning circle both front and rear.

Turntable This gives an excellent means of utilizing what would otherwise be a dead corner, or may permit the spray booth to be sited in an otherwise impossible position (Figure 17.71).

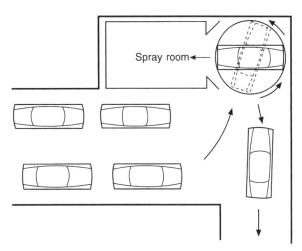

Figure 17.71 Suggested use of turntable

Rail and bogie system The vehicle is moved along railway-like tracks with bogies under each of its four wheels. This offers the greatest advantages of all sideways movement of the vehicles. Normally four sets of parallel rails are involved. The first pair constitutes the track on which the vehicles are first received into the paint shop, and along which the preparatory work is usually carried out, i.e. discing, flatting and stopping. The second pair is kept clear to allow movement of selected bodies to the spraying and/or stoving position (Figure 17.72). Vehicles are either driven or manhandled from one track to another. It is possible to have a layout utilizing only one track, but movements are more restricted than with two tracks. Two types of track are available: (a) those with sets of two rails, and (b) those with sets of three rails. With type (a) bogies long enough to accommodate vehicles of differing wheel bases must be used (Figure 17.73). Type (b) permits the use of smaller, lighter bogies which are on selected pairs of rails from each set (Figure 17.74).

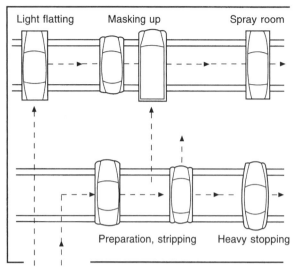

Figure 17.72 Lateral movement of vehicles using two sets of tracks

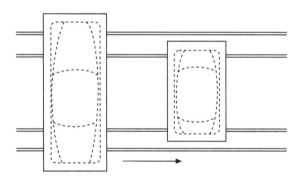

Figure 17.73 Two-track rails requiring long bogies

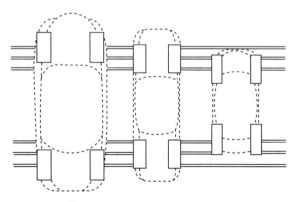

Figure 17.74 Smaller bogies can be used with three-track rails

17.19 Common spray painting defects

17.19.1 Blistering

Blisters are hemispherical projections on the painted surface, which will contain either air or water. Blisters can appear on surfaces which have been freshly painted or which have been painted for some time. Some of the causes of blistering are:

1 Moisture on the surface prior to painting
2 Excessive heat on article during service
3 Salts on surface prior to painting.

The third cause can lead to the defect known as 'contamination blistering', which appears mainly on those panels of a vehicle concealing the engine, and which are subject to fluctuating temperatures. It can be avoided by using demineralized water for the final rinse prior to painting. Blistering can occur on old paint films where the metal substrate has corroded from the reverse side, e.g. car wings, sills and lower door edges. Moisture penetrates through the metal to the paint film and causes it to swell at localized spots.

17.19.2 Blooming

This defect can be slight or severe. A slight case of blooming appears as a surface haze on a freshly dried paint film caused by moisture precipitation. It can be easily removed with a damp washleather. If moisture is deposited on to a tacky paint film it can be absorbed by the binder or medium, which will swell and give a cloudy, milky appearance with loss of gloss. This type of bloom cannot normally be cured, except by flatting and recoating. Where there is a high degree of humidity in the paint shop, it is best to use high-boiling (slow) thinner to avoid blooming.

17.19.3 Blushing

Blushing is similar in appearance to blooming but without loss of gloss. It is confined to wet paint films which are subjected to moisture precipitation, forming an emulsion with the binder. Blushing appears mainly with lacquers and can usually be cured by overspraying with a mist coat of anti-blush solvent.

17.19.4 Bridging

This occurs when the paint film contracts too much on drying so that it pulls away from recesses and corners. Too fast a thinner and too heavy a coating will produce this defect. Nitrocellulose lacquers particularly tend to exhibit this type of behaviour. Should the defect be discovered shortly after it has occurred, it can often be cured by slitting with a sharp knife or razor blade and overspraying with a slow solvent, or by adding slow solvent to any subsequent coats.

17.19.5 Cobwebbing

This is a defect of faulty atomization in which the paint leaves the spray gun in strings rather than small atomized particles. It is caused by incorrect thinning and is quite common in the spraying of acrylic enamels. When cobwebbing occurs, spraying should be stopped and solvent added to the paint. Though undesirable in the normal painting of motor vehicles, cobwebbing can be used to advantage to obtain decorative effects.

17.19.6 Dry spray

Though dry spraying can be an advantage to the spray painter, it is more often regarded as an undesirable defect as it gives the work surface a sandy appearance. It is caused by:

1   Too high an air pressure
2   Using solvents which are too fast
3   Moving the gun too rapidly or too far from the job
4   Too high a paint viscosity
5   In hot spraying, too high a paint temperature or surrounding air too hot.

17.19.7 Excessive overspray

Overspray can give a dry spray appearance and is invariably the result of faulty spray gun manipulation. However, some articles are so shaped that oversprayed areas are difficult to avoid. The spray painter should attempt to keep these areas ahead of the gun, so that with subsequent passes of the gun he can coat over the dry spray and dissolve it into the wet paint film. When this is not possible, overspray can be redissolved by mist coating the affected areas with a high-boiling (slow drying) thinner, or by flatting and burnishing.

17.19.8 Lifting

In this defect the existing paint on a surface wrinkles and then can be scraped or peeled off following the application of a fresh coating on to it. It is usually caused by the action of the solvent in the top coating attacking and softening the binder of the previous coat, such as when cellulose materials are applied over synthetic paints. When lifting occurs there is little alternative but to strip and recoat.

17.19.9 Orange peel

This is the most common of all spray painting defects. As the name implies, the dried paint film has the pebbly appearance of an orange skin, and is caused by the failure of the wet paint film to flow out smoothly after application. This defect can be caused by any, or all, of the following reasons:

1   Paint of too high a viscosity (too thick)
2   Too fast a thinner (thinner evaporating too quickly)
3   Too low an air pressure, causing poor atomization
4   Too high a temperature in the spray room, causing the paint film to 'set up' before it can flow out
5   A sudden draught in the spray room (also causing the paint film to set up prematurely)
6   In hot spraying, too high a paint temperature.

Though orange peel is a defect of faulty flow-out, it can occasionally occur after the paint film has flowed out. This is the result of uneven solvent evaporation and the formation of vortices in the film, caused by improper solvent balance.

17.19.10 Pin-holing

This is the sudden appearance of small vortices (pin-holes) over the painted surface during drying. It is caused by using a solvent in the paint mixture which is too slow to evaporate, and occurs when the topmost skin of the paint film has dried out. The slow solvent is trapped beneath the skin and ruptures it when escaping. Pin-holing can also be caused by too high a room temperature, and in hot spraying too high a fluid temperature.

17.19.11 Runs, sags, curtains

Each of these defects is the result of applying too heavy a coating thickness. Some of the factors which contribute to this are:

1 Paint which is too thin or too thick
2 Paint solvent which is too slow, retarding drying
3 Spray room temperature too low, also retarding drying
4 Spray gun held too close to the surface, or spray gun movement too slow
5 Fluid nozzle too large for material being sprayed.

17.19.12 Shelving

This term, though not widely used, is an extremely apt one to describe the defect in which the paint films tend to lie in separate layers or shelves one on top of the other. This is caused by failure of the top coat to adhere properly to those previously applied, so that one coat can be separated from the other by scratching with the fingernail or a knife. Shelving usually shows during sanding and can seldom be repaired. It is better to strip and refinish a surface rather than risk any subsequent coatings peeling off. The main causes of shelving are:

1 The solvent in the top coat is too weak.
2 The first coat is too dry before the second coat is applied.
3 There is grease or dirt between coats.
4 The top coat is too brittle.
5 There is poor natural adhesion in the top coat.

17.20 Colour mixing and matching

17.20.1 Colour mixing systems

Surveys have shown that, between them, 40 car manufacturers offer approximately 600 colours on current models. These are new colours only, and the total number of colours that the refinisher may have to match can run into thousands.

The development of a new car colour can be broken down into four stages:

1 Initiated by clothing and other fashions, new colours are reproduced in a range of paints. The car body stylist then selects a colour by trial and error, to suit a new car model.

2 Paint manufacturers then reproduce this colour and carry out the usual tests for colour stability, durability, etc.
3 The vehicle manufacturer paints a number of car bodies with the colour to ensure that the paint is suitable for mass production application.
4 Finally, the paint manufacturers are required to produce a coloured paint to suit the vehicle refinisher and to provide colour mixing formulae as well as basic colours and tinters.

There are two colour mixing systems available to the refinisher:

Gravimetric   This system involves the use of a set of scales. The base colour and tinters are mixed according to their weight, the quantity of each being obtained from a given formula.
Volumetric   The various quantities of each ingredient are added by volume. An adjustable measuring rod is the main tool involved (see Figure 17.75).

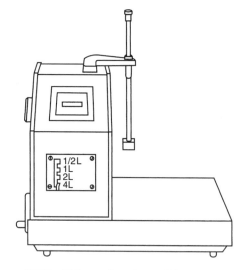

Figure 17.75   Volumetric colour mixing machine

In both systems the materials required to mix and match colours are base colours, full-strength tinters, and reduced-strength tinters. The last are necessary in order to provide the refinisher with a measurable quantity. For example, instead of adding 0.5 grams of blue-black full-strength tinter, he can more easily add 5 grams of reduced-strength tinter.

A complete set of colour formulae is provided by the paint supplier and is updated at regular intervals. Colour samples are also provided which are added to as new colours are produced. There can be several slight variations of a colour as vehicles leave the production line, and these can provide the refinisher with real problems when trying to match a colour. In order to assist with solving these problems, the paint manufacturers produce samples of these colour variants and literature with hints on how to match them.

Colour mixing systems should not be installed in the workshop if at all possible. They constitute a fire hazard, and can collect dust and overspray which may affect the accuracy of the scales. A separate colour mixing room should be provided where possible, which has plenty of natural light and also has 'daylight' fluorescent lighting.

It should be fully understood that the vehicle manufacturer uses high-bake materials, whereas the refinisher cannot. In order to match a colour, different pigments may have to be used in the refinishing material in order for them to be compatible with the binder. Consequently, a perfect match may be obtained in daylight conditions, but the colour may alter drastically when viewed under sodium or mercury street lights. There is nothing that the refinisher can do about this, and it is advisable to explain this to the customer when only part of the car, such as a wing or door, is to be painted.

The modern system of supplying the refinisher with colour formulae is in the form of microfiches, which require a microfiche viewer or reader.

17.20.2 Procedure for identifying a particular colour

1 Note the make and year of manufacture of the vehicle.
2 Find the colour identification plate on the vehicle (the positioning of this varies from one vehicle manufacturer to another) and note the colour coding.
3 Select the appropriate microfiche, place it in the viewer and find the mixing formula.
4 Double check that the correct colour has been found by finding the coloured sample in the books provided by the paint manufacturer, and placing this against a cleaned-up part of the car.
5 If this does not match perfectly, check the book on colour variants and consult the manufacturer's hints on tinting.

17.20.3 Hints on colour mixing and matching

Never try to match a colour by dabbing a sample of the colour on to the vehicle and making adjustments. Colours alter during the drying process. Always spray a sample card with the appropriate number of coats and allow it to dry before comparing it with the colour to be matched.

On a volumetric system, never use a dented tin or one with a concaved or convexed bottom to mix the colour in. On a gravimetric system, always remember to place the tin on the scales before zeroing them.

When comparing the painted sample against the vehicle, daylight, perferably from a northern source, is best. Bright sunlight or artificial lighting can sometimes exaggerate or diminish a colour effect.

Questions

1 State three of the advantages to be gained from the method used by vehicle manufacturers for applying the priming coat.

2 State two advantages to be gained by using a guide coat when sanding down a surface.

3 Explain two of the functions of a spray gun air cap.

4 Explain how a suctionfed (or syphonfed) spray gun operates.

5 Determine the difference between thermoplastic and thermosetting paint coatings.

6 What is the basic difference between those paints classed as lacquers and those classed as enamels?

7 What is the function of a ground coat?

8 Under what circumstances would a bleeder-type spray gun be selected in preference to a non-bleeder?

9 Why are bleeder-type colours a problem to the painter?

10 Explain the working principle of a single-acting, two-stage, piston-type compressor.

11 What is meant by the pot life of a two-pack paint?

12 Give two reasons why cellulose-based paints were replaced by stoving synthetics for mass production finishing on new cars.

13 What type of vehicle finish is liable to become viscid when dry sanded with a sanding tool?

14   Reduced tinters have one-tenth of the tinting strength of a full-strength tinter. Why are they necessary for inclusion in colour mixing schemes?

15   What is the best type of priming paint to use on an aluminium surface?

16   What would be the most likely result of using two-pack polyester stopper as an intercoat stopper in a lacquer-type paint system?

17   State the main disadvantage of a combined spray booth and low-bake oven unit.

18   What is the main health hazard associated with the spraying of two-pack paints?

19   Which method of heat transfer is used in (a) infrared and (b) low-bake heating units?

20   State two reasons why door handles should not be 'bandaged' when masking up.

21   In which way do thermoplastic vehicle finishes present problems to the vehicle refinisher when carrying out spot repairs?

22   What is an HVLP spray gun?

23   It is estimated that 80–90 per cent of corrosion on motor vehicles is classified as inside-out corrosion. What steps do vehicle manufacturers take to minimize this problem?

24   The spray pattern of a spray gun is top heavy. Describe a simple process to determine whether it is the air cap or fluid tip which is at fault.

25   State what is meant by 'VOCs'.

26   How do HVLP spray guns and VOCs relate to one another?

27   What is the main disadvantage of a combi-unit (spray booth and low-bake oven)?

28   What is the difference between baking and force-drying a paint film?

29   Describe a simple test to determine whether or not an existing paint finish is of the non-convertible type.

30   State three problems which can arise if cellulose stopper is applied too heavily.

Index